Bacillus thuringiensis:
Biology, Ecology and Safety

Bacillus thuringiensis: Biology, Ecology and Safety

TRAVIS R. GLARE
and
MAUREEN O'CALLAGHAN

Biocontrol & Biosecurity
AgResearch
PO Box 60, Lincoln
New Zealand

JOHN WILEY & SONS, LTD
Chichester • New York • Weinheim • Brisbane • Singapore • Toronto

Baffins Lane, Chichester,
West Sussex PO19 1UD, UK

National 01243 779777
International (+44) 1243 779777
e-mail (for orders and customer service enquiries): cs-books@wiley.co.uk
Visit our Home Page on: http://www.wiley.co.uk or http://www.wiley.com

Other Wiley Editorial Offices

John Wiley & Sons, Inc., 605 Third Avenue,
New York, NY 10158-0012, USA

WILEY-VCH Verlag GmbH, Pappelallee 3,
D-69469 Weinheim, Germany

Jacaranda Wiley, Ltd., 33 Park Road, Milton,
Queensland 4064, Australia

John Wiley & Sons (Asia) Pte, Ltd., 2 Clementi Loop #02-01,
Jin Xing Distripark, Singapore 129809

John Wiley & Sons (Canada), Ltd., 22 Worcester Road,
Rexdale, Ontario M9W 1L1, Canada

British Library Cataloguing in Publication Data

A catalogue record for this book is available from the British Library

ISBN 0-471-49630-8

Printed from camera ready copy provided by the authors
and bound in Great Britain by Antony Rowe Ltd, Chippenham, Wiltshire.
This book is printed on acid-free paper responsibly manufactured from sustainable forestry, in which at least two trees are planted for each one used for paper production.

Contents

Foreword

Bacillus thuringiensis (*Bt*) insecticides have proved effective and selective enough to eliminate outbreaks of some Lepidoptera pests. This involves aerial sprays over major cities. Their active ingredient is ideal - harmless at high dose to man, mammals and vertebrates by mouth, inhalation and injection; also harmless in most beneficial insects or having only a small effect. The active crystalline toxin component has been genetically engineered into plants used as food, thus producing a safe systemic insecticide that specifically kills caterpillars as they feed. *Bt* has jumped into the limelight at a time when the public has become more sensitive about the environment and the food they eat. The present account of the biology of *Bt* and the penetrating analysis of safety make a fascinating story, authored by two scientists who have closely followed the aerial spraying programme over the city of Auckland, New Zealand.

Written in very readable style, this book is mainly for scientists, but will interest the thinking public and environmentalists. It has wide practical and fundamental value for biologists, bacteriologists, entomologists, microbiologists, pathologists, geneticists, ecologists, chemists and pest control experts in industry, academia and government laboratories. The book is useful at graduate teaching level and is a valuable reference volume for post graduate and research work. It is the first book-sized review of the environmental consequences and mammalian safety of *Bt*. The first part is an excellent, condensed, general account of the biology and ecology of *Bt* as background knowledge. The present vast amount of literature has been assessed critically.

The text is well ordered in 15 sections. The authors reveal *Bt* as an interesting but complex organism in its own right, that has taken a large stake in the subject of bacteriology with:

- production of a wide range of toxins, particularly proteins
- survival by means of spores
- living as a facultative parasite in insects and as a saprophyte both in soil and on foliage
- showing intriguing relationships with other important pathogenic bacteria.

Study of characterisation, toxins, mode of action and natural occurrence has involved sophisticated techniques. Production for pest control occupies a whole industry: transgenic use plays a prominent role in the new industry that produces seed for transgenic crops. These general subjects are followed by comprehensive environmental sections, describing host range, effects on non-target organisms including safety tests for vertebrates and man, risk analysis, environmental persistence and incidental problems such as gene transfer and the development of pest resistance to *Bt*. The text culminates in suggestions for future research.

Text occupies nearly half the book and literature references account for about a quarter. Extensive data are partitioned into a series of long tables and even longer appendices, totalling the remaining quarter of the book, making it a must for the reference bookshelf.

H.D. Burges

Preface

Bacillus thuringiensis (*Bt*) is a fascinating bacterium, producing a number of insecticidal toxins which have spawned a whole industry. *Bt* has been in commercial production for over 60 years. For many years, it was regarded as a "safe" biopesticide, with toxicity limited to a few Lepidoptera and no indications of mammalian or non-target impacts. But knowledge of *Bt* has rapidly advanced; each year new strains and toxins have been identified and the whole picture of what constitutes *Bt* has become more confused. With the isolation of new *Bt* strains came the discovery of *Bt* activity against a greater range of organisms, which worked against one of the original safety tenets, that *Bt* was specific to Lepidoptera. *Bt* toxin genes are now expressed in other organisms and the rapid advances in transgenics have led to changes in the public and scientific perception of risk and safety. *Bt* has attracted the attention of the media on a few occasions, with incomplete reports of mammalian toxicity. Where previously community groups had been calling for *Bt* to replace chemical pesticides, more recently there have been suggestions that safety of *Bt* should be investigated further.

An interesting article "Keeping an eye on *B. thuringiensis*" by Dixon (1994) is indicative of the current climate of caution surrounding use of pesticides, both biological and chemical. In this short article, published in Bio/Technology, Dixon extols the virtues of *Bt* as a safe biopesticide, but calls for closer monitoring for potential negative health impacts. "Odd items have been appearing in the literature that point to a significantly less robust view" he states. He raised issues such as the role of *Bacillus cereus* and "nonpathogenic" *Bacillus* in human disease, even if just as opportunitist pathogens. He also drew attention to a number of isolated reports of mammalian toxicity (these are reviewed in Chapter 7), raising questions about the safety of *Bt*.

Increasing public concern over *Bt* safety, isolation of new strains of *Bt* and the discovery of new toxins, coupled with a blurring of the distinction between *Bt* and the closely related species *B. cereus,* make it timely to review the safety of *Bt.* In this book we endeavour to review environmental, non-target and mammalian safety of *Bt,* in the context of *Bt* biology and ecology. While our original intention was to examine the safety of the invertebrate active δ-endotoxins, it is impossible to assess the safety of *Bt*-based products and formulations without considering the range of toxins produced by strains of the bacterium. *Bt* forms the basis of over 90% of all biopesticides currently sold around the world. Most products, especially those sold in the developed markets of USA and Europe, do not contain the β-exotoxins of *Bt,* as early concerns over mammalian toxicity resulted in easier registration for *Bt* products free of exotoxins. Therefore, we have focused mainly on safety of δ-endotoxins, particularly the environmental impacts of *Bt* cells and spores, and toxins released from these cells. We have briefly reviewed the extensive literature on transgenic expression of individual endotoxin genes. Since 1985, a growing body of literature involves the use and safety of transgenic *Bt* expression and the topic is worthy of another book.

In our review of safety of *Bt*, we have examined vast amounts of literature that has been produced over the last 60 years, while *Bt* has been used as a commercially available pesticide. *Bt* has been applied directly over cities; it is hard to imagine another pesticide for which this would be permitted. The majority of the studies have found *Bt* to be more specific than chemical insecticides, with low mammalian toxicity. There are areas which require more study, such as the relationship between *Bt* and *B. cereus*. In Chapter 15, we evaluate what is known of *Bt* in terms of safety.

In attempting to cover the literature, we have examined over 8000 abstracts, articles and unpublished reports. However, this review cannot claim to be exhaustive because so much research is conducted on *Bt* in many countries around the world and some of the data are not widely available. Also, because of commercial interest in *Bt*, some of the more recent research, especially on toxicity to novel groups of organisms or new toxin genes, remains unpublished. We have tried to be comprehensive, but it is not possible to access all publications and reports. Therefore, there will be a number of omissions, despite our best intentions.

As in any endeavour of this size, we have had the assistance of a number of people, and the perseverance of others. We thank the AgResearch librarians (Invermay), Sue Weddell and Karen Cousins, for the mountain of work in tracking down and obtaining papers and reports. Melanie Davidson assisted in assembling Appendix 2 and Emily Gerard, Tracey Nelson, Sandra Young and Nicola Richards provided a good deal of assistance in compiling references, assembling tables and checking data. Aaron Knight and Ian Laurenson were of invaluable assistance in formatting the final document. We thank Abbott Laboratories, Australia, for partly funding this project and are especially indebted to Dr Andrew Rath and Greg Rappo of Abbotts (now with Valent BioScience) for their support. At John Wiley's, we thank Drs Sally Betteridge and Lewis Derrick for assistance in the final stages of preparing this publication.

We are grateful to Drs Joel Siegel, Bob Smith and Jean-Claude Ballaux for the constructive criticisms on early drafts of this work and especially to Drs H. Denis Burges and Wendy Gelernter for extensively reviewing the manuscript and providing additional information and suggestions.

Finally, we gratefully acknowledge the patience and support of our partners, Mary Christey and Steve Kennedy, and our neglected offspring, Nyssa, Caleb, Caitlin and Tom.

Travis R. Glare
Maureen O'Callaghan
May 2000
Lincoln, New Zealand

Summary

Bacillus thuringiensis (*Bt*) (Bacillaceae) is a Gram-positive, rod-shaped, sporeforming bacterium capable of producing large crystal protein inclusions during sporulation, which are mainly responsible for the insecticidal properties of the species. *Bt* occurs naturally in the environment and has been isolated from a range of habitats including soil, phylloplane, dust, plant material and insects throughout the world. The efficacy and specificity of *Bt* strains and individual toxins produced by *Bt* isolates are such that a large number of insecticidal products are based on this bacterium and/or its toxins. Since the launch of the first product in France in 1938, over 100 *Bt*-based products have been produced around the world and *Bt* products currently constitute over 90% of all biopesticides sold.

Because of its specificity and mode of action, *Bt* has historically been considered a safe option for pest control and has often been the preferred insect control method in IPM programmes. However, with greater understanding of the many and varied toxins produced by individual strains, there is a need to examine closely the potential non-target impacts of widespread application of *Bt*-based products. The use of *Bt* toxins is increasing rapidly with *Bt* gene expression in plants and the isolation of novel *Bt* isolates active against an increasingly wider range of organisms, for example nematodes. Consequently, there are questions about environmental impacts and, as *Bt* is often applied in close proximity to urban populations, mammalian toxicity remains of concern to the public. There is also greater recognition of the close genetic relatedness between *Bt* and other *Bacillus* species, some of which are pathogenic to humans.

Bt is not a single entity. It is a collection of isolates varying in ability to produce a range of toxins which determine toxicity to organisms. Assessment of safety of *Bt* is greatly complicated by the large number of strains identified with varying activity which are grouped together within the species designation. Serotyping is one of a number of methods used to categorise *Bt* isolates. Currently 82 serovars (based on flagellar antigens) are recognised, but some strains can't be typed by this method (such as autoagglutinating or non-flagellated strains) or represent new, uncharacterised serotypes. However, the insecticidal activity of *Bt* is related not primarily to serotype but mainly to the range of insecticidal δ-endotoxins produced by each isolate. At present, over 170 distinct δ-endotoxin genes, encoding mainly insecticidal proteins, have been described and the discovery of new toxins is increasing as new *Bt* strains are isolated. The occurrence of these toxins in the *Bt* subspecies/serovar/variety is not constant and more than one δ-endotoxin is produced by most *Bt* strains. In addition to the δ-endotoxins, some *Bt* strains can produce a number of other toxic substances, including exotoxins, enzymes, haemolysins and enterotoxins, some of which have activity against mammals. Not all *Bt* strains produce these other substances and amounts produced by individual strains vary.

Bt has been used for many years in agriculture and forestry with few reported adverse effects on humans. Concerns about the mammalian safety of the genus *Bacillus* have been raised, in part, because several other species within the genus cause various infections in humans. *Bacillus cereus,* once thought to be a totally innocuous bacterium, has more recently been recognised as a cause of food poisoning. The only consistent difference between *Bt* and *B. cereus* is the occurrence in *Bt* of the often extra-chromosomal DNA encoded insecticidal crystal proteins. Some *Bt* strains have been shown to produce toxins similar to those produced by *B. cereus*, but at much lower titres. Despite being used extensively for over 50 years, *Bt* isolates have only rarely been implicated in clinical infections and gastroenteritis in humans. Large-scale studies on human populations exposed to *Bt* have not reported significant increases in any disorders or health concerns.

Extensive mammalian safety tests against a range of small mammals have demonstrated a very low safety risk from direct exposure to *Bt* spores and δ-endotoxins. This may be partly because alkaline conditions are required to release from the crystals the large inert protoxins and suitable proteases are needed for activation of these inert protoxins into smaller proteins, the δ-endotoxins; these conditions are found in the insect gut, but are not encountered in the mammalian gut. No effects have been found after ingestion of unsolubilised endotoxin by mammals. However, there is potential for activation of the toxin under specific conditions, for example, in the presence of proteases. Solubilised δ-endotoxin from *Bt israelensis* can be toxic to mice when administered by injection but not orally; the solubilised toxin is also cytolytic to human erythrocytes. Conditions under which activation of toxin can occur outside the insect have not been fully researched, but appear to be rare.

A rare exception to the favourable mammalian safety data on *Bt* is a French study, which has recently been reported in the media. Very high levels of *Bt* caused mortality in mice following pulmonary exposure. The level of inoculum used in the experiment greatly exceeded the field application rate, but since no δ-endotoxin would be present in the active state, such studies provide information on mode of action of other groups of toxins produced by *Bt*.

Some strains produce β-exotoxins, which have demonstrated activity against a wider range of insect hosts, including some beneficial insects, fish and mammals. Regulatory authorities generally require that *Bt* products are free of β-exotoxin and the use of standardised production methods is designed to minimise the presence of unwanted toxins in the final product.

Review of the literature has shown that more than three thousand insect and mite species from over 15 orders are recorded as susceptible to at least one serovar of *Bt*, with 61 serovars recorded as toxic to at least one insect species. Susceptible insects are primarily herbivores, rarely carnivores. The range of insects susceptible to any one *Bt* isolate, toxin or serovar is difficult to assess from published studies because of incomplete information regarding the toxin composition which varies between formulated products and even within serovars. Review of the extensive studies on some of the commercially produced strains has shown that each *Bt* isolate has a limited number of species to which it is toxic, especially at recommended field rates. Generally *Bt kurstaki* and *aizawai* serovars are most effective against Lepidoptera, while *Bt israelensis* acts against Diptera and *Bt tenebrionis* against Coleoptera. However, not all species within a given order are susceptible to any one *Bt* strain, nor is activity restricted to an insect order, because of the expression of multiple δ-endotoxins in most isolates.

Reports of susceptibility in the laboratory often do not reflect potential impacts in the field. Moderate toxicity in the laboratory combined with feeding behaviour and the spatial/temporal distribution of non-target insects, factors which influence the extent of contact between *Bt* and non-target organisms, can result in minimal impact of *Bt* on field populations. Species in over 200 genera of Lepidoptera can be killed by *Bt kurstaki* and formulated *Bt kurstaki* products in the laboratory; however, it is often ineffective against many species of caterpillars in field situations.

A number of ecosystem studies have been conducted on non-target populations, such as lepidopterans in the forest canopy which were most likely to be affected after broadcast application of *Bt kurstaki*. No lepidopteran species were eliminated from the study areas, although some macrolepidoptera were reduced in numbers for up to two years. In general, *Bt* had little impact against non-target insects outside the year of application. Similarly, widespread use of *Bt israelensis* against mosquitoes and biting flies in waterways caused few non-target impacts, with only temporary changes in populations. Aquatic communities, including benthic organisms, are generally insensitive to *Bt*, especially *Bt kurstaki*, which is the most commonly applied serovar. Impacts on selected Diptera have been recorded for the mosquito-active *Bt israelensis* subspecies; however studies on non-target communities have not shown enduring impacts. *Bt* subspecies are, in the words of a review by the US Environmental Protection Agency, "practically non-toxic" to fish, with doses far in excess of the recommended field rate required for toxicity. The indirect effects of *Bt* application on ecosystems can include dietary changes in birds and insect predators, such as small mammals. Only a few studies have reported effects on bird populations and these have found small changes in bird nesting numbers or size. Overall, indirect impacts have not led to significant problems in birds and insect predatory mammals.

Bt spores and δ-endotoxins have rarely been found to impact on beneficial species, such as insect predators, pollinators or soil dwelling organisms such as microbes or worms. In general, *Bt* is compatible with other naturally occurring control agents. Review of the literature covering use of *Bt* with predators, parasitoids and pathogens showed a variety of responses ranging from synergistic to antagonistic. *Bt* application was reported to have no effect on over 50 species of insect predators and no direct effect on over 80 species of parasitoids.

However, there were occasional reports of competition between *Bt* and parasitoids for insects, which led to overall reductions in rate of parasitism. Similarly, combinations with microbial insect pathogens could lead to antagonism, probably through competition for hosts. Combined use of *Bt* with chemical pesticides resulted in a range of reported interactions, from synergism to antagonism.

The environmental fate of the various *Bt* products is of relevance as the accumulation and persistence of *Bt* and its endotoxins could result in increased exposure of non-target species and the selection of toxin-resistant target species. Persistence of *Bt* spores and toxins in water has been extensively examined only for the dipteran-active *Bt israelensis.* Persistence of toxins was measured in days rather than weeks. A few studies have examined *Bt kurstaki* persistence in water; estimates of persistence ranged from less than 24 hours (in running water) up to 12 days. Persistence of *Bt* is affected by a number of environmental factors, including temperature, rain (which washes *Bt* from leaves), soil types and UV radiation. Biotic factors influencing the effect of *Bt* in the environment include insect factors such as feeding behaviour, distribution and age. Formulation, application techniques and rates, and substrate to which *Bt* is applied also affect persistence. In general, spores are inactivated within days in the phyllosphere and the crystals survive rather longer. Conversely, the crystals are rapidly inactivated, while the spores remain viable for extended periods in the protected environment of soil. Experimentally determined half-lives of spores are usually in the range of 100-200 days, but *Bt* does not grow in natural soil.

Transmission of *Bt* between insects is not common, making it an unusual insect pathogen. Recycling of *Bt* in a target population is not expected after commercial application, with *Bt* acting primarily as a toxin rather than as an infective pathogen. *Bt* can be transmitted by cannibalism and parasitoids, and dispersed by rain, scavengers, predators, and non-susceptible organisms, including small mammals. However, poor survival of *Bt* and limited ability to sustain infections in target populations has contributed to poor horizontal and vertical transmission of *Bt* after application.

Insect resistance to *Bt* isolates and specific toxins has arisen, especially in Lepidoptera. Reports of resistance are usually from laboratory-reared insects repeatedly exposed to *Bt,* with only one insect species, the diamondback moth *Plutella xylostella,* with serious incidents of resistance to *Bt kurstaki* and *Bt aizawai* δ-endotoxins in field populations from several countries. Four other lepidopterans have shown some tolerance to *Bt* strains in some field populations, but this may reflect natural variation in susceptibility rather than resistance. Resistance has also developed to toxins of *Bt kurstaki, Bt aizawai, Bt entomocidus, Bt san diego* (=*tenebrionis*) and at very low levels to *Bt israelensis* in the laboratory. The low rate of resistance development to *Bt israelensis* may relate to the multiple toxins expressed, including both Cry and Cyt toxins. Cross-resistance between *Bt* Cry toxins has been found, but does not automatically occur. *Bt* has not been used extensively against many pest species in the field, which may explain the low frequency of field-developed resistance. The lack of field-developed resistance to *Bt,* with the exception of a few Lepidoptera, suggests that, with proper management and conservation of *Bt* as a resource, the risk of resistance developing can be kept low at current use levels.

There has been a change in public perception to application of biopesticides in recent years. Whereas the 1980s and early 90s were characterised by calls for use of *Bt* to replace chemicals, there is now often public resistance to any pesticide application, including *Bt*. Comparisons between *Bt* and chemical insecticides in terms of non-target impacts have always found *Bt* to have the least impact, for a number of target pests. There will be continued scrutiny of *Bt* safety and more studies can be expected. However, the wealth of data currently available and experience of many years of broadscale applications would suggest that *Bt* is one of the safest pesticides currently available. The continued research and identification of new toxins produced by *Bt* will result in improved screening methods for toxins and the selection of safer strains.

After examining the literature on aspects of environmental and mammalian safety of *Bt* strains and subspecies, we view *Bt*-based products used at recommended field rates as safe to use, in terms of minimal non-target impacts, little residual activity and lack of mammalian toxicity. There are a number of questions that require further research such as the relationship between *Bt* and *B. cereus* and the production by some strains of toxins other than endotoxin. It seems prudent to continue to monitor *Bt* applications closely for potential negative impacts and the techniques for monitoring are improving constantly. However, the information available at this time suggests there are few risks associated with use of *Bt*.

1 Introduction

Bacillus thuringiensis (*Bt*) is a sporeforming bacterium capable of producing a number of toxins, including insecticidal endotoxins, exotoxins, haemolysins and enterotoxins. The level of diversity between isolates within the species is high, with 80 recognised serotypes and over 170 endotoxin-encoding genes identified. The susceptibility of many insects to these various toxins has led to the use of *Bt* in biopesticides. *Bt* forms the basis of over 90% of commercially available biopesticides. In this book we aim to review published information on the biology, ecology of *Bt*, and the environmental, non-target and mammalian safety of the insecticidal endotoxins produced by *Bt*. With increasing public concern over all aspects of pesticide safety and the increasing use of *Bt* as an applied pesticide, it is timely to review again in detail, the likely impacts of *Bt* application.

Bacillus thuringiensis was first isolated around 1901 in Japan, but is named after the later discovery of a bacterium killing flour moths in Thuringia, Germany. The potential of *Bt* as control agent of insects was only slowly realised; the first product was released in France in 1938 and commercial interest did not develop in most other countries until the late 1950s. Discovery of the link between toxicity and the parasporal body was not confirmed until the 1950s. There are currently an estimated 60,000 isolates of *Bt* held in public and private collections throughout the world, indicating the level of interest in applications of this remarkable bacterium.

1.1 WHAT IS *BACILLUS THURINGIENSIS?*

"What is *Bacillus thuringiensis*?" is a question that has occupied many researchers working with this remarkable bacterial species. *Bacillus thuringiensis* (*Bt*) is a sporeforming bacterium that has insecticidal and sometimes wider toxicity. The species comprises numerous very diverse strains with widely different toxin profiles and hence activity against an extensive range of insects and some other organisms. However, the ecological niche and role of *Bt* in the environment have been debated and discussed constantly since its development as the predominant biopesticide of the late 20th century.

Bt often produces a number of insecticidal toxins in the form of a protein crystal, without being a highly effective insect pathogen. It is also a species made up of a number of distinct subspecies, varieties and pathotypes. There are over 170 individual endotoxin-encoding genes, most of which are active against organisms in a few orders at most, and each of which can be expressed in one to several of the 82 characterised serotypes (Table 2.1). In addition, *Bt* can produce other toxins, such as exotoxins, haemolysins and enterotoxins (see Chapter 2) which have both insecticidal and mammalian toxicity. *Bt* is difficult to separate from *Bacillus cereus* genetically, which has safety implications as some *B. cereus* strains produce toxins that cause gastroenteritis in humans (see Chapter 7).

The complexity resulting from the combination of a number of insecticidal proteins expressed in strains identified through numerous biochemical, genetic and serological methods and formulated into a vast array of products throughout the world makes the review of *Bt* difficult. The use of the name "*Bacillus thuringiensis*" gives no indication of the variety of toxins, cells, spores, crystals or formulations that have been used in any particular study. We have tried to reiterate constantly in this book that *Bt* is a collection of individual isolates and the formulations currently available do not only contain *Bt* spores and crystals, but also other ingredients which may also affect toxicity. However, *Bt* characterisation is not an exact science and in many cases, the specific toxicity cannot be attributed to anything but a subspecies or product in general.

1.2 NOMENCLATURE

Discussion of *Bt* is made more difficult by the bewildering array of names used and the many changes in nomenclature which have been made in the literature. Strains (whole cells) have been identified by various means (Chapter 2), but the method used most commonly is serotyping by H-antigens. Subgroups such as serotypes were originally classified as varieties, however the bacterial nomenclature code requires such subgroups to be recognised as subspecies (Sneath *et al.* 1986). Therefore, what was previously referred to as *B. thuringiensis* var. *kurstaki* is now known as *B. thuringiensis kurstaki.* The toxin genes have also been reclassified several times as more individual genes and toxic proteins have been identified. Originally a haphazard system of arbitrary designation by each discoverer was used for naming each toxin. Höfte and Whiteley (1989) introduced the first systematic classification and nomenclature for toxin proteins, based on insecticidal activity. However, exceptions to the classification began to arise as more insecticidal genes were identified. Crickmore *et al.* (1998) proposed a new nomenclature, based on hierarchical clustering using amino acid sequence identity. This system exchanged Roman numerals for Arabic numerals in the primary rank (e.g. Cry1Aa) to better accommodate the large number of expected new sequences. Herein, we have used the new nomenclature (see Appendix 1 for relationship between old and new designations).

Delta-endotoxins are not the only toxins produced by *Bt* strains and subspecies. We refer throughout to exotoxins, particularly the β-exotoxin, although more than one type has been described. Finally, the role of spores, crystal toxins and other toxins interactively or separately is not always clearly defined in studies and the use of the term "*Bt*" may mean any combination thereof, unless otherwise stated.

1.3 BRIEF HISTORY

Bacillus thuringiensis (*Bt*) was first described by Berliner (1911) when he isolated a bacillus from the Mediterranean flour moth, *Anagasta kuehniella.* He named it after the province Thuringia in Germany where the infested flour was found. Although this was the first description under the name *B. thuringiensis,* it was not the first isolation. In 1901, a Japanese biologist, S. Ishiwata isolated a bacterium as the causal agent of disease of silkworms, "sotto disease" (sudden-collapse bacillus). In 1908, Iwabuchi called the bacterium *Bacillus sotto* Ishiwata, which was later ruled an invalid name (Tanada and Kaya 1993) and the more recently ascribed name, *B. thuringiensis,* has been maintained.

Originally, *Bt* was considered a risk to silkworms and it was several decades before serious research was aimed at utilising *Bt* for insect control. In 1928, a project on *Bt* for control of the cornborer (*Ostrinia nubilalis*) was initiated and the first field use was reported by Husz (1929) against this pest. The earliest commercial production began in France in 1938, under the name Sporeine (Lambert and Peferoen 1992). However, for various reasons interest declined over the subsequent decade. A resurgence of interest in *Bt* has been attributed to Edward Steinhaus, the "father of insect pathology", who obtained a culture in 1942 and attracted attention to the potential of *Bt* through his subsequent studies (Tanada and Kaya 1993).

The presence of a paraporal body in sporulating *Bt* cells was noted as early as 1915 by Berliner (van Frankenhuyzen 1993), but it was much later that some properties of the parasporal body were identified, in particular solubility of the parasporal body under alkaline but not acidic conditions (Hannay 1953). In 1955, Hannay and Fitz-James reported on the proteinaceous nature of the crystal and Angus (1954, 1956) demonstrated that the parasporal crystal was the cause of toxicity to silkworm. Research into *Bt* crystal proteins flourished with studies on crystal structure, biochemistry and general mode of action during the next 20 years (van Frankenhuyzen 1993).

Research and development of *Bt* was strengthened considerably by Dulmage and Beegle of the USDA, who accumulated the first significant collection of strains (Nakamura and Dulmage 1988). Dulmage isolated the HD-1 strain that was used as a standard for subsequent *Bt* studies (Chapter 5) and is still used in several products today.

The first commercial use of *Bt* in the USA occurred in 1958. During the 1960s, several trial formulations were developed and some success was achieved. Thuricide, now based on *Bt kurstaki,* was commercialised with *Bt thuringiensis* in 1957 and Abbott Laboratories released Dipel (*Bt kurstaki*) in 1970 (Starnes *et al.* 1993). The first Environmental Protection Agency (EPA), USA registration of *Bt* occurred in 1961 (Starnes *et al.* 1993). In the 1970s it was believed that *Bt* was active only against lepidopterans. However, the discovery of *Bt israelensis* with activity against mosquitoes and blackflies (Goldberg and Margalit 1977), followed by the isolation of *Bt tenebrionis* in 1983 with activity against some coleopterans (Krieg *et al.* 1983), showed the broad potential of *Bt* strains. Significantly, each new strain demonstrated a certain level of specificity, suggesting that *Bt* was ideal for integrated pest management (IPM) and environmentally sensitive uses.

The search for novel *Bt* strains continues and with the new understanding of the genetics of toxic protein production, many new and unusual *Bt* toxins have been discovered. In 1992, Lambert and Peferoen estimated that 40,000 *Bt* isolates were held in collections throughout the world; more recently Boucias and Pendland (1998) estimated the figure at 60,000. In addition to strains active against Lepidoptera, Diptera and Coleoptera, *Bt* strains with activity against Hymenoptera, Hemiptera, Mallophaga, Nematoda and Protozoa have been discovered (Rajamohan *et al.* 1998).

The variety of *Bt* strains and insecticidal toxins continues to grow at an amazing rate. Lambert and Peferoen (1992) mention more than 25 different but related insecticidal crystal proteins produced by *Bt*, while the latest information from the *Bt* toxin web page (Crickmore *et al.* 1999) lists 170 *cry* and *cyt* genes, mainly from *Bt* (Appendix 1). In 1998, Rajamohan *et al.* reported that there were 58 recognised flagellar serotypes while the latest publication (Lecadet *et al.* 1999) reported identification of 69 serotypes and 13 sub-antigenic groups, giving a total of 82 serovars (listed in Table 2.1).

1.4 COMMERCIAL USE OF *BT*

1.4.1 USE OF *BT* PRODUCTS

Bt is the most widely used biopesticide in the world, accounting for over 90% of all commercial sales. In 1990, sales of *Bt* products constituted $110M of a $120M worldwide biopesticide market (Powell and Jutsum 1993). A range of products is available for different pests; more than 100 products based on over 10 serotypes have been recorded in the literature (Table 4.1). *Bt kurstaki* has been used most frequently in commercial products, some of which are still based on the HD-1 strain from the Dulmage collection (Nakamura and Dulmage 1988).

Bt has been used against a number of forestry pests. As forests usually require aerial, broadcast application, *Bt* has been the pesticide of choice to replace chemicals because of non-target and health concerns (e.g. Ecobichon 1990). *Bt kurstaki* has been applied for the control of gypsy moth, (*Lymantria dispar*) and spruce budworm (*Choristoneura fumiferana*) in North America. From 1980 to 1995, Reardon and Wagner (1995) estimated approximately 1.7 million hectares were treated with *Bt kurstaki* in the eastern USA as part of gypsy moth suppression programmes. Aerial application of *Bt* from 1990-1995 resulted in total eradication of gypsy moth in the Wasatch mountains in Utah, USA (Smith and Barry 1998). In Ontario between 1985 and 1994, approximately 250,000 ha were treated with *Bt kurstaki* for gypsy moth control. *Bt* is one of the few pesticides registered for widespread aerial application in Canada (Charest 1996).

Bt kurstaki is also sold widely for crop and horticultural pest control. For example, in Australia, *Bt kurstaki*-based products have been used against a range of Lepidoptera on cotton (*Helicoverpa* spp.), cole crops (cabbage moth and white butterfly), fruit trees (light brown apple moth, pear looper), ornamentals (loopers) and other caterpillars on crops such as kiwifruit, strawberries and tobacco (Teakle and Milner 1994). In Australia in

1990, the cotton *Bt* market was 100,000 ha, representing about 5% of total cotton pesticide market, and approximately 1% of horticulture crops (10-20,000 ha).

Bt israelensis has also been sold commercially on a large scale for mosquito and blackfly control. There are a number of successful products, including Teknar and Vectobac. *Bt israelensis* has been used in extensive control campaigns in USA and Germany for mosquitoes and biting flies, and in Africa against Simuliidae vectors in the Onchocerciasis Control Programme.

In the USSR, a number of *Bt* products have been used (see Table 4.1, Chapter 4). Rimmington (1989) reviewed production of biopesticides in the former USSR and included BIP (*Bt caucasicus*) and Gomelin (*Bt thuringiensis*) against gypsy moth and related species; Bitoksibatsillin (sic) (*Bt thuringiensis*) against the Colorado potato beetle (*Leptinotarsa decemlineata*), leafrollers, silkworms and other pests; Enterobakterin (sic) (*Bt galleriae*) against more than 50 species; and Insektin (sic) (*Bt thuringiensis*) against forest pests. In the former USSR, over 2M ha were treated annually in the early 1990s, with Dendrobacillin (840,000 ha) and Bitoxibacillin (820,000 ha) accounting for most of the area treated (Khloptseva 1992).

Bt tenebrionis has been used for control of Colorado potato beetle in the USA and elsewhere (Keller and Langenbruch 1993). In Tasmania, Australia, *Bt tenebrionis* has become the control product of choice for the chrysomelid beetle, *Chrysophtharta bimaculata* on *Eucalyptus* plantations (Elek 1997).

Bt is often promoted for use in IPM systems, because the specificity of products minimises the likelihood of non-target effects on other natural enemies of pest insects. In New Zealand, for example, a prototype IPM system for kiwifruit production, called "Kiwigreen", incorporates the use of *Bt kurstaki* and mineral oil with monitoring of pest problems, into an integrated management programme for armoured scale insects (Diaspididae) and leafrollers (Tortricidae) (Steven *et al.* 1997).

1.4.2 BROADCAST APPLICATIONS OVER URBAN POPULATIONS

One of the most contentious uses of *Bt* has been the broadcast application over urban populations, usually for the eradication of invasive pests soon after discovery. Three recent examples are the eradication of Asian gypsy moth in Vancouver (1988) and in North Carolina (1993), and the eradication of *Orgyia thyellina* (white spotted tussock moth) in Auckland, New Zealand ("Operation Evergreen" 1996). In all three examples, the use of *Bt kurstaki* was promoted as a safe alternative to chemical insecticides which was also efficacious enough to achieve eradication. Public concerns over safety resulted in closer post-spray monitoring of the human populations than would be normally associated with *Bt* use.

In New Brunswick, Canada, pesticides applied for control of the forestry pest, *Choristoneura fumiferana* had buffer zones to allow for aerial spray drift for the various pesticides used. *Bt kurstaki* could be sprayed to within 155 m of human habitation, compared to 300 m for aqueous chemical controls (Ecobichon 1990). However, in the eradication programmes, human populations were directly exposed and, in the case of the Auckland programme, were subjected to multiple sprayings. Human health impacts of aerial application over large populations are reviewed in Chapter 7.

2 Characterisation

Bacillus thuringiensis (Bacillaceae) is a Gram-positive, sporeforming bacterium that is closely related to several other *Bacillus* species, including *B. cereus*. Although usually referred to in the singular as *Bt, B. thuringiensis* is currently recognised as a complex of subspecies, almost all of which produce crystalline inclusions during sporulation. These crystalline inclusions contain the insecticidal δ-endotoxins that make *Bt* such a valuable resource in insect pest management. The similarity of *Bt* to *B. cereus* has caused concern as *B. cereus* strains can produce mammalian active enterotoxins which have been implicated in gastro-enteritis. The two species share phenotypic and biochemical characteristics but, by definition, *Bt* can be discriminated by the presence of crystalline inclusions that are formed during sporulation. DNA and biochemical analyses suggest that *Bt* and *B. cereus* could constitute one species, as the differences between the species are small and mainly plasmid-based. Many of the δ-endotoxin-encoding genes are plasmid-borne, indicating that this phenotype is susceptible to loss and capable of transmission to related species, including *B. cereus*. Thus, the distinction between the species is not clear and continues to be debated by bacterial taxonomists. However, maintaining *Bt* as a separate species still has some practical value in defining this particular group of bacteria toxic to insects.

Serotyping is the most commonly used method for differentiation of groups within the species *Bt,* but serovar does not always relate directly to insecticidal activity. Currently, 82 serotypes are recognised but not all isolates can be typed by this method. In addition, there are problems with cross reactivity with some *B. cereus* isolates, again reflecting the close relationship between the two species. Alternative, more discriminatory methods have been developed to separate and identify isolates of *Bt*. Various molecular techniques have been used to detect toxin genes in individual isolates and commercial products containing *Bt kurstaki* can be discriminated from other serovars using some of these new technologies. These techniques were also capable of discriminating between *Bt* and *B. cereus* at the strain level (but not as species) and will be powerful tools in tracking the environmental fate of *Bt* strains used for pest control.

Individual *Bt* strains vary in the number and type of toxins they produce. Over 170 specific δ-endotoxin genes are known. Endotoxins bind specifically to receptor sites in the membranes of cells in the midgut of susceptible insects, causing perforations in the membranes. This leads to loss of control of ion exchange between epithelial cells and the gut lumen, resulting in rapid death at high doses. Lower dosages result in the reduction of the pH of the gut lumen allowing spore germination, rapid vegetative multiplication, penetration into the haemocoel, gross septicaemia and eventual death. The role of spores in the infection and death of insects is often minor, although in some cases spores can synergise the activity of the endotoxins without being highly toxic when used alone.

Bt strains are capable of producing a number of toxins in addition to the insecticidal δ-endotoxins. Strains of some serovars produce β-exotoxins, which have more broad spectrum activity than endotoxins and may also have some mammalian toxicity at high doses. Strains of *Bt kurstaki* have been shown to produce a haemolysin, which is identical to a haemolysin found in *B. cereus.* Haemolysins are important virulence factors in some vertebrate bacterial pathogens. In addition, some *Bt* isolates have been shown to produce small amounts of *B. cereus*-diarrhoeal-type enterotoxins, however *Bt* has not been implicated conclusively in *B. cereus* type conditions in humans.

2.1 THE *BACILLUS CEREUS* COMPLEX

B. thuringiensis (Bacillaceae) is a Gram-positive, rod-shaped, sporeforming bacterium. Bergey's Manual of Systematic Bacteriology (Sneath *et al.* 1986) listed both *B. cereus* and *Bt* as species. However, they noted that on the basis of phenotypic tests, Logan and Berkley (1984) suggested *Bt* was a subspecies of *B. cereus.* Classification based on 16S rRNA sequence data clusters *Bt* with *B. cereus*, *B. mycoides* and *B. anthracis* within the *B. subtilis* group (see Section 2.3.2). While *Bt* is considered mainly an insecticidal bacterium, the other members of the group have strains with mammalian toxicity. Some *B. cereus* strains produce toxins that cause gastrointestinal problems in humans, while *B. anthracis* causes the mammalian disease anthrax. Therefore, the distinction between these species and *Bt* is important from a mammalian safety perspective.

During sporulation, *Bt* produces both an endospore and crystalline inclusions or parasporal bodies within a sporangium. The parasporal body is often toxic to specific insect groups and many different insecticidal crystal proteins (δ-endotoxins) can be found in different *B. thuringiensis* subspecies and strains (Appendix 1). The presence or absence of crystalline inclusions is the main characteristic used to differentiate between *Bt* and *B. cereus* strains, both being widely distributed and commonly found in soil. However, the fact that the crystal encoding genes (Cry^+) are often plasmid-borne indicates that the crystal phenotype is susceptible to loss and capable of transmission to related species. It has been shown that plasmid transfer occurring *in vitro* could give rise to Cry^+ strains of *B. cereus* and *B. anthracis* (Gonzalez *et al.* 1982). Helgason *et al.* (1998) also found that several natural isolates which did not produce parasporal crystals were toxic to the lepidopteran *Trichoplusia ni*. They suggested soluble endotoxin or vegetative insecticidal proteins (VIPs) could be involved (see Chapter 2.4).

The taxonomic relationships between members of the group are not clear (Drobniewski 1994) and the cause of some concern as the differences between *B. cereus* and *Bt* are small and may be mainly plasmid-based. Flagellar antibodies are also cross-reactive between some *Bt* and *B. cereus* (see Section 2.2.2), indicating the close homology between these species. Biochemical and physiological tests do not always distinguish between the species (see below). Carlson *et al.* (1994) suggested that *B. cereus* and *Bt* should be considered as one species based on analysis of chromosomal DNA (by pulsed field gel electrophoresis) and multilocus enzyme electrophoresis (MEE). Some DNA sequencing studies of conserved gene regions have also suggested the two species may be strains of a single species. Helgason *et al.* (1998), also using MEE, examined the genetic diversity and relationships between *B. cereus* and *Bt* from natural sources. They concluded that *B. cereus* and *Bt* from soil exhibited a high degree of recombination and little clonality.

As detailed below, these studies and others have failed to consistently distinguish between *B. cereus* and *Bt*. The status of these two species requires further study using large numbers of isolates.

2.2 CHARACTERISATION BY CLASSICAL METHODS

2.2.1 PHENOTYPIC METHODS

Differential identification of the species within the *B. cereus* complex has been problematic as the species share many phenotypic and biochemical characteristics. Methods used include flagellar serotyping, description of crystal morphology, biochemical reactions and bioassays. Such methods are very laborious, making them inappropriate for screening large collections of *Bt* isolates. However, some of the methods continue to be used for characterisation and species differentiation of *Bt,* in combination with more recently developed molecular methods (e.g. Hansen *et al.* 1998).

Biochemical characterization has proved difficult, as variations in growth responses occur and biochemical characterization does not always correlate with serotyping results. Selective media have been developed for isolation of *Bt*, and some of these methods are helpful in distinguishing between *Bacillus* spp. For example, primary isolation methods have been developed to allow the isolation of *Bt* spores from environmental samples.

The acetate selection procedure (Martin and Travers 1989) was based on the observation that the spores of most *Bt* strains do not germinate in the presence of high concentrations of buffered sodium acetate and this characteristic, combined with pasteurization, could be used to separate *Bt* from other spore-formers. However, this method does not appear to provide any greater recovery than other, more simple methods. More recently, de Vasconcellos and Rabinovitch (1995) developed a medium for selective isolation of *B. cereus* from food, on which colonies of *Bt* and *B. cereus* could be distinguished on the basis of morphology and colour. Johnson and Bishop (1996) claimed efficient recovery of *Bt* isolates from environmental samples by exploiting the apparent intrinsic resistance of *Bt* to penicillin. Use of commercial identification kits (API and VITEK) based on biochemical tests often failed to differentiate between *Bacillus* spp. correctly (Ibert *et al.* 1996). A variety of methods for the recovery of *Bt* from the environment is probably the best approach to ensure the greatest genetic diversity.

Lawrence *et al.* (1991) had difficulty in distinguishing between *Bt, B. cereus* and *B. anthracis* using fatty acid gas chromatography. Similarly, in a comparative study of enzyme variation, Zahner *et al.* (1989) obtained results that grouped *Bt* and *B. cereus* isolates in a single species.

2.2.2 SEROTYPING

Classification of subspecies or varieties based on serotyping using H-serovars (flagellar serotyping) has resulted in identification of 69 serotypes and 13 sub-antigenic groups, giving 82 serovars (Lecadet *et al.* 1999). While serotyping only reflects one characteristic of the species, it is the most common classification method used throughout the world, as it is well standardised. Rapid and economical micro-methods have been developed to facilitate serotyping of *Bt* strains (Laurent *et al.* 1996). While flagellar serotyping has greatly aided isolate classification, the technique has limitations; in particular serotyping has proved unreliable as a predictor of insecticidal activity (see Chapter 5). For example, *Bt morrisoni* (H8a8b) includes isolates active against lepidoteran, coleopteran and dipteran insects.

There are additional practical problems with serotyping using H-antigens. Firstly, all strains within a serotype do not have the same toxin profile or complement. For example, the *Spodoptera*-active Cry1C toxin is most commonly found in serotypes H6 and H7. However, a strain of H1 was also found to have the gene encoding this toxin (Lecadet *et al.* 1999). Secondly, the cross-reactivity of *B. cereus* and *B. thuringiensis* antisera causes difficulties. Lecadet *et al.* (1999) reported that 92 of 194 *B. cereus* tested were agglutinated by *Bt* antisera (mostly H14, 5, 10, 6 and 27). The number of cross-reacting H-antigens among *B. cereus* strains appears to be increasing (Lecadet *et al.* 1999), poosibly due to the increase in defined serovars. In addition, approximately 3% of *Bt* isolates are auto-agglutinating and cannot be serotyped, while some *Bt* strains do not have flagella, which also precludes serotyping (e.g. *Bt wuhanensis,* Table 2.2).

Confusingly, while serotyping is the most commonly used grouping of *Bt* isolates, subspecies are also grouped according to insecticidal activity. This can lead to several "subspecies" sharing a serotype, such as *Bt thuringiensis*, *Bt gelechiae* and *Bt berliner.* Neither method appears to be a reliable indicator of similarity when compared to genetic relatedness (Lövgren *et al.* 1998).

2.3 DNA BASED METHODS

Characterisation methods based on phenotypic characters have not proved adequate for use in studies on the environmental ecology and fate of *Bt*, as these methods do not provide unambiguous identification, often failing to distinguish between closely related strains. Recent advances in molecular biology have allowed the development of specific DNA based methods capable of interspecific and intraspecific differentiation. These methods can distinguish individual strains and isolates, allowing the tracking of the environmental fate of strains used for pest control. Such methods can also be used to identify the presence/absence of specific endotoxin genes, which means it is possible to establish whether a particular strain has lost or acquired specific δ-endotoxin genes in the environment.

Table 2.1. Classification of *Bacillus thuringiensis* strains according to the H- serotype (Reproduced from Lecadet *et al.* 1999, with permission of Blackwell Science).

H-Serotype	Serovar	First mention and first valid description
1	*thuringiensis*[a]	Berliner 1915; Heimpel & Angus 1958
2	*finitimus*	Heimpel & Angus 1958
3a3c	*alesti*	Toumanoff & Vago 1951; Heimpel & Angus 1958
3a3b3c	*kurstaki*	de Barjac & Lemille 1970
3a3d	*sumiyoshiensis*	Ohba & Aizawa 1989b
3a3d3e	*fukuokaensis*	Ohba & Aizawa 1989b
4a4b	*sotto(dendrolimus)*	Ishawata 1901; Heimpel & Angus 1958
4a4c	*kenyae*	Bonnefoi & de Barjac 1963
5a5b	*galleriae*	Shvetsova 1959[d]; de Barjac & Bonnefoi 1962
5a5c	*canadensis*	de Barjac & Bonnefoi 1972
6	*entomocidus (subtoxicus)*	Heimpel & Angus 1958
7	*aizawai*	Bonnefoi & de Barjac 1963
8a8b	*morrisoni (tenebrionsis*[b]*)*	Bonnefoi & de Barjac 1963
8a8c	*ostriniae*	Ren *et al.* 1975
8b8d	*nigeriensis*	Weiser & Prasertphon 1984
9	*tolworthi*	Norris 1964; de Barjac & Bonnefoi 1968
10a10b	*darmstadiensis*[c]	Krieg *et al.* 1968
10a10c	*londrina*	Arantes unpubl.
11a11b	*toumanoffi*	Krieg 1969
11a11c	*kyushuensis*	Ohba & Aizawa 1979b
12	*thompsoni*	de Barjac & Thompson 1970
13	*pakistani*	de Barjac *et al.* 1977
14	*israelensis*	de Barjac 1978
15	*dakota*	DeLucca *et al.* 1979
16	*indiana*	DeLucca *et al.* 1979
17	*tohokuensis*	Ohba *et al.* 1981a
18a 18b	*kumanotoensis*	Ohba *et al.* 1981b
18a 18c	*yosoo*	Lee *et al.* 1995a
19	*tochigiensis*	Ohba *et al.* 1981b
20a20b	*yunnanensis*	Wan-yu *et al.* 1979[d]
20a20c	*pondicheriensis*	Rajagopalan *et al.* unpubl.
21	*colmeri*	DeLucca *et al.* 1984
22	*shandongiensis*	Wang Ying *et al.* 1986[d]
23	*japonensis*	Ohba & Aizawa 1986a
24a 24b	*neoleonensis*	Rodriguez-Padilla *et al.* 1990
24a 24c	*novosibirsk*	Burtseva *et al.* 1995
25	*coreanensis*	Lee *et al.* 1994
26	*silo*	de Barjac & Lecadet unpubl.
27	*mexicanensis*	Rodriguez-Padilla & Galan-Wong unpubl.
28a 28b	*monterrey*	Rodriguez-Padilla *et al.* unpubl.
28a 28c	*jegathesan*	Seleena *et al.* 1995
29	*amagiensis*	Ohba unpubl.
30	*medellin*	Orduz *et al.* 1992
31	*toguchini*	Hodirev unpubl.
32	*cameroun*	Jacquemard 1990[d]; Juarez-Perez *et al.* 1994
33	*leesis*	Lee *et al.* 1994
34	*konkukian*	Lee *et al.* 1994
35	*seoulensis*	Lee *et al.* 1995a
36	*malaysiensis*	Ho unpubl.
37	*andaluciensis*	Aldebis *et al.* 1996
38	*oswaldocruzi*	Rabinovitch *et al.* 1995
39	*brasiliensis*	Rabinovitch *et al.* 1995
40	*huazhongensis*	Dia Jingyuan *et al.* 1996

Table 2.1. *(Continued)*

H-Serotype	Serovar	First mention and first valid description
41	*sooncheon*	Lee *et al.* 1995a
42	*jinghongiensis*	Li Rong Sen *et al.* in press
43	*guiyaniensis*	Li Rong Sen *et al.* in press
44	*higo*	Ohba *et al.* 1995
45	*roskildiensis*	Hinrinschen *et al.* unpubl.
46	*chanpaisis*	Chanpaisaeng unpubl.
47	*wratislaviensis*	Lonc *et al.* 1997
48	*balearica*	Caballero *et al.* unpubl.
49	*muju*	Seung Hwan Park *et al.* unpubl.
50	*navarrensis*	Caballero *et al.* unpubl.
51	*xiaguangiensis*	Jian Pin Yan unpubl.
52	*kim*	Kim *et al.* unpubl.
53	*asturiensis*	Aldebis *et al.* 1996
54	*poloniensis*	Damgaard *et al.* unpubl.
55	*palmanyolensis*	Santiago-Alvarez unpubl.
56	*rongseni*	Li Rong Sen in press
57	*pirenaica*	Caballero *et al.* unpubl.
58	*argentinensis*	Campos-Dias *et al.* unpubl.
59	*iberica*	Caballero *et al.* unpubl.
60	*pingluonsis*	Li Rong Sen in press
61	*sylvestriensis*	Damgaard unpubl.
62	*zhaodongensis*	Li Rong Sen in press
63	*bolivia*	Ferre-Manzanero *et al.* unpubl.
64	*azorensis*	Santiago-Alvarez *et al.* unpubl.
65	*pulsiensis*	Khalique & Khalique unpubl.
66	*graciosensis*	Santiago-Alvarez *et al.* unpubl.
67	*vazensis*	Santiago-Alvarez *et al.* unpubl.
68	*thailandensis*	Chanpaisaeng *et al.* unpubl.
69	*pahangi*	Seleena and Lee unpubl.

[a] includes *Bt gelechiae*, *Bt insectus* and *Bt berliner*.
[b] *tenebrionis* is identical to *Bt san diego* (Krieg *et al.* 1987).
[c] *caucasicus* was also described as H10a according to Afrikyan & Chil Akopyan 1980. This variety was first isolated in Armenia in 1961 from dead larvae of *Bombyx mori*.
[d] in Lecadet *et al.* 1999, full details not given.

2.3.1 PCR TECHNIQUES

Polymerase chain reaction (PCR) is a powerful technique that allows the multiplication of target sequences of DNA many million times. This technique has many uses and can be used to amplify known regions of DNA, or to genetically compare isolates about which little is known. PCR uses primers, short sequences of DNA, which are at either end of the fragment to be amplified. Many *Bt* toxin genes have been sequenced and specific primed PCR techniques have subsequently been used for detection and characterisation of toxin genes in individual *Bt* isolates. Carozzi *et al.* (1991) performed PCR analysis to identify different δ-endotoxins and reported the sequences of 12 primers that distinguished between three major classes of toxin genes. Restriction fragment length polymorphism (RFLP) was used in combination with PCR for identification of new variants of toxin genes (Kuo and Chak 1996). Primers with homology to variable regions have been used to identify subgroups within the *cryI* gene family (Ceron *et al.* 1994).

Arbitrary primer polymerase chain reaction (AP-PCR) is a random priming method that uses small, non-specific primers to amplify DNA fragments. The sizes of the fragments are compared and the higher the number of bands of the same size, the more homologous are the strains. The method could be used to distinguish commercial products containing *Bt kurstaki*. AP-PCR techniques have been applied to the identification of commercial strains of *Bt kurstaki* by using total DNAs extracted from single bacterial colonies as templates

(Brousseau *et al.* 1993). When a single primer was used, this method was capable of producing discriminating DNA fingerprints for 33 known serovars. Differentiation between *Bt* and *B. cereus* was also possible by this method. This technique should be a powerful tool for identification and discrimination of individual *Bt* strains, but is too discriminatory to be useful in separating *Bt* and *B. cereus* as species.

2.3.2 RIBOSOMAL RNA

Priest *et al.* (1994) examined ribosomal RNA gene restriction patterns in *Bt* strains and closely related *Bacillus* spp. Strains within a *Bt* serotype produced highly related or identical ribotype patterns. Te Giffel *et al.* (1997) developed DNA probes based on the V1 region of the 16S rRNA for the identification of *B. cereus* and *Bt*. These probes were highly specific: the *B. cereus* probe did not give a signal with the *Bt* strains that differed in only three nucleotides from *B. cereus* and only *Bt* strains were detected by the *Bt* probe. *Bt* identified strains were all also positive for primers which targeted *cry* genes. However no international standard strains have been compared on the V1 region.

Daffonchio *et al.* (1998) used the internal transcribed spacers (ITS) between the 16S and the 23S ribosomal RNA genes as molecular markers to attempt to discriminate species of *Bacillus*. Using five *Bt* isolates only, they found the ITS fingerprints failed to discriminate between *B. cereus, B. thuringiensis* and *B. mycoides*. By examining the shortest ITS in more detail, *B. mycoides* was differentiated from *B. cereus/B. thuringiensis* by restriction analysis but *B. cereus* and *Bt* could not be separated.

In practice, a combination of techniques will need to be used. For example, Hansen *et al.* (1998) found that a molecular method like colony hybridisation was suitable for screening large collections of bacteria. When colony hybridisation data were combined with RAPD analyses (random amplified polymorphic DNA), isolates could be further grouped based on genetic potential and DNA fingerprint, whereby further characterisations by PCR and more laborious phenotypic methods could be performed more efficiently.

2.3.3 COMPARISON OF CHROMOSOMES

A recent approach to comparing *Bt* isolates, subspecies and *B. cereus* is chromosome mapping, where the location of known genes is compared for different isolates. A recent paper by Lövgren *et al.* (1998) compared the location of virulence genes for strains of *Bt gelechiae, Bt thuringiensis* and *Bt berliner,* which are all serotype 1. They found that *Bt berliner* and *Bt gelechiae* were similar in gene location, but there were more differences between *Bt thuringiensis* and *Bt gelechiae*. Quoting unpublished results, Lövgren *et al.* (1998) have also found that *Bt alesti* and *Bt kurstaki* have indistinguishable chromosomes.

Comparison of physical maps of chromosomes has also been used to compare *B. cereus* and *Bt* using pulse field gel electrophoresis (PGFE) and restriction enzyme digestions. Carlson and Kolsto (1993) constructed a physical map of a *Bt* chromosome using the restriction enzyme *Not*1. From examination of *Bt* and *B. cereus* strains, they concluded the two species were very closely related. Carlson *et al.* (1996) showed that the physical map of the chromosome of the type strain of *B. cereus* was almost identical to *Bt canadensis*. Other studies (Carlson *et al.* 1994; Helgason *et al.* 1998) have also suggested that, based on PFGE and MEE, the two species overlap. Carlson *et al.* (1994) used a combination of restriction analysis by rare cutting enzymes and PFGE to compare the genomes of a number of *B. cereus* and *Bt* strains. A group of 24 *B. cereus* and 12 *Bt* strains could be identified, but the techniques did not separate the species. Carlson *et al.* (1994) concluded that the results strongly suggest *B. cereus* and *Bt* are one species.

2.4 TOXINS

Bt strains are known to produce a number of different toxins, some of which pose health risks. Therefore, consideration of the safety and non-target impacts of the various *Bt* varieties requires an understanding of the many and varied toxins produced by strains within the species. Toxins produced by *Bt* include: δ-endotoxins that exhibit insecticidal activity; exotoxins that may have mammalian toxicity at high doses; haemolysins; and enterotoxins (diarrhoeal type). Enterotoxins in *B. cereus* are responsible for food poisoning and *Bt* can produce similar toxins. The impacts of the various toxins on non-target organisms, particularly mammals, are discussed in Chapter 7.

2.4.1 EXOENZYMES AND EXOTOXINS

2.4.1.1 Beta-exotoxin

The β-exotoxin (also called thuringiensin, thermostable toxin, fly toxin) has been detected in strains from several subspecies (Table 2.2). This low molecular weight toxin is thermostable (70°C, 15 min) and has a broad spectrum of activity, killing various lepidopterans, dipterans, hymenopterans, hemipterans, isopterans, orthopterans, nematodes and mites. Beta-exotoxin preferentially inhibits biosynthesis of RNA (Beebee and Bond 1973; Sebesta *et al.* 1981). The β-exotoxin is more toxic when administered parenterally than when ingested, as it does not pass through the gut wall and is degraded by gut phosphatases (Vankova *et al.* 1974; Sebesta *et al.* 1981). Type I β-exotoxin has been detected in strains from *Bt* serotypes 1, 9 and 10, while a second β-exotoxin, type II, has been isolated from *Bt morrisoni* serotype 8ab by Levinson *et al.* (1990). They found that β-exotoxin was encoded by a δ-endotoxin *cry*-encoding plasmid in the five *Bt* strains examined. In at least one case, the plasmid was readily transmissible by the natural conjugative transfer system present in *Bt*. The authors were successful in transmitting the 75 MDa Cry+ Exo+ plasmid from isolate HD-2 to a Cry- *Bt kurstaki* background.

While exotoxin is deactivated in the digestive tract of insects, most insects seem susceptible if a sufficiently high dose is used (Tanada and Kaya 1993). Insects challenged with β-exotoxin usually stop feeding and die within several days of exposure. Beta-exotoxin can cause teratogenic effects and disrupt larval or pupal moulting. In mites, exotoxin has been reported as having a gonadotropic and morphogenetic action similar to that of juvenoids (Petrova 1987). Adult insects that survive β-exotoxin treatment may exhibit reduced longevity and fecundity. *Bt tolworthi* exotoxin caused sublethal effects on blowflies (*Lucilia cuprina*) which included interference in larval moulting (Cooper *et al.* 1990) while in *Spodoptera littoralis,* exotoxin caused a decrease in the larval digestive enzymes (El-Ghar *et al.* 1995). Of significance with respect to safety of non target organisms is the fact that β-exotoxin can synergise the activity of δ-endotoxin against normally resistant insects. This synergy may result from the inhibitory effect of β-exotoxin on the regeneration of damaged midgut cells.

Exotoxin is able to cross cell and nuclear membranes in mammalian systems and exposure of vertebrates to β-exotoxin can result in formation of lesions in liver, kidney and adrenal glands. Beta-exotoxin treated chickens lost vigour and produced undersized eggs (Boucias and Pendland 1998). Meretoja *et al.* (1977) examined the mutagenicity of *Bt* exotoxin in different mammalian systems. In human blood cultures, very high toxic concentrations of exotoxin induced a significant increase in chromosomal aberrations.

Because of vertebrate toxicity, most commercial preparations of *Bt* are composed of subspecies or isolates that do not produce β-exotoxin. As a condition for registration for pesticide use on food in the USA, *Bt* active ingredients must be tested to show the absence of β-exotoxin (McClintock *et al.* 1995). Screening methods for the presence of exotoxin were originally based on time consuming and laborious bioassays against flies (Ignoffo and Gard 1970). Several laboratories are now using high performance liquid chromotography (HPLC) methods to detect and quantify β-exotoxin in fermentation beers and formulation samples. Campbell *et al.* (1987) developed a rapid specific HPLC assay for β-exotoxin, which correlated well with fly bioassay results.

Exotoxin production was initially thought to be limited to certain flagellar serotypes, even though conjugation can occur between serotypes. Historically, serotype had been used implicitly to predict the presence or absence of exotoxin. For example, there were no reports of exotoxin production in *Bt kurstaki* until Ohba *et al.* (1981c) reported the isolation of a single Exo+ *Bt kurstaki* strain. However, Levinson *et al.* (1990) tested the type strains of 10 serotypes for exotoxin production and found that serotypes 1, 9, and 10 were type I exotoxin producers while serotype 8ab produced type II exotoxin. Exotoxin producing strains are also found in several other serotypes (Table 2.2). Thus, although exotoxin-producing strains are still unknown for many serotypes, exotoxin production is fairly broadly distributed among the serotypes. Serotyping alone is therefore not sufficient to predict whether a strain may be an exotoxin producer.

A new variant of β-exotoxin has been described from *Bt israelensis* by Weiser *et al.* (1992) and Horak *et al.* (1996). They found activity of water-soluble metabolites of *Bt israelensis* toxic to aquatic molluscs and Trematoda. An active toxin belonged to the same class as β-exotoxin but did not show toxicity in animal tests (details of species tested were not given). The toxin was termed "M-exotoxin" for mollusc-active exotoxin.

2.4.1.2 Haemolysins

Haemolysins, which lyse vertebrate erythrocytes, are important virulence factors in several vertebrate bacterial pathogens and are generally thought to be important factors for the establishment of systemic diseases in humans. A haemolysin purified from *Bt kurstaki* has been found to be identical to a haemolysin found in *B. cereus* (Honda *et al.* 1991; Matsuyama *et al.* 1995).

2.4.1.3 Enterotoxins

Bt isolates have been found to produce *B. cereus*-diarrhoeal-type enterotoxins (Carlson *et al.* 1994; Abdel-Hameed and Landen 1994). *B. cereus* enterotoxins are responsible for symptoms of food poisoning following ingestion of *B. cereus*. Abdel-Hameed and Landen (1994) found 23 out of 40 *Bt* isolates collected from soil in Sweden (typed as *Bt thuringiensis, kurstaki, galleriae, dendrolimus, darmstadiensis* and *israelensis*) produced *B. cereus*-diarrhoeal-type enterotoxin. Enterotoxins were assayed using a *B. cereus*-diarrhoeal-type enterotoxin reversed passive latex agglutination kit (BCET-RPLA) produced by Oxoid. Damgaard *et al.* (1996a) isolated enterotoxin producing strains of *Bt* from various foods. Strains were examined for cytotoxic effects on Vero cells, as an indicator of enterotoxin activity. Te Giffel *et al.* (1997) also reported that two enterotoxin producing strains that had previously been identified as *B. cereus* and implicated in incidents of food poisoning, typed as *Bt* using specific DNA probes to 16S rRNA. The implications for *Bt* as a potential agent of gastroenteritis are discussed further in Chapter 7.

2.4.1.4 Exoenzymes

Bt produces a number of exoenzymes, which play a role in its pathogenicity to insects. For example, the release of chitinase and protease by *Bt* has been proposed to disrupt the peritrophic membrane, providing access to the gut epithelium (e.g. Smirnoff and Valero 1977; Kumar and Venkateswerlu 1998; Reddy *et al.* 1998; Sampson and Gooday 1998). Two phospholipases have been implicated in virulence of *Bt gelechiae* isolate Bt 13 that lacks crystals but is still fully virulent when injected into insect larvae.

2.4.1.5 Vegetative Insecticidal Proteins

Recently, a new class of insecticidal toxins, vegetative insecticidal proteins (VIPs) have been isolated from *Bt*. Vip3A, a novel protein with a wide spectrum of activities against lepidopteran insects was first reported by Estruch *et al.* (1996). Activity was shown against black cutworm (*Agrotis ipsilon*), fall armyworm (*Spodoptera frugiperda*), beet armyworm (*Spodoptera exigua*), tobacco budworm (*Heliothis virescens*) and corn earworm (*Helicoverpa zea*). Approximately 15% of *Bt* strains examined had Vip3 homologs. The predicted 791 amino acid proteins contained no homology to known proteins and were expressed in the vegetative growth stage (Estruch *et al.* 1996). Susceptibility of tested insects varied. The midgut epithelium cells of susceptible insects are the primary target for the Vip3A insecticidal protein and their subsequent lysis appears to be the primary mechanism of lethality (Yu *et al.* 1997b). The importance of VIPs in toxicity has not yet been fully determined.

Table 2.2. The various toxins produced by strains of *Bacillus thuringiensis* (modified from Krieg and Langenbruch 1981).

Serovar	H-serotype	β-exotoxin	*B. cereus* type enterotoxin[l]	Haemolysin
thuringiensis	1	+ (or 0)		+
finitimus	2	0[b]		
alesti	3a	0		
kurstaki	3a, 3b	0 (or +[c,d,e])		+
sotto	4a, 4b	0	+ (or +[f])	
dendrolimus	4a, 4b	0 (or +[f])		+
kenyae	4a, 4c	+ (or 0)		
galleriae	5a, 5b	0 (or +[f,g,h])		+
canadensis	5a, 5c	0		
entomocidus	6	0		
aizawai	7	0 (or +[e])		
morrisoni	8a, 8b	+[e] (or 0)		
ostriniae	8a, 8c	0		
tolworthi	9	+ (or 0)		
darmstadiensis	10	+ (or 0)		+
toumanoffi	11a, 11b	0 (or +[i])		
kyushuensis	11a, 11c	0		
thompsoni	12	0 (or +[i])		
pakistani	13	0		
israelensis	14	0[j] (or +[d, k])	+	+
indiana	16	+[h]		
kumanotoensis	18a 18b	+ (or 0)		
tochigiensis	19	0		
japonensis (buibui)	23	0		
medellin	30	0		
wuhanensis	-[a]	+		

[a] non-flagellated
[b] parasporal crystals are non-toxic
[c] Ohba *et al.* 1981c
[d] Maciejewska *et al.* 1988
[e] Levinson *et al.* 1990
[f] Sebesta *et al.* 1981
[g] Wang *et al.* 1990
[h] Abdel-Hameed and Landen 1994
[i] trace, Sebesta *et al.* 1981
[j] Weiser *et al.* 1992 and Horak *et al.* 1996 describe an M-exotoxin similar to β-exotoxin produced by *Bt israelensis*
[k] Rady *et al.* 1991
[l] see also section 7.3.4.2

2.4.2 DELTA-ENDOTOXINS

At sporulation, the majority of *Bt* strains produce crystalline inclusions that contain the insecticidal δ-endotoxins. The proteins comprising these crystals account for 20-30% of the total bacterial protein at sporulation (Boucias and Pendland 1998). The composite crystalline inclusions are comprised of monomeric protoxins. Degradation of the inclusions by proteolytic enzymes releases the smaller toxic proteins (δ-endotoxins). These vary between strains but in most cases *Bt* strains produce inclusions that contain a mixture of δ-endotoxins. For example, *Bt kurstaki* HD-1 contains three Cry1 (130 kDa) and two Cry2 (70 kDa) endotoxins in the same crystal. Other *Bt* strains can contain as few as a single δ-endotoxin, as is the case for *Bt tenebrionis.* These endotoxins, which differ in their activity towards insects, form inclusions. More than 170 different genes encoding δ-endotoxins have been identified (see Appendix 1). The activity of the δ-endotoxins is confined to the insects' digestive tracts (see Section 2.5 for mode of action).

Many *Bt* toxin genes, including some exotoxins, are encoded by extra-chromosomal DNA, on plasmids of varying size and number per cell (Levinson *et al.* 1990; Lereclus *et al.* 1993). The toxin genes are normally located on large plasmids (40-150 MDa) and the presence of more than one toxin gene on a single replicon has been demonstrated, such as *Bt israelensis* with four toxin genes on the same 72 MDa plasmid (Lereclus *et al.* 1993). Not all *Bt* endotoxin genes have been located on plasmids; some are chromosomally located, although the existence of mega-plasmids cannot be ruled out (Lereclus *et al.* 1993). Carlson and Kolsto (1993) found that the *cry1A*-type toxin gene was present on the chromosome in four strains, was extrachromosomal in four strains, and was both chromosomal and extrachromosomal in two strains of *Bt* examined.

In addition to producing Cry toxins, several *Bt* strains also produce other, smaller (25-28 kDa) cytolytic endotoxins. These include the Cyt1A toxins from *Bt israelensis* and *Bt morrisoni* PG14 and the Cyt2A from *Bt kyushuensis*. Unlike the Cry δ-endotoxins, the Cyt endotoxins display broad unspecific activity *in vitro* and *in vivo*. The Cyt endotoxins are deposited together with the Cry δ-endotoxins in the crystalline inclusion bodies. The Cyt1A endotoxin from *Bt israelensis* was the first Cyt endotoxin to be characterised. The Cyt1A endotoxin (27.3 kDa) comprises 40% of the total crystal protein and is the most abundant protein in the *Bt israelensis* inclusion. The genes coding for the Cyt1A protein are carried on a large 125 kb plasmid, which also contains the genes coding for the large Cry4A and B, Cry10A and Cry11A (previously CryIVA-D). Crystals solubilised under alkaline conditions release protoxins that can be activated by proteolytic digestion. Solubilised Cyt1A toxins have been found to be cytolytic to both erythrocytes and insect cell lines, while unsolubilised Cyt1A toxins showed no cytolytic activity.

The crystals can have different morphologies as determined by light or electron microscopy. Several crystal morphologies have been described; bipyramidal crystals are most often associated with activity against Lepidoptera, while ovoidal and cuboidal crystals are active against both Lepidoptera and Diptera (Tanada and Kaya 1993). Bernhard *et al.* (1997) surveyed the distribution of the different morphologies in 2793 isolates and occurrence in different ecological types of source material. Historically, the specific δ-endotoxins (Cry proteins) have been identified with anti-Cry antibodies but more recently molecular methods have been used.

The first crystal protein gene sequences were published in 1985 (van Frankenhuyzen 1993). With the improvements in both DNA and *Bt* isolation techniques, a large number of unique toxin genes have been identified and sequenced (for a full list of toxin genes, see Appendix 1). Under the old nomenclature (see Appendix 1), the common classification of the toxins specified that CryI toxin activity was specific to Lepidoptera, CryII to Lepidoptera and Diptera, CryIII to Coleoptera and CryIV to Diptera (Crickmore *et al.* 1998), although there were exceptions within those groups. Because of confusion, several groups used the CryV designation. Some CryV were originally described by Tailor *et al.* (1992) as active against Coleoptera and Lepidoptera and the toxin is now classified as Cry1Ia, and another CryV as Cry1Ib (Shin *et al.* 1995). Two CryVAs (now Cry5As) active against nematodes were also described (K.E. Narva *et al.* Eur. Patent 0462721A2 1991; A.J. Sick *et al.* US Patent 5281530 1994). As shown in Appendix 1, Crickmore *et al.* (1998) revised the nomenclature of toxin encoding genes to reflect the phylogeny of the encoding sequences, grouping toxin genes in superfamilies such as *cry1*. While this more logical system of classification is well overdue, it does pose problems in aligning previously published names with current names. The new system replaces Roman numerals with Arabic and where possible the new system will be adhered to in this book.

2.5 MODE OF ACTION

2.5.1 DELTA-ENDOTOXINS

Consumption of food treated with endotoxins generally results in cessation of feeding of lepidopteran larvae and paralysis of the gut that retards the passage of ingested plant material and allows spores to geminate and undergo vegetative growth. Larvae eating high doses of toxin suffer a general paralysis followed by death. Histological studies have shown that toxins released from inclusions and proteolytically activated in the gut lumen bind to specific receptor sites on the membranes of midgut columnar cells, creating pores which interfere with ion transport systems across the midgut wall. At high doses of toxin, this causes lysis of the midgut epithelium, resulting in rapid death. At lower doses or in less susceptible insects, damage to the gut cells is just sufficient to stop normal gut secretion which lowers the lumen pH and allows the spores to germinate. Vegetative cells then penetrate and multiply in the rich environment of the haemolymph, eventually causing septicaemia and death. *Bt israelensis*-treated mosquito larvae generally cease feeding within 1 hour after treatment, show reduced activity by two hours, extreme sluggishness by four hours and general paralysis by six hours after ingestion (Chilcott *et al.* 1990). Most other *Bt* isolates do not have as rapid an effect as *Bt israelensis*. Feeding inhibition can occur soon after ingesting *Bt* spores and toxins, however in beetles and caterpillars, death is rapid (1-2 days) for neonate larvae only and can take longer than one week for larger larvae. Insects challenged with sublethal concentrations of *Bt* can recover within one to two days of exposure. A number of sublethal effects have been reported (Section 2.5.3).

The specificity of the receptors on the midgut epithelium is important in determining insecticidal activity. Studies using immunochemical probes have shown that both the Cry and Cyt toxins bind to the brush border of the midgut cells (Eschriche *et al.* 1995). *In vitro* binding assays have demonstrated the presence of specific high affinity binding sites on the brush border membrane vesicles (BBMV). In several cases, high affinity binding to these sites has been correlated to the susceptibility of the insect host (Hoffman *et al.* 1988; van Rie *et al.* 1990a).

2.5.2 CONTRIBUTION OF SPORES TO TOXICITY

The role of spores in the toxicity and infectivity of *Bt* has long been debated. Navon (1993) states that in most lepidopterans, the crystal protein is the major cause of mortality and the spores play a minor role. The isolation and use of specific Cry toxins supports this statement. While this may be true of many associations of insect species and *Bt* subspecies or strains, there have also been many reports where spores have been shown to synergise toxin activity or cause mortality directly (see also Chapter 5.5.7). Activity of purified toxins of *Bt kurstaki* was enhanced (146x) by addition of toxin-free spores against larvae of *Plutella xylostella* (Miyasono *et al.* 1994). Similarly, Johnson and McGaughey (1996) reported that addition of spores synergised activity of δ-endotoxins against larvae of the Indian meal moth *Plodia interpunctella*.

In much research on this subject, physical methods were used to extract spores from preparations of "pure crystals" and *vice versa*. The 1% or less of impurities left behind in these so-called "pure" preparations often significantly affected the results. Li *et al.* (1987) used genuinely pure crystal preparations, obtained by killing spores with gamma irradiation, and with spores containing no endotoxin in their coats (obtained from a crystal free mutant). They reported that pure spores or crystals *Bt aizawai* were virtually inactive in wax moth larvae (*Galleria mellonella*; LC_{50} of 10^{10} crystals /g insect food). Addition of 0.001% spores to crystals resulted in 36% mortality in larvae, 0.01% spores killed 64% of larvae, while the LC_{50} of a 1:1 mixture was 4.7×10^6 spores and crystals /g food.

In gypsy moth larvae, Dubois and Dean (1995) found that spores or vegetative cells of a number of bacteria significantly enhanced the toxicity of Cry1Aa and Cry1Ac insecticidal crystal proteins, including spores from *B. cereus, B. megaterium* and *B. subtilis*, although spores and cells were not toxic alone. Studies on *Bt kurstaki* against gypsy moth found that addition of a few spores to low toxin concentrations significantly increased mortality as a result of a lethal septicaemia, although spores alone were not pathogenic (Reardon and Wagner 1995).

The synergistic effect of spores on toxicity of *Bt* may be caused partly by additional toxin in the spore coat but is more likely to result from the outgrowth of vegetative cells from ingested spores. Spores can germinate as the pH in the insect falls, penetrating the gut wall to cause septicaemia.

2.5.3 SUBLETHAL EFFECTS

Impacts on non-target organisms are usually considered in terms of mortality. However, sublethal effects that are much more difficult to detect almost certainly occur. There are few reports of sublethal effects on non-target organisms; most sublethal effects are reported in studies on pest species.

Sublethal effects of *Bt*, mostly reported for *Bt kurstaki* against lepidopteran pests, include delayed development (Abdallah *et al.* 1970; Salama *et al.* 1981b; Ali and Watson 1982b; Biswas *et al.* 1996b; Pedersen and van Frankenhuyzen 1996; Nyouki *et al.* 1996; Tollo and Chougourou 1997), reductions in larval and pupal weight and size (Ali and Watson 1982b; Ramachandran *et al.* 1993b; Pedersen and van Frankenhuyzen 1996; Biswas *et al.* 1996b; Nyouki *et al.* 1996), reduction in pupation or adult emergence (Soliman *et al.* 1970), occasional loss or reduction of adult fecundity (Abdallah *et al.* 1970; Soliman *et al.* 1970; Salama *et al.* 1981b; Yang *et al.* 1985) and reduced egg viability.

Sublethal doses of *Bt galleriae* ingested by larvae of *Agrotis ipsilon* retarded larval and pupal development, reduced adult egg production, percentage adult emergence, egg fertility, adult lifespan and pupal weight, and caused deformities in the pupae and adults (Salama and Sharaby 1988). The effects could even be carried through to the next generation in terms of increasing generation time.

Reductions in egg production and fertility have been reported in several species following exposure of insects to sublethal doses of *Bt*. With *Choristoneura fumiferana* dosed with *Bt kurstaki*, a higher proportion of infertile eggs were produced (Pedersen and van Frankenhuyzen 1996; Pedersen *et al.* 1997). Salama *et al.* (1981) found that for three noctuid pests treated sublethally with *Bt entomocidus,* larval development was retarded and reductions in egg production of the moths and fertility of the eggs were observed.

The feeding behaviour of some insects has been affected by exposure to sublethal amounts of *Bt* (Yang *et al.* 1985; Tollo and Chougourou 1997). Larvae of *Spodoptera littoralis* dosed with sub-lethal amounts of Biotrol BTB-25W (*Bt kurstaki*) initially ate less than untreated larvae, although they eventually consumed the same amount (Abdallah and Abul Nasr 1970). Retnakaran *et al.* (1983) demonstrated feeding inhibition of larvae of *C. fumiferana* by *Bt kurstaki* ingestion, although complete recovery from feeding inhibition occurred in < 8h. Similarly, the feeding of larvae of *Lobesia botrana* treated with Thuricide (*Bt kurstaki*) diminished for a short time before increasing (El-Hakim *et al.* 1988).

3 Natural Occurrence and Role in the Environment

Bacillus thuringiensis occurs naturally and ubiquitously in the environment. It is a common component of the soil microflora and has been isolated from most terrestrial habitats. *Bt* has been isolated from a number of insects, but has only rarely been reported to cause epizootics outside laboratory-maintained insect cultures. The role of *Bt* in nature is not clear and several hypotheses exist. *Bt* could be a natural insect pathogen, although it rarely recycles in insect populations, making it a relatively transitory pathogen. Also, many *Bt* isolates are not toxic to any known component of the insect fauna from the site of isolation. It could be a natural soil bacterium with unassociated insecticidal properties, however *Bt* grows and competes relatively poorly in soil. One the theories on the natural role of *Bt,* is that *Bt* has evolved a symbiotic or mutualistic association with plants to provide protection against herbivores.

3.1 NATURAL OCCURRENCE IN THE ENVIRONMENT

Historically, *Bt* has been isolated from environments associated with insect populations and/or plant material. For example, *Bt* was initially discovered in silkworm farms and numerous isolations have been made from various insectaries, stored product environments and grain processing facilities (e.g. Ohba 1996; Bernhard *et al.* 1997; Kim *et al.* 1998a). More recently, numerous surveys have been conducted in which *Bt* has been isolated from a range of habitats in many different countries (e.g. Pakistan, Khan *et al.* 1995; China, Dai *et al.* 1994; Yao *et al.* 1998; Sweden, Landen *et al.* 1994; Chile, Vasquez *et al.* 1995; Korea, Kim *et al.* 1998a & b; Spain, Iriarte *et al* 1998; tropics, Zelazny and Welling 1994). Bernhard *et al.* (1997) conducted one of the more extensive surveys in which they isolated *Bt* from natural samples collected from eighty countries. The majority, 45% of the 5303 isolates, originated from stored products, with 25% originating from soil. The materials richest in isolates active against insects were mushroom compost and stored products. Activity against representative Lepidoptera, Diptera and Coleoptera did not appear to be correlated with origin, indicating a relatively ubiquitous distribution of insecticidal activities and of *Bt* in general. A high proportion of isolates was inactive against all test insects, as has been found in other surveys (e.g. Martin and Travers 1989). The results of the numerous surveys indicate that *Bt*, possessing minimal growth requirements, is a ubiquitous soil microbe as well as a common inhabitant of the phylloplane. Whether *Bt* actually grows in soil is debatable. *Bt* is capable of vegetative growth in sterile soil and possibly in the presence of nutrients, but most studies have not demonstrated growth and multiplication in non-sterile soil (see Chapter 8), although A. Meikle (pers. comm.) found seasonal increases in the spring and early summer in a non-acid field soil.

3.1.1 PRESENCE IN SOIL

While originally recovered mainly from insects, recent surveys indicate that *Bt* is distributed in soil sparsely but frequently and its distribution is widespread, both locally and worldwide. Martin and Travers (1989) reached this conclusion globally, as did a number of authors of regional studies, for example, Landen *et al.* (1994) in Sweden, Hossain *et al.* (1997) in Bangladesh and Iriarte *et al.* (1998) in Spain. Using selective media methods, Martin and Travers (1989) analysed soil samples collected from around the world. An analysis of 27,000 isolates collected from 1,100 soil samples demonstrated that *Bt* could be collected almost everywhere, including beach, desert and tundra habitats. The incidence of *Bt* was not associated with insects; certain environments with no insects contained high levels of this bacterium. Similarly, DeLucca *et al.* (1981) recovered *Bt* from numerous field soils remote from any large-scale aggregations of lepidopterous insects in insect rearing facilities or grain storage areas in the USA. In New Zealand, Chilcott and Wigley (1993) found that between 60-100% of soils sampled contained *Bt*, depending on source (urban, horticultural etc.).

3.1.2 PRESENCE IN THE PHYLLOPLANE

Fewer studies have focussed on the natural occurrence of *Bt* in the phylloplane. Smith and Couche (1991) surveyed the foliage of various conifers and deciduous trees and found that *Bt* isolates accounted for 30-100% of the spore-formers detected on leaf surfaces. Ohba (1996) isolated *Bt* from mulberry leaves, while Damgaard *et al.* (1997a) isolated *Bt* from cabbage foliage. They all found that a range of isolates exhibited insecticidal activity against insects from the orders Lepidoptera, Diptera and Coleoptera and proposed that *Bt* should be considered part of the common microflora of many plants. More recently Damgaard *et al.* (1998) showed that *Bt* isolates were naturally present in the phylloplane of grass foliage collected from pasture.

3.1.3 NATURAL OCCURRENCE AS AN INSECT PATHOGEN

Bt has rarely been found to cause natural epizootics in an insect population. Vankova and Purrini (1979) reported natural epizootics of *Ephestia kuhniella, E. elutella* and *Plutella maculipennis* in Yugoslavia were caused by *Bt* where *Bt* had never been applied. They isolated 18 strains, including *B. cereus* and strains of *Bt* belonging to serotypes *kurstaki, morrisoni* and *thuringiensis.* However, no data were provided on incidence of insects infected. In India, Rajagopal *et al.* (1988) reported 5.6-36.4% of larvae of *Aproaerema modicella,* the lepidopteran groundnut leaf miner, were infected with *Bt*. Epizootics are more likely to occur in crowded, stressed insects which is reflected by reports of epizootics of *Bt* in silkworms (such as *sotto* in Japan). Burges and Hurst (1977) found that two-thirds of the Lepidoptera populations occurring in stored product premises are infected with *Bt* at a low level. In 20/38 of samples of collected larvae taken into strict quarantine in the laboratory, *Bt* infection did not become apparent until the second or third generation grown under intense crowding in laboratory culture. Periodic natural epizootics caused by *Bt aizawai* in the navel orangeworm, *Amyelois transitella* (Lepidoptera; Pyralidae), reared on wheat grain in the laboratory, have been reported by Federici (1994). Unusually, larvae killed during such epizootics were completely colonised by *Bt* after death and containined spores and crystals. These larvae could contain as many as 10^8 spores per cadaver (Federici 1999). In this insect at least, cadavers were an excellent medium for *Bt* (Federici 1994).

While epizootics caused by *Bt* are rare, isolation from insects is more commonly reported. For example, *Bt dendrolimus* has been isolated from natural infection in *Dendromilus superans* and *D. sibricus* (Talalaev 1976; Krieg and Langenbruch 1981), while *Bt kurstaki* was recorded from a number of insects including the lepidopteran, *Arctia caja phaeosoma* (Arctiidae) (Kikuta and Iizuka 1990), *Homoeosoma nebulella* (Phycitidae) (Itoua Apoyolo *et al.* 1995), eggs of *Ostrinia nubilalis* (Pyralidae) (Lynch *et al.* 1976), *Papilio ploytes* (Papilionidae) (Gao *et al.* 1985), *Ephestia kuhniella, E. elutella* and *Plutella maculipennis* (Pyralidae) (Vankova and Purrini 1979) and the Orthoptera, *Calliptamus italicus* (Acridiae) (Deseo and Montermini 1990).

In some cases, *Bt* has been isolated as one of many bacterial species found within cadavers, with no indication given as to whether they were toxic to that host. For example, Panizzi and Pinzauti (1988) isolated a number of bacteria, including *Bt,* from honey bees stressed by mite attack. General surveys (e.g. Iriarte *et al.* 1998; Chilcott and Wigley 1993) have found *Bt* in dead insects, although fewer isolates were recovered from insects than other environments.

3.1.4 RECYCLING IN THE HOST POPULATION

Most insect pathogens kill a host, multiply in the cadaver and are transmitted to new hosts to repeat the cycle. However, *Bt* is a poor infectious agent and rarely recycles. While vegetative cells and spores will be produced in cadavers, *Bt* has rarely been recorded causing natural epizootics. The mode of action, which is more dependent upon toxic action than infection, probably accounts for the lack of recycling. *Bt* spores can remain viable for years in soil (Chapter 8) and *Bt* is often isolated from insect cadavers. However, re-infection in the field after application is not expected and yearly consecutive spraying may be required because of lack of persistence of toxin.

3.2 ROLE IN THE ENVIRONMENT

There are several theories on the ecological niche filled by *Bt*. Unlike most insect pathogenic microbes, *Bt* generally recycles poorly and rarely causes natural epizootics in insects, leading to speculation that *Bt* is essentially a soil micro-organism that possesses incidental insecticidal activity (Martin and Travers 1989). The fact that *Bt* is commonly found in the environment independent of insects supports this view (van Frankenhuyzen 1993). Meadows (1993) suggested four possible explanations for the presence of *Bt* in soil: 1) *Bt* rarely grows in soil but is deposited there by insects; 2) *Bt* may be infective to soil-dwelling insects (as yet undiscovered); 3) *Bt* may grow in soil when nutrients are available; and 4) an affinity with *B. cereus*. Alternatively, Smith and Couche (1991) proposed that *Bt* is a natural component of the phylloplane microflora and has evolved in a symbiotic or mutualistic association with plants to provide protection against herbivores. Speculation is confounded by the transport of *Bt* in the atmosphere as evidenced by its presence in samples taken deep in the polar ice cap (Martin and Travers 1989), long before spraying for pest control commenced.

4 Production and Formulation

Bt products usually contain a mixture of spores, endotoxin crystals, some vegetative cells, cell debris and some carryover of materials from the fermentation, in addition to a number of formulation components aimed at increasing spreading ability over leaves, persistence, shelf life and/or ease of application. Over 100 product names have been used in the literature since the launch of the first *Bt* commercial product in 1938. The majority of products have been based on *Bt* serovars *kurstaki, thuringiensis, israelensis* and *aizawai*. The components of any *Bt* product need to be independently assessed for environmental safety, however these constituents are product specific and formulations are commercially sensitive. It is unlikely that formulation components differ significantly from published experimental formulations and few problems have been reported with components. The US Environmental Protection Agency and similar regulatory authorities in other countries have cleared a number of *Bt* formulations on a case-by-case basis. However, few studies on the safety of formulation components have been published.

4.1 PRODUCTION

The technical grade active ingredient of each *Bt* product is generally manufactured using a standard fermentation batch process as described in Burges and Jones (1998). During fermentation, *Bt* first propagates in a vegetative phase. When a critical nutrient becomes depleted, *Bt* begins to sporulate, after which cells lyse to release spores and the crystal proteins which are produced during the sporulation phase. The material is then concentrated and either dried or mixed with inerts in a liquid form and packaged. The US Environmental Protection Agency (EPA) and other regulatory agencies around the world have expressed concerns about the potential for production of various undesirable *Bacillus* exotoxins because their synthesis appears to depend on unpredictable aspects of the fermentation process, either with respect to the composition of the fermentation media or growth conditions used for production. Mohd Salleh *et al.* (1980) showed that activities of exotoxins from three varieties of *Bt* grown in six different fermentation media varied between media and there was even variability when they were grown in the same medium. The toxins may be inducible toxins, with synthesis dependent on the presence of certain chemicals; they may be toxic metabolites, requiring the presence of certain chemicals for their synthesis; or their synthesis may depend on physical growth parameters such as temperature. Therefore, production batch testing is required to detect these toxins and contamination by mammalian pathogenic bacteria.

In considering the environmental safety of formulation components and products, reference to the systems and judgements of the EPA in the United States is useful, partly because many of their evaluations are publically available and because many countries have modelled their assessment systems on the EPA framework and/or accept data passed by the EPA. For example, the EPA sought to minimise the risk of inadvertent exotoxin production during production by implementing new requirements in their most recent "Reregistration Eligibility Decision" documents (EPA March 1998). Companies wishing to register their products must standardise their manufacturing process sufficiently to prevent production of significant amounts of exotoxin. Safety of products must be demonstrated by testing production batches using established procedures e.g. subcutaneous injection of at least one million spores into each of five mice and fly larvae toxicity test or HPLC detection of β-exotoxin (EPA 1998, pp. 38-39).

There is potential for strains of *Bt* to produce β-exotoxin during subsequent growth in formulated products, despite none being detected in production batches. However, the strain must have the potential to produce β-exotoxin (contain exotoxin-encoding genes). If the strain used is capable of producing β-exotoxin, the EPA requires the producer to ensure that no toxin is present in the technical grade active ingredient. In addition, the product should not be put in a medium, including formulated end products, that allows germination and/or growth at any time prior to use. End use product testing for β-exotoxin is usually conducted by testing against *Daphnia,* using a maximum hazard dose, but other methods have also been developed (Section 8.7).

4.2 ENVIRONMENTAL SAFETY OF FORMULATION COMPONENTS

4.2.1 FORMULATION COMPONENTS

Products based on *Bt* contain a large percentage of bacteria and fermentation medium, however additives are often used to improve product stability and provide other desirable characteristics such as flowability. Formulations of pesticides are closely guarded trade secrets, but for registration approval, these must be provided to the appropriate regulatory authorities. As there are differences between formulations of *Bt* products, the safety of one product does not imply safety for all. Many companies use ingredients that appear on a list known in the USA as GRAS (Generally Recognised As Safe). These are ingredients recognised by the Food and Drug Administration (FDA, USA) as safe when used as direct food additives, which allows companies to avoid extensive toxicity testing on formulated products. Toxicity testing on technical grade unformulated products is still required.

Additives and formulation components can be included at several stages during the process and for a variety of reasons. Burges and Jones (1998) listed a number of additives used with *Bt,* including dispersants and drying protectants, like gum arabic and lactose. The addition of Ca^{2+} ions has been used to precipitate β-exotoxins. After fermentation, additives can also be used in products to improve storage, persistence, spreading ability and application. Various formulations of *Bt* have been used to target insects in different habitats or where target species have different feeding habits. For example, granules, capsules and liquid products based on several subspecies including *Bt kurstaki* and *Bt israelensis,* have been produced. Specific formulations have been developed for application to water, for example, where target mosquito larvae can be bottom feeders, surface feeders or feed throughout the water column. *Bt* has been formulated as dry dusts with talcum powder (carrier) and silca powder (free-flow agent), clay granules with stickers, water-dispersible granules (corn-starch, alginate) or liquids in a gel matrix. For spray application, oils and thickeners have been added. About half the present *Bt* products are oil-in-water emulsions (Burges and Jones 1998). Typically, a water-based flowable *Bt* concentrate includes fermentation solids (with *Bt* spores and crystals), dispersants, suspenders (e.g. gum), fungistatic and bacteriostatic agents and water, while an oil-based flowable concentrate may contain the technical powder, a carrier (oil) suspender (e.g. Bentone) and activator (e.g. propylene carbonate) (Burges and Jones 1998).

Formulation additives have been used to improve the efficacy of *Bt* products with respect to coverage of sprayed material on leaves and rainfastness. Surfactants improve the coverage on hydrophobic leaves, facilitate mixing of hydrophobic spores and toxin crystals into water and formation of an emulsion between oil and water by reducing interfacial tension. Stevens *et al.* (1994) found that Silwet L-77, an organosilicone surfactant, enhanced the efficacy of *Bt* in horticultural crop protection. Stickers improve rainfastness, as toxins and spores washing off leaves have been shown to contribute significantly to lack of persistence (Chapter 8). Sunscreens are often added to formulations as lack of persistence has also been related to UV degradation.

The phytotoxicity of some formulation additives is an area of some concern. For example, non-ionic surfactants Agral, Citowett, DX and Tap in aqueous solution at 10 ppm affected seed germination and growth of sorghum (Horowitz 1977). Vegetable oils are preferable to mineral oils as they were less phytotoxic (Burges and Jones 1998).

Feeding stimulants have been trialed in tank mixes for use with *Bt* products to increase the intake of toxins and spores. Molasses, sugars, corn, wheat bran and some commercially available phagostimulants (e.g. Pheast, Coax, Gusto and Entice) have been used (Farrar and Ridgway 1995a). Patti and Carner (1974) found that Dipel was more effective against *Helicoverpa* when added to feeding stimulant baits. Salama *et al.* (1985) examined potential feeding stimulants with *Bt entomocidus* against *S. littoralis* using extracts from some food plants (molasses, soybean flour, cottonseed flour and sucrose). Several compounds, including molasses and sucrose, increased the effectiveness of *Bt* by increasing ingestion of toxins. Many compounds have been shown to synergise *Bt*. Burges and Jones (1998) listed amino acids, surfactants, inorganic salts, organic acids and their salts, phenolic compounds, protease inhibitors, protein solubilizing reagents, ascorbic acid, caffeine, dimethyl sulphoxide, doaecyclamine, dipicolinic acid, enhancin protein from granulosis virus, Neemazaal-T, salicylic acid, amino salicylic acid, sodium salicylate and sorbic acid. The extent of the synergistic effect was generally 1-5× improvement in activity, but was up to 40× in some cases. Chemical pesticides have also been shown to increase the efficacy of *Bt* (Table 12.2).

4.2.2 ENVIRONMENTAL SAFETY

The environmental impact of formulation components in *Bt* products is difficult to assess, largely because of the proprietary nature of the ingredients and lack of published accounts of the independent action of the formulation components, as described above. Few studies have separated toxicity of formulation components from the toxicity of *Bt* toxins and spores. In New Zealand, this became an issue when *Bt kurstaki* (Dipel) was proposed for use in an urban eradication campaign of an outbreak of the lymantrid, *Orgyia thyellina.* As a massive aerial application campaign over New Zealand's largest city was intended, the need for public disclosure of formulation components was considered necessary. However, the producer was reluctant to disclose commercially sensitive formulation components. A compromise was eventually reached (disclosure to a limited group of officials), however the problem with public scrutiny of formulation component safety was highlighted. For registration, formulation components are declared under confidentiality agreements and safety data are supplied.

Some reports of toxicity to bees and mice were partly attributed to carriers in early formulations (Forsberg *et al.* 1976), although in the reports reviewed by those authors, β-exotoxin was present in the formulations and it was not possible to separate the effects. Several studies have reported on safety of formulation components of Dipel. In a study by Haverty (1982), the toxicity of the carrier of Dipel 4L against predators and a parasite was tested. Dipel 4L is a non-aqueous, emulsifiable suspension of *Bt kurstaki* and there was concern about the oil toxicity after a change from the use of aqueous to oil-based formulations. Mortality (corrected for control results) caused by the carrier did not exceed 2.1% for any of the non-target species at 9.4 l/ha; when used at 18.7 l/ha rate, mortality of adult predators *Chrysopa carnea* and *Hippodamia convergens* was higher than controls, but this did not occur for the parasite *Aphytis melinus*. Mortality did not exceed 13.4% for any species. In another study, Addison and Holmes (1995) found that while unformulated *Bt kurstaki* and aqueous Dipel had no effect on the collembolan, *Folsomia candida*, an oil-based Dipel formulation did have a negative impact, suggesting direct or indirect toxicity of the formulation components. Similarly, Addison and Holmes (1996) found that 1000× effective environmental concentration of Dipel 8L reduced survival, growth and coccoon production of the forest earthworm *Dendrobaena octahedra*, whereas unformulated *Bt kurstaki* or aqueous Dipel 8AF did not.

4.3 PRODUCTS

Bt-based products make up over 90% of all biopesticides sold in the world today (Swaneder 1994). Many products (e.g. Dipel, Thuricide and Biobit) are based on the *Bt kurstaki* HD-1 standard strain, principally because of activity against over 100 species of Lepidoptera (Navon 1993). We list over 167 species of Lepidoptera as susceptible to Dipel (Appendix 3). Other products based on *Bt* strains with more specific host ranges have more restricted markets. Because of concerns about mammalian toxicity, formulations in western European and North America generally contain little or no β-exotoxin. However, several products from the former USSR and eastern bloc countries based on serotype 1 (*thuringiensis*) often contained up to 20% exotoxin (Navon 1993).

To understand the literature, it is helpful to be able to associate a product name with a subspecies or variety. However, there has been a bewildering array of *Bt*-based "products" formulated over the years and the literature is full of names of products which were in many cases short-lived or available in a single country. In Table 4.1, we have endeavoured to tabulate all product names encountered in the literature. In the USA, *Bt* products comprise one of five subspecies: *Bt kurstaki, aizawai* and *morrisoni* (lepidopteran active), *israelensis* (dipteran active) and *tenebrionis* (=*san diego*; coleopteran active; Swadener 1994). The most recent publication including current *Bt* products was a directory of microbial control products and services edited by Shah and Goettel (1999). The products listed therein are included in Table 4.1, but many more products are available than are listed in that one publication. However, exactly which products are currently available in the various countries is difficult to assess. The products listed in Shah and Goettel (1999) are likely to be available at least within the country mentioned in that publication. In addition, Abbott Laboratories (North Chicago, IL, USA) produce a number of *Bt* products, which are widely sold. According to Crop Data Management Systems (www.cdms.net), these products currently include Bactimos (granule, pellet, primary powder, wettable powder), Biobit (HP and XL), Dipel (10G, 2X, 6AF, DF, ES, ES-CPI, ES-NT, FMU, LDM and SG plus), Foray (48B, 48F and 76B), Gnatrol, Vectobac (12AS, CG and G) and Xentari. These different formulations of the same products include wettable powder, emulsifiable liquid, water dispersible granule, dry flowable, aqueous suspension, granule, technical powder, dust, wettable powder, aqueous flowable, baits and oil flowable formulations.

Table 4.1. Commercial *Bacillus thuringiensis*-based pest control agents mentioned in the literature.

Subspecies/ Product	Target pest	Company, country of origin or reference
kurstaki		
Astur (asporogenic)	Lepidoptera	Bushkovskaya *et al.* 1994
Bacilan	Lepidoptera	Lonc *et al.* 1986
Bactec Bertan Bt	Lepidoptera	Bactec Corporation
Bactospeine[1]	Lepidoptera	Abbott; Starnes *et al.* 1993
Bactucide	Lepidoptera	Switzerland; Triggiani & Sidor 1982
Baktur	Lepidoptera	Scalco *et al.* 1997
Bathurin	Lepidoptera	Slovakia; Novotny & Svestka 1986
Bioasp (asporogenous)	Lepidoptera	India; Satapathy & Panda 1997
Biobit	Lepidoptera	Abbott; Starnes *et al.* 1993
Biodart	Lepidoptera	ICI Canada; Bernier *et al.* 1990
Biolep (sporogenous)	Lepidoptera	India; Satapathy & Panda 1997
Biotrol[2]	Various	Bishop *et al.* 1973
BT Turex	Lepidoptera	Spain; Cortes & Borrero 1998
Condor	Lepidoptera	Ecogen; Starnes *et al.* 1993
CoStar	Lepidoptera	Thermo Trilogy; Shah & Goettel 1999
Crymax	Lepidoptera	Ecogen; Shah & Goettel 1999
Cutlass	Lepidoptera	Ecogen; Starnes *et al.* 1993
Delfin	Lepidoptera	Thermo Trilogy; Shah & Goettel 1999
Dipel[3]	Lepidoptera	Abbott; Shah & Goettel 1999
Foray	Lepidoptera	Abbott; Shah & Goettel 1999
Futura	Lepidoptera	Abbott; Starnes *et al.* 1993
Javelin	Lepidoptera	Thermo Trilogy; Shah & Goettel 1999
Lepidocide	Lepidoptera	Goral *et al.* 1984
Lepinox	Lepidoptera	Ecogen; Shah & Goettel 1999
Manapel	Lepidoptera	Brazil; Cruz 1977
Mattch	Lepidoptera	Mycogen
MVP[4] (Cry 1Ac)	Lepidoptera	Mycogen; Nyouki *et al.* 1996
MYX[4] (Cry 1Ab)	Lepidoptera	Mycogen; Nyouki *et al.* 1996
Novabac	Lepidoptera	Randall *et al.* 1979
Novosol	Lepidoptera	Abbott; Teakle & Milner 1994

Table 4.1. *(Continued)*

Subspecies/ Product	Target pest	Company, country of origin or reference
Rapax	Lepidoptera	D'-Offria *et al.* 1996
Raven[5]	Coleoptera	Ecogen; Shah & Goettel 1999
Safer	Lepidoptera	Home Harvest, http://homeharvest.com/bt.htm
SAN 415	Lepidoptera	Hegazy & Antounious 1987
Sok-*Bt*	Lepidoptera	Reigart & Roberts 1999
Toarow[6]	Lepidoptera	Japan; Wysoki *et al.* 1986; Navon 1993
Thuricide[3, 7]	Lepidoptera	Thermo Trilogy; Shah & Goettel 1999
Tribactur	Lepidoptera	Netherlands; Nef 1972; Reigart & Roberts 1999
Wormbuster	Lepidoptera	Bactec; Starnes *et al.* 1993
Zoocamp	Lepidoptera	Brazil; Habib 1986
Fluidosoufre BTR (+ sulfur)	Lepidoptera & mildew	France; Goebel *et al.* 1997
Bacilex (+ *aizawai*)	diamondback moth	Kimura 1991; Asano & Seki 1994
Agree (+ *aizawai*)	diamondback moth	CIBA-GEIGY; Ivey & Johnson 1997, 1998
Biocillis (+ *aizawai*)	tortricids	France; Fougeroux & Lacroze 1996
aizawai		
Xentari/Zentari/Centari	Lepidoptera	Abbott; Shah & Goettel 1999; most of the world/Japan/Thailand
Certan	wax moth/Lepidoptera	Sandoz; Starnes *et al.* 1993
Clorbac	Lepidoptera	Federici 1999
Design WSP	Lepidoptera	Mascarenhas *et al.* 1998
Florbac	diamondback moth	Abbott; Starnes *et al.* 1993
Quark		Abbott
Selectzin	Lepidoptera	Poland; Lipa *et al.* 1977; Takaki 1975
Turex	Lepidoptera	Thermo Trilogy; Shah & Goettel 1999
israelensis		
Acrobe	Mosquitoes/ Blackflies	American Cyanamid; Starnes *et al.* 1993
Aquabac	Mosquitoes/ Blackflies	Reigart & Roberts 1999
Bactimos	Mosquitoes/ Blackflies	Abbott; Shah & Goettel 1999
Bactoculicide	Mosquitoes/ Blackflies	USSR; Kanybin 1991
BMC WP	Mosquitoes/ Blackflies	Reuter
Bulmoscide		Halkova *et al.* 1993
Cybate (Aust. label)	Mosquitoes/ Blackflies	Cyanamid
Duplex (+methoprene)	Mosquitoes/ Blackflies	Zoecon - PPM
Gnatrol	Fungus gnats	Abbott; Starnes *et al.* 1993
JieJueLing preparation	Mosquitoes/ Blackflies	China; Shah & Goettel 1999
MieJueLing preparation (+ strain CS8)	Mosquitoes/ Blackflies	China; Shah & Goettel 1999
Skeetal	Mosquitoes/ Blackflies	Abbott; Chang *et al.* 1990
Teknar	Mosquitoes/ Blackflies	Thermo Trilogy; Shah & Goettel 1999
Vectobac	Mosquitoes/ Blackflies	Abbott; Shah & Goettel 1999
tenebrionis		
Foil	Coleoptera	Ecogen; Starnes *et al.* 1993
M-Trak[5]	Coleoptera	Mycogen; Starnes *et al.* 1993
M-One	Coleoptera	Mycogen; Starnes *et al.* 1993
Novodor	Coleoptera	Abbott; Shah & Goettel 1999
NovoBtt	Coleoptera	Abbott; Panetta 1989
Trident	Coleoptera	Sandoz; Starnes *et al.* 1993
thuringiensis		
Bakthane (+exo[8])	Diptera/Lepidoptera	Egypt; Farghal *et al.* 1987
Biospore	Diptera/Lepidoptera	Afify & Matter 1970a; Merdan 1977
Bitoxibacillin (BTB) (+exo)	Various	Romasheva *et al.* 1977; Rimmington 1989
Exobacillin	Coleoptera	Bulgaria; Kuzmanova & Sapundjieva 1997
Gomelin	Lepidoptera	USSR; Supatashvili 1990; Gamayunova 1987; Pishokha 1985
IMC 10,001.1 (+exo)	Diptera	Lam & Webster 1972
Insectin	Fleas & Lepiodptera	USSR; Ershova *et al.* 1976

Table 4.1. *(Continued)*

Subspecies/ Product	Target pest	Company, country of origin or reference
Muscabac (+exo)	Diptera	Finland, USSR; Powell *et al.* 1990; Pinnock & Milner 1994
Porasporin	Lepidoptera	Grain Processing Corp., USA; Forsberg *et al.* 1976
Tarmik	Lepidoptera	Turkey; Tutkun *et al.* 1987
Thiodane	Diptera	Farghal *et al.* 1987
Thuridan	Lepidoptera	Poland; Glowacka-Pilot 1984
Thuringin (+ exo)	Various	Tonkonozhenko 1981
Toxobacterin ("Insectin enriched with Enzotoxin [sic]")	Lepidoptera	USSR; Spektor & Vykhovets 1975; Malokvasova 1979
alesti		
BIP[9]	Lepidoptera	Daricheva *et al.* 1983; Goral *et al.* 1984
Alestin	Lepidoptera	USSR; Mitrofanov & Novozhilov 1982
Amdal	Lepidoptera	Creighton *et al.* 1972
galleriae		
Entobakterin	Lepidoptera	USSR; Rimmington 1989
Qing-Chong-Jun No. 6 (+cypermethrin)	mites	China; Dong & Niu 1991
Spicturin	*Plutella* spp.	India; Shah & Goettel 1999
caucasicus [10]		
BIP[9]	Fleas, Lepidoptera	USSR; Afrikyan & Chil Akopyan 1980
chinesensis		
Shuangdu preparation	Lepidoptera, Coleoptera & Diptera	China; Shah & Goettel 1999
darmstadiensis		
Bacicol	Coleoptera	USSR; Kandybin & Smirnov 1996
dendrolimus		
Dendrobacillin	Various	USSR; Rimmington 1989
sotto		
Cellstart	Lepidoptera?	Japan; Takaki 1975
Subspecies not stated		
Bitayon	Lepidoptera	Israel; Shah & Goettel 1998
Bt8010, Rijin	Lepidoptera	China & Malaysia; Shah & Goettel 1999
Cajrab	Lepidoptera	India; Srivastava & Nayak 1978
Dibeta/SAN (+exo)	Acari & Hemiptera	Grau 1987; Williams *et al.* 1987
Disparin	Lepidoptera	Bulgaria; G'onev 1977
Halt	Lepidoptera	Wockhardt, India; Lal & Lal 1996
Junduwei (+HaNPV)	*Helicoverpa armigera*	China; Zhang *et al.* 1996
Miazol (+exo)	Myiasis in animals	Tonkonozhenko *et al.* 1977
Plantibac	Lepidoptera	Mineo 1970
Shachongjin	Coleoptera, Lepidoptera & Diptera	China; Lin *et al.* 1990a,b
Sporeine	Lepidoptera	France 1938 (first *Bt* product) Lambert & Peferoen 1992
Subdu (+ *B. subtilis*)	Diptera	Campbell & Wright 1976
Tarsol	Lepidoptera	Krieg 1978
Thurintox	Mites & Lepidoptera	Baicu 1980
Turingin	Diptera	Yarnykh & Tonkonojenko 1975
Ybt-1520	Lepidoptera	China; Shah & Goettel 1999

[1] Bactospeine listed as *Bt kurstaki* in some publications but more often *Bt thuringiensis* (and could contain exotoxin, Vlayen *et al.* 1978).
[2] also listed as *Bt thuringiensis* (e.g. Hamed 1978; Rajamohan & Jayaraj 1978) in various publications.
[3] early publications listed Dipel and Thuricide as *Bt alesti*, due to the misclassification of strain HD-1.
[4] MVP, MYX and M-Trak contain no viable *Bt* cells but only δ endotoxin and dead *Pseudomonas fluorescens*.
[5] Raven is a recombinant *Bt kurstaki* strain expressing Cry3 and Cry1 genes (Johnson *et al.* 1994).
[6] Toarow CT was based on an inactivated spore product (Navon 1993).
[7] Thuricide originally contained *Bt thuringiensis* (H.D. Burges, pers. comm.), then serotype 5a5b *Bt galleriae* (Burges 1976b; Jarrett & Burges 1982b) but has also been listed as containing *Bt aizawai* (Asano & Seki 1994).
[8] (+exo) = contains b-exotoxin
[9] Biological Insecticide Preparation (BIP) is referred to as both *Bt caucasicus* and *Bt alesti* in different publications (see Appendix 2).
[10] *Bt caucasicus* is a serotype 10a, according to Afrikyan & Chil Akopyan (1980). Only mentioned in the USSR literature.

5 Toxicity to Insects

Differences between strains, assay methods, difficulty in measuring active components of *Bt* formulations and environmental and host influences on toxicity all combine to complicate assessment of *Bt* toxicity to insects and possible non-target effects. No serotype is active against only one insect order. Early reports on *Bt* stated that it was specific to Lepidoptera, but continued discovery of new strains has revealed that *Bt* has more wide ranging activity, with strains showing toxicity to over 3000 species from more than 16 orders of insects. Some generalisations can be made regarding host range: *Bt kurstaki, aizawai, thuringiensis, alesti* and *dendrolimus/sotto* are most often reported as toxic to Lepidoptera, *Bt israelensis* and *darmstadiensis* are generally toxic to Diptera, and *Bt tenebrionis* and *japonesis* are active against Coleoptera. However, *Bt* subspecies definitions such as serovar cannot be used to predict the toxin profile of strains so host susceptibility cannot be approximated by use of serovar as a grouping.

Specificity is exhibited by individual δ-endotoxins and the combination of endotoxins within a single isolate can also interact to increase (synergise) and occasionally decrease toxicity to a given insect. In some cases, spores also increase the potency of *Bt* preparations. The type of formulation can influence the spectrum of activity, with variuos formulations of the same strain showing up to 10-fold differences in toxicity in comparative assays. Comparing toxicity of *Bt* strains has been a concern for many scientists, to the extent that international standard preparations have been designated for *Bt kurstaki* and *Bt israelensis*. However, there are still difficulties when comparing between different bioassay techniques, various test insects and the many units used to express active ingredients.

Individual *Bt* strains are active against a limited number of insect species and are often active against only a few orders under field conditions. In the laboratory, some insects and other organisms may be susceptible only at very high doses. Reports on susceptibility to Dipel, useful for comparison because it has been based on a single strain for many years, show little field toxicity outside members of the Lepidoptera and relative tolerance of a number of predators and parasites of Lepidoptera.

Records of non-susceptible insects occur in 11 of the orders also containing susceptible insect species. Intrageneric differences in insect susceptibility to *Bt* are found even using a single product. While many insects appear susceptible in the laboratory, the same insect may not be affected in field situations. The difference between laboratory and field susceptibility may be related to dose, level of exposure, feeding behaviour, insect ecology or combinations of these factors. Other factors, including *Bt* persistence, may also reduce the effect of *Bt* in the field, as described in later chapters.

5.1 ASSESSING TOXICITY TO INSECTS

5.1.1 UNDERSTANDING TOXICITY OF *BT*

As described in previous Chapters, *Bt* toxicity is dependent on the variety of toxins and spores expressed in a multitude of isolates, distinguished by numerous methods including antigens (serovars). Recognition of the multiple toxicity of various subspecies and toxins produced by *Bt* is essential in understanding the limits of insecticidal activity and assessing risk to non-target organisms.

One of the difficulties with evaluating environmental safety, or even understanding *Bt* toxicology, is that the terms *Bt*, varieties, subspecies, serovars, toxins and products have all been used without strict definition to describe the virulent unit. In many cases, when a particular product or subspecies has been used, the actual toxins which caused the insecticidal activity have not been identified. It is therefore extremely difficult to compare specific activity of "*Bt*" within the literature.

An additional difficulty is that a wide range of methods have been used to evaluate potency. The role of spores in potency is variable, with some studies demonstrating higher insecticidal activity when spores were present than when pure crystal preparations were used (Section 5.6). However, in many cases spores do not contribute significantly to the potency of *Bt*.

5.1.2 USE OF REFERENCE STRAINS

The difficulties in comparing between isolates and toxins of *Bt* became apparent in the 1960-1970s. As Dulmage *et al.* (1981) states, it is practically impossible to reproduce exactly the conditions of bioassays to estimate LC_{50}s. This led to the proposal that a number of "standard" strains be designated for use in all *Bt* efficacy experimentation. The first generally accepted standard that was agreed between countries, different industrial firms, government laboratories and academia, was a *Bt thuringiensis* strain E-61 prepared in France (most early industrial products contained this subspecies/variety), with an assigned potency of 1000 International Toxic Units (ITU) or International Units (IU)/mg (Vago and Burges 1964; Burges 1967a, b; Burges *et al.* 1967; Dulmage *et al.* 1981). Dulmage (1975) proposed another standard strain HD-1-S-1971, based on *Bt kurstaki*, when this more effective subspecies replaced *Bt thuringiensis* in most products. "HD" indicates isolates from the USDA Howard Dulmage collection of *Bt*, currently housed in the ARS/USDA collection at Peoria, Illinois (Nakamura and Dulmage 1988). HD-1-S-1971 was assigned a potency of 18,000 IU/mg on the basis of assays against E-61 (Dulmage *et al.* 1981). However this was a comparison between two subspecies with differing host ranges and gives only an indication of relative potency to that insect (Burgerjon and Dulmage 1977). The supply of HD-1-S-1971 was later exhausted and HD-1-S-1980 was adopted as the primary standard with an activity determined as 16,000 IU/mg (Beegle *et al.* 1986). Three standards have been used for *Bt israelensis*, IPS-78, HD-968-1983 (Dulmage *et al.* 1985) and IPS-82 (Skovmand *et al.* 1997). HD-968-1983 was assigned a potency of 4740 IU/mg against *Aedes aegypyti* larvae (Dulmage *et al.* 1985).

5.1.3 MEASURING TOXICITY

Potency is established using concentration-mortality data, transforming to a log-probit scale, and comparing an estimated LC_{50} (lethal concentration required to kill 50% of targets) of a test substance with that of the standard material. As stated by Skovmand *et al.* (1997), the rationale is that daily within-laboratory or inter-laboratory variations can be accounted for using relative estimates. In addition to the standard strains, standardised protocols have also been suggested (McLaughlin *et al.* 1984), but variation is still found between laboratories (Skovmand *et al.* 1997).

Efficacy of a formulation is generally determined by measurement against a standard formulation and expressed in terms of ITU/mg (Dulmage *et al.* 1981). Potency is calculated by:

$$\frac{LC_{50} \text{ of a standard}}{LC_{50} \text{ of a sample}} \times \text{potency of the standard} = \text{Potency of the sample (ITU/mg)}$$

The difficulties in comparing toxicity are highlighted by the study of Skovmand *et al.* (1997) who compared 11 products or formulations of *Bt israelensis* over >1500 bioassays, to assess if the slopes of the probit lines were parallel, a necessary condition for comparison of such results. They found that, in general, particle size and product type affected the slope of probit lines, which could result in incorrect comparisons with the standard.

Robertson *et al.* (1995) compared bioassay results of *Bt tenebrionis* against *Leptinotarsa decemlineata* and *Bt kurstaki* against *Plutella xylostella* and found that LC_{50}, LC_{90} and LC_{99} values varied among strains tested on the same species. Robertson *et al.* (1995) concluded that natural variation could account for much of the variation found when comparing LC/LD values between strains or targets, leading to erroneous conclusions, such as resistance developing in populations where none existed.

Variation in effect of *Bt* can be measured by a number of methods, including the concentration/dose required to kill (LC/LD_{50}) and time taken to kill (LT_{50}). LC_{50}s, while a common measure, are often difficult to compare between studies because of the various bioassay techniques. It is difficult to compare the toxicity of *Bt* strains without considering the natural variation in susceptibility of insect populations, differences between the bacterial isolates and the interactions between these factors (Kinsinger *et al.* 1980).

Measurement and evaluation of toxicity can be complicated by the mode of action. Estimates of the actual amount of δ-endotoxins per dose or ingested are sometimes used rather than the dose supplied. Van Frankenhuyzen and Gringorten (1991) suggested alternative measures to LC and LD_{50} which might be more appropriate to some lepidopteran species which can take several days to die and where death may be caused by secondary infection rather than rapid action of the *Bt* toxin. They suggested comparing failure to produce frass (FFD_{50}) and failure to reach pupation (PFD_{50}) with untreated pupating controls. Similarly, de Leon and Ibarra (1995) proposed using an ED_{50} using weight gain as a measure of effect. They found that concentration of *Bt* and weight gain were inversely related, at least for their test insect, *Leptinotarsa texana*.

Although these are valuable methods for use with particular insects at the research level, the LD_{50} is the most common method used to measure potency of industrial products. LD_{50} is a quantitative measure so that if product A has twice the mortality-based potency units of product B, then in field use, half the quantity of A should give the same result as that obtained with B. The other measures of potency are qualitative and many detect differences in assays with better confidence limits, but without this direct porportionality.

5.2 SENSITIVITY OF INSECTS TO *BT*

We have attempted to compile a list of organisms against which *Bt* has been directly or indirectly tested, covering much of the accessible literature (Appendix 2, summarised in Table 5.1). These tables remain incomplete, however, as in some cases it was impossible for us to access more than an abstract and many references are in the form of unpublished reports, limited distribution journals or company records. We have included several previously published lists: Krieg and Langenbruch (1981) listed insects susceptible to *Bt* up to 1980; Keller and Langenbruch (1993) reported species susceptible to *Bt tenebrionis* and Glare and O'Callaghan (1998) reported on *Bt israelensis*. However, very little information is available for some serovars. In fact, for most of the recently described serovars no susceptible organisms are known as many of these new isolates have been isolated from the environment (Lecadet *et al.* 1999). A useful database for comparing *Bt* toxins efficacy against different insects, with descriptions of bioassay methods and references is maintained by van Frankenhuyzen and Nystrom (1998).

Listing organisms sensitive to *Bt* is a difficult task, made more confusing by the multiple toxins (including exotoxins) which may be produced by strains within a given grouping (serovar, variety, subspecies etc.), the number of products based on *Bt* (Chapter 4, Table 4.1), the range of bioassay methods, field application methods and dosages used. As an aim of this review was to examine possible non-target effects, we have taken a conservative approach and listed all species (by serovar/variety where possible) to which susceptibility to *Bt* has been recorded (Appendix 2.1). In the tables and appendices, insects are described as susceptible, where insect death following laboratory treatment with *Bt* was significantly greater than in the untreated controls or where, in the researchers' opinion, *Bt* was toxic following field application.

Comprehensive data are available for only a few serotypes, particularly *Bt kurstaki, aizawai, thuringiensis, galleriae, alesti, israelensis, dendrolimus/sotto* and *tenebrionis*. *Bt kurstaki* has been reported to be effective against insects from 318 genera, while 166 genera contained species reported to be non-susceptible either in the laboratory or field (Table 5.1). Some genera, which contained both susceptible and non-susceptible species are counted in both groups. Although *Bt kurstaki* is generally regarded as lepidopteran active, it has been recorded as toxic to a number of Coleoptera, Diptera and Hymenoptera, and there are isolated reports of toxicity in other orders (Table 5.1; Appendix 2). Interestingly, no serotype is restricted in its activity to one order of insects. However, *Bt kurstaki, thuringiensis, alesti, dendrolimus/sotto* are more often reported as toxic to Lepidoptera, while *Bt israelensis* and *darmstadiensis* show toxicity to Diptera, and *Bt tenebrionis* is active against Coleoptera. *Bt tenebrionis* was more toxic to Chrysomelidae than other Coleoptera (Keller and Langenbruch 1993).

In published reviews, it has been common practice to list the serovars/subspecies with insecticidal properties assigned to insect order. For example, Powell *et al.* (1990) listed serovars from H-1 (*Bt thuringiensis*) to H-16 (*Bt indiana*) with specificities to insect order given for each. While this is a good method for indicating broad host range, it is biased by the number of experiments conducted with any one serovar. We have attempted to give an indication of the limitations of the data by listing where non-susceptibility was recorded, in addition to studies which have reported toxicity. For example, Powell *et al.* (1990) reported *Bt pakistani* as lepidopteran active; subsequent review of the literature has shown a single dipteran species was sensitive in addition to tested lepidopterans but, as the number of reported studies using *Bt pakistani* was very small (6), the spectrum of activity is unlikely to have been fully established.

Comparison of Appendices 2.1 (susceptibles) and 2.2 (non-susceptibles or lack of field efficacy) shows that a number of species occur in both lists for the same serovar. There are a number of reasons for this. When compiling their list of susceptible organisms, Krieg and Langenbruch (1981) attributed species being recorded as both susceptible and non-susceptible to differences between strains, which is a likely explanation in many cases, considering the expression of toxins varies within a single serovar. In addition, a number of the studies used different doses of *Bt*, whether as spores, crystal toxins, mixtures or formulated products. Of course, in laboratory assessments, mortality decreases as the dose rate is decreased so in studies at a single rate, the designation 'susceptible' depends on the rate selected. Also, a single insect may appear non-susceptible at a field dosage but susceptible at rates evaluated in the laboratory, as has been demonstrated many times. Even the reverse has been demonstrated, where an insect appears susceptible in the field but was non-susceptible in the laboratory (Kreutzweiser *et al.* 1994). Misidentification of strains is also possible and in some cases, more than one name has been attributed to the same *Bt* subspecies (e.g. *Bt tenebrionis* = *san diego*).

There are a number of variables which need to be understood in evaluating both toxicity and non-target effects of *Bt*-based products. There is considerable evidence that the δ-endotoxins produced by different strains of *Bt* differ in the spectra of their insecticidal activity, most likely due to differential expression of individual toxins between strains. In addition, even formulations based on the same isolate of *Bt* may differ in their toxicity to the same insect. The expression of β-exotoxin is also a variable factor. Certainly, the broad spectrum of activity exhibited by *Bt thuringiensis* can be partly attributed to the expression of β-exotoxin, but to what extent particular strains produce both this toxin and other less well defined toxins is often unclear. Some of these aspects are examined in more detail in the following sections.

5.2.1 INTRAGENERIC DIFFERENCES IN SUSCEPTIBILITY TO *BT*

While we have summarised the sensitivity of insects to *Bt* in Table 5.1, it has been demonstrated that a single *Bt* strain or product is not necessarily toxic to all species within a genus. For example, Peacock *et al.* (1998) found that of 42 species (in seven families) of native Lepidoptera tested for susceptibility to Foray 48B (equivalent of 89 BIU/ha), significant mortality was recorded in 27 of the species. These included the only Papilionidae tested, all 3 Nymphalidae and all 3 Saturniidae, but only 5 of 7 Geometridae (the only Lasicampidae), and 14 of 26 Noctuidae. The only Lymantriidae tested was not susceptible. They also found intrageneric differences were recorded in *Lithophane* and *Catocala*, with only 6 of 8 species of *Catocala* showing susceptibility to Foray. These authors concluded that susceptibility to *Bt* needed to be assessed on a case by case basis.

Similarly, *Spodoptera litura* was sensitive to *Bt kurstaki*, *Bt aizawai* and *kenyae* but not to *Bt tolworthi* or *thuringiensis* (Amonkar *et al.* 1985), whereas larvae of the *Spodoptera frugiperda* were most sensitive to strains of *Bt kenyae* and *tolworthi*, followed by other strains from *Bt kenyae, aizawai* and *kurstaki* serovars. The less toxic strains belonged to the *Bt alesti, dendrolimus, sotto* and *colmeri* serovars (Hernandez 1988). In another study by Salama (1991), the most active cultures against *S. littoralis* were HD-593 (*aizawai*), HD-110 (*entomocidus*), HD-263 (*Bt kurstaki*) and HD-554 (*canadensis*) with LD_{50}s of 22, 27, 36 and 38 µg endotoxin per ml, respectively, as compared with 2160 µg/ml for HD-1-1980 standard preparation. Kalfon and de Barjac (1985) also tested a number of serovars and strains against 1st instar *S. littoralis* and found that the highest toxicity was for strains belonging to *Bt aizawai, kenyae* and *entomocidus*. In contrast, there were also many reports that *Spodoptera* spp. were not susceptible to a wide range of serovars (Krieg and Langenbruch 1981; Appendix 2). These studies have demonstrated that *Bt* toxicity to a particular insect species cannot be accurately predicted by examining toxicity to related species.

5.2.2 BETWEEN SEROVAR VARIATION

A single insect species is often susceptible to more than one toxin, serovar or subspecies. Appendix 2 provides a number of examples of this. In many cases, several *Bt* subspecies are active against one order, in particular Lepidoptera where *Bt aizawai, alesti, anduze, berliner, canadensis, caucasicus, cereus, colmeri, darmstadiensis, entomocidus/subtoxicus, finitimus, fukuokaensis, galleriae, gelechiae, kenyae, kurstaki, kyushuensis, japonensis, morrisoni/tenebrionis, ostriniae, pondicheriensis, shandongiensis, sotto/dendrolimus, subtoxicus, sumiyoshiensis, thompsoni, thuringiensis, tianmensis, tohokuensis, tolworthi, toumanoff, wuhanensis* and *yunnanensis* have all exhibited toxicity to various species.

In many other studies, isolates of various serovars showed differing toxicity to the same insect. For example, larvae of *Achaea janata* were most effectively controlled by *Bt kurstaki* (most toxic to larvae), followed by *Bt galleriae* and *Bt thuringiensis* (Deshpande and Ramakrishnan 1982). However, *Bt sotto, Bt entomocidus* and *Bt aizawai* or their endotoxins alone were not toxic to the larvae. *Bt kurstaki* is the most common subspecies applied against gypsy moth, *Lymantria dispar*, but among other serovars toxic to this species, *Bt galleriae*, *Bt subtoxicus* and *Bt entomocidus* were the most effective in terms of mortality, while *Bt thuringiensis*, *Bt alesti*, *Bt sotto* and *dendrolimus* gave lower mortality rates (Ridet 1973). The LC_{50} values of purified parasporal inclusions of *Bt sumiyoshiensis*, *fukuokaensis*, *darmstadiensis* and *japonensis*, for the silkworm (*Bombyx mori*) larvae were 7.35, 6.45, 3.08 and 2.63 µg/g diet, respectively, showing that their toxicity levels were 5-15 times lower than that of the strain HD-1 (*Bt kurstaki*) (0.49 µg/g diet)(Wasano *et al.* 1998). Where *Bt* was used against *Helicoverpa armigera* and *H. punctigera* in Australia, *Bt kurstaki* HD-1 and a strain of *Bt kenyae* were the most toxic (LC_{50} 40-70 µg/ml); *Bt aizawai, kurstaki* (HD-336), *galleriae* and *wuhanensis* had intermediate toxicity, *Bt thuringiensis* was less toxic (LC_{50} 2100-11,000 µg/ml) and *Bt canadensis* was almost non-toxic (LC_{50} 22,000-44,000 µg/ml) (Teakle *et al.* 1992).

Of 12 H-serotypes tested against the wax moth, *Galleria mellonella*, none was more toxic than *Bt galleriae* (serotype 5a5b). Serotypes 3a3b (*Bt kurstaki*), 3a (*alesti*?), 4a4b (*sotto*) and 4a4c (*kenyae*) all contained some strains which were moderately active against *G. mellonella* but only 5a5b (*Bt galleriae*) and 7 (*Bt aizawai*) contained highly active strains (Jarrett and Burges 1982).

Although *Bt israelensis* is commonly used against mosquitoes, it is not the only subspecies with mosquitocidal toxins. Some *Bt kurstaki* strains produce mosquitocidal toxins, although these are less active than *Bt israelensis* (Ignoffo *et al.* 1981). The LC_{50}s for *Bt israelensis* and *Bt kurstaki* against *A. aegypti* were 0.054 mg/ml and 210.0 mg/ml, respectively. Similarly, Ignoffo *et al.* (1981) reported that *Bt israelensis* was effective against the lepidopterans *Trichoplusia ni, Helicoverpa zea* and *Heliothis virescens* with LC_{50}s of 109.6, 19.3 and 27.6 mg/ml, respectively compared to 15.9, 2.0 and 7.8 mg/ml for *Bt kurstaki*. Ibarra and Federici (1986) assayed parasporal crystals of *Bt morrisoni* and *Bt israelensis* isolates against *Aedes aegpyti* and found their LC_{50}s similar to each other, even though *Bt morrisoni* was generally considered to be toxic to Lepidoptera. Other serovars found toxic to mosquitoes include *Bt thompsoni, malaysiensis, canadensis, jegathesan, darmstadiensis, medellin, kyushuensis* and *fukuokaensis* (Ragni *et al.* 1996). In addition, the bioassays of Dulmage *et al.* (1981) showed

Table 5.1. Number of insect genera (species) susceptible to serovars[1] of *Bacillus thuringiensis* (top line, outside box) and genera (species) recorded as not susceptible (bottom line, inside box). Summary of Appendix 2.

Serovar[1]	Serotype/ subspecies	Coleoptera	Diptera	Lepidoptera	Hymenoptera	Hemiptera	Isoptera	Orthoptera	Siphonaptera	Thysanoptera	Neuroptera	Plecoptera	Phthiraptera & Ephemeroptera	Trichoptera	Dictyoptera	Total no. genera
1	*thuringiensis*	20(25)	20(31)	190(266)	14(14)	9(11)	2(2)	1	2(2)	1			5(9)		2	266(362)
		26(27)	10(17)	20(22)	22(30)	8(9)	2(4)	7(8)		1	1(2)			1		98(121)
2	*finitimus*		1		10(10)									1		12(12)
			2(5)	6(6)												8(11)
3a3c	*alesti*		3(3)	57(75)	4(4)			1					1			66(84)
			5(9)	9(10)	6(6)											20(25)
3a3d	*sumiyoshiensis*		1	4(4)												5(5)
			4(4)													4(4)
3a3b3c	*kurstaki*	14(15)	19(33)	234(330)	22(25)	17(18)	2(2)		1	1	1(4)	2(2)	5(6)			318(437)
		28(35)	24(28)	41(55)	37(65)	12(17)	2(2)	1			1	7(7)	12(13)		1	166(224)
3a3d3e	*fukuokaensis*		5(7)	1												6(8)
		1	5(7)	4(4)												10(12)
4a4b	*dendrolimus*	4(4)	6(9)	57(77)	6(6)	2(2)			1	1			2(2)			79(102)
		6(7)	4(7)	9(9)	5(8)	2(3)				1	1					28(36)
4a4b	*sotto*		1(2)	45(56)	3(3)											49(61)
			5(9)	6(6)	3(6)											14(21)
4a4c	*kenyae*	1	3(6)	21(27)									2(2)			25(34)
			3(7)	5(5)												8(12)
5a5b	*galleriae*	4(4)	12(20)	119(175)	6(6)	5(5)	1	1	3(5)		2(4)		3(3)			156(224)
		7(10)	9(14)	10(13)	19(31)	4(5)	1	1		1	1					53(77)
5a5c	*canadensis*		3(4)	3(3)									1			7(8)
			3(6)	3(3)												6(9)
6†	*entomocidus*	2(3)	4(8)	47(56)	3(3)						1		1			58(72)
		1(1)	4(7)	3(3)	2(2)											10(13)
6†	*subtoxicus*			12(14)									1			13(15)
			1	1												2(2)
7	*aizawai*		2(5)	41(60)		1							2(2)			45(68)
			5(8)	6(6)	1								1			13(16)
8a8b*	*tenebrionis*	42(50)	1	4(5)		3(5)	2(2)	2(2)							2	56(67)
		20(21)	4(4)	8(10)	2(2)	1	1	3(3)			1				1	41(44)
8a8b*	*morrisoni*	1	7(13)	11(12)		1							2(2)			22(29)
		3(3)	4(9)	12(14)												19(26)
8a8c	*ostriniae*			1									1			2(2)
			2(2)	2(2)												4(4)
8b8d	*nigeriensis*			1												1
																0
9	*tolworthi*	2(2)	5(7)	16(18)									2(2)			22(29)
		1	4(7)	4(4)												9(12)
10a"	*caucasicus*			9(10)					1							10(11)
																0
10a10b	*darmstadiensis*	2(2)	9(14)	10(11)									1			22(28)
		2(2)	5(8)	7(7)												14(17)

10a10c	*londrina*											0
		1	1	3(3)								5(5)
11a11b	*toumanoffi*		3(3)	4(4)						1		8(8)
			2(6)									2(6)
11a11c	*kyushuensis*		4(7)	3(3)						1		8(11)
			4(5)	4(4)								8(9)
12	*thompsoni*		3(3)	8(8)						1		12(12)
				3(3)								3(3)
13	*pakistani*		1	5(5)						2(2)		8(8)
		1	4(4)	4(4)								9(9)
14	*israelensis*		47(103)	5(5)	8(9)					2(2)	2(2)	64(121)
		21(27)	18(21)	12(12)	1	11(15)		1	1	7(9)	5(5)	77(92)
15	*dakota*											0
			2(2)	2(2)								4(4)
16	*indiana*			3(3)						1		4(4)
		1	1	2(2)								4(4)
17	*tohokuensis*		2(2)									2(2)
			3(3)	1								4(4)
18a18b	*kumamotoensis*	1	1	2(2)								4(4)
		1	2(2)	3(3)								6(6)
19	*tochigiensis*		2(2)									2(2)
			1	1								2(2)
20a20b	*yunnanensis*		1									1
			2(3)	1			1					4(5)
20a20c	*pondicheriensis*		1	1								2
			2(3)	1								3(4)
21	*colmeri*			1								1
			2(2)	1								3(3)
22	*shandongiensis*		1									1
		1	4(5)	4(5)								9(11)
23	*japonensis*	6(12)		6(6)								12(18)
			3(4)									3(4)
24a24b	*neoleonensis*		3(3)	2(2)								5(5)
			3(5)	3(3)								6(8)
24a24c	*novosibirsk*			1								1
												0
25	*coreanensis*		1									1
			2(2)	1								3(3)
26	*silo*											0
		1	1	3(3)								5(5)
27	*mexicanus*		2(2)									2(2)
		1	3(3)	3(3)								7(7)
28a28c	*jegathesan*	4(8)										4(8)
												0
29	*amagiensis*		2(3)	1			1					4(5)
			2(2)	2(2)								4(4)
30	*medellin*		3(5)									3(5)
												0
31	*toguchini*			1								1
												0
33	*leesis*		1	2(2)								3(3)
			4(4)	1								5(5)
34	*konkukian*			1								1
		1	2(2)	3(3)								6(6)

Table 5.1. *(Continued)*

Serovar[1]	Serotype/ subspecies	Coleoptera	Diptera	Lepidoptera	Hymenoptera	Hemiptera	Isoptera	Orthoptera	Siphonaptera	Thysanoptera	Neuroptera	Plecoptera	Phthiraptera & Ephemeroptera	Trichoptera	Dictyoptera	Total no. genera
36	*malaysiensis*		3(5)													3(5)
																0
37	*andaluciensis*															0
		1	1	3(3)												5(5)
38	*oswaldocruzi*			1									1			2(2)
			3(3)													3(3)
39	*brasiliensis*		2(2)													2(2)
				1												1
40	*huazhongensis*		2(2)													2(2)
			2(2)	3(4)												5(6)
43	*guiyangiensis*			1												1
		1	1	3(3)												5(5)
44	*higo*		3(4)													3(4)
		1		2(2)												3(3)
47	*wratislaviensis*		2(2)													2(2)
															2(2)	2(2)
	wenguanensis			1												1
		1		2(2)												3(3)
	amagiensis		2(3)	1			1									4(5)
			2(2)	2(2)												4(4)
	tianmensis			3(3)												3(3)
																0
	sooncheon															0
		1	1	3(3)												5(5)
	wuhanensis		1	13(15)									1			15(17)
				1			1									2(2)
	oyamensis	1	2(2)	1				1								5(5)
		1	4(5)	4(4)	1											10(11)

[1] serovars as listed in Lecadet *et al.* 1999
[*] *morrisoni* and *tenebrionis* are both H8a8b serotype
† *subtoxicus* and *entomocidus* both serotype 6
# *caucasicus* is H10a (see Chapter 2)

activity against mosquitoes of the same strains of *Bt* subspecies *thuringiensis, kurstaki, alesti, kenyae, galleriae, canadensis, entomocidus, aizawai* and *tolworthi* (Appendix 2.3). Four day old larvae of the mosquito *Ae. aegypti* exposed to four *Bt* varieties had vastly different LT_{50} values (0.01 days for *Bt israelensis*, 1.3 for *Bt kurstaki*, 3.9 for *Bt galleriae* and 3.4 for *Bt thuringiensis* (Ignoffo *et al.* 1980a) and caused 100% mortality, except *Bt thuringiensis* which killed 83% of larvae. Larget and de Barjac (1981a) also examined whole cultures of 15 serotypes against mosquitoes and found only *Bt israelensis* to be toxic, being active at 1000× lower rates than other serovars. Some mortality of at least one mosquito species was obtained with more concentrated suspensions of *Bt aizawai, entomocidus, galleriae, kyushuensis* and *tolworthi.*

5.2.3 WITHIN SEROVAR VARIATION

The variety/subspecies cannot be used to predict the potency of a strain (Jarrett and Burges 1982b; Park *et al.* 1998) as isolates show considerable variation within a single serovar/subspecies. One of the most comprehensive studies of toxicity of isolates within serovars to a variety of insects was the work of Dulmage and cooperators, first reported in a preliminary report by Dulmage *et al.* (1981). The full study has never been published, but a summary of the data can be found in Appendix 2.3. The study found that grouping by serovar partially classifies the spectra of toxicity of the isolates, but there was great variation in activity, often ranging from almost zero to high in each serovar against a particular insect. Dulmage *et al.* found each serovar could be subgrouped by using activity ratio (dividing toxicity to one insect by toxicity to another insect) into statistically homogeneous groups, giving further partial classification within serovar.

Many other studies have shown similar variability in toxicity within serovars. For example, the biting louse *Damalinia ovis* was susceptible to many subspecies (Appendix 2.1), but individual isolates within some serovars showed little or no activity (Drummond *et al.* 1992). Various isolates of *Bt kurstaki* caused high (>75%) mortality, moderate (25-75%) and low to nil mortality to this insect. Saitoh *et al.* (1996) examined 1449 isolates, representing 31 serovars for activity against the mosquito *Culex pipiens molestus* and the moth-fly *Telmatoscopus albipunctatus* (Appendix 2). They found 152 of 1297 isolates were toxic to *C. p. molestus* and 30 of 1419 were toxic to *T. albipunctatus,* with no serovar containing isolates which were all toxic to either insect. Another example of serovar not reflecting toxicity is isolate NTB-88, an isolate from soil in Korea which belongs to H8a8b (*Bt morrisoni*) but was not toxic to a range of Lepidoptera, Coleoptera and Diptera including two mosquito species (Park *et al.* 1998).

5.2.4 TOXICITY OF DELTA-ENDOTOXINS

Specificity of subspecies of *Bt* is dependent on the toxins produced, although separating activity of individual toxins from variables in the bioassay system, insects and type of formulation or preparation of *Bt* is difficult. The spectra of toxins produced by isolates of *Bt* varies greatly, even for the δ-endotoxins on which most *Bt* products are currently based. Toxicity of *Bt tenebrionis*, for example, is based on a single Cry3A toxin, while *Bt israelensis* produces up to 5 Cry and Cyt toxins in one strain. *Bt kurstaki* HD-1 strain contains genes for several crystal proteins: Cry1Aa, Cry1Ab, Cry1Ac, Cry2A and Cry2B (Moar *et al.* 1995a; Lee *et al.* 1996). Despite the high amino acid similarity often found between some of these toxins, they frequently exhibit different insecticidal effects (e.g. Ge *et al.* 1989,1991).

As most reports involve the use of a *Bt* isolate, strain or subspecies, the actual individual toxin effects are often unknown, although the increasing interest in transgenic plants expressing *Bt* insecticidal toxins has resulted in several recent studies focussing on the effect of single toxins. Van Frankenhuyzen *et al.* (1991) tested individually expressed Cry1Aa, Cry1Ab and Cry1Ac toxins against larvae of the tortricids *Choristoneura fumiferana, C. occidentalis* and *C. pinus*, the lymantriids *Lymantria dispar* and *Orgyia leucostigma,* the lasiocampid *Malacosoma disstria* and the noctuid *Actebia fennica* (*Ochropleura fennica*). The effects varied between different insects, with all 3 Cry1A toxins equally effective against *M. disstria* but Cry1Aa and Cry1Ab up to 5-fold more toxic than Cry1Ac to the tortricids and 100-fold more to lymantriids. No Cry1A toxin had any effect against *A. fennica*. The toxicity of HD-1 was similar to that of the individual Cry1Aa or Cry1Ab toxins to all the species. Against *B. mori*, Cry1Aa was much more toxic than Cry1Ab or Cry1Ac (Table 5.2).

Table 5.2. Examples of the activity spectra of Cry1 toxins against Lepidoptera (modified from van Frankenhuyzen 1993, reproduced with permission of John Wiley & Sons; additional references therein).

\+ = active; ± = low activity; - = no activity; * = not tested

Species	Cry gene							
	1Aa	1Ab	1Ac	1B	1C	1D	1E	1F
Spodoptera exigua	±	±	-	-	+	+	±	+
S. littoralis	-	-	-	-	+	-	+	*
S. exempta[1]	±	±	-	±	+	±	+	*
Actebia fennica	-	-	-	-	-	-	-	*
Trichoplusia ni	+	+	+	*	+	+	+	+
Heliothis virescens	+	+	+	-	±	-	±	±
Ostrinia nubilalis	+	+	+	*	*	*	+	+
Mamestra brassicae	+	+	-	-	+	-	-	*
Pieris brassicae	+	+	+	+	+	-	-	*
Manducta sexta	+	+	+	-	+	+	+	*
Choristoneura fumiferana	+	+	+	+	+	+	+	+
Lymantria dispar	+	+	±	-	±	+	-	+
Orgyia leucostigma	+	+	±	-	+	±	-	+
Malacosoma disstria	+	+	+	+	+	+	+	*
Bombyx mori	+	±	-	-	+	+	+	*
Plutella xylostella	+	+	+	+	+	-	-	*
Lambina fiscellaria	+	+	+	-	+	+	+	*
Perlieucoptera coffeella[2]	-	-	+	+	-	*	-	*

[1] Bai *et al.* (1993)
[2] Filho *et al.* (1998)

MacIntosh *et al.* (1990c) examined the effect of purified Cry 3A from *Bt tenebrionis*, and Cry1Ab and Cry1Ac from *Bt kurstaki* on a number of pests. Only Lepidoptera were sensitive to Cry1Aa and Cry 1Ac, while only Colorado potato beetle was sensitive to Cry3A. There was no synergism found when all three toxins were used. Interestingly, while Herrnstadt *et al.* (1986) found that corn rootworm and cotton boll weevil were sensitive to *Bt tenebrionis*, MacIntosh *et al.* (1990c) found no effect of the purified Cry3A, suggesting some other metabolites or toxins could have caused the sensitivity.

Cry2A has an unusual dual toxicity to lepidopteran and dipteran insects, although considerable variation in response to purified toxinis exhibited within those orders. Purified Cry2A from *Bt kurstaki* was toxic only to Lepidoptera and Diptera when tested against 35 insect species (Coleoptera, Collembola, Diptera, Hemiptera, Hymenoptera, Isoptera, Lepidoptera, Neuroptera and Orthoptera) and an isopod (Sims 1997). While the mosquitoes *Aedes aegypti, Ae. triseriatus* and *Anopheles quadrimaculatus* had LC_{50}s in the range of 0.37-37 µg/ml of purified Cry2A, other Diptera such as *Musca domestica, Drosophila melanogaster* and *Culex pipens* required over 100 µg/ml (Sims 1997).

In some cases, Cry or Cyt proteins identified have no known toxicity (Appendix 1; Table 2) or the toxicity is expressed only after an *in vitro* activation and solubilisation step. For example, the Cry3C protein is not toxic to any lepidopteran or coleopteran tested as a spore-crystal or solubilized crystal state, but is toxic to *Leptinotarsa decemlineata* (Chrysomelidae) after trypsin activation (Lambert *et al.* 1992).

Binding of endotoxins to receptors in the gut is crucial to toxicity. However, gut binding assays do not always predict toxicity. Filho *et al.* (1998) found that although Cry1Ac and Cry1E were both positive for binding assay tested, Cry1E was negative in toxicity tests against *Perlieucoptera coffeella* caterpillars.

The ratio of endotoxin present in a single cell is relevant to the comparative toxicity, but is rarely published. Liu *et al.* (1996) analysed two commercial formulations by HPLC and found that CryIC accounted for 26% of the CryI protein in the *Bt aizawai* formulation, but did not occur in the *Bt kurstaki* formulation. CryIAb was the most abundant CryI protein in the commercial formulations of *Bt aizawai* and *kurstaki* (Liu *et al.* 1996).

5.3 SYNERGY BETWEEN DELTA-ENDOTOXINS

Several studies have examined interactions between the Cry1 toxins from HD-1. Van Frankenhuyzen *et al.* (1991) reported that Cry1A toxins had a slightly synergistic effect against lymantrids, especially with *Lymantria dispar* (28× of that of toxins used alone, when used in combination). However, Tabashnik (1992) re-evaluated the data and concluded that the study by van Frankenhuyzen *et al.* did not demonstrate sufficient effect to indicate synergy for two of the insects and no synergy was shown between the 3 Cry1A toxins against 7 species. Lee *et al.* (1996) examined synergism between CryIA toxins against gypsy moth and silkworms and synergism was observed between Cry1Aa and Cry1Ac for gypsy moth (about 4-7× above predicted value from individual toxin effects), but Cry1Aa and Cry1Ab were antagonistic. Laboratory studies showed that when Cry1Ac and Cry1F δ-endotoxins from *Bt* were mixed, toxicity (EC_{50}) to larvae of *Helicoverpa armigera* was increased 26 times (Chakrabarti *et al.* 1998).

Tabashnik (1992), using the data of Chilcott and Ellar (1988), found synergy between a 27-kDa (Cyt1A) toxin and 130- or 65-kDa (Cry4) toxins from *Bt israelensis* against larvae of *Aedes aegypti* (up to 10 fold increase in potency) and possibly weak antagonism between the 65 and 130 kDa proteins. Poncet *et al.* (1995) tested Cry4A (=CryIVA), Cry4B (=CryIVB) and Cry11A (=CryIVD) toxins from *Bt israelensis* alone and in combinations against three mosquito species (*Aedes aegypti, Anopheles stephensi* and *Culex pipiens*) larvae and found clear evidence of synergy, particularly between Cry4A and both Cry4B and Cry11A . Wirth *et al.* (1997) also found that tolerance to CryIV proteins (from *Bt israelensis* and a transformed acrystalliferous *Bt kurstaki* strain expressing CryIVD) in larvae of *Cx. quinquefasciatus* was markedly reduced by sublethal quantities of Cyt1A.

The use of multiple formulations can also result in additive or antagonistic effects, although effects may not only relate to toxicity of endotoxins. Ameen *et al.* (1998) found that effects of formulations of *Bt aizawai* (Xentari As R) and *kurstaki* (Dipel 6AF R) used together were additive against *Heliothis virescens*, but Xentari and Dipel ES R were antagonistic against the same host. For *H. zea,* Xentari was significantly antagonistic to three formulations of Dipel.

Asano and Hori (1995) and Asano *et al.* (1995) used culture supernatants to synergise δ-endotoxins against *Spodoptera litura*. This effect was not found against *Plutella xylostella. Bt* active against *S. litura* could be synergised by culture supernatants from non-lepidopteran active strains such as *Bt japonensis* and *Bt morrisoni*, suggesting the effect was not due to solubised δ-endotoxins but some other factor not yet identified.

5.4 TOXICITY OF BETA-EXOTOXIN

The toxicity of some strains against insects appears to result from β-exotoxin (Maciejewska *et al.* 1988). The toxicity of exotoxin to *Musca domestica*, coupled with lack of activity of *Bt* endotoxins or with retention of activity after autoclaving (exotoxins are heat-stable and endotoxins are heat-labile), was used as a primary screen for the occurrence of exotoxin for many years. Beta-exotoxins are produced by a number of subspecies (see Table 2.2). Krieg and Langenbruch (1981) listed over 70 species of insects as susceptible to β-exotoxin. In addition to a number of pest species, exotoxin has been shown to be harmful to beneficial insects such as bees (Krieg and Langenbruch 1981). As detailed in Chapter 7, exotoxin also poses a risk to mammals, with an intraperitoneal LD_{50} for mice of 10-20 μg/g body weight and *per os* toxicity of some salts of beta-exotoxin considerable in hens, even at low doses (Sebesta *et al.* 1981).

The risk to non-target organisms and mammals has led to the requirement that products in most countries must be free from β-exotoxin. However, in the former USSR and other countries of Eastern Europe, *Bt* preparations containing exotoxins have been used, especially those based on *Bt thuringiensis,* providing some data on the effect of beta-exotoxin against some insects. Beta-exotoxins have shown toxicity to numerous insects, especially against Diptera, mites (Inogamov and Sharafutdinov 1975; Hoy and Ouyang 1987; McKeen *et al.* 1988) and fleas (Yakunin *et al.* 1974). Beta-exotoxins from *Bt darmstadiensis* and *Bt thuringiensis* were toxic to *Ostrinia nubilalis* (Lepidoptera: Pyralidae) and only *Bt tolworthi* produced both endo and exotoxins active against this insect (Mohd Salleh and Lewis 1983). Against 1st, 3rd and 4th instar larvae of *Helicoverpa zea*, β-exotoxin LC_{50}s were 4.9, 134.6 and 286.2 μg a.i./ml diet after 7 days (Herbert and Harper 1985). The "type II" β-exotoxin from *Bt morrisoni* isolated by Levinson *et al.* (1990) was lethal to the Colorado potato beetle, *Leptinotarsa decemlineata*. Beta-exotoxin from *Bt thuringiensis* was toxic to the nematode *Meloidogyne* sp., killing 85-95% of 2nd stage juveniles after 7 days (Rai and Rana 1979). Herbert and Harper (1986) reported that nymphs of *Geocoris punctipes* were susceptibility to β-exotoxin, whereas adults of the same species were not.

Exotoxin can also act as a feeding deterrent in some insects (e.g. Thomsen *et al.* 1998; Gardner *et al.* 1986).

5.4.1 SYNERGIES BETWEEN BETA-EXOTOXIN AND DELTA-ENDOTOXINS/SPORES

Several studies have examined the effect of using both exotoxin and δ-endotoxin and/or spores from *Bt*. Use of a commercially produced exotoxin product, Thuringiensin, with a spore-crystal mixture of *Bt kurstaki* could cause potentiation effects against *Spodoptera exigua* neonate larvae (Moar *et al.* 1986). Mueller and Harper (1987) found synergism between exotoxin and Dipel against neonate (3- to 6-h-old) larvae of *S. frugiperda*, especially at LC_{30} concentrations of both where 100% mortality was achieved, compared to a predicted mortality of 51%. Subsequently, Gardner (1988) found that against *S. frugiperda,* the use of combinations of exotoxin (preparation ABG-6162 used at 120-360 μg/ml) and LC_{50} rates of *Bt kurstaki* (Dipel) could be synergistic, but only from 3-7 days, as after that β-exotoxin-related mortality was too high. Navon (1993) thought this result reflected the slow action of β-exotoxin, which is expressed as moulting failure. Dubois (1986) showed synergism between *Bt kurstaki* HD-1 and exotoxin against 2nd instar gypsy moth (*Lymantria dispar*). Against the Larch bud moth, *Zeiraphera diniana*, exotoxin from *Bt thuringiensis* increased mortality when Thuricide 90T was used from 53% to 80%, but had very little effect by itself (Benz 1975).

5.5 COMPARATIVE EFFICACY OF FORMULATED *BT*

It is extremely difficult to compare *Bt* products for activity against the same insect. The main problem is lack of standardisation in expressing the active ingredients in each product. The application rates vary, active ingredients can be expressed as spores/ml (meaningless in most cases), dry weight (often includes fermentation materials, and therefore not measuring only the active ingredient) and international units (only available for some subspecies and measured against a single insect species - see Section 5.1.3). Because of this problem of standardisation, the comparisons below must be considered a general guide only.

Several studies have failed to find differences between various formulations in the laboratory or field. For example, Patti and Carner (1974) used Bactospeine, Biotrol, Dipel and Thuricide against *Heliothis virescens* and found no significant differences in effectiveness against larvae caged on cotton plants. However, in many cases the various types of *Bt* formulations have differing effects on target species and the environment. In some cases, the type of formulation (ie. pellet, liquid) has a major impact on efficacy. For example, Grant *et al.* (1985) tested several commercial formulations of *Bt israelensis* for mosquito control in the field in California. They showed VectoBac sand granules, which are heavy with low content of active ingredient, were ineffective against *Cx. tarsalis* at application rates of less than 28 kg/ha. VectoBac aqueous suspension, which was more concentrated, gave 98% control at 0.5 kg/ha. Some variation in efficacy attributed to formulation may result from varying feeding behaviour of exposed insects (Section 9.2). Similar varying responses to different formulations were reported by Moar and Trumble (1990), who tested five *Bt kurstaki* formulations against neonate *Spodoptera exigua.* They reported that the wettable powder/granule formulations of SAN 415, WG 354 and ABG-6218 had lower (7 and 3.35x) LC_{50}s than Dipel 2× when used at the same μg/ml diet rate. Liquid formulation (SAN 415) had an even lower LC_{50} expressed as μl/ml diet than the wettable formulations, although field rates for these products were not stated.

Formulations can also vary in efficacy between insect species. For example, Burgerjon and Martouret (1971) found that two formulations with approximately equal activity against a cabbage white butterfly, *Pieris brassicae*, can differ widely in activity against the silkworm, *Bombyx mori*, with one preparation being approximately 15 times more active than the other.

Differences in sensitivity may also be related to developmental stage of the insect tested (Section 9.7). Usually young larvae, particularly first instars, are more susceptible under the same conditions than older larvae. However, Morris (1986) found that young larvae of *Mamestra configurata* were more sensitive to Dipel 132 than older larvae, but older larvae were more susceptible to Thuricide 48 LV than younger larvae. Against young instars of *Hyphantria cunea*, Thuricide had a 48h LC_{50} of 541 μg/ml while Biobit (*Bt kurstaki)* had a LC_{50} of only 444 μg/ml. Against older instars, the difference between Thuricide (2133 μg/ml) and Biobit (400 μg/ml) was more pronounced (Ecevit *et al.* 1994).

Products may contain mixtures of *Bt* serotypes or toxin genes usually present in two serotypes may be combined in a single isolate by conjugational gene transfer, which will also impact on the environmental risk profiles. A number of studies have suggested there are advantages in using endotoxins from more than one serotype or variety of *Bt* against some pests. The noctuid cotton pests, *Earias biplaga* and *E. insulana*, are highly sensitive to *Bt galleriae* and *aizawai*, with *aizawai* 10× more effective than *Bt kurstaki* and 100× more effective than *Bt berliner* (*Bt thuringiensis*), both of which have been used commercially against these pests (Frutos *et al.* 1987). Although *Bt kurstaki* is often the base for biopesticides used against lepidopteran pests, it is not always the most toxic to moths. Of six strains evaluated against larvae of the pine moth (*Dendrolimus pini*), nun moth (*Lymantria monacha*) and winter moth (*Operophtera brumata*), the most toxic were *Bt sotto* (biotype *dendrolimus*), *thuringiensis* and *aizawai* respectively, rather than *Bt kurstaki* (Sierpinska 1997). It is common to find isolates of higher toxicity than the commercially available *Bt* products, especially when testing strains from a number of subspecies (e.g. Jarrett and Burges 1986).

The composition of the toxins and spores also has a marked influence on the efficacy of products against *Plutella xylostella* larvae. Asano and Seki (1994) used four commercially available wettable powder formulations (Bacilex [*Bt aizawai* + *kurstaki*], Thuricide [*Bt aizawai* in this publication, usually *Bt kurstaki*], Toarow-*BT* and Dipel [*Bt kurstaki*]) and three experimental wettable powder formulations (KM301- crystal toxin Cry1Ac and killed spores, KM202 and KM302 - Cry1Ac toxin in killed *Pseudomonas fluorescens* cells). Formulations based on multiple toxins and killed spores were more potent against 3rd instar larvae than encapsulated single crystal toxin, and Thuricide was the most toxic commercial formulation.

Often there is no obvious reason for reported differences between products. For example, it is surprising that when results for Dipel and Thuricide are compared, (at various times, both these products have been based on *Bt kurstaki* HD-1 strain), differences in toxicity to the same insect under similar conditions were evident (e.g. Mathur *et al.* 1994). In the laboratory, ABG 6105 (*Bt kurstaki*) was more toxic than Dipel WP (*Bt kurstaki*) and Bactospeine 1 (*Bt thuringiensis*) against 1st-5th instar *Spodoptera littoralis*, but Thuricide WP (*Bt kurstaki*) was "ineffective" (Jansens *et al.* 1984). Takaki (1975) also found differences in efficacy between a number of *Bt*-based products against insects in Japan. For example, Takaki recorded Dipol (sic) (*Bt kurstaki*) and Selectozin (*Bt aizawai*) as effective against the vegetable pest, *Prodenia litura*, but Cellstart (*Bt sotto*) and Thuricide A ("*Bt aizawai*?" in Takaki 1975) as ineffective.

5.5.1 DIPEL: SPECTRUM OF ACTIVITY

One of the most widely used *Bt* products over the years is Dipel, produced by Abbott Laboratories, Chicago. Dipel is based on the *Bt kurstaki* strain, HD-1 and, while formulations have changed over the years, it has been used in many countries against numerous pest insects and a few mites in both experimental and commercial control efforts. A list of insects and other organisms recorded as susceptible or as tolerant to Dipel formulations (Appendix 3) illustrates the specificity of a single strain of *Bt kurstaki* under field situations. Dipel was toxic to only a few species of Acari, Coleoptera, Diptera and Hemiptera, but was recorded as toxic to 170 species of Lepidoptera.

The majority of insects listed as not susceptible to Dipel are parasitoids and predators, reflecting the concerns of researchers, rather than the full range of activity of *Bt kurstaki* (Appendix 3). A number of species of Lepidoptera are listed as both susceptible and non-susceptible. Such contradictory reports may result from field studies where target pests were recorded as non-susceptible because Dipel exhibited low efficacy. However, poor efficacy may also have resulted from factors such as lack of persistence rather than inherent non-susceptibility of insect species (e.g. *Cydia* spp. - Zivanovic and Stamenkovic 1976).

5.6 ROLE OF SPORES IN MORTALITY

The amount of insect mortality which can be attributed to spores as opposed to toxins produced by *Bt* strains has been investigated in a number of studies. Some studies (e.g. Mohd-Salleh and Lewis 1982b) have found that while pure toxins or crystals are toxic to insects, a combination of spores and crystals often gives the highest mortality. Schesser and Bulla (1978) also showed toxicity of *Bt* spores to the tobacco hornworm, *Manduca sexta*. Against *Galleria mellonella*, separate use of spores (with <0.01% crystals) and crystals (with <0.02% spores) of *Bt galleriae*, *aizawai* and *wuhanensis* were only moderately toxic compared to 1:1 mixtures. In addition, an acrystalliferous *Bt aizawai* mutant was not toxic, suggesting the spore activity may have been caused by the small amount of crystals contaminating the spore preparation (Li *et al.* 1987). Conversely, crystals without live spores had little effect. The addition of 0.01% spores to crystals increased the kill to 64% of larvae, from 36% mortality at 0.001% spores to crystals. These results with *G. mellonella,* demonstrating the importance of spores and crystals combined, were not found with a number of other lepidopteran pests (Li *et al.* 1987; Navon 1993). Li and Luo (1989) found that while high concentrations of spores were toxic to *G. mellonella* and *Pieris brassicae*, there was a large protein present in both spores and crystals which was not present in acrystalliferous mutants and strains not virulent to the two insects, suggesting that it was the presence of specific toxins in the spore coats which was related to virulence.

In the study of Moar *et al.* (1989), whole cultures were more toxic than either P1 (135 KDa, = Cry 1) or P2 (65 KDa, = Cry 2) crystal proteins from two *Bt kurstaki* strains, except for the P2 endotoxins of strain HD-1 (Table 5.3). Preparations containing spores alone were the least toxic, but did have activity, showing some role for spores in activity against *S. exigua.* Nyouki *et al.* (1996) examined spore and toxins interactions against *S.*

Table 5.3. LC_{50}s (μg/ml diet) against *Spodoptera exigua* neonate larvae of *Bt kurstaki* strains HD-1 and NRD-12 whole cultures and δ-endotoxins (from the data of Moar *et al.* 1989).

Strain	Sporulating culture	Spore preparation	P1 (Cry 1) endotoxins (135 kDa)	P2 (Cry 2) endotoxins (65 kDa)	P1 and P2
NRD-12	20.8	166	63.0	72.4	82.0
HD-1	49.3	117	153.0	34.2	157

frugiperda. Using formulations based on single toxins (MVP - Cry1Ac and MYX - Cry1Ab) and a toxin free formulation of spores (MYD) they found a non-dose related low level mortality using toxin-free spores against 3rd instar larvae. Interactions where spores and toxins were used in combination varied and were dose dependent, with Cry1Ac and spores additive at lower doses, but synergistic at higher doses, while Cry1Ab and spores ranged from antagonistic at low concentrations to additive at high doses.

With *Bt tenebrionis* against 1st instar larvae of the Colorado potato beetle, *Leptinotarsa decemlineata*, spores were not involved in the efficacy of spore-crystal mixtures (Riethmuller and Langenbruch 1989; Meyers 1989 in Keller and Langenbruch 1993). In contrast, studies with the same subspecies against *Agelastica alni* (adler leaf beetle), found that ultraviolet irradiation of spores reduced the efficacy of a spore-crystal mixture. In general, however, studies suggest the role of spores in toxicity to insects is minor.

6 Effects on Non-target Microbes and Invertebrates

Laboratory tests against invertebrates indicate that *Bt* has limited impacts on non-target organisms. However, given the wide host range of many *Bt* strains, there can be no doubt that some non-target impacts of *Bt* application will occur. For example, more than 300 lepidopteran species have been reported as susceptible to at least one strain of *Bt kurstaki.* However, dose rates, targeting of applications, inherent variability in susceptibility of insect populations, various environmental factors and insect feeding behaviour all contribute to reduce non-target impacts in the field. Ecosystem studies on Lepidoptera have shown some non-target impacts of *Bt kurstaki* strains but these were insufficient to endanger any species assessed and effects were rarely seen beyond one season. *Bt* is safe for earthworms, spiders and bees at field application rates, unless excess exotoxin is included in the product. *Bt* toxins are not phytotoxic, although it is possible that some formulations may contain phytotoxic components.

Following aerial application, there is potential for *Bt* to enter waterways and some subspecies (e.g. *Bt israelensis*) are used against mosquito and blackfly larvae in water. Only a limited number of studies have examined effects in aquatic communities, but these have found only minimal impacts of *Bt israelensis, Bt tenebrionis* and *Bt kurstaki* on aquatic communities.

Review of a number of studies on the effect of *Bt* application on insect predators and parasitoids indicates that *Bt* is rarely harmful. On the contrary, when *Bt* replaced routine use of chemical insecticides, the numbers of beneficial arthropods usually increased. While competition for hosts and reduced host quality can indirectly retard development or cause premature death of adult predators and parasitoids, *Bt* is generally considered valuable in IPM systems.

6.1 HISTORY AND PROBLEMS WITH NON-TARGET STUDIES

Early reviews on the non-target impacts of *Bt* application often contained sweeping statements on the safety of *Bt* sprays and it was generally considered safe for use in most situations and unlikely to influence non-target populations. This view possibly reflects the broad spectrum of activity of chemical pesticides in use at the time, and was based on a relatively low level of knowledge of the impacts on non-target organisms. It was also considered that *Bt* was restricted in specificity to Lepidoptera. In the 1990s, concern was expressed about the specificity of *Bt*, particularly as related to the impacts on organisms exposed to the two most commonly applied subspecies, *Bt kurstaki* and *Bt israelensis*. Non-target impacts are directly related to the range of susceptible insects (Chapter 5). This chapter reports on studies where impact of *Bt* has been examined against non-target species or beneficial species such as predators, parasitoids, bees and earthworms. Non-target impacts have been measured directly by bioassay or indirectly at an ecosystem/community level.

There are several problems or considerations in reviewing non-target impacts reported in the literature. As discussed in the previous chapter, dose of *Bt* is directly related to toxicity and some organisms that are susceptible at very high doses would not be susceptible at field rates. In other cases, high doses result in feeding inhibition, even in relatively tolerant organisms. Therefore, studies need to be evaluated in terms of field dosage. This is not always possible, as some studies have used purified toxins or experimental formulations, or have given no indication of field dose. Another problem in the interpretation of non-target impacts is the likelihood of an organism coming into contact with *Bt*. In some cases, organisms which have been shown to be susceptible to a particular *Bt* may be unlikely to be present in the same area as a pest that *Bt* is used to treat. Also, *Bt* entering environments such as waterways from runoff would, in most cases, be relatively dilute compared to the levels used in tests. Therefore, defining non-target impacts can be difficult from laboratory and even field studies. These factors need to be considered when assessing non-target impacts.

6.2 PHYTOTOXICITY

Phytotoxic effects by *Bt* have rarely been reported. A single report (Sharma *et al.* 1977) using the β-exotoxin thuringiensin A and the protein subunit of the δ-endotoxin from *Bt* found a depressive effect on mitosis in root-meristem cells of *Allium cepa,* possibly by prolonging the cell-generation time. However, there have been no other reports of detrimental effects on plants from *Bt* or *Bt*-based product application. Phytotoxicity is more likely to occur with formulations than *Bt* toxins or cells alone, as some formulation components could be phytotoxic (see Chapter 4). No detrimental effect of *Bt* treatment was reported on tobacco (Cheng and Hanlon 1990), fibre quality of cotton (Satpute *et al.* 1988), seed yield or seed oil content of sunflowers (Rogers *et al.* 1984), avocado fruits or leaves (Izhar *et al.* 1979), on sprouting of potatoes (Das *et al.* 1998) or poinsettia growth (Schuster and Harbaugh 1977) or against a variety of plants in over 100 greenhouse trials (Zuckerman *et al.* 1994; US Patent 05378460). Falcon (1973 in Forsberg *et al.* 1976) found that *Bt* application on cotton did not interfere with plant physiology. The USA EPA Reregistration Eligibility Decision (RED) documents on *Bt* (EPA 1998) reported no detrimental effects on plant life, including terrestrial, semi-aquatic and aquatic plant life.

6.3 MICROORGANISMS

Several studies have found antagonistic effects of *Bt* against other microorganisms. Yudina and Burtseva (1997) found that the endotoxins of *Bt tenebrionis* produced an antibiotic effect on three out of four species of micrococci tested. Yudina and Burtseva (1997) also reported antibiotic activity of *Bt* subspecies *amagiensis, novosibirsk, wuhanensis* and *yunnanensis* against the bacteria *Micrococcus aurantiacus, Erwinia carotovora* and *Streptomyces chrysomallus.* Visser *et al.* (1994) examined the impact of Dipel 176 (*Bt kurstaki*) on soil microflora and decomposition processes. There was no effect on microbial respiration, microbial biomass and processes such as cellulose decay at the recommended field rate. When applied at 1000× field rate, substrate induced respiration and biomass were increased over levels in the control and ammonification and nitrification were reduced after eight weeks incubation. The experiments were of relatively short duration and it seems unlikely that the changes in microbial processes would be other than short-lived in field soils. The increase in microbial biomass after *Bt kurstaki* application could have been caused by germination of *Bt* but could also have resulted from formulation components of this petroleum-based product (Visser *et al.* 1994).

Bernier *et al.* (1990) reported no effect of the A20 strain of *Bt kurstaki* on soil microorganisms even at 100 times the concentration recommended for aerial spraying. There was also no significant build-up of spores over a period of 11 months. Thuringiensin was reported to have no effect against the fungi *Aspergillus* sp., *Fusarium* sp., *Pythium* sp., *Phytophthora* sp. and *Rhizoctonia* sp. (Bolanos *et al.* 1991).

Bt israelensis can be encapsulated in vacuoles of the protozoan *Tetrahymena pyriformis,* which, at least in an experimental system, can then act as a delivery vehicle to mosquitoes (Zaritsky *et al.* 1991; Manasherob *et al.* 1998). In an earlier study, soil dwelling protozoa were unaffected by the addition of *Bt* (Casida 1989).

6.4 INVERTEBRATES

6.4.1 COLLEMBOLA

Addison and Holmes (1995) examined the effect of *Bt kurstaki* (Dipel 8L and 8AF) on the soil collembolan *Folsomia candida* in a microcosm study. Collembolan populations initially increased in untreated soil and soil amended with the aqueous formulation of *Bt* (Dipel 8AF). Populations in microcosms treated with 1000× effective environmental concentration of the oil-based formulation (Dipel 8L) remained extremely low throughout the experiment, although the population did not drop to zero in any of the replicates. By comparison with an oil-based formulation blank, the authors concluded that the negative effects at this very high level were attributed to the oil, rather than *Bt kurstaki* itself. There are several mechanisms whereby an oil-based formulation might adversely affect collembolan populations, including direct toxicity or suffocation.

Studies with purified toxins showed that low doses of four *Bt* toxins (Cry1Ab, Cry1Ac, Cry2A, and Cry3A) did not reduce survival or reproduction of the Collembola species, *F. candida* and *Xenylla grisea* in laboratory bioassays (Sims and Martin 1997).

6.4.2 MOLLUSCS AND CRUSTACEA

Generally molluscs and crustaceans are insensitive to *Bt* at field rates, with the exception of larvae of the coot clam, *Mulinia lateralis*. For larvae of *M. lateralis,* a commercial preparation of *Bt israelensis* caused significantly higher mortality than heat-killed controls, making this a very sensitive test (Gormly *et al.* 1996). Other studies which have included molluscs, have not found toxicity of *Bt israelensis* on snails, mussels, oysters and other molluscs (Garcia *et al.* 1980; Larget and de Barjac 1981b; Beck 1982; Becker and Margalit 1993). In the lists of susceptible organisms compiled by Krieg and Langenbruch (1981), the bivalve *Ostrea edulis* and several Gastropoda were listed as not sensitive to *Bt israelensis*, mainly from an unpublished study of Morawczik and Schnetter. Purified Cry2A insecticidal protein from *Bt kurstaki* was tested against one isopod (Crustacea, Isopoda) without noticeable effect (Sims 1997). Several *Bt* products, including Bitoxibacillin which contained β-exotoxins, were tested against three slug species, *Deroceras reticulatum, Arion distinctus* and *Limax valentianus* (Kienlen *et al.* 1996). No strain of *Bt* tested was toxic to the slug species.

6.4.3 NEMATODES

Entomopathogenic nematodes have not been found to be directly susceptible to *Bt*. Infective-stage juveniles of the entomopathogenic nematodes in the genera *Neoaplectana, Steinernema* and *Heterorhabditis* were unharmed by applications of *Bt san diego (= tenebrionis), israelensis* and *kurstaki* (e.g. Lam and Webster 1972; Bari and Kaya 1984; Koppenhofer and Kaya 1997; Baur *et al.* 1998b; Koppenhofer *et al.* 1999; see also Chapter 11), although co-inoculation into hosts resulted in less nematode development in one case (Poinar *et al.* 1990).

It has been known for over 30 years that some endoparasitic nematodes are susceptible to *Bt thuringiensis* (Pinnock 1994). Representatives of thirty *Bt* serotypes were toxic to eggs of the ruminant nematode, *Trichostrongylus colubriformis*, with LD_{50}s ranging from 0.09 (*Bt tochigiensis*) to 71 (*Bt toumanoffi*) μg total protein/ml, but not toxic to adults (Bottjer *et al.* 1985). In the same study, eggs of six other nematode species (free living and zooparasitic) were reported to be susceptible to *Bt israelensis* with LD_{50}s ranging from 0.38-7.1 μg total protein/ml.

Among plant parasitic nematodes, emergence of juveniles of the nematode *Meloidogyne incognita* was almost zero when egg masses were maintained in 0.25-1% Thuricide (*Bt kurstaki*), even after transfer to water (Chahal and Chahal 1993). While an asporogenic strain of *Bt kurstaki* was not active against free-living nematodes (*Panagrellus* sp.), it was highly toxic towards the parasitic phytonematodes *Aphelenchus avenae* and *Meloidogyne hapla,* and the animal parasitic nematode *Nippostrongylus braziliensis*. The formulations tested may have contained exotoxin (Shevtsov *et al.* 1996) and exotoxin has been demonstrated to impact on several nematode species. Multiplication of the mushroom nematode, *Ditylenchus myceliophagus*, was significantly reduced after exposure to exotoxin of *Bt* in acetone (Cayrol 1974). Exotoxin was listed as toxic to *Aphelenchus avenae, Meloidogyne incognita* and *Panagrellus redivivus,* but not the entomopathogenic nematode *Neoaplectana carpocapsae* (Krieg and Langenbruch 1981). *Bt israelensis* has been found to affect population growth of the free-living nematode *Turbatrix aceti* (Meadows *et al.* 1990) and weakens nematode eggshells (Wharton and Bone 1989).

Nematocidal *Bt* toxins have been patented recently (e.g. Narva *et al.* 1991 Eur. Patent 0462721A2; A.J. Sick *et al.* 1994 US Patent 5281530; Payne *et al.* 1996 patent US Patent 5589382; Feitelson 1997 US Patent 5670365) and the action of some of these toxins described in detail against *Caenorhabditis elegans* (Borgonie *et al.* 1996a, b, c). However, no nematocidal *Bt* is currently commercially available and to our knowledge no products are under development. The requirement to pre-dissolve the toxin crystal due to the small size of stylets on nematodes may preclude the use of *Bt* against nematodes.

6.4.4 SPIDERS

No negative impacts have been reported in studies which included spiders. The effect of two *Bt* formulations (Thuricide 16B and Dipel 4L) on spiders was studied in spruce/fir forests in Maine in 1980 (Hilburn and Jennings 1988). Following aerial spraying with Dipel and Thuricide at 9.35 and 5.8 litre/ha, 12 families, 42 genera and 62 species of Araneae were captured in linear-pitfall traps. Comparison of mean pre-spray and post-spray trap catches indicated no significant reduction following *Bt* treatments. In glasshouse experiments in which the usual chemical insecticide treatment to control Lepidoptera pests was replaced by *Bt,* increases in spider populations were always noticed (H.D. Burges, pers. comm.)

Several other studies have failed to show any effects on spiders after *Bt* applications in the former USSR. Bitoxibacillin (*Bt thuringiensis*) applied against the forestry pest *Dichomeris marginella* in the Ukraine had “hardly any adverse effects” against linyphiid spiders (Zelenev 1982). Several *Bt* products (and subspecies) used in Moldova caused no reported adverse effects on spiders (Sklyarov 1983).

Spiders are predominantly obligate predators that feed chiefly on insects, and therefore could be sensitive to changes in pest/prey numbers resulting from *Bt* applications. Hilburn and Jennings (1988) studied the spider populations in spruce-fir forests of west central Maine, which had previously been sprayed with chemical insecticides for control of spruce budworm. After Dipel and Thuricide application, the abundance of species of webspinning spiders increased while hunter spiders remained unchanged. Total numbers of spiders post-spray were not significantly different to pre-spray populations.

6.4.5 INSECTS AND MITES

6.4.5.1 Mites

There are a few reports of non-target terrestrial organisms showing susceptibility in the laboratory to *Bt israelensis.* Mite species have shown susceptibility to *Bt israelensis* when inoculated in the laboratory (Temeyer 1984). Krieg and Langenbruch (1981) listed a number of studies where mite species were not susceptible to *Bt thuringiensis* and in Appendix 2.2, further studies are listed which showed predatory mites were not susceptible to *Bt* subspecies. One exception was the predatory mite, *Metaseiulus occidentalis,* which was sensitive to a wettable powder formulation of *Bt tenebrionis* tested at 0.1, 0.5 and 1.0 times the recommended field rate (0.9 kg a.i./75.7 litres per acre (Chapman and Hoy 1991).

6.4.5.2 Aquatic insects

Several studies have reported on susceptibility of non-target aquatic insects to Dipel formulations. Dipel application at field rates caused no mortality to larvae of the detritivorous aquatic insect *Hydatophylax argus* (Kreutzweiser and Capell 1996). Kreutzweiser *et al.* (1992) reported that aquatic insects in Ephemeroptera, Plecoptera and Trichoptera are unlikely to be directly affected by Dipel application, even at high (100×

recommended rate) concentrations. A significant effect (30% mortality) was observed in only 1 of the 12 species exposed to 600 IU/ml, (the stonefly *Taeniopteryx nivalis*), but this concentration represents a considerable margin of safety as it was approximately 100× the expected environmental concentration in 50 cm of water. Populations of another stonefly, *Leuctra tenuis*, were significantly reduced (70%) after *Bt kurstaki* application to a stream in Canada (Kreutzweiser *et al.* 1994), although laboratory bioassay using leaf material did not find *Bt kurstaki* toxic to *L. tenius*, suggesting another cause.

Again in Canada, Eidt (1984) tested *Bt kurstaki* (Thuricide) applied for the control of spruce budworm, *Choristoneura fumiferana*, on laboratory-reared aquatic insects. Species of blackflies, caddisflies, midges, stoneflies, mayflies and dobsonflies were exposed, with only the larvae of one species of blackfly (*Prosimulium fuscum/mixtum*) significantly affected. Eidt concluded there was no cause for concern about the effects of *Bt* sprays on aquatic insects.

The most extensive tests against aquatic insects have been using *Bt israelensis*, although most studies on the effects of this serovar have examined impacts on non-target organisms (Section 6.6.3). Studies have shown that toxicity of *Bt israelensis* is limited mainly to Diptera, including Culicidae, Simuliidae, Dixiidae, Chironomidae and some Ceratopogonidae (Garcia *et al.* 1980). Wipfli and Merrit (1994) also found few effects of *Bt israelensis* on Ephemeroptera, Plecoptera and Trichoptera.

6.4.5.3 Laboratory studies on Lepidoptera

Because the largest number of studies with *Bt* have been conducted with lepidopteran-active strains, the number of caterpillar species listed as sensitive to *Bt* is large. Despite this, over 200 species of Lepidoptera have been reported as not sensitive to at least one serovar of *Bt* (Appendix 2), compared with around 1500 reports of lepidopteran species susceptible to a serovar of *Bt*. Strain/subspecies variation is an important factor in consideration of the non-target effects of *Bt,* as toxicity is dependent upon the endotoxins and exotoxins. Even within a single genus of insect, interpecific variation in susceptibility has been recorded. Intrageneric differences in susceptibility were found in *Catocala* and *Lithophane* sp., with some species reported as susceptible, while others were not. In the laboratory, *Bt kurstaki* (Foray 48B) caused significant mortality in 27 of 42 Lepidoptera, which would be present as larvae at the time of gypsy moth spraying (Peacock *et al.* 1998). Peacock *et al.* (1998) also found that all four butterfly species tested were highly susceptible, compared to only 10 of 38 moth species. In some cases, they found that mortality was delayed, with little larval mortality but no insects surviving to adulthood. Such delayed mortality may not be detected in routine laboratory screening, which could lead to *Bt* impacts being underestimated.

Non-target effects of *Bt* applications are of most concern where they impact on endangered species. The endangered butterfly, *Lycaeides melissa samuelis,* was susceptible to *Bt kurstaki* in the laboratory, especially at high field application rates (Herms *et al.* 1997). Another much-loved butterfly, the monarch *Danaus plexippus*, showed low susceptibility to *Bt* in the laboratory even at twice the recommended field rate (Leong *et al.* 1992). However, it was the monarch butterfly which was the centre of recent publicity when Losey *et al.* (1999) demonstrated that transgenic corn pollen expressing *Bt* toxin on milkweed leaves killed about 50% of larvae after 4 days feeding. The study did not quantify the amount of toxin used and has been criticised for lack of detail. The study by Leong *et al.* (1992), which reported little susceptibility of adult monarchs to Javelin (*Bt kurstaki*) and MVP (encapsulated toxin), suggested that spraying of overwintering populations should be avoided because a small mortality was to be expected. Hansen and Obrycki (1999) also examined the effects of transgenic *Bt* pollen on monarch larvae by placing potted milkweeds near the edge of a transgenic corn field. Leaf samples were removed and neonate larvae of *D. plexippus* exposed in the laboratory. After 48h, larvae on leaves exposed to *Bt* pollen had 19% mortality compared to no mortality in untransformed crops. This study, together with work by Losey *et al.* (1999), indicated that monarchs may be adversely affected by *Bt,* especially when the insect is in the larval stage. However, it is not known whether insect behaviour in the field, such as movement away from contaminated leaves, would mitigate the effect.

6.4.5.4 Toxicity of purified toxins to insects and mites

Toxicities of purified toxins differs from whole products based on spores and crystals, as indicated in Table 5.2. Individual endotoxins vary in toxicity to a group of lepidopteran species and a number of toxins can occur in a single bacterial cell and crystal. Therefore, the toxicity established for a strain or isolate can be wider than that of the specific toxins individually, although the amount of toxin actually ingested may be higher where a single toxin is produced transgenically than in products.

Tests on a number of non-target organisms using purified proteins are becoming more common, as much of the focus on *Bt* use is in transgenic plants. For example, purified Cry2A protein from *Bt kurstaki* tested against 35 insect species representing the orders Coleoptera, Collembola, Diptera, Hemiptera, Hymenoptera, Isoptera, Lepidoptera, Neuroptera and Orthoptera found only Lepidoptera and Diptera (Culicidae) were sensitive (Sims 1997). MacIntosh *et al.* (1990a) used single gene products from *Bt kurstaki,* (HD-1: Cry1Ab and HD-73: Cry1Ac) and *Bt tenebrionis* (Cry3A) against pest insects (17 species covering 5 orders) and a mite. Only the chrysomelid, *Leptinotarsa decemlineata,* was sensitive to Cry3A protein with an LC_{50} of 6.5 µg/ml. Lepidopterans, *Manduca sexta, Trichoplusia ni, Heliothis virescens, Agrotis ipsilon, Spodoptera exigua, Heliothis zea* (= *Helicoverpa zea*) and *Ostrinia nubilalis* were sensitive to both *Bt kurstaki* proteins (LC_{50}s of 0.036, 0.09, 1, 18, 44, 10 and 37 µg/ml, respectively, for HD-73, and 0.04, 0.19, 1.7, >80, 34, 33 and 3.6 µg/ml for HD-1) but the cockroach *Blattella germanica* and mosquito *Aedes aegypti* were not. Interestingly, a number of studies using *Bt kurstaki* in the laboratory have found *A. aegypti* to be susceptible (e.g. Panbangred *et al.* 1979; Ignoffo *et al.* 1980a, 1981), although sometimes with only low susceptibility (e.g. Johnson and Davidson 1984). Panbangred *et al.* (1979) established a LC_{50} for HD-1 of 5.6×10^4 spores/ml for *A. aegypti*, although comparative toxicities determined by Ignoffo *et al.* (1981) for this mosquito species for *Bt kurstaki* and *Bt israelensis* were LC_{50}s of 210.0 µg/ml (139.0-281.4) and 0.054 µg/ml, respectively. *Blattella germanica* has shown a low degree sensitivity to *Bt kurstaki* in other studies (Sandhu and Varma 1980; Zukowski 1995).

6.5 BENEFICIAL SPECIES

6.5.1 NATURAL ENEMIES

It is important to assess the impact of *Bt* strains and toxins on natural enemies. *Bt* is often promoted as an environmentally friendly and safe alternative to chemical insecticides, partly because it is considered compatible with naturally occurring or introduced biocontrol agents such as parasitoids and predators (e.g. Sarmiento and Razuri 1978; Panis 1979; Farias *et al.* 1980; Boller and Remund 1981; Zelenev 1982; Lorenzato and Corseuil 1982; Snodgrass and Stadelbacher 1994).

Below and in Appendix 2.4, studies on the effect of *Bt* subspecies and products on predators and parasites are reported. An early review of studies before 1976, where *Bt kurstaki*-based products had been applied in the field in the presence of predators and/or parasitoids, concluded that there was "an incomplete picture" of the impact of *Bt*, particularly for parasites (Forsberg *et al.* 1976). Since then, a number of studies have demonstrated the relative safety of *Bt* for predators and parasitoids, except for competition-related effects. The level of specificity of *Bt* strains generally supports these conclusions, although some non-target mortality of natural enemies has been reported occasionally.

Several studies have examined large numbers of predators and parasitoids, or reported field studies that included general observations of predator/parasitoid numbers after *Bt* application. The working group "Pesticides and Beneficial Arthropods" of the International Organization for Biological Control examined non-target effects of 40 pesticides in 6 countries using standard test methods against *Amblyseius potentillae, Anthocoris nemorum, Chrysopa* (= *Chrysoperla*) *carnea, Pimpla turionellae, Cryptolaemus montrouzieri, Palexorista inconspicua, Encarsia formosa, Leptomastix dactylopii, Opius* sp., *Phygadeuon trichops, Phytoseiulus persimilis, Syrphus vitripennis* and *Trichogramma cacoeciae. Bt kurstaki* was harmless to most of the arthropods and was recommended for use in integrated control programmes (Hassan *et al.* 1983). In many different pest control situations, *Bt* application has not resulted in obvious impacts on predators and pathogens. Use of *Bt thuringiensis* against *Leptinotarsa decemlineata* in Moldova apparently did not harm the complex of natural enemies (including 26 carabids, 11 coccinellids and 3 chrysopids) (Novozhilov *et al.* 1991). Presumably after years of intense spraying with chemical insecticides, orchard spraying with several *Bt* products (*Bt thuringiensis* - Bitoxibacillin, *Bt dendrolimus* - Dendrobacillin, *Bt galleriae* - Entobacterin and *Bt alesti* - BIP) in Moldova led to an increase in natural enemies such as coccinellids, chrysopids, predacious beetles and spiders, with a subsequent increase in the natural control of the pest species (Sklyarov 1983). Similarly, when *Bt* (including *Bt kurstaki*) was first used for pest control in vegetable crops in China, the number of aphid natural enemies increased while use of chemical insecticides led to a marked decrease (Lu *et al.* 1986). Predators and parasitoids of *Opsiphanes tamarindi*, a defoliator of bananas in Venezuela, were unaffected by *Bt* application (Briceno 1997). Application of Dipel (*Bt kurstaki*) in Connecticut forests for gypsy moth control had "good compatibility" with naturally occurring nuclear polyhedrosis virus and parasitism by larval and pupal parasites (Andreadis *et al.* 1983). Against gypsy moth in cork oak, *Bt kurstaki* did not have any direct effect on beneficial insects (Lentini and Luciano 1995). In most cases, use of *Bt* appears to be compatible with natural enemies.

6.5.1.1 Insect predators

Insect predators are important natural control agents of insect pests and insecticides which do not harm predators have a great advantage. Predators are often generalists and will attack several insect pests. Therefore, control measures which reduce predator numbers may result in outbreaks in non-target pests.

Overall, *Bt* has rarely been found to be directly toxic to predator species at field dosages. A summary of records of *Bt* and predator species (Appendix 4.1) revealed the most frequent response was little or no toxicity. Of the 87 individual records of predator species response to *Bt* strains and products (Appendix 4.1), over 70 could be classified as showing no or little effect. The remainder reported responses ranging from reduced egg predation and food intake (Giroux *et al.* 1994a, b; Hafez *et al.* 1997a), effects on predator growth and oviposition of adults (Zhou 1992), delayed mortality in the next generation of predator nymphs (El-Husseini *et al.* 1987) and reduced fertility and temporary sterility (Petrova and Khrameeva 1989).

A direct toxic effect of *Bt tenebrionis* has been reported against the mite predator, *Metaseiulus occidentalis* (Chapman and Hoy 1991). Treatment with *Bt tenebrionis* as a wettable powder at the equivalent of a recommended field rate of 0.9 kg a.i./75.7 litres/acre or less was more toxic to the predator than to the mite host. Toxicity was higher when female predators were starved for 24h before treatment, possibly as a result of drinking the spray. Female predators, which survived treatment as eggs and developed in the presence of residues, did not show a significant decrease in fecundity. The prey species *Tetranychus urticae* is not a major pest of crops where *Bt tenebrionis* has traditionally been used, which would limit exposure of the predator to sprays but it is a ubiquitous plant feeder and would occasionally be found on treated crops.

There are several instances where different studies on the same predator have yielded contradictory results. Lacewings, *Chrysopa* (= *Chrysoperla*) spp. (Neuroptera: Chrysopidae), are sometimes used in tests for environmental contamination as they can be particularly sensitive to insecticidal contamination. Several studies found that *Bt* could be toxic to *Chrysopa* spp. The feeding of Cry1Ab toxin to *Chrysopa carnea* in the laboratory was toxic at 100 µg/ml of diet by using encapsulated artificial diet, although immature mortality was only 57% compared to control mortality of 30% (Hilbeck *et al.* 1998b). Babrikova *et al.* (1982) found three products (*Bt galleriae*-based Entobacterin-3, *Bt thuringiensis*-based Bactospeine and *Bt kurstaki*-based Dipel) toxic to adult *C. carnea* in the laboratory, but they were less toxic to larvae. Babrikova and Kuzmanova (1984) also found *C. septempunctata, C. formosa* and *C. perla* adult mortality of 60-90% to the same *Bt* products in the laboratory. Unfortunately, the dosage used was not reported. In cotton crops in the Tashkent region of Uzbekistan, USSR, Lepidocide (*Bt kurstaki*) or Dendrobacillin (*Bt dendrolimus*) applications had no adverse effect on the numbers of adults and immature stages of the common lacewing *C. carnea,* although Bitoxibacillin (*Bt thuringiensis*) reduced populations within 15 days but not subsequently (Adylov *et al.* 1990). Other studies on *Chrysopa* spp., including those using exotoxin-containing products (Zuo *et al.* 1994; Slabospitskii and Yakulov 1979), *Bt kurstaki*-

based Dipel (Salama and Zaki 1984), *Bt dendrolimus* (Dendrobacillin), *Bt galleriae* (Entobakterin) (Umarov Sh *et al.* 1975) and *Bt tenebrionis* (Krieg *et al.* 1984; Langenbruch 1992), did not report significant effects.

Chrysopa spp. is not the only predator for which studies have yielded contradictory reports of sensitivity to *Bt*. While Greener and Candy (1994) reported *Bt tenebrionis* affected survival of Australian plague soldier beetle, *Chauliognathus lugubris*, Beveridge and Elek (1999) reported no toxicity with the same subspecies of *Bt*. The work by Greener and Candy (1994) caused some debate in Australia. Milner and Akhurst (1995) were concerned that as the work by Greener and Candy (1994) was the first to show a direct adverse effect, it could lead to inhibition of the development of *Bt* as part of integrated control programmes. Milner and Akhurst (1995) claimed that direct effects of *Bt tenebrionis* and formulation components of Novodor were not determined, that control mortality was too high (reaching 100% after 26 days), adults of the target pest species *Chrysophtharta bimaculata* were not as susceptible as would normally be expected and unusually, *Bt*-treatment mortality occurred after 7 days, which is extended for *Bt* at the temperatures used. In reply, Greener and Candy (1995) stood by their study, stating that mortality was caused by Novodor but added that impacts in the field were likely to be minimal compared with a pyrethroid.

In some cases, there can be an indirect effect of *Bt* on predators. The seven spot ladybird, *Coccinella septempunctata,* suffered 63% mortality when placed for 6 days on potato shoots with aphids dipped in 1% Novodor (*Bt tenebrionis*)(Langenbruch, in Keller and Langenbruch 1993). However, Langenbruch (1992) found no direct effect on ladybird larvae. Keller and Langenbruch (1993) thought that *Bt* would not influence ladybirds populations at field concentrations, as was found by Krieg *et al.* (1984). Reports on the use of *Bt kurstaki* (Mohamed 1993; Young *et al.* 1997), *Bt tenebrionis* (Krieg *et al.* 1984; Greener and Candy 1994), *Bt thuringiensis* (Slabospitskii and Yakulov 1979), *Bt dendrolimus* and *Bt galleriae* (Umarov *et al.* 1975) have reported no or little effect on Coccinellidae following *Bt* application.

6.5.1.2 Parasitoids

Generally, where the host insect was not susceptible to *Bt*, no effects on associated parasitoids were observed (e.g. Wysoki *et al.* 1975; Johnson *et al.* 1980). *Bt* applications, especially *Bt kurstaki* against Lepidoptera, have not usually reduced field parasitism rates, making *Bt* application compatible with most naturally occurring or introduced parasitoid controls. Occasionally, laboratory bioassays have demonstrated low level susceptibility to *Bt*, but this is rarely observed in the field. Direct toxicity has been reported only by Hamed (1979) for *Diadegma armillata, Pimpla turionellae, Ageniaspis fuscicollis* and *Tetrastichus evonymellae,* where ingestion of 10^8 spores/ml and associated crystals in food sap killed the parasitoids. Large numbers of vegetative cells of *Bt* were detected microscopically in the haemolymph and gut of dead parasitoids. Fewer parasitoids emerged from larvae and pupae of the caterpillar host, *Yponomeuta evonymellus,* when larvae had fed on *Prunus* leaves contaminated with *Bt kurstaki* than when they had fed on uninfected leaves. A single report of 100% mortality of *Trichogramma* spp. caused by Dendrobacillin (Tolstova and Ionova 1976) is in contrast to the numerous studies of *Trichogramma* spp. associated with a number of hosts, which have not reported any adverse effects (Appendix 4.2).

In some cases, parasitoid parasitism rates have increased after *Bt* application when compared with untreated plots as reported for parasitism of spruce budworm in Canada (Hamel 1977; Nealis *et al.* 1992). Ticehurst *et al.* (1982) also reported enhanced parasitism by *Apanteles melanoscelus* of *Lymantria dispar* after Dipel 4L application at 19.8 BIU/ha. Hamel (1977) speculated that the increase in parasitism by species such as *A. fumiferanae* and *Glypta fumiferanae* was possibly caused by both parasitoids attacking pre-diapausing 1st instar budworm and the progeny beginning development along with 2nd instar budworm, pre-spraying. Parasitized larvae are photonegative and do not come into contact with *Bt kurstaki* as readily as the photopositive unparasitized larvae, meaning their hosts remain unexposed to *Bt kurstaki.*

As with predators, a range of effects on parasitoids has been recorded with many studies finding no impact. However, treatment of hosts with *Bt* may adversely affect larval parasitoids by increasing mortality of larval parasitoids still within hosts which survived the treatment, and by reducing the body size of adult parasitoids (e.g. Monnerat and Bordat 1998; Varlez *et al.* 1993). If *Bt* kills the host too quickly, the parasitoid does not have sufficient resources to develop. For example, in a parasitoid of spruce budworm, *Apanteles fumiferanae*, populations were reduced by 50-60% because of lack of parasitoid emergence before host death (Nealis and van Frankenhuyzen 1990).

Reduced egg production, cocoon formation and longevity of parasitoid adults resulted from feeding *Bt* to *Plodia interpunctella* just prior to attack by the parasitoid (Salama *et al.* 1991). Similarly, *Apanteles glomeratus, A. melanoscelus* and *A. litae* showed reduced adult life-span and/or a developmental lag after *Bt kurstaki* applications against the host (Marchal-Segault 1975; Weseloh and Andreadis 1982; Salama *et al.* 1996c). Effects on development or adult life span were also reported for some other braconids (Appendix 4.2). In contrast, among tachinid flies, most studies found no impact from *Bt* application against the host.

6.5.2 POLLINATORS

Many studies have found that *Bt* was safe for use around bees (Cantwell *et al.* 1972; Ozino-Marletto *et al.* 1972). Numerous subspecies and formulations of *Bt* have been tested and found harmless to bees: Bactospeine, Dipel, Tarsol and Thuricide (Krieg 1978; Verma 1995); Lepidocide (Kuteev *et al.* 1983); Bitoxibacillin and Astur (Kandybin 1991; Bushkovskaya *et al.* 1994). Studies with various individual strains have yielded similar results. *Bt dendrolimus* was harmless to workers of the honeybee, *Apis mellifera ligustica,* even at high doses (Ozino-Marletto *et al.* 1972). Krieg *et al.* (1980) found that several serotypes (*kurstaki, thuringiensis* and *israelensis*) were relatively non-toxic to bees at field concentrations. An unidentified mosquito active strain of *Bt* was non-toxic to bees at 10× the concentration required to kill mosquito larvae (Balaraman *et al.* 1981).

In the field, no adverse effects on bees have been reported in the treated area (e.g. Malyi *et al.* 1978). Thuricide (*Bt kurstaki*) sprays of 4 billion IU in 5.6 litre/ha had no detectable effect on honeybee hives (Buckner *et al.* 1975). Panizzi and Pinzauti (1988) reported on the natural occurrence of *Bt* and several other common bacteria in brood of *Apis mellifera* in a mite infested colony in Italy, but this appears to be the only report of natural occurrence in hives.

Bt has been used as a control agent of wax moth, *Galleria mellonella,* in honeybee hives (e.g. Verma 1995, using Dipel). Moran-Rodriguez and Sandovaly y Trujillo (1991) used *Bt thuringiensis, aizawai* and *galleriae* against larvae and adults of *G. mellonella* and *A. mellifera,* with no effect on bees recorded. Specific studies have found *Bt* (as Thuricide-HP) did not affect adult honeybees when used at a rate sufficient to control larvae of *G. mellonella* (Ali *et al.* 1973). One product, Certan (*Bt aizawai*), has been specifically developed for use in wax combs (McKilling and Brown 1991) and had no detrimental effect on adult honeybee mortality, brood rearing and honey production (Cantwell and Shieh 1981). Spores of *Bt galleriae* from Thuricide 90TS which were impregnated into foundation wax, used by bees to construct honeycomb, leached into honey where no loss of viability was detected over 49 weeks (Burges 1976c).

However, adverse effects on bees have occasionally been reported and these have generally involved exotoxins rather than endotoxins. Early studies by Cantwell and Franklin (1966), Cantwell *et al.* (1972), Krieg and Herfs (1963) and Martouret and Euverte (1964) found that exotoxin producing strains of *Bt thuringiensis* could kill bees. Krieg (1973b) found that both an α- and β-exotoxin from *Bt* were toxic when administered orally to honeybees. Kazakova and Dzhunusov (1977) reported the Russian product Bitoxibacillin-202 (containing spores, endotoxin and exotoxin of *Bt thuringiensis*) caused "complete mortality" of honeybees. Oral feeding of two commercial preparations of Thuringiensin (β–exotoxin) to adults of *A. mellifera* had significant effects on life expectancy, although only high doses reduced lifespan when spiked sucrose was sprinkled on caged bees (Vandenberg and Shimanuki 1986). The authors considered the likelihood of foraging bees being affected was very low. *Bt tenebrionis* reduced bee longevity at the highest concentration tested (10^8 spores and crystals/ml sucrose syrup), but did not cause disease and Vandenberg (1990) thought it unlikely to cause undesirable effects on honeybee colonies under field application conditions.

6.5.3 WEED BIOCONTROL AGENTS

As insects are used as biological control agents of many weeds, it is possible *Bt* application could affect their efficacy. Few studies have reported directly on impacts on weed biocontrol agents. In Oregon, the biological control agent of weeds, the arctiid *Tyria jacobaeae,* was tested for susceptibility to *Bt kurstaki*. In laboratory feeding tests, early instars were not susceptible but 4th and 5th instars had LC_{50}s of 0.31 and 0.22 mg formulation/ml (Dipel HG; potency 4320 IU/mg). Laboratory and field results suggested *Bt kurstaki* could reduce the impact of *T. jacobaeae,* although use of *Bt kurstaki* against target pests does not coincide with the susceptible stages of the biocontrol agent (James *et al.* 1993b). The curculionid *Neochetina eichhorniae*, a biological control agent of water hyacinth, was not susceptible to 21 strains of *Bt* (subspecies *Bt aizawai, alesti, darmstadiensis, entomocidus, israelensis, kenyae, kurstaki, san diego, tenebrionis, thuringiensis* and two unidentified). Beta-exotoxin was a feeding deterrent but was not significantly toxic (Haag and Boucias 1991). Susceptibility of biocontrol insects needs to be determined on a case by case basis, but there is no reason to suggest that biocontrol agents would be likely to be more sensitive than other non-target insects.

6.5.4 EARTHWORMS

There are few reports directly recording the effect of *Bt* on earthworms, although they are included as a non-target invertebrate in tests for most registration packages, presumably without effect (see EPA review, Section 6.8). Smirnoff and Heimpel (1961) applied Thuricide to soil at very high doses (10^4-$10^5 \times$ higher than normal levels after a commercial application according to Benz and Altwegg 1975), and found this *Bt* product fatal for earthworms, *Lumbricus terrestris*. A subsequent study showed that commercial preparations of *Bt kurstaki* (Dipel) applied at 60, 600 and 6000 mg/m^2 (recommended field rate of 60 mg/m^2) and Bactospeine applied at 30 g/m^2 did not reduce the numbers of earthworms (*L. terrestris*) in an ash/maple forest compared with numbers in a plot which received water alone (Benz and Altwegg 1975). The experiments of Smirnoff and Heimpel (1961) used an early formulation of Thuricide which was almost certainly based on *Bt thuringiensis* and may have contained β-exotoxin as many *Bt* products in the USA contained exotoxin prior to 1970 (Forsberg *et al.* 1976).
Addison and Lolmes (1996) examined the effects of formulated (aqueous and oil) and unformulated *Bt kurstaki* on the forest earthworm *Pendrobaena octaedra* in soil-litter microcosms. At 1000× the expected environmental concentration (EEC), unformulated *Bt kurstaki* and aqueous Dipel 8AF had no adverse effects. However, Dipel 8L at 1000× EEC reduced survival, growth and coccoon production, an effect attributed the oil by comparison with an oil based formulation blank. Dipel 8L had no effect at 100× EEC.

6.6 ECOSYSTEM STUDIES

6.6.1 MICROBIAL AQUATIC COMMUNITIES

Kreutzweiser *et al.* (1996) examined the effect of *Bt kurstaki* (Dipel) on activity of microbial communities on leaves in aquatic microcosms. They used two formulations of Dipel (64AF and 8AF), at an expected environmental concentration (EEC) after a spray application as well as 100-1000× the EEC. Measurements of bacterial and protozoa cell density, microbial respiration and decomposition were used to examine the effect of *Bt kurstaki* application. Not surprisingly, bacterial cell numbers increased even at the 1× EEC, but protozoan numbers were not affected. Microbial respiration increased and decomposition decreased, but the decrease was significant at only 1000× EEC treatment. The authors concluded that *Bt kurstaki* contamination of waterways is unlikely to have any significant effects on microbial communities.

6.6.2 INVERTEBRATES

Toxicity of *Bt* to several invertebrate taxa (Acarina, Nematoda, Collembola, Annelida, Hymenoptera) inhabiting the soil has been demonstrated, but only rarely is it possible to relate dosage information to field situations and, in many cases, the *Bt* subspecies tested are not currently used for pest control in North America (Addison 1993). Use of Dipel (*Bt kurstaki*) in a spruce stand in Germany resulted in a large increase in emergence of adult Diptera with no negative effects on non-target soil arthropods (Wernicke and Funke 1995). Traps placed next to elms in Dakota detected little effect of *Bt* application on most families caught, such as sarcophagids, coccinellids, calliphorids and staphylinids (Frye *et al.* 1988). In the Philippines, *Bt kurstaki* (Dipel) had no residual effect on 15 families of mites, including soil mites and predatory mites, in decomposing rice stubble (Sayaboc *et al.* 1973), although in other studies some mites have been shown to be susceptible in the laboratory (Section 6.4.5). In Europe, Sklodowski (1996) studied the effect of various pesticides (Foray, Trebon, Decis and Dimilin) used for control of nun moth on carabid beetles. He found that while carabids were reduced in the autumn following spraying of all agents compared to a non-sprayed control, Foray (*Bt kurstaki*) had little impact on carabid communities overall. Diversity of species was greatest in the Foray sprayed areas. Similarly, Riddick and Mills (1995) examined effects of *Bt kurstaki* and oil on carabids in apple orchards in California and found few changes in populations of several species.

6.6.2.1 Lepidoptera

With *Bt kurstaki*, several studies have shown impacts on Lepidoptera after broadcast application. The use of *Bt* δ-endotoxins can result in a temporary reduction in susceptible insect populations. Following *Bt* application in forests, several studies have reported a significant decrease in numbers of adult and larval Lepidoptera in the year of application, with some reductions extending into the following year in species whose susceptible life stage occurs in the year previous to the appearance of adults. Reported effects are usually of limited duration, which has generally been attributed to the lack of persistence and replication of *Bt* in the field.

Bt kurstaki applied in West Virginia for control of gypsy moth, *Lymantria dispar*, caused a significant decline in Macrolepidoptera in the year of application (Butler *et al.* 1995). This study included about 90 species of Macrolepidoptera and nearly 200 other arthropod taxa and found that effects were limited to those Lepidoptera present in the canopy at the time of spraying. Butler *et al.* (1995) found no significant impact on species which were present after the spray period and no effect was detected on non-lepidopteran species. In another study in West Virginia, where *Bt kurstaki* was applied for gypsy moth control application, detectable effects were generally restricted to Lepidoptera (Sample *et al.* 1996).

Similarly, Wagner *et al.* (1996) examined the effects of a single application of *Bt* (90 BIU/ha) on native Lepidoptera in West Virginia in the treatment year and the following two recovery years. Larval abundance was lower on foliage from treated plots but differences were small and mostly not significant. Nineteen of the twenty common lepidopteran species (Gelechiidae, Tortricidae and Noctuidae) on foliage decreased slightly in relative abundance after *Bt* application. Most affected species recovered in the first year after spraying. Johnson *et al.* (1995) examined the toxicity and persistence of *Bt kurstaki* towards larvae of tree-feeding Lepidoptera (*Papilio glaucus, P. canadensis* and *Callosamia primethea*) on seven of their natural host plants in the field. *Bt kurstaki* was toxic to the test species, and the effect persisted for up to 30 days against *P. glaucus*. Scriber (1998) attributed the extended risk to *Papilio* spp. (swallowtail butterflies) partly to the higher sensitivity than the target pest, gypsy moth. This ranged from several hundred to several thousand times lower concentrations for LD_{50}s than required for gypsy moth control (Scriber 1998).

After a year of *Bt kurstaki* application for a year in Oregon oak forests, Miller (1990) monitored 35 lepidopteran species for three years and found leaf-feeding lepidopterans were significantly reduced in treated areas, as were total numbers of lepidopterans in the first two years. Two years after treatment, no significant effects were found on numbers, but species diversity was reduced. A similar study of lepidopterans in tobacco brush forests (*Ceanothus velutinus*) found significantly higher total numbers in untreated populations in the two weeks following treatment with a similar difference found one year later, although the effect disappeared in the subsequent summer (Miller 1992). Species diversity was not significantly affected by *Bt kurstaki* spraying of tobacco brush.

The extent of drift of *Bt kurstaki* after application was measured in Utah during 1989-93, using the lepidopterans, *Incisalia fotis* and *Callophrys sheridanii* (Whaley *et al.* 1998). An aerial application of a water based formulation, Dipel 6AF, was made by helicopter at 59 BIU/ha. Mortality was found when the lepidopterans were fed on host plants exposed up to 3000m from the application site, indicating non-target effects need to be considered well outside the spray area.

6.6.3 BENTHIC AND AQUATIC COMMUNITIES

The effect of *Bt* contamination of waterways is a major concern in areas where pests attack crops that are in the vicinity of, or occasionally immersed in water. While safety of non-target organisms has been studied for *Bt israelensis*, which is applied to water (reviewed in Glare and O'Callaghan 1998), fewer studies report on safety of other *Bt* to non-target organisms in water.

The impact of *Bt israelensis* on non-target arthropods in water is minimal. Long-term application of *Bt israelensis* against mosquitoes (Germany) (Schnetter *et al.* 1981) and simuliids (Africa) (Onchoncerciasis Control Programme) showed no impacts on non-target insects. Several other studies have examined the non-target effects of *Bt israelensis* (e.g. Colbo and Undeen 1980; Miura *et al.* 1980; Sinegre *et al.* 1980; Ali 1981; Molloy and Jamnback 1981; Mulligan and Schaefer 1981; Mulla *et al.* 1982; Rettich 1983; Gharib and Hilsenhoff 1988; Yameogo *et al.* 1988; Merritt *et al.* 1989; Becker and Margalit 1993; Jackson *et al.* 1994; Hershey *et al.* 1995; McCracken and Matthews 1997) without reporting any significant impacts. Laboratory studies have similarly found few concerns (e.g. Garcia *et al.* 1980; Wipfli and Merritt 1994). After reviewing a number of studies, we found that the risks of *Bt israelensis* were limited to Nematocera, including mosquitoes, Simuliidae and some chironomid midge larvae (Glare and O'Callaghan 1998). Not all chironomids were susceptible and the safety margin for use was quite large in some cases (Molloy 1992). *Chironomus plumosus* was sensitive to *Bt israelensis* only at doses much higher than those used in the field (Larget and de Barjac 1981b). Chironomid midge larvae showed only low level susceptibility and, in some cases, no effect on chironomid populations was observed under field conditions (Miura *et al.* 1980; Lebrun and Vlayen 1981) and in New Zealand by Chilcott *et al.* (1983). However, in other cases chironomids (Sinegre *et al.* 1980b; Pistrang and Burger 1984; Back *et al.* 1985; de Moor and Car 1986; Charbonneau *et al.* 1994) and some Blephariceridae (Back *et al.* 1985) were susceptible in the field.

As *Bt israelensis* forms the basis of a number of dipteran-active water-applied products, safety of *Bt israelensis* products in water is a primary consideration. With other *Bt* products, such as *Bt kurstaki* and *Bt aizawai*, the impact of application on aquatic insects has not been considered in many studies. Surgeoner and Farkas (1989 in Kreutzweiser *et al.* 1992) reviewed available literature on the effects of *Bt kurstaki* in aquatic systems and found no reports of adverse effects. Laboratory assays of *Bt kurstaki* against aquatic insects have rarely identified sensitive species (Eidt 1984; Kreutzweiser *et al.* 1992). In one of the few studies conducted in water, Kreutzweiser *et al.* (1994) examined the effect of *Bt kurstaki* at 10× the expected environmental concentration on macroinvertebrates in a forest stream. They found an increase in drift density of approximately two-fold over pre-treatment densities. There were short-term alterations in the community structure but no significant change in taxomonic diversity after *Bt kurstaki* application. Only the stonefly (*Leuctra tenuis*) was reduced significantly (around 70%), although subsequent bioassays did not find *Bt kurstaki* on leaves directly toxic to stoneflies. Kreutzweiser *et al.* (1994) hypothesized that the stonefly reduction in the field when it was not sensitive in the laboratory may indicate an interaction with other agents, which increased toxicity to this insect, or some unknown factor not related to *Bt kurstaki* application. Caged caddisfly larvae in a stream were unaffected by *Bt kurstaki*.

Two earlier studies reported in Kingsbury (1975) found no significant adverse effects on aquatic insects following aerial application of *Bt kurstaki* in Ontario. Again in Canada, commercial *Bt kurstaki* formulations at 50-5000 BIU/ha were used to examine the non-target effects on invertebrate water colonists (Richardson and Perrin 1994). *Bt kurstaki* affected neither the density nor composition of benthos sampled 7 days after application.

The emergence rate of adult Baetidae, Chironomidae, Hydroptilidae, Simuliidae and Nemouridae was unchanged. The mayfly, *Baetis* sp., exhibited increased drift rates when exposed to *Bt kurstaki* in the first 2.5 h, but this effect disappeared subsequently. There were no significant differences in the numbers of macro-invertebrates on leaf packs and Richardson and Perrin (1994) concluded there was no evidence that addition of *Bt kurstaki* to stream mesocosms created adverse effects for this benthic community, even at >100 times expected exposure rates.

In a study by Eidt (1984) on aquatic insects susceptible to Thuricide (*Bt kurstaki*), it was concluded there was no cause for concern about the effects of *Bt* sprays on aquatic insects. Morris (1980) also found no evidence of undue hazard to aquatic non-target organisms in field studies during spruce budworm control with *Bt kurstaki.*

A Chinese *Bt* emulsion '187' had no adverse effects on the diversity and quantity of a number of non-target organisms (including: Copepoda, Cladocera, Rotifera, Algae and Tubifex) when applied for mosquito control (Huang *et al.* 1995).

Bt application against chrysomelids in Tasmania, Australia, was considered highly likely to result in stream contamination. Therefore, Davies (1994) evaluated the possible effects on stream macroinvertebrates. Artificial streams were dosed every two hours with Novodor (*Bt tenebrionis*) at 5 and 10 l/ha, and colonising fauna from a natural stream fed into the system were assessed. There was a slight effect of Novodor on drift of chironomid pupae and simuliid larvae, but not for other taxa. However, emergence rates of chironomids, which constituted greater than 99% of Diptera emerging, showed no effect of *Bt tenebrionis* exposure. The abundance of benthic organisms (mainly chironomid larvae, ostacods, mayflies, simuliids and beetles) showed no significant effects of *Bt tenebrionis* treatment. Despite *Bt tenebrionis* being coleopteran active, Davies (1994) found no effect on aquatic coleopteran larvae or adults.

6.7 US ENVIRONMENTAL PROTECTION AGENCY REVIEW

Non-target insect susceptibility data from numerous company-supplied studies have been submitted to the US Environmental Protection Agency (EPA) in support of registration packages, including many studies not available for review here. Some of these studies were summarised in Reregistration Eligibility Decision documents (EPA 1998). According to the EPA summaries, *Bt kurstaki, Bt israelensis, Bt tenebrionis* and *Bt aizawai* showed little to no toxicity to the tested Neuroptera, Hymenoptera, Coleoptera, "arthropoda" and annelida group indicator species (Table 6.1).

Bt δ-endotoxins were reported to have "no appreciable effect on aquatic (freshwater) invertebrates" (p 29, EPA 1998). The EPA report listed immature and adult stages of mayflies, caddisflies, dragonflies, damselflies, beetles, midges and dobsonflies as unaffected by *Bt* application. However, the EPA report did include reference to moderate to high levels of toxicity to *Daphnia (*a freshwater invertebrate), especially by *Bt aizawai* (Table 6.1). The LC_{50} estimates of between 5 and 50 ppm indicated that *Bt kurstaki* and *Bt israelensis* were moderately toxic to *Daphnia*, while *Bt aizawai* was highly toxic. The toxicity was attributed to factors other than δ-endotoxins (presumably exotoxins), as the report goes on to state in the same paragraph that as long as *Bt* is used to label specification and the production process is carefully managed "to prevent higher levels of non-target toxicity due to the exotoxins" then "no freshwater aquatic invertebrate hazards [are] expected" (p 30, EPA 1998). Marine invertebrates were also considered to be unaffected by *Bt* application (EPA 1998).

The EPA document highlights one problem in comparing non-target impacts of *Bt* formulations: the lack of consistency in expression of active ingredient in the products. Rates of *Bt* used in experiments are listed in Table 6.1 as: spores/g diet; spores/ml; ppm; ppm of food; multiple of field rate; cfu/ g food; cfu/l; mg or g/l and μg/bee. Some of these measures are relatively meaningless, in particular spores/ml (normally with accompanying crystals) as in many cases spores do not influence toxicity. Most of the terms do not relate directly to amount of toxin ingested by test organisms nor are phrases such as "practically non-toxic" explained.

Table 6.1. Non-target susceptibility: summary of unpublished studies submitted to EPA in support of registration of products - Reregistration Eligibility Decision documents, EPA 1998. (NOEL = no observed effect level; NOEC = no observable effects concentration). No experimental details or species names were presented in the report.

Organism	*Bt* subspecies	Results
Predaceous Neuroptera	*kurstaki*	NOEL = 3000 ppm
		Practically nontoxic at $1x10^8$ cfu/g food for 9 days; NOEL = $1x10^8$ spores /g diet
		Practically nontoxic at $1.2x10^8$ cfu/g food for 9 days; NOEL = $1.2x10^8$ spores /g diet
		Slightly toxic; 10x field rate resulted in 18% mortality
Parasitic Hymenoptera	*kurstaki*	Practically nontoxic at 3000 ppm of food for 15 days; NOEL = 3000 ppm
		Practically nontoxic at $2.4x10^8$ spores/ml diet for 23 days; NOEL > $2.4x10^8$ spores/ml diet
		Practically nontoxic at $1x10^8$ spores/ml diet for 30 days; NOEL > $1x10^8$ spores/ml diet
	israelensis	30-day LC_{50} > $7.9x10^7$ cfu/g diet
	tenebrionis	NOEL = 100 ppm
Predaceous Coleoptera	*kurstaki*	NOEL = 1500 ppm, slightly toxic
		Practically nontoxic at $2.4x10^8$ spores/ml diet for 28 days; NOEL > $2.4x10^8$ spores
	israelensis	9 day LC_{50} $1.8x10^8$ cfu/g diet
	aizawai	NOEL = 10,000 ppm
		NOEL = 1566 ppm
Arthropod predators and parasites	*kurstaki*	Slightly toxic (6.2 g/l resulted in 12-21% mortality) Toxic; 10x field rate resulted in 100% mortality within 6 days
Honeybee	*kurstaki*	48-hour LD_{50} > 23.2 mg/bee; NOEL = 7.7 mg/bee
		48-hour LD_{50} > 23.2 mg/bee; NOEL = 7.7 mg/bee
		10-day LC_{50} 118 mg/bee (consumed)
		No significant effects noted at 10x field rate
	aizawai	Highly toxic; LC_{50} = 15 ppm
	israelensis	5 day LC_{50} > 7.0 x 10^7 cfu/g diet
	tenebrionis	NOEL = > 100 ppm (100x field conc.) in 18 day test
Green lace-wing larvae	*israelensis*	16 day LC_{50} > 1.5x 10^8 cfu/g diet; 16 day NOEL = 2.5 x 10^4 cfu/g
	aizawai	NOEL = 10,000 ppm
		Toxic to larvae at 10x field rate
Earthworm	*tenebrionis*	NOEC => 100 ppm (120x in 1 kg soil) in 14 day test
Predatory mite	*aizawai*	1x field rate resulted in 24% corrected mortalty
Daphnia (freshwater invertebrate)	*kurstaki*	Moderately toxic; 21 day LC_{50} is between 5 ppm and 50 ppm Aqueous LC_{50} >4.9 ml/l
	israelensis	Moderately toxic; 21 day LC_{50} is between 5 ppm and 50 ppm
	tenebrionis	48 hour EC_{50} > 100 mg/l
	aizawai	Highly toxic; 21 day NOEC = $6.4x10^8$ cfu/l

6.8 NON-TARGET SAFETY OF EXOTOXINS

Toxicity to non-target organisms has not been found with the δ-endotoxins when they are separated from the bacterial growth medium (EPA 1998). However, a number of *Bt* fermentation based products tested at high dosages have shown intrinsic toxicity to non-target organisms. Some investigations conducted to determine what is responsible for the non-target activity have implicated heat-labile soluble substances contaminating the technical material (β-exotoxin is heat stable). Toxic effects have been seen in the water flea *Daphnia magna* (microcrustacean), the honeybee, some beneficial insects and fish (rainbow trout, bluegill) and mammal (mouse and rat) studies, with *Daphnia* being the most sensitive indicator of toxicity. *Daphnia* studies are a useful toxicity screen and may indicate that additional non-target testing is required.

The toxic components were found in the supernatant fluids separated from the δ-endotoxins and toxicity did not appear to be caused by heat stable β-exotoxin, as toxic effects were removed by autoclaving. The heat-labile, soluble toxic impurities have been seen in *Bt kurstaki*, *aizawai* and *israelensis* but may also be present in other varieties. Damgaard (1995) reported on the presence of at least one soluble exotoxin in all commercial *Bt* products tested. *Bt aizawai*-based products showed the greatest negative effects on non-target organisms. With *Bt kurstaki*, the manifestation of the toxins appears to be at least partly related to production methodology, especially the composition of the growth media used in industrial fermentation (see Chapter 4).

A β-exotoxin-like "M-exotoxin" ("M" for mollusca-active) produced by strains of *Bt israelensis* was tested against the final larval stage of the Trematoda *Schistosoma mansoni, Trichobilharzia szidati, Apatemon minor, Echinostoma revolutum, Diplostomum pseudospathaceum, Hypoderaeum conoideum* and *Paryphostomum* sp. Treatment resulted in separation of the tegument from the rest of the body and other morphological abnormalities (Horak *et al.* 1996).

7 Effects on Vertebrates

Laboratory studies on vertebrate safety indicate very little safety risk from direct exposure. *Bt* has very low toxicity to fish and amphibians. No direct effects of *Bt* have been reported for a number of bird species, both in the laboratory and after field applications. Indirect effects on some insectivorous birds and small mammals include temporary switches in diet. Reduction in lepidopteran populations following field application of *Bt* has led to fewer nesting attempts by some bird species but this did not necessarily impact on the number of fledglings per breeding territory in the year of application or subsequent years.

Review of mammalian toxicity of *Bt* by the US Environmental Protection Agency concluded no adverse effects have been demonstrated in acute toxicity studies with small mammals. Following oral and pulmonary dosing, spores were generally cleared from animals rapidly, with no adverse effects. Some mortality was observed in mice following intraperitoneal injection of high doses (10^8 cfu/animal), but the same isolates were shown to be non-toxic at doses of 10^7 cfu and below. By comparison, high doses of bacteria generally regarded as non-toxic, such as *Bacillus subtilis*, can cause mortality if used at rates exceeding 10^8 cfu/animal.

Despite many years of broadscale application of *Bt*-based products, very few associated human health problems have been reported and proven cases of *Bt* causing clinical infections in humans remain extremely rare. There have been two occupational health incidents where accidental exposure to *Bt* resulted in conditions requiring treatment with antibiotics. *Bt* isolates have been recovered from patients with severe burn wounds but these isolates were different from those used in commercial preparations. Recent toxicity studies on a new serotype, *Bt konkukian*, suggest that some isolates, at least in an immunocompromised mouse model system, could produce infection and necrosis similar to that caused by *B. cereus*. Monitoring of public health and bacterial cultures collected for routine purposes in communities exposed to *Bt* spraying has raised few concerns. Where *Bt* was recovered from patients, it could not be determined as a cause of illness.

The production of *B. cereus*-type-diarrhoeal enterotoxins by *Bt* isolates has raised concerns that *Bt* may cause gastroenteritis in humans. The few studies conducted to date show that some strains of *Bt*, including those found in many commonly used commercial products, are capable of producing diarrhoeal enterotoxins, but at a level around 10 fold less than *B. cereus* strains. Evidence to date suggests food poisoning would be unlikely to occur due to *Bt* contamination as conditions for toxin production are not prevalent in commercial products and the titre of enterotoxin produced by most *Bt* strains is low. As yet there is no valid evidence to link usage of *Bt* with episodes of diarrhoea following ingestion of food. However, it may be prudent to screen strains with potential commercial value for ability to produce enterotoxin. Several assay systems have already been developed for detecting enterotoxins in *B. cereus*.

In general, the recorded effects on vertebrates indicate very little safety risk. Considering the extensive use of *Bt* already made for pest control and the ubiquitous natural occurrence in the environment, the reports of human infections are remarkably rare, suggesting a corresponding low risk.

7.1 FISH AND AMPHIBIANS

The few studies in the literature examining effects of *Bt* against fish have mainly been performed using *Bt israelensis*-based products, which are applied to water directly. No effect has been found on fish present during *Bt israelensis* application, nor in direct exposure experiments (Table 7.1). *Bt israelensis* application against blackflies in Michigan had no effect on mortality or weight gain of caged rock bass (*Ambloplites rupestris*) or fish numbers and species composition (Merritt *et al.* 1989). In laboratory studies, Christensen (1990 a, b and c) found no effect of *Bt israelensis* on bluegill sunfish (*Lepomis macrochirus*), sheephead minnow (*Cyprinodon variegatus*) and rainbow trout (*Oncorhynchus mykiss*) when exposed to $1.3\text{-}1.7 \times 10^{10}$ cfu/g of diet. In a comparison of toxicity of various mosquitocidal agents to the mummichog, *Fundulus heteroclitus*, *Bt israelensis* was reported to be the least toxic of the five preparations (Lee and Scott 1989; Table 7.2). They found that fenoxycarb and *Bt israelensis* together were more toxic than fenoxycarb alone, suggesting combined use of control agents should be examined carefully. *Bt tenebrionis* was not toxic to the fish, *Lebistes reticulata* (Meyer 1989 in Keller and Langenbruch 1993).

In the Reregistration Eligibility Decision (RED) documents (1998) submitted to the US Environmental Protection Agency (EPA), a number of fish studies (trout and bluegill) with subspecies *Bt kurstaki*, *tenebrionis*, *aizawai* and *israelensis* were included (Table 7.3). With aqueous LC_{50}s ranging from $>8.7 \times 10^9$ to $>4.6 \times 10^{10}$ cfu/l, no toxicity or pathogenicity was evident in the bluegill or rainbow trout (Table 7.3).

Table 7.1. Fish species reported as non-sensitive to *Bt israelensis* in laboratory bioassays or after field exposure (from Glare and O'Callaghan 1998).

Fish species	Common name	Reference
Gambusia affinis	mosquito fish	Abbott Laboratories; Garcia *et al.* 1980
Lucania parva	rainwater killifish	Abbott Laboratories; Garcia *et al.* 1980
Gasterosteus wheatlandi	twospine stickleback	Abbott Laboratories; Garcia *et al.* 1980
Lepomis macrochirus	bluegill	Christensen 1990a
Salvelinus fontinalis	brook trout	Wipfli *et al.* 1994
Salmo trutta	brown trout	Wipfli *et al.* 1994
Oncorhynchus mykiss	rainbow trout	Wipfli *et al.* 1994; Christensen 1990b
Pseudomugil signifer	Pacific blue-eye fish	Brown *et al.* 1998
Poecilia reticulata	larvivorous fish	Mittal *et al.* 1994
Tilapia nilotica		Lebrun and Vlayen 1981
Esox lucius		Becker and Margalit 1993
Cyprinus carpio		Becker and Margalit 1993
Perca fluviatilis		Becker and Margalit 1993
Ambloplites rupestris	rock bass	Merritt *et al.* 1989
Epiplatys sp.	killifish	Beck 1982
Cyprinoidei	goldfish	Beck 1982
Cyprinodon variegatus	sheephead minnow	Christensen 1990c

Table 7.2. Acute toxicity of mosquito larvicides to *Fundulus heteroclitus* (reproduced from Lee and Scott 1989, with permission of Springer-Verlag GmbH & Co.).

Insecticide	Mean 96h LC_{50} (mg/l)	95% confidence limits (mg/l)
Temephos	0.04	0.02-0.05
Fenoxycarb	2.14	2.01-2.27
Diflubenzuron	32.99	29.01-37.52
Methoprene	124.95	90.01-171.64
VectoBac (*Bt israelensis*)	980.00	730-1330
Fenoxycarb/VectoBac	1.55	1.40-1.72

Table 7.3. Non-target susceptibility of fish: summary of unpublished studies submitted to EPA in support of product registration - Reregistration Eligibility Decision documents, EPA 1998 (NOEC = no observable effects concentration). No experimental details or species names were presented in the report.

Fish	*Bt* subspecies	Result
Trout	*kurstaki*	LC_{50} > 1.5×10^{10} cfu/l
		Practically non-toxic; aqueous LC_{50}>4.9 ml/l and oral LC_{50} > 2.5 nl/g of food
		Practically non-toxic with an aqueous LC_{50} > 4.6×10^{10} cfu/l of dilution water
	israelensis	Aqueous LC_{50} >8.7×10^{9} cfu/l; oral LC_{50}>1.7×10^{10} cfu/g food. Slightly toxic
		Aqueous LC_{50}>1.4×10^{10} cfu/l; oral LC_{50}>5.3×10^{9} cfu/g food
	tenebrionis	Aqueous NOEC = 100 mg/l
	aizawai	Aqueous LC_{50}>3.9×10^{7} cfu/ml; oral LC_{50} > 1.5×10^{10} cfu/g food
		96 hour LC_{50} >100 mg/l
Bluegill	*kurstaki*	Practically non-toxic at 1.5×10^{10} cfu/l of dilution water and at 1.2×10^{10} cfu/g of food for 32 days
	israelensis	Aqueous LC_{50} >8.9×10^{9} cfu/l; oral LC50>1.3×10^{10} cfu/g food
		Aqueous LC_{50}>1.6×10^{10} cfu/l; oral LC50 >4.3×10^{9} cfu/g food

The World Health Organisation (1992) reviewed a number of laboratory and field studies examining the impact of *Bt* on frogs, newts, salamanders and toads in which no adverse effects were recorded. No effect of *Bt israelensis* on a range of amphibian species has been reported (Glare and O'Callaghan 1998); similarly *Xenopus laevis* was insensitive to *Bt tenebrionis* (Meyer 1989, in Keller and Langenbruch 1993).

7.2 BIRDS

7.2.1 DIRECT EFFECTS

No significant toxicity of *Bt* strains to any species of bird has been recorded. Extensive studies on birds with high doses of unsolubilised *Bt israelensis* spores and crystals have shown no adverse effects (Meadows 1993). Lattin *et al.* (1990a and b) found no toxicity, pathogenicity or weight loss when *Bt israelensis* was administered orally in large doses ($3.4\text{-}6.2 \times 10^{11}$ cfu/kg/day for 5 days) to young brown quail (*Colinus virginianus*) and young mallards (*Anas platyrhynchus*). Smirnoff and MacLeod (1961) orally administered *Bt thuringiensis* to starling (*Sturnus vulgaris*), white-crowned sparrow (*Zonotrichia leucophrys*) and slate coloured junco (*Junco hyemalis*). No adverse effects on birds were noted and viable undigested bacteria could be recovered from the faeces. Several laboratory studies on toxicity of *Bt kurstaki*, *Bt israelensis*, *Bt tenebrionis* and *Bt aizawai* to mallard duck and bobwhite quail were summarised in RED documents (EPA 1998). The EPA concluded that these *Bt* subspecies were not toxic or pathogenic to the mallard and bobwhite quail after acute or subacute testing. LC_{50} values for *Bt aizawai* ranged from 8-16 g/kg for the two bird species. *Bt kurstaki* and *Bt israelensis* were described as "practically non-toxic" after doses ranging from 2.5 - 3.1 g/kg/day were administered for 5 days. No mortality was found after injection of 3 mg/kg *Bt tenebrionis* into mallard. No information on clinical effects was provided in the report, nor was the meaning of "practically non-toxic" explained.

Kachekova and Frolov (1977) examined the effect of oral application of Thuringin to fowls. The active ingredient of the preparation was a heat-stable exotoxin. LD_{50}s of Thuringin were 676.1 mg/kg body weight for 2 week old fowls, 691.8 mg/kg for month-old fowls, and 912 and 1148 mg/kg for young and mature adult fowls, respectively, demonstrating lower tolerance to exotoxin than endotoxin.

The effect of *Bt israelensis, Bt thuringiensis* and *Bt kurstaki* on the health of laying hens (and albino mice) was studied by Hassanian *et al.* (1997a). No gross lesions were found in postmortem tests and no differences in the rates of feeding and in blood analyses could be discerned between the treated and untreated groups.

Bird mortality was monitored following application of several pesticides applied over several years to control eight outbreaks of the nun moth, *Lymantria monacha,* in Denmark. While endosulfan gave good levels of control (60-97% mortality), it also caused some bird mortality; in comparison Dipel (*Bt kurstaki*) gave 76% control of nun moth larvae and did not cause bird mortality (Bejer 1986).

7.2.2 INDIRECT EFFECTS

Any effect of *Bt* on insectivorous birds is caused by a reduction in food supply, as direct toxicity has not been found (Section 7.2.1). Birds that feed on caterpillars in the spring will have a reduced number of prey on which to feed for a short time, which may force a switch in diet. The number of nesting attempts per year may be reduced but this will not necessarily impact on the number of fledglings per breeding territory in the year of application or subsequent years (RED, EPA 1998).

In specific studies, few indirect effects have been found. The Tennessee warbler behaviour and reproduction was not affected by application of lepidopteran-active *Bt* although nests in treated blocks had non-significantly smaller clutches, smaller broods and lower hatch rates (Holmes 1998). The researcher concluded that the indirect effects of forest spraying with Lepidoptera-specific *Bt* products posed little risk to forest songbirds. In Arkansas, *Bt kurstaki* application for the eradication of gypsy moth was examined for effect on reproduction of hooded warblers (*Wilsonia citrina*). Despite some statistically significant differences, including higher nestling masses at day 5 in the treated zone, only minimal effects on reproduction of warblers were caused by reduction of Lepidoptera (Nagy and Smith 1997).

The application of Dendrobacillin (*Bt dendrolimus*) at 2.5 kg/ha to an oak stand did not significantly affect the diet of nesting starlings (*Sturnus vulgaris*) and great tits (*Parus major*), despite serious pests constituting 55% and 72% of the diet of these birds respectively (Bezvesil' nyu 1983). Fernandez de Cordova and Cabezuelo Perez (1993) studied bird populations in Cordoba, Spain, after application of a number of pesticides, including Dipel, Foray and Bactospeine. They found small, temporary population reductions only.

Bird communities in wetlands in Minnesota, USA, were compared before and several years after application of *Bt israelensis* (VectoBac-G granules) for mosquito control. No effects on 19 species of birds were observed, with the few differences detected more likely caused by natural variation in response to changes in habitat and water levels (Hanowski *et al.* 1997).

However, impacts have been noted in some studies. Rodenhouse and Holmes (1992) showed that a reduction in biomass of lepidopteran larvae after application of *Bt* led to significantly fewer nesting attempts by certain birds. Scriber (1998) reported that the Kirtland's Warbler in Michigan was threatened by extensive spraying to control gypsy moth in late 1980s - early 1990s, because of non-target impacts on some Lepidoptera which are part of that bird's diet. However, Scriber (1998) goes on to point out that not spraying may actually have greater adverse effects on non-target organisms. Generalist parasites living off a gypsy moth outbreak in Michigan attacked other Lepidoptera which were 50-100 miles from the outbreak front, reducing food supply for birds. Gypsy moth directly impacted on butterfly populations by defoliating host plants, eggs were eaten by gypsy moth, or butterflies were forced to shift host plant.

7.3 MAMMALS

7.3.1 EFFECT OF *BT* ON SMALL MAMMALS

7.3.1.1 Summary of US Environmental Protection Agency mammalian safety assessment (1998)

The US EPA maintains a historical database for *Bt* and a summary review of mammalian toxicity studies was published by Agency reviewers (J.T. McClintock, C.R. Schaffer and R.D. Sjoblad 1995). The most recent RED documents released by the EPA (March 1998) state that "no known mammalian health effects have been demonstrated in any infectivity/pathogenicity study". Acute toxicity studies (predominantly from unpublished safety assessments) reflecting five routes of exposure, including the so called "maximum hazard" intravenous route, are summarised below. The individual studies summarised below are not referenced but further details are available from the EPA (contact details in McClintock *et al.* 1995).

Acute oral toxicity studies: Acute oral toxicity (large single dose) data submitted have not demonstrated any adverse effects or signs of infectivity to mammals with any of the *Bt* subspecies tested at the specified dose level. Oral ingestion of 4.7×10^{11} spores/kg body weight and associated crystals of *Bt kurstaki* failed to produce adverse toxic effects or any signs of infectivity in rats. Microbial clearance data showed that *Bt* passed through the digestive tract with complete clearance demonstrated, in some instances, as early as day 2 post-dosing. Low numbers of *Bt* have occasionally been detected in the spleen, caecum and small intestine, but these numbers decreased to below the limit of detection during the study. McClintock *et al.* (1995) concluded that after oral dosing, *Bt* remains confined to the gastrointestinal tract and does not become systemically distributed in the test animal.

Acute pulmonary toxicity studies: Lack of adverse effects have been demonstrated following inhalation exposure. For example, when *Bt kurstaki* was administered to rats by the pulmonary route at a concentration of 2.6×10^{7} spores/kg, no adverse effects were observed. A general pattern of clearance from the lung was observed in most studies, with numbers of *Bt* recovered decreasing rapidly following dosing.

Intravenous toxicity studies: Acute IV toxicity tests provide "maximum-hazard" information on potential health effects when the skin is bypassed as a barrier. After IV injection, there was a general pattern of clearance from organs and tissues. When rats were challenged with *Bt israelensis*, viable bacteria were recovered on Day 1 from blood and all organs examined. On Day 4, a general decline of the bacterium was observed in the lungs, kidneys, lymph, blood and brain. The animal cleared the majority of the inoculum from the spleen and liver by Day 57.

Intraperitoneal toxicity studies: The intraperitoneal injection assay of *Bt* in mice is used as a quality control measure by the EPA to demonstrate the lack of mammalian toxicity of the technical grade active ingredient. When mice were injected with a technical grade preparation of *Bt aizawai, israelensis, kurstaki,* or *tenebrionis* at 10^6 and 10^7 cfu/animal, no toxicity or pathogenicity was observed. No mortality was observed with either of two *Bt aizawai* isolates following a single IP challenge at a dose of 10^8 cfu/animal. The only toxic effects reported at 10^8 dose rate, included slightly enlarged spleens in three of the five treated mice. One *Bt israelensis* isolate out of five tested caused mortality at 10^8 level, while the remaining four isolates had no adverse effects. The EPA has found that high dose levels of 10^8 cfu/animal often caused mortality, even for bacteria generally regarded as non-pathogenic and non-toxic, such as *Bacillus subtilis*. Some of the assays submitted to the EPA showed mortality in mice at these high doses, but the same studies supported lack of toxicity at doses of 10^7 and below for these isolates of *Bt*.

When mice were injected IP with *Bt kurstaki* (dose not stated), various mortality rates were observed. Three of six isolates tested caused 70-100% mortality, with most deaths occurring from the day of dosing up to 72 h after challenge. Clinical observations included abdominal sensitivity and eyes crusted closed. Treated mice showed small abdominal abscesses, enlarged spleens, pale kidneys and haemorrhagic lungs. The remaining *Bt kurstaki* isolates showed no toxicity.

Dermal/ocular exposure: *Bt* subspecies have been applied to intact and abraded rat and rabbit skin in a number of studies and no mortality or adverse effects on normal body weight gain were reported. The isolates tested were not infectious, pathogenic or toxic to mammals. Slight to moderate skin irritation has been observed following ocular exposure of rabbits to various *Bt* subspecies. In most cases, the only observed sign of irritation was reversible within 7 days without any noted effects on the cornea. The effects were thought to be caused by the physical nature of the formulation, often dry anhydrous formulations.

7.3.1.2 Additional laboratory studies

The potential pathogenicity of the strain of *Bt konkukian* isolated from severe human tissue necrosis (Hernandez *et al.* 1998; see Section 7.3.4.1) was investigated in immunocompetent mice following pulmonary infection (Hernandez *et al.* 1999). Mice infected intranasally with a suspension containing 10^8 spores/mouse died within 8 hours in a clinical toxic-shock syndrome. Autopsy revealed lesions of acute bronchitis associated with ulceration, injuries of the mucociliary apparatus, oedema and alveolar damage. Blood cultures obtained after intracardiac puncture yielded pure cultures of *Bt*. Lower inocula (10^5 - 10^7 spores/mouse) induced only a local inflammatory reaction with bacterial persistence observed for at least 10 days. Mortality rates were lower when mice were treated with commercial strains obtained from Abbott Laboratories; mice treated with 10^8 *Bt kurstaki* suffered 80% mortality, while *Bt israelensis* caused 40% mortality. Histological examinations revealed identical lesions to those observed with *Bt konkukian*. This is the first report of any serotype able to kill mice when applied at high concentration by the pulmonary route, however the concentrations used for infection under this protocol were very high compared to field use rates. The authors suggested that death was caused by a *B. cereus*-like haemolytic toxin as described by Honda *et al.* (1991). A concentration of 10^7 bacteria/mouse did not produce enough toxin to kill the mice, indicating a high level of haemolysin was required for lethality.

Subcutaneous injection of rats by Bishop *et al.* (1999) with *Bt kurstaki* did not result in any necrosis or bacterial growth in organs. No organisms were recovered from blood three weeks after injection. As no effect of enterotoxin producing *Bt* strains was detected in this study, Bishop *et al.* (1999) suggested rodents may not be a good model system for detection of potential effects of *Bt* on humans. However, no evidence was presented that the amount of enterotoxin produced would be sufficient to cause human illness (see Section 7.3.4.2).

Tsai *et al.* (1997) examined the clearance and distribution of spores of *Bt* (strain not specified) in rats after single intra-tracheal administrations of 1×10^8 spores. The spores penetrated into the body beyond the lungs, a process commonly seen with this route of infection, with spores isolated from not only the lung but also from the liver, spleen, kidney and mesenteric lymph nodes. Inflammatory changes were observed in the lungs. The recovery culture examination and histological studies revealed that the total number of colony forming units recovered was less than the initial inoculation. While spores of *Bt* survived in tissues for up to 21 days, no spore germination was found.

A number of other studies in several countries have found no effect on small mammals. Halkova *et al.* (1993) found no effect of subacute oral, dermal and inhalation exposure of Bulmoscide (*Bt israelensis*) on rats. Siegel *et al.* (1987) tested six commercially available preparations of *Bt israelensis* for toxicity and infectivity to mice, rats and rabbits. Oral (4×10^7 bacteria/animal) or pulmonary (2.05×10^6 bacteria/animal) exposure to rats caused no mortality. While intraperitoneal injection of one preparation to immuno-compromised mice produced substantial mortality, a subsequent experiment with a different preparation did not cause mortality. Colony forming units of *Bt israelensis* seemed to be cleared from the host as inert particles and multiplication in the mammals was not evident. They concluded that "*Bt israelensis* is not a virulent, invasive mammalian pathogen and it can be used safely in environments where human exposure is likely to occur". Further work by Siegel and Shadduck (1990) produced similar results. Following oral, intraperitoneal, subcutaneous, aerosol, intra-cerebral and ocular application of *Bt israelensis* to rats and rabbits, only intraperitoneal injection of *Bt israelensis* led to significant mortality when animals were treated with high concentrations (10^8 cfu/rat). Studies by Siegel and Shadduck (1990) included immuno-suppressed mice and showed that an intact immune system was not necessary for successful clearance of *Bt israelensis*.

7.3.1.3 Field studies on small mammals

While numerous studies have been conducted on the impact of *Bt* in the laboratory, few researchers have monitored the effect of *Bt* applications on small mammals under natural conditions. As part of the wider study monitoring the impact of several insecticides applied for control of jack pine budworm in Ontario, Innes and Bendell (1989) monitored populations of rodents and shrew in treated (20 billion international units /ha Thuricide 48 LV) and untreated forest plots. Four species of shrews and eight rodent species were captured by trapping after application of insecticides. *Bt* appeared to have little detectable impact on the abundance of small mammals, supporting previous work by Buckner *et al.* (1973), who found no effect of *Bt* on five species of small mammals in Manitoba.

The effects of a simulated operational spray with *Bt kurstaki* (Dipel 8L at 30 BIU and 1.8 litres/ha) on masked shrew (*Sorex cinereus*) in a jack pine (*Pinus banksiana*) plantation in Ontario were studied during May-July

1989 (Bellocq *et al.* 1992). The diet of masked shrews includes large numbers of insect larvae. During the pre-treatment period, the abundance and population structure of *S. cinereus* was similar in the control and experimental areas. Although the total abundance of shrews was also similar after spraying, there were fewer adult males and more juveniles in the treated than the untreated area. The emigration of adult males apparently increased after spraying. Larvae of Lepidoptera and Araneae were the most abundant diet items. After spraying, more lepidopteran larvae were eaten in untreated than treated areas. Juveniles and adult females but not adult male shrews shifted from lepidopteran larvae to alternative prey in the treated area. Generalist insectivores such as *S. cinereus* are more likely to control the abundance of arthropods and less likely to be negatively affected by selective insecticides such as *Bt*.

Mammals, including bats, which feed on susceptible insects, might be affected indirectly by reductions in food abundance due to *Bt* sprays. This may trigger a switch in diet. However, unlike the situation following treatment with many conventional pesticides, insect feeding mammals are not adversely affected by ingestion of moribund insects killed by *Bt*.

7.3.2 EFFECT OF *BT* ON LARGE MAMMALS

Bt has been isolated from a fatal case of bovine mastitis (Gordon 1977, in Samples and Buettner 1983) but few studies have been conducted on large mammals. Hadley *et al.* (1987) conducted a five month oral toxicity study of *Bt* insecticides in sheep. *Bt* products Dipel and Thuricide were administered in the diet for 5 months at a dose of 500 mg/kg/day (approximately 10^{12} spores/day). No treatment related effect was seen on weight gain or clinical chemical parameters, nor were gross clinical changes observed. Blood and tissue samples taken prior to the time the animals were killed were found to be positive for *Bt*.

7.3.3 EFFECT OF *BT* ON HUMANS

More recently, concerns about the mammalian pathogenicity of the genus *Bacillus* have been raised, in part because several other species within the genus cause various infections in humans. *B. anthracis* has long been recognised as a virulent mammalian pathogen and *B. cereus,* previously thought to be a totally innocuous relative of the anthrax bacillus, has more recently been identified as a cause of mild food poisoning and has been incriminated in several very severe types of eye infection that can lead to blindness. It has also been isolated from burn wounds, insect bites and meningitis patients (Damgaard *et al.* 1997b). Microbiologists have increasingly realised that other "non-pathogenic" members of the genus *Bacillus* can trigger human disease too, if only in special situations such as in immuno-compromised individuals.

Although *Bt* has been used for over 60 years for insect pest control, there have been few reports in the literature of clinical infections caused by this organism. Some workers have suggested that the low number of reported cases may be an underestimate due to inadequate diagnostic laboratory facilities, failure to identify *Bacillus* isolates to species, the mixed microbiological nature of some clinical specimens and the rejection of clinically significant isolates as contaminants. For example, Jackson *et al.* (1995) reported that *Bt* isolates recovered from patients with burn wounds were initially incorrectly identified as *B. cereus*. There is a need for simple accurate identification systems capable of differentiating *Bt* isolates from closely related species, in particular *B. cereus*.

Direct human exposure experiments would now be considered unethical, but have been performed in the past. Human volunteers exposed to *Bt* by oral ingestion (3-10 $\times$ 10^9 spores /day for 3-5 days) and inhalation (100 mg Thuricide for 5 days) showed no ill effects (Fisher and Rosner 1959 in Drobniewski 1994). Fisher and Rosner (1959) also reported normal health records for eight employees of a *Bt* product manufacturer. They had been exposed for 7 months to *Bt* as a whole fermentation broth, moist bacterial cake, effluent and final powder preparation.

In contrast, three out of four human volunteers who ate food artificially contaminated with a Russian strain of *Bt galleriae* (10^5 - 10^9 cells/g) showed signs typical of food poisoning following an incubation time of eight hours (Ray 1991). The rapid onset of symptoms, which included nausea, vomiting, dairrhea and fever, suggests the toxicity may have been caused by β-exotoxin.

7.3.3.1 Infections

Two occupational health incidents have occurred during the handling of highly concentrated *Bt* fluids. Sample and Buettner (1983) reported a corneal ulcer attributed to *Bt* after accidental splashing of a commercial preparation (Dipel, *Bt kurstaki*) into the eye of a previously healthy 18-year old farmer. Treatment with gentamycin cured the ulcer. Siegel and Shadduck (1990) suggested the possibility that while the spores might persist in the eye and be recovered, they may not have actually caused the ulcer. No vegetative cells were looked for, or found, in the ulcer. Sample and Buettner (1983) suggested that some form of ocular protection may be advisable while using active biological insecticides, which have previously been considered harmless to humans.

In the second incident, a research student in England accidentally punctured his hand with a hypodermic needle containing a suspension of spores and endotoxin crystals of *Bt israelensis* (6×10^8 /ml), together with a contaminating vegetative bacterium *Acinetobacter calcoaceticus* var. *anitratus,* the latter a normal human skin bacterium which seldom causes serious soft-tissue infection (Warren *et al.* 1984). Within 2 hours, the finger became painful and later became swollen and discoloured. The patient recovered after 5 days treatment with antibiotics. Further studies were conducted on the isolates recovered from the wound. Cell free culture filtrates of the *Bt* strain showed intermediate levels of exotoxin activity, whereas the *Acinetobacter* strain, alone or in mixed culture with *Bt* produced a strong reaction with necrosis in an exotoxin assay system. Culture filtrates of neither strain were lethal on intravenous injection in mice, but a filtrate from a mixed culture was consistently lethal. The authors found that the crystalline *Bt israelensis* protoxin could be activated *in vitro* by proteases in aged cultures of *Bt* or in culture filtrates from unrelated bacteria.

Damgaard *et al.* (1997b) isolated *Bt* from infections in burn wounds and from water used in the treatment of the wounds. The patients had deep burn wounds on 30-70% of their body surface and were therefore considered to be highly immuno-suppressed. The source of contamination of the water supply was unknown, however, it was clear that the contaminating *Bt* isolates did not originate from pesticide residues. The observed molecular size of the protoxins comprising the δ-endotoxin was different from those used in commercial preparations and showed no activity against the caterpillar, *Pieris brassicae,* or the mosquito, *Aedes aegypti*. In addition, strains were non-flagellated *Bt;* all strains used in pesticides are flagellate.

More recently, Hernandez *et al.* (1998) presented a case of a French soldier with wounds infected by *Bt konkukian* serotype H34, a serotype originally described from Korea. As it was unclear whether the *Bt* was the causative agent of the infection or simply a contaminant, the potential pathogenicity of this strain to mammals was examined in a mouse model of cutaneous challenge using both immunosuppressed mice and a control group of mice which were not immunosuppressed. Cutaneous inflammatory lesions occurred in all animals when 10^7 cfu inoculum was used. Lesions healed spontaneously after 48 h in non-immunosuppressed mice but increased in the other animals. *Bt* was recovered from the samples taken from immunosuppressed mice only. The animal model used demonstrated the ability of *Bt konkukian* to produce infection and myonecrosis in the absence of an effective immune response. The clinical data suggested that the infection could be produced in patients with tissue devitalisation caused by massive tissue destruction. The authors speculated that soil and dust particles projected by a landmine explosion was the source of the organism in this unusual case of infection. This reported episode of soft-tissue infection and necrosis by *Bt* was reminiscent of *B. cereus* infection after predisposing tissue necrosis caused by severe trauma.

7.3.3.2 Gastric illnesses

B. cereus is recognized as a food contaminant involved in some cases of food poisoning each year (Johnson 1984). The illness occurs in two distinct syndromes, characterized by toxin-induced emetic or diarrhoeal syndromes caused by *Bacillus* diarrhoeal enterotoxin. For this reason, the presence and quantity of *B. cereus* is routinely monitored in certain foods. Andersson *et al.* (1998) suggested that the infective dose could be as low as 5×10^4 spores (200 spores /g food). Techniques currently used for monitoring *B. cereus* in food and clinical samples do not allow discrimination between *Bt* and *B. cereus*. Samples are plated onto *B. cereus*-selective medium and colonies are then spread onto blood agar. *B. cereus* strains are identified by their irregular-edges and by the production of lecithinase and inability to ferment mannitol on selective media. This method is completely unable to differentiate *Bt* from *B. cereus*.

Jackson *et al.* (1995) reported the isolation of *Bt* strains from four hospitalized patients suffering gastroenteritis. The isolates were originally incorrectly identified as *B. cereus* using traditional morphological and biochemical characteristics. While it was not confirmed that *Bt* caused the gastroenteritis, the failure to find other enteric pathogens suggests a possible role for *Bt*. *Bt* isolates recovered showed cytotoxic effects characteristic of enterotoxin-producing *B. cereus*, which may have implications for *Bt* as an agent of gastroenteritis.

While Jackson *et al.* (1995) reported the only case where *Bt* was associated with gastroenteric illness in humans, Damgaard (1995) suggested that *Bt* may have been involved in more cases but went undetected because the specialised techniques required for differentiation between *Bt* and *B. cereus* were not used when identifying causal agents of food poisoning (cultures are not normally held and stained for toxin crystal formation).

The close relationship between *Bt* and *B. cereus* led Damgaard (1995) to investigate the levels of diarrhoeal enterotoxin produced by strains isolated from various *Bt* products (Bactimos, Dipel, Florbac FC, Foray 48B, Novodor FC, Turex, VectoBac, Xentari). He found that all strains produced diarrhoeal enterotoxin, including the *Bt israelensis* based products, Bactimos and VectoBac. These two products had titres of 242 and 120, respectively, compared to *B. cereus* standard of 1629. The product MVP in which δ-endotoxin is expressed in a *Pseudomonas* cell does not contain viable *Bt* spores and did not contain diarrhoeal enterotoxin. Damgaard (1995) considered the ability to produce enterotoxin indicated a potential risk for gastroenteritis outbreak caused by *Bt*. Pathogenicity of *B. cereus*, in terms of food poisoning by diarrhoeal enterotoxin is caused by ingestion of vegetative cells rather than pre-formed enterotoxin. Therefore, the product contents of *Bt* insecticides, spores and d-endotoxins, are not at all toxic to mammals. However, because the spores are viable, they may pose a risk of causing gastroenteritis when the right conditions occur.

More recently, Bishop *et al.* (1999) tested several strains of *Bt*, including three which had been used world-wide for many years to control pest insects, for production of *B. cereus*-type enterotoxin, using commercially available kits, BCET-RPLA and VIA ELISA. They detected enterotoxin in all six strains tested, consisting of *Bt kurstaki* (HD-1 and 3 others), *Bt israelensis* and *Bt tenebrionis*. However, high dose challenge to rats over three weeks did not result in detectable effects. Bishop *et al.* (1999) suggested that rodents may not be a sensitive test organism for investigating the potential for food poisoning.

Several recent publications have reported the isolation of enterotoxin-producing *Bt* strains in food. Damgaard *et al.* (1996c) isolated *Bt* strains from various food items (pasta, pitta bread and milk); strains belonged to either H-serotype *kurstaki* or *neoleonensis*. Nearly all strains were found to express cytotoxic effects on Vero cells as an indicator of enterotoxin activity. Further, the *Bt* strains HD-1 (*Bt kurstaki*), NB-125 (*Bt tenebrionis*) and HD-567 (*Bt israelensis*), which are used commercially for insect pest management, were also found to have cytotoxic effects on Vero cells. It was not possible to determine the origin of the strains from food items; i.e. whether they were naturally occurring *Bt* or from pesticide residues in the raw agricultural commodities. The serotype of some of the isolated strains was found to be *kurstaki*, which is the most widely used serotype for lepidopteran pest control. However, this serotype also has a wide natural distribution.

Isolation of *Bt* from different types of food has been reported previously. Rusul and Yaacob (1995) recovered *Bt* isolates while surveying a range of foods (pasta, spices, grain and legumes) for enterotoxin-producing *B. cereus*. Phillips and Griffiths (1986) isolated *Bt* from farm bulk tank and creamery silo milk. Bidochka *et al.* (1987) isolated a strain of *Bt kurstaki* from grapes for human consumption and hypothesized that the strain could have originated from pesticide residues. Burges (1976a) reported on the leaching of *Bt* spores from foundation beeswax into honey. Spores survived for extended periods but were unable to multiply in neat honey. It is conceivable that bacteria in honey could cause such food preparations as sauces to spoil if the honey was well diluted and the preparations kept improperly, e.g. stored at warm temperatures for extended periods. While it seems unlikely that the presence of *Bt* in honey would cause food poisoning, a precautionary limit of 1% w/w of *Bt* in foundation wax was recommended by Ministry of Agriculture, Fisheries and Food in Britain (Burges 1976c). In unpublished work, cited in Damgaard *et al.* (1996c), the authors reported isolating 10^3-10^4 *Bt* spores/g of pasta, indicating either contamination of the raw materials, e.g. flour, or substantial germination and growth during food processing. They further suggested that insect control products containing live spores which can germinate and produce enterotoxins may constitute a potential health hazard if food items are contaminated, emphasising the caution required in assessing the safety of *Bt* as an insecticide applied to crops close to consumption.

Recent work by Bishop *et al.* (1999) also suggests a potential food poisoning hazard from food sprayed with commercial preparations of *Bt*. Spinach plants sprayed with Bactospeine were harvested 24 hours after spraying and subjected to various washing and food preparation routines. Viable spores were enumerated on solid growth medium and colonies were examined for the presence of crystals. Washing of leaves in cold water resulted in a reduction of only about 50% of the spore load; 1.8×10^6 colonies/g wet weight of leaves remained after washing. While a credit to the efficacy of the formulation, the authors suggested that if a commercial product contained a strongly enterotoxic strain, then sufficiently high doses of spores could be ingested, possibly causing food poisoning. The authors suggested there was potential for replication of the bacterium if, for example, washed salad leaves were supplemented with carbohydrate and protein rich foods and kept unrefrigerated.

The most recent RED documents released by the EPA (March 1998), state that the agency has no valid evidence to link actual usage of *Bt* insecticides with episodes of diarrhoea following ingestion of food. *B. cereus* and other toxigenic microorganisms can normally be found on many kinds of food but must multiply in the foods in order to produce the toxins responsible for the symptoms. A review of morbidity data for food-borne diseases compiled by the Centre for Disease Control and Prevention and cited in RED documents (EPA 1998) showed that from 1988 to 1992, the average number of reported outbreaks/year attributed to *B. cereus* was 4.2 (0.64% of total). No deaths were attributed to these outbreaks. For these reasons, the EPA takes the view that the current uses of *Bt* are not likely to contribute to the prevalence of diarrhoea induced by microbial toxins in food that has been improperly handled or stored.

Damgaard (1995) recommended that strains used in commercial *Bt*-based insecticides be tested in specific assays to determine the amount of enterotoxin produced, as part of the product registration procedure. He also suggested that only diarrhoeal enterotoxin-negative strains be used in insecticides, as is the case with β-exotoxin producing strains, where toxin-positive strains have been banned in USA since 1971 because of toxicity to humans. Bishop *et al.* (1999) reiterated the need for enterotoxin-free strains: "To ensure that biopesticides containing *B. thuringiensis* continue to enjoy the excellent safety record that they have to date it would seem to be advisable to ensure that existing and newly registered products are tested reliably for their potential as enterotoxigenic food poisoning agents."

As discussed in Chapter 2, one problem is that the taxonomic separation of *B. cereus* and *Bt* is unclear; there is also a need for techniques to discriminate between these species in clinical samples, and studies to determine prevalence of each species in gastroenteritis cases. Assays for enterotoxins have already been developed for *B. cereus* (Rusul and Yaacob 1995; Tan *et al.* 1997) and new strains should be tested for production early in the selection process for new biopesticides.

7.3.3.3 Hypersensitivity incidents

Despite the widespread use of *Bt*-based products, only two incidents of reported allergic reaction have been reported to the EPA (reported in McClintock *et al.* 1995). In the first incident, it was concluded that the exposed individual was most likely suffering from a previously diagnosed disease. The second incident involved an individual with a previous history of life-threatening food allergies. Following exposure to a *Bt* product, the person experienced breathing difficulties and nasal congestion. The formulation to which the individual was exposed contained a specific carbohydrate and preservatives which have been implicated in food allergy. Investigation of the case concluded that *Bt* was not the causative agent.

7.3.3.4 Community exposure

Bt has been used for many years in agriculture and forestry situations with no reported problems. Most researchers in the field of safety of *Bt* have found no reason to discontinue the use of *Bt* on grounds of risk to human health (e.g. Siegel and Shadduck 1990; World Health Organisation WHO - Becker 1992; Becker and Margalit 1993; Drobniewski 1994; McClintock *et al.* 1995).

Green *et al.* (1990) monitored cultures obtained for routine clinical purposes from people who lived in a spray zone of *Bt kurstaki* used against gypsy moth in Oregon, USA. Over a 2 year period, 55 *Bt*-positive cultures were identified (around 120,000 people lived in the sprayed area), 95% of which were thought to be probable contaminants and not the cause of clinical illness. Of the remaining three patients, *Bt* could not be determined to be a pathogen, nor could it be ruled out as having no role in the illness. All three *Bt*-positive patients had pre-existing medical problems. The authors suggested that *Bt* may have some risk for immuno-compromised individuals, and this should be considered by agencies responsible for regulating the use of microorganisms for pest control. They recommended that immuno-compromised individuals should be advised on how to use biopesticides and how to protect themselves from undue exposure in areas where they are used.

Ecobichon (1990) noted one further case of detrimental effects on humans following aerial overspray. Two elderly people were oversprayed by a *Bt* formulation in New Brunswick. No details of the case were given, but Ecobichon (1990) suggested that the case raised concerns about post-exposure, non-specific health effects including dermal rash, hive-like wheals, increased incidence of respiratory infections and general malaise which might have been associated with the bioinsecticide.

Proven cases of *Bt* causing clinical disease in mammals remain extremely rare and after reviewing available data, Drobniewski (1994) concluded that the risk to public health from *Bt* was extremely small. Risks to immuno-suppressed individuals may be higher than to the general public, as is the case with all environmental bacteria. During a spray programme, monitoring which incorporates specific identification methods for *Bt*, capable of identification to the level of isolate, would greatly assist evaluation and monitoring the effects of use. Such a monitoring programme should be conducted in conjunction with health monitoring in the spray areas to detect any possible symptoms resulting from spraying.

Similar safety data to that of Drobniewski (1994) have been presented in New Zealand following the use of *Bt kurstaki* against tussock moth during the aerial application programme, "Operation Evergreen". The Ministries of Forestry and Health commissioned a health risk assessment following "Operation Evergreen". The regional health authority found that there were: no miscarriages or premature baby deliveries associated with *Bt kurstaki* spraying; no allergies caused by *Bt*; no increased attendance at local health centres as a result of spraying; and no increased incidence of measles or meningococcal disease. The New Zealand Government has commissioned a health surveillance programme to run until 1999, monitoring any health effects, which might have arisen following tussock moth spraying.

7.3.4 CYTOTOXICITY OF SOLUBILISED TOXIN

It is generally concluded that the δ-endotoxin is not toxic to mammals when administered *per os*, and as the mammalian gut is not alkaline, the toxin is not activated following ingestion. The δ-endotoxin can be solubilised (i.e. activated) under alkaline conditions (pH >12), or by a combination of alkaline conditions (pH 9-10) and reducing conditions (Huber *et al.* 1981).

With *Bt israelensis*, Thomas and Ellar (1983) showed that while intact δ-endotoxin crystals showed no detectable toxicity in the *in vitro* and *in vivo* systems used, alkaline-dissolved δ-endotoxin was cytotoxic to five mammalian cultured cell lines (mouse fibroblasts, primary pig lymphocytes and three mouse epithelial carcinoma cell types). Soluble crystal protein also caused haemolysis of rat, mouse, sheep, horse and human erythrocytes. Intravenous administration of the solubilised toxin crystals to mice at a dose rate of 15-30 mg of protein/gram body weight resulted in rapid paralysis followed by death in 12 h. Subcutaneous injection caused death of suckling mice in 2-3 h. Cheung *et al.* (1985) examined the effects of alkaline-dissolved δ-endotoxin from *Bt israelensis*, when they were injected into mice. High *in vivo* toxicity at 1-5 ppm (mg toxin/g body wet weight) was found. Symptoms in mice were suggestive of those caused by neuromuscular toxins and included loss of alertness, shallow breathing and some animals lost activity of the hind legs. Dead animals showed signs of diaphragm arrest. Results from tests on human erythrocytes *in vitro* have concluded that *Bt israelensis* insecticidal crystal proteins are safe to humans when they are intact and not when solubilised (Rani and Balaraman 1996).

The cause of this toxicity in *Bt israelensis* was shown in a study by Mayes *et al.* (1989), in which the mammalian effects of the major *Bt israelensis* parasporal crystal components were examined. A 28-kDa Cyt endotoxin protein was confirmed to be haemolytic and was more potent than the solubilised *Bt israelensis* crystals in male mice. Acute administration of the 28-kDa protein (mg/kg, IP) produced severe hypothermia and bradycardia in the mouse. Preliminary histological examination of the hearts of mice exposed to the 28-kDa protein (IP) did not reveal any specific lesion, suggesting that the deficient cardiac performance might be a secondary physiological response. Gross pathological examination of mice as well as rats acutely treated with equivalent doses of solubilised *Bt israelensis* crystal preparations revealed focal to segmental reddened and oedematous areas within the small intestine. Histopathology indicated that the major lesion was in the jejunum. Contrary to expectations from *in vitro* haemolysis assays, cytolysis of mouse red and white blood cells was not detectable after *in vivo* exposure to the *Bt israelensis* solubilised proteins. Results indicated that the 28-kDa protein is the mammalian toxic component of *Bt israelensis* crystals.

However, only some *Bt* species and varieties cause these effects when the toxin is solubilised. For example, the alkali-soluble fraction from the δ-endotoxin of *Bt kurstaki,* which has no Cyt protein, showed no *in vitro* or *in vivo* toxicity to mammalian cultured cell lines (Thomas and Ellar 1983).

8 Persistence and Activity in the Environment

The accumulation and persistence of *Bt* and its toxins may result in environmental hazards, such as increased exposure of non-target species and the selection of toxin-resistant pest species. The fate of *Bt* in outdoor environments is known only in relatively general terms: the toxins are usually inactivated within days in the phyllosphere and soil, while endospores can remain in the soil for months to years. The bacteria can potentially multiply in favourable microhabitats, one of which is known to be the targeted larvae. As natural epizootics almost never occur outdoors, in contrast to tropical commodity stores, it is difficult to understand how *Bt* is maintained in the outdoor environment. One possibility could be growth and sporulation outside the insect host such as in nutrient-rich soil. The balance of evidence suggests that artificially boosted *Bt* soil populations decline to a fluctuating, non-hazardous level.

The survival of *Bt* on plant material and in water must also be considered when evaluating the risk of toxicity to non-target organisms. In water, *Bt* lacks persistence with little activity remaining just days after application. The rate of decline of various *Bt* products is influenced by many factors, including type of formulation, application techniques and rates, and substrate to which they are applied. Environmental factors, such as UV radiation, also impact on persistence.

8.1 PERSISTENCE IN SOIL

8.1.1 SURVIVAL AS SPORES

Bt spores can survive for long periods in the relatively protected environment of soil. Experimentally determined half-lives are usually in the range of 100-200 days (Hansen *et al.* 1996). In a field study in Utah, spores of *Bt*, previously applied to control *Lymantria dispar* (gypsy moth) in a forest, did not proliferate but remained viable for about 24 months (Smith and Barry 1998). No significant loss of viability of Dipel 176 (*Bt kurstaki*) occurred in acidic soils over an 8 week period when a recommended field application rate was applied (equivalent to 1.8 l/ha) (Visser *et al.* 1994). Pedersen *et al.* (1995) studying the population dynamics of *Bt kurstaki* in a cabbage crop found a half-life exceeding 100 days in topsoil.

8.1.2 GERMINATION OF SPORES AND GROWTH IN SOIL

While vegetative growth in sterile soils has been demonstrated several times, in the presence of other microorganisms *Bt* sporulated rather than grew vegetatively (Yara *et al.* 1997). Petras and Casida (1985) incubated commercially available *Bt* in natural soil and found a decrease of approximately 1 log in the first two weeks, but relatively constant spore numbers over the subsequent 8 months. However, if vegetative cells were added to soil, cells survived for only 1-2 days (Petras and Casida 1985; Akiba 1986a). Several authors (West *et al.* 1985; Petras and Casida 1985) attributed good survival in soil to lack of spore germination.

For *Bt,* nutrient availability appears to be a major factor limiting growth in soil (West *et al.* 1985). While *Bt* cannot grow under most soil conditions, in some cases *Bt* appears to germinate and grow in soil, although direct evidence for this is limited. Saleh *et al.* (1969) showed that spores germinated, grew and sporulated in soils of neutral pH amended with alfalfa or casein. In another study, vegetative cells did not survive in acidic soils. *Bt* spores were capable of germinating in sterilized soils but not unsterilized soils (Akiba 1986a). Growth occurred in sterile soil between pH 5.5-7.1 but not below 5.4 (Akiba *et al.* 1980). Similarly, West *et al.* (1985) found that growth was reduced at pH 5.2 compared to 7.3. West and Burges (1985) showed that amending soil with different organic fractions influenced growth and survival. Dried grass-supplemented soil resulted in 22× increase in *Bt* cells in soil over 64 days, while chicken manure reduced viability to 0.22× the original level.

8.1.3 PERSISTENCE OF *BT* TOXIN ACTIVITY

Information about the fate of toxins from subspecies of *Bt* in soil is limited and estimates on persistence of *Bt* toxin activity in the environment vary widely. However, this is an area where new studies on transgenic organisms expressing *Bt* toxins are making a contribution to knowledge (Chapter 13). There is some evidence that binding of *Bt* toxins to humic acids, organic supplements or on soil particles protects toxins from degradation by microbes, without eliminating their insecticidal activity (West 1984; Koskella and Stotzky 1997; Crecchio and Stotzky 1998).

Toxins from *Bt kurstaki* and *Bt tenebrionis* were rapidly adsorbed and tightly bound on the clay minerals, montmorillonite and kaolinite, on the clay-size fractions separated from soil (Tapp and Stotzky 1995) and on humic acids extracted from soil (Crecchio and Stotzky 1998). Insecticidal activity was retained when *Bt kurstaki* toxins were incubated in soil, but the amount of retention varied with the type of soil. Toxins added to soil amended with kaolinite remained toxic to *Manduca sexta* larvae for more than 6 months, while the same soil amended with montmorillonite showed reduced insecticidal activity after only 35 days (Tapp and Stotzky 1998). The pH of soils in which insecticidal activity was reduced was higher (5.8 to 7.3) than that of soils in which insecticidal activity was retained (4.9 to 5.1). As microbial activity is greater at higher pH values, microbes in soils with higher pH values may have degraded more of the toxins. Tapp and Stotzky (1998) concluded that there was potential for *Bt* toxins to accumulate in soil and retain insecticidal activity, especially when bound on clays or other particles and particularly in soils with a low pH.

Earlier work provided a range of estimates of persistence of toxin activity in soil, although distinctions between toxins and spores were not always made. Pruett *et al.* (1980) examined the persistence of *Bt galleriae* spores and toxins in laboratory microcosms at 25°C. Viability of spores reduced over 135 days, decreasing to 24% of the original numbers while toxicity to *Galleria mellonella* fell to <1%. This suggests a more rapid degradation of pure toxin than spores, which was confirmed by West *et al.* (1984). Residual activity of toxins from *Bt japonensis* (as measured by bioassay against larvae of *Anomala cuprea*) was prolonged in soils from sweet potato fields as well as in fields of turfgrass, lasting up to 40 days (Suzuki *et al.* 1994). In a study on *Bt kurstaki* and *Bt tenebrionis* in clay soil, insecticidal activity persisted for >40 days in non-sterile soil (Koskella and Stotzky 1997). West and Burges (1985) found that insecticidal activity of *Bt* parasporal crystals decreased exponentially in grass- and manure-supplemented soils, with half-lives of approximately 9.5 and 8.5 days, respectively. It must be remembered that in most experiments, *Bt* spores and crystals were initially added to soil at concentrations far greater than would be needed to control larvae in the phyllosphere in order to measure degradation over time. In the field, however, *Bt* aimed at the phyllosphere which reaches the soil will become diluted well below insecticidal levels. Degradation in the soil would probably exceed the rate of acquisition from repeated foliar sprays, so populations of applied *Bt* reaching the soil should decline over time to a fluctuating natural level.

Survival of toxicity of single *Bt* δ-endotoxins expressed in plants has been reported for the Cry2A protein in cotton, with a half-life of growth inhibition of *Heliothis virescens* of around 15-30 days (Sims and Ream 1997). Palm *et al.* (1996) found little purified *Bt kurstaki* toxin or transgenic cotton leaves-encapsulated toxin present in soil at 140 days.

8.2 PERSISTENCE ON FOLIAGE

8.2.1 SURVIVAL AS SPORES

The half-life of *Bt* spores in the phyllosphere is much shorter than in soil. Hansen *et al.* (1996) reported that experimentally determined half-lives were usually in the range of less than one to three days and these estimates are supported by results of field studies. Pedersen *et al.* (1995) showed an initial half-life (1st week) was 16 hours for *Bt kurstaki* applied to cabbage foliage, with numbers declining by five log units during the first four weeks after spraying. Similarly, *Bt kurstaki* (Foray 48) spores applied to needles of *Pinus* in Lithuania decreased rapidly in the first 1-2 days; bacterial mortality reached 68% after two days on needles and 88% by five days (Bartninkaite and Ziogas 1996). Morris (1977a) reported that when Dipel 36B formulations were applied to white spruce and balsam fir foliage, the viability of spores was greatly reduced after one day of "weathering". On cotton in Egypt, the half-life of spores varied from 75 to 256 h, depending on formulation (Salama *et al.* 1983c). Smith and Barry (1998) examined persistence of *Bt kurstaki* after applications in Utah (72 billion international units per acre per year for five years for eradication of gypsy moth). They found that while spores persisted in soil for over 2 years, there was no difference between numbers of *Bt* spores on sprayed and unsprayed leaves sampled 12 months after application. A range of *Bt* genotypes, including the applied strain, was found on leaves.

Niccoli and Pelagatti (1986) tested spore viability of Dipel, Bactucide (*Bt kurstaki*) and Bactimos (*Bt israelensis*) formulations. On foliage of *Quercus cerris, Crataegus oxyacantha* and *Pinus nigra,* spore viability declined sharply in 8 days. Pishokha (1985) recorded *Bt* survival after aerial application on oak leaves (*Quercus robur*) and found bacterial survival of between 20.4 - 56.5% for four *Bt* subspecies. This dropped to 1.1-19.2% after 20 days and <0.28% by 3 months. In contrast, Scriber (1998) mentioned several reports where the toxicity of *Bt* formulations containing live spores extended for 30-40 days. Although toxicity would normally reflect primarily persistence of crystals, Scriber thought reproduction and re-sporulation could partly explain these results. Johnson *et al.* (1995) found toxicity to *Papilio glaucus* persisted for up to 30 days after application of *Bt kurstaki* in the field in Michigan.

The chemistry of the leaf surface (high pH or the presence of proteases and allelochemicals) may lead to degradation of the crystal protein. Ultrastructure and the chemical and physical microenvironment of leaf surfaces differs widely between plant species (reviewed by Burges and Jones 1998). Therefore, differences in *Bt* persistence on different plant species are not unexpected. Pinnock *et al.* (1975) found that initial deposition rates and subsequent rate of decay of viable spores varied significantly between four tree species. On pine needles, several studies have shown more extended survival than on leaves of other trees. Where *Bt kurstaki* (Foray 48B) was applied to needles of Scots pine *Pinus sylvestris* for control of *Lymantria monacha* in Poland, spores were detected for over 121 days after application (Damgaard *et al.* 1996b) at concentrations significantly higher than the natural background population of *Bt* on needles prior to spraying. Bioassays using *L. monacha* showed that no measurable insecticidal activity was left on the needles 21 days after spraying, although spores could still be detected.

Position on leaves also has an effect on persistence; Thuricide HPC decayed more slowly on the underside of oak (*Q. agrifolia*) than the upper (Milstead *et al.* 1980). Prolonged survival has been recorded occasionally, such as when viable endospores of the pathogen were recovered from spruce foliage collected one year after treatment (Reardon and Haissig 1984).

8.2.2 PERSISTENCE OF *BT* TOXIN ACTIVITY

Persistence of larvicidal activity at operational doses in the phylloplane has been well studied. Since crystals survive damage by UV radiation in sunlight much better than spores, recorded larvicidal activity reflects activity of crystals because viable spore numbers can be considerably reduced without significantly lowering activity against most insects. In the phyllosphere, the half-life of toxins can be as short as less than one day (Hansen *et al.* 1996) and several field studies have detected short residual activity on foliage, while other studies have demonstrated quite lengthly survival. Obviously, there could be many reasons for variability in persistence of toxin activity. These studies used different formulations and improving survival and persistence on foliage is one of the aims of formulation. Also, as mentioned above, substrate directly affects survival. Formulation is reviewed in detail by Burges and Jones (1998).

8.2.2.1 Trees

Van Frankenhuyzen and Nystrom (1989) found an aqueous high-potency formulation of *Bt* (Thuricide 48LV) had residual toxicity to *Choristoneura fumiferana* of <2 days on white spruce (*Picea glauca*) foliage. However, other field studies have indicated that insecticidal activity can be maintained on foliage for longer periods. In Corvalis, USA, *Bt* spraying of *Quercus garryana* and Douglas fir (*Pseudotsuga menziesii*) was effective against gypsy moth for at least 64 days (Miller and West 1987). On white spruce, Reardon and Haissig (1984) reported that *Bt* toxins (from Thuricide 16B and Dipel 4L) were detected on foliage up to 16 days after treatment, and were still capable of killing *C. fumiferana.* On Austrian black pine (*Pinus nigra*) in southern France, Dipel 8L and Foray 48B remained active for 6 and 8 days, respectively against *Thaumetopoea pityocampa* (Demolin and Martin 1998). Sundaram and Sundaram (1996) also found persistence of *Bt kurstaki* (Foray 76B) on mixed spruce-fir foliage in Canada after aerial application with foliar deposits remaining active for up to 18 days (mortality in bioassays of 33%). Sundaram *et al.* (1996) reported the average half-life of Foray 48B toxin on *Picea glauca* foliage was 20.9 h and persistence of up to 10 days, while comparative values for Foray 76B were 22.2 h and 7 days.

8.2.2.2 Crops

Duration of viability and insecticidal activity has been measured on a number of crops. Half-lives of insecticidal activity have often been reported as less than 24 h. Ignoffo *et al.* (1974) found 65% of insecticidal activity and 90% of spore viability were lost on soybean leaves on the first day after application, although some insecticidal activity remained 7 days after application. On tomato foliage, *Bt* was toxic to *Heliothis* (=*Helicoverpa*) *zea* for <48 h (Walgenbach *et al.* 1991) and on potato foliage, mortality caused by *Bt tenebrionis* (M-One) reduced from 85% in early instar *Leptinotarsa decemlineata* 1 h after application to <5% mortality at 48 h (Ferro *et al.* 1993). Similarly, *Bt tenebrionis* (M-Trak) caused 90% mortality of *L. decemlineata* 1 h after application and 20% mortality at 72 h (Ferro *et al.* 1993). The δ-endotoxin from *Bt kurstaki* when aerially and ground applied to pecan plots was found to be short-lived (Sundaram *et al.* 1997). The persistence of *Bt* was <4 days in phyllospheres of mulberry, peanut, chickpea, tomato and rice (Sudarsan *et al.* 1994).

In Australia, activity against *Helicoverpa punctigera* in the laboratory on cotton leaves removed after field application of Dipel (*Bt kurstaki*) was reduced by 50% within 36 h (Wilson *et al.* 1983). In Texas, insecticidal activity of *Bt kurstaki* (HD-1) on cotton leaves in the field had a half-life of between 1.5 and 2 days, measured against *Heliothis virescens* and *Trichoplusia ni* (Beegle *et al.* 1981). Investigating the difficulty of *Bt* control of *H. virescens* on cotton compared with soya or tobacco, Luthy *et al.* (1985) showed that cotton leaf extracts could inactivate insecticidal toxins of *Bt* when incubated for 30 mins at 23°C.

The crops above showed half-lives of only around 1 day for insecticidal activity. In contrast, Hamed (1978) found that on leaves of *Prunus padus,* viable bacterial spores from Dipel were detected 10 and 18 days after treatment, while 100% mortality of third-instar larvae of *Yponomeuta* was recorded 12 days after field treatment with Dipel and Thuricide. On rice, Thuricide HP activity against borers was still detected 10 days after application (Israel and Padmanabhan 1978).

8.2.2.3 Stored grain

Bt has often been used against stored grain pests. Stored grain is a relatively protected environment and persistence of *Bt* activity is generally improved compared to crops. Dipel added to wheat grain did not appreciably decrease in activity against *Plodia interpunctella* over one year in a farm bin under fluctuating temperatures (Kinsinger and McGaughey 1976). Thuricide (*Bt kurstaki*) added to grains of rye remained effective against *P. interpunctella* for up to 5 months at temperatures of 12-28°C and a relative humidity of 75% (Schmidt 1979). *Bt kurstaki* protected potato tubers from attack by *Phthorimaea operculella* when applied inside stores for periods of up to 255 days (Salama *et al.* 1995b). Salama *et al.* (1996a) reported *Bt kurstaki* δ-endotoxin and β-exotoxin decreased in activity against larvae of *P. interpunctella* on stored grains slowly over 150 days, with β-exotoxin decreasing more rapidly. *Bt* survives virtually indefinitely in dry residues of products in stores, together with Lepidoptera infestation left behind at unloading, only to be carried by larvae moving back on to the next new load of product, causing epizootics of *Bt* if substantial insect infestation develops (Burges and Hurst 1977).

8.3 PERSISTENCE IN WATER

Most reports on *Bt* persistence in water are for the mosquitocidal and blackfly toxic subspecies, such as *Bt israelensis,* although a few studies have followed the fate of *Bt kurstaki* in water. Following aerial application of *Bt kurstaki* (Thuricide 16B) for control of the tortricid *Choristoneura fumiferana* in eastern Canadian forests, the bacterium was recovered from rivers up to 13 days after spraying (Menon and de Mestral 1985), presumably as a result of continual leaching into the water. Laboratory studies indicated that *Bt kurstaki* spores could survive in fresh and seawater for more than 70 and 40 days, respectively, at 20°C. Buckner (1974 in Forsberg *et al.* 1976) reported levels of 1730 spores/ml of river water after *Bt* application but no spores were found in the river 1 month later. Clams tested at the same location contained viable *Bt* spores at 2 days after application but no spores were found after one month.

Bt israelensis is considered to lack persistence in water when compared to other available mosquito controls (e.g. methoprene, temephos). In water, the toxicity of *Bt israelensis* has generally been found to persist in larval feeding zones for only days, rather than weeks. Early formulations of *Bt israelensis* had virtually no residual effect against mosquito larvae beyond day of application, although the δ-endotoxin remained chemically stable in neutral and acid waters (Sinegre *et al.* 1981a). Sedimentation can decrease the persistence of *Bt israelensis*. Sheeran and Fisher (1992) found agitation to be the most important factor in maintaining persistence of *Bt israelensis* cells and the bioavailability of the *Bt israelensis* toxin. In their study, sediment acted to decrease efficacy against 3rd instar larvae of *Ae. aegypti* by increasing settling of the toxic particles, but did not decrease the persistence of the *Bt israelensis* spores themselves.

In most studies, larvicidal activity of *Bt israelensis* disappears within 1-4 weeks (Mulla *et al.* 1985; Beehler *et al.* 1991; Hougard *et al.* 1985). Laboratory microcosm studies in France with 4th instar larvae of *Ae. aegypti* showed that suspensions prepared from 2 preparations of *Bt israelensis* retained their activity against the larvae for only 3-5 days, but this period could be extended by leaving dead larvae in the water so that they became the victims of cannibalism (Larget 1981). In contrast, the *Bt israelensis* product Acrobe provided significant control for 47 days in Florida in tyres in which larval cadavers would accumulate (Becnel *et al.* 1996).

As studies on *Bt* growth outside the insect indicate little or no multiplication, *Bt* rarely spreads from the point of inoculation on land, although non-target organisms such as fish (Snarski 1990) have been shown to disseminate *Bt israelensis* in water. The lack of recycling of *Bt* in insect populations, or growth in water, reduces the persistence.

The presence of free chlorine in the water can inhibit or destroy the endotoxin and a clear inverse correlation has been observed between the amount of chlorine in the water and larval mortality due to *Bt israelensis* (Sinegre *et al.* 1981a). However, chlorine concentrations in *Bt* solutions between 0-100 ppm had little impact on activity (Neisess 1980). The amount of residual chlorine normally applied to a standard water purification system does not appear to be sufficient to destroy *Bt kurstaki* spores. The isolation of *Bt* spores from the public water distribution system raised questions about the effectiveness of chlorination for the inactivation of *Bt kurstaki* in the water purification system. A total chlorine residual of at least 1.5 mg/litre and a 60 mins contact time was needed to inactivate 99% of the *Bt kurstaki* population in tap water at pH 7.2 and 20°C (Menon and de Mestral 1985).

8.4 FACTORS AFFECTING PERSISTENCE

8.4.1 BIOTIC FACTORS

Microbial degradation of *Bt* has been reported/suggested by several authors, in response to differences in growth and survival of *Bt* in sterilized versus natural soils (Akiba *et al.* 1977, 1980; West *et al.* 1984, 1985; Akiba 1986a). However, on leaves Haas and Scriber (1998) found no difference in survival of *Bt kurstaki* on bleached (5% chlorox bleach) compared to unbleached treated leaves. The bleach pre-treatment was designed to remove microbial colonies from leaves. Sheeran and Fisher (1992) found that competitive effects from other microorganisms did not influence persistence of *Bt israelensis* in water, but efficacy of *Bt israelensis* endotoxin was adversely affected.

Bt was occasionally reported as passing through animals or birds, without inactivation. Smirnoff and MacLeod (1961) fed *Bt* spores to birds and mice and found that considerable proportions of the spores were still active after passage. Vegetative cells of *Bt* (ATCC 10792) decreased by 90% in 4 h in cattle rumens while the spore count did not decrease even after 24 h (Adams and Hartman 1965). The presence of tadpole shrimps, *Triops longicaudatus*, in water slowed the natural decline of *Bt israelensis* activity, by constantly stirring up the infective particles (Fry-O'Brien and Mulla 1996). The presence of some crustaceans, such as the fairy shrimp and an ostracod, decreased the effectiveness of *Bt israelensis* in water under some conditions (Blaustein and Margalit 1991).

8.4.2 ABIOTIC FACTORS

A number of abiotic factors including solar radiation, temperature and humidity (moisture) may impact on persistence of *Bt* (Leong *et al.* 1980). Temperature can affect *Bt* survival, but generally extremes are needed to inactivate spores. Decreased *Bt* survival in stored grains was directly proportional to storage temperature (Kinsinger and McGaughey 1976).

Rain has been shown to cause a sharp decline in *Bt* activity on leaves (Kiselek 1974; Svestka and Vankova 1976; Pishokha and Timchenko 1981; Behle *et al.* 1997a; Ferro *et al.* 1997; reviewed by Burges and Jones 1998). With Dipel, only 31% of viable spores remained on leaves after 10 mm of artificial rain, reducing to 12% for 50 mm and 6% for 100 mm (Svestka and Vankova 1976). Van Frankenhuyzen and Nystrom (1989) reported that wash-off by rain (as little as 6 mm) was the primary cause of loss of residual toxicity when Thuricide 48LV was applied to white spruce. Simulated rainfall of 3 mm removed over 50% of *Bt kurstaki* protein from foliage in a study by Sundaram *et al.* (1993). Wash-off can be reduced by adding stickers to sprays (Burges and Jones 1998).

Microcosm studies have shown that suspended particles in water greatly reduced the activity of *Bt israelensis* products towards mosquito larvae. Some authors found that toxins bound to particulate material became irreversibly reduced in toxicity (Ramoska *et al.* 1982). Other factors affecting persistence of *Bt israelensis* include agitation, water quality and constituents such as pollutants, also environmental conditions such as pH and to a minor extent temperature (e.g. Standaert 1981; Cokmus and Elciin 1995). Sheeran and Fisher (1992) found agitation to be the most important factor in maintaining persistence of *Bt israelensis* spores and the bioavailability of the *Bt israelensis* toxin. In their study, sediment acted to decrease efficacy against 3rd instar larvae of *Ae. aegypti* by increasing settling of the toxic particles, but did not decrease the persistence of the *Bt israelensis* spores themselves.

Sunlight causes a significant reduction in spore viability (reviewed by Burges and Jones 1998). Morris and Moore (1975) showed that sunlight can inactivate more than 90% of *Bt kurstaki* spores on white spruce needles in one day. However, in the dark, only 78% of the spores were inactivated in 14 days.

Several studies have shown rapid (hours-days) loss of activity against insects due to UV degradation. Formulations of *Bt* spores and crystals, encapsulated together within a starch matrix containing no ultraviolet screens, lost all spore viability and insecticidal activity against *Ostrinia nubilalis* within 4 days even though the starch itself was a partial screen (Dunkle and Shasha 1989). Bagci and Shareef (1989) examined the effect of UV on 27 subspecies of *Bt* and found extreme sensitivity in most, with >99% inactivation at approximately 1 J/m^2. Exceptions requiring a UV dose between 1 and 2 J/m^2 for 99% inactivation were *Bt canadensis, Bt colmeri, Bt galleriae, Bt indiana, Bt kenyae* and *Bt thompsoni,* although differences might be expected between different isolates/strains of the same subspecies. Laboratory investigations into the photostability of Dipel 76AF deposits on spruce foliage after different periods of exposure to two radiation intensities showed that bioactivity (measured by insect bioassay) decreased with an increasing duration of radiation exposure and increasing radiation intensity (Sundaram and Sundaram 1996). The half-life (DT_{50}) was 5.1 days for the low intensity and 3.9 days for the higher intensity.

While a number of reports have shown that UV radiation has a significant negative impact on the survival of *Bt* spores, the effect on toxins is not as clearly understood. The influence of light on inactivation depends on the wavelength and intensity of light, the presence of oxygen and the length of exposure.

Several studies failed to find any impact of exposure to the shorter bands in the UV range on toxin persistence or found only minor degradation. Burges *et al.* (1975) found no reduction of insecticidal activity in crystals of *Bt* against larvae of *Pieris brassicae* after gamma - or short wave ultraviolet irradiation. According to Pusztai *et al.* (1991) the toxin-destructive spectrum of sunlight is in the range of 300-380 nm. Sundaram and Sundaram (1996) also found that total protein levels showed no decrease with an increasing duration of radiation exposure, or with increasing radiation intensity on spruce foliage, although bioassay against insects did show such a decrease, indicating that the protein is inactivated but not destroyed.

UV protectants have been used to improve persistence with some success where *Bt* must be applied under conditions exposed to sunlight. Griego and Spence (1978) examined which sunlight wavelengths were most damaging to *Bt.* They found that inactivation of *Bt* spores by sunlight was due in part to wavelengths of the visible spectrum, especially by those wavelengths (near 400 nm) that are readily absorbed by spores. Liu *et al.* (1993) studied the effect of adding UV light absorbing melanin to a spore-crystal preparation of *Bt israelensis*. Live cell counts and bioassay of residual mosquitocidal activity indicated that melanin was an excellent photoprotective agent.

8.5 EFFECTS OF FORMULATION ON PERSISTENCE

Companies producing *Bt* as a biopesticide usually view lack of persistence as a problem with the use of *Bt* and several attempts have been made to improve persistence through novel formulations or use of various biological carriers (reviewed by Burges and Jones 1998). For example, several workers have reported on the use of starch matrices to encapsulate *Bt kurstaki* toxins (McGuire and Shasha 1989; Dunkle and Shasha 1988; McGuire *et al.* 1996). Starch matrix formulations showed good storage properties (Dunkle and Shasha 1988) and extended persistence (> 2 weeks) on cotton leaves in glasshouses, probably by giving protection against the highly adverse chemical environment of the cotton leaf surface (McGuire and Shasha 1989). Outdoors toxicity to neonate *Ostrinia nubilalis* was maintained longer when cornstarch encapsulated *Bt kurstaki* granules were applied to whorls of corn than for commercial *Bt kurstaki* in wet years, but not dry years (McGuire *et al.* 1994). Behle *et al.* (1997b) used solubilized wheat-gluten, added to sprayable formulations of *Bt kurstaki*, to form a film that resisted wash-off upon drying. Adding 1% (wt:vol) gluten to formulations resulted in greater tolerance to high leaf pH than traditional formulations and resisted wash-off of *Bt kurstaki* by 5 cm of simulated rain. Bryant (1994) reported that *Bt* deposited with a sticker might withstand up to 10 mm of rain without serious loss of activity. Stickers are spray additives, which form a coating that bonds the dried *Bt* deposit to the surface of the foliage. Rainfastness contributes significantly to persistence of many *Bt kurstaki* products, for example, Sundaram *et al.* (1993) tested eight formulations for rainfastness and found between 47-100% loss of protein.

Formulations based on casein have also been evaluated (Behle *et al.* 1996). When applied to cotton leaves and subjected to simulated rain and artificial sunlight, casein formulations resisted wash-off and retained >60% insecticidal activity against neonate *O. nubilalis* compared to <20% of the original activity for unformulated and commercially formulated *Bt* preparations. The casein formulations also provided some protection from light and allelochemical-induced degradation compared with unformulated *Bt*, although the amount of protection was less than that provided by other experimental formulations (Behle *et al.* 1996).

Another approach to improving persistence of *Bt* has been the expression of *Bt* genes and toxins in other bacteria, an approach that has not always been successful. A commercial formulation MVP (Table 4.1) was developed, consisting of *Pseudomonas fluorescens* genetically engineered to express a δ-endotoxin gene of *Bt kurstaki* so that the bacterial cell encapsulated the toxin. Some work suggested improved persistence, but sprayed foliage bioassayed with 3rd instar larvae of *Pseudoplusia includens* (*Chrysodeixis includens*) indicated that overall persistence of Dipel was significantly better than that of MVP (Nyouki and Fuxa 1994).

With *Bt israelensis*, lack of persistence in the mosquito feeding zones in water has been a widely perceived problem and therefore many formulations have been developed in an effort to extend persistence. Slow release briquettes improve the persistence of *Bt israelensis* above liquid formulations, extending control from days to weeks in some cases. For example, Bactimos (*Bt israelensis*) briquettes provided complete control of *Ae. aegypti* adult emergence in Malaysia in plastic containers of water for up to 75 days post-treatment (Sulaiman *et al.* 1991).

UV radiation is a principal factor in the lack of persistence of *Bt* in some situations and many formulations contain UV protectants. These may be in the form of UV absorbing compounds or encapsulation. Researchers continue to look for new UV protectants, such as the recent example of Ragaei (1998) who showed that shellac is an excellent photo-protective agent of residual insecticidal activity of *Bt entomocidus*.

8.6 APPLICATION METHODS AND RATES

Choice of application method on land and in water has implications for environmental risk as it determines the spread of *Bt*, the range of non-target organisms exposed and even the deposition of dose. Applications can be made by pouring into water, spraying (including ultra-low volume), aerial techniques and by specialist formulation such as briquettes. Aerial application has been used in several control efforts (e.g. Yates 1985; Palmisano 1987; Schmidt 1988; Knepper *et al.* 1991) and often ultra-low volume technology is used in conjunction with aerial application.

Sundaram and Sundaram (1996) showed that droplet size affected *Bt* survival, with large drops (250 μm) having a mean half life of about 26 h against gypsy moth, *Lymantria dispar*, while the smallest droplet (42 μm) had a half life of active *Bt* of only 12 h. Above 130-160 μm, persistence on oak leaves was independent of droplet size.

The initial inoculum level applied may affect the persistence of *Bt*. Higher inoculum levels can take longer to degrade, as found by Beckwith and Stelzer (1987) who applied 20 and 30 billion IU/ha by helicopter against *Choristoneura occidentalis* in Oregon. However, these authors concluded that inactivation rate was less important than initial larval mortality in the success of *Bt* application. In contrast, excessive VectoBac 12AS dosages did not significantly extend the duration of effectiveness for controlling mosquito larvae, although treated populations never recovered to the same level because of increased predation (Mulla *et al.* 1993).

8.7 DETECTING *BT* AND TOXINS IN THE ENVIRONMENT

The accumulation and persistence in soil and other natural habitats of the insecticidal toxins of *Bt* may result in environmental hazards, such as toxicity to non-target species and the selection of resistant target species. Therefore, sensitive methods for the detection of *Bt* and/or toxins in the environment are required.

There are several methods used for detecting *Bt* in the environment. Direct plating using selective media has been widely used (e.g. Martin and Travers 1989). Akiba and Katoh (1986) developed a selective medium for the growth of vegetative cells of *Bt morrisoni* and *Bt thuringiensis*, although spores would not germinate on it. Yousten *et al.* (1982) reported on selective media for *Bt israelensis*.

Insecticidal toxins have often been measured indirectly, using activity against selected pests such as *Galleria mellonella*. Other methods for the detection of endotoxins include enzyme-linked immunosorbent assay (ELISA) (e.g. Takahashi *et al.* 1998) and the use of immunofluorescence (e.g. West *et al.* 1984), but these methods do not reveal whether toxins have undergone any inactivation, which can be detected only by bioassay.

Tapp and Stotzky (1997) developed the use of flow cytometry as a method for detecting and tracking the fate of *Bt* toxins in soil that does not require their extraction and purification. Toxins from *Bt kurstaki* and *Bt tenebrionis* were bound on clay or silt sized particles. Toxins were detected by measuring fluorescence after particle-toxin complexes were incubated with rabbit antibody to the *Bt tenebrionis* toxin, followed by incubation with fluorescently labelled anti-rabbit antibody. This method proved more sensitive than a previously developed dot blot ELISA.

Molecular methods are also being applied to the detection of *Bt*. Damgaard *et al.* (1996a) used specific polymerase chain reaction (PCR) primers targeting the phosphatidylinositol-specific phospholipase C gene of *Bt* to detect bacteria in soil.

9 Insect and Environmental Factors Affecting Toxicity

As a live bacterium with associated endotoxin crystals, *Bt* is subject to a number of environmental and insect influences. Many environmental factors reduce persistence and can therefore result in reduced risk to non-target organisms. Efficacy of *Bt* can be affected by abiotic factors, for example, efficacy in water is reduced by low temperatures and poor water quality. However, abiotic factors do not influence the efficacy of *Bt* to the same extent as non-toxin producing pathogens. Physical factors in waterways, for example the amount of vegetation present, can also influence efficacy.

Various other factors, such as plant diet of insects and stage of insect development have more impact on toxicity. Generally *Bt* is more effective against younger larvae than older larvae and is not effective against all developmental stages. The species of leaf on which caterpillars feed has been shown to have a strong impact on toxicity with some plants enhancing and others inhibiting *Bt* toxicity. The feeding behaviour of insects determines the amount of inoculum ingested, which in turn influences susceptibility. Temperature and humidity can affect insect feeding, influencing the amount of inoculum consumed. Other factors, such as distribution by non-susceptible insects, can influence the area and range of organisms at risk after *Bt* application.

9.1 ENVIRONMENTAL FACTORS

In Chapter 8, the effect of abiotic factors on persistence of toxins and spores was discussed, with solar radiation and rainfall being two key factors that reduced persistence on plants and non-protected surfaces. Temperature and humidity have a greater influence on *Bt* effectiveness than on its persistence. Higher temperature increases toxicity just as it increases the effectivenss of toxins in general. Temperature has a more important influence on *Bt* effectiveness through its action on the insect, as raising the temperature increases the insect metabolic rate and also the rate at which insects feed and consume *Bt*. Raising the temperature also accelerates bacterial development inside the insect.

Generally, *Bt* is more toxic at higher temperatures and is not effective in the field at temperatures below around 15°C (e.g. Lacey and Oldacre 1983; Barreiro and Santiego Alvarez 1985; van Frankenhuyzen 1990b), possibly because of reduced feeding by the target insect. The Russian product Entobakterin (based on *Bt galleriae*) has been reported to lose effectiveness against a number of insects below 18°C (Sikura 1976) and it is not effective against gypsy moth under a mean daily temperature of 13°C (Khryanina 1974; Korchagin 1980). The optimum temperature for toxicity to gypsy moth larvae is around 25°C (Galani 1978). *Bt kurstaki* did not kill larvae of *Spodoptera litura* at 18°C on diet in the laboratory (Palanichamy and Arunachalam 1995). Against the sorghum stem borer, *Chilo partellus*, a *Bt* (unspecified) caused increasing mortality as temperature increased over 26°C, with highest mortality at 36°C (Sharma *et al.* 1998).

Increasing temperatures (up to 25°C) increased the larval mortality of *Choristoneura rosaceana* fed on raspberry leaves treated with *Bt kurstaki* in the laboratory (Li *et al.* 1995). With larvae of *Choristoneura fumiferana* on balsam fir, temperatures between 13 and 25°C affected progression but not the level of cumulative mortality during 14 days of feeding on sprayed foliage. The LT_{50} decreased from 12-17 days at 13°C to 2-4 days at 25°C, depending on droplet density (van Frankenhuyzen 1990b). Van Frankenhuyzen (1994) found that the temperature effect on disease progression was due to its effect on the proliferation of vegetative cells in the infected larvae. Another serovar, *Bt san diego* (= *tenebrionis*) was less effective against 1st instar *Leptinotarsa decemlineata* at 18°C than 28-33°C (Ferro and Lyon 1991).

Temperature of water has the same effect as air temperature, with toxicity increasing with higher temperatures (e.g. Lacoursiere and Charpentier 1988). Most published data on water temperature are for the mosquito and blackfly active subspecies, *Bt israelensis*, where low temperature (5°C) yielded 10-fold higher LC_{50} and LC_{90} values against mosquitoes in bioassays compared with those conducted at a high temperature (25°C)(Becker *et al.* 1992). Although mosquito toxicity was reduced below 19°C for *Bt israelensis* (Sinegre *et al.* 1981a), some toxicity has been recorded as low as 0°C (Walker 1995). There are few reports on the effect of water temperature on other *Bt* strains. The activity of *Bt darmstadiensis* against 4th instar larvae of the mosquito, *Culex quinquefasciatus,* increased with increasing temperature between 18 and 31°C (Lacey and Oldacre 1983). In laboratory tests in the former Soviet Union, the toxicity of Entobakterin (*Bt galleriae*) against larvae of *Aedes caspius* in water increased with increasing temperature, with mortalities of 22.6, 43.0 and 72.7%, at 15, 19 and 25°C respectively after 4 days (Saubenova 1973).

A low temperature reduces the amount of feeding by insect targets. For example, low water temperature reduced feeding by mosquito and blackfly larvae, which is probably directly related to lower levels of *Bt israelensis* virulence (Walker 1995). Lepidopteran larvae do not feed much below 17-18°C, which may explain the commonly reported lack of field effectiveness of *Bt* preparations below 18°C (Videnova *et al.* 1980).

Moisture can have a detrimental effect on the toxicity of *Bt*. Rain has been shown to wash spores and crystal from leaves, although in the laboratory spraying wet leaves of balsam fir with *Bt* did not reduce its toxicity to larvae of *Choristoneura fumiferana* confined on foliage the following day, as compared with dry foliage (van Frankenhuyzen 1987). In contrast, Afify and Matter (1970b) found that decreasing moisture in the rearing medium of *Ephestia kuehniella* larvae increased mortality due to *Bt* (Biospor 2802), but this could have been due to direct effect of changing moisture on the larvae. In general, humidity alters *Bt* effectiveness less than temperature and probably acts primarily on the rate at which an insect feeds and develops.

9.2 INSECT FEEDING BEHAVIOUR

Several insects have been shown to be more likely to move from *Bt* contaminated leaves than untreated or control treated leaves and in some cases, *Bt* formulations or crystals have been shown to be repellent to feeding insects. Neonates and 3rd instar *Spodoptera exigua* larvae showed significant avoidance of *Bt* (CryIC) treated diet (Berdegue *et al.* 1996). Several other studies have reported *S. exigua* avoided food contaminated with *Bt* products (Berdegue and Trumble 1997; Stapel *et al.* 1998). In a choice test, *Lymantria dispar* larvae avoided diet or oak leaves treated with *Bt* (Yendol *et al.* 1975; Farrar and Ridgway 1995b). However when *L. dispar* was offered oak leaf discs treated with crystals alone, little difference could be detected from water treated discs. *Bt* significantly reduced feeding rate on cabbage leaves by larvae of *Plutella xylostella* and *L. decemlineata* (Hoy and Hall 1993). Neonate *Heliothis virescens* larvae avoided moderate to high concentrations of toxins in a choice test with *Bt kurstaki* (Gould and Anderson 1991). *Mamestra brassicae* avoided treated leaves when offered cabbage leaves treated with *Bt* (Bactospeine) (Espinel 1981). On plants, larvae of many species lose coordination on eating leaf sprayed with *Bt* crystals, become hyperactive and tend to fall off the plants (H.D. Burges, pers. comm.), which could lead to sublethal dosing.

In contrast, 4th instar larvae of Colorado potato beetle, *L. decemlineata* did not avoid *Bt tenebrionis*-treated potato foliage in a choice test, nor did they leave treated foliage when untreated foliage was in contact with the treated foliage (Ferro and Lyon 1991). *Bt tenebrionis* (Novodor FC) did not deter *L. decemlineata* from feeding in laboratory and field trials in New Jersey (Ghidiu *et al.* 1996). Larvae and adults of *L. decemlineata* were not behaviourally deterred from, and in some cases may have been stimulated to increase, feeding initially by the presence of M-One (*Bt tenebrionis* encapsulated in *Pseudomonas* cells) on foliage (Hough-Goldstein *et al.* 1991).

In water, the zone where insects feed can result in exposure to differing levels of *Bt* inoculum. For example, *Culex* spp. usually feed throughout the water column, while *Aedes* spp. are substrate feeders, which means the two genera will ingest different rates of settling particles of *Bt israelensis*. *Anopheles* spp. feed primarily at the surface where *Bt israelensis* remains for only a short time (Mulla 1990 and references therein) so are exposed for shorter periods than species feeding through the water profile. In addition, filtering rates vary between genera and species. For example, Aly (1988) measured the filtration rates of three mosquito species: 632 ml/larva/h for *Ae. aegypti*; 515 for *Cx. quinquefasciatus* or 83.9 for *An. albimanus*. This has the effect that *Anopheles* spp. are generally less susceptible to *Bt israelensis* than *Culex* or *Aedes* spp. Mahmood (1998) suggested that larvae of *Ae. aegypti* were more susceptible to *Bt israelensis* than *An. albimanus* because of faster feeding rates. *Ae. aegypti* ingested 11.5 times more spores than did *An. albimanus*, resulting in lower LT_{50} values.

In field studies on *Bt kurstaki* non-target impacts, Wagner *et al.* (1996) found few effects on some microlepidoptera, which they attributed to feeding behaviour and habitat rather than natural low susceptibility to *Bt*. The unaffected larvae rolled or folded leaves and others bored into plants, reducing the likelihood of encountering *Bt* spores and crystals.

9.3 SOIL FACTORS

As discussed in Chapter 8, *Bt* can bind to soil particles, reducing the amount of toxin degradation and increasing the residual toxicity. For example, *Bt kurstaki* and *Bt tenebrionis* were rapidly (30 mins for 70% of eventual adsorption) adsorbed onto clay particles and washing with water failed to remove much of the adsorbed material (Tapp *et al.* 1994). Bound toxins of *Bt kurstaki* and *Bt tenebrionis* were toxic to larvae of *Manduca sexta* and *L. decemlineata* and sometimes binding to clay soils enhanced the insecticidal activity (Tapp and Stotzky 1995). Adsorption of the *Bt kurstaki* toxin was higher in the clay loam soil than in sandy loam (Sundaram 1996).

Kleiner *et al.* (1998) found a relationship between low soil nitrogen and increasing efficacy of Cry1Aa toxin against gypsy moth on poplar leaves. Low nitrogen availability increased larval mortality and prolonged the development times of 2nd instar larvae feeding on leaves treated with the *Bt* Cry1Aa endotoxin.

9.4 WATER FACTORS

In addition to temperature, density of insects and inoculum level (all discussed elsewhere in this Chapter), factors found to affect *Bt* efficacy in water are pH, salinity and water quality. Reported effects of pH on efficacy vary but some workers have shown that increasing the pH of water increased mortality of the target insect. For example, mortality of *Simulium* caused by *Bt israelensis* increased with pH (Lacoursiere and Charpentier 1988). Response to pH also varies between insect species. Entobakterin (*Bt galleriae*) against larvae of *Ae. caspius* was more toxic at pH 7 than pH 9, although for larvae of *Culicoides* sp. and *Culiseta alaskaensis,* pH had little effect on susceptibility (Saubenova 1973). In field tests against *Culex tarsalis,* all the formulations of *Bt israelensis* were less effective in water where the pH was higher than 8 (Mulla *et al.* 1980).

Varying responses to salinity have also been reported. At 14.4, 135.6 and 215.2 meq sodium chloride/litre, mortality of *Ae. caspius* larvae was 98.0, 80.6 and 43.6% respectively, 7 days after treatment with Entobakterin. In contrast, salinity had little effect on the susceptibility of larvae of *Ae. flavescens (*Saubenova 1973). Balaraman *et al.* (1983b) found *Bt israelensis* activity against *Culex* spp. and *Anopheles stephensi* stable at different salinity ranges. One percent salinity did not alter the character of the correlation between dose and mortality of *Ae. aegypti* and *Cx. pipiens* when exposed to *Bt israelensis* (Rasnitsyn *et al.* 1993).

The quality of water, in terms of pollutants and organic and inorganic particles can affect toxicity. There appears to be a direct correlation between organic pollution and dosage required to kill mosquitoes for *Bt israelensis,* with extraneous material apparently resulting in less *Bt israelensis* ingested and consequently reduced efficacy (Bakr *et al.* 1986; Mulla *et al.* 1990b; Becker and Margalit 1993). *Bt israelensis* is less toxic to mosquito larvae in the presence of soil particles, possibly because of increased sedimentation and binding to particles (Sinegre *et al.* 1981b; van Essen and Hembree 1982; Ramoska *et al.* 1982). Sludge from soil decreased the effectiveness of *Bt israelensis* more than decomposing organic matter, inorganic mud or silica gel (Margalit and Bobroglo 1984). As with plants, the presence of tannic acid in water can reduce *Bt israelensis* efficacy at concentrations as low as 0.25 mM (425 mg/litre) (Lord and Undeen 1990).

Garcia *et al.* (1982) found that *Bti* control of *Culex peus* in a primary oxidation pond with a dense cover of water hyacinth was unsuccessful and the presence of dense vegetation reduced the effectiveness of *Bt israelensis* against *Odagmia ornata* in the Czech Republic (Mata *et al.* 1986). Skovmand and Eriksen (1993), using a fizzy tablet formulation of *Bt israelensis,* found that a full tablet dose per 5 m^2 pond killed 93-100% of 2nd- to 3rd and early 4th-instar *Ae. cataphylla* and *Ae. cantans* larvae in ponds without dense vegetation but only 75-82% of the same instars in ponds with dense vegetation. Conversely, the presence of extensive aquatic vegetative growth had little effect in reducing the activity of *Bt israelensis* against larvae of *Simulium vittatum* in a stream in Tennessee (Frommer *et al.* 1981a).

9.5 PLANT/DIET FACTORS

Host plant of insects can strongly affect their susceptibility to *Bt* (e.g. Kudler and Lysenko 1975; Merdan *et al.* 1975; Moldenke *et al.* 1994; Guerra 1995). It has been demonstrated many times that changing the plant diet of Lepidoptera alters the susceptibility of the larvae.

An example of the variation in response to *Bt kurstaki* found within a single insect species is well illustrated by gypsy moth, *L. dispar.* Several *Bt kurstaki* formulations have been shown to alter in toxicity to gypsy moth, depending on host plant (Appel and Schultz 1994; Moldenke *et al.* 1994; Farrar *et al.* 1996b). Oaks were 2-5 times more inhibitory to *Bt kurstaki* toxicity to gypsy moth larvae than foliage of aspen (Appel and Schultz 1994). Hwang *et al.* (1995) showed that *L. dispar* fed on different aspen (*Populus*) trees varied in susceptibility to *Bt kurstaki* (Foray 48B) by up to 100 fold.

Laboratory-reared larvae of another lymantrid *Orgyia antiqua* fed on blackberry, pear, *Crataegus* sp. and *Picea abies* differed in their susceptibility to Thuricide 90TS (Skatulla 1973). Within-season variation in celery quality affected the efficacy of *Bt kurstaki* and the interaction between *Bt kurstaki* and host plant cultivar for *S. exigua* but not for *Trichoplusia ni* (Meade and Hare 1994). Similar variation depending on diet was demonstrated with stored grain pests *Plodia interpunctella* and *Sitotroga cerealella* reared on artificial diet or grains (Salama and Abdel-Razek 1992).

Testing *Bt morrisoni* for control of *L. dispar* and *Euproctis chrysorrhoea* (Cabral 1977) indicated that the leaves of *Quercus* contain some substance that inhibits the action of the bacterium. Several authors (Appel and Schultz 1994; Hwang *et al.* 1995) have suggested that condensed tannins and phenolic glycosides are involved in such variations. Purified tannins from the oaks were highly inhibitory to *Bt kurstaki* (Appel and Schultz 1994). The inhibitory effect of forest trees on *Bt* in other cases has been linked to terpenes (Morris 1972a). In some cases, plant phenolics have been found to enhance endotoxin toxicity (e.g. Sivamani *et al.* 1992). As phenolics from sunflowers increased the effectiveness of *Bt* against larvae of *Homoeosoma electellum,* high-phenolic producing sunflowers might be more effective in the field (Brewer and Anderson 1990).

Smirnoff and Hutchison (1965) tested the effect of 74 plant species foliage on *Bt thuringiensis.* Effects of plant juices and extracts ranged from completely inhibitory to growth to no effect, with conifers the most inhibitory. While juices and water extracts were generally not inhibitory, ethyl acetate, alcohol and ether extracts were more inhibitory.

There was a marked difference in *Bt israelensis* susceptibility between two larval populations of a euryphagous species, *Pentapedilum tigrinum*, when reared on different foods prior to treatment; the larvae reared on aquatic plant tissue were 100 times more susceptible than the larvae fed on detritus (Kondo *et al.* 1995b).

Plant allelochemicals affect efficacy of *Bt* (reviewed by Burges and Jones 1998). The role of plant allelochemicals in reducing the ability of *Bt kurstaki* to kill larvae of the sphingid *Manduca sexta* was studied by Krisc (1988), with both alkaloid nicotine (from tobacco) and the flavonoid rutin (rutoside from tomato plants) the toxicity of *Bt* to the insect. Tannic acid has been shown to increase the efficacy of *Bt kurstaki,* res higher mortality than where *Bt* was used alone (Gibson *et al.* 1995). Caffeine has also been shown to enhance the activity of Dipel against larvae of *Mamestra configurata* (Noctuidae) (Morris *et al.* 1994).

9.6 INOCULUM AND INSECT DENSITY

Inoculum and insect densities also play a role in efficacy. As has been demonstrated with many insect species, the higher the larval density, the greater the *Bt israelensis* dose required to kill (e.g. Farghal *et al.* 1983; Chen *et al.* 1984; Becker *et al.* 1992). Mulla (1990) estimated that 1.5-2× more inoculum would be required against 50-100 larvae/dip than 5-20 larvae/dip to give the same mortality. Use of VectoBac 12AS against *Ae. albopictus* was less effective against high population densities (200-250 larvae/litre medium)(Chui *et al.* 1993) and increasing densities of 2nd instar larvae of *Ae. aegypti* population led to a decrease in the proportion of dead individuals after treatment with *Bt israelensis* (Alekseev *et al.* 1983). The same effect was shown for *Bt israelensis* against *Culex* spp. and *Culiseta longiareolata* (Farghal *et al.* 1983). Presumably at the higher larval densities, there is insufficient *Bt* to enable all larvae to filter out a lethal dose. A similar effect could be expected against plant feeding insects at sufficiently high densities.

9.7 DEVELOPMENTAL STAGE OF LARVAE

Generally, insects become more resistant to *Bt* as larvae age. However, adults are not consistently more or less susceptible than larvae. For example, larval mortality after feeding on *Bt* decreased with increasing age for *L. decemlineata* with *Bt san diego* (= *tenebrionis*)(Stewart *et al.* 1991), *Archips cerasivoranus* (Smirnoff 1972), *Helicoverpa zea* with *Bt kurstaki* (Ali and Young 1996), *S. litura* (Sareen *et al.* 1983), *Malacosoma neustria* with *Bt galleriae* (Bartninkaite 1983) and larvae of the stored-product pests *Ephestia cautella, Plodia interpunctella* (McGaughey 1978), *Ephestia kuehniella* (Afify and Matter 1970a) and with *Bt kurstaki* against larvae of *Boarmia selenaria* (Izhar *et al.* 1979). With *Bt israelensis* against the chironomid *Chironomus tepperi* in Australia, the activity of a wettable powder formulation of Bactimos against 4th instar larvae had an LC_{90} of 0.79 ppm, about 4 times that for the 1st instar (Treverrow 1985). Second-instar larvae of *Cx. quinquefasciatus* and *Ae. aegypti* were about 10 times as susceptible as 4th instar larvae to *Bt darmstadiensis* spore-crystal suspension (Lacey and Oldacre 1983).

Some studies have found susceptibility did not simply decrease with increasing larval age. With silkworms *Bt* was more effective against 2-4th instar than 1st or 5th (Asano *et al.* 1997). In the field, Fast and Dimond (1984) found that 3rd instar larvae of *Choristoneura fumiferana* were controlled less effectively than 4th or 5th instar larvae (which were controlled equally as well as each other). Sixth-instar larvae were also less effectively controlled, probably because a significant proportion had reached the non-feeding prepupal stage before the full effect of treatment was obtained. Similar lack of susceptibility in the final instar of mosquitoes (e.g. Molloy *et al.* 1981) could be attributed to lack of feeding prior to pupation. Pupae and pre-pupae do not feed and are therefore not affected directly by *Bt*.

For almost all species tested against *Bt israelensis*, increasing age of the larvae resulted in reduced susceptibility in mosquito and blackflies (e.g. Mulla *et al.* 1980; Wraight *et al.* 1981; Chen *et al.* 1984). *Bt galleriae* toxicity against larvae of *Ae. caspius* lessened with increasing age; mortality after a single dosage was 91% in the 1st instar, 55.5% in the 2nd, 42.2% in the 3rd and 20% in the 4th after exposure to treated water for 5 days (Saubenova 1973).

Comparisons of susceptibility of different insect development stages (larvae, pupae and adults) generally show different susceptibilities. Larvae of *S. litura* were more susceptible to *Bt* than adults, with pupae non-susceptible (Zaz 1989) and while *Bt. galleriae* (Entobakterin), *Bt thuringiensis* (Bactospeine) and *Bt kurstaki* (Dipel) were practically non-toxic to larvae of *Chrysoperla carnea*, but highly toxic to adults, causing up to 75% mortality (Babrikova *et al.* 1982).

Eggs are not considered directly susceptible to *Bt*, although larvae hatching from eggs of *Heliothis virescens* treated with *Bt kurstaki* had significantly higher mortality than controls (Ali and Watson 1982a). Eggs of *Boarmia selenaria* treated with *Bt kurstaki* hatched larvae but only 2.2% survived (Izhar *et al.* 1979). Neonate larvae of some species eat their shells immediately after hatching, which may explain this mortality.

9.8 OTHER ORGANISMS

In water, co-inhabiting organisms can reduce the number of *Bt* spores and crystals available and therefore reduce efficacy to target insects. For example, competition in food intake by filter feeding *Daphnia curvirostris* resulted in lower mortality of mosquito larvae after *Bt israelensis* applications (Becker *et al.* 1992). In some cases, non-susceptible insects act as transmission agents for *Bt* infections, such as transfer on the ovipositors of parasitoids or dispersal by scavengers which consume cadavers. While these may act to disperse *Bt* spores, it is likely to be at a reduced inoculum level (Chapter 11).

Both in its role as a naturally occurring and an artifically applied bacterium, *Bt* may interact with other microbes. In some cases *in vitro*, *Bt* has been found to have antibacterial properties. Favret and Yousten (1989) reported that a strain of *Bt thuringiensis* demonstrated antibacterial activity against 48 of 56 strains of *Bt* and against some other gram-positive species, but not against gram-negative species. In a large field experiment in Sweden, a streptomycin resistant strain of *Bt israelensis* was released into a swampland and effects on other bacteria monitored. A minor and transient increase in *Bt*-like bacteria occurred and the released strain established at a low number and outside the release site. Seven weeks after release, no remaining effects on the total bacterial population could be detected (Eskils and Lovgren 1997). Laboratory experiments showed that the released strain could grow in non-sterilised as well as sterilised soil, but that growth was very slow and transient in non-sterilized soil.

Some soil microbes are antagonistic to *Bt*. The microflora antagonistic to *Bt* depends on the crop, but in one study (Galan-Wong *et al.* 1993) bacteria were the most antagonistic group.

9.9 BACTERIOPHAGE

Bacteriophages are viruses that attack bacteria and can therefore reduce the effectiveness of a bacterial control agent. A number of studies have reported bacteriophage from *Bt* strains (e.g. Skvortsova *et al.* 1976; Bal'-man *et al.* 1988; Kanda and Aizawa 1989; Inal *et al.* 1992; Azizbekyan *et al.* 1996, 1997). Bacteriophage which inhibited *Bt israelensis* were found in 9 of 12 habitats surveyed in Egypt (Ali *et al.* 1993). Bacteriophage infections have been reported during fermentation production of *Bt* (Wu *et al.* 1998). In the USA, infection in *Bt galleriae* decimated growth in industrial fermentation and a switch to another *Bt* subspecies was necessary to enable orders for the product to be fulfilled (H.D. Burges, pers. comm.). Azizbekyan *et al.* (1996) phage-typed strains of *Bt*, using 12 phages, resulting in 37 different phagotypes among 45 serovars tested. This and early work showed that classification of *Bt* strains according to susceptibility to the different bacteriophages cut across accepted classification by serotyping.

Few reports mention the effect of bacteriophages on *Bt* field performance. Skvortsova *et al.* (1976) reported that contamination of Entobakterin with the phages decreased its biological activity to 25-33% of the normal. Hussein *et al.* (1990, 1991) also reported that some phages of *Bt israelensis* decreased activity against *Cx. pipiens*. In addition, phages of *B. cereus* and *B. sphaericus* can attack *Bt israelensis*, with an inhibitory effect on insecticidal activity (Hussein *et al.* 1991) and phages first located in *Bt aizawai* were able to form a stable lysogen with *B. cereus* (Reynolds *et al.* 1988). Phages are unlikely to have a direct effect on efficacy after application, as toxicity of *Bt* is dependent on the protein crystal and not the bacterial cell. Bacteriophages may have an impact on any recycling of the bacterium and fermentation of *Bt* preparations.

10 Transmission and Dispersal

Transmission and dispersal are usually crucial to success of insect pathogens. However, insect to insect transmission is usually poor to non-existent for *Bt* and its success as a biopesticide is not dependent on reinfection of insects. Recycling of *Bt* in a target population is not expected after commercial application, with *Bt* acting more as a toxin than an infective pathogen in most cases. Transmission and dispersal factors impacting on ecology of the bacterium include rain, cannibalism, presence of scavengers, predators, parasitoids and non-susceptible organisms, including birds and small mammals. Wind can carry *Bt* up to 3 km from the site of application. However, poor survival and limited ability to sustain infections in populations limits the temporal and spatial distribution of *Bt*. With poor transmission between susceptible insects, dispersal inevitably leads to reductions in inoculum density resulting in little downstream toxicity.

10.1 HORIZONTAL TRANSMISSION

Bt is not generally considered a very infectious insect pathogen and does not always sporulate in insects before or after death. This reduces recycling of the bacterium, which in turn reduces the environmental risks from *Bt*, as it does not persist in most populations after application. Therefore, horizontal transmission is considerably limited compared to most insect pathogens. For example, Takatsuka and Kunimi (1998) studied horizontal transmission of *Bt aizawai* using 5th instar larvae of *Ephestia kuehniella*. No transmission was detected after infected and healthy larvae were introduced into containers in various ratios. This is in contrast to epizootics in *Plodia interpunctella* caused by introduction of 10^4 spores in pieces of larval cadaver (Burges and Hurst 1977). The difference may result from the observation that in stored products Lepidoptera do not cannibalise live larvae, but they do eat cadavers. Burges and Hurst (1977) considered cannibalism of infected cadavers was the most potent means of natural spread of *Bt* in a study on *Plodia interpunctella, Ephestia cautella, E. kuehniella, E. elutella* and *Galleria mellonella*. Natural epizootics caused by *Bt* are discussed in Section 3.1.4.

Many of the insects killed by *Bt* applications do not support sporulation. Failure to sporulate is caused either by rapid drying of neonate larvae and small cadavers, or to the overgrowth of aerobic vegetative *Bt* developing in the body cavity by competing gut flora, which is permitted by the rapid onset of anaerobic conditions when insect respiration ceases at death. There have been some reports of sporulation in cadavers (e.g. Prasertphon *et al.* 1973; Federici 1994; Katagiri and Shimazu 1974). Higher temperatures (34°C compared to 10-19°C) favoured sporulation in lepidopteran cadavers (Katagiri and Shimazu 1974) and it appears that conditions favouring sporulation do not always occur. In *P. interpunctella* and other stored product Lepidoptera, natural infections are usually light and occur mainly in protected environments such as food stores. Occasionally high mortality rates are observed and sporulation is usual in these insects (Burges and Hurst 1977).

10.2 DISPERSAL

Dispersal of *Bt* has been linked to wind, rain, cannibalism of infected insects, transport by non-hosts, infected mobile insects prior to death, and birds and mammals. Rain may be one of the most important factors (Akiba 1991). Rain splash moved *Bt* spores from cabbage leaves to the topsoil (Pedersen *et al.* 1995) and is one of the most common factors affecting persistence on plant foliage (see Chapter 8.2). *Bt* can also be moved down the soil profile by irrigation or rain. After commercial application, *Bt* spores can be carried by spray drift and air currents. In one study, wind currents along valleys carried *Bt* for up to 15 km from the point of application and winds permitted spores to pass over ridges up to 300 m higher than the point at which they were applied (Grison *et al.* 1976). As discussed in Section 6.6.2.1., Whaley *et al.* (1998) studied drift of *Bt kurstaki* following application in Utah and found *Bt*-related insect mortality up to 3000 m from the application site.

Transportation by predators and parasitoids of infected insects or on non-susceptible insects has been demonstrated. Pitfall sampling demonstrated dispersal of *Bt kurstaki* by means of carabid beetles (up to 135 m) and other insects active on the soil surface (Pedersen *et al.* 1995). Feeding *Bt* to predacious ants (*Formica polyctena*) and then letting the ants bite larvae of *Chrysomela populi* resulted in 80-100% mortality, whereas non-*Bt* fed ant bites resulted in only 6.7-16.7% mortality. A high percentage of cadavers from *Bt*-fed ant bite victims contained viable *Bt*, showing some ability to recycle from this direct introduction to the insect haemocoel (Orekhov *et al.* 1978). It is also possible such an effect would be seen with most contaminating bacteria, whether an insect pathogen or not. A study by Cheng *et al.* (1983) showed transmission of *Bt galleriae* by predaceous mites of scale in pine stands at Nanjing College, China. Spraying *Bt* onto evenly spaced trees in a stand that already contained some predators caused very rapid spread of the bacteria and significantly reduced the scale population. In laboratory studies by Temerak (1982), ovipositors of female *Bracon brevicornis* were contaminated with *Bt* before parasitism of *Sesamia cretica*, a pest of sorghum. *Bt* was readily transmitted to healthy *Sesamia* larvae by oviposition; the hosts were immobilised much more rapidly by contaminated than by uncontaminated parasites, possibly owing to the combined action of the bacterium and the venom of the parasite. Kustak (1964 in Forsberg *et al.* 1976) also found that parasitoids *Nemeritis canescens* exposed to *Bt thuringiensis* could transmit *Bt* to the moth *Ephestria kuehniella*. *Apanteles glomeratus* was also reported to transmit *Bt alesti* to *Pieris brassicae* (Toumanoff 1959 in Forsberg *et al.* 1976).

Bt kurstaki application against the stored grain pests, showed that non-susceptible insects *Rhyzopertha dominica, Tribolium castaneum, T. confusum* and *Sitophilus oryzae* transferred viable spores from treated to untreated wheat by carrying them externally or voiding them with their excreta (McGaughey *et al.* 1975). In field studies, Baur *et al.* (1998a) observed the Argentine ant (*Linepithema humile*) scavenging on *Bt*-killed insect cadavers, possibly contributing to dispersal. *Bt kurstaki* (Dipel) inoculated common cabbageworm, *Pieris rapae*, fed to ray starlings resulted in starling faeces toxic to silkworms in bioassays (Fushimi *et al.* 1995).

In water, dispersal of *Bt* is dependent on water movement and sedimentation of the *Bt* spores and crystals. A study was made of the spatial and temporal distribution of spores of *Bt kurstaki* in four streams, after aerial spraying with Thuricide to control spruce budworm, *Choristoneura fumiferana*, in Quebec. In this study, the spores were dispersed rapidly and uniformly along watercourses (Dostie *et al.* 1988). In contrast, Undeen *et al.* (1981) reported that downstream movement of *Bt israelensis* was often poor. In Tennessee, application of *Bt israelensis* to a stream resulted in almost no *Bt* detected in stream eddies near the point of application 24 and 48 h after treatment (Frommer *et al.* 1981b).

The effectiveness of spread of *Bt* in insect populations depends on the mechanism of spread. Infection by the oral route is poor when spores and crystals are distributed by the physical factors such as rain, wind and in soil because concentrations become progressively diluted below infective dosage levels. However, infection is greater where conditions allow the concentration of *Bt* to increase, i.e. through filter-feeding, and collection in faeces of insectivores and infected cadavers. Wounding by the biting of predators and oviposition of parasitoids spreads *Bt* efficiently because it bypasses the natural barrier presented by the gut wall and introduces the bacteria directly into the vulnerable body tissues. By this route infective doses can be very low. Spread of *Bt* infection is demonstrated only by experiments that record insect mortality, as observation of spores, visually or by plating, does not prove the presence of *Bt* at infectious levels.

11 Effects in Combination with Other Insecticidal Agents

Bt-based insecticides are sometimes used in combination with other insecticidal agents such as pathogens or chemicals or in situations where residual activity of previously applied agents is likely. It is therefore important to understand potential interactions when recommending *Bt* as an environmentally safe biopesticide. Measurement of combination effects is often difficult, as true synergistic responses are rare and synergy is often confused with additive effects, even in the laboratory. It is even more difficult to determine synergies and incompatibilities in the field. Few studies have examined impacts of combined use of *Bt* with pesticides or pathogens against non-target insects, therefore results need to be extrapolated from studies on pest species. Responses often vary with dose; a synergistic or additive effect may be found at one dose, while the response may be antagonistic at a different dose.

Generally, chemical pesticides enhanced *Bt* activity, although there were some exceptions. Interactions of *Bt* with pathogens were variable and rarely synergistic, *Bt* competed with other pathogens for the resources provided by the insect host.

11.1 INTERACTIONS WITH PATHOGENS

Microbial insect pathogens (as well as entomopathogenic nematodes) occur naturally in insect populations and are sometimes used as introduced biocontrol agents. As pathogens occur in both pest and non-pest populations, their interactions with *Bt* are important in assessing the environmental impact of *Bt*. Because pathogens compete with *Bt* for hosts, it would be expected that effects would be less than additive, or even antagonistic. Some reports on the use of nematodes, protozoans, viruses, and fungi with *Bt* applications are listed in Table 11.1. In general, the interactions tended to be slightly antagonistic, with a few reports of more positive interactions including possible synergies.

The entomopathogenic nematodes *Heterorhabditis* and *Steinernema* spp. have been used with *Bt kurstaki, aizawai* and *thuringiensis* β-exotoxin. When used against scarabs in soil, results were additive or better (Koppenhofer and Kaya 1997; Koppenhofer *et al.* 1999), whereas impacts on Lepidoptera were generally negative (Table 11.1), possibly because the comparatively greater effect of *Bt* on lepidopteran larvae in the phyllosphere obscurred the relatively low impact of the nematodes in that habitat. In the one test with a dipteran species (*Tipula paludosa*), *Neoaplectana carpocapsae* and the β-exotoxin were possibly synergistic (Lam and Webster 1972). The use of *Bt* and the entomopathogenic nematode *Steinernema feltiae* was antagonistic in the control of *Spodoptera exigua*, as cadavers did not support nematode reproduction (Kaya and Burlando 1989).

Lepidoptera often suffer from nucleopolyhedroviruses (NPVs) and these viruses have been used as biocontrol agents for many years. The reports on the use of *Bt kurstaki* with NPVs against lepidopterans are conflicting, even for the same host. For example, Luttrell *et al.* (1982) found that the impact of NPV with *Bt kurstaki* on *Heliothis virescens* feeding on cotton, was less than additive, while Bell and Romine (1980) reported the effect of NPV/*Bt kurstaki* on the same host on cotton was additive. Incompatibility between NPV and *Bt* has been recorded for many lepidopterans, and has been attributed to feeding arrestment caused by *Bt,* which prevents ingestion of the virus (Navon 1993). NPVs against lepidopterans sometimes have an additive effect, such as when *Bt* and virus (AcNPV) were used against *H. virescens* (Bell and Romine 1980). Sublethal amounts of *Bt kurstaki* (half field rate of Dipel WP) with the lowest dose of NPV (3×10^{11} polyhedral inclusion bodies (PIB) /ha compared to the highest dose of 1.5×10^{12} PIB/ha) was better than either agent alone in controlling the brassica pest, *Trichoplusia ni* (Tompkins *et al.* 1986), perhaps because sublethal doses of *Bt* did not reduce feeding (Navon 1993). With cytoplasmic polyhedroviruses (CPV) and *Bt* against the pine caterpillar *Dendrolimus spectabilis,* the effect was additive or higher after field application (Katagiri and Iwata 1976; Katagiri *et al.* 1977). When *Bt* (Thuricide or Dipel) was used with a granulosis virus of *Zeiraphera diniana*, *Bt* was reported to be antagonistic to the virus (Schmid 1975). In contrast to this result, a synergistic effect was observed when *Bt* was applied to larvae already infected with the virus (Schmid 1975), indicating the range of responses possible, depending on experimental conditions, timing of application etc. The activation of latent insect viruses is also possible by *Bt* treatment. Manchev (1980) reported silkworm mortality caused by virus increased to 6-13% after

Table 11.1. Examples of interactions between *Bacillus thuringiensis* and pathogens.

Microbial pathogen	*Bt* subspecies/ serovar	Target insect	Assay	Interaction/ joint action	Reference
Nematodes					
Heterorhabditis bacteriophora, Steinernema glaseri or *S. kushidai*	*japonensis*	*Cyclocephala hirta, C. pasadenae, Anomala orientalis*	3rd instar, laboratory in soil, greenhouse and field	additive or synergistic for all but *S. kushidai*	Koppenhofer & Kaya 1997; Koppenhofer *et al.* 1999
Heterorhabditis bacteriophora	*kurstaki*	*Agrotis ipsilon*	laboratory	< additive	Shamseldean and Ismail 1997
Steinernema feltiae (Neoaplectana carpocapsae)	*kurstaki*	*Platyptilia carduidactyla*	field, artichoke sprays	no increased mortality	Bari & Kaya 1984
Steinernema carpocapsae	*kurstaki*	*Plutella xylostella*	laboratory	no interaction against susceptible strain of *P. xylostella*, antagonistic interaction *Bt* resistant strain	Baur *et al.* 1998b
Steinernema carpocapsae	*aizawai*	*Plutella xylostella*	field	slightly additive	Baur *et al.* 1998b
Neoaplectana carpocapsae	*thuringiensis* (exotoxin)	*Tipula paludosa*	laboratory, 3rd and 4th instar	possibly synergistic	Lam & Webster 1972
Steinernema feltiae (=*Neoaplectana carpocapsae*)	*kurstaki*	*Spodoptera exigua*	*Bt* infected larvae	antagonistic	Kaya & Burlando 1989
Neoaplectana carpocapsae and *Heterorhabditis heliothidis*	*san diego, israelensis* and *kurstaki*	*Pyrrhalta luteda* and *Galleria mellonella*	laboratory	applied simultaneously, reduce nematode development, no effect when *Bt* applied after 24h	Poinar *et al.* 1990
Protozoa					
Nosema pyrausta	*kurstaki*	*Ostrinia nubilalis*	field	additive	Lublinkhof *et al.* 1979
Nosema pyrausta	*kurstaki*	*Ostrinia nubilalis*	laboratory and field	no effect or antagonistic to protozoan infection intensity	Lublinkhof & Lewis 1980
Nosema sp.	*alesti* and *kurstaki*	*Loxagrotis albicosta*	laboratory	reduced infectivity of *Nosema*	Helms & Wedberg 1976
Vairimorpha necatrix	*kurstaki* (Dipel)	*Heliothis zea*	laboratory	additive	Fuxa 1979
Vairimorpha nectrix	*kurstaki*? (Thuricide)	*Anticarsia gemmatalis*	laboratory	antagonistic	Richter & Fuxa 1984
Vairimorpha nectrix	?	*Spodoptera litura*	laboratory	synergistic?	Liang *et al.* 1983
Viruses					
Nucleopolyhedrovirus (NPV)	*kurstaki*? (Thuricide)	*Anticarsia gemmatalis*	laboratory	additive	Richter & Fuxa 1984

Table 11.1. *(Continued)*

Microbial pathogen	*Bt* subspecies/ serovar	Target insect	Assay	Interaction/ joint action	Reference
NPV	unspec.	*Spodoptera exigua*	laboratory, 3rd instar larvae	synergistic?	Lipa *et al.* 1975
NPV	*aizawai*	*Spodoptera exigua*	laboratory, 2nd instar	mainly antagonistic, except at low doses	Geervliet *et al.* 1991
NPV	*galleriae*	*Spodoptera littoralis*	laboratory, 2nd instar	antagonistic	Salama *et al.* 1987
NPV	*entomocidus*	*Spodoptera littoralis*	laboratory, 2nd instar	additive	Salama *et al.* 1987
NPV	?	*Spodoptera litura*	laboratory, larvae	< additive	Hwang & Ding 1975
NPV	Bactospeine	*Spodoptera litura*	dipped leaves	< additive	Komolpith & Ramakrishnan 1978
NPV	*kurstaki* (HD-1)	*Trichoplusia ni*	diet	additive	McVay *et al.* 1977
NPV	*kurstaki*	*Trichoplusia ni*	laboratory	additive at $>LC_{45}$, antagonistic at lower doses	Young *et al.* 1980
NPV	*kurstaki* (half field rate of Dipel)	*Trichoplusia ni*	field	better control than either alone	Tompkins *et al.* 1986
NPV	*kurstaki*	*Heliothis virescens*	cotton field/neonate	<additive	Luttrell *et al.* 1982
NPV	*kurstaki*	*Heliothis virescens*	cotton field	additive	Bell & Romine 1980
NPV	*kurstaki* (Dipel)	*Heliothis virescens* and *H. zea*	laboratory, diet	antagonistic at high doses, synergistic at sublethal doses	Bell & Romine 1986
NPV	*kurstaki*	*Lymantria dispar*	lab, 3rd instar	synergy?	Triggiani 1980
NPV	*kurstaki*	*Lymantria monacha*	diet	> additive?	Schonherr & Ketterer 1979
NPV + adjuvants	*kurstaki*?	*Heliothis* spp.	field, cotton sprays	< additive	Johnson 1982
NPV + adjuants	*kurstaki*	*Heliothis* spp.	laboratory	< additive	Luttrell *et al.* 1982
NPV + feeding adjuvant	*kurstaki*	*Bucculatrix thurberiella*	field sprays	< additive	Bell & Romine 1982
Cytoplasmic polyhedrosis virus (CPV)	unspecified	*Dendrolimus spectabilis*	field sprays	additive	Katagiri *et al.* 1977
CPV	*aizawai*	*Dendrolimus spectabilis*	field sprays	additive to <additive	Katagiri *et al.* 1978
CPV	Bactospeine	*Dendrolimus spectabilis*	sprays	additive or >	Katagiri & Iwata 1976
Granulosis virus	?	*Zeiraphera diniana*	laboratory, larvae	antagonistic	Schmid 1975
Fungi					
Beauveria bassiana	*kurstaki*	*Ostrinia nubilalis*	field	independent action	Lewis & Bing 1991
Beauveria bassiana	*kurstaki* (Dipel)	*Spodoptera littoralis*	laboratory	antagonistic	El-Maghraby *et al.* 1988
Destruxins from *Metarhizium anisopliae*	*kurstaki* (Thuricide)	*Choristoneura fumiferana*	5th instar larvae	synergistic at lower doses	Brousseau *et al.* 1998
Nomuraea rileyi	*kurstaki*	*Trichoplusia ni*	laboratory, 1-2 day old larvae	< additive	Ignoffo *et al.* 1980b

application of sublethal doses of Dipel, whereas untreated silkworm larvae had virus-related mortality of only 1.7%.

The two most commonly encountered entomopathogenic fungal species, *Beauveria bassiana* and *Metarhizium anisopliae,* have shown varying results in combination with *Bt kurstaki.* The combination with *B. bassiana* against *Spodoptera littoralis* was antagonistic in the laboratory while the actions were "independent" (additive?) on *Ostrinia nubilalis* (Lewis and Bing 1991). Only the toxins from *M. anisopliae* have been used in combination with *Bt kurstaki* and the toxins were synergistic at low doses (Brouseau *et al.* 1998). The microsporidian protozoan pathogens, *Nosema pyrausta* and *Vairimorpha nectrix* again showed interactions ranging from antagonistic to possibly synergistic with *Bt* against lepidopterans (Table 14.1). Inoculation with *Bt* reduced the infectivity of the protozoan insect pathogen, *Nosema* sp. to the lepidopteran, *Loxagrotis albicosta,* larvae by causing epithelial cells to 'shed' the microsporidian spores and vegetative cells into the midgut lumen (Helms and Wedberg 1976).

11.2 INTERACTIONS WITH CHEMICAL PESTICIDES

Combined use of chemical pesticides and *Bt* has generally increased efficacy against target pests, with most responses either additive or synergistic (Table 11.2). Benz (1971) reviewed studies of the interactions between microorganisms and chemical insecticides. In the 1970s, several studies demonstrated that sublethal doses of chemical insecticides had an additive or synergistic effect with *Bt* (Morris 1972). However, reports on combination effects in the literature are varied. For example, the effect of diflubenzuron with *Bt* ranged from antagonistic, when used against *H. virescens,* (Mohamed *et al.* 1983) to synergistic against *S. litura* (Rao and Krishnaya 1996) and possibly *L. dispar* (Novotny 1988a). In the former USSR, combinations of *Bt* products, such as either Gomelin (*Bt thuringiensis*) or Dendrobacillin (*Bt dendrolimus*) with Dimilin (diflubenzuron) have been used against larvae of *Malacosoma neustria, Aporia crataegi* and *Hyponomeuta malinellus* (*Yponomeuta malinellus*), achieving up to 100% mortality of young larvae at lower doses than when used alone (Grigoryan *et al.* 1988). Pyrethroids were generally synergistic (Table 11.2) and chitinases ranged from no effect to synergistic against three different species. The few studies reporting on use of combinations against insects other than Lepidoptera reported effects ranging from additive to synergistic where chemical pesticides were used in conjunction

Table 11.2. Examples of interactions of *Bacillus thuringiensis* with pesticides.

Target insect	*Bt* subspecies/ serovar	Assay	Pesticide	Interaction/ joint action	Reference
Diptera: Curculidae					
Culex tarsalis	*kurstaki*	laboratory, 1st instar	mon-585 (IGR)	additive	Goldberg & Ford 1974
Culex pipiens pallens	*israeliensis*	laboratory	mieyoubenzuron	synergy	Chen *et al.* 1984b
Culex pipiens pallens	*Israeliensis*	laboratory	trichlorfon, malathion, temephos, fenthion or fenitrothion	additive	Chen *et al.* 1984b
Coleoptera: Tenebrionidae					
Tenebrio molitor	*thuringiensis*	laboratory, larvae	DDT	>additive to synergistic at different doses	Benz & Lebrun 1976
Tribolium castaneum	*kurstaki* (Thuricide)	6th instar	pyrethroids	up to synergistic	Saleem *et al.* 1995
Lepidoptera: Arctiidae					
Halisidota argentata	*galleriae* (Thuricide 90TS)	3rd instar larvae	pyrethrum	supplementary	Morris 1972b
Hyphantria cunea	*galleriae* (Thuricide 90TS)		malathion, phosphamidon or pyrethrum	supplementary to antagonistic at higher doses	Morris 1972b

Table 11.2. *(Continued)*

Target insect	*Bt* subspecies/ serovar	Assay	Pesticide	Interaction/ joint action	Reference
Spilosoma obliqua	*kurstaki* (Dipel 0.05%)	7 day old larvae	endosulfan, BHC, malathion, quinalphos, fenvalerate & cypermethrin	additive to synergistc	Biswas *et al.* 1996c
Geometridae					
Lycia (Biston) hirtaria	*kurstaki* (Dipel) and *galleriae* (Entobacterin)	laboratory & field	trichlorphon (Dipterex), quinalphos (Ekalux), tetrachlorvinphos (Gardona) and diazinon (Basudin)	additive, possibly synergistic	Lecheva & Kuzmanova 1980
			phosalone (Zolone)	antagonistic	
Operophtera brumata	*kurstaki* (Thuricide)	spray	pyrethroid	synergistic	Svestka & Vankova 1980
Gelechiidae					
Pieris rapae	*kurstaki* (Dipel)	field, cabbage & collards spray	chlordimeform hydrochloride	synergistic	Creighton & McFadden 1974
Lasiocampidae					
Malacosoma disstria	*galleriae* (Thuricide 90TS)	3rd instar larvae	malathion, phosphamidon or pyrethrum	subadditive to antagonistic	Morris 1972b
Lymantriidae					
Euproctis chrysorrhoea	*kurstaki* (Dipel)	laboratory, sprayed leaves	DDT, lindane, dieldrin diflubenzuron pyrethrins, azinphos-methyl, tetrachlorvinphos, chlorfenvinphos, carbaryl and propoxur	antagonistic to synergistic	Canivet *et al.* 1978
Lymantria dispar	*kurstaki* (Bathurin 82 'S')	laboratory and field	teflubenzuron and diflubenzuron	synergistic?	Novotny 1988a
Orgyia pseudotsugata	*galleriae* (Thuricide 90TS)	3rd instar larvae	malathion, zectran, piperonyl butoxide or pyrethrum	supplemental to antagonistic	Morris 1972b
Noctuidae					
Anticarsia gemmatalis	*kurstaki*?	laboratory	carbaryl	synergy	Richter & Fuxa 1984
Anticarsia gemmatalis		laboratory	parathion-methyl and permethrin	additive	Richter & Fuxa 1984
Helicoverpa spp.	*kurstaki*	field, cotton spray	thiodicarb or cyhalothrin	no effect	Young *et al.* 1998
Helicoverpa (=*Heliothis*) spp.	*kurstaki*	field, cotton spray	chlordimeform	< additive	Yearian *et al.* 1980
Helicoverpa armigera	*kurstaki* (Dipel)	laboratory, 5th instar	carbaryl, endosulfan, malathion, monocrotophos and phenthoate	synergistic	Dabi *et al.* 1988
Helicoverpa armigera	*kurstaki* (Dipel)	laboratory, neonate larvae	endosulfan	synergistic	Pree & Daly 1996
Helicoverpa armigera and *Spodoptera litura*	*thuringiensis* (Bactospeine)	laboratory, 2nd instar	endosulfan, monocrotophos, fenvalerate and cypermethrin	increased pesticide susceptibility	Justin *et al.* 1989
Helicoverpa virescens	*kurstaki* (Dipel)	laboratory 1st or 3rd instar	chlordimeform, methoprene, thiabenazole and cyhexatin	additive to synergistic	Mohamed *et al.* 1983

Table 11.2. *(Continued)*

Target insect	*Bt* subspecies/ serovar	Assay	Pesticide	Interaction/ joint action	Reference
Helicoverpa virescens	*kurstaki* (Dipel)	laboratory, 1st or 3rd instar	diflubenzuron and fentin hydroxide	antagonistic	Mohamed *et al.* 1983
Helicoverpa virescens 1983	*kurstaki* (Dipel)	laboratory, 1st or 3rd instar	chlorothalonil	antagonistic to 1st instar, additive to 3rd	Mohamed *et al.*
Helicoverpa virescens 1983	*kurstaki* (Dipel)	laboratory, 1st or 3rd instar	benomyl and chlorothalonil	no effect	Mohamed *et al.*
Helicoverpa virescens	Biotrol XK	diet, 5th instar larvae	carbaryl	synergistic	Chen *et al.* 1974
Helicoverpa virescens	Biotrol XK	diet, 5th instar larvae	methomyl, phosmet and carbofuran	synergistic ?	Chen *et al.* 1974
Helicoverpa virescens	Biotrol XK	diet, 5th instar larvae	trichlorphon and propoxur	no effect	Chen *et al.* 1974
Helicoverpa virescens	Biotrol XK	diet, 5th instar larvae	tetrachlorvinphos	antagonistic	Chen *et al.* 1974
Helicoverpa virescens		cotton spray	chlordimeform	synergistic	Mohamed *et al.* 1983
Helicoverpa virescens		cotton spray	diflubenzuron	antagonistic	Mohamed *et al.* 1983
Panolis flammea	Thuricide	laboratory, 3rd or 4th instar	dichlorvos (Nogos)	no effect	Swiezynska & Glowacka-Pilot 1973
Spodoptera exigua		diet	neem	antagonistic	Moar & Trumble 1987b
Spodoptera littoralis	*thuringiensis* (Thuricide 90 TS)	4th instar	trichlorphon (Dipterex) monocrotophos (Nuvacron)	synergism	Altahtawy & Abaless 1973
Spodoptera littoralis		diet/larvae	pyrethroids	synergistic	Salama *et al.* 1984
Spodoptera littoralis	Biotrol	cotton leaves, 5th instar	calcium arsenate, DDT or methyl-parathion	no synergism	Abdallah 1969
Spodoptera littoralis	Cry1C	laboratory, larvae	endochitinase	synergistic	Regev *et al.* 1996
Spodoptera littoralis	*kurstaki* (Dipel)	laboratory, larvae	fenvalerate	potentiation at low doses	El-Zemaity *et al.* 1987
Spodoptera litura	*kurstaki*	larvae	diflubenzuron	synergistic	Rao & Krishnayya 1996
Spodoptera litura	*kurstaki*?	lab, dipping, 4th instar	boric acid, dicrotophos, fenitrothion, dichlorvos	enhanced mortality and shorter lethality	Govindarajan *et al.* 1976
Trichoplusia ni	*kurstaki* (Dipel)	field, cabbage & collards spray	chlordimeform hydrochloride	synergistic	Creighton & McFadden 1974
Trichoplusia ni	*kurstaki*	laboratory	permethrin	> additive	Jaques *et al.* 1989
Pieridae					
Pieris rapae	*kurstaki*	laboratory	permethrin	< additive	Jaques *et al.* 1989
Tortricidae					
Choristoneura fumiferana	*kurstaki* (Thuricide)	foliage of fir, 5th instar	chitinase	no effect	Smirnoff & Fast 1978
Tortrix viridana	*kurstaki* (Thuricide)	spray	pyrethroid	synergistic	Svestka & Vankova 1980
Thaumetopoeidae					
Thaumetopoea pityocampa	Thuricide	laboratory	chitinase	no significant effect	Cavalcaselle 1978
Trichostrongylidae					
Trichostrongylus colubriformis	*israelensis*	eggs	benzimidazoles	synergy	Bone & Coles 1987
Yponomeutidae					
Plutella xylostella		laboratory, larvae	diazinon	additive to synergistic	Zhou *et al.* 1994

with *Bt israelensis* in water. Similar effects were found with *Bt kurstaki* and *Bt thuringiensis* against Tenebrionidae.

The evidence from a number of field studies indicates that combined use of chemical pesticides and *Bt* led to higher impacts on non-target predators and parasites than using *Bt* alone (e.g. Kennedy and Oatman 1976; Morris and Armstrong 1975). While this is not an unexpected result, further studies on some of the more recently registered pesticides would be useful.

11.3 DIRECT EFFECT OF INSECTICIDES ON *BT* SPORES AND TOXINS

Several studies have reported on the effect of common insecticides, herbicides and other common pest control agents directly on growth, germination and viability of *Bt*. In India and some other countries, Neem tree extracts are used as pesticides. Coventry and Allan (1997) screened Neem preparations for antibacterial activity against several species including *Bt*. A solution containing 1% active ingredient inhibited growth of *Bt in vitro,* but sensitivity to commercial preparations varied significantly. Moar and Trumble (1987b) reported use of *Bt* with Neem against *Spodoptera exigua* on artificial diet was antagonistic to toxicity.

Bt has been used in conjunction with organophosphates and other insecticides and herbicides, either intentionally or through exposure to spray residues. The effect of a number of organophosphates on the viability of spores from Biotrol-XK and Biotrol BTB on an inert surface showed carbaryl and tetrachlorvinphos (stirofos) decreased the viability of Biotrol XK spores, propoxur increased it, and trichlorphon, phosmet, methomyl and carbofuran had no effect (Chen *et al.* 1974). No compound had a significant effect on the viability of Biotrol BTB spores, presumably because of differences in formulation.

Growth of *Bt thuringiensis* on nutrient agar was inhibited by the insecticides thanite, naled and parathion, but rotenone, pyrethrin, allethrin, DDT, methoxy-DDT (methoxychlor), gamma BHC (lindane), dieldrin and malathion caused minor inhibition while coumaphos (Co-Ral), dichlorvos (DDVP), carbaryl (Sevin) and propoxur (Baygon) caused no inhibition (Dougherty *et al.* 1971).

Thuricide 16B compatibility with 27 chemical insecticides (organophosphates, carbamates, pyrethrins, chlordimeform, urea derivatives and antifeedants) in terms of germination of bacterial spores and replication of vegetative cells was investigated by Morris (1977b). The most compatible insecticides were acephate, trichlorphon, methomyl, carbaryl, mexacarbate and diflubenzuron (as Dimilin) (Morris 1977b), while Morris (1975) found that fenitrothion at 2 ppm or resmethrin and tetrachlorvinphos at 1 ppm inhibited *Bt kurstaki* replication after 2 h growth in liquid broth culture. Acephate at 10,000 ppm for 2 h had no significant inhibitory effect on replication, nor on spore germination and crystal size (Morris 1975). Sixty minutes mixing of *Bt galleriae* (Thuricide 90TS) vegetative cells with Zectran, phosphamidon, piperonyl butoxide, prythrum and malathion, did not affect fermentation of sugars, lecithinase production or spore viability (Morris 1972b).

Morris (1975, 1977a) investigated the effect of types of formulation of *Bt* on spore viability. Technical formulations were less harmful to the bacteria than wettable powders, which were less harmful than emulsifiable concentrates. The commonly used emulsifiers, Atlox and Triton X-100, at 1000 ppm totally inhibited germination of *Bt* spores in broth, impaired vegetative cell growth and reduced crystal size. In contrast, Triton X-100 was a commonly used *Bt* spore wetting agent in the preparation of suspensions for viable spore plate counting and plant sprays in the laboratory and field for many years, without reported adverse effects. Similarly, there was no effect on larval mortality in bioassay after soaking a spore-crystal suspension for 24 hours in Triton X-100 aqueous solution (reviewed by Burges and Jones 1998).

12 Gene Transfer

Many toxin-encoding genes in *Bt* are carried on plasmids, extra-chromosomal DNA which can be transferred from cell to cell relatively easily by conjugation. There have been few studies on transfer of genetic elements in the environment, but it has been demonstrated in the laboratory that *Bt* strains can interchange toxin-encoding plasmids with other *Bt* strains and with other bacterial species. Based on present knowledge, transfer of the highly mobile plasmid-encoded δ-endotoxin genes is most likely to occur in dead insect larvae, but it may also occur in other environments conducive to growth of *Bt*, such as nutrient-rich soil or decaying organic material. The possible effects of gene transfer to or from an introduced *Bt* include establishment of toxin genes in bacteria that colonize specific habitats more efficiently than *Bt* and altered host range of *Bt* or other transcipient bacteria. Both of these possibilities could lead to increased risk of developing resistance to *Bt* in target insects and greater risk to sensitive non-target insects. However, despite the widespread use of *Bt* and its ubiquitous natural occurrence, no examples of *Bacillus* with unexpected properties that could be attributed to gene transfer have been reported.

12.1 EXTRA-CHROMOSOMAL DNA ENCODING *BT* TOXINS

A relatively new area of environmental concern regarding the use of *Bt* is the mobility of toxin genes. Many *Bt* toxins, including some exotoxins, are encoded by extra-chromosomal DNA (e.g. plasmids) (Lereclus *et al.* 1993; Levinson *et al.* 1990) and such DNA is known to be exchanged between bacteria by conjugation and transformation.

In an evaluation of the possible risks of gene transfer resulting from release of *Bt,* both transfer from *Bt* and transfer into *Bt* should be considered. Many of the δ-endotoxin genes presently known are located on large plasmids in *Bt* strains. *Bt* strains carry an array of plasmids, both small (<10 MDa) and large (>30 MDa). The number per cell and size of the plasmids varies greatly between strains. Not all *Bt* endotoxin genes are located on plasmids and some are presently thought to be chromosomally located although the existence of mega-plasmids cannot be ignored (Lereclus *et al.* 1993). The toxin genes are normally located on large plasmids (40-150 MDa), and the presence of more than one toxin gene on a single replicon has been demonstrated. For example, *Bt israelensis* has four toxin genes on the same 72 MDa plasmid (Lereclus *et al.* 1993).

Beta-exotoxin produced by *Bt* can also be encoded on plasmids. Levinson *et al.* (1990) correlated exotoxin production in some strains of *Bt thuringiensis, Bt tolworthi* and *Bt darmstadiensis* with the presence of plasmids (65-110 MDa), through the use of plasmid curing (loss of plasmid) and plasmid transfer into a *cry-* strain. Exotoxin positive *Bt kustaki* strains were generated by conjugation *in vitro* with strains carrying plasmids encoding exotoxin (Levinson *et al.* 1990).

Transfer of DNA between organisms can occur by several mechanisms. Horizontal gene transfer can occur either directly by transformation with naked DNA, transduction with phages, or by uptake of chromosomal fragments, or with the greatest frequency, through transfer of plasmids by conjugation.

12.2 CONJUGATION

Under laboratory conditions, plasmids can be shuttled between and within *B. thuringiensis*, *B. cereus* and *B. anthracis* (Battisti *et al.* 1985; Wiwat *et al.* 1990). Among the plasmids transferred frequently were those encoding toxin genes or those involved in regulation of protoxin synthesis. Few *Bt* subspecies transfer plasmids at a sufficiently high rate to be detectable without selectable markers, although in some cases colonies of derivatives that do or do not produce protein inclusions can be distinguished on certain media (Gonzalez *et al.* 1982). Aronson and Beckman (1987) demonstrated a low frequency of chromosomal gene transfer from *Bt* to *B. cereus* during cell mating but plasmid transfer between the two species occurred frequently in the mating experiments.

Muller-Cohn *et al.* (1994) examined the frequency of conjugation between *Bt kurstaki* and *Bt aizawai* under different growth conditions. Filter mating using 1:1 donor to recipient cells resulted in a high frequency of conjugation (2×10^{-3} transconjugants/recipient cell). The effect of soil texture, pH and water holding capacity was examined in sterile soil microcosms. Frequency of conjugation was highest in an amended sandy soil (pH 6.3) resulting in $5\text{-}30 \times 10^{-6}$ transconjugants/recipient/g soil after 7 days.

There is increasing evidence that phyllosphere animals are involved in bacterial gene exchange in terrestrial environments. Jarrett and Stephenson (1990) observed plasmid transfer between strains of *Bt* infecting larvae of *Galleria mellonella* and *Spodoptera littoralis* at rates similar to those obtained by *in vitro* plasmid transfer experiments, in which rates of up to 60% were achieved. The ability of *Bt* to multiply to populations of 10^8 cells per larva may contribute to the high rates of transfer achieved, although *Bt* does not regularly colonise cadavers to this level. Muller-Cohn *et al.* (1994) also demonstrated conjugation among *Bt* in insect larvae. In this study, conjugation was observed only if at least one of the *Bt* strains was infective in the insect larvae. More recently, Vilas-Boas *et al.* (1998) monitored conjugal transfer of a genetically marked plasmid carrying the gene *cry*1Ac between *Bt* strains in broth culture, soil microcosms and infected larvae of *Anticarsia gemmatalis.* The lowest conjugation frequency was in soil (8×10^{-8} - 2.7×10^{-5}) while high rates were recorded in larvae ($1.0\text{-}8.4 \times 10^{-1}$). Laboratory studies of this type indicate the potential for gene transfer to occur in the environment when the conditions favouring conjugation occur. The insect appears to be the most favourable environment for transfer of genetic material but as natural *Bt* contamination of insects is infrequent (see Chapter 3), conjugal transfer in soil may be more important.

The potential for *Bt* to acquire plasmids from indigenous bacteria other than *Bacillus* in soil has been shown. Transfer of a plasmid encoding erythromycin resistance from *Streptococcus faecalis* into *Bt israelensis* occurred in compost prepared from cow manure (Byzov *et al.* 1999). Other studies have demonstrated that there is potential for conjugative DNA transfer from *Bt* to indigenous soil bacteria. For example, Jarrett and Stephenson (1990) isolated spore-forming soil bacteria and used the vegetative cells as recipients for *in vitro* conjugation experiments. Up to 3% of the transconjugants (cells receiving DNA) gained the ability to produce protein crystals.

Bora *et al.* (1994) studied a *Bacillus megaterium* which had received a plasmid carrying a lepidopteran active toxin gene from *Bt kurstaki* HD-1 by *in vitro* congugation. The authors concluded that a *Bt* plasmid encoding a toxin gene could potentially be introduced and maintained in a bacterium with a much higher persistence in the phylloplane. *B. megaterium* was able to produce insecticidal toxins on the leaf surface. Evaluation of the risk of this type of transfer would require determination of the likelihood of the donor and recipient proliferating sufficiently in the same habitat.

The likelihood of gene transfer is directly related to the number of cells capable of participating in conjugation. *Bt* is primarily present in the environment as spores, which can not participate in gene transfer. This greatly reduces the occurrence of gene transfer. The likelihood of transfer between *Bt* and other *Bacillus* species such as *B. cereus* or *B. anthracis* depends on the other species being present in the same environment in a suitable growth phase. These requirements greatly reduce the likelihood of transfer events taking place.

12.3 TRANSPOSONS

Conjugation may also be mediated by transposons, which are DNA structures surrounded by repeats that are able to move around in the genome. The δ-endotoxin genes are often found within a composite transposon-like

structure (Hallet *et al.* 1991; Malvar and Baum 1994). No investigations of transposon mediated conju
of genes involving *Bt* in natural environments have been described, however horizontal tranfer of large
encoding regions is a growing area of interest (Hacker *et al.* 1997).

12.4 TRANSFORMATION AND TRANSDUCTION

Possible transfer of genetic information from and into *Bt* strains may not be limited to the mechanism of conjugational plasmid transfer. Transduction and transformation might also play a role in the movement of DNA, but recorded rates of transfer have consistently been much lower than for conjugation. Transformation, where a piece of naked DNA is taken into a foreign cell and incorporated into the genome, has been demonstrated for *Bacillus* (Reanney *et al.* 1982). Bacteriophages are common in *Bt*, but no data are available on phage-mediated gene transfer among *Bt* in the natural environment. The ability of phages to transfer DNA between *Bt* strains has been used intensively *in vitro* for mapping chromosomal *Bt* genes and for transfer of plasmids among *B. anthracis*, *B. cereus* and *Bt*.

12.5 POSSIBLE EFFECTS OF GENE TRANSFER

While studies suggest that gene transfer to and from *Bt* is a natural event in the environment, gene transfer will occur only in the presence of metabolically active bacteria. *Bt* is primarily present in the environment as spores that cannot participate in gene transfer. Therefore, gene transfer could be expected to occur in habitats with actively growing *Bt*, primarily insect cadavers. In addition, genetic transfer is unlikely to take place at high rates in soil or water because of the low population levels of *Bt* (Meadows 1993). There may also be environmental barriers between such transfer in nature. For example, with other bacteria, plasmid transfer has been difficult to achieve in non-sterile soil while possible in sterile soils (e.g. Lebaron *et al.* 1997). Thus natural gene transfer may not be common.

As the frequency of gene transfer is directly proportional to the density of the donor and/or recipient organisms, the release of large amounts of *Bt* during pest control efforts may result in increased frequency of gene transfer. Hansen *et al.* (1996) suggested a number of possible unwanted effects resulting from gene transfer:

- Establishment of *Bt* toxin genes in bacteria with ecology different from *Bt*.
- Any transferred gene giving the recipient bacteria a better chance for survival could affect the biodiversity of the particular environment.
- New genes originating from recombination events could alter the host range of the recipient bacteria.
- All these events could result in prolonged exposure of the toxin genes to the environment, giving rise to increased resistance in target insects. Furthermore, sensitive non-target insects could be more exposed. Improved molecular methods will allow the specific tracking of toxin encoding plasmids and more accurate assessment of transfer frequency in the environment.

To put these conclusions into perspective, it must be remembered that the relatively high number of *Bt* spores applied sporadically and locally for pest control - when summed globally over time - is miniscule compared to the long term natural occurrence of *Bt* spores, which are found in low numbers almost universally in soil and commonly in the phyllosphere. Furthermore, local relatively high densities of *Bt* spores immediately following application soon revert to natural population levels.

13 Transgenic Use of *Bt* Toxin Genes

The use of *Bt* toxin genes expressed in other organisms, especially plants, supports a growing industry. A number of plant species which contain *Bt* toxins are now either commercially available or under field trials and other organisms, such as viruses and bacteria, have been modified to express *Bt* toxins. Public concerns about transgenics have brought environmental and mammalian safety research on impacts and fate of transgenic *Bt* toxins to the fore. However, the majority of published research on *Bt*-expressing plants has been on the production and stability of toxins in plants and ecological studies are in their infancy in comparison with the body of literature on impacts of *Bt*-based insecticides. The currently available studies on non-target impacts and persistence of toxic proteins in plants have not indicated any significant divergence from results reported in studies on *Bt* toxins not expressed in transgenics.

13.1 TRANSGENIC EXPRESSION

The use of *Bt* endotoxin genes for transgenic pest control has been an established field of research since the mid-1980s. Some early efforts to genetically enhance *Bt* efficacy included plasmid curing of *Bt* strains (deleting plasmids which did not encode toxin genes) and expression of novel toxin combinations in a single *Bt* strain (e.g. Lereclus *et al.* 1992). As understanding of the genetics of endotoxin production increased, so did potential for novel application of these toxins that were encoded by single genes. The expression of individual *Bt* toxin genes in plants (e.g. corn, potatoes, cotton, brassicas, rice), other bacteria (e.g. Xu *et al.* 1998; Selinger *et al.* 1998), viruses (e.g. Ribeiro and Crook 1998) and other organisms (e.g. Gelernter and Schwab 1993) has been published extensively. Cotton varieties containing *Bt* toxin genes have been on sale since 1996 (Roush 1997) and a number of other crops with *Bt* genes are undergoing field trials or are at the early commercial stage. However, the costs of developing *Bt* crops will probably limit the number of species eventually produced (Peferoen 1997) and many unresolved issues still surround the use of transgenic organisms. The main issues at present are those of transient expression, public acceptability and safety.

Generally, publications on transgenic use of *Bt* genes have been on the techniques for production, primary testing of transgenic organisms for toxicity to target pests and persistence of gene expression, as well as an increasing volume of reports on the expanding agricultural use of transgenic plant varieties. Comparatively few studies have looked at environmental and mammalian safety, but this is rapidly becoming a very topical issue. We do not intend to extensively review the environmental safety of transgenic expression of *Bt* genes and most of the issues are similar to those involved in the application of *Bt* insecticidal products.

13.2 RESISTANCE

Development of resistance to singly and continuously expressed *Bt* genes is considered to be more frequent and more severe than to strains containing multiple toxins, which has made resistance and its management a major issue for transgenics. Resistance management strategies suggested include the use of refugia of non-transgenic crops, augmented where possible with high toxin expression in the plants (Gelernter 1997; Roush 1997). Currently in the USA, refugia are the strategy of choice and there are specified percentages of total acreage that are required to be planted in non-transgenic crops, such as 20% for corn and 25% for cotton.

There are concerns about refuge size however, as practical size of refuge in cropping situations is often considered too small to delay resistance. The use of gene pyramiding of two or more insecticidal genes in the same plant is another resistance management strategy under consideration.

13.3 NON-TARGET IMPACTS

Non-target impacts of transgenic *Bt* plants has received media attention recently following publication of a study showing transgenic pollen from corn plants was toxic to monarch butterfly caterpillars (Losey *et al.* 1999). As discussed in Section 6.4.5.3., there has been discussion about the lack of detail in the study, unmeasured levels of toxin used, uncertainty as to whether monarchs would encounter large quantities of transgenic pollen in natural situations and whether monarchs could avoid *Bt*-containing pollen. However, this study and others are demonstrating some potential problems of *Bt* toxin use. There have been few studies of non-target effects of continually expressed insecticidal toxin genes in plants but, in many ways, transgenic expression of toxins poses the same problems as application of formulated *Bt* products.

Non-target impacts of transgenic *Bt* plants on predators and parasites have been examined in a few studies (e.g. Riddick and Mills 1995; Orr and Landis 1997; Hilbeck *et al.* 1998a; Riddick *et al.* 1998), with similar results as found for *Bt* insecticidal products (see Chapter 7.5). There is little direct impact, but there may be some competition for host resources. Johnson and Gould (1992) found some synergism between a parasitoid, *Campoletis sonorensis*, and transgenic tobacco against larvae of *H. virescens*. No detrimental effect of transgenic maize pollen expressing a Cry1Ab protein was found on predators of *Ostrinia nubilalis* (Pilcher *et al.* 1997). In another study, a diamondback moth parasitoid was unaffected by transgenic *Bt* crops (Riggin-Bucci and Gould 1997). In a Chinese study, no effect was found on the total number of natural enemies of cotton pests recorded in transgenic *Bt* cotton and unmodified cotton (Cui and Xia 1997a).

Other studies have begun examining impacts on a range of other beneficials and non-target organisms. Single endotoxin protein toxicity for honeybees was assessed by Arpaia (1996) for the Cry3B protein, to mimic expected toxin concentrations in pollen. No toxic effect on larvae or pupae was observed and the author concluded transgenic crops with this gene would be unlikely to affect pollinators.

Studies on effects of plants expressing transgenic *kurstaki* and *tenebrionis* proteins on soil arthropods, (collembolan, *Folsomia candida* and an oribatid mite, *Oppia nitens*), found no effect on several indices examined (Yu *et al.* 1997a). Potato plants expressing a *Bt tenebrionis* protein showed no persistent significant differences in phylloplane microflora (including plant pathogenic fungi and bacteria) compared to unmodified plants (Donegan *et al.* 1996). It would not be expected that single or even dual *Bt* gene expression in plants would significantly alter arthropod susceptibility from the tests on strains containing those genes. However, if the level of toxin expression differed greatly, significant changes in susceptibility could occur. A study by Sims (1995) found no significant mortality of non-lepidopteran species from transgenically expressed Cry1Ac and Cry1A among 14 species and concluded that risks to beneficial non-lepidopteran insect species were negligible. Total arthropod numbers on transgenic cotton were less than on untreated plants in a 1992-94 study in Mississippi but more than on insecticidal treated plots (Luttrell *et al.* 1995). To date, there has been little examination of sublethal effects on non-targets, but these will also need to be considered.

13.4 ENVIRONMENTAL IMPACTS

Environmental risks which can be attributed to genetically modified plants might include expression of undesirable phenotypic traits as well as the putative transfer of the new genes to wild relatives or to microorganisms (Dietz 1993; Snow 1997). As an example of an unexpected change in characteristics, introduction of the *Bt tenebrionis cry3A* gene into *Rhizobium* resulted in changes not associated with the intended function of the transgene; the transgenic strain exhibited a greater ability to compete for nodule occupancy (Giddings *et al.* 1997). It is conceivable such an increased competitive ability could lead to alterations in the distribution of *Rhizobium* strains if released.

Because the toxins are always present in transgenic plants, the fate of toxic proteins after plant death is of considerable interest. Sims and Ream (1997) studied the loss of activity of Cry2A proteins in transgenic cotton in the soil over 120 days. They found, using an insect bioassay, that half-life of bioactivity (based on growth inhibition of *H. virescens*) was 15.5 days in the laboratory and 31.7 days in the field. Less than 25% of activity remained after 120 days. In another study by Palm *et al.* (1996), purified *Bt kurstaki* toxin added to soil microcosms showed a rapid decline in concentration and little detectable activity after 140 days, with the decline being attributed largely to biotic influences. Such studies emphasize the lack of soil persistence of *Bt* toxins outside the spore.

13.5 MAMMALIAN TOXICITY

The registration requirements for transgenic crops in relation to human and animal health impacts are currently under consideration. In most countries *Bt* transgenic plants will be required to undergo primary screening for novel toxicity products, in case the process of gene insertion has resulted in new mammalian or animal toxins being produced. The presently established mamalian safety of *Bt* insecticidal products is unlikely to be sufficient grounds for exclusion from screening. Published studies in this area are relatively few, although companies developing transgenic *Bt* plants have conducted studies. Noteborn *et al.* (1996) determined that oral exposure to Cry1Aa in tomatoes posed no additional risk to human and animal health than the toxin alone. Similarly, Noteborn *et al.* (1994) and Kuiper and Noteborn (1996) found no changes in the mammalian safety of Cry1Ab when expressed in tomatoes.

13.6 FUTURE

Regulatory authorities in some countries are under pressure to restrict transgenic field use. However, the EPA (USA) found no significant risk of gene expression beyond the spatial and temporal limits of the proposed field tests for transgenic cotton, potato and maize (LaSota *et al.* 1996). One problem with transgenics is the lack of independent studies assessing safety, with much of the information coming from companies directly benefiting from the sale of transgenic plants. While this is not necessarily a scientific problem, it has led to claims of bias from some environmental action groups. There is a great need for independent environmental impact assessment. There are additional implications arising from the continual advances made in transgenic expression, such as higher protein production in plants and site-specific expression of toxins.

A potential negative aspect of transgenic organisms (specifically food plants) is that the *Bt* toxin(s) is continually expressed, which increases exposure to the toxin and includes many perceived risks of eating such toxins in food. There are also concerns over gene transfer, but such concerns apply as much or more to *Bt* cells, as many toxin genes reside on the extrachromosomal bacterial plasmids. Objective consideration suggests that transgenic plants expressing one or a few known toxin genes may have an advantage of less complexity in risk assessment over *Bt* formulations as formulations include spores, other toxins and enzymes which may not have been fully characterised. With transgenic expression, usually only one *Bt* toxin-encoding gene is expressed and, in theory at least, only one foreign toxin is expressed in the plant. There have been many press articles on the lack of knowledge about the effect of expressing foreign genes in cells and possible mutations which could result in unexpected toxic products, but to our knowledge there is no evidence for this from transgenic use of *Bt* genes. However, additional research is needed both on the effect of gene insertion as well as on environmental effects of each toxin.

14 Resistance

Development of insect resistance to biological insecticides was once considered unlikely. However, the evolution of resistance to both *Bt* subspecies and individual toxins has been demonstrated in the laboratory for several insects and found in field populations of *Plutella xylostella.* In the laboratory, resistance has developed to *Bt* subspecies *kurstaki, aizawai, entomocidus* and *tenebrionis* (*san diego*) and individual Cry toxins from *Bt kurstaki, aizawai, entomocidus* and *israelensis*. Despite repeated attempts, significant resistance to whole cultures of *Bt israelensis* has not developed. The difficulty of generating resistance to *Bt israelensis* is because of the involvement of Cyt toxins, which appear capable of overcoming resistance generated to individual or multiple Cry toxins. Cross-resistance between *Bt* toxins has been found, but does not automatically occur. For example, development of resistance to *Bt aizawai* in the Indian mealmoth, *Plodia interpunctella,* resulted in resistance to *Bt kurstaki*, but not *vice versa*. In most cases, with the exception of *Bt israelensis,* resistance to *Bt* was developed in the laboratory in less than 20 generations.

Detecting resistance in field populations is complicated by natural variation within populations of the same species in their susceptibility to *Bt*. Further, as discussed earlier, bioassays are not precise indicators of relative toxicity. Therefore, small variations in relative toxicity (i.e. <10x) may not indicate development of resistance but reflect the natural variation in insect populations or difficulties in bioassay technique. Resistance can decrease when *Bt* selection pressure is removed, however studies on several insects found that resistance may reach a stabilized level for long periods where reversion to susceptibility does not occur. In some cases, resistance may be accompanied by reduced biotic fitness.

Risk assessment for resistance development in most pest populations is a complicated issue. The relatively restricted use of *Bt* (around 1% of the insecticide market worldwide) may contribute to the low rates of *Bt* resistance in field populations currently observed. If the present application levels were continued, it should be possible to minimise resistance through careful management. However, as *Bt* use increases, so does the risk of resistance developing. In addition, the widespread use of plants expressing single *Bt* toxins increases the risk of resistance. Use of some of the more recently discovered *Bt* toxins, especially those with a different action from Cry toxins (such as Cyt1A), may be of some help in reducing the risk of resistance, as well as management strategies recommended for *Bt* transgenic crops, such as use of refugia and high doses of toxins. However, because little is known about the occurrence and inheritance of resistance, cross-resistance and aspects of pest behaviour that may influence development of resistance, it is premature to assume that any resistance management plan will be successful.

14.1 DEVELOPMENT OF RESISTANCE

14.1.1 LABORATORY

Resistance has developed in a number of insect species in the laboratory after exposure to *Bt* or individual toxins (Table 14.1). In most of the species tested resistance can develop rapidly to *Bt kurstaki, entomocidus, tenebrionis* or *aizawai*, but has been more difficult to induce experimentally with *Bt israelensis*.

When the difference between resistant and susceptible insect populations is low (<10×), it is not always clear whether a population has developed true resistance or is displaying a range of tolerances among subgroups or genotypes of that species (see Chapter 5). As a rule, differences in toxicity of 10× or greater indicate resistance strains of a species, while those below 10× may indicate only a natural tolerance to *Bt.* In addition, bioassay systems are unlikely to be sensitive enough to consistently detect differences below 10× (see Section 5.1.3).

The first demonstration of resistance in the laboratory was in *Musca domestica* using *Bt thuringiensis* (Harvey and Howell 1964; Wilson and Burns 1968), with resistance presumably developing against β-exotoxin. Similarly, a 10× resistance to β-exotoxin in *Drosophila melanogaster* was developed in 30 generations in the laboratory (Carlberg and Lindstrom 1987).

The first laboratory demonstration of resistance to δ-endotoxin producing strains was in the Indian mealmoth, *Plodia interpunctella.* Resistance reached >250 times that of susceptible populations under high selection pressure in some lines (McGaughey and Beeman 1988; McGaughey 1990). Since these studies, resistance has been developed in the laboratory in more than a dozen insect species (Table 14.1). In some cases resistance has been demonstrated to the whole bacterial preparation, while in other cases, resistance is to a single toxin. McGaughey and Johnson (1994) demonstrated that selecting for resistance to spores and toxins of strains of *Bt kurstaki, aizawai* or *entomocidus* did not result in resistance to the individual toxins expressed by those strains in equal levels. For example, a *Bt kurstaki* resistant line of *P. interpunctella* was highly resistant to Cry1Ac (2,816x), less resistant to Cry1Ab (263x) and not resistant to singly used toxins Cry1Aa, Cry1B, Cry1C or Cry2A. This may partly reflect the differing levels of expression of the various toxins in the *Bt kurstaki* strain.

Resistance in the laboratory has been demonstrated for three species of *Spodoptera*: *S. frugiperda* (Fonseca and De Polania 1998), *S. littoralis* (Muller-Cohn *et al.* 1996) and *S. exigua* (Moar *et al.* 1995a). *Heliothis virescens* strains have been selected with over 10,000× resistance to Cry1Ac toxin relative to the susceptible strain (Gould *et al.* 1995; Lee *et al.* 1995b).

In some cases, such as *Plodia* and *Plutella* spp., resistance can be achieved after only a few generations of high selection pressure. Resistance of colonies of *P. interpunctella* increased from 2- to 29× within 3 generations to *Bt kurstaki* (McGaughey and Beeman 1988). In the laboratory, resistance in the diamondback moth, *Plutella xylostella,* has also been raised after 6 generations to *Bt aizawai* Cry1C, increasing the level of resistance, 22 to 62× (Liu and Tabashnik 1997a, b). The Colorado potato beetle, *Leptinotarsa decemlineata,* was selected for resistance to the Cry3A coleopteran specific δ-endotoxin of *Bt san diego/tenebrionis* and reached a level of resistance >700× the susceptible population after 35 generations (Trisyono and Whalon 1997). Stone *et al.* (1989) found that *Pseudomonas* expressing a single 130 KDa endotoxin could lead to laboratory resistance of *H. virescens* in as little as 3 generations, reaching 24× after 7 generations. After 7 generations, a laboratory colony of *Ostrinia nubilalis* was selected for resistance of 73× greater than a susceptible baseline to Dipel ES (*Bt kurstaki*), indicating that *O. nubilalis* populations can respond rapidly to intense selection pressure with a commercial formulation of *Bt* (Huang *et al.* 1997). Resistance in the laboratory has also been selected for a number of other coleopteran, lepidopteran and dipteran species (Table 14.1).

Laboratory attempts to induce resistance by continual exposure of mosquitoes to the full spectra of *Bt israelensis* toxins have generally failed (e.g. Lee and Cheong 1985; Goldman *et al.* 1986; Gharib and Szalay-Marzso 1986) or selected for only low level resistance (tolerance?)(Georghiou 1984; Georghiou and Wirth 1997). However, resistance to individual toxins and even multiple toxins has been shown in the laboratory. Cheong *et al.* (1997) selected for *Culex quinquefasciatus* larvae resistant to Cry11A toxin. Wirth *et al.* (1998) reported on a number of multiple resistant lines of *Cx. quinquefasciatus* (Table 14.1). The lack of resistance development to whole *Bt israelensis* could be because of its complex mode of action, involving synergistic interaction between up to four proteins (Becker and Margalit 1993). Georghiou and Wirth (1997) also showed that resistance could be raised in only a few generations when a single *Bt israelensis* toxin was used, and was progressively more difficult to raise in mosquitoes with combinations of two and three toxins. When all four *Bt israelensis* toxins were used, including the Cyt toxin, incidence of resistance was remarkably low (Table 14.1).

The genetic basis for resistance is incompletely understood and some of the results seem contradictory. Resistance in *Plutella* to Cry1Ab has been shown to be a recessive trait primarily controlled by an incompletely recessive, autosomal single allele (Hama *et al.* 1992), although it has not always been attributed to the same loci. Tabashnik *et al.* (1997) reported that populations in Hawaii and Pennsylvania share a genetic locus at which a recessive mutation associated with reduced toxin binding confers extremely high resistance to four *Bt* toxins. However, resistance in a population from the Philippines showed multilocus control, a narrower spectrum, and, for some *Bt* toxins, inheritance that was not recessive and not associated with reduced binding. A major *Bt*-resistance locus in a strain of *Heliothis virescens* exhibiting up to 10,000× resistance to Cry1Ac toxin has been identified and mapped (Heckel *et al.* 1997a, b). Tabashnik *et al.* (1997) found that a recessive gene which conferred resistance in *Plutella* to four *Bt* toxins (Cry1Aa, Cry1Ab, Cry1Ac, and Cry1F) was present in 21% of individuals from a susceptible strain, suggesting the potential of relatively rapid evolution of resistance under field selection.

14.1.2 FIELD

The first report of field developed resistance to *Bt* was in the diamondback moth, *Plutella xylostella,* to *Bt kurstaki* (Tabashnik *et al.* 1997). Despite the demonstration of resistance potential in a number of insects, *P. xylostella* is presently the only insect where resistance poses serious problems for field control using *Bt* as a biopesticide (Kennedy and Whalon 1995). Four other lepidopterans (Table 14.2.) have been suggested as showing resistance or the potential for resistance to *Bt* under field selection, although only low-level differences in toxicity have been demonstrated. In the field, *S. frugiperda* was 1.48× more "resistant" than a baseline level (Fonseca and De Polania 1998) and natural occurrence of resistance in *H. virescens* to MVP II R (Cry1Ac in dead *Pseudomonas*) was termed "tolerance" rather than resistance by Hasty *et al.* (1997). In China, field populations of *Helicoverpa armigera* have also been reported resistant to *Bt kurstaki* (Shen *et al.* 1998). Levels of difference in the four species between susceptible and tolerant populations may not be sufficient to indicate resistance, given the low sensitivity of bioassays and natural variations in populations in response to *Bt.*

Although *P. xylostella* was the first insect to show high level resistance to *Bt* in the field, early laboratory investigations did not reveal the potential of *P. xylostella* to develop resistance to *Bt*. Devriendt and Martouret (1976) held a strain of *P. xylostella* under high selection pressure (LD_{90}) from the spore-crystal complex of *Bt thuringiensis* E-61 for 10 generations in the laboratory and found no significant change in susceptibility. Subsequently, field resistance in *P. xylostella* to *Bt kurstaki* has been recorded in Thailand (Zoebelein 1990), Hawaii (Tabashnik *et al.* 1990; Tabashnik 1994), Japan (Tanaka and Kimura 1991), Korea (Song 1991), China (Zhao *et al.* 1993), the mainland USA (Shelton 1993; Tang and Shelton 1995) and Central America (Perez and Shelton 1997). Commercial application of *Bt kurstaki* in Hawaii resulted in resistance 25-33× greater than for a susceptible laboratory colony. Shelton *et al.* (1993) reported resistance levels of >300× to *Bt kurstaki*, but only 3-4× to *Bt aizawai* in field populations of *P. xylostella* from the USA. In the Philippines, a field population was >200 times more resistance to Cry1Ab than a laboratory strain (Ferre *et al.* 1991).

The species of insect influences the likelihood of resistance developing, both in terms of predisposition and in other ways such as likelihood of encountering inoculum. Forrester (1994a) suggests that *Helicoverpa punctigera* in Australian cotton is unlikely to become resistant to *Bt* because of its ecology is different from the more resistance-prone *H. armigera.* The highly migratory and polyphagous nature of *H. punctigera* results in large natural populations of unsprayed *H. punctigera*, which can dilute any resistant populations which may develop (Forrester 1994b). *H. punctigera* has never developed resistance to synthetic insecticides and would be less likely to develop *Bt* resistance than *H. armigera* for this reason. The diamondback moth, *P. xylostella*, the only insect to show high levels of resistance to *Bt* in the field, is not a mobile insect. Gelernter (1997) suggests that this may be a contributing factor to *P. xylostella* developing resistance as lack of mobility reduces the likelihood of mating with individuals from other locations, thereby diluting the resistance genepool.

One group of insects which have been exposed to high and constant applications of *Bt* products in the field are mosquitoes. Most of these applications have been using *Bt israelensis* and, despite many studies, no true resistance to *Bt israelensis* products has been detected in field populations (reviewed in Glare and O'Callaghan 1998). Constant exposure of *Ae. vexans* in the field for 10 years in Germany resulted in no difference in the level of resistance in exposed and unexposed populations (Becker and Margalit 1993). Similar results were reported for blackflies (Kurtak *et al.* 1989). These findings confirm the laboratory findings that Cyt toxin may reduce the likelihood of resistance developing to *Bt israelensis* (see above).

14.1. Reports of resistance developing to *Bacillus thuringiensis* varieties or toxins from laboratory studies.

Insect	Subspecies/ variety	Levels achieved compared to susceptible population	Toxin(s)	Cross resistance	References
Lepidoptera: Gelechiidae					
Pectinophora gossypiella			Cry1Ac		Bartlett *et al.* 1997
Lepidoptera: Noctuidae					
Heliothis virescens	*Pseudomonas fluorescens* expressing 130-kDa δ-endotoxin of *kurstaki* HD-1	up to 24×	Cry1		Stone *et al.* 1989
	Pseudomonas fluorescens expressing 130-kDa δ-endotoxin of *kurstaki* HD-1	12-69×	Cry1Ab		Sims & Stone 1991
		50-53×	Cry1Ac	Cry1Ab and Cry2A	Gould *et al.* 1992
		10,000×	Cry1Ac	Cry1Aa, Cry1Ab, Cry1F, Cry1B[1], Cry1C[1] and Cry2A[1])	Gould *et al.* 1995
Spodoptera exigua	*kurstaki*	1-2×[1]			Moar *et al.* 1995b
		850×	Cry1C (from *aizawai*)	trypsinized Cry1Ab, Cry1C (from *entomocidus*) Cry1E/Cry1C fusion protein, Cry1H and Cry2A	
		11×	Cry1C (in *E. coli)*		
Spodoptera frugiperda	*kurstaki*	4.9×			Fonseca & De Polania 1998
Spodoptera littoralis	*aizawai*	up to >500×	Cry1C	partial to Cry1D, Cry1E, Cry1Ab	Muller-Cohn *et al.* 1996; Chaufaux *et al.* 1997
Trichoplusia ni		31×	Cry1Ab	(not cross-resistant to CryIAa or CryIAc)	Estada & Ferre 1994
Lepidoptera: Phycitidae					
Homoeosoma electellum	*kurstaki*	1.7×[1]			Brewer 1991
Lepidoptera: Pyralidae					
Ephestia cautella	*kurstaki*	7×[1]			McGaughey & Beeman 1988
Ostrinia nubilalis		low level only[1, 2]	Cry1Ab		Lang *et al.* 1996
	kurstaki	up to 73×			Huang *et al.* 1997
Plodia interpunctella	*kurstaki*	>250x			McGaughey & Beeman 1988
	kurstaki (Dipel)	>250×		*thuringiensis, kurstaki* (isolates other than HD-1) and *galleriae* (not	McGaughey & Johnson 1987; McGaughey 1990
				cross resistant to *kenyae, entomocidus, aizawai, tolworthi* and *darmstadiensis*	
	kurstaki (Dipel)	>25×			van Rie *et al.* 1990b
		>106×	Cry1Ab protoxin		van Rie *et al.* 1990b
		~875×	Cry1Ab toxin		van Rie *et al.* 1990b
	kurstaki	140×			McGaughey & Johnson 1992
	aizawai	28-61×			
	entomocidus	21×			

	kurstaki+aizawai	15×			
	kurstaki HD-1	307×		not cross-resistant to *aizawai* or *entomocidus*	McGaughey & Johnson 1994
	aizawai HD-112, HD-133	28× 97×		cross resistant to HD-1,-198 & -133 cross resistant to HD-1, -112 & -198	
	entomocidus HD-198	32×		cross resistant to HD-1, -112 & -133	
	kurstaki + *aizawai*	164× to *kurstaki* 62-100 × to *aizawai*		resistant to *entomocidus*	
Lepidoptera: Tortricidae					
Choristoneura fumiferana	*sotto*	3.8×[1]			van Frankenhuyzen & Milne 1993 in Tabashnik 1994
Lepidoptera: Yponomeutidae					
Plutella xylostella	*kurstaki*	15-66×			Tabashnik *et al.* 1991
		22-62×	Cry1C		Liu & Tabashnik 1997a, b
			Cry1Aa, Cry1Ab, Cry1Ac, and Cry1F		Tabashnik *et al.* 1997
			Cry1Aa, Cry1Ab and Cry1Ac	Cry1F, Cry1J and H04[3]	Tabashnik *et al*. 1996
	kurstaki		Cry1A and Cry2	Cry1F from *aizawai*	Tabashnik *et al.* 1994
Coleoptera: Chrysomelidae					
Chrysomela scripta		>5,000×	Cry3Aa	Cry1Ba	Federici & Bauer 1998
	tenebrionis (san diego)	59×	Cry3A		Whalon *et al.* 1993
			Cry3A		Trisyono & Whalon 1997
Diptera: Drosophilidae					
Drosophila melanogaster		6-10×[1]	thuringiensin		Carlberg & Lindstrom 1987
Diptera: Culicidae					
Culex quinquefasciatus	*israelensis*	2.3 -5.1×?[1]	Cry11A Cry4A plus Cry4B Cry4A plus Cry4B	*jegathesan* Cry11B in some strains	Wirth *et al.* 1998
			Cry11A	marginally to *israelensis*, *fukuokaensis*, *jegathesan* and *kyushuensis*	Cheong *et al.* 1997
	israelensis[2]	-			Georghiou 1984
		>1000× >122× 13×	Cry11A Cry4A + Cry4B Cry4A + Cry4B + Cry11A	reduced to 8x by addition of CytlA suppressed by CytlA suppressed by CytlA	Georghiou & Wirth 1997; Wirth *et al.* 1997
		3.2×[1]	Cry4A + Cry4B + Cry11A + CytlA		
Culex pipens pallens	*israelensis*	6-11×[1, 2]			Han 1990
Diptera: Muscidae					
Musca domestica	*thuringiensis*	6×[1]			Harvey & Howell 1964 Wilson & Burns 1968

[1] low levels which may not indicate resistance.
[2] not sustained.
[3] a hybrid with domains I and II of Cry1Ab and domain III of Cry1C (Tabashnik *et al.* 1996).

14.2 NATURAL VARIATION IN SUSCEPTIBILITY TO *BT*

Considerations of resistance are influenced by the natural variation of sensitivity to a single *Bt* strain or toxin. While such variation has been suggested as providing potential to develop resistance (Rossiter *et al.* 1990), it also indicates a difficulty in distinguishing low level resistance from naturally occurring tolerance. Several studies have provided evidence of natural variation in susceptibility to *Bt*. For example, Rossiter *et al.* (1990) found significant variation (LC_{50}s ranging from 76-180 IU/ml diet) in susceptibility among four populations of gypsy moth, *Lymantria dispar*. Field populations of the maize pest, *Ostrinia nubilalis* (European corn borer), without significant exposure to the toxins, were found to differ up to 5× in susceptibility to Cry1Ab protein (Siegfried *et al.* 1995).

14.3 REVERSION OF RESISTANCE

Laboarory resistance to *Bt* can be readily selected in some insects, but is often unstable. In published work on seven strains of *P. xylostella*, level of resistance to *Bt* declined when exposure to insecticide ceased (Tabashnik *et al.* 1994). Resistance to *Bt kurstaki* in *P. xylostella* in a laboratory colony started from a Florida population was initially >1500 times compared to susceptible insects, but declined and stabilised at 300× after 3 generations without selection pressure (Tang *et al.* 1997).

In four other pests (*Heliothis virescens, Leptinotarsa decemlineata, Musca domestica* and *Plodia interpunctella*), resistance to *Bt* declined slowly or not at all in the absence of exposure of *Bt* in the laboratory (Tabashnik *et al.* 1994). In the Indian mealmoth, *P. interpunctella,* resistance was stable after the resistance levels reached a plateau, but declined if selection was discontinued earlier (McGaughey and Beeman 1988).

S. littoralis developed resistance to Cry1C toxin of >500× after 12-14 generations (Chaufaux *et al.* 1997; Muller-Cohn *et al.* 1996), but resistance declined from >500× to >74× after 1 generation without selection pressure (Muller-Cohn *et al.* 1996). With *L. decemlineata*, the stability of resistance was also studied by testing a colony over 17 generations after the selection pressure was removed. Resistance level of the colony decreased after 5 generations. The resistance level did not decrease further when the selection was removed for >12 generations (Utami *et al.* 1995), suggesting it might never have returned to susceptible levels.

Reduced biotic fitness associated with resistance is the most likely cause of instability of resistance in *P. xylostella*. Groeters *et al.* (1994) demonstrated such a reduction in fitness. After 5 generations, egg hatch and fecundity were reduced by 10% in the resistant strain compared to a susceptible strain. Understanding the instability of resistance may clarify why resistance to *Bt* has been relatively rare and may also help to devise strategies for extending the usefulness of this environmentally benign insecticide (Tabashnik *et al.* 1994).

14.4 CROSS RESISTANCE

The issue of cross-resistance to other control agents or *Bt* toxins has been investigated. Generally the incidence of cross-resistance is low but has been reported for a number of toxins (Table 14.1). Cross-resistance between subspecies appears related to the toxin composition of the isolate used. One of the most detailed studies on cross resistance was the work of McGaughey and Johnson (1994), who examined the resistance to individual toxins after selection for resistance to isolates of *Bt kurstaki* (HD-1), *Bt aizawai* (HD-112, HD-133) and *Bt entomocidus* (HD-198). For *P. interpunctella, Bt kurstaki* HD-1 resulted in resistance to Cry1Ab and Cry1Ac, but not Cry1Aa, Cry 1B, Cry1C or Cry2A. Resistance to *Bt aizawai* (HD-133) and *Bt entomocidus* (HD-198) strains involve a broader spectrum resistance to Cry1Aa, Cry1Ab, Cry1Ac, Cry1B, Cry1C, and Cry2A, despite neither isolate producing Cry2A (McGaughey and Johnson 1994). McGaughey and Johnson (1994) suggested that the relatively narrow spectrum of resistance to different toxins in *Bt kurstaki* meant that cross-resistance to other subspecies was less likely to develop, therefore *Bt kurstaki*-based products should be used first in the field. If resistance developed to either *Bt aizawai* and *Bt entomocidus* it would result in cross-resistance to *Bt kurstaki,* at least in the Indian mealmoth.

Bt resistance in *H. virescens* is usually reported for *Bt kurstaki* Cry1Ac (Gould *et al.*1992; Heckel *et al.* 1997a, b). However, cross resistance to other toxins, such as Cry1Aa, Cry1Ab, Cry1F and Cry2A has also been reported (Gould *et al.* 1992; Gould *et al.* 1995; Lee *et al.* 1995b). In one case, some cross-resistance to Cry1B, Cry1C and Cry2A was found, but the level of resistance to these toxins was moderate (Gould *et al.* 1995). In another study, *P. xylostella* showed high cross resistance to Thuricide, based on a *Bt kurstaki* strain but, surprisingly, not to Dipel (also *Bt kurstaki*) or, as expected, not to Centari (*Bt aizawai*) (Sarnthoy *et al.* 1997). Selection of *P. xylostella* with *Bt kurstaki* (containing Cry1A and Cry2 toxins) caused a >200× cross-resistance to Cry1F toxin from *aizawai* (Tabashnik *et al.* 1994).

Some cross resistance between mosquitocidal toxins of *Bt jegathesan* and *Bt israelensis* has been demonstrated (Table 14.1), but the mixture of Cry and Cyt proteins produced by these two subspecies makes field occurrence of resistance unlikely (Wirth *et al.* 1997; 1998). Resistance to Cry11A resulted in marginal cross-resistance to the multiple Cry toxin crystals from *Bt israelensis* and also to toxin crystals from three other mosquitocidal strains, *Bt fukuokaensis*, *jegathesan* and *kyushuensis* (Cheong *et al.* 1997).

Several studies have reported no link between chemical pesticide susceptibility and *Bt*. For *Bt* resistant *P. xylostella*, there was no cross-resistance to fenthoate (phenthoate), benfuracarb, fenvalerate, chlorfluazuron, cartap, and abamectin (Sarnthoy *et al.* 1997). Similarly, permethrin- and methomyl-resistant strains of *H. virescens* had no resistance to *Bt* (Ignoffo and Roush 1986). *Spodoptera littoralis, P. xylostella* and *Pseudoplusia includens* resistant to chemical control measures were no less susceptible to *Bt* (Matter and Keddis 1988; Jespersen and Keiding 1990; Mink and Boethel 1993). These results are to be expected because of the modes of action of the chemicals differ from that of *Bt*, but there was an unexpected result in another study in which *Bt kurstaki*-resistance was shown to be inversely related to pyrethroid resistance in *Bovicola* (*Damalinia*) *ovis,* with pyrethroid-resistant strains significantly more susceptible to *Bt* than a permethrin-susceptible line (Drummond *et al.* 1995).

14.5 MANAGEMENT OF RESISTANCE

Resistance to *Bt* can evolve in the field in less than two years with sufficient selection pressure (Liu and Tabashnik 1996). Gelernter (1997) has attributed the lack of field resistance in insects other than *P. xylostella* to the relatively small amounts of *Bt* currently applied, especially in comparison to chemical pesticides. With the increasing use of single *Bt* genes in transgenic plants, the risk of resistance would be expected to be much higher and many strategies are under discussion to combat this threat. Resistance has led to the withdrawal of a number of chemical pesticides and *Bt* could be lost as a resource for the same reason.

Efforts to delay resistance by the use of two or more *Bt* toxins assume that independent mutations are required to counter each toxin and that resistance alleles are rare in susceptible populations (Tabashnik *et al.* 1997). However, several studies have found recessive resistance genes to be common in susceptible populations. Tabashnik *et al.* (1997) found that a recessive gene that conferred resistance in *Plutella* to four *Bt* toxins (Cry1Aa, Cry1Ab, Cry1Ac, and Cry1F) was present in 21% of individuals from a susceptible strain, suggesting resistance could evolve relatively rapidly under field selection. Cross-resistance among toxins and the ability of insects to develop resistance to multiple toxins, as demonstrated by resistance in a single insect to both *Bt kurstaki* and *aizawai*, also suggests expression of multiple toxins may not delay resistance in all situations. Cross resistance and broad spectrum resistance to *Bt* may be insect species specific (McGaughey 1994a). In *Plutella*, resistance reduced only slowly when *Bt* selection pressure was removed, which may work against the use of rotation strategies of different *Bt* toxins if the rotation is too rapid (McGaughey 1994b).

Use of transgenic plants expressing *Bt* toxins will not automatically result in resistant populations. For example, resistant Colorado potato beetle (*L. decemlineata*) which survived on transgenic potatoes expressing Cry3A δ-endotoxin did not survive continuous or successive exposure over three generations (DiCosty and Whalon 1997).

The risk of insects developing resistance from sublethal exposure to repeated *Bt* applications is well recognised, if still hotly debated as to the significance for many insects. However, most researchers recognise the need to develop management strategies for transgenic crops expressing *Bt* toxins as well as biopesticide use. Many strategies have been suggested for overcoming resistance problems, including the use of multiple toxins, non-sprayed refugia areas or non-*Bt* plants (in the case of transgenics) (see also Section 14.2), high doses of *Bt* toxin, mixtures of normal and transgenic seeds and toxin rotation (Gelernter 1997). The interplanting of resistant and susceptible lines has been suggested as refugia for transgenic crops, with seed mixtures of *Bt* and non-*Bt* plants a variation on the same idea. Refugia and interplantings are aimed at maintaining susceptible populations in numbers which will allow sufficient dilution of any resistance in target populations. The amount and size of such refugia will differ depending on the mobility and ecology of the insect, crop and geographical area. Overspray with pesticides in conjunction with suitable refugia has been suggested for cotton (Harris *et al.* 1998), in addition to tissue-specific expression of toxins, such as in boll-tissue of cotton where early generations of *Helicoverpa* would not be exposed (Gould 1988). However, in cabbages, Perez *et al.* (1997) found that the deliberate inclusion of a refuge may reduce the proportion of marketable produce, which may affect use of this resistance management strategy in both sprayed *Bt* and transgenic crops expressing *Bt* toxins. Also evidence that some insects move from *Bt* expressing plants to non-*Bt* plants (Hoy and Head 1995; Gould 1997) would reduce the value of interplanting or seed mixtures. Refugia are not the strategy of choice for sprayed plants, because imperfect spray cover normally results in a proportion of the larval population not being exposed to *Bt* anyway.

A strategy which has received support from researchers and the Envrionmental Protection Agency in the USA is a combination of refugia and high doses of *Bt* (Gelernter 1997). Because resistance is thought to be a recessive trait and possible rare, high dose levels are hypotheses to kill most of those with the resistance genes and the remaining to mate the resistance genes out with insects in refugia. However, there is some evidence that resistance genes are not rare in populations, and what is a high dose for one insect may not be for another insect on the same crop. Whatever strategy is adopted, it is presently based on incomplete data and it is too soon to conclude that current strategies have been successful. More information is required on mechanisms of resistance, inheritance and various other aspects before a complete management plan can be developed for pest insects.

Table 14.2. Reports of possible resistance developing to *Bacillus thuringiensis* varieties or toxins in the field.

Insect	Subspecies/variety	Level of resistance	Toxin(s)	Representative references
Lepidoptera: Noctuidae				
Helicoverpa armigera	*kurstaki*	?		Shen *et al.* 1998
Heliothis virescens	MVP II R (Cry 1Ac encapsulated from *Bt*)	"tolerance"		Hasty *et al.* 1997
Spodoptera frugiperda	*kurstaki*	1.48×		Fonseca & De Polania 1998
Lepidoptera: Pyralidae				
Plodia interpunctella	*kurstaki*	low levels		McGaughey 1990
Lepidoptera: Yponomeutidae				
Plutella xylostella	*kurstaki*	25-33×		Tabashnik *et al.* 1990
	kurstaki	321-426×		Shelton *et al.* 1993
	aizawai	3-4.1×		
	kurstaki		Cry1Ab	Liu *et al.* 1996
	aizawai	>20×	Cry1C	

15 Conclusions: Safety and Risks

15.1 FACTORS IN ASSESSING SAFETY OF *BT*

Bt is a naturally occurring bacterium, found in a wide range of habitats world-wide. The strains used as microbial control agents are present as constituents of soil, leaf and insect microflora. However, where *Bt* populations occur naturally, their density is relatively low. In comparison with naturally occurring *Bt* populations, pest control efforts result in the periodic localised release of high numbers of one particular *Bt* strain into the environment. It is the environmental and health impacts of large scale application of *Bt* spores and crystals that are the focus of this review. Extensive use of *Bt* may impact on the environment in many ways. There may be increased exposure of sensitive non-target organisms, seen at worst during exposure for prolonged periods following repeat applications for pest eradication. In addition, the use of various formulations that extend persistence may also lead to more organisms or different groups of non-target species being exposed locally to *Bt* than would occur naturally. People and animals are exposed by drift from sprays or from intentional application over cities. Two recent examples of such applications are the eradication of Asian gypsy moth in Vancouver, Canada, and the white-spotted tussock moth in Auckland, New Zealand, both of which involved aerial application of *Bt kurstaki*. Development of resistance to *Bt* may reduce its effectiveness as a pest control agent. Use of *Bt* toxins is also increasing through the use of transgenic crops, raising new concerns over development of resistance, non-target effects and potential gene transfer.

We have reviewed much of the extensive literature on *Bt*, with a view to assessing the environmental and health impacts possible from broadcast application of various strains in formulated products. Our intention was to draw some conclusions on the safety of *Bt* but it became quickly apparent that there are many complicating issues. Firstly, the so-called species "*Bt*" comprises a diverse range of strains, each having unique toxin profiles, differing in the amounts and types of toxins produced. It is therefore difficult to make broad declarations on the safety of *Bt* as it is a mistake to treat the multiple toxins and spores from the >60,000 isolates presently held in collections (Boucias and Pendland 1998) as a single biopesticide. Results on the toxicity and safety of one isolate cannot be extrapolated to other isolates, as they may produce different toxins (such as exotoxins).

Secondly, there are major limitations in the methods used to assess toxicity, both in terms of the accuracy of bioassays and the measurement of dose. Insect strains and species differ from one study to the next and application methods and rates differ, resulting in a number of species listed as both non-susceptible and susceptible in the Appendices. The vast range of strains and products used are not always properly defined in the literature and formulations can affect toxicity even of the same strain. There is major difficulty in expressing the units of toxicity as most of the units used (e.g. estimated toxin content, international toxicity units, spores/ml, dry weight, vegetative cells) do not directly relate to the amount of toxin produced.

There are several additional problems in assessing non-target impacts from records in the literature. Toxicity of *Bt* is directly related to dose and some organisms recorded as susceptible at very high doses in the laboratory may not prove to be susceptible at field rates. In other cases, high doses result in feeding inhibition, even in relatively tolerant organisms. Therefore, studies need to be evaluated in terms of field dosage. This is not always possible, as some studies have used purified toxins or experimental formulations, or given no indication of field dosage. In addition, limited presence of susceptible stages and feeding behaviour limit the non-target impacts.

Another problem in the interpretation of non-target impacts is assessing the likelihood of an organism coming into contact with *Bt*. In some cases, organisms which have been shown to be susceptible to a particular *Bt* may be unlikely ever to be in the same area as a pest that *Bt* is used to treat. Also, *Bt* entering environments such as waterways from runoff would, in most cases, be relatively dilute compared to the levels used in tests.

The rapid increase in *Bt* toxin expressing transgenic plants may further complicate the situation, as the toxins are expressed continuously. Therefore, defining non-target impacts from laboratory and even field studies can be difficult. These limitations need to be kept in mind when considering non-target impacts.

Within the above limitations, we can make some generalisations on safety of *Bt*.

15.2 NON-TARGET AND ENVIRONMENTAL SAFETY

15.2.1 IMPACTS ON NON-TARGET INVERTEBRATES

Bt strains are not specific to genera, families or even orders of insects. For example, *Bt kurstaki* is reported to be toxic to over 300 species of Lepidoptera and various other species. While this is an advantage in terms that a *Bt* based pesticide can be used against more than one pest, it poses a risk in terms of potential non-target impacts. Most specific ecosystem studies on non-target impacts of *Bt* have been conducted with *Bt kurstaki* and *Bt israelensis.* In summary, while some non-target toxicity has been reported (certain caterpillars in the case of *Bt kurstaki* and chironomids for *Bt israelensis*), no non-target species has been shown to be at risk of eradication through use of *Bt.* As detailed below (Section 15.3), *Bt* still rates very highly in comparison to other pesticides in terms of impacts on non-target invertebrates.

One concern with assessing non-target impacts is whether studies are currently underestimating impacts by assessing larval mortality, but not adult emergence or sublethal effects. In a study on non-target toxicity to Lepidoptera, mortality was delayed in some species, with little larval mortality but no insects surviving to adulthood (Peacock *et al.* 1998) . Such delayed mortality may not be detected in routine laboratory screening, which could lead to an underestimation of *Bt* impacts. In addition, few studies have examined sublethal effects, which can include shortened lifespan and reduced fecundity. Sublethal effects may be the only effects in non-target organisms, in situations where the field dose applied fails to cause mortality. Some sublethal effects, for example reduced fecundity, would be difficult to detect in the short term unless specifically measured.

Consideration of non-target impacts, even when detected, must be compared with the impact on ecosystems of "doing nothing", of not controlling pest outbreaks. For example, are the impacts on birds greater if a single pest reduces the available foliage for other caterpillars and insects, or if that pest is controlled by *Bt* application and its numbers are reduced and the spray impacts on some non-target insects? In the case of *Bt* application against *Lymantria dispar* (gypsy moth), Scriber (1998) considered strong non-target impacts would occur whether gypsy moth was sprayed or not.

15.2.2 IMPACTS ON BENEFICIAL SPECIES

Review of a number of studies on the effect of *Bt* application on insect predators, parasitoids and pathogens indicates that *Bt* is rarely directly harmful to beneficial invertebrates. While competition for hosts and reduced host food quality can indirectly retard development or cause premature death of mature stages of predators and parasitoids, *Bt* has been shown to be useful in IPM systems in most situations. Remarkably, *Bt* is a pathogen almost entirely of herbivores, with virtually no carnivorous hosts, resulting in little effect on predator species, which contribute to natural insect control. No adverse effects on bees and earthworms have been reported from non-exotoxin producing strains.

15.2.3 MAMMALIAN TOXICITY

As discussed in Chapter 7, exceptionally few problems associated with human health have been reported and proven cases of *Bt* causing clinical disease in humans remain extremely rare, despite large scale use of *Bt* for over 60 years. Extensive safety tests against small mammals have rarely shown toxicity; the few reports of toxicity were when artificially high levels of inoculum were used.

Mammalian safety issues arise partly from the close genetic relationship between *Bt* and *B. cereus,* a species which can cause gastro-enteritis in humans. The production of *B. cereus*-type-diarrhoeal enterotoxins by *Bt* isolates has raised concerns that *Bt* may cause gastroenteritis outbreaks in humans. The few studies conducted to date show that strains of *Bt*, including those found in many commonly used commercial products, are capable of producing diarrhoeal enterotoxins. No valid evidence has been found to link usage of *Bt* with episodes of diarrhoea following ingestion of food and enterotoxin titres produced by *Bt* strains were significantly lower than those for *B. cereus*. However, this is a area which requires significantly more research.

Studies such as that by Hernandez *et al.* (1998) have drawn the attention of the public and raised concerns about mammalian toxicity. The authors suggested that mortality of mice following pulmonary exposure to *Bt* in the laboratory resulted from activity of a *B. cereus*-like haemolytic toxin. While the doses used in the study were artificially high, the fact that a toxin produced by a *Bt* strain could cause effects in mice after inhalation of spores requires more detailed investigation. Overall, however, it can be concluded that *Bt* poses little risk to mammals at dosages equivalent to field level exposure.

15.2.4 ENVIRONMENTAL PERSISTENCE

The persistence of *Bt* in the environment has a significant impact on its environmental risk profile. *Bt* spores can survive in environments protected from sunlight, such as soil, for many months but half-lives of endotoxins are generally measured in days rather than months in most situations. In water, *Bt* has rarely been detected more than a few days after application. This lack of residual activity is often been perceived as a disadvantage of *Bt* as a pesticide but short persistence actually reduces the risk of non-target impacts and increases environmental safety. *Bt* does not cause insect epizootics in most circumstances, so does not pose a threat of persistent toxicity to non-target populations after application. *Bt* soon degrades on foliage and the lack of persistence of toxins reduces the risk of contamination of food after crop application of *Bt*-based biopesticides.

One area of concern with *Bt* application is contamination of waterways. In Canada, the increasing use of *Bt* as the biopesticide efficacy increased has led to concerns over the fate of aerially applied *Bt kurstaki* sprays and the effect of *Bt* entering aquatic ecosystems (e.g. Kreutzweiser *et al.* 1996). In Australia, the use of *Bt kurstaki* against cotton pests is restricted near waterways because of concerns about aquatic organisms (A. Rath pers. comm.). *Bt israelensis* is applied directly to water and therefore has the greatest potential impact against non-target aquatic organisms. Review of non-target impacts of *Bt israelensis* (Glare and O'Callaghan 1998) suggests minimal impacts. Given the limited number of insect species susceptible to *Bt kurstaki* and *Bt aizawai,* it is unlikely these subspecies would have greater impacts than *Bt israelensis*. In addition, a number of specific studies on aquatic communities after application of *Bt kurstaki* and *Bt tenebrionis* (see Section 6.6.3) found minimal disruption, even using high dosages (e.g. Eidt 1984; Kreutzweiser *et al.* 1992, 1996). No *Bt* has been shown to be toxic to fish or macro-invertebrates in aquatic communities.

15.2.5 GENE TRANSFER

Many *Bt* toxins are encoded on extrachromosmally located plasmids, which increases the likelihood of movement of toxin-producing ability between *Bt* strains. There is also the possibility of transfer between *Bacillus* species and other genera. Although there have been relatively few studies on gene transfer in the environment, it has been demonstrated in the laboratory that *Bt* strains can transfer toxin-encoding plasmids to other bacterial species, including *B. cereus*. Studies have shown that dead insects are one of the most likely sites for plasmid transfer events, among other high nutrient environments (organic material, nutrient rich soils).

The risk posed by plasmid or gene transfer is that the recipient organism might then colonise specific habitats (probably only insect populations) more efficiently and outcompete natural microbiota, or alter the spectrum of toxicity of *Bt* or other species. An increase of *Bt* toxin expression in the environment could possibly lead to greater resistance in target insects or greater non-target effects. There is also the possibility that genetic material transferred may encode toxins other than endotoxins, such as the enterotoxins of *B. cereus,* leading to *Bt* strains with unwanted characteristics.

Currently, gene transfer is an area of increasing interest, partly generated by concern over transgenic organisms. While plasmid transfer has been shown to occur in the laboratory, it has not been quantified in the natural environment. Despite the ubiquitous occurrence of *Bt* in the environment and its widespread use as a pesticide, no examples of *Bacillus* with unexpected properties that are readily attributable to gene transfer have been reported, but it is an area of current research interest.

15.3 SAFETY IN COMPARISON WITH OTHER INSECTICIDES

While *Bt* sales make up less than 1% of all pesticide sales world-wide, there has been a move towards use of pesticides, such as *Bt,* with low perceived environmental risk (e.g. van Frankenhuyzen 1990a). It is pertinent to consider the relative safety, environmental and mammalian, of *Bt* strains in comparison to other pesticides used against the same pests. Although there is a strong movement towards reduced or even nil pesticide use in some developed countries, it will still be imperative to control insect pests. The world population is increasing rapidly and feeding the large populations of some countries is already a major problem. Pest outbreaks can devastate crops. Is *Bt* a safer option than other commonly used pesticides? In the absence of total guarantees of safety (and in reality there are never absolute guarantees), judgements need to be made on comparative safety.

In a number of studies, *Bt* was the least toxic to non-target organisms of a range of insecticides. For example, in Tasmania, the use of pyrethroid insecticides such as alpha- and cypermethrin has caused mortality to non-target fauna, especially stream dwelling insects (Davies and Cook 1993; Barton and Davies 1993, in Davies 1994). Typical LC_{50} values for these insecticides are 0.1-1 μg/l for aquatic macro-invertebrates and 1-10 μg/l for fish (Coats and O'Donnell-Jeffrey 1979 and other references in Davies 1994). Davies (1994) concluded that *Bt* was a safer option for control of *Chrysophtharta* sp. Similarly, several studies on pesticide effects on fish (reviewed in Chapter 7), found *Bt israelensis* to have LD_{50}s between 100-25,000× lower than other pesticides (e.g. Lee and Scott 1989).

An IOBC/WPRS working group conducted a series of bioassays of 40 pesticides, including *Bt* (Dipel), on beneficial arthropods (Hassan *et al.* 1983). Dipel rated "harmless" to 8 of 9 test insects, which made it the least harmful of the 20 insecticides tested. The working group has continued its role in testing the non-target impacts of pesticides on beneficials and Hassan (1992) provided a list of pesticides found to be "less hazardous" to relevant natural enemies on certain crops. *Bt* was listed for use on cereals, corn, forage and root, vegetable, fruit, vineyard and forest crops. Generally, only one or two insecticides were listed for each crop, from over 65 tested in total, indicating a low comparative risk from *Bt.* This result is common to a number of other studies which report *Bt* as a safer insecticide than many currently in use. At present, the only agents that appear safer are specific pathogens, such as the gypsy moth nucleopolyhedrovirus, which are generally not as effective and may not commercially viable.

15.4 FUTURE RESEARCH REQUIREMENTS

15.4.1 SPECIES AND STRAIN DIFFERENTIATION

Are *Bt* and *B. cereus* distinct species, or do are they strains of the same species, with varying abilities to produce the various toxins and enzymes known (or even unknown)? Phenotypic and biochemical methods have not succeeded in answering this question and DNA level analyses with a large number of strains are needed. Studies conducted to date have examined relatively few isolates of each species and, given the large amount of strain variability, have been inconclusive. A multi-laboratory study, employing multiple strains and applying many phenotypic, biochemical and molecular techniques would elucidate the species definition. Simple discriminatory methods would allow closer monitoring of clinical and food samples, providing more information on the currently speculative role of *Bt* in gastroenteritis. Methods which allow discrimination of specific strains would allow closer monitoring of the environmental fate of commercially applied strains and would allow definitive differentiation from environmental and clinical isolates.

15.4.2 TOXINS AND THEIR MODE OF ACTION

Definition of the types and concentrations of the various toxins from *Bt* is crucial to assessment of safety. New toxins are being described regularly from *Bt* and the complement of toxins produced by any strain needs to be fully understood before its widespread application. It may be that some toxins, for example β-exotoxin, do not pose the level of mammalian risk once thought, but as strains are known that do not produce this toxin, it has been prudent to use such strains. The production by *Bt* of vegetative insecticidal proteins (VIPs), and *B. cereus*-like enterotoxins and haemolysins needs to be better understood and the toxins fully characterised in terms of mammalian and non-target safety. Are the toxin(s) involved produced in all *Bt*? What level is harmful? The development of better assay systems for toxins will be essential for screening of new isolates and products for undesirable toxins.

As the search for new isolates with specific insecticidal properties based on δ-endotoxins continues, it will be necessary to screen potential biocontrol strains for several other toxic substances of concern to human health. As outlined in Chapter 7, of particular concern are the non-specific β-exotoxin and *B. cereus*-type enterotoxins. The possession of either is undesirable in a microbial insecticide to be used in the field, but their prevalence is generally not known. An example of the problem with enterotoxin is the study of Perani *et al.* (1998) which examined the frequency with which the β-exotoxin and diarrhoeal toxin was produced in a collection of natural isolates of *Bt*. Thirty five percent of the isolates possessed the *cry1B* gene; of these 83% also produced enterotoxin and 58% produced β-exotoxin. Methods for removing the ability to produce unwanted exotoxins and enterotoxins from valuable strains (e.g. deletion of specific genes) require urgent study.

15.4.3 NON-TARGET IMPACTS AND ECOSYSTEM DISRUPTION

Reviewing the literature on *Bt* subspecies and products has shown that, while any particular *Bt* isolate can have a broad host range in the laboratory at high doses, possibly involving several orders of insect, this does not translate into obvious impacts on non-target invertebrates in the field. While laboratory toxicity studies are unlikely to provide further illumination for the more studied products and subspecies, ecosystem studies would be useful as very few have been completed. Of particular interest would be studies on non-target impacts on aquatic organisms, by species other than *Bt israelensis*. Ecosystem studies which examined sublethal and delayed impacts would also fill a gap in current knowledge.

15.4.4 RESISTANCE

Resistance to *Bt* has been selected in numerous insects in the laboratory, but rarely found in the field. There are a number of questions which, if answered, may assist in determining both the likelihood of resistance developing in the field and suggest management strategies to avoid resistance developing. One of the main questions is how is resistance inherited and which insect populations have the genetic ability to develop resistance rapidly?

15.5 IS *BT* SAFE?

We have covered much of the literature on environmental and mammalian safety of *Bt* strains and subspecies. After covering this vast amount of literature, our view is a qualified verdict of "safe to use". Certainly, with the amount already applied and the widespread natural occurrence of *Bt*, it must rate as safer than most insecticides currently in use. In terms of non-target impacts, it is specific enough not to eradicate even the most "at risk" non-target insect populations, but is sufficiently broadacting to allow development of commercially viable products in a market demanding insecticides active against a number of pests on a single crop. At field levels used in practical pest control, it is not harmful to mammals, fish or birds.

Much of the recent controversy over safety of *Bt* reflects the public's growing sense of caution over the use of any pesticide. In "Keeping an eye on *B. thuringiensis*", Benard Dixon (1994) does not advocate removing *Bt* from the market, pointing out any organism so widely disseminated with virtually no adverse incidents over several decades is likely to be relatively safe. It seems prudent to continue to monitor *Bt* applications closely for potential negative impacts and the techniques for monitoring are improving constantly. There are a number of issues that require further research, the most obvious being the relationship between the bacilli species and the safety of some of the toxins recently discovered from some *Bt* strains. However, the information available at this time does not suggest there is danger from use of *Bt*.

References

Abai, M. (1976) *Porthesia melania* Stgr. (Lep. Lymantriidae) in Iran. *Entomol. Phytopath. Appl.*, **41**, 7–15.

Abai, M. (1981) A contribution to the knowledge of *Leucoma wiltshirei* Coll. (Lep., Lymantriidae), a new pest of Iranian oak forests. 2. Biology, population dynamics and control. *Z. Angew. Entomol.*, **91**, 86–99.

Abai, M. & Faseli, G. (1986) Morphology, biology and control of the fig tree defoliator *Ocnerogyia amanda* Stgr. (Lep., Lymantriidae). *J. Entomol. Soc. Iran*, **8**, 31–44.

Abbott Laboratories (1993) VectoBac. Tomorrow's answer, Today. Product Information, 7 p. AG-4864/R3.

Abdallah, M.D. (1969) The joint action of microbial and chemical insecticide in the cotton leafworm, *Spodoptera littoralis* (Boisd.) (Lepidoptera:Noctuidae). *Bull. Entomol. Soc. Egypt Econ. Ser.*, **3**, 209–217.

Abdallah, M.D. & Abul Nasr, S. (1970) Effect of *Bacillus thuringiensis* Berliner on reproductive potential of the cotton leafworm (Lepidoptera: Noctuidae). Lethal and sublethal action of *Bacillus thuringiensis* Berliner on the cotton leafworm *Spodoptera littoralis* (Boisd.). *Bull. Entomol. Soc. Egypt Econ. Ser.*, **4**, 171–176.

Abdel Hameed, A. & Landen, R. (1994) Studies on *Bacillus thuringiensis* strains isolated from Swedish soils: insect toxicity and production of *B. cereus*-diarrhoeal-type enterotoxin. *World J. Microbiol. Biotech.*, **10**, 406–409.

Abdel Megeed, K.N., Abdel Rahman, E.H. & Hassanain, M.A. (1997) Field application of *Bacillus thuringiensis* var. *kurstaki* (Dipel–2X) against soft and hard ticks. *Vet. Med. J. Giza*, **45**, 389–395.

Abe, Y. (1987) Culture of *Steinernema feltiae* (DD-136) on bran media. *Jap. J. Nematol.*, **17**, 31–34.

Aboul Ela, R.A., Salama, H. & Ragaei, M. (1993) Assay of *Bacillus thuringiensis* (Berl.) isolates against the greasy cutworm *Agrotis ypsilon* (Rott.) (Lep., Noctuidae). *J. Appl. Entomol.*, **116**, 151–155.

Abrahamson, L.P. & Harper, J.D. (1973) Microbial insecticides control forest tent caterpillar in southwestern Alabama. *USDA For. Ser. Res. Note, Southern For. Ext. St.*, SO–157, 3 p.

Abul Nasr, S. & Abdallah, M.D. (1970) Lethal and sublethal action of *Bacillus thuringiensis* Berliner on the cotton leafworm, *Spodoptera littoralis* (Boisd.). *Bull. Entomol. Soc. Egypt Econ. Ser.*, **4**, 151–160.

Abul Nasr, S.E., El Nahal, A.K.M. & Shahoudah, S.K. (1968) Field experiments on the chemical control of the corn borer, *Sesamia cretica* Led. (Lepidoptera: Agrotidae-Zenobiinae). *Bull. Entomol. Soc. Egypt Econ. Ser.*, **2**, 131–142.

Abul Nasr, S.E., Ammar, E.D. & Merdan, A.I. (1978) Field application of two strains of *Bacillus thuringiensis* for the control of the cotton bollworms, *Pectinophora gossypiella* (Saund.) and *Earias insulana* (Boisd.). *Bull. Entomol. Soc. Egypt Econ.*, **11**, 35–39.

Abul Nasr, S.E., Ammar, E.D., Merdan, A.I. & Farrag, S.M. (1979) Infectivity tests on *Bacillus thuringiensis* and *B. cereus* isolated from resting larvae of *Pectinophora gossypiella* (Lepidoptera, Gelechiidae). *Z. Angew. Entomol.*, **88**, 60–69.

Adams, J.C. & Hartman, P.A. (1965) Longevity of *Bacillus thuringiensis* Berliner in the rumen. *J. Invertebr. Pathol.*, **7**, 245–247.

Adashkevich, B.P. & Rashidov, M.I. (1986) Biological control of the cotton bollworm on vegetable crops. *Zaschc. Rast.*, **6**, 51–52.

Addison, J.A. (1993) Persistence and nontarget effects of *Bacillus thuringiensis* in soil: a review. *Canad. J. For. Res.*, **23**, 2329–2342.

Addison, J.A. & Holmes, S.B. (1995) Effect of two commercial formulations of *Bacillus thuringiensis* subsp. *kurstaki* (Dipel R 8L and Dipel R 8AF) on the collembolan species *Folsomia candida* in a soil microcosm study. *Bull. Environ. Contam. Toxic.*, **55**, 771–778.

Addison, J.A. & Holmes, S.B. (1996) Effect of two commercial formulations of *Bacillus thuringiensis* subsp. *kurstaki* on the forest earthworm *Dendrobaena octaedra*. *Canad. J. For. Res.*, **26**, 1594–1601.

Adeli, E. (1980) The oak tree pest '*Leucoma wiltshirei* Collen' in Iran. *Proc. Intern. Symp. IOBC-WPRS Integr. Con. Agr. For.* (eds. K. Russ & H. Berger), Vienna, 8th-12th Oct.1979, 514.

Adomas, J. (1994) Pest control in Taborskie Forests in 1994. *Ochr. Rosl.*, **38**, 5–6.

Adylov, Z.K., Khakmov, A., Babebekov, K. & Agzamova Kh, K. (1990) Influence of microbiological preparations on the entomophages of the cotton agrocoenosis. *Zaschc. Rast.*, **7**, 34.

Aeschlimann, J.P. (1978) Recent developments in biological control: the example of the grey larch tortrix. *Ann. Soc. Hort. Hist. Nat. Herault*, **118**, 17–20.

Afify, A.M. & Matter, M.M. (1970a) Increasing tolerance (LT value) of *Anagasta kuhniella* Z. towards *Bacillus thuringiensis* with the stage of larval development. *Anz. Schaedlingskd. Pflanzenschutz*, **43**, 97–100.

Afify, A.M. & Matter, M.M. (1970b) Interacting effect of *Bacillus thuringiensis* and moisture content in diet on duration and mortality of immature stages of *Anagasta kuhniella* Z. *Z. Angew. Entomol.*, **66**, 284–291.

Afrikyan, E.K. & Chil Akopyan, L.A. (1980) The bacterial insecticidal preparation BIP and the biological characteristics of cultures of *Bacillus thuringiensis* var. *caucasicus*. *Biol. Z. Armenii,* **33**, 355–365.

Aguda, R.M., Rombach, M.C. & Shepard, B.M. (1988) Effect of Dimilin and Dipel on leaffolder (LF) larvae. *Intern. Rice Res. Newsl.,* **13**, 34–35.

Aguilera, A.P., Vargas, H.C. & Bobadilla, D.G. (1992) Selective control of the chief olive pests in northern Chile. *Olivae,* **41**, 24–30.

Agzamova Kh, K., Bababekov, K., Inogamov, R.U. & Sharafutdinov Sh, A. (1988) Microbiological pest control in cotton. *Zaschc. Rast. Moskva,* **5**, 16–17.

Ahmad, S., O.Neill, J.R., Mague, D.L. & Nowalk, R.K. (1978) Toxicity of *Bacillus thuringiensis* to gypsy moth larvae parasitized by *Apanteles melanoscelus*. *Environ. Entomol.,* **7**, 73–76.

Ahmed, K., Raja, N.U., Afza, M., Khalique, F. & Malik, B.A. (1988) Bio-efficacy of four strains of *Bacillus thuringiensis* Berliner against hairy caterpillar *Diacrisia obliqua* Wlk. *Pakistan J. Sci. Ind. Res.,* **31**, 637–639.

Ahmed, S.M., Nagamma, M.V. & Majumder, S.K. (1973) Studies on granular formulations of *Bacillus thuringiensis* Berliner. *Pest. Sci.,* **4**, 19–23.

Aizawa, K. (1975) Selection and strain improvement of insect pathogenic micro-organisms for microbial control. In: *Approaches to Biological Control* (eds. K. Yasumatsu & H. Mori), pp. 99–105. University of Tokyo Press, Tokyo.

Akhmedov, A.M. (1982) The Turkestan vapourer - a serious pest of orchards in the Zeravshan Plain. *Iz. Akad. Nauk Tadzhik. SSR Biol. Nauk,* **2**, 79–80.

Akhurst, R.J., Lyness, E.W., Zhang, Q.Y., Cooper, D.J. & Pinnock, D.E. (1997) A 16S rRNA gene oligonucleotide probe for identification of *Bacillus thuringiensis* isolates from sheep fleece. *J. Invertebr. Pathol.,* **69**, 24–31.

Akiba, Y. (1986a) Microbial ecology of *Bacillus thuringiensis* VI. Germination of *Bacillus thuringiensis* spores in the soil. *Appl. Entomol. Zool.,* **21**, 76–80.

Akiba, Y. (1986b) Microbial ecology of *Bacillus thuringiensis*. VII. Fate of *Bacillus thuringiensis* in larvae of the silkworm, *Bombyx mori*, and the fall webworm, *Hyphantria cunea*. *Jap. J. Appl. Entomol. Zool.,* **30**, 99–105.

Akiba, Y. (1991) Assessment of rainwater-mediated dispersion of field-sprayed *Bacillus thuringiensis* in the soil. *Appl. Entomol. Zool.,* **26**, 477–483.

Akiba, Y. & Katoh, K. (1986) Microbial ecology of *Bacillus thuringiensis*. V. Selective medium for *Bacillus thuringiensis* vegetative cells. *Appl. Entomol. Zool.,* **21**, 210–215.

Akiba, Y., Sekijima, Y., Aizawa, K. & Fujiyoshi, N. (1977) Microbial ecological studies on *Bacillus thuringiensis*. II. Dynamics of *Bacillus thuringiensis* in sterilised soil. *Jap. J. Appl. Entomol. Zool.,* **21**, 41–46.

Akiba, Y., Sekijima, Y., Aizawa, K. & Fujiyoshi, N. (1980) Microbial ecological studies on *Bacillus thuringiensis*. IV. Growth of the bacterium in soils of mulberry plantations. *Jap. J. Appl. Entomol. Zool.,* **24**, 13–17.

Al Badry, M.S., Hensley, S.D.& Henderson, M.T. (1972) Comparison of *Bacillus thuringiensis* Berliner and azinphos-methyl for control of the sugarcane borer *Diatraea saccharalis* (F.). *Proc. Intern Soc. Sugar Cane Techn.,* New Orleans, Louisiana, October 22-November 5, 1971, 526–531.

Al Hafidh, E.M.T. (1985) The integration of *Nosema whitei* and some insecticides on *Tribolium castaneum*. *Thesis, University of Newcastle upon Tyne; UK,* 241pp.

Al Shayji, Y., Shaheen, N., Saleem, M. & Ibrahim, M. (1998) The efficacy of some *Bacillus thuringiensis* formulations against the whitefly *Bemisia tabaci* (Homoptera: Aleyrodidae). *Kuwait J. Sci. Eng.,* **25**, 223–229.

Aldebis, H.K., Vargus-Osuna, E. & Santiago-Alvarez, C. (1996) Ecological study of *Bacillus thuringiensis* on soils all over Spain. *Abst. Soc. Invertebr. Pathol. 29th Ann. Meet. 3rd Intern. Coll.* Bacillus thuringiensi*s, Universidad de Cordoba.* 2.

Alekseev, A.N., Sokolova, E.I., Kosovskykh, V.L., Khorkhordin, E.G., Rasnitsyn, S.P., Ganushkina, L.A. & Bikunova, A.N. (1983) The influence of larval density in the mosquito *Aedes aegypti* on their mortality under the action of preparations of *Bacillus thuringiensis* Berl. *Med. Parazitol. Parazit. Bolezni,* **52**, 78–80.

Alfonso, J., Coll, Y., Armas, R., Pujol, M., Ayala, J.L., De la Riva, G. & Selman Housein, G. (1994) Identification of a strain of *Bacillus thuringiensis* with high insecticidal activity against the maize stem borer *Spodoptera frugiperda* (J.E. Smith). *Centro Agricola,* **21**, 19–25.

Ali, A. (1981) *Bacillus thuringiensis* serovar *israelensis* (ABG-6108) against chironomids and some nontarget aquatic invertebrates. *J. Invertebr. Pathol.,* **38**, 264–272.

Ali, A. & Young, S.Y. (1996) Activity of *Bacillus thuringiensis* Berliner against different ages and stages of *Helicoverpa zea* (Lepidoptera: Noctuidae) on cotton. *J. Entomol. Sci.,* **31**, 1–8.

Ali, A., Baggs, R.D. & Stewart, J.P. (1981) Susceptibility of some Florida chironomids and mosquitoes to various formulations of *Bacillus thuringiensis* serovar *israelensis*. *J. Econ. Entomol.,* **74**, 672–677.

Ali, A., Abdellatif, M.A., Bakry, N.M. & El Sawaf, S.K. (1973) Studies on biological control of the greater wax moth, *Galleria mellonella*. I. Susceptibility of wax moth larvae and adult honeybee workers to *Bacillus thuringiensis*. *J. Apicult. Res.,* **12**, 117–123.

Ali, A., Weaver, M.S. & Cotsenmayer, E. (1989) Effectiveness of *Bacillus thuringiensis* serovar *israelensis* (Vectobac 12 AS) and *Bacillus sphaericus* 2362 (ABG-6232) against *Culex* spp. mosquitoes in a dairy lagoon in central Florida. *Florida Entomol.*, **72**, 585–591.

Ali, A.S.A. & Watson, T.F. (1982a) Effects of *Bacillus thuringiensis* var. *kurstaki* on tobacco budworm (Lepidoptera: Noctuidae) adult and egg stages. *J. Econ. Entomol.*, **75**, 596–598.

Ali, A.S.A. & Watson, T.F. (1982b) Survival of tobacco budworm (Lepidoptera: Noctuidae) larvae after short-term feeding periods on cotton treated with *Bacillus thuringiensis*. *J. Econ. Entomol.*, **75**, 630–632.

Ali, S.M., Saleh, M.B. & Merdan, A.I. (1993) A field survey of bacteriophage contamination of mosquito breeding places, inhibiting bacterial insecticide. *J. Egypt. Soc. Parasit.*, **23**, 389–397.

Aliev, D.G. & Yakubov, Z.B. (1971) Against the oriental fruit moth. *Zaschc. Rast.*, **16**, 44.

AliNiazee, M.T. (1974) Evaluation of *Bacillus thuringiensis* against *Archips rosanus* (Lepidoptera: Tortricidae). *Canad. Entomol.*, **106**, 393–398.

AliNiazee, M.T. (1986) The European winter moth as a pest of filberts: damage and chemical control. *J. Entomol. Soc. Br. Col.*, **83**, 6–12.

AliNiazee, M.T. & Jensen, F.L. (1973) Microbial control of the grape leaffolder with different formulations of *Bacillus thuringiensis*. *J. Econ. Entomol.*, **66**, 157–158.

Alizadeh, M.H.S. (1977) Study of the pathological effect of the bacterium *Bacillus thuringiensis* on the larvae of *Leucoma wiltshirei* Collen. *Entomol. Phytopathol. App.*, **43**, 58–65.

Allsopp, P.G., Chilcott, C.N. & McGhie, T.K. (1996) Activity of proteins from two New Zealand strains of *Bacillus thuringiensis* against larvae of *Antitrogus consanguineus* (Blackburn) (Coleoptera: Scarabaeidae). *Aust. J. Entomol.*, **35**, 107–112.

Alm, S.R., Villani, M.G., Yeh, T. & Shutter, R. (1997) *Bacillus thuringiensis* serovar *japonensis* strain Buibui for control of Japanese and oriental beetle larvae (Coleoptera: Scarabaeidae). *Appl. Entomol. Zool.*, **32**, 477–484.

Almela Pons, G.R., Domenech, H. & Martini, N.U. (1972) Action of *Bacillus thuringiensis* Berliner on *Oiketicus kirbyi* Guild. in a laboratory test. *Rev. Fac. Cienc. Agrar. Univ. Nac. Cuyo*, **18**, 55–60.

Alrubeai, H.F. (1988) Susceptibility of *Ectomyelois ceratoniae* to *Bacillus thuringiensis* isolates under laboratory and field conditions. *J. Agr. Water Resource Res. Plant Prod.*, **7**, 125–136.

Altahtawy, M.M. & Abaless, I.M. (1973) Signal and symptomatological responses of *Spodoptera littoralis* (Boisd.) to Thuricide 90 TS Flowable alone and associated with Dipterex or Nuvacron. *Z. Angew. Entomol.*, **74**, 373–383.

Alten, B. & Bosgelmez, A. (1990) Effectiveness of several *Bacillus* isolates on *Culex laticinctus* Edwards (Diptera: Culicidae) larvae under natural conditions. *Doga Turk Zooloji Dergisi*, **14**, 252–262.

Alves, S.B., Laranjeiro, A.J. & Alves, J.E.M. (1988) Forest vigilance and the study of suppressing agents of pests. *IPEF Inst. Pesquis. Estud. Florestais*, **38**, 50–52.

Aly, C. (1988) Filter feeding of mosquito larvae (Dipt., Culicidae) in the presence of the bacterial pathogen *Bacillus thuringiensis* var. *israelensis*. *J. Appl. Entomol.*, **105**, 160–166.

Aly, C. & Mulla, M.S. (1987) Effect of two microbial insecticides on aquatic predators of mosquitoes. *J. Appl. Entomol.*, **103**, 113–118.

Ameen, A.O., Fuxa, J.R. & Richter, A.R. (1998) Antagonism between formulations of different *Bacillus thuringiensis* subspecies in *Heliothis virescens* and *Helicoverpa zea* (Lepidoptera: Noctuidae). *J. Entomol. Sci.*, **33**, 129–134.

Amend, J. & Basedow, T. (1997) Combining release/establishment of *Diadegma semiclausum* (Hellen) (Hym., Ichneumonidae) and *Bacillus thuringiensis* Berl. for control of *Plutella xylostella* (L.) (Lep., Yponomeutidae) and other lepidopteran pests in the Cordillera Region of Luzon (Philippines). *J. Appl. Entomol.*, **121**, 337–342.

Ames, B.N., Cantwell, G.E., Gingrich, R.E. & Kunz, S.E. (1982) Activity of a "thermostable exotoxin" of *Bacillus thuringiensis* subsp. *morrisoni* in the *Salmonella*/microsomal assay for bacterial mutagenicity. *J. Invertebr. Pathol.*, **40**, 350–358.

Amonkar, S.V., Kulkarni, U. & Anand, A. (1985) Comparative toxicity of *Bacillus thuringiensis* subspecies to *Spodoptera litura* (F.). *Curr. Sci.*, **54**, 475–478.

Amonkar, S.V., Pal, A.K., Vijayalakshmi, L. & Rao, A.S. (1979) Microbial control of potato tuber moth (*Phthorimaea operculella* Zell.). *Indian J. Exp. Biol.*, **17**, 1127–1133.

Amsheev, R.M. (1991) Ecological problems of the maintenance and protection of sea buckthorn in the Buryat SSR. *Sibirskii Biol. Zh.*, **2**, 42–45.

Andermatt, M., Mani, E., Wildbolz, T. & Luthy, P. (1988) Susceptibility of *Cydia pomonella* to *Bacillus thuringiensis* under laboratory and field conditions. *Entomol. Exp. Appl.*, **49**, 291–295.

Anderson, J.F. & Kaya, H.K. (1975) Biological control of the elm spanworm, *Ennomos subsignarius*. *VIII Intern. Plant Prot. Cong., Moscow*, 8–15.

Andersson A., Einer Granum, P. & Ronner, U. (1998). The adhesion of *Bacillus cereus* spores to epithelial cells might be an additional virulence mechanism. *Intern. J. Food Micr.*, **39**, 93–99.

Andrashchuk, V.V. (1981) Susceptibility of the adult sea-buckthorn fly to entomopathogenic microorganisms. *Izv. Sibir. Otdel. Akad. Nauk SSSR, Seriya Biol. Nauk*, **15**, 119–125.

Andreadis, T.G., Dubois, N.R., Moore, R.E.B., Anderson, J.F. & Lewis, F.B. (1983) Single applications of high concentrations of *Bacillus thuringiensis* for control of gypsy moth (Lepidoptera: Lymantriidae) populations and their impact on parasitism and disease. *J. Econ. Entomol.*, **76**, 1417–1422.

Angeli, G., Forti, D. & Cappelletti, C. (1998) Side effects on *Orius* of some pesticides used on garden flowers. *Colture Protette*, **27**, 73–77.

Angelini, A. & Couilloud, R. (1972) Biological control measures against certain pests of cotton and a view of integrated control on the Ivory Coast. *Coton Fibres Trop.*, **27**, 283–289.

Anglade, P., Stengel, M., Russ, K.F & Berger, H. (1980) Integrated protection in maize culture. *Proc. Intern. Symp. IOBC-WPRS Integr. Cont. Agri. For.* Vienna, 8-12 October 1979, 231–234.

Angus, T.A. (1954) A bacterial toxin paralysing silkworm larvae. *Nature*, **173**, 545–546.

Angus, T.A. (1956) General characteristics of certain insect pathogens related to *Bacillus cereus*. *Canad. J. Microbiol.*, **2**, 111–121.

Anishchenko, B.I., Torchik, M.V. & Fleisher, O.G. (1982) Against the nun moth. *Zaschc. Rast.*, **4**, 17.

Anonymous (1972) INTA studied the biological control of *Hylesia nigricans* (Berg.) with an entomopathogenic bacterium. *Idia*, **290**, 40.

Anonymous (1976) Fresh market tomato research 1975. *Res. Rep. Univ. Calif. Veg. Crops Ser.*, **176**, 125 pp.

Anonymous (1980) A new variety of *Bacillus thuringiensis* Berliner. *Acta Microbiol. Sinica*, **20**, 1–5.

Anonymous (1985a) Demonstration of air spraying technique in control of *Dendrolimus punctatus* with Liquid agent 6. *For. Sci. Tech. Linye Keji Tongxun*, **3**, 22–25.

Anonymous (1987a) Aerial application of *Bacillus thuringiensis* var. *galleriae* for the control of *Dendrolimus punctatus*. *Chin. J. Biol. Cont.*, **3**, 61–65.

Anonymous, (1987b) Screening insecticides for cabbage webworm control. Progress Report, Asian Vegetable Research and Development Center, Shanhua, Taiwan.

Antonelli, A. & Collman, S.J. (1993) Pest management options for silver-spotted tiger moth. *Ext. Bull. Coop. Ext., Coll. Agri. Home Econ., Washington State Univ.*, EB1718, 2 p.

Antonelli, A.L., LaGasa, E. & Bay, E.C. (1989) Apple ermine moth. *Ext. Bull. Coop. Ext., Coll. Agri. Home Econ., Washington State Univ.*, EB1526, 2 p.

Appel, H.M. & Schultz, J.C. (1994) Oak tannins reduce effectiveness of Thuricide (*Bacillus thuringiensis*) in the gypsy moth (Lepidoptera: Lymantriidae). *J. Econ. Entomol.*, **87**, 1736–1742.

Appleby, J.E., Bristol, P. & Eickhorst, W.E. (1975) Control of the fall cankerworm. *J. Econ. Entomol.*, **68**, 233–234.

Araujo Coutinho, C. & Lacey, L.A. (1990) Control of Simuliidae with flowable concentrate formulations of *Bacillus thuringiensis*. *Bol. Ofic. Sanit. Panamer.*, **108**, 213–219.

Arellano, A., Cooper, D.J., Smart, M. & Pinnock, D.E. (1990) Evidence of a new *Bacillus thuringiensis* toxin active against the Australian sheep blowfly *Lucilia cuprina*. *Proc. Abst. Vth Intern. Coll. Invertebr. Pathol. Microb. Con.*, Adelaide, Australia, 291.

Aronson, A.I. (1993) The two faces of *Bacillus thuringiensis*: insecticidal proteins and post-exponential survival. *Mol. Microbiol.*, **7**, 489–496.

Aronson, A.I. & Beckman, W. (1987) Transfer of chromosomal genes and plasmids in *Bacillus thuringiensis*. *Appl. Environ. Microbiol.*, **53**, 1525–1530.

Arpaia, S. (1996) Ecological impact of *Bt*-transgenic plants: 1. Assessing possible effects of CryIIIB toxin on honey bee (*Apis mellifera* L.) colonies. *J. Genet. Breed.*, **50**, 315–319.

Arthur, F.H. & Brown, S.L. (1994) Evaluation of diatomaceous earth (Insecto) and *Bacillus thuringiensis* formulations for insect control in stored peanuts. *J. Entomol. Sci.*, **29**, 176–182.

Asano, S. & Hori, H. (1995) Enhancing effects of supernatants from various cultures of *Bacillus thuringiensis* on larvicidal activity of delta-endotoxin against the common cutworm, *Spodoptera litura*. *Appl. Entomol. Zool.*, **30**, 369–374.

Asano, S. & Seki, A. (1994) Comparative bioactivity of several formulations of *Bacillus thuringiensis* toxins against diamondback moth, *Plutella xylostella* (Linnaeus) (Lepidoptera: Yponomeutidae) using a diet incorporation method. *Appl. Entomol. Zool.*, **29**, 285–288.

Asano, S., Indrasith, L.S. & Hori, H. (1995) Synergism in larvicidal activity between supernatant and pellet from *Bacillus thuringiensis* culture. *Appl. Entomol. Zool.*, **30**, 153–158.

Asano, S., Sakakibara, H., Kitagaki, T., Nakamura, K. & Matsushita, Y. (1973a) Laboratory and field evaluation of the effectiveness of *Bacillus thuringiensis* products 'Thuricide' on some lepiopterous pests of crucifers. *Jap. J. Appl. Entomol. Zool.*, **17**, 91–96.

Asano, S., Nakamura, K. & Matsushita, Y. (1973b) Some biological effects of a *Bacillus thuringiensis* product on the gypsy moth larvae. *Jap. J. Appl. Entomol. Zool.*, **17**, 141–146.

Asano, S., Sakakibara, H. & Nakamura, K. (1976) Susceptibility of the pine caterpillar, *Dendrolimus spectabilis* Butler (Lepidoptera, Lasiocampidae) to *Bacillus thuringiensis*. *Kontyu*, **44**, 217–227.

Asano, S., Hori, H. & Cui, Y.L. (1994) A unique insecticidal activity in *Bacillus thuringiensis* growth medium. *Appl. Entomol. Zool.*, **29**, 39–45.
Asano, S., Iwasa, T. & Seki, A. (1997) Silkworm assay for *Bacillus thuringiensis* formulations using the diet incorporation method. 1. Evaluation based on larval mortality. *Jap. J. Appl. Entomol. Zool.*, **41**, 187–194.
Asokan, R. & Mohan, K.S. (1996) Safety of sporeless mutant of *Bacillus thuringiensis* Berliner subsp. *kurstaki* to *Cotesia plutellae* Kurdj., a larval parasitoid of *Plutella xylostella* (L.). *Pest Manag. Hort. Ecosyst.*, **2**, 45–48.
Asylbaeva, N.S., Fedorova, S.Z. & Romasheva, L.F. (1977) The effect of bacterial preparations of the group of *Bacillus thuringiensis* on the fowl parasite *Goniocotes hologaster* and the reaction of the organism to the introduction of cultures. *Biologicheskie Metody Bor'by s Krovososushchimi Nasekomymi i Kleshchami*, 38–46.
Attathom, T., Chongrattanmeteekul, W., Chanpaisang, J. & Siriyan, R. (1995) Morphological diversity and toxicity of delta-endotoxin produced by various strains of *Bacillus thuringiensis*. *Bull. Entomol. Res.*, **85**, 167–173.
Atwood, D.W., Robinson, J.V., Meisch, M.V., Olson, J.K. & Johnson, D.R. (1992) Efficacy of *Bacillus thuringiensis* var. *israelensis* against larvae of the southern buffalo gnat, *Cnephia pecuarum* (Diptera: Simuliidae), and the influence of water temperature. *J. Amer. Mosq. Con. Assoc.*, **8**, 126–130.
Atwood, D.W., Young, S.Y., III & Kring, T.J. (1995) Development of *Cotesia marginiventris* in *Bacillus thuringiensis* exposed *Heliothis virescens* larvae. *1995 Proc. Beltwide Cotton Conf.*, USA, January 4-7, 1995: Volume 2., 853–855.
Atwood, D.W., Young, S.Y., III & Kring, T.J. (1996) Interactions of *Cotesia marginiventris* parasitization and field applied *Bacillus thuringiensis*, Thiodicarb, and their combination on tobacco budworm mortality and parasitoid emergence. *1996 Proc. Beltwide Cotton Conf.*, USA, January 9-12, 1996: Volume 2., 905–908.
Atwood, D.W., Young, S.Y., III & Kring, T.J. (1997a) Development of *Cotesia marginiventris* (Hymenoptera: Braconidae) in tobacco budworm (Lepidoptera: Noctuidae) larvae treated with *Bacillus thuringiensis* and thiodicarb. *J. Econ. Entomol.*, **90**, 751–756.
Atwood, D.W., Young, S.Y., III & Kring, T.J. (1997b) Impact of *Bt* and thiodicarb alone and in combination on tobacco budworm, mortality and emergence of the parasitoid *Microplitis croceipes*. *1997 Proc. Beltwide Cotton Conf.*, USA, January 6-10, 1997: Volume 2., 1305–1310.
Atwood, D.W., Young, S.Y., III & Kring, T.J. (1998) Mortality of tobacco budworm larvae (Lepidoptera: Noctuidae) and emergence of *Cotesia marginiventris* (Hymenoptera: Braconidae) exposed to *Bacillus thuringiensis* and thiodicarb alone and in combination. *J. Entomol. Sci.*, **33**, 136–141.
Aukshtikal' nene, A.M. & Imnadze, T.S. (1981) The use of new microbiological preparations against leaf-gnawing forest pests in Georgia. *Noveishie dostizheniya lesnoi entomologii po materialam USh s"ezda VEO, Vil'nyus,* 9-13 Oktyabrya 1979, 62–64.
Aukshtikal' nene, A.M., Shcherbakova, L.N. & Ovcharov, D.V. (1984) The susceptibility of lackey moths to bacterial preparations in relation to food-plant. *Noveishie dostizheniya lesnoi entomologii po materialam USh s"ezda VEO, Vil'nyus, 9-13 Oktyabrya 1979,* 174–178.
Averill, B. & Vandre, W. (eds.)(1982) *Proc. 2nd Alaska Integr. Pest Manage. Conf.*, January 21-22, 1982, Matanuska Susitna Community College.
Azizbekyan, K.R., Kuzin, A.I. & Dobrzhanskaya, E.O. (1996) A study of the range of lytic spectrum of the phages isolated from the type strains of *Bacillus thuringiensis* and the use of phagotyping for identification of the *Bacillus thuringiensis* strains. *Biotekhnologiya,* **12**, 9–13.
Azizbekyan, K.R., Kuzin, A.I., Shamshina, T.N. & Dobrzhanskaya, E.O. (1997) Morphology and protein pattern of bacteriophages isolated from type strains of *Bacillus thuringiensis*. *Microbiol. New York,* **66**, 199–202.
Babrikova, T. & Kuzmanova, I. (1984) The toxicity of biological preparations based on *Bacillus thuringiensis* to some stages of *Chrysopa septempunctata* Wesm., *Chrysopa formosa* Br. and *Chrysopa perla* L. *Gradinar. Lozar. Nauka,* **21**, 55–59.
Babrikova, T. & Lecheva, I. (1986) The effect of synthetic pyrethroids alone and in combination with Dipel on the seven-spotted ladybird (*Coccinella septempunctata* L., Coleoptera: Coccinellidae). *Pochvoz. Agrokhim. Rast. Zashc.*, **21**, 107–110.
Babrikova, T., Kuzmanova, I. & Lai, N.T. (1982) The effect of biological preparations based on *Bacillus thuringiensis* on some stages of the lacewing *Chrysopa carnea* Steph. *Gradinar. Lozar. Nauka,* **19**, 40–45.
Babu, B.R. & Krishnayya, P.V. (1998) Bioinsecticides to mitigate the load of quinalphos against cauliflower caterpillars. *Pestic. Res. J.*, **10**, 231–234.
Babu, P.C.S. & Subramaniam, T.R. (1973) Studies with *Bacillus thuringiensis* Berliner on *Spodoptera litura* Fabricius. *Madras Agr. J.*, **60**, 487–491.
Babu, P.C.S., Lakshmanan, P. & Subramaniam, T.R. (1971) Preliminary study on the efficacy of certain bacterial insecticides on the rhinoceros beetle. *Madras Agr. J.*, **58**, 511–513.
Back, C., Boisvert, J., Lacoursiere, J.O. & Charpentier, G. (1985) High-dosage treatment of a Quebec stream with *Bacillus thuringiensis* serovar *israelensis*: efficacy against black fly larvae (Diptera: Simuliidae) and impact on non-target insects. *Canad. Entomol.*, **117**, 1523–1534.
Baganich, M.I. (1976) The yellow noctuid on oak. *Zaschc. Rast.*, **9**, 38.

Bagci, H. & Shareef, S.R. (1989) Photoreactivation in *Bacillus thuringiensis*. *Doga Turk Biyoloji Dergisi,* **13**, 123–131.

Baggiolini, M., Descoins, C., Baillod, M., Touzeau, J., Simon, J.L. & Schmid, A. (1976) Integrated control in viticulture. *Rev. Suisse Vitic. Arboric. Hortic.*, **8**, 147–160.

Bai, C. & Degheele, D. (1993) Larvicidal activity of spores, crystals and mixtures of spores and crystals of *Bacillus thuringiensis* to *Spodoptera exempta* and hydrolysis of the crystals. *Parasitica,* **49**, 17–25.

Bai, C., Yi, S.X. & Degheele, D. (1992) The comparative potency of commercial *Bacillus thuringiensis* formulations to larvae of *Spodoptera exempta* (Walker) (Lepidoptera: Noctuidae). *Parasitica,* **48**, 35–42.

Bai, C., Degheele, D., Jansens, S. & Lambert, B. (1993) Activity of insecticidal crystal proteins and strains of *Bacillus thuringiensis* against *Spodoptera exempta* (Walker). *J. Invertebr. Pathol.*, **62**, 211–215.

Baicu, T. (1980) Current problems in the integrated control of pests and diseases of field crops. *Cereale si Plante Tehnice, Productia Vegetala,* **32**, 21–23.

Baicu, T. & Hussein, S.M. (1984) Joint action of mixtures of insecticides with preparations of *Bacillus thuringiensis* Berliner against different insect pests. *Bulet. Prot. Plantelor,* **4**, 55–62.

Bailey, J.B. & Olsen, K.N. (1990) Supplemental chemical control for omnivorous looper on avocados. *Calif. Agr.*, **44**, 8–9.

Bailey, P., Baker, G. & Caon, G. (1996) Field efficacy and persistence of *Bacillus thuringiensis* var. *kurstaki* against *Epiphyas postvittana* (Walker) (Lepidoptera: Tortricidae) in relation to larval behaviour on grapevine leaves. *Aust. J. Entomol.*, **35**, 297–302.

Baklanova, O.V., Lappa, N.V. & Doroshenko, N.N. (1990) Biological investigation on the potato moth and its sensitivity to microbial pesticides. *Zaschc. Rast. Kiev,* **37**, 38–42.

Baki, A., Ali, M.H. & Al Jubury, A.R.Y. (1988) Bioassay of certain insecticides against larvae of beet armyworm *Spodoptera exigua* (Hub.) (Lepidoptera: Noctuidae). *Mesopotamia J. Agr.*, **20**, 345–363.

Bakr, H.A., El Husseini, M.M., Merdan, A.I. (1986) Breeding water and mosquito strain as factors influencing susceptibility of *Culex pipiens* L. to *Bacillus thuringiensis* serotype H-14. *J. Egypt. Soc. Parasit.*, **16**, 235–241.

Balaraman, K., Hoti, S.L. & Manonmani, L.M. (1981) An indigenous virulent strain of *Bacillus thuringiensis*, highly pathogenic and specific to mosquitoes. *Curr. Sci.*, **50**, 199–200.

Balaraman, K., Balasubramanian, M. & Jambulingam, P. (1983a) Field trial of *Bacillus thuringiensis* H-14 (VCRC B-17) against *Culex* and *Anopheles* larvae. *Indian J. Med. Res.*, **77**, 38–43.

Balaraman, K., Balasubramanian, M. & Manonmani, L.M. (1983b) *Bacillus thuringiensis* H-14 (VCRC B-17) formulation as mosquito larvicide. *Indian J. Med. Res.*, **77**, 33–37.

Balazs, K., Bujaki, G. & Farkas, K. (1996) Incorporation of apple clearwing (*Synanthedon myopaeformis* Bork.) control into the IPM system of apple. *Bull. OILB-SROP,* **19**, 134–139.

Balazs, K., Molnar, M., Bujaki, G., Gonda, I., Karacsony, D. & Bartha, J. (1997) Possibility and problems of organic apple growing in Hungary. *Entomol. Res. Organic Agri,* **15**, 223–232.

Balevski, A. & Ivanov, S. (1979) Control of leafrollers - damagers of fruit in apple orchards. *Rastit. Zashc.*, **27**, 32–35.

Bal'man, R.A., Smirnova, T.A. & Azizbekyan, R.R. (1988) Phages of *Bacillus thuringiensis* H1 *insectus*. *Biotekhnologiya,* **4**, 46–52.

Banaszak, R. & Szmidt, A. (1987) Control of *Larix decidua* cone and seed pests in seed orchards. *Sylwan,* **131**, 13–20.

Band, R.J., Pinnock, D.E., Jackson, K.L. & Milstead, J.E. (1976) Viable spore count as an index of effective dose of *Bacillus thuringiensis*. *J. Invertebr. Pathol.*, **27**, 141–148.

Baranovskii, V.I., Zurabova, E.R., Larionov, G.V., Omelenchuk, S.V. & Bakhvalov, S.A. (1986) Lepidocide for control of *Dendrolimus sibiricus*. *Lesnoe Khoz.*, **1**, 62–64.

Baranovskii, V.I., Remorov, V.V. & Lamikhov, K.L. (1988) Ecological aspects of using micro-organisms against *Dendrolimus sibiricus*. *Lesnoe Khoz.*, **8**, 54–55.

Bari, M.A. & Kaya, H.K. (1984) Evaluation of the entomogenous nematode *Neoaplectana carpocapsae* (=*Steinernema feltiae*) Weiser (Rhabditida: Steinernematidae) and the bacterium *Bacillus thuringiensis* Berliner var. *kurstaki* for suppression of the artichoke plume moth (Lepidoptera: Pterophoridae). *J. Econ. Entomol.*, **77**, 225–229.

Barker, J.F. (1998) Effect of *Bacillus thuringiensis* subsp. *kurstaki* toxin on the mortality and development of the larval stages of the banded sunflower moth (Lepidoptera: Cochylidae). *J. Econ. Entomol.*, **91**, 1084–1088.

Barreiro, J.M. & Santiago Alvarez, C. (1985) Laboratory study of the susceptibility of *Ocnogyna baetica* (Lepidoptera: Arctiidae) to *Bacillus thuringiensis*. *Bol. Ser. Def. contra Plagas Inspeccion Fitopatol.*, **11**, 173–177.

Barry, J.W. & Ekblad, R.B. (1978) Deposition of insecticide drops on coniferous foliage. *Trans. ASAE,* **21**, 438–441.

Bartlett, A.C., Dennehy, T.J. & Antilla, L. (1997) An evaluation of resistance to *Bt* toxins in native populations of the pink bollworm. *1997 Proc. Beltwide Cotton Conf.*, USA, January 6-10, 1997: Volume 2., 885–888.

Bartninkaite, I.S. (1983) Susceptibility of larvae of the lackey moth to Entobakterin-3 according to age. *Acta Entomol. Lit.*, **6**, 46–54.

Bartninkaite, I.S. & Babonas, I.L. (1985) Susceptibility of the Colorado beetle to microbial preparations. *Acta Entomol. Lit.*, **8**, 78–86.

Bartninkaite, I. & Ziogas, A. (1996) Dynamics of elimination of entomopathogenic bacteria included in the composition of the preparation Foray 48B in the forest following its industrial application. *Ekologija,* **2**, 8–16.

Barton, W.E., Noblet, R. & Kurtak, D.C. (1991) A simple technique for determining relative toxicities of *Bacillus thuringiensis* var. *israelensis* formulations against larval blackflies (Diptera: Simuliidae). *J. Amer. Mosq. Con. Assoc.,* **7**, 313–315.

Baskaran, P. & Kumar, A. (1980) Further studies on Dipel-insecticide combinations against the insect pests of brinjal. *Pesticides,* **14**, 9–11.

Bastian, R.A. & Hart, E.R. (1989) First-generation parasitism of the mimosa webworm (Lepidoptera: Plutellidae) by *Elasmus albizziae* (Hymenoptera: Eulophidae) in an urban forest. *Environ. Entomol.,* **19**, 409–414.

Batalova, T.S. (1970) The biological method of control of the codling moth. *Zaschc. Rast.,* **15**, 22–23.

Battisti, L., Green, B.D. & Thorne, C.B. (1985) Mating system for transfer of plasmids among *Bacillus anthracis*, *Bacillus cereus*, and *Bacillus thuringiensis*. *J. Bacteriol.,* **162**, 543–50.

Bauer, L.S. (1992) Response of the imported willow leaf beetle to *Bacillus thuringiensis* var. *san diego* on poplar and willow. *J. Invertebr. Pathol.,* **59**, 330–331.

Bauernfeind, R.J. & Wilde, G.E. (1993) Control of army cutworm (Lepidoptera: Noctuidae) affects wheat yields. *J. Econ. Entomol.,* **86**, 159–163.

Baum, D. (1986) Field trials for controlling the European grape berry moth (*Lobesia botrana* Schiff.) and the honeydew moth (*Cryptoblabes gnidiella* Mill.) in vineyards. *Alon Hanotea,* **40**, 795–799.

Baur, M.E., Kaya, H.K. & Strong, D.R. (1998a) Foraging ants as scavengers on entomopathogenic nematode-killed insects. *Biological Control,* **12**, 231–236.

Baur, M.E., Kaya, H.K., Tabashnik, B.E. & Chilcutt, C.F. (1998b) Suppression of diamondback moth (Lepidoptera: Plutellidae) with an entomopathogenic nematode (Rhabditida: Steinernematidae) and *Bacillus thuringiensis* Berliner. *J. Econ. Entomol.,* **91**, 1089–1095.

Beck, W.R. (1982) Current status of the biological larvicide, Teknar. *Proc. 69th Ann. Meet. New Jersey Mosq. Cont. Assoc.,* Atlantic City, New Jersey, 17-19 March 1982, 83–90.

Becker, N. (1992) Community participation in the operational use of microbial control agents in mosquito control programs. *Bull. Soc. Vector Ecol.,* **17**, 114–118.

Becker, N. & Margalit, J. (1993) Use of *Bacillus thuringiensis israelensis* against mosquitoes and blackflies. In: *Bacillus thuringiensis, an Environmental Biopesticide: Theory and Practice* (eds. P.F. Entwistle, J.S. Cory, M.J. Bailey & S. Higgs), pp. 147–170. John Wiley and Sons, Chichester.

Becker, N., Zgomba, M., Ludwig, M., Petric, D. & Rettich, F. (1992) Factors influencing the activity of *Bacillus thuringiensis* var. *israelensis* treatments. *J. Amer. Mosq. Con. Assoc.,* **8**, 285–289.

Beckwith, R.C. & Stelzer, M.J. (1987) Persistence of *Bacillus thuringiensis* in two formulations applied by helicopter against the western spruce budworm (Lepidoptera: Tortricidae) in north central Oregon. *J. Econ. Entomol.,* **80**, 204–207.

Becnel, J.J., Garcia, J. & Johnson, M. (1996) Effects of three larvicides on the production of *Aedes albopictus* based on removal of pupal exuviae. *J. Amer. Mosq. Con. Assoc.,* **12**, Part 1, 499–502.

Beebee, T.J.C. & Bond, R.P.M. (1973) Effect of the exotoxin of *Bacillus thuringiensis* on normal and ecdysone-stimulated ribonucleic acid polymerase activity in intact nuclei from the fat-body of *Sarcophaga bullata* larvae. *Biochem. J.,* **136**, 1–7.

Beegle, C.C. (1996) Efficacy of *Bacillus thuringiensis* against lesser grain borer, *Rhyzopertha dominica* (Coleoptera: Bostrichidae). *Biocontr. Sci. Technol.,* **6**, 15–21.

Beegle, C.C., Pedigo, L.P., Poston, F.L. & Stone, J.D. (1973) Field effectiveness of the granulosis virus of the green cloverworm as compared with *Bacillus thuringiensis* and selected chemical insecticides on soybean. *J. Econ. Entomol.,* **66**, 1137–1138.

Beegle, C.C., Dulmage, H.T., Wolfenbarger, D.A. & Martinez, E. (1981) Persistence of *Bacillus thuringiensis* Berliner insecticidal activity on cotton foliage. *Environ. Entomol.,* **10**, 400–401.

Beegle, C.C., Couch, T.L., Alls, R.T., Versoi, P.L. & Lee, B.L. (1986) Standardization of HD-1-S-1980: U.S. standard for assay of lepidopterous-active *Bacillus thuringiensis*. *Bull. Entomol. Soc. Amer.,* **32**, 44–45.

Beehler, J.W., Quick, T.C. & DeFoliart, G.R. (1991) Residual toxicity of four insecticides to *Aedes triseriatus* in scrap tires. *J. Amer. Mosq. Con. Assoc.,* **7**, 121–122.

Beglyarov, G.A. & Maslienko, L.V. (1978) The toxicity of certain pesticides to *Encarsia*. *Zaschc. Rast.,* **11**, 36–37.

Begunov, V.I. & Storozhkov Yu, V. (1986) We are developing the biological method. *Zaschc. Rast. Moskva,* **9**, 8–9.

Behle, R.W., McGuire, M.R. & Shasha, B.S. (1996) Extending the residual toxicity of *Bacillus thuringiensis* with casein-based formulations. *J. Econ. Entomol.,* **89**, 1399–1405.

Behle, R.W., McGuire, M.R. & Shasha, B.S. (1997a) Effects of sunlight and simulated rain on residual activity of *Bacillus thuringiensis* formulations. *J. Econ. Entomol.,* **90**, 1560–1566.

Behle, R.W., McGuire, M.R., Gillespie, R.L. & Shasha, B.S. (1997b) Effects of alkaline gluten on the insecticidal activity of *Bacillus thuringiensis*. *J. Econ. Entomol.,* **90**, 354–360.

Beitia, F., Garrido, A. & Castaner, M. (1991) Mortality produced by various pesticides applied to eggs of *Diglyphus isaea* (Walker) (Hym.: Eulophidae) in laboratory tests. *Ann. Appl. Biol.,* **118**, 16–17.

Bejer, B. (1986) Outbreaks of nun moth (*Lymantria monacha* L.) in Denmark with remarks on their control. *Anz. Schaedlingskd. Pflanzenschutz Umweltschutz,* **59**, 86–89.

Bejer Petersen, B. (1974) Outbreaks of *Lymantria monacha* in Denmark. *Dansk Skovforen. Tidsskr.,* **59**, 59–80.

Bekheit, H.K.M., Moawad, G.M., El Bedawy, R.A., Mabrouk, M.A., El Halim, S.M.A. & Mahgoub, M.M. (1997) Control of the potato tuber moth, *Phthorimaea operculella* (Zeller) in potato crop. *Egypt. J. Agri. Res.,* **75**, 923–938.

Bekheit, S. S. (1984). Laboratory trials with *Bacillus thuringiensis* serotype H-14 in controlling mosquito larvae. *J. Egypt. Soc. Parasit.* **14,** 71–76.

Bell, M.R. & Romine, C.L. (1980) Tobacco budworm field evaluation of microbial control in cotton using *Bacillus thuringiensis* and a nuclear polyhedrosis virus with a feeding adjuvant. *J. Econ. Entomol.,* **73**, 427–430.

Bell, M.R. & Romine, C.L. (1982) Cotton leafperforator (Lepidoptera: Lyonetiidae): effect of two microbial insecticides on field populations. *J. Econ. Entomol.,* **75**, 1140–1142.

Bell, M.R. & Romine, C.L. (1986) *Heliothis virescens* and *H. zea* (Lepidoptera: Noctuidae): dosage effects of feeding mixtures of *Bacillus thuringiensis* and a nuclear polyhedrosis virus on mortality and growth. *Environ. Entomol.,* **15**, 1161–1165.

Bellocq, M.I., Bendell, J.F. & Cadogan, B.L. (1992) Effects of the insecticide *Bacillus thuringiensis* on *Sorex cinereus* (masked shrew) populations, diet and prey selection in a jack pine plantation in northern Ontario. *Canad. J. Zool.,* **70**, 505–510.

Bellotti, A., Arias, B. (1978) Biology, ecology and biological control of the cassava hornworm (*Erinnyis ello*). *Proceedings cassava protection workshop CIAT* (eds. T. Brekelbaum, A. Bellotti & J.C. Lozano), Cali, Colombia, 7-12 Nov., 1977, 227–232.

Bellows, T.S., Jr. & Morse, J.G. (1993) Toxicity of insecticides used in citrus to *Aphytis melinus* DeBach (Hymenoptera: Aphelinidae) and *Rhyzobius lophanthae* (Blaisd.) (Coleoptera: Coccinellidae). *Canad. Entomol.,* **125**, 987–994.

Bellows, T.S., Jr., Morse, J.G. & Gaston, L.K. (1992) Residual toxicity of pesticides used for control of lepidopteran insects in citrus to the predaceous mite *Euseius stipulatus* Athias-Henriot (Acarina, Phytoseiidae). *J. Appl. Entomol.,* **113**, 493–501.

Bellows, T.S., Jr., Morse, J.G. & Gaston, L.K. (1993) Residual toxicity of pesticides used for lepidopteran insect control on citrus to *Aphytis melinus* Debach (Hymenoptera: Aphelinidae). *Canad. Entomol.,* **125**, 995–1001.

Benavides Gomez, M. & Cardenas Murillo, R. (1975) Effect of various insecticides on the control of the coffee leaf-miner, *Leucoptera coffeella* (Guerin-Meneville) (Lepidoptera: Lyonetiidae). *Cenicafe,* **26**, 151–160.

Benfatto, D. (1981) Necessity for control against the citrus shoot leaf-roller and use of chemical products. *Inf. Fitopatol.,* **31**, 23–27.

Benuzzi, M. & Antoniacci, L. (1995) Recent successes in biological and integrated control strategies on strawberry. *Riv. Fruttic. Ortofloric.,* **57**, 63–65.

Benz, G. (1971) Synergism of micro-organisms and chemical insecticides. *Microbial Control of Insects and Mites* (eds. H.D. Burges & N.W. Hussey), pp. 327–355. Academic Press, London.

Benz, G. (1975) Action of *Bacillus thuringiensis* preparation against Larch bud moth, *Zeiraphera diniana* (Gn) enhanced by beta-exotoxin and DDT. *Experientia,* **31**, 1288–1290.

Benz, G. & Altwegg, A. (1975) Safety of *Bacillus thuringiensis* for earthworms. *J. Invertebr. Pathol.,* **26**, 125–126.

Benz, G. & Joeressen, H.J. (1994) A new pathotype of *Bacillus thuringiensis* with pathogenic action against sawflies (Hymenoptera, Symphyta). *Bull. OILB-SROP,* **17**, 35–38.

Benz, G. & Lebrun, P. (1976) Study of the synergistic action of chemical and biological insecticides on an experimental population of *Tenebrio molitor* (Col.: Tenebrionidae). *Entomophaga,* **21**, 141–150.

Berdegue, M. & Trumble, J.T. (1997) Interaction between linear furanocoumarins found in celery and a commercial *Bacillus thuringiensis* formulation on *Spodoptera exigua* (Lepidoptera: Noctuidae) larval feeding behavior. *J. Econ. Entomol.,* **90**, 961–966.

Berdegue, M., Trumble, J.T. & Moar, W.J. (1996) Effect of CryIC toxin from *Bacillus thuringiensis* on larval feeding behavior of *Spodoptera exigua*. *Entomol. Exp. Appl.,* **80**, 389–401.

Berendt, O., Stenseth, C., Svensson, G. & Tittanen, K. (1973) The biological control of the glasshouse spider mite, *Tetranychus urticae* Koch (Acarina: Tetranychidae). 2. The use of pathogens and incompatible genes. *Ugeskr. Agron. Hortonomer,* **21**, 360–364.

Berliner, E. (1911) Uber die Schlaffsucht der Mehlmottenraupe. *Z. Gesamte Getreidewe. (Berlin),* **3**, 63–70.

Berliner, E. (1915) Über die Schaffsucht der Mehlmottenraupe (*Ephestia kuhniella* Zell.). *Z. Angew. Entomol.,* **2**, 29–56.

Berndt, K.P., Schmied, G. & Nitschmann, J. (1974) Evaluation of *Bacillus thuringiensis* for the control of Pharaoh's ant. *Angew. Parasit.,* **15**, 223–224.

Bernhard, K., Jarrett, P., Meadows, M., Butt, J., Ellis, D.J., Roberts, G.M., Pauli, S., Rodgers, P. & Burges, H.D. (1997) Natural isolates of *Bacillus thuringiensis*: worldwide distribution, characterization, and activity against insect pests. *J. Invertebr. Pathol.,* **70**, 59–68.

Bernier, R.L., Jr., Gannon, D.J., Moser, G.P., Mazzarello, M., Griffiths, M.M. & Guest, P.J. (1990) Development of a novel *Bt* strain for the control of forestry pests. *Brighton Crop Protection Conference, Pests and Diseases,* Vol. 1, 245–251.

Berti Filho, E. & Gallo, D. (1977) The use of *Bacillus thuringiensis* Berliner for the control of the palm worm *Brassolis astyra astyra* Godart, 1765 (Lepidoptera, Brassolidae). *An. Soc. Entomol. Bras.*, **6**, 85–91.

Bertoldo, N.G. & Morosini, S. (1982) Use of three commercial formulations of *Bacillus thuringiensis* for the control of the velvetbean caterpillar *Anticarsia gemmatalis* Hubner, 1818. *Agron. Sulriogr.*, **18**, 19–23.

Beveridge, N. & Elek, J.A. (1999) *Bacillus thuringiensis* var. *tenebrionis* shows no toxicity to the predator *Chauliognathus lugubris* (F.) (Coleoptera: Cantharidae). *Aust. J. Entomol.*, **38**, 34–39.

Bezvesil' nyu, V.A. (1983) Feeding of hole-nesting birds in an oak stand sprayed with Dendrobacillin for the control of leaf-eating pests. *Lesovods. Agrolesomelior.*, **66**, 63–65.

Bhateshwar, N.R. & Mandal, B.C. (1991) Efficacy of newer microbial formulations against different mosquito larvae. *Proc. Symp. Entomol. Defence Services.* (eds. P.K. Ramachandran, D. Sukumaran & S.S. Rao), pp. 169–174.

Bidochka, M.J., Selinger, L.B. & Khachatourians, G.G. (1987) A *Bacillus thuringiensis* isolate found on grapes imported from California. *J. Food Prot.*, **50**, 857–858.

Biever, K.D. & Hostetter, D.L. (1975) *Bacillus thuringiensis*: against lepidopterous pests of wine grapes in Missouri. *J. Econ. Entomol.*, **68**, 66–68.

Bishop, A.H., Johnson, C. & Perani, M. (1999) The safety of *Bacillus thuringiensis* to mammals investigated by oral and subcutaneous dosage. *World J. Microbiol. Biotech.*, **15**, 375–380.

Bishop, E.J., Helms, T.J. & Ludwig, K.A. (1973) Control of bagworm with *Bacillus thuringiensis*. *J. Econ. Entomol.*, **66**, 675–676.

Biswas, S., Upadhyay, K.D. & Kumar, A. (1994) Bio-efficacy of various *Bacillus thuringiensis* formulations and dosages against hairy caterpillar, *Spilosoma* (*Diacrisia*) *obliqua*. *J. Ecotoxicol. Environ. Monitor.*, **4**, 185–188.

Biswas, S., Upadhyay, K.D. & Dubey, R.K. (1996a) Effect of Dipel (*Bacillus thuringiensis* var. *kurstaki*) on different larval instars of *Spilosoma obliqua* Walker. *Ann. Plant Prot. Sci.*, **4**, 165–166.

Biswas, S., Ashok, K., Upadhyay, K.D. & Kumar, A. (1996b) Effect of sub-lethal concentration of Dipel on the post embryonic development of *Spilosoma obliqua*. *Indian J. Entomol.*, **58**, 359–363.

Biswas, S., Upadhyay, K.D., Ashok, K. & Kumar, A. (1996c) Efficacy of some insecticides alone and in combination with Dipel (a bacterial formulation) against *Spilosoma obliqua* Walker. *Z. Angew. Zool.*, **81**, 227–235.

Blackshaw, R.P. (1984a) Studies on the chemical control of vine weevil larvae. *1984 British Crop Protection Conference,* England, November 19-22, Volume 3, 1093–1098.

Blackshaw, R.P. (1984b) An evaluation of three insecticides as compost treatments for the control of vine weevil larvae in hardy-annual nursery stock. *J. Hort. Sci.*, **59**, 559–562.

Blaustein, L. & Margalit, J. (1991) Indirect effects of the fairy shrimp, *Branchipus schaefferi* and two ostracod species on *Bacillus thuringiensis* var. *israelensis*-induced mortality in mosquito larvae. *Hydrobiologia,* **212**, 67–76.

Bleicher, E., de Jesus, F.M.M. & de Sousa, S.L. (1990) Use of selective insecticides in the control of the cotton leaf worm. *Pesqu. Agropecu. Bras.*, **25**, 277–280.

Blumberg, D., Navon, A., Keren, S., Goldenberg, S. & Ferkovich, S.M. (1997) Interactions among *Helicoverpa armigera* (Lepidoptera: Noctuidae), its larval endoparasitoid *Microplitis croceipes* (Hymenoptera: Braconidae), and *Bacillus thuringiensis*. *J. Econ. Entomol.*, **90**, 1181–1186.

Bohorova, N., Maciel, A.M., Brito, R.M., Aguilart, L., Ibarra, J.E. & Hoisington, D. (1996) Selection and characterization of Mexican strains of *Bacillus thuringiensis* active against four major lepidopteran maize pests. *Entomophaga,* **41**, 153–165.

Bolanos, A.E., Mendoza Zamora, C. & Resendiz Garcia, B. (1991) Identification and control of mites contaminating fungal cultures. *Rev. Mex. Fitopatol.*, **9**, 14–15.

Boldyrev, M.I. (1970) Integrated control of apple leaf miner. *Sb. Nauchn. Rab. Vses. Nauchno-issled. Inst. Sadovod. I.V. Michurina,* **14**, 323–326.

Boller, E. & Remund, U. (1981) Grape moth in East Switzerland in 1981. Forecasts and hints for its chemical and biological control. *Schweiz. Z. Obst Weinb.*, **117**, 424–430.

Bolotnikova, V.V. (1984) Geometrids. *Zaschc. Rast.*, **10**, 52.

Boman, H.G., Faye, I., Pye, A., Rasmuson, T., Bulla, L.A., Jr. & Cheng, T.C. (1978) The inducible immunity system of giant silk moths. *Compar. Pathobiol.*, **4**, 145–163.

Bone, L.W. (1989) Activity of commercial *Bacillus thuringiensis* preparations against *Trichostrongylus colubriformis* and *Nippostrongylus brasiliensis*. *J. Invertebr. Pathol.*, **53**, 276–277.

Bone, L.W. & Coles, G.C. (1987) Lethal synergism between benzimidazoles and *Bacillus thuringiensis israelensis* toxin for *Trichostrongylus colubriformis* eggs. *J. Parasit.*, **73**, 1059–1060.

Bonnefoi, A. & de Barjac, H. (1963) Classification des souches du groupe *Bacillus thuringiensis* par la determination de l'antigene flagellaire. *Entomophaga,* **8**, 223–229.

Bora, R.S., Murty, M.G., Shenbagarathai, R., Vaithilingam, S. & Sekar, V. (1994) Introduction of a Lepidoptera-specific insecticidal crystal protein gene of *Bacillus thuringiensis* subsp. *kurstaki* by conjugal transfer into a *Bacillus megaterium* strain that persists in the cotton phyllosphere. *Appl. Environ. Microbiol.*, **60**, 214–222.

Borah, M. & Basit, A. (1996) Effect of certain insecticides on the emergence of *Trichogramma japonicum* Ashmead. *J. Agr. Sci. Soc. North East India*, **9**, 224–225.

Borgonie, G., Claeys, M., Leyns, F., Arnaut, G., de Waele, D. & Coomans, A. (1996a) Effect of nematicidal *Bacillus thuringiensis* strains on free-living nematodes. 1. Light microscopic observations, species and biological stage specificity and identification of resistant mutants of *Caenorhabditis elegans*. *Fund. Appl. Nematol.*, **19**, 391–398.

Borgonie, G., Claeys, M., Leyns, F., Arnaut, G., de Waele, D. & Coomans, A. (1996b) Effect of a nematicidal *Bacillus thuringiensis* strain on free-living nematodes. 3. Characterization of the intoxication process. *Fund. Appl. Nematol.*, **19**, 523–528.

Borgonie, G., Claeys, M., Leyns, F., Arnaut, G., de Waele, D. & Coomans, A. (1996c) Effect of nematicidal *Bacillus thuringiensis* strains on free-living nematodes. 2. Ultrastructural analysis of the intoxication process in *Caenorhabditis elegans*. *Fund. Appl. Nematol.*, **19**, 407–414.

Borisevich, L.V. (1998) The effectiveness of biopreparations against the tobacco thrips. *Zashc. Karantin Rast.*, **6**, 28.

Borthakur, M.C. & Raghunathan, A.N. (1987) Biological control of tea looper with *Bacillus thuringiensis*. *J. Coffee Res.*, **17**, 120–121.

Botelho, P.S.M., Nakano, O. & Rodella, R.J. (1975) Action of some insecticides on the cucurbit borer *Margaronia nitidalis* (Cramer, 1782). *Rev. Agr. Piracicaba*, **50**, 157–161.

Botha, J.H., Du Plessis, D. & Calitz, F.J. (1994) Contact toxicity of some pesticides to *Oligota fageli* (Bernhauer), a predatory beetle in deciduous fruit orchards. *J. South. African Soc. Hort. Sci.*, **4**, 50–51.

Bottjer, K.P., Bone, L.W. & Gill, S.S. (1985) Nematoda: susceptibility of the egg to *Bacillus thuringiensis* toxins. *Exp. Parasit.* **60**, 239–244.

Boucias, D.G. & Pendland, J.C. (1998) *Principles of Insect Pathology*. Kluwer Academic Publishers, Norwell, Massachusetts.

Boyd, M.L. & Boethel, D.J. (1998) Susceptibility of predaceous hemipteran species to selected insecticides on soybean in Louisiana. *J. Econ. Entomol.*, **91**, 401–409.

Bradley, R.S., Stuart, G.S., Stiles, B. & Hapner, K.D. (1989) Grasshopper haemagglutinin: immunochemical localization in haemocytes and investigation of opsonic properties. *J. Insect Physiol.*, **35**, 353–361.

Brewer, G.J. (1991) Resistance to *Bacillus thuringiensis* subsp. *kurstaki* in the sunflower moth (Lepidoptera: Pyralidae). *Environ. Entomol.*, **20**, 316–322.

Brewer, G.J. & Anderson, M.D. (1990) Modification of the effect of *Bacillus thuringiensis* on sunflower moth (Lepidoptera: Pyralidae) by dietary phenols. *J. Econ. Entomol.*, **83**, 2219–2224.

Briceno, A.J. (1997) Perspectives of an integrated pest management system against the banana green worm, *Opsiphanes tamarindi* Felder (Lepidoptera: Brassolidae). *Revta. Fac. Agron. Uni. Zulia*, **14**, 487–495.

Broadley, R.H., Giles, J.E., Adams, G.D. & Halfpapp, K.H. (1979) Control of *Heliothis* spp. (Lepidoptera: Noctuidae) larvae on flue-cured tobacco in north Queensland. *Queensland J. Agr. Animal Sci.*, **36**, 125–132.

Brousseau, C., Charpentier, G. & Belloncik, S. (1998) Effects of *Bacillus thuringiensis* and destruxins (*Metarhizium anisopliae* mycotoxins) combinations on spruce budworm (Lepidoptera: Tortricidae). *J. Invertebr. Pathol.*, **72**, 262–268.

Brousseau, R., Saint Onge, A., Prefontaine, G., Masson, L. & Cabana, J. (1993) Arbitrary primer polymerase chain reaction, a powerful method to identify *Bacillus thuringiensis* serovars and strains. *Appl. Environ. Microbiol.*, **59**, 114–119.

Brown, K.L. & Whiteley, H.R. (1992) Molecular characterization of two novel crystal protein genes from *Bacillus thuringiensis* subsp. *thompsoni*. *J. Bacteriol.*, **174**, 549–557.

Brown, M.D., Thomas, D., Watson, K., Greenwood, J.G. & Kay, B.H. (1996) Acute toxicity of selected pesticides to the estuarine shrimp *Leander tenuicornis* (Decapoda: Palaemonidae). *J. Amer. Mosq. Con. Assoc.*, **12**, 721–724.

Brown, M.D., Thomas, D. & Kay, B.H. (1998) Acute toxicity of selected pesticides to the Pacific Blue-eye, *Pseudomugil signifer* (Pisces). *J. Amer. Mosq. Con. Assoc.*, **14**, 463–466.

Brownbridge, M. (1990) Evaluation of *Bacillus thuringiensis* for the control of cereal stem borers. *Proc. Abst. Vth Intern. Coll. Invertebr. Pathol. Microb. Con.*, Adelaide, Australia, 495.

Brownbridge, M. (1991) Native *Bacillus thuringiensis* isolates for the management of lepidopteran cereal pests. *Insect Sci. Its Applic.*, **12**, 57–61.

Brownbridge, M. & Onyango, T. (1992a) Laboratory evaluation of four commercial preparations of *Bacillus thuringiensis* (Berliner) against the spotted stem borer, *Chilo partellus* (Swinhoe) (Lep., Pyralidae). *J. Appl. Entomol.*, **113**, 159–167.

Brownbridge, M. & Onyango, T. (1992b) Screening of exotic and locally isolated *Bacillus thuringiensis* (Berliner) strains in Kenya for toxicity to the spotted stem borer, *Chilo partellus* (Swinhoe). *Trop. Pest Manag.*, **38**, 77–81.

Broza, M. & Sneh, B. (1994) *Bacillus thuringiensis* spp. *kurstaki* as an effective control agent of lepidopteran pests in tomato fields in Israel. *J. Econ. Entomol.*, **87**, 923–928.

Broza, M., Brownbridge, M. & Sneh, B. (1991a) Monitoring secondary outbreaks of the African armyworm in Kenya using pheromone traps for timing of *Bacillus thuringiensis* application. *Crop Prot.*, **10**, 229–233.

Broza, M., Brownbridge, M., Hamal, M. & Sneh, B. (1991b) Control of the African armyworm *Spodoptera exempta* Walker (Lepidoptera: Noctuidae) in Kenyan fields with highly effective strains of *Bacillus thuringiensis* Berliner. *Biocontr. Sci. Technol.*, **1**, 127–135.

Bryant, J.E. (1994) Application strategies for *Bacillus thuringiensis*. *Agric. Ecosys. Environ.*, **49**, 65–75.

Buchanan, G.A. (1977) The seasonal abundance and control of light brown apple moth, *Epiphyas postvittana* (Walker) (Lepidoptera: Tortricidae), on grapevines in Victoria. *Aust. J. Agr. Res.*, **28**, 125–132.

Buckner, C.H., Ray, D.G.H. & McLeod, B.B. (1973) The effects of pesticides on small forest vertebrates of the Spruce Woods Provincial Forest, Manitoba. *Manitoba Entomol.*, **7**, 37–45.

Buckner, C.H., Gochnauer, T.A. & McLeod, B.B. (1975) The impact of aerial spraying of insecticides on bees. *Aerial Control of Forest Insects in Canada*, (ed. M.L. Prebble), pp. 276–279, Department of the Environment; Ottawa; Canada.

Buess, U. & Bassand, D. (1976) Experiments with Thuricide HP as a measure against grapevine tortricids. *Meded. Fac. Landbouwwet. Rijksuniv. Gent*, **41**, 919–925

Bull, D.L., House, V.S., Ables, J.R. & Morrison, R.K. (1979) Selective methods for managing insect pests of cotton. *J. Econ. Entomol.*, **72**, 841–846.

Bulukhto, N.P. & Korotkova, A.A. (1988) The yellow redcurrant sawfly. *Zaschc. Rast. Moskva*, **3**, 35.

Burgerjon, A. (1972) Some physiological effects of the thermostable toxin of *Bacillus thuringiensis* on the Colorado potato beetle *Leptinotarsa decemlineata*. *Entomol. Exp. Appl.*, **15**, 112–127.

Burgerjon, A. & Dulmage, H. (1977) Industrial and international standardization of microbial pesticides - I. *Bacillus thuringiensis*. *Entomophaga*, **22**, 121–129.

Burgerjon, A. & Martouret, D. (1971) Determination and significance of the host spectrum of *Bacillus thuringiensis*. *Microbial Control of Insects and Mites* (eds. H.D. Burges & N.W. Hussey), pp. 305–325. Academic Press, London.

Burgerjon, A., Yamvrias, C., Charmoille, L., Vincent, B., D'Oultremont, P., Debacq, J.J. & Prunet, P. (1974) The bioassay of commercial preparations of *Bacillus thuringiensis* Berliner with the aid of *Anagasta kuhniella* Z. and *Musca domestica* L. *Phytiat. Phytopharm.*, **23**, 223–234.

Burges, H.D. (1967a) Standardization of *Bacillus thuringiensis* products: homology of the standard. *Nature* **215 (5101)**, 664–665.

Burges, H.D. (1967b). The standardization of products based on Bacillus thuringiensis. *Proc. Int. Colloq. Insect Path., Microbial Control*, Wageningen, 1966, 306–314.

Burges, H.D. *et al.* (1967). The standardization of *Bacillus thuringiensis*: tests on three candidate reference materials. *Proc. Int. Colloq. Insect Pathol. Microbial Control*, Wageningen, 1966, 314–338.

Burges, H.D. (1976a) Persistence of *Bacillus thuringiensis* in foundation beeswax and beecomb in beehives for the control of *Galleria mellonella*. *J. Invertebr. Pathol.*, **28**, 217–222.

Burges, H.D. (1976b) Techniques for the bioassay of *Bacillus thuringiensis* with *Galleria mellonella*. *Entomol. Exp. Appl.*, **19**, 243–254.

Burges, H.D. (1976c) Leaching of *Bacillus thuringiensis* spores from foundation beeswax into honey and their subsequent survival. *J. Invertebr. Pathol.*, **28**, 393–394.

Burges, H.D. (1978) Control of wax moths: physical, chemical and biological methods. *Bee World*, **59**, 129–138.

Burges, H.D. (ed.)(1981) *Microbial Control of Pests and Plant Diseases 1970-80*. Academic Press, New York, 949 p.

Burges, H.D. (2000) Techniques for testing microbials for control of arthropod pests in greenhouses. *Field Manual of Techniques for Evaluation and Application of Insect Pathogens*. Kluwer Academic Publishers, Dordrecht (in press).

Burges, H.D. & Hurst, J.A. (1977) Ecology of *Bacillus thuringiensis* in storage moths. *J. Invertebr. Pathol.*, **30**, 131–139.

Burges, H.D. & Jarrett, P. (1976) Adult behaviour and oviposition of five noctuid and tortricid moth pests and their control in glasshouses. *Bull. Entomol. Res.*, **65**, 501–510.

Burges, H.D. & Jarrett, P. (1978) Caterpillar control with *Bacillus thuringiensis*. *Grower*, **90**, 589–590, 593–595.

Burges, H.D. & Jarrett, P. (1980) Application and distribution of *Bacillus thuringiensis* for control of tomato moth in glasshouses. *Proc. 1979 Brit. Crop Prot. Conf. Pests and Diseases*, 19-22 November 1979, Brighton, England, 433–439.

Burges, H.D. & Jones, K.A. (1998) Formulation of Bacteria, Viruses and Protozoa to Control Insects. *Formulation of Microbial Biopesticides; Beneficial Microorganisms, Nematodes and Seed Treatments* (ed. H.D. Burges), pp. 33–127. Kluwer Academic Publishers, Dordrecht, The Netherlands.

Burges, H.D., Hillyer, S. & Chanter, D.O. (1975) Effect of ultraviolet and gamma rays on the activity of delta -endotoxin protein crystals of *Bacillus thuringiensis*. *J. Invertebr. Pathol.*, **25**, 5–9.

Burgio, G., Ferrari, R. & Maini, S. (1992) Laboratory trials with a *Bacillus thuringiensis* Berliner ssp. *tenebrionis*-based formulation against *Gonioctena fornicata* (Brugg.). *Inf. Fitopatol.*, **427**, 45–47.

Burtseva, L.I., Burlak, V.A., Kalmikova, G.V., de Barjac, H. & Lecadet, M.M. (1995) *Bacillus thuringiensis novosibirsk* (serovar H24a24c) a new subspecies from the West Siberian plain. *J. Invertebr. Pathol.*, **66**, 92–93.

Buryi, V.F., Buraya, P.K. & Kosenkova, T.V. (1991) Biopreparations in rice-growing in the Maritime Territory. *Zaschc. Rast.*, **6**, 9.

Bushkovskaya, L.M., Shevtsov, V.V. & Shchelokova, E.V. (1994) Astur for protection of medicinal plants. *Zaschc. Rast. Moskva,* **5**, 9.

Butani, D.K. (1979) Insect pests of citrus and their control. *Pesticides,* **13**, 15–21, 31–33.

Butler, L., Zivkovich, C. & Sample, B.E. (1995) Richness and abundance of arthropods in the oak canopy of West Virginia's Eastern Ridge and Valley Section during a study of impact of *Bacillus thuringiensis* with emphasis on macrolepidoptera larvae. *Bull. Agr. For. Exp. Station West Virginia Univ.,* **711**, 19 p.

Butter, N.S., Battu, G.S., Kular, J.S., Singh, T.H. & Brar, J.S. (1995) Integrated use of *Bacillus thuringiensis* Berliner with some insecticides for the management of bollworms on cotton. *J. Entomol. Res.,* **19**, 255–263.

Byzov, B.A., Claus, H. Tretyakova, E.B., Ryabchenko, N.F., Mozgovaya, I.N. Zvyagintsev, D.G. & Filip, Z. (1999). Plasmid transfer between introduced and indigenous bacteria in leaf litter, soil and vermicompost as affected by soil invertebrates. *Biology and Fertility of Soils,* **28**, 169–176.

Cabral, M.T. (1973) Activity of nine *Bacillus thuringiensis* strains on *Lymantria dispar* compared. *Zast. Bilja,* **24**, 197–203.

Cabral, M. (1977) The influence of the food of *Lymantria dispar* L. and of *Euproctis chrysorrhoea* L. on the effect of *Bacillus thuringiensis*. *An. Inst. Super. Agron.,* **37**, 179–221.

Cabral, M. (1978) Possibility of using microbiological control of *Lymantria dispar* L. and *Euproctis chrysorrhoea* L., two cork-oak pests by means of *Bacillus thuringiensis*. *Cienc. Biol.,* **3**, 20.

Cabral, M. (1980) The influence of the food of *Lymantria dispar* L. and of *Euproctis chrysorrhoea* L. on the effect of *Bacillus thuringiensis*. *An. Inst. Super. Agron.,* **37**, 179–221.

Cadapan, E.P. & Gabriel, B.P. (1972) Field evaluation of Dipel in comparison with other commercial *Bacillus thuringiensis* and chemical insecticides against *Plutella xylostella* (L.) and other insect pests of cabbage. *Philipp. Entomol.,* **2**, 297–305.

Cadogan, B.L. (1993) Field weights of jack pine budworm larvae (Lepidoptera: Tortricidae) surviving aerial applications of *Bacillus thuringiensis* and two chemical insecticides. *Proc. Entomol. Soc. Ontario,* **124**, 189–196.

Cadogan, B.L., Zylstra, B.F., Nystrom, C., Ebling, P.M. & Pollock, L.B. (1986) Evaluation of a new Futura formulation of *Bacillus thuringiensis* on populations of jack pine budworm, *Choristoneura pinus pinus* (Lepidoptera: Tortricidae). *Proc. Entomol. Soc. Ontario,* **117**, 59–64.

Cadogan, B.L., Nealis, V.G. & Van Frankenhuyzen, K. (1995) Control of spruce budworm (Lepidoptera: Tortricidae) with *Bacillus thuringiensis* applications timed to conserve a larval parasitoid. *Crop Prot.,* **14**, 31–36.

Cameron, E.A. & Reeves, R.M. (1990) Carabidae (Coleoptera) associated with gypsy moth, *Lymantria dispar* (L.) (Lepidoptera: Lymantriidae), populations subjected to *Bacillus thuringiensis* Berliner treatments in Pennsylvania. *Canad. Entomol.,* **122**, 123–129.

Cameron, R.S. (1989) Promising new pesticides for cone and seed insect control in the southern United States. *Proc. 3rd Cone Seed Insects Working Party Conf.,* British Columbia, Canada, on 26-30 June 1988, 193–202.

Campbell, C.D., Walgenbach, J.F. & Kennedy, G.G. (1991) Effect of parasitoids on lepidopterous pests in insecticide-treated and untreated tomatoes in western North Carolina. *J. Econ. Entomol.,* **84**, 1662–1667.

Campbell, D.P., Dieball, D.E. & Brackett, J.M. (1987) Rapid HPLC assay for the beta-exotoxin of *Bacillus thuringiensis*. *J. Agric. Food Chem.,* **35**, 156–158.

Campbell, J.B. & Wright, J.E. (1976) Field evaluations on insect growth regulators, insecticides, and a bacterial agent for stable fly control in feedlot breeding areas. *J. Econ. Entomol.,* **69**, 566–568.

Campos, A.R. & Gravena, S. (1984) Insecticides, *Bacillus thuringiensis* and predacious arthropods for the control of budworms on cotton. *An. Soc. Entomol. Bras.,* **13**, 95–105.

Cane, J.H., Cox, H.E. & Moar, W.J. (1995) Susceptibility of *Ips calligraphus* (Germar) and *Dendroctonus frontalis* Zimmermann (Coleoptera: Scolytidae) to coleopteran-active *Bacillus thuringiensis*, a *Bacillus* metabolite, and avermectin B1. *Canad. Entomol.,* **127**, 831–837.

Cangardel, H. (1971) Research on the protection of maize ears intended for seed production against the second-generation caterpillars of the European corn borer, *Ostrinia nubilalis* Hbn. *Rev. Zool. Agri. Pathol. Veg.,* **70**, 7–19.

Canivet, J.P., Nef, L. & Lebrun, P. (1978) Mixtures of *Bacillus thuringiensis* with reduced doses of chemical insecticides against *Euproctis chrysorrhoea*. *Z. Angew. Entomol.,* **86**, 85–97.

Cantwell, G.E. & Cantelo, W.W. (1982) Potential of *Bacillus thuringiensis* as a microbial agent against the Mexican bean beetle. *J. Econ. Entomol.,* **75**, 348–350.

Cantwell, G.E. & Cantelo, W.W. (1984) Effectiveness of *Bacillus thuringiensis* var. *israelensis* in controlling a sciarid fly, *Lycoriella mali*, in mushroom compost. *J. Econ. Entomol.,* **77**, 473–475.

Cantwell, G.E. & Franklin, B.A. (1966) Inactivation by irradiation of spores of *Bacillus thuringiensis* var. *thuringiensis*. *J. Invertebr. Pathol.,* **8**, 256–258.

Cantwell, G.E. & Shieh, T.R. (1981) Certan - a new bacterial insecticide against the greater wax moth, *Galleria mellonella* L. *Amer. Bee J.,* **121**, 424–426, 430–431.

Cantwell, G.E., Lehnert, T. & Fowler, J. (1972) Are biological insecticides harmful to the honey bee? *Amer. Bee J.,* **112**, 255–258.

Cantwell, G.E., Cantelo, W.W. & Schroder, R.F.W. (1985) The integration of a bacterium and parasites to control the Colorado potato beetle and the Mexican bean beetle. *J. Entomol. Sci.*, **20**, 98–103.
Car, M. & Kutzer, E. (1988) Field studies on the control of Simuliidae with *Bacillus thuringiensis* var. *israelensis* in the Austrian Alpenvorland. *Mitt. Dtsch. Ges. Allg. Angew. Entomol.*, **6**, 207–210.
Carballo, M., Hernandez, M. & Rutilio Quezada, J. (1989) Effect of insecticides and weeds on *Plutella xylostella* (L) and its parasitoid *Diadegma insulare* (Cress) in cabbage crop. *Manejo Integrado de Plagas,* **11**, 1–20.
Cardei, E. & Rominger, E. (1997) Research concerning phytosanitary protection of cherry tree plantation at the Iasi Fruit-growing Research Station. *Cercet. Agron. Moldova,* **30**, 133–139.
Cardona, C. (1995) Management of *Trialeurodes vaporariorum* (Westwood) on beans in the Andean zone: technical aspects, farmers attitude and technology transfer. *Ceiba,* **36**, 53–64.
Carlberg, G. & Lindstrom, R. (1987) Testing fly resistance to Thuringiensin produced by *Bacillus thuringiensis*, serotype H-1. *J. Invertebr. Pathol.*, **49**, 194–197.
Carles, L. (1984) The citrus flower moth. *Arboricult. Fruit.*, **31**, 42–43.
Carlson, C.R. & Kolsto, A.B. (1993) A complete physical map of a *Bacillus thuringiensis* chromosome. *J. Bacteriol.*, **175**, 1053–60.
Carlson, C.R., Caugant, D.A. & Kolsto, A.B. (1994) Genotypic diversity among *Bacillus cereus* and *Bacillus thuringiensis* strains. *Appl. Environ. Microbiol.*, **60**, 1719–1725.
Carlson, C.R., Johansen, T. & Kolsto, A.B. (1996) The chromosome map of *Bacillus thuringiensis* subsp. *canadensis* HD224 is highly similar to that of the *Bacillus cereus* type strain ATCC 14579. *FEMS Microbiol. Lett.*, **141**, 163–167.
Carlson, E.C. (1975) Pesticides for controlling sunflower moth larvae. *Calif. Agr.*, **29**, 12–13.
Carnegie, A.J.M. & Dick, J. (1972) Notes on sugarcane trash caterpillars (Noctuidae) and effects of defoliation on the crop. *Proc. South African Sugar Technol. Assoc.*, **46**, 160–167.
Carolin, V.M. & Thompson, C.G. (1967) Field testing of *Bacillus thuringiensis* for control of western hemlock looper. *Research Paper, Pacific Northwest Forest and Range Experiment Station, U.S. Forest Service, USDA,* **38**, 24 p.
Carozzi, N.B., Kramer, V.C., Warren, G.W., Evola, S. & Koziel, M.G. (1991) Prediction of insecticidal activity of *Bacillus thuringiensis* strains by polymerase chain reaction product profiles. *Appl. Environ. Microbiol.*, **57**, 3057–3061.
Carrillo, L.R. & Mundaca, B.N. (1975) Evaluation of *Bacillus thuringiensis* Berliner and insecticides for the control of larvae of *Maculella* spp. (Lep. Hepialidae) in pastures. *Agro. Sur.*, **3**, 54–62.
Carroll, J., Li, J. & Ellar, D.J. (1989) Proteolytic processing of a coleopteran-specific delta-endotoxin produced by *Bacillus thuringiensis* var. *tenebrionis*. *Biochem. J.*, **261**, 99–105.
Casida, L.E., Jr. (1989) Protozoan response to the addition of bacterial predators and other bacteria to soil. *Appl. Environ. Microbiol.*, **55**, 1857–1859.
Castaner, M. & Garrido, A. (1995) Contact toxicity and persistence of various insecticides to three insects used in biological control: *Cryptolaemus montrouzieri*, *Lysiphlebus testaceipes* and *Encarsia formosa*. *Invest. Agrar. Prod. Prot. Vegetales,* **10**, 139–147.
Castello Branco, A., Jr. & Andrade, C.F.S. (1992) Susceptibility of *Simulium (Chirostilbia) pertinax* Kollar, 1832 (Diptera, Simuliidae) to *Bacillus thuringiensis* var. *israelensis* in an atypical breeding habitat. *Mem. Inst. Oswaldo Cruz,* **87**, 317–318.
Castelo Branco, M. & Franca, F.H. (1995) Impact of insecticides and bioinsecticides on adults of *Trichogramma pretiosum*. *Hort. Bras.*, **13**, 199–201.
Castineiras, A. & Calderon, A. (1985) Susceptibility of *Pheidole megacephala* to three microbial insecticides: Dipel, Bitoxobacillin 202 and *Beauveria bassiana* under laboratory conditions. *Cienc. Tec. Agricult. Prot. Plantas,* Suppl., 61–66.
Castineiras, A., Neyra, M., Fernandez Larrea, O. & Ponce, E. (1991) Evaluation of strains of *Beauveria bassiana* (Bals.) Vuill., *Metarhizium anisopliae* (Metsch.) Sor. and *Bacillus thuringiensis* Berl. for biological control of *Wasmannia auropunctata* (Roger). *Rev. Proteccion Vegetal,* **6**, 21–26.
Castro Franco, R., Garcia Alvarado, J.S. & Galan Wong, L.J. (1998) An alternative bioinsecticide formulation to encapsulate *Bacillus thuringiensis* delta toxin and extracts of *Agave lecheguilla* Torr. for the control of *Spodoptera frugiperda* Smith. *Phyton Buenos Aires,* **62**, 71–77.
Cate, P.C. (1997) Biology and biological control of the caraway moth (*Depressaria nervosa* Hw.) in Austria. *Pflanzenschutz Wien,* **13**, 2–5.
Cavalcaselle, B. (1976) Evaluation of the effectiveness of two commercial preparations of *Bacillus thuringiensis* in control of *Thaumetopoea pityocampa*. *Cellulosa e Carta,* **27**, 21–26.
Cavalcaselle, B. (1978) Tests of the use of a compound based on *Bacillus thuringiensis* Berliner with the addition of chitinase for the control of the pine processionary. *Cellulosa e Carta,* **29**, 3–7.
Cayrol, J.C. (1974) Action of *Bacillus thuringiensis* toxins on the mycophagous nematode *Ditylenchus myceliophagus*. *Simposio Internacional (XII) DE Nematologia, Sociedad Europea de Nematologos,* 1-7 Sept, 1974, Granada, Spain, 21–22.
Ceber, K. (1992) Biological control of larvae of *Culex pipiens* and *Anopheles sacharovi* in Cukurova District. *Turk Hij. Deneysel Biyol. Derg.*, **49**, 1–12.

Celli, G. (1974) Conditions for the survival of the honeybee in present agricultural systems. *Annali Accad. Naz. Agric. Bologna*, **94**, 395–411.

Ceron, J., Covarrubias, L., Quintero, R., Ortiz, A., Ortiz, M., Aranda, E., Lina, L. & Bravo, A. (1994) PCR analysis of the *cryI* insecticidal crystal family genes from *Bacillus thuringiensis*. *Appl. Environ. Microbiol.*, **60**, 353–356.

Chabanenko, A.A., Bogdanova, E.N., Ermishev Yu, V. & Dremova, V.P. (1992) Efficacy of a combined preparation based on *Bacillus sphaericus* and *B. thuringiensis* H-14 against larvae of blood-sucking mosquitoes. *Med. Parazitol. Parazit. Bolezni*, **1**, 23–25.

Chadhar, S.K. (1996) Field evaluation of *Bacillus thuringiensis* (a biopesticide) in relation to control of teak skeletoniser. *Vaniki Sandesh*, **20**, 1–6.

Chahal, P.P.K. & Chahal, V.P.S. (1993) Effect of Thuricide on the hatching of eggs of root-knot nematode (*Meloidogyne incognita*). *Curr. Nematol.*, **4**, 247.

Chakrabarti, S.K., Mandaokar, A.D., Kumar, P.A. & Sharma, R.P. (1998) Synergistic effect of Cry1Ac and Cry1F delta-endotoxins of *Bacillus thuringiensis* on cotton bollworm, *Helicoverpa armigera*. *Curr. Sci.*, **75**, 663–664.

Chalfant, R.B. (1978) Chemical control of insect pests of greens in Georgia. *Res. Bull. Georgia Exp. St.*, **216**, 21.

Chan, T. (1973) *Mamestra oleracea* (L.) and measures for its control. *Rast. Zashc.*, **21**, 28–31.

Chang, D.C. & Chen, C.C. (1993) Population fluctuation of major insect pests on kidney bean and the proper time of control. *Bull. Taichung District Agric. Improv. St.*, **38**, 11–22.

Chang, L.C. (1972) Insect control on vegetables with Thuricide and other insecticides. *Taiwan Agric. Quart.*, **8**, 164–169.

Chang, M.S., Ho, B.C. & Chan, K.L. (1990) The effect of *Bacillus thuringiensis israelensis* (H-14) on emergence of *Mansonia* mosquitoes from natural breeding habitat. *Southeast Asian J. Trop. Med. Public Health*, **21**, 430–436.

Chang, S.L. & Pegn, C.C. (1971) An investigation of chemical control of some important insect pests on Cruciferae in Singapore. *Plant Prot. Bull. Taiwan*, **13**, 110–120.

Chapman, M.H. & Hoy, M.A. (1991) Relative toxicity of *Bacillus thuringiensis* var. *tenebrionis* to the two-spotted spider mite (*Tetranychus urticae* Koch) and its predator *Metaseiulus occidentalis* (Nesbitt) (Acari, Tetranychidae and Phytoseiidae). *J. Appl. Entomol.*, **111**, 147–154.

Charbonneau, C.S., Drobney, R.D. & Rabeni, C.F. (1994) Effects of *Bacillus thuringiensis* var. *israelensis* on nontarget benthic organisms in a lenthic habitat and factors affecting the efficacy of the larvicide. *Environ. Toxicol. Chem.*, **13**, 267–279.

Charest, P.J. (1996) Biotechnology in forestry: examples from the Canadian Forest Service. *For. Chron.*, **72**, 37–42.

Chari, M.S., Sreedhar, U., Rao, R.S.N. & Reddy, S.A.N. (1996) Studies on compatibility of botanical and microbial insecticides to the natural enemies of *Spodoptera litura* F. *Tob. Res.*, **22**, 32–35.

Charles, J.G., Collyer, E. & White, V. (1985) Integrated control of *Tetranychus urticae* with *Phytoseiulus persimilis* and *Stethorus bifidus* in commercial raspberry gardens. *NZ J. Exp. Agri.*, **13**, 385–393.

Chaudhry, M.I. & Shah, B.H. (1977) Laboratory trials of Dipel against poplar defoliator - *Ichthyura anastomosis* Steph. (Notodontidae: Lepidoptera). *Pakistan J. For.*, **27**, 29–32.

Chaufaux, J., Muller Cohn, J., Buisson, C., Sanchis, V., Lereclus, D. & Pasteur, N. (1997) Inheritance of resistance to the *Bacillus thuringiensis* CryIC toxin in *Spodoptera littoralis* (Lepidoptera: Noctuidae). *J. Econ. Entomol.*, **90**, 873–878.

Chen, K.S., Funke, B.R., Schulz, J.T., Carlson, R.B. & Proshold, F.I. (1974) Effects of certain organophosphate and carbamate insecticides on *Bacillus thuringiensis*. *J. Econ. Entomol.*, **67**, 471–473.

Chen, L., Hang, S. & Zhang, F. (1991a) A field experiment on the control effect of "Mie-e-ling", a *Bacillus thuringiensis* preparation against *Plutella xylostella* (in China). *Chin. J. Biol. Cont.*, **7**, 140.

Chen, H.F., Zhang, H.F., Xue, S.D., Deng, P.J. & Li, G.Q. (1991b) Occurrence and control of *Anomis flava* Fabr. *China's Fiber Crops*, **3**, 41–44.

Chen, S.F., Xiao, T.C. & Lu, J.F. (1984a) A study of the toxicity of *Bacillus thuringiensis* var. *israelensis* to mosquito larvae and factors affecting it. *Natural Enemies of Insects*, **6**, 115–117.

Chen, S.F., Li, J.F. & Xiao, Y.C. (1984b) Observations on the toxicity of *Bacillus thuringiensis* var. *israelensis* mixed with chemical insecticides to larvae of *Culex pipiens pallens* Coquillett. *Natural Enemies of Insects*, **6**, 153–154.

Cheng, C.C, Tang, L.C & Hou, R.F. (1998) Efficacy of the entomopathogenic nematode, *Steinernema carpocapsae* (Rhabditida: Steinernematidae), against the Asian corn borer, *Ostrinia furnacalis* (Lepidoptera: Pyralidae). *Chin. J. Entomol.*, **18**, 51–60.

Cheng, H.H. (1973) Laboratory and field tests with *Bacillus thuringiensis* against the dark-sided cutworm, *Euxoa messoria* (Lepidoptera: Noctuidae), on tobacco. *Canad. Entomol.*, **105**, 941–945.

Cheng, H.H. & Hanlon, J.J. (1990) A note on the use of microbial insecticides for the control of tomato hornworms on flue-cured tobacco. *Phytoprotection*, **71**, 101–103.

Cheng, H.Y., Ming, W.J. & Ge, Q.J. (1983) Experiments on integrated pest control of *Matsucoccus matsumurae* Kuwana. *J. Nanjing Technolog. Coll. For. Prod.*, **1**, 11–30.

Chenot, A.B. & Raffa, K.F. (1998) Effects of parasitoid strain and host instar on the interaction of *Bacillus thuringiensis* subsp. *kurstaki* with the gypsy moth (Lepidoptera: Lymantriidae) larval parasitoid *Cotesia melanoscela* (Hymenoptera: Braconidae). *Environ. Entomol.*, **27**, 137–147.

Cheong, H., Dhesi, R.K. & Gill, S.S. (1997) Marginal cross-resistance to mosquitocidal *Bacillus thuringiensis* strains in Cry11A-resistant larvae: presence of Cry11A-like toxins in these strains. *FEMS Microbiol. Lett.*, **153**, 419–424.

Chepurnaya, V.I. (1994) Formulations against pests of peach trees. *Zaschc. Rast. Moskva,* **5**, 14.

Chernev, T. (1976) Biological control of forest pests. *Rast. Zashc.*, **24**, 5–8.

Cheung, P.Y.K., Roe, R.M., Hammock, B.D., Judson, C.L. & Montague, M.A. (1985) The apparent *in vivo* neuromuscular effects of the delta-endotoxin of *Bacillus thuringiensis* var. *israelensis* in mice and insects of four orders. *Pestic. Biochem. Physiol.*, **23**, 85–94.

Cheung, W.W. & Lam, Y. (1993) Histochemical studies on normal and *Bacillus thuringiensis*-infected *Pieris canidia* larval midgut. *Bull. Inst. Zool. Acad. Sin.*, **32**, 12–22.

Cheung, W.W.K., Lam, Y.F. & Chang, S.T. (1990) Histopathological effects of *Bacillus thuringiensis* on *Pieris canidia* larva: a transmission electron microscope study. *Acta Biol. Exp. Sin.*, **23**, 177–191.

Chiang, A.S., Yen, D.F. & Peng, W.K. (1986) Germination and proliferation of *Bacillus thuringiensis* in the gut of rice moth larva, *Corcyra cephalonica*. *J. Invertebr. Pathol.*, **48**, 96–99.

Chilcott, C.N. & Ellar, D.J. (1988) Comparative toxicity of *Bacillus thuringiensis* var. *israelensis* crystal proteins in vivo and in vitro. *J. Gen. Microbiol.*, **134**, 2551–2558.

Chilcott, C.F. & Tabashnik, B.E. (1997) Independent and combined effects of *Bacillus thuringiensis* and the parasitoid *Cotesia plutellae* (Hymenoptera: Braconidae) on susceptible and resistant diamondback moth (Lepidoptera: Plutellidae). *J. Econ. Entomol.*, **90**, 397–403.

Chilcott, C. & Wigley, P. (1990) Toxicity of *Bacillus thuringiensis* against grass grub (Coleoptera: Scarabaeidae). *Proc. Abst. Vth Intern. Coll. Invertebr. Pathol. Microb. Con., Adelaide, Australia,* 342.

Chilcott, C.N. & Wigley, P.J. (1993) Isolation and toxicity of *Bacillus thuringiensis* from soil and insect habitats in New Zealand. *J. Invertebr. Pathol.*, **61**, 244–247.

Chilcott, C.N., Pillai, J.S. & Kalmakoff, J. (1983) Efficacy of *Bacillus thuringiensis* var. *israelensis* as a biocontrol agent against larvae of Simuliidae (Diptera) in New Zealand. *NZ J. Zool.*, **10**, 319–325.

Chilcott, C.N., Knowles, B.H., Ellar, D.J. & Drobniewski, F.A. (1990) Mechanism of action of *Bacillus thuringiensis israelensis* parasporal body. *Bacterial Control of Mosquitoes and Black Flies: Biochemistry, Genetics and Applications of* Bacillus thuringiensis israelensis *and* Bacillus sphaericus, (eds. H. de Barjac, & D.J. Sutherland), Unwin Hyman, London, 45–65.

Chilcutt, C.F. & Tabashnik, B.E. (1997) Host-mediated competition between the pathogen *Bacillus thuringiensis* and the parasitoid *Cotesia plutellae* of the diamondback moth (Lepidoptera: Plutellidae). *Environ. Entomol.*, **26**, 38–45.

Chilingaryan, V.A., Kazaryan, B.K., Ormanyan Zh, K. & Pogosyan, A.G. (1977) Acaricidal effectiveness of Bitoxibacillin. *Zaschc. Rast.*, **5**, 24.

Christensen, K.P. (1990a) Vectobac technical material (*Bacillus thuringiensis* variety *israelensis*): Infectivity and pathogenicity to bluegill sunfish (*Lepomis mrochirus*) during a 32 day static renewal test. Springborne Laboratories Inc. Wareham, Massachusetts. Report No. 90-2-3-228 (unpublished), pp. 1–55.

Christensen, K.P. (1990b) Vectobac technical material (*Bacillus thuringiensis* variety *israelensis*): Infectivity and pathogenicity to rainbow trout (*Oncorhyncus mykiss*) during a 32 day static renewal test. Springborne Laboratories Inc. Wareham, Massachusetts. Report No. 90-2-342 (unpublished), pp. 1–55.

Christensen, K.P. (1990c) Vectobac technical material (*Bacillus thuringiensis* variety *israelensis*): Infectivity and pathogenicity to sheephead minnow (*Cyprindon variegatus*) during a 32 day static renewal test. Springborne Laboratories Inc. Wareham, Massachusetts. Report No. 90-2-3288 (unpublished), pp. 1–57.

Christie, G.D. (1990) Salt marsh mosquito control in Portsmouth, Rhode Island. *J. Amer. Mosq. Con. Assoc.*, **6**, 144–147.

Chui, V.W.D., Koo, C.W., Lo, W.M., Qiu, X. & Qiu, X.J. (1993) Laboratory evaluation of Vectobac-12AS and teflubenzuron against *Culex* and *Aedes* mosquito larvae under different physical conditions. *Environ. Int.*, **19**, 193–202.

Chung, C.S. & Hyun, J.S. (1980) The effects of temperatures on the development of Oriental tobacco budmoth, *Heliothis assulta* Guenee, and control effects of Thuricide HP. *Korean J. Plant Prot.*, **19**, 57–65.

Chung, S.B. & Shin, S.C. (1986) Control of the black tipped-sawfly, *Acantholyda posticalis posticalis* Matsumura (Hymenoptera: Pamphilidae), with some insecticides. *Res. Rep. For. Res. Inst. Seoul,* **33**, 126–131.

Cilek, J.E. & Knapp, F.W. (1992) Distribution and control of *Chironomus riparius* (Diptera: Chironomidae) in a polluted creek. *J. Amer. Mosq. Con. Assoc.*, **8**, 181–183.

Clarke, A.R., Zalucki, M.P., Madden, J.L., Patel, V.S. & Paterson, S.C. (1997) Local dispersion of the Eucalyptus leaf-beetle *Chrysophtharta bimaculata* (Coleoptera: Chrysomelidae), and implications for forest protection. *J. Appl. Ecol.*, **34**, 807–816.

Cloutier, C. & Jean, C. (1998) Synergism between natural enemies and biopesticides: a test case using the stinkbug *Perillus bioculatus* (Hemiptera: Pentatomidae) and *Bacillus thuringiensis tenebrionis* against Colorado potato beetle (Coleoptera: Chrysomelidae). *J. Econ. Entomol.*, **91**, 1096–1108.

Coates, J.R. & O'Donnell-Jeffrey, N.L. (1979) Toxicity of four synthetic pyrethroid insecticides to rainbow trout. *Bull. Environ. Contam. Toxicol.*, **23**, 250–255.

Cohen, M., Wysoki, M. & Sneh, B. (1983) The effect of different strains of *Bacillus thuringiensis* Berliner on larvae of the giant looper, *Boarmia* (*Ascotis*) *selenaria* Schiffermuller (Lepidoptera, Geometridae). *Z. Angew. Entomol.*, **96**, 68–74.

Cokmus, C. & Elciin, Y.M. (1995) Stability and controlled release properties of carboxymethylcellulose-encapsulated *Bacillus thuringiensis* var. *israelensis*. *Pestic. Sci.*, **45**, 351–355.

Colbo, M.H. & O'Brien, H. (1984) A pilot black fly (Diptera: Simuliidae) control program using *Bacillus thuringiensis* var. *israelensis* in Newfoundland. *Canad. Entomol.*, **116**, 1085–1096.

Colbo, M.H. & Undeen, A.H. (1980) Effect of *Bacillus thuringiensis* var. *israelensis* on non-target insects in stream trials for control of Simuliidae. *Mosq. News*, **40**, 368–371.

Collingwood, E.F. & Bourdouxhe, L. (1980) Trials with decamethrin for the control of *Heliothis armigera* on tomatoes in Senegal. *Trop. Pest Manag.*, **26**, 3–7.

Connell, J.H., Zalom, F.G., Bentley, W.J., Ferguson, L. & Kester, D. (1998) Navel orangeworm control in almond with *Bacillus thuringiensis*. *Proc. 2nd Intern. Symp. Pistachios Almonds, Davis, California, USA, 24-29 August 1997*, 470, 547–552.

Coombs, R.M., Dancer, B.N., Davies, D.H., Houston, J. & Learner, M.A. (1991) The use of *Bacillus thuringiensis* var. *israelensis* to control the nuisance fly *Sylvicola fenestralis* (Anisopodidae) in sewage filter beds. *Water Res. Oxford*, **25**, 605–611.

Cooper, D.J. (1983) The susceptibility of *Etiella behrii* to *Bacillus thuringiensis*. *J. Aust. Entomol. Soc.*, **22**, 93–95.

Cooper, D.J. (1984) The application of a model to achieve predicted mortality in a field trial using *Bacillus thuringiensis* to control *Heliothis punctiger*. *Entomol. Exp. Appl.*, **36**, 253–259.

Cooper, D.J., Pinnock, D.E., Were, S.T. & Chapman, R.B. (1985) Bacterial pathogens of the Australian sheep blowfly *Lucilia cuprina*. *Proc. 4th Australasian Conf. Grassl. Invertebr. Ecol.*, Canterbury, New Zealand, 13-17 May, 1985, 236–243.

Cooper, D.J., Zhang, Q.Y., Arellano, A. & Pinnock, D.E. (1990) The effects of *Bacillus thuringiensis* var. *israelensis* on *Lucilia cuprina* larval tissue - an ultrastructural study. *Proc. Abst. Vth Intern. Coll. Invertebr. Pathol. Microb. Con.*, Adelaide, Australia, 357.

Cordero, R.J. & Cave, R.D. (1990) Parasitism of *Plutella xylostella* L. (Lepidoptera: Plutellidae) by *Diadegma insulare* (Cresson) (Hymenoptera: Ichneumonidae) in cabbages *Brassica oleracea* var. *capitata* in Honduras. *Manejo Integrado de Plagas*, **16**, 19–22.

Corino, L., Accotto, G.P. & Cattaneo, E. (1983) Infestation by *Panonychus ulmi* (Koch) following the use of insecticides against vine moths. *Atti XIII Congresso Nazionale Italiano di Entomologia*, 607–614.

Corliaa, J. (1994) Pheromone monitoring tracks borers. *Agr. Res. Washington*, **42**, 20.

Cortes, J.A. & Borrero, S. (1998) Control trials for olive moth. *Agric. Rev. Agropecuaria*, **67**, 904–906.

Cortes, M.L., Cardona, C. & Trujillo, F. (1990) Effectivity of three insecticides for the control of the bean tentiform leaf miner *Phyllonorycter* sp. (Lepidoptera: Gracillariidae) in Narino. *Rev. Colombiana Entomol.*, **16**, 12–15.

Coupland, J.B. (1993) Factors affecting the efficacy of three commercial formulations of *Bacillus thuringiensis* var. *israelensis* against species of European black flies. *Biocontr. Sci. Technol.*, **3**, 199–210.

Coventry, E. and Allan, E.J. (1997) The effect of neem-based products on bacterial and fungal growth. *Practice Oriented Results on Use and Production of Neem-Ingredients and Pheromones. Proceedings 5th Workshop* (eds. Kleeberg, H. & Zebitz, C.P.W.), Wetzlar, Germany, 22-25 Jan. 1996, 237–242

Cowan, C.B., Jr. & Davis, J.W. (1972) Insecticides evaluated in field tests against cotton insects in central Texas, 1970. *J. Econ. Entomol.*, **65**, 1111–1112.

Cranshaw, W.S., Day, S.J., Gritzmacher, T.J. & Zimmermann, R.J. (1989) Field and laboratory evaluations of *Bacillus thuringiensis* strains for control of elm leaf beetle. *J. Arboric.*, **15**, 31–34.

Crecchio, C. & Stotzky, G. (1998) Insecticidal activity and biodegradation of the toxin from *Bacillus thuringiensis* subsp. *kurstaki* bound to humic acids from soil. *Soil Biol. Biochem.*, **30**, 463–470.

Creighton, C.S. & McFadden, T.L. (1974) Complementary actions of low rates of *Bacillus thuringiensis* and chlordimeform hydrochloride for control of caterpillars on cole crops. *J. Econ. Entomol.*, **67**, 102–104.

Creighton, C.S., McFadden, T.L. & Bell, J.V. (1970) Pathogens and chemicals tested against caterpillars on cabbage. *Production Research Report, ARS/USDA*, 114, 10 pp.

Creighton, C.S., McFadden, T.L. & Cuthbert, R.B. (1973) Tomato fruitworm: control in South Carolina with chemical and microbial insecticides 1970-1971. *J. Econ. Entomol.*, **66**, 473–475.

Creighton, C.S., McFadden, T.L., Cuthbert, R.B. & Onsager, J.A. (1972) Control of four species of caterpillars on cabbage with *Bacillus thuringiensis* var. *alesti*, 1969-70. *J. Econ. Entomol.*, **65**, 1399–1402.

Creighton, C.S., McFadden, T.L. & Robbins, M.L. (1981) Comparative control of caterpillars on cabbage cultivars treated with *Bacillus thuringiensis*. *J. Georgia Entomol. Soc.*, **16**, 361–367.

Crickmore, N., Zeigler, D.R., Feitelson, J., Schnepf, E., van Rie, J., Lereclus, D., Baum, J. & Dean, D.H. (1998) Revision of the nomenclature for the *Bacillus thuringiensis* pesticidal crystal proteins. *Microbiol. Mol. Biol. Rev.*, **62**, 807–813.

Crickmore, N., Zeigler, D.R., Schnepf, E., van Rie, J., Lereclus, D., Baum, J, Bravo, A. and Dean, D.H. "*Bacillus thuringiensis* toxin nomenclature" (2000). http://www.biols.susx.ac.uk/ Home/Neil _Crickmore/Bt/index.html (February 2000).

Cristescu, A., Duport, M., Tacu, V., Durbaca, S. & Iancu, L. (1975) Contributions to the study on the biology of the *Anopheles hyrcanus* species from the Danube Delta. *Arch. Roum. Pathol. Exp. Microbiol.*, **34**, 277–284.

Crooks, M. (1975) Bacterial control of insect pests. *NZ J. Agr.*, **130**, 24–27.

Crosswhite, C.D. (1985) Damage to mescal bean (*Sophora secundiflora*) by a pyralid moth (*Uresiphita reversalis*). *Desert Plants*, **7**, 32.

Cruz, P.F.N. (1977) Preliminary results on the effectiveness of *Bacillus thuringiensis* Berliner for the control of the rubber hawkmoth (*Erinnyis ello* L.), (Lepidoptera: Sphingidae), in Bahia. *Rev. Theobroma*, **7**, 93–98.

Cui, J. and Xia, J. (1997a) Effects of transgenic *Bt* cotton on population dynamics of the main pests and their natural enemies. *Acta Agricult. Univ. Henanensis*, **31**, 351–356.

Cui, J. and Xia, J. (1997b) The effect of *Bt* transgenic cotton on the feeding function of major predators. *China Cottons*, **24**, 19.

Culbert, D.F. (1995) An IPM approach for the control of Atala (*Eumaeus atala*) on Florida coonties (*Zamia floridana*). *107th Annual meeting of the Florida State Horticultural Society*, Orlando, Florida, USA, 30 October 1 November 1994, 427–430.

Currado, I. & Brussino, G. (1985) Control experiments with *Bacillus thuringiensis* Berl. in the forests of Piedmont. *Difesa delle Piante*, **8**, 339–343.

Dabi, R.K., Gupta, H.C. & Sharma, S.K. (1980a) Bio-efficacy of *Bacillus thuringiensis* Berliner against *Euproctis lunata* Walker on pearlmillet. *Indian J. Agric. Sci.*, **50**, 356–358.

Dabi, R.K., Mehrotra, P. & Shinde, V.K.R. (1980b) Bio-efficacy of different levels of *Bacillus thuringiensis* Berliner against *Diacrisia obliqua* Walker. *J. Entomol. Res.*, **4**, 231–233.

Dabi, R.K., Puri, M.K., Gupta, H.C. & Sharma, S.K. (1988) Synergistic response of low rate of *Bacillus thuringiensis* Berliner with sub-lethal dose of insecticides against *Heliothis armigera* Hubner. *Indian J. Entomol.*, **50**, 28–31.

Daffonchio, D., Borin, S., Consolandi, A., Mora, D., Manachini, P.L. & Sorlini, C. (1998) 16S-23S rRNA internal transcribed spacers as molecular markers for the species of the 16S rRNA group I of the genus *Bacillus*. *FEMS Microbiol. Lett.*, **163**, 229–236.

Dahlsten, D.L. (1994) Elm leaf beetle management - the California experience. *Plant Prot. Quart.*, **9**, 42.

D'Aguilar, J., Large, M., Moussion, G. & Riom, J. (1978) The introduction of the webworm into France. *Phytoma*, **30**, 27–30.

Dai J.Y., Yu L., Wang B., Luo X., Yu Z. & Lecadet, M.M. (1996) *Bacillus thuringiensis* subspecies *huazonghensis*, serotype H-40, isolated from soils in the People's Republic of China. *Lett. Appl. Microbiol.*, **22**, 44–45.

Dai, J.Y., Wang, B., Luo, X.X., Yu, L., Zan, X.M. & Yu, Z.N. (1994) A study on 410 strains of *Bacillus thuringiensis* isolated from soils. *J. Huazhong Agric. Uni.*, **13**, 144–152.

Dai, L. & Wang, X. (1988) A new subspecies of *Bacillus thuringiensis*. *Microbiol. Sin.*, **28**, 301–306.

Dai, L.Y., Wang, X.P. & Ma, Z.L. (1989) The toxicity of three strains of *Bacillus thuringiensis* with different forms of parasporal crystal. *Sci. Silvae Sin.*, **25**, 29–32.

Damgaard, P.H. (1995) Diarrhoeal enterotoxin production by strains of *Bacillus thuringiensis* isolated from commercial *Bacillus thuringiensis*-based insecticides. *FEMS Immun. Med. Microbiol.*, **12**, 245–50.

Damgaard, P.H., Jacobsen, C.S. & Sorensen, J. (1996a) Development and application of a primer set for specific detection of *Bacillus thuringiensis* and *Bacillus cereus* in soil using magnetic capture hybridization and PCR amplification. *Sys. Appl. Microbiol.*, **19**, 436–441.

Damgaard, P.H., Malinowski, H., Glowacka, B. and Eilenberg, J. (1996b) Degradation of *Bacillus thuringiensis* serovar *kurstaki* after aerial application to a Polish pine stand. *Bull. OILB-SROP* **19**, 61–65.

Damgaard, P.H., Larsen, H.D., Hansen, B.M., Bresciani, J. & Jorgensen, K. (1996c) Enterotoxin-producing strains of *Bacillus thuringiensis* isolated from food. *Lett. Appl. Microbiol.*, **23**, 146–150.

Damgaard, P.H., Hansen, B.M., Pedersen, J.C. & Eilenberg, J. (1997a) Natural occurrence of *Bacillus thuringiensis* on cabbage foliage and in insects associated with cabbage crops. *J. Appl. Microbiol.*, **82**, 253–258.

Damgaard, P.H., Granum, P.E., Bresciani, J., Torregrossa, M.V., Eilenberg, J. & Valentino, L. (1997b) Characterization of *Bacillus thuringiensis* isolated from infections in burn wounds. *FEMS Immun. Med. Microbiol.*, **18**, 47–53.

Damgaard, P.H., Abdel Hameed, A., Eilenberg, J. & Smits, P.H. (1998) Natural occurrence of *Bacillus thuringiensis* on grass foliage. *World J. Microbiol. Biotech.*, **14**, 239–242.

Danguy, R. (1971) On the subject of a proliferation of the brown-tail moth in Ariege. *Phytoma*, **23**, 25–26.

Daricheva, M.A., Klyuchko, Z.F. & Sakchiev, A. (1983) A pest of tomatoes. *Zaschc. Rast.*, **7**, 44–45.

Dariichuk, Z.S. (1981) Dendrobacillin in oak forests. *Zaschc. Rast.*, **6**, 34.

Das, L.K. & Singh, B. (1998) Integrated management of jute pests. *Environ. Ecol.*, **16**, 218–219.

Das, G.P., Magallona, E.D., Raman, K.V. & Adalla, C.B. (1992) Effects of different components of IPM in the management of the potato tuber moth, in storage. *Agric. Ecosys. Environ.*, **41**, 321–325.

Das, G.P., Lagnaoui, A., Salah, H.B. & Souibgui, M. (1998) The control of the potato tuber moth in storage in Tunisia. *Trop. Sci.*, **38**, 78–80.

da Silva, M.T.B. (1987) Bioecology, damage and control of *Rachiplusia nu* (Guenee, 1852) in flax. *Rev. Cen. Cienc. Rurais,* **17**, 351–367.

da Silva, M.T.B. (1995) *Baculovirus anticarsia* associated with dosage reduction of insecticides for the control of larvae of *Anticarsia gemmatalis* (Hubner, 1818). *Cienc. Rural,* **25**, 353–358.

da Silva, R.F.P. & Heinrichs, E.A. (1975) Influence on defoliation and yield of soy bean of attack by *Anticarsia gemmatalis* Hubner, 1818 and its control with *Bacillus thuringiensis* Berliner and chlordimeform. *An. Soc. Entomol. Bras.*, **4**, 53–60.

Davidson, E.W., Sweeney, A.W. & Cooper, R. (1981) Comparative field trials of *Bacillus sphaericus* strain 1593 and *B. thuringiensis* var. *israelensis* commercial powder formulations. *J. Econ. Entomol.*, **74**, 350–354.

Davidson, E.W., Patron, R.B.R., Lacey, L.A., Frutos, R., Vey, A. & Hendrix, D.L. (1996) Activity of natural toxins against the silverleaf whitefly, *Bemisia argentifolii*, using a novel feeding bioassay system. *Entomol. Exp. Appl.*, **79**, 25–32.

Davies, P.E., (1994) Effect of BT (*Bacillus thuringiensis* var. *tenebrionis*) on stream macroinvertebrate abundance, emergence and drift. Report for the Division of Silviculture, Forestry Commission, Tasmania, 19 p.

Davies, P.E. & Cook, L.S.J. (1993) Catastrophic macroinvertebrate drift and sublethal effects on brown trout, *Salmo trutta*, caused by cypermethrin spraying on a Tasmanian stream. *Aquat. Toxicol.*, **27**, 201–224.

Davies, R.A.H. & McLaren, I.W. (1977) Tolerance of *Aphytis melinus* DeBach (Hymenoptera: Aphelinidae) to 20 orchard chemical treatments in relation to integrated control of red scale, *Aonidiella aurantii* (Maskell) (Homoptera: Diaspididae). *Aust. J. Exp. Agr. Animal Husb.*, **17**, 323–328.

de Abreu, J.M. (1974) Pathogenicity of *Bacillus thuringiensis* Berliner against the rubber hawkmoth (*Erinnyis ello* L.)(Lepidoptera: Sphingidae). *Rev. Theobroma,* **4**, 33–36.

de Abreu, J.M. (1982) Investigations on the rubber leaf caterpillar *Erinnyis ello* in Bahia, Brazil. *Revista Theobroma,* **12**, 85–99.

de Barjac, H. (1978) Une nouvelle variete de *Bacillus thuringiensis* tres toxique pour les moustiques: *Bacillus thuringiensis* var. *israelensis* serotype H14. *Compt. Rend. Acad. Sci. Paris,* **286**, 797–800.

de Barjac, H. & Bonnefoi, A. (1962) Essai de classification biochemique et serologique de 24 souches de *Bacillus* du type *B. thuringiensis*. *Entomophaga,* **7**, 5–31.

de Barjac, H. & Bonnefoi, A. (1968) A classification of strains of *Bacillus thuringiensis* with a key to their differentiation. *J. Invertebr. Pathol.*, **11**, 335–347.

de Barjac, H. & Bonnefoi, A. (1972) Presence of H-antigenic subfactors in serotype 5 of *Bacillus thuringiensis* var. *canadensis*. *J. Invertebr. Pathol.*, **20**, 212–213.

de Barjac, H. & Coz, J. (1979) Comparative susceptibility of six different species of mosquitoes to *Bacillus thuringiensis* var. *israelensis*. *Bull. World Health Organ.*, **57**, 139–141.

de Barjac, H. & Lemille, F. (1970) Presence of flagellar antigenic subfactors in serotype 3 of *Bacillus thuringiensis*. *J. Invertebr. Pathol.*, **15**, 139–140.

de Barjac, H. & Thompson, J.V. (1970) A new serotype of *Bacillus thuringiensis* var. *thompsoni* (serotype 12). *J. Invertebr. Pathol.*, **15**, 141–144.

de Barjac, H., Larget, I. & Killick Kendrick, R. (1981) Toxicity of *Bacillus thuringiensis* var. *israelensis*, serotype H14, to the larvae of phlebotomine vectors of leishmaniasis. *Bull. Soc. Pathol. Exotiq.*, **74**, 485–489.

de Barjac, H., Cosmao Dumanoir, V., Shaik, R. & Viviani, G. (1977) *Bacillus thuringiensis* var. *pakistani*; une nouvelle sousepece correspondant au serotype 13. *Compt. Rend. Acad. Sci. Paris,* **248D**, 2051–2053.

de Barjac, H., Lee, H.L. & Seleena, P. (1990) Isolation of indigenous larvicidal microbial control agents of mosquitos (sic): the Malaysian experience. *Southeast Asian J. Trop. Med. Public Health,* **21**, 281–287.

de Bellis E. & Cavalcaselle, B. (1970) Control tests against the pine processionary with aerial treatments with *Bacillus thuringiensis*. *Bol. Serv. Plagas For.*, **13**, 145–149.

de Chenon, R.D. (1982) *Latoia* (*Parasa*) *lepida* (Cramer) Lepidoptera Limacodidae, a coconut pest in Indonesia. *Oleagineux,* **37**, 177–183.

de Leon, T. & Ibarra, J.E. (1995) Alternative bioassay technique to measure activity of Cry III proteins of *Bacillus thuringiensis*. *J. Econ. Entomol.*, **88**, 1596–1601.

de Lucca, A.J., Simonson, J.L. & Larson, A.D. (1979) Two new serovars of *Bacillus thuringiensis*: serovars *dakota* and *indiana* (serovars 15 and 16). *J. Invertebr. Pathol.*, **34**, 323–324.

de Moor, F.C. & Car, M. (1986) A field evaluation of *Bacillus thuringiensis* var. *israelensis* as a biological control agent for *Simulium chutteri* (Diptera:Nematocera) in the middle Orange River. *Onderstepoort J. Vet. Res.*, **53**, 43–50.

de Reede, R.H. (1985) Integrated pest management in apple orchards in the Netherlands: a solution for selective control of tortricids. *Meded. Laborat. Entomol. Wageningen,* No. 493, 105 p.

de Vasconcellos, F.J.M. & Rabinovitch, L. (1995) A new formula for an alternative culture medium, without antibiotics, for isolation and presumptive quantification of *Bacillus cereus* in foods. *J. Food Prot.*, **58**, 235–238.

del Bene, G. (1981) Effects of *Bacillus thuringiensis* Berliner on *Epichoristodes acerbella* (Walk.) Diak. *Redia,* **64**, 229–236.

del Bene, G. & Melis Porcinai, G. (1980) Histopathology of *Bacillus thuringiensis* Berliner on larvae of *Anagasta kuhniella* Zeller. *Redia,* **63**, 19–23.

Degaspari, N. & Gomez, S.A. (1982) Chemical control of the soyabean caterpillar under field conditions in Mato Grosso do Sul. *Pesqu. Agropecu. Bras.*, **17**, 513–517.

Delattre, R. (1974) Base data for integrated control in cotton crops in Africa. *Food and Agriculture Organization: Proceedings of the FAO Conference on Ecology in relation to Plant Pest Control,* Italy, 11-15 December 1972, 161–181.

Delplanque, A. & Gruner, L. (1975) The use of *Bacillus thuringiensis* Berliner against some Lepidoptera injurious to vegetable crops in the Antilles. *Nouvelles Agronomiques des Antilles et de la Guyane,* **1**, 71–82.

Delplanque, A., Gruner, L., Brathwaite, C.W.D., Phelps, R.H. & Bennett, F.D. (1974) Some results and prospects of the use of heat stable toxin of *Bacillus thuringiensis* on entomological pests of horticultural crops. *Crop Prot. in the Caribbean,* The University of the West Indies, St. Augustine, Trinidad. April 8-11, 1974, 245–251.

DeLucca, A.J., II, Palmgren, M.S. and de Barjac, H. (1984) A new serovar of *Bacillus thuringiensis* from grain dust: *Bacillus thuringiensis* serovar *colmeri* (serovar 21). *J. Invertebr. Pathol.*, **43**, 437–438.

DeLucca, A.J., Simonson, J.G. & Larson, A.D. (1981) *Bacillus thuringiensis* distribution in soils of the United States. *Canad. J. Microbiol.*, **27**, 865–870.

Delrio, G. (1995) Integrated control of olive pests. *Inf. Fitopatol.*, **45**, 9–15.

Demolin, G. & Martin, J.C. (1998) Control of pine-tree processionary caterpillar. Effectiveness and persistence of the activity of two formulations based on *Bacillus thuringiensis*. *Phytoma,* **507**, 11–14.

Denver, C. (1974) The use of biological control in an Arkansas farming operation. *Tall Timbers Research Station: Proceedings Tall Timbers Conference on Ecological Animal Control by Habitat Management, No,* March 1-2, 1973, Tallahassee, Florida.

Derecha, Z.A., Stetsenko, V.A., Emel' yanchuk, A.M. & Taran, F.I. (1981) Potato moth on hops. *Zaschc. Rast.*, **1**, 51.

Deschle, W.E., Hagen, H.E., Rutschke, J., Stamer, M. & Meyer, T. (1988) Black fly (Diptera: Simuliidae) control in West Germany with the biological larvicide *Bacillus thuringiensis* var. *israelensis*. *Bull. Soc. Vector Ecol.*, **13**, 280–286.

Deseo, K.V. & Docci, R. (1985) Microbiological control against *Zeuzera pyrina* L. (Lepidoptera: Cossidae). *Difesa delle Piante,* **8**, 285–291.

Deseo, K.V. & Montermini, A. (1990) Three year long monitoring of pathogens in *Calliptamus italicus* L. (Orthoptera: Acrididae) in central-northern Italy. *Proc. Abst. Vth Intern. Coll. Invertebr. Pathol. Microb. Con., Adelaide, Australia,* 249.

Deseo, K.V., Grassi, S., Foschi, F. & Rovesti, L. (1984) A system of biological control against the leopard moth (*Zeuzera pyrina* L.; Lepidoptera, Cossidae). *Atti Giornate Fitopatol.*, **2**, 403–414.

Deseo, K.V., Balbiani, A., Sanino, L. & Zampelli, G. (1993) Observations on the biology and biological control of the tobacco flea beetle, *Epitrix hirtipennis* Melsh. (Col., Chrysomelidae) in Italy. *Anz. Schaedlingskd. Pflanzenschutz Umweltschutz,* **66**, 26–29.

Deshmukh, A.D.D. & Deshpande, A.D. (1989) Bioefficacy of *Bacillus thuringiensis* Berliner against *Achoea janata* L. and *Bombyx mori* L. *Entomon,* **14**, 91–94.

Deshpande, A.D. & Ramakrishnan, N. (1982) Pathogenicity of certain serotypes of *Bacillus thuringiensis* Berliner against *Achaea janata* L. *Entomon,* **7**, 239–245.

Devriendt, M. & Martouret, D. (1976) Absence of resistance to *Bacillus thuringiensis* in the diamondback moth, *Plutella maculipennis* (Lep.: Hyponomeutidae). *Entomophaga,* **21**, 189–199.

Dhouibi, M.H. (1992) Effect of Bactospeine XLV on the date pyralid *Ectomyelois ceratoniae* Zeller (Lepidoptera: Pyralidae). *Meded. Fac. Landbouwwet. Uni. Gent,* **57**, 505–514.

Dhouibi, M.H. & Jemmazi, A. (1996) Warehouse biological control of the pyralid date pest *Ectomyelois ceratoniae*. *Fruits Paris,* **51**, 39–46.

Diaz, T., Restrepo, N., Orduz, S. & Rojas, W. (1993) Distribution and isolation of *Bacillus thuringiensis* in Colombia. *Rev. Colombiana Entomol.*, **19**, 35–40.

DiCosty, U.R. & Whalon, M.E. (1997) Selection of Colorado potato beetle resistant to CryIIIA on transgenic potato plants. *Resist. Pest Manag.*, **9**, 33–34.

Dietz, A. (1993) Risk assessment of genetically modified plants introduced into the environment. In *Transgenic organisms: Risk Assessment of Deliberate Release,* (eds. K. Wohrmann & J. Tomiuk) Advances in Life Sciences, Birkhauser Verlag AG; Basel; Germany, 209–227.

Dimitriadis, V.K. & Domouhtsidou, G.P. (1996) Effects of *Bacillus thuringiensis* strain *ormylia* spore-crystal complex on midgut cells of *Dacus* (*Bactrocera*) *oleae* larvae. *Cytobios,* **87**, 19–30.

Dirimanov, M. & Lecheva, I. (1980) The effects of some microbial preparations on larvae of the cherry spinner. *Rast. Zashc.*, **28**, 23–25.

Dirimanov, M., Angelova, R. & Babrikova, T. (1980) State of the harmful entomo- and acarofauna and predacious species of insects in apple orchards with some plant protection technologies. *Nauchni Tr. Entomol. Mikrobiol. Fitopatol.*, **25**, 15–30.

Dixon, B. (1994) Keeping an eye on *B. thuringiensis*. *BioTechnology,* **12**, 435.

Dixon, W.N. (1982) *Anacamptodes pergracilis* (Hulst), a cypress looper (Lepidoptera: Geometridae). *Entomology Circular, Division of Plant Industry, Florida Department of Agriculture and Consumer Services,* **244**, 2 p.

do Amaral Filho, B.F. & de Carvalho, G.A. (1996) Effect of *Bacillus thuringiensis* var. *kurstaki* on the 1st and 4th instar larvae of *Corcyra cephalonica* (Stainton, 1865) (Lep., Pyralidae). *Rev. Agric. Piracicaba,* **71**, 119–125

Dobrivojevic, K. & Injac, M. (1975) Laboratory investigation of the effectiveness of the microbial insecticides Dipel™ and Bactospeine for the control of the woolly moth (*Hypogymna morio* L.). *Zast. Bilja,* **26**, 365–369.

Dobzhenok, N.V. (1976) The effect of ascorbic acid on the physiological condition of the codling moth and its resistance to fungus and bacterial infection. *Zakhist Rosl.*, **19**, 3–7.

D'-Offria, E., Galbero, G. & Furlan, G. (1996) Biological control against pine processionary moth. *Inform. Agr.,* **52**, 23, 57–58.

Dolidze, G.V. (1983) Minimal application of chemical treatments. *Zaschc. Rast.*, **6**, 26.

Donaubauer, E. (1976) The use of bacteria and viruses in the control of forest pests. *Allg. Forstz.*, **87**, 112–114.

Donaubauer, E. & Schmutzenhofer, H. (1973) Aerial application of *Bacillus thuringiensis* (Dipel) against *Colotois* (*Himera*) *pennaria* L. in comparison with ULV-application of various insecticides. *Bulletin, Organisation Europeenne et Mediterraneenne pour la Protection des Plantes,* **3**, 111–115.

Donegan, K.K., Schaller, D.L., Stone, J.K., Ganio, L.M., Reed, G., Hamm, P.B. & Seidler, R.J. (1996) Microbial populations, fungal species diversity and plant pathogen levels in field plots of potato plants expressing the *Bacillus thuringiensis* var. *tenebrionis* endotoxin. *Transgenic Res.*, **5**, 25–35.

Dong, H.F. & Niu, L.P. (1991) Effect of commercial preparations of *Bacillus thuringiensis* on predacious mites. *Chin. J. Biol. Cont.*, **7**, 10–12.

Dostie, R., Delisle, S. & Marotte, P.M. (1988) Distribution of spores of *Bacillus thuringiensis* in running water. *Ministere de l'Energie et des Ressources (Forets); Quebec; Canada* 15 pp.

Dougherty, D.E. & Schuster, D.J. (1985) Compatibility of fungicide-insecticide combinations for disease and pickleworm control on honeydew melon. *Proc. Florida State Hort. Soc.*, **97**, 205–208.

Dougherty, E.M., Reichelderfer, C.F. & Faust, R.M. (1971) Sensitivity of *Bacillus thuringiensis* var. *thuringiensis* to various insecticides and herbicides. *J. Invertebr. Pathol.*, **17**, 292–293.

Drobniewski, F.A. (1994) The safety of *Bacillus* species as insect vector control agents. *J. Appl. Bacteriol.*, **76**, 101–109.

Dronka, K., Niemczyk, E. & Dadaj, J. (1976) The effectiveness of the bacterial preparation Dipel and certain chemical insecticides in controlling the apple leaf skeletoniser (*Simaethis pariana* Clerk) in apple orchards). *Rocz. Nauk Roln. E,* **6**, 159–164.

Drummond, J., Miller, D.K. & Pinnock, D.E. (1992) Toxicity of *Bacillus thuringiensis* against *Damalinia ovis* (Phthiraptera: Mallophaga). *J. Invertebr. Pathol.*, **60**, 102–103.

Drummond, J., Kotze, A.C., Levot, G.W. & Pinnock, D.E. (1995) Increased susceptibility to *Bacillus thuringiensis* associated with pyrethroid resistance in *Bovicola* (*Damalinia*) *ovis* (Phthiraptera: Mallophaga) possible role of monooxygenases. *J. Econ. Entomol.*, **88**, 1607–1610.

Dua, V.K., Sharma, S.K. & Sharma, V.P. (1993) Application of Bactoculicide (*Bacillus thuringiensis* H-14) for controlling mosquito breeding in industrial scrap at BHEL, Hardwar (U.P.). *Indian J. Malariol.*, **30**, 17–21.

Dubitskii, A.M., Saubenova, O.G. & Cherkashin, A.N. (1981) A new effective biological pathogen for the control of mosquitoes. *Vestn. Akad. Nauk Kazakh. SSR,* recd. 1983, **5**, 38–43.

Dubois, N.R. (1986) Synergism between beta-exotoxin and *Bacillus thuringiensis* subspecies *kurstaki* (HD-1) in gypsy moth, *Lymantria dispar*, larvae. *J. Invertebr. Pathol.*, **48**, 146–151.

Dubois, N.R. & Dean, D.H. (1995) Synergism between CryIA insecticidal crystal proteins and spores of *Bacillus thuringiensis*, other bacterial spores, and vegetative cells against *Lymantria dispar* (Lepidoptera: Lymantriidae) larvae. *Environ. Entomol.*, **24**, 1741–1747.

Dubois, N.R, Huntley, P.J. & Newman, D. (1989) Potency of *Bacillus thuringiensis* strains and formulations against gypsy moth and spruce budworm larvae: 1980-96. *USDA For. Ser. Northeastern For. Exp. St. Gen. Tech. Rep. NE-131,* 25 p.

Dulmage, H.T. (1975) The standardization of formulations of the delta -endotoxins produced by *Bacillus thuringiensis*. *J. Invertebr. Pathol.*, **25**, 279–281.

Dulmage, H.T. *et al.* (1981) Insecticidal activity of isolates of *Bacillus thuringiensis* and their potential for pest control. *Microbial Control of Pests and Diseases 1970-1980* (ed H.D. Burges), pp. 193–222. Academic Press, London.

Dulmage, H.T. & Martinez, E. (1973) The effects of continuous exposure to low concentrations of the delta -endotoxin of *Bacillus thuringiensis* on the development of the tobacco budworm, *Heliothis virescens*. *J. Invertebr. Pathol.*, **22**, 14–22.

Dulmage, H.T., McLaughlin, R.E., Lacey, L.A., Couch, T.L., Alls, R.T. & Rose, R.I. (1985) HD-968-S-1983, a proposed U.S. standard for bioassays of preparations of *Bacillus thuringiensis* subsp. *israelensis*-H-14. *Bull. Entomol. Soc. Amer.*, **31**, 31–34.

Dunbar, J.P. & Johnson, A.W. (1975) *Bacillus thuringiensis*: effects on the survival of a tobacco budworm parasitoid and predator in the laboratory. *Environ. Entomol.*, **4**, 352–354.

Dunbar, D.M. & Kaya, H.K. (1972) *Bacillus thuringiensis*: control of the gypsy moth and elm spanworm with three new commercial formulations. *J. Econ. Entomol.*, **65**, 1119–1121.

Dunbar, D., Kaya, H.K., Doane, C.C., Anderson, J.F. & Weseloh, R.M. (1973) Aerial application of *Bacillus thuringiensis* against larvae of the Elm spanworm and gypsy moth and effects on parasitoids of the gypsy moth. *Bull. Connecticut Agric. Exp. St.*, **735**, 23 pp.

Dunkle, R.L. & Shasha, B.S. (1988) Starch-encapsulated *Bacillus thuringiensis*: a potential new method for increasing environmental stability of entomopathogens. *Environ. Entomol.*, **17**, 120–126.

Dunkle, R.L. & Shasha, B.S. (1989) Response of starch-encapsulated *Bacillus thuringiensis* containing ultraviolet screens to sunlight. *Environ. Entomol.*, **18**, 1035–1041.

Dusenko, E.A. (1986) Biological control. *Zaschc. Rast.*, **6**, 34.

Duvlea, I., Rusu, L. & Palagesiu, I. (1969) Research on some aspects of integrated control of cabbage moth, *Mamestra brassicae*, under field conditions. *Lucr. Sti. Inst. Agron. Timisoara Agronomie,* **12**, 459–467.

Dyadechko, N.P., Tsybul' skaya, G.N., Chizhik, R.I. & Venger, V.M. (1976) Biological agents reducing the numbers of the meadow moth. *Zaschc. Rast.*, **7**, 43–44.

Dzhivladze, K.N. (1979) Use of biological methods in Georgia. *Zaschc. Rast.*, **5**, 28.

Eberhardt, H.J. (1997) Alternative forms of storage protection: biological insecticides for the control of the cigarette beetle (*Lasioderma serricorne*) and the tobacco moth (*Ephestia elutella*). *Beitr. Tabakforsch. Int.*, **17**, 31–47.

Ebersold, H.R., Luethy, P. & Mueller, M. (1977) Changes in the fine structure of the gut epithelium of *Pieris brassicae* induced by the delta-endotoxin of *Bacillus thuringiensis*. *Mitt. Schweiz. Entomol. Ges.*, **50**, 269–276.

Ecevit, O., Tuncer, C., Hatat, G. & Kececi, S. (1994) Studies on the efficiency of two *Bacillus thuringiensis* formulations (Thuricide HP and Biobit), azinphos-methyl and triflumuron against fall webworm (*Hyphantria cunea* Drury Lepidoptera: Arctiidae). *Turkiye III,* 25-28 Ocak 1994, Ege Universitesi Ziraat Fakultesi, Bitki Koruma Bolumu, Izmir. 1994, 519–528.

Eckberg, T.B. & Cranshaw, W.S. (1994) Larval biology and control of the rabbitbrush beetle, *Trirhabda nitidicollis* Leconte (Coleoptera: Chrysomelidae). *Southwestern Entomol.*, **19**, 249–256.

Eckenrode, C.J., Andaloro, J.T. & Shelton, A.M. (1981) Suppression of lepidopterous larvae in commercial sauerkraut cabbage fields and research plots. *J. Econ. Entomol.*, **74**, 276–279.

Ecobichon, D.J. (1990) Chemical management of forest pest epidemics: a case study. *Biomed. Environ. Sci.*, **3**, 217–239.

Edwards, B.A.B. & Hodgson, P.J. (1973) The toxicity of commonly used orchard chemicals to *Stethorus nigripes* (Coleoptera: Coccinellidae). *J. Aust. Entomol. Soc.*, **12**, 222–224.

Effremova, T.G. (1976) Pests of cabbage. *Zaschc. Rast.*, **8**, 58–59.

Eidt, D.C. (1984) *B.t.* budworm spray is innocuous to aquatic insects. *Tech. Note, Maritimes For. Res. Centre Canad.*, **No. 114**, 4 p.

Eidt, D.C. & Dunphy, G.B. (1991) Control of spruce budmoth, *Zeiraphera canadensis* Mut. and Free., in white spruce plantations with entomopathogenic nematodes, *Steinernema* spp. *Canad. Entomol.*, **123**, 379–385.

El Bahrawi, A., Saad, A.S.A. & Soliman, S.S. (1979) Dimilin, urea growth regulator for control of brown tail moth *Euproctis melania* Staud in Iraq. *Med. Fac. Landbouww. Rijk. Gent,* **44**, 31–37.

El Ghar, G., Radwan, H.S.A., El Bermawy, Z.A. & Zidan, L.T.M. (1995) Inhibitory effect of thuringiensin and abamectin on digestive enzymes and non-specific esterases of *Spodoptera littoralis* (Boisd.) (Lep., Noctuidae) larvae. *J. Appl. Entomol.*, **119**, 355–359.

El Hakim, A.M. & Hanna, S. (1982) Evaluation of *Bacillus thuringiensis* Berliner against the olive-leaf moth *Palpita unionalis* Hb. (Pyralidae, Lepidoptera). *Agric. Res. Rev.*, **60**, 17–28.

El Hakim, A.M., El Ela, R.G.A. & Hanna, S.K. (1988) Effect of sublethal concentrations of Thuricide, Gardona and Sumicidin on the larvae of the grape moth, *Lobesia botrana* Den and Schiff Tortricidae: Lepidoptera. *Ann. Agric. Sci. Ain Shams Univ.*, **33**, 609–619.

El Husseini, M.M. (1980) Effect of two *Bacillus thuringiensis* Berl. preparations on early larval instars of the European corn borer, *Ostrinia nubilalis* Hbn. *Bull. Soc. Entomol. Egypt,* **63**, 175–179.

El Husseini, M.M. (1984) New approach to control the cotton leaf-worm, *Spodoptera littoralis* (Boisd.) by *Bacillus thuringiensis* Berl. in clover fields. *Bull. Entomol. Soc. Egypt,* **12**, 1–4.

El Husseini, M.M. & Afifi, A.I. (1984a) Effect of Entobakterin-3 on the spiny bollworm, *Earias insulana* (Boisd.). *Bull. Entomol. Soc. Egypt,* **12**, 59–70.

El Husseini, M.M. & Afify, A.I. (1984b) Increasing the efficacy of *Bacillus thuringiensis* Berliner against the spiny bollworm, *Earias insulana* Boisd., by adding a feeding stimulant. *Bull. Soc. Entomol. Egypt,* **63**, 37–41.

El Husseini, M.M., Tawfik, M.F.S., Awadallah, K.T. & Afifi, A.E. (1987) Effect of *Bacillus thuringiensis*-diseased prey on the development and adult activity of the anthocorid *Xylocoris flavipes* (Reut.). *Bull. Soc. Entomol. Egypt,* **65**, 205–213.

El Maghraby, M.M.A., Hegab, A. & Yousif Khalil, S.I. (1988) Interactions between *Bacillus thuringiensis* Berl., *Beauveria bassiana* (Bals.) Vuill. and the host/parasitoid system *Spodoptera littoralis* (Boisd.)/*Microplitis rufiventris* Kok. *J. Appl. Entomol.*, **106**, 417–421.

El Moursy, A.A., Sharaby, A. & Awad, H.H. (1996) Some chemical additives to increase the activity spectrum of *Bacillus thuringiensis* var. *kurstaki* (Dipel 2x) against the rice moth *Corcyra cephalonica*. *Bull. Nat. Res. Centre Cairo,* **21**, 161–172.

El Sayed, G.N., Moawad, G.M. & Ahmed, G.S. (1979) The effect of certain bacterial and chemical pesticides on the potato tuber worm, *Phthorimaea operculella* (Zeller), infesting tomatoes in Saudi Arabia (Lepidoptera: Gelechiidae). *Agric. Res. Rev. Plant Prot.*, **57**, 223–231.

El Sebae, A.H. & Komeil, A.M. (1990) Interaction of conventional insecticides with *Bacillus thuringiensis* against the Mediterranean fruit fly, *Ceratitis capitata* (Wied.). *Brighton Crop Prot. Conference, Pests and Diseases,* Volume 1, 241–244.

El Zemaity, M.S., El Refai, S.A., Zemaity, M.S.E. & Refai, S.A.E. (1987) Joint action of fenvalerate mixtures against the cotton leafworm, *Spodoptera littoralis* (Boisd). *Ann. Agric. Sci. Ain Shams Univ.,* **32**, 1741–1749.

Eldridge, B.F. & Callicrate, J. (1982) Efficacy of *Bacillus thuringiensis* var. *israelensis* de Barjac for mosquito control in a western Oregon log pond. *Pestic. Sci.,* **42**, 102–105.

Eldridge, B.F., Washino, R.K. & Henneberger, D. (1985) Control of snow pool mosquitoes with *Bacillus thuringiensis* serotype H-14 in mountain environments in California and Oregon. *J. Amer. Mosq. Con. Assoc.,* **1**, 69–75.

Elek, J.A. (1997) Assessing the impact of leaf beetles in eucalypt plantations and exploring options for their management. *Tasforests,* **9**, 139–154.

Elsey, K.D. (1973) *Jalysus spinosus*: effect of insecticide treatments on this predator of tobacco pests. *Environ. Entomol.,* **2**, 240–243.

Endakov, E. (1979) On the fields of the Altai. *Zaschc. Rast.,* **4**, 6–7.

Englert, W.D. & Kettner, J. (1983) Side-effects of plant protection materials on spider mites and predacious mites. *Mitt. Dtsch. Ges. Allg. Angew. Entomol.,* **4**, 89–91.

EPA, Schneider, W.R. & others (1998) EPA Reregistration Eligibility Decision (RED) *Bacillus thuringiensis.* US EPA, Prevention, Pesticides and toxic substances, EPA738-R-98-004, 91 p.

Erfurth, P. & Motte, G. (1971) First report of experiments on gold-tail moth control in the year 1971. *Nachrbl. Pflschutzdienst. DDR,* **25**, 234–235.

Ershova, L.C., Afanas' eva, O.V. & Barbashova, H.M. (1980) The effect of the bacterial preparation Bitoxibacillin on fleas. *Parazitologiya,* **14**, 35–38.

Ershova, L.S., Afanas' eva, O.V. & Bundzhe, Z.F. (1976) The effect of the bacterial preparations Insectin on *Xenopsylla cheopis. Parazitologiya,* **10**, 552–554.

Ershova, L.S., Afanas' eva, O.V. & Bundzhe, Z.F. (1982) On the action of Dendrobacillin on the fleas *Xenopsylla cheopis* (Siphonaptera). *Parazitologiya,* **16**, 165–168.

Escobar, A.L., Millan, J.M., & de la Cruz, J. (1979) Contribution to knowledge of the population dynamics and control of the bagworm *Oiketicus kirbyi* (Lepidoptera: Psychidae) in banana plantations at Rozo, Valle. *Acta Agron.,* **29**, 29–39.

Escriche, B.B. Tabashnik, N., Finson, N. & Ferre, J. (1995). Immunochmical detection of binding of CryIA crystal proteins of *Baciilus thuringiensis* in highly resistant strains of *Plutella xylostella* from Hawaii. *Biochem. Biophys. Res. Comm.,* **212**, 388–395.

Eskils, K. & Lovgren, A. (1997) Release of *Bacillus thuringiensis* subsp. *israelensis* in Swedish soil. *FEMS Microbiol. Ecol.,* **23**, 229–237.

Espinel, R. (1981) Laboratory study of the feeding behaviour of *Mamestra* (*Barathra*) *brassicae* L. in the presence of foliage treated with *Bacillus thuringiensis* B. *Z. Angew. Entomol.,* **91**, 383–388.

Espino, E., Barroso, J. & Carnero, A. (1988) Preliminary results of integrated control in cucumber in the Canaries. *Boletin de Sanidad Vegetal, Plagas,* **14**, 55–66.

Estada, U. & Ferre, J. (1994) Binding of insecticidal crystal proteins of *Bacillus thuringiensis* to the midgut brush border of the cabbage looper, *Trichoplusia ni* (Hubner) (Lepidoptera: Noctuidae), and selection for resistance to one of the crystal proteins. *Appl. Environ. Microbiol.,* **60**, 3840–3846.

Ester, A. & Nijenstein, J.H. (1995) Control of the field slug *Deroceras reticulatum* (Muller) (Pulmonata: Limacidae) by pesticides applied to winter wheat seed. *Crop Prot.,* **14**, 409–413.

Estruch, J.J., Warren, G.W., Mullins, M.A., Nye, G.J., Craig, J.A. & Koziel, M.G. (1996) Vip3A, a novel *Bacillus thuringiensis* vegetative insecticidal protein with a wide spectrum of activities against lepidopteran insects. *Proc. Nat. Acad. Sc. USA,* **93**, 5389–5394.

Fadeev, A.V. (1974) Biological preparations against *Bupalus piniarius. Lesnoe Khoz.,* **2**, 64–65.

Farghal, A.I. (1982a) Effect of temperature on the effectiveness of *Bacillus thuringiensis* var. *israelensis* against *Culex pipiens molestus* Forsk larvae. *Z. Angew. Entomol.,* **94**, 408–412.

Farghal, A.I. (1982b) On the combined effect of a juvenile hormone analogue (Altosid) and a preparation of *Bacillus thuringiensis* (Bactimos) against the larvae of the mosquitoes *Culex pipiens molestus* Forsk. and *Theobaldia longiareolata* Macq. (Dipt., Culicidae). *Anz. Schadlingskd. Pflanzenschutz Umweltschutz,* **55**, 164–167.

Farghal, A.I., Ahmed, S.A. & Salit, A.M. (1983) Effect of overcrowding on the biolethal efficiency of *Bacillus thuringiensis* var. *israelensis* against *Culex pipiens molestus* and *Theobaldia longiareolata* larvae. *Z. Angew. Entomol.,* **95**, 457–460.

Farghal, A.I., Darwish, Y.A. & Abdel Galil, F.A. (1987) Relative susceptibility of four insect species to some bacterial insecticides. *Assiut J. Agric. Sci.,* **18**, 21–30.

Farias, A.R.N., Ezeta, F.N. & Dantas, J.L.L. (1980) The cassava hawkmoth. *Circular Tecnica, EMBRAPA, CNPMF,* **5**, 11 p.

Farrar, R.R., Jr. & Ridgway, R.L. (1995a) Enhancement of activity of *Bacillus thuringiensis* Berliner against four lepidopterous insect pests by nutrient-based phagostimulants. *J. Entomol. Sci.,* **30**, 29–42.

Farrar, R.R., Jr. & Ridgway, R.L. (1995b) Feeding behavior of gypsy moth (Lepidoptera: Lymantriidae) larvae on artificial diet containing *Bacillus thuringiensis*. *Environ. Entomol.*, **24**, 755–761.

Farrar, R.R., Jr., Martin, P.A.W. & Ridgway, R.L. (1996) Host plant effects on activity of *Bacillus thuringiensis* against gypsy moth (Lepidoptera: Lymantriidae) larvae. *Environ. Entomol.*, **25**, 1215–1223.

Farris, M.E. & Appleby, J.E. (1980) Chemical control studies of the walnut caterpillar. *J. Arboric.*, **6**, 150–152.

Faruki, S.I. & Khan, A.R. (1996) The pathogenicity of *Bacillus thuringiensis* var. *kurstaki* on the tropical warehouse moth, *Cadra cautella* (Walker), as influenced by groundnut varieties. *J. Invertebr. Pathol.*, **68**, 113–117.

Fast, P.G. & Dimond, J.B. (1984) Susceptibility of larval instars of spruce budworm, *Choristoneura fumiferana* (Lepidoptera: Tortricidae), to *Bacillus thuringiensis*. *Canad. Entomol.*, **116**, 131–137.

Fast, P.G. & Donaghue, T.P. (1971) The delta -endotoxin of *Bacillus thuringiensis*. II. On the mode of action. *J. Invertebr. Pathol.*, **18**, 135–138.

Fast, P.G. & Morrison, I.K. (1972) The delta -endotoxin of *Bacillus thuringiensis*. 4. The effect of delta -endotoxin on ion regulation by midgut tissue of *Bombyx mori* larvae. *J. Invertebr. Pathol.*, **20**, 208–211.

Fatzinger, C.W., Yates, H.O., III & Barber, L.R. (1992) Evaluation of aerial applications of acephate and other insecticides for control of cone and seed insects in southern pine seed orchards. *J. Entomol. Sci.*, **27**, 172–184.

Favret, M.E. & Yousten, A.A. (1989) Thuricin: the bacteriocin produced by *Bacillus thuringiensis*. *J. Invertebr. Pathol.*, **53**, 206–216.

Federici, B.A. (1994) *Bacillus thuringiensis*: Biology, application, and prospects for further development. *Proceedings of the Second Canberra Meeting on* Bacillus thuringiensis (ed. R. J. Akhurst). pp. 1–15. CPN Publications Pty, Ltd., Canberra, Australia.

Federici, B.A. (1999) *Bacillus thuringiensis*. Handbook of Biological Control (eds. T.S. Bellows and T. W. Fisher), pp. 575–593. Academic Press, San Diego.

Federici, B.A. & Bauer, L.S. (1998) Cyt1Aa protein of *Bacillus thuringiensis* is toxic to the cottonwood leaf beetle, *Chrysomela scripta*, and suppresses high levels of resistance to Cry3Aa. *Appl. Environ. Microbiol.*, **64**, 4368–4371.

Fedorinchik, N.S. & Sogoyan, R.S. (1975) Microbial preparations for the control of tortricids in orchards on the Black Sea coast of the Krasnodar region. *Tr. Vses. Nauchno issled. Inst. Zashch. Rast.*, **42**, 110–121.

Felip, J., Aguilar, L., Serra, J. & Muntaner, A. (1987) Trial of insecticides against corn borers *Ostrinia nubilalis* Hbn and *Sesamia nonagrioides* Lef. *Fulls d'Informacio Tecnica*, **135**, 8 pp.

Feng, G.K., Chou, Y.L., Chang, K.S. & Nieh, S.H. (1977) Studies on the control of European corn borers by using trichogrammatid egg parasites. *Acta Entomol. Sin.*, **20**, 253–258.

Feng, L., Liu, H., Gao, X., Chi, X. (1997) Study on the regularity of occurrence of *Herse convolvuli* Linnaeus. *J. Shandong Agric. Uni.*, **28**, 465–470.

Feng, S., Fu, Y., Fan, X., Wang, R. (1995) Toxic correlation between the indicator insect *Ostrinia furnacalis* for the toxicity of *Bacillus thuringiensis* to *Heliothis armigera*, *Mythimna separata* and *Euxoa segetum*. *Acta Agric. Boreali Sin.*, **10**, 91–94.

Feng, X.X. & Xing, X.Q. (1982) A preliminary study on a polycrystal strain of *Bacillus thuringiensis*. *Kunchong Zhishi*, **19**, 18–19.

Fernandez de Cordova, J. & Cabezuelo Perez, P. (1993) Effect of various pesticides used on oak on the bird fauna. *Boletin de Sanidad Vegetal, Plagas*, **19**, 687–705.

Fernandez, B.R. & Clavijo, A.S. (1984) Effects of two insecticides (one chemical and the other biological) on the parasitism observed in larvae of *Spodoptera frugiperda* (S.) from experimental plots of maize. *Rev. Fac. Agron. Univ. Cent. Venezuela*, **13**, 101–109.

Ferrari, R. & Maini, S. (1992) *Bacillus thuringiensis* ssp. *tenebrionis*. *Inf. Fitopatol.*, **5**, 45–49.

Ferre, J., Real, M.D., Van Rie, J., Jansens, S. & Peferoen, M. (1991) Resistance to the *Bacillus thuringiensis* bioinsecticide in a field population of *Plutella xylostella* is due to a change in a midgut membrane receptor. *Proc. Nat. Acad. Sci. USA*, **88**, 5119–5123.

Ferro, D.N. & Lyon, S.M. (1991) Colorado potato beetle (Coleoptera: Chrysomelidae) larval mortality: operative effects of *Bacillus thuringiensis* subsp. *san diego*. *J. Econ. Entomol.*, **84**, 806–809.

Ferro, D.N., Slocombe, A.C. & Mercier, C.T. (1997) Colorado potato beetle (Coleoptera: Chrysomelidae): residual mortality and artificial weathering of formulated *Bacillus thuringiensis* subsp. *tenebrionis*. *J. Econ. Entomol.*, **90**, 574–582.

Ferro, D.N., Yuan, Q.C., Slocombe, A. & Tuttle, A.F. (1993) Residual activity of insecticides under field conditions for controlling the Colorado potato beetle (Coleoptera: Chrysomelidae). *J. Econ. Entomol.*, **86**, 511–516.

Filho, O.G., Denolf, P., Peferoen, M., Decazy, B., Eskes, A.B. & Frutos, R. (1998) Susceptibility of the coffee leaf miner (*Perileucoptera* spp.) to *Bacillus thuringiensis* delta-endotoxins: a model for transgenic perennial crops resistant to endocarpic insects. *Curr. Microbiol.*, **36**, 175–179.

Filip, I. & Blaise, P. (1998) Research on the biology and integrated control system of *Peribatodes rhomboidaria* Den. et Schiff. in Romanian viticulture. *Integ. Con. Viticult.*, **21**, 83–84.

Finlayson, D.G. (1979) Combined effects of soil-incorporated and foliar–applied insecticides in bed-system production of brassica crops. *Canad. J. Plant Sci.*, **59**, 399–410.

Finnegan, R.J. (1979) Predacious red wood ants in Quebec forests. *Bulletin-SROP*, **2**, 65–74.

Finney, J.R. & Harding, J.B. (1982) The susceptibility of *Simulium verecundum* (Diptera: Simuliidae) to three isolates of *Bacillus thuringiensis* serotype 10 (*darmstadiensis*). *Mosq. News*, **42**, 434–435.

Fisher, R. & Rosner, L. (1959) Toxicology of the microbial insecticide Thuricide. *J. Agr. Food Chem.*, **7**, 686–688.

Fodale, A.S. & Mule, R. (1990) Bioethological observations on *Palpita unionalis* Hb. in Sicily and trials of defence. *Acta Hort.*, **286**, 351–353.

Fonseca, F.B. & De Polania, I.Z. (1998) Resistance potential of *Spodoptera frugiperda* to *Bacillus thuringiensis* subsp. *kurstaki*. *Manejo Integrado de Plagas*, **47**, 18–23.

Foo, A.E.S. & Yap, H.H. (1982) Comparative bioassays of *Bacillus thuringiensis* H-14 formulations against four species of mosquitoes in Malaysia. *Southeast Asian J. Trop. Med. Public Health*, **13**, 206–210.

Forrester, N.W. (1994a) Use of *Bacillus thuringiensis* in integrated control, especially on cotton pests. *Agric. Ecosys. Environ.* **49**, 77–83.

Forrester, N.W. (1994b) Resistance management options for conventional *Bacillus thuringiensis* and transgenic plants in Australian summer field crops. *Biocontr. Sci. Technol.* **4**, 549–553.

Forsberg, C.W., Henderson, M., Henry, E. & Roberts, J.R. (1976) *Bacillus thuringiensis*: its effects on environmental quality. *Publication, National Research Council Canada*, No. 15383, 135pp.

Forti, D. & Ioriatti, C. (1990) *Pammene rhediella* makes its appearance in Trentino. *Inf. Fitopatol.*, **40**, 111–113.

Forti, D. & Ioriatti, D. (1992) Plum fruit moth (*Pammene rhediella* Cl.): life cycle, damage and control methods. *Acta Phytopath. Entomol. Hungarica*, **27**, 239–244.

Foster, G.N. & Crook, N.E. (1983) A granulosis disease of the tomato moth *Lacanobia oleracea* (L.). *Bulletin SROP*, **6**, 163–166.

Fougeroux, A. & Lacroze, T. (1996) The preventive grape insecticide Biocillis, regarding the control of *Botrytis*. *Phytoma*, **489**, 61–62.

Francardi, V. (1990) Results in the use of a new microbiological preparation based on *Bacillus thuringiensis* var. *tenebrionis* in the control of the larvae of *Pyrrhalta luteola* (Mull.). *Redia*, **73**, 463–472.

Franz, J.M., Bogenschutz, H., Hassan, S.A., Huang, P., Nation, E., Suter, H. & Viggiani, G. (1980) Results of a joint pesticide test programme by the Working Group: Pesticides and Beneficial Arthropods. *Entomophaga*, **25**, 231–236.

Frolov, B.A. (1974) Chemical and biological methods of controlling ectoparasites of birds. *Veterinariya, Moscow, USSR*, **12**, 66–68.

Frolov, B.A. (1977) External parasites of birds and their control. *Arakhnozy i protozoinye bolezni sel'skokhozyaistvennykh zhivotnykh*, 70–82.

Frolov, B.A., Tonkonozhenko, A.P. & Kachekova, S. (1979) Study of the effectiveness of thuringin mixed with low doses of chlorophos and Sevin against bird lice (Mallophaga) on poultry. *Problemy Veterinarnoi Sanitarii*, **58**, 92–96.

Frolov, B.A., Li, R.A., Tonkonozhenko, A.P. & Molochaeva, T.S. (1984) Effectiveness of 'Thuringin' against Mallophaga of chickens. *Sanitarno toksikologicheskaya otsenka veterinarnykh preparatov i voprosy mikologii i sanitarii kormov*, 102–106.

Frommer, R.L., Nelson, J.H., Remington, M.P. & Gibbs, P.H. (1981a) The influence of extensive aquatic vegetative growth on the larvicidal activity of *Bacillus thuringiensis* var. *israelensis* in reducing *Simulium vittatum* (Diptera: Simuliidae) larvae in their natural habitat. *Mosq. News*, **41**, 707–712.

Frommer, R.L., Hembree, S.C., Nelson, J.H., Remington, M.P. & Gibbs, P.H. (1981b) The distribution of *Bacillus thuringiensis* var. *israelensis* in flowing water with no extensive aquatic vegetative growth. *Mosq. News*, **41**, 331–338.

Frutos, R., Jacquemard, P. & Amargier, A. (1987) Compared activity of various varieties of *Bacillus thuringiensis* Berl. on two lepidopterous cotton pests, *Earias biplaga* Wlk. and *Earias insulana* (Boisd.). *Coton Fibres Trop.*, **42**, 5–21.

Fry O' Brien, L.L. & Mulla, M.S. (1996) Effect of tadpole shrimp, *Triops longicaudatus*, (Notostraca: Triopsidae), on the efficacy of the microbial control agent *Bacillus thuringiensis* var. *israelensis* in experimental microcosms. *J. Amer. Mosq. Con. Assoc.*, **12**, 33–38.

Frye, R.D., Scholl, C.G., Scholz, E.W. & Funke, B.R. (1973) Effect of weather on a microbial insecticide. *J. Invertebr. Pathol.*, **22**, 50–54.

Frye, R.D., Elichuk, T.L. & Stein, J.D. (1976) Dispersing *Bacillus thuringiensis* for control of cankerworm in shelterbelts. *Research Note, Rocky Mountain Forest and Range Experiment Station, Forest Service, USDA*, RM-315, 7 p.

Frye, R.D., McBride, D.K., Carey, D.R., Elichuk, T.L. & Dregseth, R.L. (1977) Cankerworm control in shelterbelts. *North Dakota Farm Res.*, **34**, 3–7.

Frye, R.D., Hard, J., Carey, D. & Dix, M.E. (1983) Day and night application of *Bacillus thuringiensis* for cankerworm control. *Res. Rep. Agr. Exp. St. North Dakota St. Univ.*, **94**, 10.

Frye, R.D., Dix, M.E. & Carey, D.R. (1988) Effect of two insecticides on abundance of insect families associated with Siberian elm windbreaks. *J. Kansas Entomol. Soc.*, **61**, 278–284.

Fu, Z.B., Chen, D.Z. & Huang, R.X. (1983) A preliminary study on the sweet potato moth. *Fujian Agric. Sci. Tech. Fujian Nongye Keji*, **3**, 13–15.

Fushimi, S., Iseki, S. & Nakamura, K. (1995) Feeding on Dipel (*Bacillus thuringiensis*)-infested common cabbage worm, *Pieris rapae*, by the gray starling; and toxicity of the gray starling's feces to the silkworm, *Bombyx mori*. *Jap. J. Appl. Entomol. Zool.*, **39**, 338–340.

Fuxa, J.R. (1979) Interactions of the microsporidium *Vairimorpha necatrix* with a bacterium virus, and fungus in *Heliothis zea*. *J. Invertebr. Pathol.*, **33**, 316–323.

Gadais, M., Mahieu, N. & Gilbert, R. (1978) Control of the pine processionary in the Loire-Atlantique and Vendee. *Phytoma*, **30**, 19–21.

Galan Wong, L.J., Rodriguez Padilla, C., Tamez Guerra, R.S., Espinosa Meade, E. & Luna Olvera, H.A. (1993) Soil microbial antagonism against *Bacillus thuringiensis*. *Rev. Latinoam. Microbiol.*, **35**, 391–398.

Galani, G. (1973) On the histology of larvae of *Orgyia antiqua* L. (Lepid., Lymantriidae) after treatment with a preparation containing *Bacillus thuringiensis*. *Anz. Schadlingskd. Pflanzen Umweltschutz*, **46**, 150–152.

Galani, G. (1978) Influence of temperature on the pathogenicity of the bacterium *Bacillus thuringiensis* to the insects *Lymantria dispar* and *Hyphantria cunea*. *Anal. Inst. Cerc. Prot. Plantelor*, **14**, 101–106.

Gamayunova, S.G. (1987) Application of bacterial preparations in foci of leaf-roller multiplication. *Vestnik Sel'skokhozy. Nauki, Moscow*, **7**, 88–91.

Gamayunova, S.G. & Timchenko, G.A. (1985) Changes in numbers of parasites of *Tortrix viridana* following application of bacterial preparations. *Lesovods. Agrolesomelior.*, **70**, 56–59.

Gambino, P.V. (1988) Interactions between yellowjackets (Hymenoptera: Vespidae) and insect pathogens. PhD Thesis. University of California. 161p.

Gangwar, S.K., Yadava, C.P.S., Mishra, R.K., Dadheech, L.N. & Saxena, R.C. (1980) Efficiency of *Bacillus thuringiensis* against lepidopterous pests. *Indian J. Plant Prot.*, **8**, 154–156.

Gaprindashvili, N.K. (1975) Biological control of the main pests on tea plantations in the Georgian SSR. *8th Intern. Plant Prot. Congr., Moscow*, 29–33.

Gao, R.X., Lin, G.X., Guan, X. & Luo, Y.F. (1985) Studies on the new strain of *Bacillus thuringensis* var. *kurstaki* 8010. *J. Fujian Agric. Coll.*, **15**, 1–10.

Garcia, R.J. (1991) Effect of diazinon, deltamethrin and *Bacillus thuringiensis* var. *kurstaki* on the control of larvae of Lepidoptera on cabbage. *Bol. Entomol. Venezolana*, **6**, 19–25.

Garcia, R., des Rochers, B., & Tozer, W. (1980) Studies on the toxicity of *Bacillus thuringiensis* var. *israelensis* against organisms found in association with mosquito larvae. *Proc. Papers 48th Ann. Conf. Calif. Mosq. Vector Cont. Assoc.*, (ed. C.D. Grant) Anaheim, California. 33–36.

Garcia, R., des Rochers, B. & Tozer, W. (1982) Studies on *Bacillus thuringiensis* var. *israelensis* against mosquito larvae and other organisms. *Proc. Papers 49th Ann. Conf. Calif. Mosq. Vector Cont. Assoc*, 25–29.

Gardner, W.A. (1988) Enhanced activity of selected combinations of *Bacillus thuringiensis* and beta-exotoxin against fall armyworm (Lepidoptera: Noctuidae) larvae. *J. Econ. Entomol.*, **81**, 463–469.

Gardner, W.A. & Fuxa, J.R. (1980) Pathogens for the suppression of the fall armyworm. *Southeastern Branch, Entomol. Soc. Amer.*, **63**, 439–447.

Gardner, W.A., Pendley, A.F. & Storey, G.K. (1986) Interactions between *Bacillus thuringiensis* and its beta-exotoxin in fall armyworm (Lepidoptera: Noctuidae) neonate larvae. *Florida Entomol.*, **69**, 531–536.

Ge, A.Z., Shivarova, N.I. & Dean, D.H. (1989) Location of the *Bombyx mori* specificity domain on a *Bacillus thuringiensis* delta-endotoxin protein. *Proc. Nat. Acad. Sci. USA*, **86**, 4037–4041.

Ge, A.Z., Rivers, D., Milne, R. & Dean, D.H. (1991) Functional domains of *Bacillus thuringiensis* insecticidal crystal proteins. Refinement of *Heliothis virescens* and *Trichoplusia ni* specificity domains on CryIA(c). *J. Biol. Chem.*, **266**, 17954–17958.

Geervliet, J.B.F., Vlak, J.M. & Smits, P.H. (1991) Effects of *Spodoptera exigua* nuclear polyhedrosis virus and *Bacillus thuringiensis* subsp. *aizawai* mixtures on mortality of beet armyworm, *Spodoptera exigua*. *Med. Fac. Landbouww. Rijk. Gent*, **56**, 305–311.

Gelernter, W.D. (1990) *Bacillus thuringiensis*, bioengineering and the future of bioinsecticides. *Proc. Brighton Crop Prot. Conf., Pests Dis.*, **2**, 617–624.

Gelernter, W.D. (1997) Resistance to microbial insecticides: the scale of the problem and how to manage it. *BCPC Symposium Proc. no. 68: Microbial Insecticides: Novelty or Neccessity?* pp. 201–213.

Gelernter, W. & Schwab, G.E. (1993) Transgenic bacteria, viruses, algae and other microorganisms as *Bacillus thuringiensis* toxin delivery systems. *Bacillus thuringiensis, an Environmental Biopesticide: Theory and Practice* (eds. P.F. Entwistle, J.S. Cory, M.J. Bailey & S. Higgs), pp. 89–104. John Wiley and Sons, Chichester.

Genty, P. (1978) Morphology and biology of a moth defoliating oil palm in Latin America, *Stenoma cecropia* Meyrick. *Oleagineux*, **33**, 421–427.

Georghiou, G.P. (1984) Insecticide resistance in mosquitoes: research on new chemicals and techniques for management. *Ann. Rep. Mosq. Cont. Res. Uni. Calif.*, 97–99.

Georghiou, G.P. & Wirth, M.C. (1997) Influence of exposure to single versus multiple toxins of *Bacillus thuringiensis* subsp. *israelensis* on development of resistance in the mosquito *Culex quinquefasciatus* (Diptera: Culicidae). *Appl. Environ. Microbiol.*, **63**, 1095–1101.

Gerasimovich, M.M. (1971) The ecological features of the apple fruit moth, *Argyresthia conjugella*, and methods of controlling it. *Izv. Timiryazev. Sel'skokhozy. Akad.*, **No. 3**, 157–167.

Gerginov, L. (1978) Measures for the control of the corn borer. *Rast. Zashc.*, **26**, 12–15.

Gerst, J.J., Gachon, C., Stengel, B. & Stengel, M. (1977) Leek moth in Alsace. A review of five years of pheromone trapping. *Pepinier. Hortic. Maraichers,* **182**, 31–39.

Gharib, A.H. & Hilsenhoff, W.L. (1988) Efficacy of two formulations of *Bacillus thuringiensis* var. *israelensis* (H-14) against *Aedes vexans* and safety to non-target macroinvertebrates. *J. Amer. Mosq. Cont. Assoc.*, **4**, 252–255.

Gharib, A.H. & Szalay-Marzso, L. (1986) Selection of resistance to *Bacillus thuringiensis* serotype H-14 in a laboratory strian of *Aedes aegyptii* L. *Fundamental and Applied Aspects of Invertebrate Pathology* (eds. R.A. Samson, J.M. Vlak & D. Peters), Wageningen, 37.

Ghidiu, G.M., Collins, D.E. & Kirfman, G.W. (1996) Laboratory and field studies of *Bacillus thuringiensis* subsp. *tenebrionis* as a feeding deterrent to Colorado potato beetle (Coleoptera: Chrysomelidae). *J. Agric. Entomol.*, **13**, 349–357.

Ghobrial, A. & Ali, A. (1978) Semi-large field cotton experiments on the late season pest-predator complex in Iraq. *Proc. 4th Conf. Pest Cont.*, 471–479.

Ghobrial, A. & Dittrich, V. (1980) Early and late pest complexes on cotton, their control by aerial and ground application of insecticides and side-effects on the predator fauna. *Z. Angew. Entomol.*, **90**, 306–313.

Gibson, D.M., Greenspan Gallo, L., Krasnoff, S.B. & Ketchum, R.E.B. (1995) Increased efficacy of *Bacillus thuringiensis* subsp. *kurstaki* in combination with tannic acid. *J. Econ. Entomol.*, **88**, 270–277.

Giddings, G., Mytton, L., Griffiths, M., McCarthy, A., Morgan, C. & Skot, L. (1997) A secondary effect of transformation in *Rhizobium leguminosarum* transgenic for *Bacillus thuringiensis* subspecies *tenebrionis* delta-endotoxin (*cry*IIIA) genes. *Theor. Appl. Genet.*, **95**, 1062–1068.

Gill, S.A. & Raupp, M.J. (1994) Using entomopathogenic nematodes and conventional and biorational pesticides for controlling bagworm. *J. Arboric.*, **20**, 318–322.

Gilreath, M.E. & Funderburk, J.E. (1987) Entomopathogens for suppression of lesser cornstalk borer. *Proc. Amer. Peanut Res. Educ. Soc.*, **19**, 18.

Gingrich, R.E. (1987) Demonstration of *Bacillus thuringiensis* as a potential control agent for the adult Mediterranean fruit fly, *Ceratitis capitata* (Wied.). *J. Appl. Entomol.*, **104**, 378–385.

Gingrich, R.E. & Eschle, J.L. (1971) Susceptibility of immature horn flies to toxins of *Bacillus thuringiensis*. *J. Econ. Entomol.*, **64**, 1183–1188.

Gingrich, R.E., Allan, N. & Hopkins, D.E. (1974) *Bacillus thuringiensis*: laboratory tests against four species of biting lice (Mallophaga: Trichodectidae). *J. Invertebr. Pathol.*, **23**, 232–236.

Gingrich, R.E. & Haufler, M. (1978) Components of *Bacillus thuringiensis* active against larvae of the horn fly *Haematobia irritans*. *Folia Entomol. Mex.*, 121–122.

Giroux, S., Coderre, D., Vincent, C. & Cote, J.C. (1994b) Effects of *Bacillus thuringiensis* var. *san diego* on predation effectiveness, development and mortality of *Coleomegilla maculata lengi* (Col.: Coccinellidae) larvae. *Entomophaga,* **39**, 61–69.

Giroux, S., Cote, J.C., Vincent, C., Martel, P. & Coderre, D. (1994a) Bacteriological insecticide M-One effects on predation efficiency and mortality of adult *Coleomegilla maculata lengi* (Coleoptera: Coccinellidae). *J. Econ. Entomol.*, **87**, 39–43.

Glare, T.R. & O'Callaghan, M. (1998) Environmental and health impacts of *Bacillus thuringiensis israelensis*. *Report for the New Zealand Ministry of Health.* 58 pp.

Gloria, B.R. (1975) Chemical control of the armyworm *Prodenia sunia* (G.) on tomatoes. *Rev. Peruana Entomol.*, **18**, 120–123.

Glowacka Pilot, B. (1974) Susceptibility of some species of forest insect pests to the biopreparation Dipel. *Sylwan,* **118**, 17–25.

Glowacka Pilot, B. (1984) Control of *Lymantria monacha* - Bactospeine and Thuridan. *Las Polski,* **11**, 20.

Glowacka Pilot, B., Swiezynska, H. & Drabarczyk Grabarczyk, S. (1974) Trial in the control of larvae of Geometridae with the preparations Thuricide and Nogos. *Prace Inst. Bad. Lesn.*, No. 427–432, 213–221.

Goarant, G., Andre, F., Glement, G. & Betbeder Matibet, M. (1995) The rice stem borer in the Camargue: (harmfulness). *Phytoma,* **479**, 26–29.

Goebel, O., Roubaud, G. & Vergnet, C. (1997) Fluidosoufre BTR: a new concept for the control of grape berry moths and powdery mildew in vine. *Association Nationale pour la Protection des Plantes (ANPP); Paris; France. Quatrième Conference Internationale sur les Ravageurs en Agriculture,* le Corum, Montpellier, France, 397–402.

Goettel, M.S., Toohey, M.K. & Pillai, J.S. (1982) Laboratory bioassays of four formulations of *Bacillus thuringiensis israelensis* against *Aedes polynesiensis*, *Ae. pseudoscutellaris* and *Ae. aegypti*. *Mosq. News,* **42**, 163–167.

Goldberg, L.J. & Ford, I. (1974) Mortality of *Culex tarsalis* mosquito larvae obtained in laboratory studies using various combinations of *Bacillus thuringiensis* (HD-1) with two growth regulators. *Proc. Papers 42nd Ann. Conf. Calif. Mosq. Cont. Assoc.* (ed. T.D. Mulhern), 24-27 Feb. 1974, Anaheim, California, 169–174.

Goldberg, L.J. & Margalit, J. (1977) A bacterial spore demonstrating rapid larvicidal activity against *Anopheles sergentii, Uranotaenia unguiculata, Culex univittatus, Aedes aegypti* and *Culex pipiens*. *Mosq. News,* **37**, 355–358.

Goldberg, L.J., Ford, I. & Mulhern, T.D. (1974) Mortality of *Culex tarsalis* mosquito larvae obtained in laboratory studies using various combinations of *Bacillus thuringiensis* (HD-1) with two growth regulators. *Proc. Papers 42nd Ann. Conf. Calif. Mosq. Cont. Assoc.,* Feb. 24-27, 1974, 169–174.

Goldberg, L.J., Margalit, J., Undeen, A.H. & Nagel, W.L. (1978) The effect of *Bacillus thuringiensis* ONR-60A strain (Goldberg) on *Simulium* larvae in the laboratory. *Mosq. News,* **38**, 524–527.

Goldman, I.F., Arnold, J. & Carlton, B.C. (1986) Selection for resistance to *Bacillus thuringiensis* subspecies *israelensis* in field and laboratory populations of the mosquito *Aedes aegypti*. *J. Invertebr. Pathol.,* **47**, 317–24.

G'onev, G. (1975) Integrated and biological control in apple and plum orchards in the Stara Zagora district. *Rast. Zashc.,* **23**, 31–34.

G'onev, G. (1977) Bacterial preparations for *Laspeyresia funebrana* control. *Ovoshcharstvo,* **56**, 9–11.

Gonzalez, J.M., Jr., Brown, B.J. & Carlton, B.C. (1982) Transfer of *Bacillus thuringiensis* plasmids coding for delta-endotoxin among strains of *B. thuringiensis* and *B. cereus*. *Proc. Nat. Acad. Sci. USA,* **79**, 6951–6955.

Goodwin, W.D. (1985) A unique method for the prevention and amelioration of greater wax moth infestations in honeycombs and wax foundations. *South African Bee J.,* **57**, 36, 38–41.

Goral, V.M., Drozda, V.F., Zurabova, E.R., Lappa, N.V. & Ustimenko, A.A. (1984) Bacterial preparations on cabbage. *Zaschc. Rast.,* **7**, 20.

Gormly, N.M., Singer, S. & Genthner, F.J. (1996) Nontarget testing of microbial pest control agents using larvae of the coot clam *Mulinia lateralis*. *Dis. Aquat. Organ.,* **26**, 229–235.

Gorokhov, V.A. & Kaplenko, V.M. (1980) Biological control of *Dendrolimus pini*. *Lesnoe Khoz.,* **5**, 50–52.

Gould, F. (1988) Genetic engineering, integrated pest management and the evolution of pests. *Planned release of genetically engineered organisms. S15-S18; in Trends in Biotechnology 6, and Trends in Ecology and Evolution 3, combined special issue.* Elsevier Publications; Cambridge; UK.

Gould, F. (1997) Integrating pesticidal engineered crops into Mesoamerican agriculture. *Transgenic plants:* Bacillus thuringiensis *in Mesoamerican Agriculture,* (eds. A.J. Hruska & M.L. Pavon) pp. 6–36. Escuela Agricola Panamericana Zamorano; Tegucigalpa, Central America.

Gould, F. & Anderson, A. (1991) Effects of *Bacillus thuringiensis* and HD-73 delta-endotoxin on growth, behavior, and fitness of susceptible and toxin-adapted strains of *Heliothis virescens* (Lepidoptera: Noctuidae). *Environ. Entomol.,* **20**, 30–38.

Gould, F., Anderson, A., Reynolds, A., Bumgarner, L. & Moar, W. (1995) Selection and genetic analysis of a *Heliothis virescens* (Lepidoptera: Noctuidae) strain with high levels of resistance to *Bacillus thuringiensis* toxins. *J. Econ. Entomol.,* **88**, 1545–1559.

Gould, F., Martinez Ramirez, A., Anderson, A., Ferre, J., Silva, F.J. & Moar, W.J. (1992) Broad-spectrum resistance to *Bacillus thuringiensis* toxins in *Heliothis virescens*. *Proc. Nat. Acad. Sci. USA,* **89**, 7986–7990.

Grace, J.K. & Ewart, D.M. (1996) Recombinant cells of *Pseudomonas fluorescens*: a highly palatable encapsulation for delivery of genetically engineered toxins to subterranean termites (Isoptera: Rhinotermitidae). *Lett. Appl. Microbiol.,* **23**, 183–186.

Grant, C.D., Combs, J.C., Coykendall, R.L., Lusk, E.E., Washino, R.K., Mulligan, F.S., III & Schaefer, C.H. (1985) Field tests of *Bacillus thuringiensis* H. 14 commercial formulations. *Proc. Papers 52nd Ann. Conf. Calif. Mosq. Vector Cont. Assoc.,* Jan. 29-Feb. 1, 1984, Long Beach, California, 80–81.

Granum, P.E., Pinnavaia, S.M. & Ellar, D.J. (1988) Comparison of the in vivo and in vitro activity of the delta-endotoxin of *Bacillus thuringiensis* var. *morrisoni* (HD-12) and two of its constituent proteins after cloning and expression in *Escherichia coli*. *Eur. J. Biochem.,* **172**, 731–738.

Grau, P. (1987) Dibeta: a promising material for lygus bug and spider mite control in cotton. *Proc. Beltwide Cotton Prod. Conf.,* Dallas, USA, 136–137.

Gravena, S. (1984) Effect of insecticides used to control the coffee leaf-miner *Perileucoptera coffeella* (Guerin-Meneville, 1842) and *Bacillus thuringiensis* Berliner on the population of predacious ants. *An. Soc. Entomol. Bras.,* **13**, 389–390.

Gravena, S., Campos, A.R., Maia, O.S. & Paulaneto, G.T. (1980) Effectiveness of *Bacillus thuringiensis* Berliner and *Bacillus thuringiensis* + methomyl for the control of Lepidoptera on tomato. *An. Soc. Entomol. Bras.,* **9**, 241–248.

Gravena, S., Araujo, C., Campos, A.R., Villani, H.C. & Yotsumoto, T. (1983) Strategies for integrated pest management on cotton in Jaboticabal, Sao Paulo, with *Bacillus thuringiensis* Berliner and natural enemies. *An. Soc. Entomol. Bras.,* **12**, 17–29.

Grechkanev, O.M. & Maksimova, N.L. (1980) Effect of Entobakterin. *Zaschc. Rast.,* **8**, 36.

Green, M., Heumann, M., Sokolow, R., Foster, L.R., Bryant, R. & Skeels, M. (1990) Public health implications of the microbial pesticide *Bacillus thuringiensis*: an epidemiological study, Oregon, 1985-86. *Amer. J. Public Health*, **80**, 848–852.

Greene, G.L., Whitcomb, W.H. & Baker, R. (1974) Minimum rates of insecticide on soybeans: Geocoris and Nabis populations following treatment. *Florida Entomol.*, **57**, 114.

Greener, A. & Candy, S.G. (1994) Effect of the biotic insecticide *Bacillus thuringiensis* and a pyrethroid on survival of predators of *Chrysophtharta bimaculata* (Olivier) (Coleoptera: Chrysomelidae). *J. Aust. Entomol. Soc.*, **33**, 321–324.

Greener, A. & Candy, S.G. (1995) Effect of the biotic insecticide *Bacillus thuringiensis* on predators of *Chrysophtharta bimaculata*: reply. *J. Aust. Entomol. Soc.*, **34**, 99–100.

Griego, V.M. & Spence, K.D. (1978) Inactivation of *Bacillus thuringiensis* spores by ultraviolet and visible light. *Appl. Environ. Microbiol.*, **35**, 906–10.

Griego, V.M., Moffett, D. & Spence, K.D. (1979) Inhibition of active K+ transport in the tobacco hornworm (*Manduca sexta*) midgut after ingestion of *Bacillus thuringiensis* endotoxin. *J. Insect Physiol.*, **25**, 283–288.

Grigoryan, E.G., Sarkisyan, M.A. & Davtyan, L.T. (1988) The use of biological insecticides and Dimilin against leaf-feeding pests. *Biol. Z. Armenii*, **41**, 503–507.

Grinberg Sh, M., Pynzar, B.V. & Boubetryn, I.N. (1992) *Trichogramma* is not effective? Let's find out the reason. *Zaschc. Rast. Moskva*, **12**, 4–8.

Grison, P., Martouret, D. & Auer, C. (1971) Microbiological control of the larch tortrix. *Ann. Zool. Ecol. Animale*, **3**, 91–121.

Grison, P., Martouret, D., Servais, B. & Devriendt, M. (1976) Pesticides microbiens et environnement. *Ann. Zool. Ecol. Animale*, **8**, 133–160.

Groeters, F.R., Tabashnik, B.E., Finson, N. & Johnson, M.W. (1994) Fitness costs of resistance to *Bacillus thuringiensis* in the diamondback moth (*Plutella xylostella*). *Evolution*, **48**, 197–201.

Groner, A. (1977) Development and testing of a virus preparation (nuclear polyhedrosis virus) for control of the cabbage moth. *Mitteilungen aus der Biologischen Bundesanstalt fur Land und Forstwirtschaft Berlin Dahlem*, 180.

Guerra, G.P. (1995) The influence of diet in the control of the nun moth (*Lymantria monacha*) (Lepidoptera: Lymantriidae) with *Bacillus thuringiensis*. *Mitt. Dtsch. Ges. Allg. Angew. Entomol.*, **10**, 147–150.

Guillet, P., Escaffre, H. & Prud' hom, J.M. (1982) Utilisation of a formulation based on *Bacillus thuringiensis* H14 in the control of Onchocerciasis in West Africa. I- Effectiveness and application methods. *Cah. ORSTOM Entomol. Med. Parasitol.*, **20**, 175–180.

Gukasyan, V.M. & Gukasyan, A.B. (1980) The physiological activity of *Bacillus thuringiensis* var. *caucasicus* against leaf-gnawing insects and pests of pine. *Biol. Z. Armenii*, **33**, 432–435.

Guo, S.J., Lin, W.C., Wang, X.G., Liu, S.S. & Song, H.M. (1998) Toxicity of insecticides to *Oomyzus sokolowskii*, a parasitoid of diamondback moth (*Plutella xylostella*). *Chin. J. Biol. Cont.*, **14**, 97–100.

Guo, Y.J., Zuo, G.S., Zhao, J.H., Wang, N.Y. & Jiang, J.W. (1993) A laboratory test on the toxicity of thuringiensin to *Tetranychus urticae* (Acari: Tetranychidae) and *Phytoseiulus persimilis* (Acari: Phytoseiidae). *Chin. J. Biol. Cont.*, **9**, 151–155.

Gupta, M. & Rana, R.S. (1991) Effect of *Bacillus thuringiensis* var. *thuringiensis* Berliner on dry matter utilization by *Spilosoma obliqua* Walker (Lepidoptera: Arctiidae). *Indian J. Entomol.*, **53**, 59–65.

Guzeev, G.F. (1986) Microbiological preparations for control of defoliating insects in the pistacio forests of Central Asia. *Lesnoe Khoz.*, **10**, 61–63.

Guzeev, G.F. & Mansurov, O.A. (1983) The Turkestan saturniid. *Zaschc. Rast.*, **11**, 42.

Gusev, G.V., Zayats Yu, V., Topashchenko, E.M. & Rzhavina, G.K. (1983) Control of the Colorado beetle on aubergines. *Zaschc. Rast.*, **8**, 34.

Gusteleva, L.A. (1980) Results of the testing of microbial preparations against the large larch bark-beetle. *Izv. Sib. Otd. Akad. Nauk SSSR Biol*, **15**, 49–54.

Gusteleva, L.A. (1982) Prospects for using microbial preparations against *Ips subelongatus*. *Lesnoe Khoz.*, **9**, 67.

Guzeev, G.F. (1986) Microbiological preparations for control of defoliating insects in the pistacio forests of Central Asia. *Lesnoe Khoz.*, **10**, 61–63.

Gyoutoku, Y. & Kastio, T. (1990) Toxicity of pesticides to *Oligota* spp. (Coleoptera: Staphylinidae). *Proc. Assoc. Plant Prot. Kyushu*, 155–159.

Haag, H. & Boucias, D.G. (1991) Infectivity of insect pathogens against *Neochetina eichhorniae*, a biological control agent of water hyacinth. *Florida Entomol.*, **74**, 128–133.

Haas, E. (1987) Side effects of plant protectants on predatory mites and leafhoppers. *Obstbau Weinbau*, **24**, 70–73.

Haas, L. & Scriber, J.M. (1998) Phylloplane sterilization with bleach does not reduce *Btk* toxicity for *Papilio glaucus* larvae (Lepidoptera: Papilionidae). *Great Lakes Entomol.*, **31**, 49–57.

Habib, M.E.M. (1983) Potency of *Bacillus thuringiensis* var. *israelensis* (H-14) against some aquatic dipterous insects. *Z. Angew. Entomol.*, **95**, 368–376.

Habib, M.E.M. (1986) Responses of larvae of *Spodoptera latifascia* (Walker) to three preparations based on *Bacillus thuringiensis* var. *kurstaki* (H-3a:3b). *Rev. Agr. Piracicaba Brazil*, **61**, 9–16.

Habib, M.E.M. & De Andrade, C.F.S. (1984) Pathogenicity of *Bacillus thuringiensis* var. *kurstaki* (H: 3a-3b) for the cotton leafworm *Alabama argillacea* (Hubner, 1818) (Lepidoptera, Noctuidae). *Rev. Agric. Bras.*, **59**, 263–282.

Habib, M.E.M., Andrade, C.F.S., Favaro Junior, A., Heimpel, A.M. & Angus, T.A. (1986) Classification, pathology and susceptibility of larvae of *Brassolis sophorae* (L., 1758) infected by *Bacillus thuringiensis* var. *kurstaki* (H:3a-3b). *Rev. Agr. Piracicaba Brazil,* **61**, 105–113.

Hachler, M., Jermini, M. & Brunetti, R. (1998) Two new harmful noctuids on tomatoes in glasshouse in South and Western Switzerland. *Rev. Suisse Vitic. Arboric. Hortic.*, **30**, 281–285.

Hacker, J., Blum-Oehler, G., Muhldorfer, I & Tschape, H. (1997) Pathogenicity islands of virulent bacteria: structure, function and impact on microbial evolution. *Mol. Microbiol.* **26**, 1089–1097.

Hadley, W.M., Burchiel, S.W., McDowell, T.D., Thilsted, J.P., Hibbs, C.M., Whorton, J.A., Day, P.W., Friedman, M.B. & Stoll, R.E. (1987) Five-month oral (diet) toxicity/infectivity study of *Bacillus thuringiensis* insecticides in sheep. *Fund. Appl. Toxicol.*, **8**, 236–242.

Hafez, M., Abol Ela, F.N., Zaki, F.N., Salama, H.S. & Ragaei, M. (1997a) The potential of the predator *Orius albidepennis* on *Agrotis ypsilon* as affected by *Bacillus thuringiensis*. *Anz. Schaedlingskd. Pflanzenschutz Umweltschutz,* **70**, 127–130.

Hafez, M., Salama, H.S., Aboul Ela, R., Zaki, F.N. & Ragaei, M. (1997b) *Bacillus thuringiensis* affecting the larval parasite *Meteorus laeviventris* Wesm. (Hym., Braconidae) associated with *Agrotis ypsilon* (Rott.) (Lep., Noctuidae) larvae. *J. Appl. Entomol.*, **121**, 535–538.

Hafez, M., Salama, H.S. & Abdel Rahman, A. (1998a) Activity of *Bacillus thuringiensis* isolates on the corn borers, *Chilo agamemnon* Bles. (Crambidae) and *Ostrinia nubilalis* Hbn. (Pyraustidae). *Anz. Schaedlingskd. Pflanzenschutz Umweltschutz,* **71**, 110–112.

Hafez, M., Salama, H.S., El Moursy, A. & Abdel Rahman, A. (1998b) A biochemical approach to potentiate the activity of *Bacillus thuringiensis* against corn borers. *Anz. Schaedlingskd. Pflanzenschutz Umweltschutz,* **71**, 100–103.

Hagstrum, D.W. & Sharp, J.E. (1975) Population studies on *Cadra cautella* in a citrus pulp warehouse with particular reference to diapause. *J. Econ. Entomol.*, **68**, 11–14.

Halkova, Z., Zaykov, C., Antov, G., Mihaylova, A., Mircheva, V., Dinoeva, S. & Chipilska, L. (1993) Experimental study of subacute oral, dermal and inhalation toxicity of Bulmoscide preparation. *Polish J. Occupat. Med. Environ. Health,* **6**, 19–25.

Hall, I.M., Hunter, D.K. & Arakawa, K.Y. (1971) The effect of the beta -exotoxin fraction of *Bacillus thuringiensis* on the citrus red mite. *J. Invertebr. Pathol.*, **18**, 359–362.

Hall, I.M., Dulmage, H.T. & Arakawa, K.Y. (1972) Laboratory tests with entomogenous bacteria and the fungus *Beauveria bassiana* against the little house fly species *Fannia canicularis* and *F. femoralis*. *Environ. Entomol.*, **1**, 105–107.

Hall, I.M., Arakawa, K.Y., Dulmage, H.T. & Correa, J.A. (1977) The pathogenicity of strains of *Bacillus thuringiensis* to larvae of *Aedes* and to *Culex* mosquitoes. *Mosq. News,* **37**, 246–251.

Hallet, B., Rezsohazy, R. & Delcour, J. (1991) IS231A from *Bacillus thuringiensis* is functional in *Escherichia coli*: transposition and insertion specificity. *J. Bacteriol.*, **173**, 4526–4529.

Halos, S.C., Quiniones, S.S., de Guzman, E.D., Lapis, E.B. & Lucero, R.M. (1985) Towards an integrated control measure against shoot moth and needle blight affecting pines. *PCARRD Monitor,* **13**, 6–8.

Hama, H., Suzuki, K. & Tanaka, H. (1992) Inheritance and stability of resistance to *Bacillus thuringiensis* formulations of the diamondback moth, *Plutella xylostella* (Linnaeus) (Lepidoptera: Yponomeutidae). *Appl. Entomol. Zool.*, **27**, 355–362.

Hamed, A.R. (1978) Effects of *Bacillus thuringiensis* on *Yponomeuta evonymellus* (L.) and *Y. padellus* (L.) (Lep., Yponomeutidae). *Z. Angew. Entomol.*, **85**, 392–412.

Hamed, A.R. (1979) Effects of *Bacillus thuringiensis* on parasites and predators of *Yponomeuta evonymellus* (Lep., Yponomeutidae). *Z. Angew. Entomol.*, **87**, 294–311.

Hamel, D.R. (1977) The effects of *Bacillus thuringiensis* on parasitoids of the western spruce budworm, *Choristoneura occidentalis* (Lepidoptera: Tortricidae), and the spruce coneworm, *Dioryctria reniculelloides* (Lepidoptera: Pyralidae), in Montana. *Canad. Entomol.*, **109**, 1409–1415.

Hamilton, J.T. & Attia, F.I. (1977) Effects of mixtures of *Bacillus thuringiensis* and pesticides on *Plutella xylostella* and the parasite *Thyraeella collaris*. *J. Econ. Entomol.*, **70**, 146–148.

Han, L.Z. (1990) Selection of resistance to biological insecticide, *Bacillus thuringiensis* var. *israelensis* in *Culex pipiens pallens*. *Contr. Shanghai Inst. Entomol.*, **8**, 153–156.

Hannay, C.L. (1953) Crystalline inclusions in aerobic spore-forming bacteria. *Nature,* **172**, 1004.

Hannay, C.L. & Fitz-James, P. (1955) The protein crystals of *Bacillus thuringiensis* Berliner. *Canad. J. Microbiol.*, **1**, 694–709.

Hanowski, J.M., Niemi, G.J., Lima, A.R. & Regal, R.R. (1997) Response of breeding birds to mosquito control treatments of wetlands. *Wetlands,* **17**, 485–492.

Hansen, B.M., Damgaard, P.H., Eilenberg, J. & Pedersen, J.C., (1996) *Bacillus thuringiensis. Ecology and Environmental Effects of its Use for Microbial Pest Control.* Ministry of Environment and Energy, Denmark, Danish Environmental Protection Agency, Environment Project no. 316, 126 pp.

Hansen, B.M., Damgaard, P.H., Eilenberg, J. & Pedersen, J.C. (1998) Molecular and phenotypic characterization of *Bacillus thuringiensis* isolated from leaves and insects. *J. Invertebr. Pathol.*, **71**, 106–114.

Hansen, L.C. & Obrycki, J.J. (1999) Field deposition of transgenic pollen: lethal effects on the monarch buttefly. *Entomol. Soc. Amer. Ann. Meet.*, Atlanta, Georgia (submitted).

Haragsim, O. & Vankova, J. (1973) Pathogenic effect on the honeybee of exotoxins of eleven varieties of the group *Bacillus thuringiensis*. *Apidologie*, **4**, 87–101.

Harcourt, R.L., Llewellyn, D., Morton, R., Dennis, E.S. & Peacock, W.J. (1996) Effectiveness of purified *Bacillus thuringiensis* Berliner insecticidal proteins in controlling three insect pests of Australian eucalypt plantations. *J. Econ. Entomol.*, **89**, 1392–1398.

Hardman, J.M., Smith, R.F. & Bent, E. (1995) Effects of different integrated pest management programs on biological control of mites on apple by predatory mites (Acari) in Nova Scotia. *Environ. Entomol.*, **24**, 125–142.

Harpaz, I. & Wysoki, M. (1984) Susceptibility of the carob moth, *Ectomyelois ceratoniae*, to *Bacillus thuringiensis*. *Phytoparasitica*, **12**, 189–191.

Harper, J.D. & Abrahamson, L.P. (1979) Forest tent caterpillar control with aerially applied formulations of *Bacillus thuringiensis* and Dimilin. *J. Econ. Entomol.*, **72**, 74–77.

Harris, J.G., Hershey, C.N. & Watkins, M.J. (1998) The usage of Karate (lambda-cyhalothrin) oversprays in combination with refugia, as a viable and sustainable resistance management strategy for B.T. cotton. *1998 Proc. Beltwide Cotton Conf.*, (eds. P. Dugger & D. Richter) California, USA, 5-9 January 1998. Volume 2, 1217–1220.

Harris, M. (1993) Fungus gnats: they're more than just a nuisance. *Grower Talks*, **56**, 49–59.

Harris, M.K. & Cutler, B. (1975) Walnut caterpillar response to insecticides. *Progress Report, Texas Agric. Exp. St.*, PR-3314, 2 p.

Harris, M.O., Mafile'o, F. & Dhana, S. (1997) Behavioural responses of lightbrown apple moth neonate larvae on diets containing *Bacillus thuringiensis* formulations or endotoxins. *Entomol. Exp. Appl.*, **84**, 207–219.

Harvey, T.L. & Howell, D.E. (1964) Resistance in the house fly to *Bacillus thuringiensis* Berliner. *J. Invertebr. Pathol.*, **7**, 92–100.

Hasan, R., Khan, A.M. & van Emden, H.F. (1994) Efficacy of *Bacillus thuringiensis* against caterpillars of *Corcyra cephalonica* Stainton. *Ann. Plant Prot. Sci.*, **2**, 5–7.

Hassan, E. & Graham Smith, S. (1995) Toxicity of endosulfan, esfenvalerate and *Bacillus thuringiensis* on adult *Microplitis demolitor* Wilkinson and *Trichogrammatoidea bactrae* Nagaraja. *Z. Pflanzenkrankheiten Pflanzenschutz* **102**, 422–428.

Hassan, S. & Krieg, A. (1975) *Bacillus thuringiensis* preparations harmless to the parasite *Trichogramma cacoeciae* (Hym.:Trichogrammatidae). *Z. Pflanzenkrankheiten Pflanzenschutz*, **82**, 515–521.

Hassan, S.A. (1983) Results of laboratory testing of a series of plant protection preparations on egg parasites of the genus *Trichogramma* (Hymenoptera, Trichogrammatidae). *Nachrichtenbl. Dtsch. Pflanzenschutzdienst*, **35**, 21–25.

Hassan, S.A. (1992) Testing methodology and the concept of the IOBC/WPRS working group. *Pesticides and Non-target Invertebrates* (ed. P.C. Jepson), pp. 1–18. Intercept, Wimborne, Dorset.

Hassan, S.A., Bigler, F., Bogenschutz, H., Brown, J.U., Firth, S.I., Huang, P., Ledieu, M.S., Naton, E., Oomen, P.A., Overmeer, W.P.J., Rieckmann, W., Samsoe Petersen, L., Viggiani, G. & van Zon, A.Q. (1983) Results of the second joint pesticide testing programme by the IOBC/WPRS-Working Group "Pesticides and Beneficial Arthropods". *Z. Angew. Entomol.*, **95**, 151–158.

Hassanian, M.A., Soufy, H., El Ghaffar, F.A.A. & El Megeed, K.N.A. (1997a) Some studies on the safety of *Bacillus thuringiensis* as a biological control agent. *Egypt. J. Comp. Pathol. Clinical Pathol.*, **10**, 141–147.

Hassanain, M.A., El Garhy, M.F., Abdel Ghaffar, F.A., El Sharaby, A., & Abdel Megeed, K.N. (1997b) Biological control studies of soft and hard ticks in Egypt. I. The effect of *Bacillus thuringiensis* varieties on soft and hard ticks (Ixodidae). *Parasit. Res.*, **83**, 209–213.

Hastowo, S., Lay, B.W. & Ohba, M. (1992) Naturally occurring *Bacillus thuringiensis* in Indonesia. *J. Appl. Bacteriol.*, **73**, 108–113.

Hasty, M., Durham, E. & Payne, G. (1997) Evaluation of the susceptibility of tobacco budworm (*Heliothis virescens*) and cotton bollworm (*Helicoverpa zea*) populations in Georgia to various insecticides. *Proc. Beltwide Cotton Conf.*, La, USA, January 6-10, 1997: Volume 2. 1292–1293.

Haufler, M. & Kunz, S.E. (1985) Laboratory evaluation of an exotoxin from *Bacillus thuringiensis* subsp. *morrisoni* to horn fly larvae (Diptera: Muscidae) and mice. *J. Econ. Entomol.*, **78**, 613–616.

Haverty, M.I. (1982) Sensitivity of selected nontarget insects to the carrier of Dipel 4L in the laboratory. *Environ. Entomol.*, **11**, 337–338.

Havukkala, I. (1982) Natural control methods against cabbage flies. *Vaxtskyddsnotiser*, **46**, 90–93.

Havukkala, I. (1988) Non-chemical control methods against cabbage root flies *Delia radicum* and *Delia floralis* (Anthomyiidae). *Ann. Agric. Fenniae*, **27**, 271–279.

Hayashi, H. (1996) Side effects of pesticides on *Encarsia formosa* Gahan. *Bull. Hiroshima Prefectural Agric. Res. Center*, **64**, 33–43.

Heckel, D.G., Gahan, L.C., Gould, F. & Anderson, A. (1997a) Identification of a linkage group with a major effect on resistance to *Bacillus thuringiensis* Cry1Ac endotoxin in the tobacco budworm (Lepidoptera: Noctuidae). *J. Econ. Entomol.*, **90**, 75–86.

Heckel, D.G., Gahan, L.J., Gould, F., Daly, J.C. & Trowell, S. (1997b) Genetics of *Heliothis* and *Helicoverpa* resistance to chemical insecticides and to *Bacillus thuringiensis*. *Pestic. Sci.* **51**, 251–258.

Hegazy, E. & Antounious, A.G. (1987) Feeding deterrent efficacy of certain microbial insecticides against the cotton leafworm, *Spodoptera littoralis* (Boisd.). *Ann. Agric. Sci. Ain Shams Univ.*, **32**, 1765–1778.

Heimpel, A.M. & Angus, T.A. (1958) The taxonomy of insect pathogens related to *Bacillus cereus* Frankland and Frankland. *Canad. J. Microbiol.*, **4**, 531–541.

Heinz, K.M., Newman, J.P. & Parrella, M.P. (1988) Biological control of leafminers on greenhouse marigolds. *Calif. Agr.*, **42**, 10–12.

Helgesen, R.G. & Zenner Polania, I. (1974) A new control program for the omnivorous leaf roller during rose production. *New York State Flower Indust. Bull.*, **44**, 2–3.

Helgason, E., Caugant, D.A., Lecadet, M.M., Chen, Y., Mahillon, J., Lovgren, A., Hegna, I., Kvaloy, K. & Kolsto, A.B. (1998) Genetic diversity of *Bacillus cereus/B. thuringiensis* isolates from natural sources. *Curr. Microbiol.*, **37**, 80–87.

Helms, T.J. & Wedberg, J.L. (1976) Effect of *Bacillus thuringiensis* on *Nosema*-infected midgut epithelium of *Loxagrotis albicosta* (Lepidoptera: Noctuidae). *J. Invertebr. Pathol.*, **28**, 383–384.

Helyer, N. (1991) Laboratory pesticide screening method for the aphid predatory midge *Aphidoletes aphidimyza* (Rondani) (Diptera: Cecidomyiidae). *Biocontr. Sci. Technol.*, **1**, 53–58.

Hendricks, L.C. & Barbera, G. (1994) Almond growing and sustainable agriculture in California. *1st Intern. Congr. Almond, Agrigento, Italy, May, 1993*, **373**, 265–267.

Heppner, J.B., Pena, J.E. & Glenn, H. (1987) The banana moth, *Opogona sacchari* (Bojer) (Lepidoptera: Tineidae), in Florida. *Entomology Circular, Division of Plant Industry, Florida Department of Agriculture and Consumer Services*, No. **293**, 4 p.

Herbert, D.A. & Harper, J.D. (1985) Bioassay of a beta-exotoxin of *Bacillus thuringiensis* against *Heliothis zea* larvae. *J. Invertebr. Pathol.*, **46**, 247–250.

Herbert, D.A. & Harper, J.D. (1986) Bioassays of a beta-exotoxin of *Bacillus thuringiensis* against *Geocoris punctipes* (Hemiptera: Lygaeidae). *J. Econ. Entomol.*, **79**, 592–595.

Herms, C.P., McCullough, D.G., Baue, L.S., Haack, R.A., Miller, D.L. & Dubois, N.R. (1997) Susceptibility of the endangered Karner blue butterfly (Lepidoptera: Lycaenidae) to *Bacillus thuringiensis* var. *kurstaki* used for gypsy moth suppression in Michigan. *Great Lakes Entomol.*, **30**, 125–141.

Hernandez, J.L.L. (1988) Evaluation of toxicity of *Bacillus thuringiensis* to *Spodoptera frugiperda* (Smith). *Entomophaga*, **33**, 163–171.

Hernandez, E., Ramisse, F., Cruel, T., le Vagueresse, R. & Cavallo, J.-D. (1999) *Bacillus thuringiensis* serotype H34 isolated from human and insecticidal strains serotypes H3a3b and H14 can lead to death of immunocompetent mice after pulmonary infection. *FEMS Immun. Med. Microbiol.*, **24**, 43–47.

Hernandez, E., Ramisse, F., Ducoureau, J.P., Cruel, T. & Cavallo, J.D. (1998) *Bacillus thuringiensis* subsp. *konkukian* (serotype H34) superinfection: case report and experimental evidence of pathogenicity in immunosuppressed mice. *J. Clin. Microbiol.*, **36**, 2138–2139.

Hernandez, J.L.L. (1988) Evaluation of toxicity of *Bacillus thuringiensis* to *Spodoptera frugiperda* (Smith). *Entomophaga*, **33**, 163–171.

Herrnstadt, C., Gaertner, F., Gelernter, W. & Edwards, D.L. (1987) *Bacillus thuringiensis* isolate with activity against Coleoptera. *Biotechnology in Invertebrate Pathology and Cell Culture*, pp. 101–113, Academic Press, Inc., San Diego, USA

Herrnstadt, C., Soares, G.G., Wilcox, E.R. & Edwards, D.L. (1986) A new strain of *Bacillus thuringiensis* with activity against coleopteran insects. *Biotechnology*, **4**, 305–308.

Hershey, A.E., Shannon, L., Axler, R., Ernst, C. & Mickelson, P. (1995) Effects of methoprene and *Bti* (*Bacillus thuringiensis* var. *israelensis*) on non-target insects. *Hydrobiologia*, **308**, 219–227.

Heungens, A. & Pelerents, C. (1977) Control of *Trialeurodes vaporariorum* on hibiscus and gerbera. *Med. Fac. Landbouww., Rijk. Gent*, **42**, 1471–1477.

Hidalgo Salvatierra, O. & Palm, J.D. (1972) Studies on the shootborer *Hypsipyla grandella* Zeller. (Lep., Pyralidae). XIV. Susceptibility of first instar larvae to *Bacillus thuringiensis*. *Grijpma P*, **22**, 467–468.

Hidalgo Salvatierra, O. & Grijpma, P. (1973) Control of *Hypsipyla grandella* (Zeller) by microbiological methods. *Proc. 1st Symp. Integr. Cont. Hypsipyla, Turrialba, Costa Rica*, 24.

Higuchi, K., Saitoh, H., Mizuki, E. & Ohba, M. (1998a) Similarity in moth-fly specific larvicidal activity between two serologically unrelated *Bacillus thuringiensis* strains. *FEMS Microbiol. Lett.*, **169**, 213–8.

Higuchi, K., Saitoh, H., Mizuki, E., Hwang, S.H. & Ohba, M. (1998b) A novel isolate of *Bacillus thuringiensis* serovar *leesis* that specifically exhibits larvicidal activity against the moth-fly, *Telmatoscopus albipunctatus*. *Syst. Appl. Microbiol.*, **21**, 144–150.

Hilbeck, A., Baumgartner, M., Fried, P.M. & Bigler, F. (1998a) Effects of transgenic *Bacillus thuringiensis* corn-fed prey on mortality and development time of immature *Chrysoperla carnea* (Neuroptera: Chrysopidae). *Environ. Entomol.*, **27**, 480–487.

Hilbeck, A., Moar, W.J., Pusztai Carey, M., Filippini, A. & Bigler, F. (1998b) Toxicity of *Bacillus thuringiensis* Cry1Ab toxin to the predator *Chrysoperla carnea* (Neuroptera: Chrysopidae). *Environ. Entomol.*, **27**, 1255–1263.

Hilburn, D.J. & Jennings, D.T. (1988) Terricolous spiders (Araneae) of insecticide-treated spruce-fir forests in west-central Maine. *Great Lakes Entomol.*, **21**, 105–114.

Hildahl, V. & Peterson, L.O.T. (1974) Fall and spring cankerworms in the prairie provinces. *Inf. Rep. Northern For. Res. Centre Canad.*, NOR-X-100, 10 p.

Hill, C.A. & Pinnock, D.E. (1998) Histopathological effects of *Bacillus thuringiensis* on the alimentary canal of the sheep louse, *Bovicola ovis*. *J. Invertebr. Pathol.*, **72**, 9–20.

Hodgkinson, R.S., Finnis, M., Shepherd, R.F. & Cunningham, R.C. (1979) Aerial applications of nuclear polyhedrosis virus and *Bacillus thuringiensis* against western spruce budworm. *Joint Report, Ministry of Forests, British Columbia Canadian Forestry Service,* **No. 10**, 19 p.

Hoffmann, D. (1980) Induction of antibacterial activity in the blood of the migratory locust *Locusta migratoria* L. *J. Insect Physiol.*, **26**, 539–549.

Hoffmann, D. & Brehelin, M. (1976) On the origin and role of an activity of the lysozyme type demonstrated in the blood of *Locusta migratoria migratorioides* R. & F. *Acrida,* **5**, 181–188.

Hoffmann, D., Brehelin, M. & Hoffmann, J.A. (1974) Modifications of the hemogram and of the hemocytopoietic tissue of male adults of *Locusta migratoria* (Orthoptera) after injection of *Bacillus thuringiensis*. *J. Invertebr. Pathol.*, **24**, 238–247.

Hoffman, C. Van der Bruggen, H., Hofte, H., Van Rie, J., Jansens, S. & Mellaert, H.V. (1988) Specificity of *Bacillus thuringiensis* d-endotoxins is correlated with the presence of high affinity binding sites in the brush border membrane of target insect midguts. *Proc. Nat. Acad. Sci. USA,* **85**, 7844–7848.

Hofte, H. & Whiteley, H.R. (1989) Insecticidal crystal proteins of *Bacillus thuringiensis*. *Microbiol Rev.*, **53**, 242–255.

Holck, A.R. & Meek, C.L. (1987) Dose-mortality responses of crawfish and mosquitoes to selected pesticides. *J. Amer. Mosq. Con. Assoc.*, **3**, 407–411.

Holden, A.V. & Bevan, D. (eds.) (1978) Control of pine beauty moth by fenitrothion in Scotland 1978. Forestry Commission, Edinburgh, 176 pp.

Holmes, S.B. (1998) Reproduction and nest behaviour of Tennessee warblers *Vermivora peregrina* in forests treated with Lepidoptera-specific insecticides. *J. Appl. Ecol.*, **35**, 185–194.

Holsten, E.H. & Hard, J. (1985) Efficacy of *Bacillus thuringiensis* Berliner for suppressing populations of large aspen tortrix in Alaska. *Canad. Entomol.*, **117**, 587–591.

Honda, T., Shiba, A., Seo, S., Yamamoto, J., Matsuyama, J. & Miwatani, T. (1991) Identity of hemolysins produced by *Bacillus thuringiensis* and *Bacillus cereus*. *FEMS Microbiol. Lett.*, **79**, 205–210.

Horák, P., Weiser, J., Mikes, L. & Kolárová, L. (1996) The effect of *Bacillus thuringiensis* M-exotoxin on Trematode *Cercariae*. *J. Invertebr. Pathol.*, **68**, 41–49.

Hori, H., Suzuki, N., Ogiwara, K., Himejima, M., Indrasith, L.S., Minami, M., Asano, S., Sato, R., Ohba, M. & Iwahana, H. (1994) Characterization of larvicidal toxin protein from *Bacillus thuringiensis* serovar *japonensis* strain Buibui specific for scarabaeid beetles. *J. Appl. Bacteriol.*, **76**, 307–313.

Horowitz, M. (1977) Specific phytotoxicity of surfactants. *Proc. EWRS Symp. Different Methods Weed Cont. Integration*, 79–86.

Hossain, M.A., Sohel, A., Sirajul, H., Ahmed, S. & Hoque, S. (1997) Abundance and distribution of *Bacillus thuringiensis* in the agricultural soil of Bangladesh. *J. Invertebr. Pathol.*, **70**, 221–225.

Hou, R.F. (1987) Microbial control of insects in Taiwan. *Tech. Bull. Food Fert. Tech. Center Asian Pacific Region Taiwan,* **102**, 5pp.

Hougard, J.M., Darriet, F. & Bakayoko, S. (1983) Evaluation in a natural environment of the larvicidal activity of *Bacillus thuringiensis* serotype H-14 on *Culex quinquefasciatus* Say, 1823 and *Anopheles gambiae* Giles, 1902 s. l. (Diptera: Culicidae) in West Africa. *Cah. ORSTOM Entomol. Med. Parasitol.*, **21**, 111–117.

Hougard, J.M., Duval, J. & Escaffre, H. (1985) Evaluation in the natural environment of the larvicidal activity of a formulation of *Bacillus thuringiensis* H-14 on *Aedes aegypti* (Linne) in a yellow fever epidemic focus in the Ivory Coast. *Cah. ORSTOM Entomol. Med. Parasitol.*, **23**, 235–240.

Hough Goldstein, J. & Keil, C.B. (1991) Prospects for integrated control of the Colorado potato beetle (Coleoptera: Chrysomelidae) using *Perillus bioculatus* (Hemiptera: Pentatomidae) and various pesticides. *J. Econ. Entomol.*, **84**, 1645--1651.

Hough Goldstein, J.A. & Whalen, J. (1993) Inundative release of predatory stink bugs for control of Colorado potato beetle. *Biological Control,* **3**, 343–347.

Hough Goldstein, J., Janis, J.A. & Ellers, C.D. (1996) Release methods for *Perillus bioculatus* (F.), a predator of the Colorado potato beetle. *Biological Control,* **6**, 114–122.

Hough Goldstein, J., Tisler, A.M., Zehnder, G.W. & Uyeda, K.A. (1991) Colorado potato beetle (Coleoptera: Chrysomelidae) consumption of foliage treated with *Bacillus thuringiensis* var. *san diego* and various feeding stimulants. *J. Econ. Entomol.,* **84**, 87–93.

Houston, J., Dancer, B.N. & Learner, M.A. (1989a) Control of sewage filter flies using *Bacillus thuringiensis* var. *israelensis* I. Acute toxicity tests and pilot scale trial. *Water Res.,* **23**, 369–378.

Houston, J., Dancer, B.N. & Learner, M.A. (1989b) Control of sewage filter flies using *Bacillus thuringiensis* var. *israelensis* II. Full scale trials. *Water Res.,* **23**, 379–385.

Howard, F.W. (1990) Population suppression of mahogany webworm, *Macalla thyrsisalis* (Lepidoptera: Pyralidae), with natural products. *Florida Entomol.,* **73**, 225–229.

Hoy, C.W. & Hall, F.R. (1993) Feeding behaviour of *Plutella xylostella* and *Leptinotarsa decemlineata* on leaves treated with *Bacillus thuringiensis* and esfenvalerate. *Pestic. Sci.,* **38**, 335–340.

Hoy, C.W. & Head, G. (1995) Correlation between behavioral and physiological responses to transgenic potatoes containing *Bacillus thuringienesis* delta-endotoxin in *Leptinotarsa decemlineata* (Coleoptera: Chrysomelidae). *J. Econ. Entomol.,* **88**, 480–486.

Hoy, M.A. & Ouyang, Y.L. (1987) Toxicity of the beta-exotoxin of *Bacillus thuringiensis* to *Tetranychus pacificus* and *Metaseiulus occidentalis* (Acari: Tetranychidae and Phytoseiidae). *J. Econ. Entomol.,* **80**, 507–511.

Huang, F., Higgins, R.A. & Buschman, L.L. (1997) Baseline susceptibility and changes in susceptibility to *Bacillus thuringiensis* subsp. *kurstaki* under selection pressure in European corn borer (Lepidoptera: Pyralidae). *J. Econ. Entomol.,* **90**, 1137–1143.

Huang, G.Q., Yuan, F.Y. & Xu, B.Z. (1995) Influence of BT-187 emulsion on nontarget organisms in a rice field. *Chin. J. Vector Biol. Cont.,* **6**, 1–3.

Huang, R.N., Lo, I.P., Ho, C.M., Huang, J.S. & Hsu, E.L. (1993) Effectiveness of two formulations of *Bacillus thuringiensis* var. israelensis on mosquito larvae. *Chin. J. Entomol.,* **13**, 177–185.

Huber, H.E., Luthy, P., Ebersold, H.R. & Cordier, J.L. (1981) The subunits of the parasporal crystal of *Bacillus thuringiensis*: size, linkage and toxicity. *Arch. Mikrobiol.,* **129**, 14–18.

Hukuhara, T., Midorikawa, M. & Iwahana, H. (1984) The effect of sigma-endotoxin of *Bacillus thuringiensis* on the gut movements of the silkworm, *Bombyx mori*. *Appl. Entomol. Zool.,* **19**, 221–226.

Hull, L.A., McPheron, B.A. & Lake, A.M. (1997) Insecticide resistance management and integrated mite management in orchards: can they coexist? *Pestic. Sci.,* **51**, 359–366.

Hurpin, B. (1974) Biological control of *Oryctes*. Testing of various diseases in the laboratory. *Oleagineux,* **29**, 135–140.

Hussain, M. & Askari, A. (1976) Field tests of *Bacillus thuringiensis* and chemical insecticides for control of *Earias insulana* on cotton. *J. Econ. Entomol.,* **69**, 343–344.

Hussein, M.E., Merdan, A., Razik, N.A.A., Morsy, S. & Botros, M. (1990) Presence of certain bacteriophages in mosquito larval habitats inhibiting the larvicidal activity of *Bacillus thuringiensis* and *B. sphaericus*. *J. Appl. Entomol.,* **109**, 513–519.

Hussein, M.E., Merdan, A., Razik, N.A.A., Morsy, S. & Botros, M. (1991) Presence of certain bacteriophages in mosquito larval habitats inhibiting the larvicidal activity of *Bacillus thuringiensis* and *B. sphaericus*. *Zentralbl. Mikrobiol.,* **146**, 271–277.

Hussey, N.W., Burges, H.D. & David, W.A.L. (1976) Annual report 1975. *Report, Glasshouse Crops Research Institute,* 160 p.

Husz, B. (1929) On the use of *Bacillus thuringiensis* in the flight against the corn borer. *Int. Corn Borer Invest. Sci. Rep.,* 2, 99–110.

Hwang, G.H. & Ding, T. (1975) Studies on the nuclear polyhedrosis-virus disease of the cotton leafworm, *Prodenia litura* F. *Acta Entomol. Sin.,* **18**, 17–24.

Hwang, S.Y., Lindroth, R.L., Montomery, M.E. & Shields, K.S. (1995) Aspen leaf quality affects gypsy moth (Lepidoptera: Lymantriidae) susceptibility to *Bacillus thuringiensis*. *J. Econ. Entomol.,* **88**, 278–282.

Iacob, M., Matei, I., Voica, E. & Vladu, S. (1981) Influence of some control treatments against *Grapholitha molesta* Busck., on the quantity and quality of peach fruits, on the irrigated sands of the left bank of the river Jiu. *Anal. Inst. Cercet. Prot. Plantelor,* **16**, 385–394.

Ibarra, J.E. & Federici, B.A. (1986) Parasporal bodies of *Bacillus thuringiensis* subsp. *morrisoni* (PG-14) and *Bacillus thuringiensis* subsp. *israelensis* are similar in protein composition and toxicity. *FEMS Microbiol. Lett.,* **34**, 79–84.

Ibarra, L.F., Araya, J.E. & Arretz, P., III (1992) Laboratory and field studies in Chile on the control of *Epinotia aporema* (Lepidoptera: Olethreutidae) and *Rachiplusia nu* (Lepidoptera: Noctuidae) on *Phaseolus vulgaris* with growth regulators, *Bacillus thuringiensis*, and avermectin. *Crop Prot.,* **11**, 185–190.

Ibert, H., Frechen, M., Schallehn, G. & Kramer, J. (1996) Identification of *Bacillus* species using the computer based analysis system VITEK. *Arch. Lebensmittelhyg.*, **47**, 122–125.

Idris, A.B. & Grafius, E. (1993) Differential toxicity of pesticides to *Diadegma insulare* (Hymenoptera: Ichneumonidae) and its host, the diamondback moth (Lepidoptera: Plutellidae). *J. Econ. Entomol.*, **86**, 529–536.

Ignatowicz, S. (1997) Combination of gamma irradiation with a bioinsecticidal treatment for controlling larvae of the Mediterranean flour moth, *Ephestia kuehniella* (Zell.). *Ann. Warsaw Agric. Univ. Agric.*, **31**, 53–58.

Ignoffo, C.M. & Gard, I. (1970) Use of an agar base diet and house fly larvae to assay b-exotoxin activity of *Bacillus thuringiensis*. *J. Econ. Entomol.* **63**, 1987–1989.

Ignoffo, C.M. & Roush, R.T. (1986) Susceptibility of permethrin- and methomyl-resistant strains of *Heliothis virescens* (Lepidoptera: Noctuidae) to representative species of entomopathogens. *J. Econ. Entomol.*, **79**, 334–337.

Ignoffo, C.M., Hostetter, D.L. & Kearby, W.H. (1973) Susceptibility of walkingstick, orangestriped Oakworm and variable Oakleaf caterpillar, to *Bacillus thuringiensis* var. *alesti*. *Environ. Entomol.*, **2**, 807–809.

Ignoffo, C.M., Hostetter, D.L. & Pinnell, R.E. (1974) Stability of *Bacillus thuringiensis* and *Baculovirus heliothis* on soybean foliage. *Environ. Entomol.*, **3**, 117–119.

Ignoffo, C.M., Garcia, C., Kroha, M.J. & Fukuda, T. (1980a) Susceptibility of *Aedes aegypti* to four varieties of *Bacillus thuringiensis*. *Mosq. News*, **40**, 290–291.

Ignoffo, C.M., Garcia, C., Kroha, M.J. & Hoffman, J.D. (1980b) Effects of bacteria and a fungus fed singly or in combination on mortality of larvae of the cabbage looper (Lepidoptera: Noctuidae). *J. Kansas Entomol. Soc.*, **53**, 797–800.

Ignoffo, C.M., Couch, T.L., Garcia, C. & Kroha, M.J. (1981) Relative activity of *Bacillus thuringiensis* var. *kurstaki* and *B. thuringiensis* var. *israelensis* against larvae of *Aedes aegypti, Culex quinquefasciatus, Trichoplusia ni, Heliothis zea* and *Heliothis virescens*. *J. Econ. Entomol.*, **74**, 218–222.

Ignoffo, C.M., Garcia, C. & Kroha, M. (1982a) Susceptibility of the Colorado potato beetle *Leptinotarsa decemlineata* to *Bacillus thuringiensis*. *J. Invertebr. Pathol.*, **39**, 244–246.

Ignoffo, C.M., Garcia, C., Kroha, M. & Couch, T.L. (1982b) High-temperature sensitivity of formulations of *Bacillus thuringiensis* var. *israelensis*. *Environ. Entomol.*, **11**, 409–411.

Ilieva, S. (1973) The results of yield trials. 1. Control of *Cydia funebrana* and *Tranzschelia pruni-spinosae*. *Rast. Zashc.*, **21**, 5 – 6.

Imenes, S.D.L., Bergmann, E.C., Takematsu, A.P., Hojo, H. & De Campos, T.B. (1990) Influence of insecticides on the population of *Myzus persicae* (Sulzer, 1776) and its parasitoids on a tomato crop (*Lycopersicum esculentum*). *An. Soc. Entomol. Bras.*, **19**, 291–299.

Inal, J.R.M., Karunakaran, V. & Burges, H.D. (1992) Generalized transduction in *Bacillus thuringiensis* var. *aizawai*. *J. Appl. Bacteriol.*, **72**, 87–90.

Indrasith, L.S., Suzuki, N., Ogiwara, K., Asano, S. & Hori, H. (1992) Activated insecticidal crystal proteins from *Bacillus thuringiensis* serovars killed adult house flies. *Lett. Appl. Microbiol.*, **14**, 174–177.

Injac, M. & Dulic, K. (1982) Activation and control of overwintered larvae of summer fruit leaf-rollers (Tortricidae: *Pandemis heparana* Den et Schiff. and *Adoxophyes orana* F. v. R.). *Zast. Bilja*, **33**, 27–37.

Innes, D.G.L. & Bendell, J.F. (1989) The effects on small-mammal populations of aerial applications of *Bacillus thuringiensis*, fenitrothion, and Matacil used against jack pine budworm in Ontario. *Canad. J. Zool.*, **67**, 1318–1323.

Inogamov, R.U. & Sharafutdinov Sh, A. (1975) Exotoxin against the common spider mite. *Zaschc. Rast.*, **7**, 27.

Inserra, S., Calabretta, C. & Garzia, G.T. (1987) Attack by *Cacoecimorpha pronubana* (Hbn.) on protected crops of gerbera and rose and possibilities of chemical and biological control. *Difesa delle Piante*, **10**, 97–100.

Intari, S.E. (1996) The control of *Mucanum* sp. causing damage to seedlings of *Shorea javanica* in Krui, West Lampung. *Bul. Penelitian Hutan*, **604**, 31–37.

Ioriatti, C., Pasqualini, E. & Delaiti, M. (1996) Effectiveness of *Bacillus thuringiensis* Berliner on three species of apple leafrollers. *Boll. Ist. Entomol. 'Guido Grandi' Univ. Studi Bologna*, **50**, 73–93.

Iriarte, J., Bel, Y., Ferrandis, M.D., Andrew, R., Murillo, J., Ferre, J. & Caballero, P. (1998) Environmental distribution and diversity of *Bacillus thuringiensis* in Spain. *Syst. Appl. Microbiol.* **21**, 97–106.

Ironside, D.A. & Giles, J. (1981) Effect of insecticides on *Homoeosoma vagella* Zellar damage to macadamia flowers. *Queensland J. Agric. Animal Sci.*, **38**, 61–64.

Irshad, M., Mirza, S. & Rahatullah (1982a) Control of the sugarcane borers *Chilo infuscatellus* and *Tryporyza nivella* by Bactospeine, a microbial pesticide. *Pertanika*, **5**, 263–264.

Irshad, M., Shah, I. & Beg, M.N. (1982b) Chemical control of the Gurdaspur borer, *Acigona steniellus* Hamps. (Lepidoptera: Pyralidae) in Pakistan. *Intern. Pest Cont.*, **24**, 108.

Isakova, N.P. & Mogilevskaya, A.B. (1975) Intestinal flora of the cabbage white butterfly (*Pieris brassicae* L.), the gamma moth (*Phytometra gamma* L.), and the cabbage moth (*Barathra brassicae* L.) and some factors influencing its composition. *Tr. Vses. Nauchno issled. Inst. Zashch. Rast.*, **42**, 80–84.

Ishawata, S. (1901) On a kind of severe flacherie (sotto disease). *Dainihon Sanshi Kaiho*, **114**, 1–5.

Ishii, T. & Ohba, M. (1993) Characterization of mosquito-specific *Bacillus thuringiensis* strains isolated from a soil population. *Syst. Appl. Microbiol.* **16**, 494–499.

Ishii, T. & Ohba, M. (1994) The 23-kilodalton CytB protein is solely responsible for mosquito larvicidal activity of *Bacillus thuringiensis* serovar *kyushuensis*. *Curr. Microbiol.*, **29**, 91–94.

Ishii, T. & Ohba, M. (1997) Investigation of mosquito-specific larvicidal activity of a soil isolate of *Bacillus thuringiensis* serovar *canadensis*. *Curr. Microbiol.*, **35**, 40–43.

Israel, P. & Padmanabhan, S.Y. (1978) Biological control of stem borers of rice in India. *Final technical report (USPL 480 Project). Central Rice Research Institute, Indian Council of Agricultural Research.* Cuttack; India, 155 p.

Itoua Apoyolo, C., Drif, L., Vassal, J.M., DeBarjac, H., Bossy, J.P., Leclant, F. & Frutos, R. (1995) Isolation of multiple subspecies of *Bacillus thuringiensis* from a population of the European sunflower moth, *Homoeosoma nebulella*. *Appl. Environ. Microbiol.*, **61**, 4343–4347.

Ivanova, L. (1984) The strawberry butterfly - a pest of apricot. *Rast. Zashc.*, **32**, 33–34.

Ivanova, T.S., Vyalykh, A.K., Cokolov, M.S. & Monastyrskii, O.A. (1996) Efficiency of biopreparations in control of gall nematode in protected soil. *Agrokhimiya*, **3**, 101–106.

Ivey, P.W. & Johnson, S.J. (1997) Efficacy of *Bacillus thuringiensis* and cabbage cultivar resistance to diamondback moth (Lepidoptera: Yponomeutidae). *Florida Entomol.*, **80**, 396–401.

Ivey, P.W. & Johnson, S.J. (1998) Using integrated pest management to manage insect pests of cabbage. *Louisiana Agric.*, **41**, 14–15.

Izhar, Y., Wysoki, M. & Gur, L. (1979) The effectiveness of *Bacillus thuringiensis* Berliner on *Boarmia* (*Ascotis*) *selenaria* Schiff. (Lepidoptera, Geometridae) in laboratory tests and field trials. *Phytoparasitica*, **7**, 65–77.

Jacas, J., Vinuela, E., Budia, F., Estal, P.d. & Marco, V. (1992a) Laboratory evaluation of selected pesticides against *Opius concolor* Szepl. (Hymenoptera, Braconidae). *Tests Agrochem. Cultivars*, **13**, 140–141.

Jacas, J., Vinuela, E., Adan, A., Budia, F., Estal, P.d. & Marco, V. (1992b) Secondary effects of some pesticides used in Spanish olive groves on adults of *Opius concolor* Szepl. (Hym. Braconidae), a parasitoid of the olive fly, *Bactrocera oleae* (Gmel.) (Dip. Tephritidae). *Boletin de Sanidad Vegetal, Plagas*, **18**, 315–321.

Jackson, J.K., Sweeney, B.W., Bott, T.L., Newbold, J.D. & Kaplan, L.A. (1994) Transport of *Bacillus thuringiensis* var. *israelensis* and its effect on drift and benthic densities of nontarget macroinvertebrates in the Susquehanna River, northern Pennsylvania. *Canad. J. Fish. Aquat. Sci.*, **51**, 295–314.

Jackson, S.G., Goodbrand, R.B., Ahmed, R. & Kasatiya, S. (1995) *Bacillus cereus* and *Bacillus thuringiensis* isolated in a gastroenteritis outbreak investigation. *Lett. Appl. Microbiol.*, **21**, 103–105.

Jackson, T.A. & Poinar, G.O., Jr. (1989) Susceptibility of eucalyptus tortoise beetle (*Paropsis charybdis*) to *Bacillus thuringiensis* var. *san diego*. *Proc. 42nd NZ Weed Pest Cont. Conf.* (ed. A.J. Popay), 140–142.

Jacobs, S.J. (1989) Micro-organisms as potential biological control agents of *Eldana saccharina* Walker (Lepidoptera: Pyralidae). *Proc. Ann. Cong. South African Sugar Technol. Assoc.*, **63**, 186–188.

Jacquemard, P. (1982) Resultats des essais de lutte microbiologique effectues en culture cotonniere au Cameroun en 1979, 1980 et 1981 (Results of microbiological control tests in cotton crops in Cameroon in 1979, 1980 and 1981). *Coton Fibres Trop.*, **37**, 279–293.

Jadhav, D.B. & Khaire, V.M. (1987) Persistent toxicity of insecticides and microbial preparations to the exotic egg-larval parasite, *Chelonus blackburni* Cameron. *Curr. Res. Rep.*, **3**, 59–64.

Jafri, R.H. & Sabiha, H. (1974) Development of *Bacillus thuringiensis* in *Galleria mellonella* larvae exposed to gamma radiation. *J. Invertebr. Pathol.*, **23**, 76–84.

Jamaa, B., Graf, P. & Bourarach, K. (1996) Factors of natural death of the forest caterpillar: *Thaumetopoea pityocampa* Schiff. (Lepidoptera: Thaumetopoeidae) at the egg stage in Morocco. *Arab J. Plant Prot.*, **14**, 86–90.

James, R.R., Miller, J.C. & Lighthart, B. (1993) *Bacillus thuringiensis* var. *kurstaki* affects a beneficial insect, the cinnabar moth (Lepidoptera: Arctiidae). *J. Econ. Entomol.*, **86**, 334–339.

Janes, M.J. (1973) Corn earworm and fall armyworm occurrence and control on sweet corn ears in south Florida. *J. Econ. Entomol.*, **66**, 973–974.

Jansens, S., Vanhaecke, M. & Degheele, D. (1984) Effect of *Bacillus thuringiensis* Berliner formulations against larval instars of *Spodoptera littoralis* Boisduval. *Parasitica*, **40**, 41–50.

Jaques, R.P. (1973) Tests on microbial and chemical insecticides for control of *Trichoplusia ni* (Lepidoptera: Noctuidae) and *Pieris rapae* (Lepidoptera: Pieridae) on cabbage. *Canad. Entomol.*, **105**, 21–27.

Jaques, R.P., Laing, D.R. & Maw, H.E.L. (1989) Efficacy of mixtures of microbial insecticides and permethrin against the cabbage looper (Lepidoptera: Noctuidae) and the imported cabbageworm (Lepidoptera: Pieridae). *Canad. Entomol.*, **121**, 809–820.

Jardak, T. & Ksantini, M. (1986) Control trials against the leaf-eating generation of *Prays oleae* by *Bacillus thuringiensis* and diflubenzuron. *Bulletin OEPP*, **16**, 403–406.

Jarrett, P. & Burges, H.D. (1982a) Control of tomato moth *Lacanobia oleracea* by *Bacillus thuringiensis* on glasshouse tomatoes and the influence of larval behaviour. *Entomol. Exp. Appl.*, **31**, 239–244.

Jarrett, P. & Burges, H.D. (1982b) Effect of bacterial varieties on the susceptibility of the greater wax moth *Galleria mellonella* to *Bacillus thuringiensis* and its significance in classification of the bacterium. *Entomol. Exp. Appl.*, **31**, 346–352.

Jarrett, P. & Burges, H.D. (1986) Isolates of *Bacillus thuringiensis* active against *Mamestra brassicae* and some other species: alternatives to the present commercial isolate HDI. *Biol. Agric. Hortic.*, **4**, 39–45.

Jarrett, P. & Stephenson, M. (1990) Plasmid transfer between strains of *Bacillus thuringiensis* infecting *Galleria mellonella* and *Spodoptera littoralis*. *Appl. Environ. Microbiol.*, **56**, 1608–1614.

Jassim, H.K., Foster, H.A. & Fairhurst, C.P. (1990) Biological control of Dutch elm disease: *Bacillus thuringiensis* as a potential control agent for *Scolytus scolytus* and *S. multistriatus*. *J. Appl. Bacteriol.*, **69**, 563–568.

Jayanthi, P.D.K. & Padmavathamma, K. (1996) Cross infectivity and safety of nuclear polyhedrosis virus, *Bacillus thuringiensis* subsp. *kurstaki* Berliner and *Beauveria bassiana* (Balsamo) Vuille to pests of groundnut (*Arachis hypogaea* Linn.) and their natural enemies. *J. Entomol. Res.*, **20**, 211–215.

Jensen, F. & Aliniazee, M.T. (1972) Microbial insecticides for grape leaf folder control. *Calif. Agr.*, **26**, 5.

Jespersen, J.B. & Keiding, J. (1990) The effect of *Bacillus thuringiensis* var. *thuringiensis* on *Musca domestica* L. larvae resistant to insecticides. *Biocontrol of Arthropods affecting Livestock and Poultry* (eds. D.A. Rutz & R.S. Patterson), Westview Press, Inc.; Boulder, Colorado; USA 215–229.

Jimenez, J. & Fernandez, R. (1985) Behaviour of *Diadegma* sp. parasitism of *Heliothis virescens* caterpillars in field plots with different microbiological and chemical insecticide treatments. *Cienc. Tec. Agricult. Prot. Plantas,* **8**, 67–79.

Jimenez, J., Karabash, Y. & Fernandez, R. (1983) Comparative effectiveness of several preparations based on *Bacillus thuringiensis*, for the control of *Heliothis virescens* in tobacco grown under cover. *Ciencia y Tecnica en la Agricultura,* **6**, 27–41.

Johansson, K. (1971) Laboratory tests with Entobakterin-3 (*Bacillus thuringiensis* var. *galleriae*) against larvae of the gypsy moth, the nun moth, the greater budroller, the cabbage moth and the cabbage sawfly. *Medd. Statens Vaxtskyddanstalt,* **15**, 111–137.

Johnson, A.W. (1974) *Bacillus thuringiensis* and tobacco budworm control on flue-cured tobacco. *J. Econ. Entomol.*, **67**, 755–759.

Johnson, K.M. (1984) *Bacillus cereus* foodborne illness - an update. *J. Food Prot.*, **47**, 145–153.

Johnson, C. & Bishop, A.H. (1996) A technique for the effective enrichment and isolation of *Bacillus thuringiensis*. *FEMS Microbiol. Lett.*, **142**, 173–177.

Johnson, D.E. & Davidson, L.I. (1984) Specificity of cultured insect tissue cells for bioassay of entomocidal protein from *Bacillus thuringiensis*. *In Vitro,* **20**, 66–70.

Johnson, D.E. & McGaughey, W.H. (1996) Contribution of *Bacillus thuringiensis* spores to toxicity of purified cry proteins towards Indianmeal moth larvae. *Curr. Microbiol.*, **33**, 54–59.

Johnson, D.E., Oppert, B. & McGaughey, W.H. (1998) Spore coat protein synergizes *Bacillus thuringiensis* crystal toxicity for the Indianmeal moth (*Plodia interpunctella*). *Curr. Microbiol.*, **36**, 278–282.

Johnson, D.R. (1982) Suppression of *Heliothis* spp. on cotton by using *Bacillus thuringiensis*, *Baculovirus heliothis*, and two feeding adjuvants. *J. Econ. Entomol.*, **75**, 207–210.

Johnson, J. (1968) On the control of *Hymenia recurvalis* (Fabricius) on amaranthus. *Agric. Res. J. Kerala,* **6**, 132–133.

Johnson, K.S., Scriber, J.M., Nitao, J.K. & Smitley, D.R. (1995) Toxicity of *Bacillus thuringiensis* var. *kurstaki* to three nontarget Lepidoptera in field studies. *Environ. Entomol.*, **24**, 288–297.

Johnson, M.T. & Gould, F. (1992) Interaction of genetically engineered host plant resistance and natural enemies of *Heliothis virescens* (Lepidoptera: Noctuidae) in tobacco. *Environ. Entomol.*, **21**, 586–597.

Johnson, M.W., Oatman, E.R. & Wyman, J.A. (1980) Effects of insecticides on populations of the vegetable leafminer and associated parasites on summer pole tomatoes. *J. Econ. Entomol.*, **73**, 61–66.

Johnson, M.W., Oatman, E.R., Toscano, N.C., Welter, S.C. & Trumble, J.T. (1984) The vegetable leafminer on fresh market tomatoes in southern California. *Calif. Agr.*, **38**, 10–11.

Johnson, T.B, Lye, B.-H. & Hannan, R.-P. (1994) Control of Colorado Potato Beetle, *Leptinotarsa decemlineata* with Raven™ Bioinsecticide. *Abst. VIth Intern. Colloq. Invertebr. Pathol. Microb. Cont. 2nd Intern. Conf.* Bacillus thuringiensis, Montpellier, France, p 62.

Johnson, T.B., Slaney, A.C., Donovan, W.P. & Rupar, M.J. (1993) Insecticidal activity of EG4961, a novel strain of *Bacillus thuringiensis* toxic to larvae and adults of southern corn rootworm (Coleoptera: Chrysomelidae) and Colorado potato beetle (Coleoptera: Chrysomelidae). *J. Econ. Entomol.*, **86**, 330–333.

Johnson, W.T. & Morris, O.N. (1981) Cold fog applications of pesticides for control of *Malacosoma disstria*. *J. Arboric.*, **7**, 246–251.

Jones, V.P., Parrella, M.P. & Hodel, D.R. (1986) Biological control of leafminers in greenhouse chrysanthemums. *Calif. Agr.*, **40**, 10–12.

Joshi, R.C., Cadapan, E.P. & Heinrichs, E.A. (1987) Natural enemies of rice leaf folder, *Cnaphalocrocis medinalis* Guenee (Pyralidae: Lepidoptera) - a critical review (1913-1983). *Agric. Rev.*, **8**, 22–34.

Joshi, K.C., Roychoudhury, N., Sambath, S., Humane, S. & Pandey, D.K. (1996) Efficacy of three varietal toxins of *Bacillus thuringiensis* against *Ailanthus defoliator*, *Atteva fabriciella* Swed. (Lepidoptera: Yponomeutidae). *Indian For.*, **122**, 1023–1027.

Jouvenaz, D.P., Lord, J.C. & Undeen, A.H. (1996) Restricted ingestion of bacteria by fire ants. *J. Invertebr. Pathol.*, **68**, 275–277.

Juarez Perez, V.M., Jacquemard, P. & Frutos, R. (1994) Characterization of the type strain of *Bacillus thuringiensis* subsp. *cameroun* serotype H32. *FEMS Microbiol. Lett.*, **122**, 43–48.

Jung, Y., Kim, S., Cote, J.C., Lecadet, M.M., Chung, Y., Bok, S., (1998) Characterization of a new *Bacillus thuringiensis* subsp. *higo* strain isolated from rice bran in Korea. *J. Invertebr. Pathol.*, **71**, 95–96.

Justin, C.G.L., Rabindra, R.J. & Jayaraj, S. (1989) Increased insecticide susceptibility in *Heliothis armigera* (Hbn.) and *Spodoptera litura* F. larvae due to *Bacillus thuringiensis* Berliner treatment. *Insect Sci. Its Applic.*, **10**, 573–576.

Kabour, K. & Sane', S. (1972) Chemical control of tomato fruitworm, *Heliothis armigera* Hb., at Deiralla Station. *Plant Protection Research Section, Ministry of Agriculture, Jordan*, 4–6 .

Kachekova, S. & Frolov, B.A. (1977) The toxicity of thuringin to fowls. *Problemy Veterinarnoi Sanitarii*, **58,** 97–99.

Kadamshoev, M. (1996) Cutworm moths as pests of Calendula. *Zaschc. Rast. Moskva*, **1**, 32–33.

Kadyrov, A.K., Kimsanbaev Kh, K. & Kuchkorov, D.K. (1994) Residual toxicity of pesticides for *Encarsia*. *Zaschc. Rast. Moskva*, **7**, 10–11.

Kadyrova, M.K., Shcherbak, V.P. & Shcherban, Z.P. (1977) The effect of bacterial preparations on larvae of blood-sucking mosquitoes and horseflies. *Dokl. Akad. Nauk Uzb. SSR*, **1**, 69–70.

Kailides, D.S., Georgebits, R., Kailidis, D.S. & et al. (1971) Control of *Thaumetopoea pityocampa* in Greece during 1969. *Dasika Hron.*, **13**, 7–12.

Kakaliev, K. & Saparliev, K. (1975) A study of the pathogenicity of Entobakterin to termites in nature. *Izv. Akad. Nauk Turkmen. Biol. Nauk*, **6**, 39–41.

Kalfon, A.R. & De Barjac, H. (1985) Screening of the insecticidal activity of *Bacillus thuringiensis* strains against the Egyptian cotton leaf worm *Spodoptera littoralis*. *Entomophaga*, **30**, 177–186.

Kalia, S. & Joshi, K.C. (1996) Efficacy of three products of *Bacillus thuringiensis* Berliner against three defoliators of *Populus deltoides*. *J. Trop. For.*, **12,** 237–241.

Kalia, S. & Joshi, K.C. (1997) Efficacy of foliar spraying of three varietal strains of *Bacillus thuringiensis* against *Moringa* defoliator *Noorda blitealis* Tanss. (Lepidoptera: Pyralidae). *Indian J. Plant Prot.*, **25**, 65–66.

Kalia, S., Joshi, K.C. & Pant, N.C. (1997) Field evaluation with three varietal products of *Bacillus thuringiensis* Berliner for the control of the larvae of *Sylepta balteata* Fab. on *Sterculia urens* Roxb. *Indian J. Plant Prot.*, **25,** 140–141.

Kal' vish, T.K. & Krivtsova, N.V. (1978) Interaction of muscardine fungi and *Bacillus thuringiensis* var. *galleriae*. *Izv. Sib. Otd. Akad. Nauk SSSR Biol*, **No. 5**, 40–46.

Kalyuzhnaya, N.S., Gorbacheva, O.V. & Didyk, L.K. (1995) *Galerucella luteola* Mull. (Coleoptera, Chrysomelidae) as a pest of plantations of trees in the southern Ergeni Hills (Kalmykia). *Entomologicheskoe Obozrenie*, **74**, 45–51.

Kamenek, L.K. (1988) Effect of *Bacillus thuringiensis* and its delta-endotoxin on adults of the mustard beetle *Phaedon cochleariae* F. *Sib. Vestn. Sel'skokhozy. Nauki*, **1**, 37–39.

Kanda, K. & Aizawa, K. (1989) Selective induction of two temperate phages in *Bacillus thuringiensis* strain AF 101. *Agric. Biol. Chem.*, **53**, 2819–2820.

Kandybin, N.V. (1991) Microbial preparations. *Zaschc. Rast. Moskva*, **8**, 54–55.

Kandybin, N.V. & Smirnov, O.V. (1996) Novel ecologically safe biopesticides against insects and mites. *Bull. OILB-SROP* **19,** 15–17

Kang, S.C. & Chen, C.S. (1986) Histopathological studies of mosquito larvae infected by *Bacillus thuringiensis* var. *israelensis*. *Chin. J. Entomol.*, **6**, 39–56.

Kapustina, O.V. (1975) The effect of certain pesticides on *Trichogramma*. *Tr. Vses. Nauchno issled. Inst. Zashch. Rast.*, **44**, 33–47.

Karabash, Y. (1974) The wax moths (*Galleria mellonella* and *Achroia grisella*) and their control. *Pchelovodstvo*, **12**, 24–25.

Karadzhov, S. (1973a) Dynamics of the phytophagous and predacious mites in a biological system of pest control on apple. *Gradinar. Lozar. Nauka*, **10**, 51–62.

Karadzhov, S. (1973b) The problem of harmful Acarina on apple. *Rast. Zashc.*, **21**, 21–26.

Karadzhov, S. (1974) Efficiency of some *Trichogramma* spp. (Hymenoptera Chalcididae) in controlling the codling moth (*Carpocapsa pomonella* L.). *Gradinar. Lozar. Nauka*, **11**, 49–56.

Karadzhov, S. (1975) The effectiveness of certain species of *Trichogramma* against *Cydia pomonella*. *Rast. Zashc.*, **23**, 27–31.

Karamanlidou, G., Lambropoulos, A.F., Koliais, S.I., Manousis, T., Ellar, D. & Kastritsis, C. (1991) Toxicity of *Bacillus thuringiensis* to laboratory populations of the olive fruit fly (*Dacus oleae*). *Appl. Environ. Microbiol.*, **57**, 2277–2282.

Karel, A.K. & Schoonhoven, A.V. (1986) Use of chemical and microbial insecticides against pests of common beans. *J. Econ. Entomol.*, **79**, 1692–1696.

Kariya, A. (1977) Control of tea pests with *Bacillus thuringiensis*. *JARQ*, **11**, 173–178.

Karnowski, W. & Labanowski, G. (1998) *Cacyreus marshalli* - a potential pest of Pelargonium in Poland. *Ochr. Rosl.*, **42**, 12–13.

Kashkarova, L.F. (1975) Stimultaneous use of pathogenic microorganisms and parasitic insects against *Agrotis segetum*. *Tr. Vses. Nauchno issled. Inst. Zashch. Rast.*, **42**, 130–136.

Katagiri, K. & Iwata, Z. (1976) Control of *Dendrolimus spectabilis* with a mixture of cytoplasmic polyhedrosis virus and *Bacillus thuringiensis*. *Appl. Entomol. Zool.*, **11**, 363–364.

Katagiri, K. & Shimazu, M. (1974) Sporulation of *Bacillus thuringiensis* in the cadavers of insect hosts. *J. Jap. For. Soc.*, **56**, 325–331.

Katagiri, K., Iwata, Z., Ochi, K. & Kobayashi, F. (1978) Aerial application of a mixture of CPV and *Bacillus thuringiensis* for the control of the pine caterpillar, *Dendrolimus spectabilis*. *J. Jap. For. Soc.*, **60**, 94–99.

Katagiri, K., Iwata, Z., Kushida, T., Fukuizumi, Y. & Ishizuka, H. (1977) Effects of application of *Bt*, CPV and a mixture of *Bt* and CPV on the survival rates in populations of the pine caterpillar, *Dendrolimus spectabilis*. *J. Jap. For. Soc.*, **59**, 442–448.

Katanyukul, W. & Bhudhasamai, T. (1983) Chemical control of azolla insects. *Intern. Rice Res. Newsl.*, **8**, 14–15.

Katayama, R.W., Cobb, C.H., Burleigh, J.G. & Robinson, W.R. (1987) Susceptibility of adult *Microplitis croceipes* (Hymenoptera: Braconidae) to insecticides used for *Heliothis* spp. (Lepidoptera: Noctuidae) control. *Florida Entomol.*, **70**, 530–532.

Kawalek, M.D., Benjamin, S., Lee, H.L. & Gill, S.S. (1995) Isolation and Identification of novel toxins from a new mosquitocidal isolate from Malaysia, *Bacillus thuringiensis* subsp. *jegathesan*. *Appl. Environ. Microbiol.*, **61**, 2965–2969.

Kaya, H.K. (1974) Laboratory and field evaluation of *Bacillus thuringiensis* var. *alesti* for control of the Orangestriped oakworm. *J. Econ. Entomol.*, **67**, 390–392.

Kaya, H.K. & Burlando, T.M. (1989) Development of *Steinernema feltiae* (Rhabditida: Steinernematidae) in diseased insect hosts. *J. Invertebr. Pathol.*, **53**, 164–168.

Kaya, H.K. & Dunbar, D.M. (1972) Effect of *Bacillus thuringiensis* and carbaryl on an elm spanworm egg parasite *Telenomus alsophilae*. *J. Econ. Entomol.*, **65**, 1132–1134.

Kaya, H.K. & Reardon, R.C. (1982) Evaluation of *Neoaplectana carpocapsae* for biological control of the western spruce budworm, *Choristoneura occidentalis*: ineffectiveness and persistence of tank mixes. *J. Nematol.*, **14**, 595–597.

Kazakova, S.B., Dzhunusov, K.K. & Protsenko, A.I. (1977) The effect of Bitoxibacillin-202 on certain orchard insects in the Issyk-Kul' depression. *Entomol. Issled. Kirgizii. Vypusk 10,* (ed. A.I. Protsenko), 49–51.

Kearby, W.H., Hostetter, D.L. & Ignoffo, C.M. (1972) Laboratory and field evaluation of *Bacillus thuringiensis* for control of the bagworm. *J. Econ. Entomol.*, **65**, 477–480.

Kearby, W.H., Hostetter, D.L. & Ignoffo, C.M. (1975) Aerial application of *Bacillus thuringiensis* for control of the bagworm in pine plantations. *J. For.*, **73**, 29–30.

Keever, D.W. (1994) Reduced adult emergence of the maize weevil, lesser grain borer, and tobacco moth due to thuringiensin. *J. Entomol. Sci.*, **29**, 183–185.

Kellen, W.R., Hunter, D.K., Lindegren, J.E., Hoffmann, D.F. & Collier, S.S. (1977) Field evaluation of *Bacillus thuringiensis* for control of navel orangeworms on almonds. *J. Econ. Entomol.*, **70**, 332–334.

Keller, B. & Langenbruch, G.A. (1993) Control of coleopteran pests by *Bacillus thuringiensis*. *Bacillus thuringiensis, an Environmental Biopesticide: Theory and Practice.* (eds. P.F. Entwistle, J.S. Cory, M.J. Bailey & S. Higgs), pp. 171–191. John Wiley & Sons Ltd., New York.

Kennedy, F.J.S. (1993) Evaluation of certain insecticides against early shoot borer (*Chilo infuscatellus* Snellen) in sugarcane. *Meded. Fac. Landbouwwet., Uni. Gent,* **58**, 635–640.

Kennedy, G.G. & Oatman, E.R. (1976) *Bacillus thuringiensis* and pirimicarb: selective insecticides for use in pest management on broccoli. *J. Econ. Entomol.*, **69**, 767–772.

Kennedy, G.G. & Whalon, M.E. (1995) Managing pest resistance to *Bacillus thuringiensis* endotoxins: constraints and incentives to implementation. *J. Econ. Entomol.*, **88**, 454–460.

Khan, M.R. & Khan, B.M. (1968) Biology and control of maize stem-borer (*Chilo partellus* Swinhoe) in Peshawar. *Sci. Indust.*, **6**, 124–130.

Khan, A.M., Rashid, H. & Nuzhat, J. (1994) Bacterial and insecticidal control of *Hieroglyphus nigrorepletus* Bolivar (Orthoptera: Acrididae). *Indian J. Entomol.*, **56**, 107–110.

Khan, E., Karim, S., Makhdoom, R. & Riazuddin, S. (1995) Abundance, distribution and diversity of *Bacillus thuringiensis* in Pakistanian environment. *Pakistan J. Sci. Indust. Res.*, **38**, 5–6.

Khan, K.I., Jafri, R.H., Ahmad, M., Idrees Khan, K. & Husain Jafri, R. (1985) The pathogenicity and development of *Bacillus thuringiensis* in termites. *Pakistan J. Zool.*, **17**, 201–209.

Khanna, V., Gupta, V.K., Kanta, U., Dhaliwal, H.S. & Sekhon, S.S. (1995) Control of maize stem borer, *Chilo partellus* (Swinhoe) by *Bacillus thuringiensis* based bioinsecticides. *J. Entomol. Res.*, **19**, 101–105.

Kharchenko, N.A. & Chelysheva, L.P. (1975) Parasites and agents of disease of the larvae of the Japanese hazel saturniid. *Zaschc. Rast.*, **5**, 45.

Kharizanov, A. & Kharizanov, S. (1983) The vine bud moth. *Rast. Zashc.*, **31**, 31–34.

Khawaja, P., Hafiz, I.A. & Chaudhry, M.I. (1983) Efficacy of *Bacillus thuringiensis* Berliner against *Piesmopoda obliquifasciella* (Hamps), Pyralidae, Lepidoptera, a leaf stitcher of *Cassia fistula*. *Pakistan J. For.*, **33**, 83–86.

Khazipov, N.Z. & Alekseev, A.A. (1982) Bactoculicides - larvicides for the control of the pre-adult stages of blood-sucking mosquitoes. *Diagnostika zaraznykh zabolevanii sel'skokhozyaistvennykh zhivotnykh*, 88–91.

Kheiri, I.M., Abdallah, M.J. & Humadi, A.K.H. (1974 - publ. 1978)) The biology and control of peach worm *Anarsia lineatella* Zeller. *Yearbook of Plant Protection Research, Iraq Ministry of Agriculture and Agrarian Reform*, **1**, 4–8.

Kheyri, M. (1977) The necessity of integrated control application against beet armyworm (*Spodoptera exigua* Hb.). *Entomol. Phytopathol. Appl.*, **45**, 5–28.

Khloptseva, R.I. (1992) The use of microbial preparations in the USSR. *Biocont. News Inf.*, **13**, 27–32.

Kholchenkov, V.A. & Galetenko, S.M. (1979) Geometrids – important pests of orchards in the Crimea. *Zaschc. Rast.*, **7**, 34.

Khrameeva, A.V. (1972) The Riga bacterial strain of the *Bacillus thuringiensis* sporuliferous bacterial group. *Patalogiya nasekomykh i kleshchei* (ed. P. Tsinovskii Ya), Riga, Izdatel'stvo Zinatne, Latvian SSR, 167–182.

Khrameeva, A.V., Toman, G.A. & Klimpinya, A.E. (1988) Ecological factors influencing the quantitative dynamics of the green peach aphid *Myzus persicae* Sulzer. *Biologicheskaya regulyatsiya chislennosti vrednykh chlenistonogikh*, 68–85.

Khryanina, R.A. (1974) Liquid Entobakterin. *Zaschc. Rast.*, **4**, 46.

Kiel, C.B. (1991) Field and laboratory evaluation of a *Bacillus thuringiensis* var. *israelensis* formulation for control of fly pests of mushrooms. *J. Econ. Entomol.*, **84**, 1180–1188.

Kienlen, J.C., Gertz, C., Briard, P., Hommay, G. & Chaufaux, J. (1996) Research on the toxicity of various *Bacillus thuringiensis* Berliner strains to three species of slugs. *Agronomie*, **16**, 347–353.

Kikuta, H. & Iizuka, T. (1990) Isolates of *Bacillus thuringiensis* from soil and dead larvae of *Arctia caja phaeosoma* at Biei-cho, Hokkaido. *Proc. Abst. Vth Intern. Coll. Invertebr. Pathol. Microb. Con.*, Adelaide, Australia, 258.

Kim Chi, C. (1978) The effect of some insecticides on the common *Trichogramma* (*Trichogramma evanescens* Westw.). *Gradinar. Lozar. Nauka*, **15**, 73–79.

Kim, H.S., Lee, D.W., Woo, S.D., Yu, Y.M. & Kang, S. (1998a) Biological, immunological, and genetic analysis of *Bacillus thuringiensis* isolated from granary in Korea. *Curr. Microbiol.*, **37**, 52–57.

Kim, H.S., Lee, D.W., Woo, S.D., Yu, Y.M. & Kang, S.K. (1998b) Distribution, serological identification, and PCR analysis of *Bacillus thuringiensis* isolated from soils of Korea. *Curr. Microbiol.*, **37**, 195–200.

Kimball, M.R., Bauer, S.D. & Kauffman, E.E. (1986) Field evaluation of *Bacillus thuringiensis* var. *israelensis* in various habitats of Sutter-Yuba Mosquito Abatement District. *Proc. Papers Ann. Conf. Calif. Mosq. Vector Cont. Assoc.*, **54**, 27–32.

Kimura, T. (1991) Evaluation of *Bacillus thuringiensis* formulations and insect growth regulators for the control of diamondback moth (Lepidoptera: Yponomeutidae) in Aomori Prefecture. *Ann. Rep. Soc. Plant Prot. North Japan*, **42**, 141–144.

Kingsbury, P.D. (1975) Monitoring aquatic insect populations in forest streams exposed to chemical and biological insecticide applications. *Proc. Entomol. Soc. Ontario*, **106**, 19–24.

Kinsinger, R.A. & McGaughey, W.H. (1976) Stability of *Bacillus thuringiensis* and a granulosis virus of *Plodia interpunctella* on stored wheat. *J. Econ. Entomol.*, **69**, 149–154.

Kinsinger, R.A., McGaughey, W.H. & Dicke, E.B. (1980) Susceptibilities of Indian meal moth and almond moth to eight *Bacillus thuringiensis* isolates (Lepidoptera: Pyralidae). *J. Kansas Entomol. Soc.*, **53**, 495–500.

Kirby, R.D. (1978) Insecticidal control of the cabbage looper and imported cabbageworm in the Rolling Plains. *Texas Agricultural Experiment Station; College Station, Texas; USA*, PR-3522, 4 p.

Kiselek, E.V. (1974) The persistence of bacterial entomopathogens in the crowns of trees and in the areas round the trunks. *Vestn. Sel'skokhozy. Nauki*, **5**, 68–71.

Kiselek, E.V. (1975) The effect of biopreparations on insect enemies. *Zaschc. Rast.*, **12**, 23.

Kiselek, E.V. & Zayats Yu, V. (1976) Microbial preparations in the control of the Colorado beetle. *Vestn. Sel'skokhozy. Nauki*, **3**, 108–112.

Kleimenova, V.A. (1970) Integrated control of apple codling moth. *Sadovod.*, **12**, 13–14.

Kleiner, K.W., Raffa, K.F., Ellis, D.D. & McCown, B.H. (1998) Effect of nitrogen availability on the growth and phytochemistry of hybrid poplar and the efficacy of the *Bacillus thuringiensis cry*1A(a) d-endotoxin on gypsy moth. *Canad. J. For. Res.*, **28**, 1055–1067.

Kneifl, V. (1977) Dipel, an effective biological preparation against caterpillars of *Euproctis phaeorrhoea*. *Ved. Pr. Ovocnarske*, **6**, 371–381.

Knepper, R.G. & Walker, E.D. (1989) Effect of *Bacillus thuringiensis* israelensis (H-14) on the isopod *Asellus forbesi* and spring *Aedes* mosquitoes in Michigan. *J. Amer. Mosq. Con. Assoc.*, **5**, 596–598.

Knepper, R.G., Wagner, S.A. & Walker, E.D. (1991) Aerially applied, liquid *Bacillus thuringiensis* var. *israelensis* (H-14) for control of spring *Aedes* mosquitoes in Michigan. *J. Amer. Mosq. Con. Assoc.*, **7**, 307–309.

Ko, J.H. & Lee, B.Y. (1972) On the virulence of Thuricide and virus for the control of pine caterpillar, *Dendrolimus spectabilis* Butler, and fall webworm, *Hyphantria cunea* Drury. *Res. Rep. For. Res. Inst., Korea*, **19**, 43–50.

Kochanov, F.I., Rybin, V.I., Marchenko Ya, I. & Kul' kov, I.D. (1976) Aerial applications of bacterial preparations against *Bupalus piniarius*. *Lesnoe Khoz.*, **10**, 73–75.

Kolmakova, V.D. (1971) *Trichogramma* and Entobakterin against the apple fruit moth. *Zaschc. Rast.*, **16**, 24.

Komolpith, U. & Ramakrishnan, N. (1978) Joint action of a baculovirus of *Spodoptera litura* (Fabricius) and insecticides. *J. Entomol. Res.*, **2**, 15–19.

Kondo, S., Ohba, M. & Ishii, T. (1992) Larvicidal activity of *Bacillus thuringiensis* serovar *israelensis* against nuisance chironomid midges (Diptera: Chironomidae) of Japan. *Lett. Appl. Microbiol.*, **15**, 207–209.

Kondo, S., Fujiwara, M., Ohba, M. & Ishii, T. (1995a) Comparative larvicidal activities of the four *Bacillus thuringiensis* serovars against a chironomid midge, *Paratanytarsus grimmii* (Diptera: Chironomidae). *Microbiol. Res.*, **150**, 425–428.

Kondo, S., Ohba, M. & Ishii, T. (1995b) Comparative susceptibility of chironomid larvae (Dipt., Chironomidae) to *Bacillus thuringiensis* serovar *israelensis* with special reference to altered susceptibility due to food difference. *J. Appl. Entomol.*, **119**, 123–125.

Kondrya, V.S., Tret' yakova, M.F., Lupulchuk, M.D. & Smelyi, V.L. (1980) The testing of biopreparations against some leaf-gnawing lepidopteran larvae in orchards. *Biol. Z. Armenii*, **33**, 439–441.

Konig, E. (1975) Control of *Pristiphora abietina* with *Bacillus thuringiensis* preparations. *Z. Angew. Entomol.*, **77**, 424–429.

Konig, E. (1970) The effect of control of *Tortrix viridana* L. on the fructification of common oak (*Quercus robur*) and sessile oak (*Quercus petraea* Liebl.). *Z. Angew. Entomol.*, **65**, 319–333.

Koppenhofer, A.M. & Kaya, H.K. (1997) Additive and synergistic interaction between entomopathogenic nematodes and *Bacillus thuringiensis* for scarab grub control. *Biological Control*, **8**, 131–137.

Koppenhofer, A.M., Choo, H.Y., Kaya, H.K., Lee, D.W. & Gelernter, W.D. (1999) Increased field and greenhouse efficacy against scarab grubs with a combination of an entomopathogenic nematode and *Bacillus thuringiensis*. *Biological Control*, **14**, 37–44.

Korchagin, V.N. (1975) The gold-tail moth. *Zaschc. Rast.*, **12**, 56–57.

Korchagin, V.N. (1980) The gypsy moth. *Zaschc. Rast.*, **11**, 64–65.

Korchagin, V.N. (1983) The thorn butterfly and the brown-tail moth. *Zaschc. Rast.*, **12**, 48–49.

Kornilov, V.G. & Ivanova, G.P. (1987) Complex system for the control of the greenhouse whitefly in the greenhouse. *Tr. Latv. Sel'skokhozy. Akad.*, **237**, 31–34.

Kornilov, V.G., Stroeva, I.A. & Novozhilov, K.V. (1982) Insecticides of chemical and biological origin for control of the meadow moth. *Proc. All Union Res. Inst. Plant Prot.*, 91–93.

Korol, I.T. (1986) Effectiveness of bacterial preparations. *Zaschc. Rast.*, **3**, 34–35.

Korostel, S.I. & Kapustina, O.V. (1975) Effect of the thermostable exotoxin of *Bacillus thuringiensis* on *Trichogramma* (*Trichogramma* sp.) and *Ageniaspis* (*Ageniaspis fuscicollis* Dalm.). *Tr. Vses. Nauchno issled. Inst. Zashch. Rast.*, **42**, 102–109.

Korzh, K.P., Tonkonozhenko, A.P., Kotlyar, V.I., Mikityuk, V.V. & Markevich, A.P. (1975) Use of larvicidal effect of thermostable exotoxin of *Bacillus insectus* in farm conditions. *Probl. Parasitol.*, 249–251.

Kosenko, G.I. & Anferov, A.N. (1996) The spotted ash looper (*Calospilos pantaria*). *Lesnoe Khoz.*, **5**, 51–52.

Koskella, J. & Stotzky, G. (1997) Microbial utilization of free and clay-bound insecticidal toxins from *Bacillus thuringiensis* and their retention of insecticidal activity after incubation with microbes. *Appl. Environ. Microbiol.*, **63**, 3561–3568.

Kostadinov, D. (1979) Susceptibility of single-sexed *Trichogramma* to some widely used pesticides. *Rast. Zashc.*, **27**, 28–30.

Kosugi, Y. (1998) Difference in efficacy of BT (*Bacillus thuringiensis*) formulation against oriental tea tortrix, *Homona magunanima* (*H. magnanima*), and smaller tea tortrix, *Adoxophyes* sp., the between tea leaf dipping method and the synthetic diet dipping method. *Proc. Kanto Tosan Plant Prot. Soc.*, **45**, 211–214.

Kouskolekas, C.A. & Harper, J.D. (1973) Control of insect defoliators of collards in Alabama. *J. Econ. Entomol.*, **66**, 1159–1161.

Koval, A.G. (1985) The Carpathian ground beetle - a natural enemy of the Colorado beetle. *Zaschc. Rast.*, **6**, 25–26.

Koval, A.G. (1986) Predatory carabids as natural enemies of the Colorado potato beetle. *Zaschc. Rast. Moskva*, **11**, 45–46.

Kovalenkov, V.G. (1983) Biological control measures as part of integrated pest management in cotton. *10th Intern. Cong. Plant Prot.*, England, 20-25 November, 1983.

Kovtun, N.N. (1984) The effect of BTB-200 and chemical treatments on the insect fauna. *Zaschc. Rast.*, **3**, 39.

Kozlov, M.P., Savel' ev, V.N. & Reitblat, A.G. (1975) Effect of Entobakterin on larvae of fleas. *Med. Parazitol. Parazit. Bolezni*, **44**, 89–92.

Kozlov, M.P., Savel' ev, V.N., Nadeina, V.P. & Markevich, A.P. (1977) Infestation of adult fleas with spores of *Bacillus thuringiensis* by non-transmissive means. *Probl. Parasitol.*, 229–231.

Kramer, K.J., Hendricks, L.H., Wojciak, J.H. & Fyler, J. (1985) Evaluation of fenoxycarb, *Bacillus thuringiensis*, and malathion as grain protectants in small bins. *J. Econ. Entomol.*, **78**, 632–636.

Kramer, V. (1984) Evaluation of *Bacillus sphaericus* and *B. thuringiensis* H-14 for mosquito control in rice fields. *Indian J. Med. Res.*, **80**, 642–648.

Kramer, V.L. (1989) The ecology and biological control of mosquitoes (*Culex tarsalis*, *Anopheles freeborni* and *A. franciscanus* (*A. pseudopunctipennis franciscanus*)) in California wild and white rice fields (using *Bacillus thuringiensis* and larvivorous fish). *Dissert. Abst. Intern. B Sci. Engineer.*, **49**, 4670.

Kreidl, H. (1974) Dipel against the fruit moth. *Obstbau Weinbau*, **11**, 307.

Kreutzweiser, D.P. & Capell, S.S. (1996) Palatability of leaf material contaminated with *Bacillus thuringiensis* var. *kurstaki* to *Hydatophylax argus*, a detritivorous aquatic insect. *Bull. Environ. Contam. Toxicol.*, **56**, 80–84.

Kreutzweiser, D.P., Holmes, S.B., Capell, S.S. & Eichenberg, D.C. (1992) Lethal and sublethal effects of *Bacillus thuringiensis* var. *kurstaki* on aquatic insects in laboratory bioassays and outdoor stream channels. *Bull. Environ. Contam. Toxicol.*, **49**, 252–258.

Kreutzweiser, D.P., Capell, S.S. & Thomas, D.R. (1994) Aquatic insect responses to *Bacillus thuringiensis* var. *kurstaki* in a forest stream. *Canad. J. For. Res.*, **24**, 2041–2049.

Kreutzweiser, D.P., Gringorten, J.L., Thomas, D.R. & Butcher, J.T. (1996) Functional effects of the bacterial insecticide *Bacillus thuringiensis* var. *kurstaki* on aquatic microbial communities. *Ecotoxicol. Environ. Safety*, **33**, 271–280.

Krieg, A. (1969) *In vitro* determination of *Bacillus thuringiensis/Bacillus cereus* and related bacilli. *J. Invertebr. Pathol.*, **15**, 313–320.

Krieg, A. (1971) Is the potential pathogenicity of bacilli for insects related to production of alpha -exotoxin? *J. Invertebr. Pathol.*, **18**, 425–426.

Krieg, A. (1972) The action of preparations of *Bacillus thuringiensis* on spider mites (Tetranychidae). *Anz. Schaedlingskd. Pflanzenschutz*, **45**, 169–171.

Krieg, A. (1973a) On the possibilities of microbiological control of wax moths. *Z. Angew. Entomol.*, **74**, 337–343.

Krieg, A. (1973b) The toxic action of cultures of *Bacillus cereus* and *Bacillus thuringiensis* on honey bees (*Apis mellifera*). *Z. Pflanzenkrankheiten Pflanzenschutz*, **80**, 483–486.

Krieg, A. (1974) Effect of ultraviolet rays on spores of *Bacillus thuringiensis* and experiments on phytoprotection. *Z. Pflanzenkrankheiten Pflanzenschutz*, **81**, 591–596.

Krieg, A. (1978) Insect control by means of *Bacillus thuringiensis* preparations and their effect on the environment. *Nachrichtenbl. Dtsch. Pflanzenschutzdienst*, **30**, 177–181.

Krieg, A. (1982) The potential pathogenicity of some spore-formers (genus: *Bacillus*) to larvae of *Galleria mellonella* and its causes. *Z. Angew. Entomol.*, **93**, 355–365.

Krieg, A. & Herfs, W. (1963) Uber die Wirkung von *Bacillus thuringiensis* auf Bienen. *Entomol. Exp. Appl.*, **6**, 1–9.

Krieg, A. & Langenbruch, G.A. (1981) Susceptibility of arthropod species to *Bacillus thuringiensis*. *Microbial Control of Pests and Diseases 1970-1980* (ed. H.D. Burges), pp. 837–896. Academic Press, London.

Krieg, A. & Miltenburger, H.G. (1984) Bioinsecticides: I. *Bacillus thuringiensis*. *Adv. Biotechnol. Processes*, **3**, 273–290.

Krieg, A., de Barjac, H. & Bonnefoi, A. (1968) A new serotype of *Bacillus thuringiensis darmstadiensis*. *J. Invertebr. Pathol.*, **10**, 428–430.

Krieg, A., Hassan, S. & Pinsdorf, W. (1980) Comparison of the effect of the variety *israelensis* with other varieties of *Bacillus thuringiensis* on nontarget organisms of the order Hymenoptera: *Trichogramma cacoeciae* and *Apis mellifera*. *Anz. Schadlingskd. Pflanzenschutz Umweltschutz*, **53**, 81–83.

Krieg, A., Huger, A.M., Langenbruch, G.A. & Schnetter, W. (1983) *Bacillus thuringiensis* var. *tenebrionis*: a new pathotype effective against larvae of Coleoptera. *Z. Angew. Entomol.*, **96**, 500–508.

Krieg, A., Huger, A.M., Langenbruch, G.A. & Schnetter, W. (1984) Neue ergebnisse uber *Bacillus thuringiensis* var. *tenebrionis* unter besonderer Berucksichtigung selner Wirkung auf den Kartoffelkafer (*Leptinotarsa decemlineata*). *Anz. Schaedlingskd. Pflanzenschutz Umweltschutz* **57**, 145–150.

Krieg, A., Huger, A.M., Schnetter, W. & Herrnstadt, C. (1987) '*Bacillus thuringiensis* var. *san diego* strain M-7 is identical to *B. thuringiensis* subsp. *tenebrionis* strain BI 256-82, which was previously isolated in Germany and is pathogenic to beetles. *J. Appl. Entomol.*, **104**, 417–424.

Kring, T.J. & Smith, T.B. (1995) *Trichogramma pretiosum* efficacy in cotton under *Bt*-insecticide combinations. *1995 Proc. Beltwide Cotton Conf.*, Texas, USA, January 4-7, 1995: Volume 2, 856–857.

Krischik, V.A., Barbosa, P. & Reichelderfer, C.F. (1988) Three trophic level interactions: allelochemicals, *Manduca sexta* (L.) and *Bacillus thuringiensis* var. *kurstaki* Berliner. *Environ. Entomol.*, **17**, 476–482.

Krishnaiah, K., Mohan, N.J. & Prasad, V.G. (1981) Efficacy of *Bacillus thuringiensis* Ber. for the control of lepidopterous pests of vegetable crops. *Entomon*, **6**, 87–93.

Krishnaiah, K., Jaganmohan, N., Prasad, V.G. & Rajendran, R. (1978) A note on chemical control of pod borers on creeper bean (*Dolichos lablab* vari. *typicus*). *Pesticides*, **12**, 60–63.

Krushev, L.T. & Marchenko Ya, I. (1981) Against the nun moth. *Zaschc. Rast.*, **11**, 35.

Krushev, L.T., Mashnina, T.I. & Entin, L.I. (1972) Addition of an exotoxin to increase the effectiveness of bacterial preparations. *Lesnoe Khoz.*, **8**, 59–61.

Kudler, J. (1984) Susceptibility of larvae of the larch leaf-roller *Zeiraphera diniana* (Gn.) to different preparations of *Bacillus thuringiensis*. *Prace VULHM*, **64**, 263–282.

Kudler, J. & Lysenko, O. (1975) The influence of food of forest insect pests on the viability of entomogenous bacteria. *Commun. Inst. For. Cechosloveniae,* **9**, 29–37.

Kudler, J. & Lysenko, O. (1976) A contribution to the use of entomogenous bacteria in combination with sublethal doses of synthetic pyrethrin for the control of forest pests. *Prace VULHM,* **48**, 111–124.

Kudler, J. & Lysenko, O. (1977) The results of bacterial tests with the pine looper moth (*Bupalus piniarius* L.). *Lesnictvi,* **23**, 117–126.

Kuiper, H.A. & Noteborn, H. (1996) Food safety assessment of transgenic insect-resistant *Bt* tomatoes. *Food safety evaluation, Proceedings of an OECD-sponsored workshop*, 12-15 September 1994, Oxford, UK. 1996, 50–57.

Kulikovskii Yu, N. (1984) Experience in aerial treatment of forests with biopreparations. *Zaschc. Rast.,* **7**, 28.

Kulkarni, U.V. & Amonkar, S.V. (1988a) Microbial control of *Heliothis armigera* (Hb): Part I - isolation and characterization of a new strain of *Bacillus thuringiensis* and comparative pathogenicity of three isolates of *B. thuringiensis* against *H. armigera. Indian J. Exp. Biol.,* **26**, 703–707.

Kulkarni, U.V. & Amonkar, S.V. (1988b) Microbial control of *Heliothis armigera* (Hb): Part II - relative toxicity of spores and crystals of *Bacillus thuringiensis* varieties to *H. armigera* and their efficacy in field control. *Indian J. Exp. Biol.,* **26**, 708–711.

Kulshreshtha, J.P., Prakasa Rao, P.S., Rajamani, S. & Rao, P.S.P. (1971/1973) Recent developments in control of insect pests of rice crop in India. *Association of Rice Research Workers: Symposium on Rice Production under Environmental Stress,* **8**, 329–336.

Kumar, N.S. & Venkateswerlu, G. (1998) Intracellular proteases in sporulated *Bacillus thuringiensis* subsp. *kurstaki* and their role in protoxin activation. *FEMS Microbiol. Lett.,* **166**, 377–382.

Kumar, S. & Jayaraj, S. (1978) Mode of action of *Bacillus thuringiensis* Berliner in *Pericallia ricini* and *Euproctis fraterna. Indian J. Exp. Biol.,* **16**, 128–131.

Kuo, W.S. & Chak, K.F. (1996) Identification of novel cry-type genes from *Bacillus thuringiensis* strains on the basis of restriction fragment length polymorphism of the PCR-amplified DNA. *Appl. Environ. Microbiol.,* **62**, 1369–1377.

Kurtak, D., Back, C., Chalifour, A., Doannio, J., Dossou Yovo, J., Duval, J., Guillet, P., Meyer, R., Ocran, M. & Wahle, B. (1989) Impact of *B.t.i.* on blackfly control in the Onchocerciasis Control Programme in West Africa. *Israel J. Entomol.,* **23**, 21–38.

Kuteev, F.S., Lyashenko, L.I., Zurabova, E.R. & Chekanov, M.I. (1983) Lepidocide concentrate against forest pests. *Lesnoe Khoz.,* **8**, 52–53.

Kuzmanova, I. & Lecheva, I. (1981) Study of the possibility of the combined use of various strains of entomopathogenic bacteria of the group *Bacillus thuringiensis* Berliner. *Gradinar. Lozar. Nauka,* **18**, 40–46.

Kuzmanova, I. & Lecheva, I. (1984) Possibilities of combined use of the preparation Dipel with lower dosages of pyrethroids. *Gradinar. Lozar. Nauka,* **21**, 35–39.

Kuzmanova, J. & Sapundjieva, K. (1997) Exobacillin - a potential control agent for the Colorado potato beetle. *Proceedings of the first Balkan symposium on vegetables and potatoes* (eds. S. Jevtic & B. Lazic) *Volume 2,* Belgrade, Yugoslavia, 4-7 June 1996. 977–981.

Lacey, L.A. & Federici, B.A. (1979) Pathogenesis and midgut histopathology of *Bacillus thuringiensis* in *Simulium vittatum* (Diptera: Simuliidae). *J. Invertebr. Pathol.,* **33**, 171–182.

Lacey, L.A. & Lacey, J.M. (1981) The larvicidal activity of *Bacillus thuringiensis* var. *israelensis* (H-14) against mosquitoes of the central Amazon basin. *Mosq. News,* **41**, 266–270.

Lacey, L.A. & Mulla, M.S. (1978) Factors that influence the pathogenicity of *Bacillus thuringiensis* in black-flies. *Folia Entomol. Mex.,* 186–187.

Lacey, L.A. & Oldacre, S.L. (1983) The effect of temperature, larval age, and species of mosquito on the activity of an isolate of *Bacillus thuringiensis* var. *darmstadiensis* toxic for mosquito larvae. *Mosq. News,* **43**, 176–180.

Lacey, L.A. & Singer, S. (1982) Larvicidal activity of new isolates of *Bacillus sphaericus* and *Bacillus thuringiensis* (H-14) against anopheline and culicine mosquitoes. *Mosq. News,* **42**, 537–543.

Lacey, L.A., Lacey, C.M. & Padua, L.E. (1988) Host range and selected factors influencing the mosquito larvicidal activity of the PG-14 isolate of *Bacillus thuringiensis* var. *morrisoni. J. Amer. Mosq. Cont. Assoc.,* **4**, 39–43.

Lacey, L.A., Mulla, M.S. & Dulmage, H.T. (1978) Some factors affecting the pathogenicity of *Bacillus thuringiensis* Berliner against blackflies. *Environ. Entomol.,* **7**, 583–588.

Lacoursiere, J.O. & Charpentier, G. (1988) Laboratory study of the influence of water temperature and pH on *Bacillus thuringiensis* var. *israelensis* efficacy against black fly larvae (Diptera: Simuliidae). *J. Amer. Mosq. Cont. Assoc.,* **4**, 64–72.

Ladoni, H., Zaim, M. & Motabar, M. (1986) The effect of *Bacillus thuringiensis* H-14 on the anopheline mosquitoes in southern Iran. *J. Entomol. Soc. Iran,* **8**, 45–52.

Lake, R.W., Weber, R.G., Donofrio, R.M. & Kintzer, W.D. (1982) Field tests of *Bacillus thuringiensis* var. *israelensis* de Barjac in micromarsh pools, Delaware, 1980. *New Jersey Mosq. Cont. Assoc.,* Cherry Hill, New Jersey, 25-27 February 1981, 113–118.

Lal, O.P. & Lal, S.K. (1996) Failure of control measures against *Heliothis armigera* (Hubner) infesting tomato in heavy pesticidal application areas in Delhi and satellite towns in western Uttar Pradesh and Haryana (India). *J. Entomol. Res.*, **20**, 355–364.

Lam, A.B.Q. & Webster, J.M. (1972) Effect of the DD-136 nematode and of a beta -exotoxin preparation of *Bacillus thuringiensis* var. *thuringiensis* on leatherjackets, *Tipula paludosa* larvae. *J. Invertebr. Pathol.*, **20**, 141–149.

Lambert, B. & Peferoen, M. (1992) Insecticidal promise of *Bacillus thuringiensis*. *BioScience*, **42**, 112–122.

Lambert, B., Hofte, H., Annys, K., Jansens, S., Soetaert, P. & Peferoen, M. (1992) Novel *Bacillus thuringiensis* insecticidal crystal protein with a silent activity against coleopteran larvae. *Appl. Environ. Micro.* **58**, 2536–2542.

Landen, R., Bryne, M. & Abdel Hameed, A. (1994) Distribution of *Bacillus thuringiensis* strains in southern Sweden. *World J. Microbiol. Biotech.*, **10**, 45–50.

Landi, S. (1990) Possibility of biological control of *Otiorhynchus* spp. on ornamental plants in nurseries. *Redia*, **73**, 261–273.

Lang, B.A., Moellenbeck, D.J., Isenhour, D.J. & Wall, S.J. (1996) Evaluating resistance to CryIA(b) in European corn borer (Lepidoptera: Pyralidae) with artificial diet. *Resist. Pest Manag.*, **8**, 29–31.

Langenbruch, G.A. (1977) Experiments on the possibility of controlling *Agrotis segetum* by means of *Bacillus thuringiensis*. *Nachrichtenbl. Dtsch. Pflanzenschutzdienst*, **29**, 133–137.

Langenbruch, G.A. (1984a) Control of the American thuja miner (*Agyresthia thuiella*) with pyrethrins, diflubenzuron or *Bacillus thuringiensis*. *Nachrichtenbl. Dtsch. Pflanzenschutzdienst*, **36**, 161–164.

Langenbruch, G.A. (1984b) On the effect of *Bacillus thuringiensis* against harmful Lepidoptera on cabbage. *Mitt. Biol. Bundesanst. Land Forstwirtsch. Berlin Dahlem*, **218**, 108–118.

Langenbruch, G.A. (1992) Experiences with *Bacillus thuringiensis* subsp. *tenebrionis* in controlling the colorado potato beetle. *Mitt. Dtsch. Ges. Allg. Angew. Entomol.*, **8**, 193–195.

Laranjeiro, A.J. & Evans, H.F. (1994) Integrated pest management at Aracruz Celulose. *For. Integ. Pest Manag. Prog.*, **65**, 45–52.

Larget, I. (1981) Study of the persistence of *Bacillus thuringiensis* var. *israelensis*. *Rev. Gen. Bot.*, **88**, 33–42.

Larget, I. & Charles, J.F. (1982) Study of the larvicidal activity of *Bacillus thuringiensis* variety *israelensis* on the larvae of Toxorhynchitinae. *Bull. Soc. Pathol. Exotiq.* **75**, 121–130.

Larget, I. & de Barjac, H. (1981a) Comparative activity of 22 varieties of *Bacillus thuringiensis* against 3 species of Culicidae. *Entomophaga*, **26**, 143–148.

Larget, I. & de Barjac, H. (1981b) Specificity and active principle of *Bacillus thuringiensis* var. *israelensis*. *Bull. Soc. Pathol. Exotiq.*, **74**, 216–227.

Larget Thiery, I., Hamon, S. & De Barjac, H. (1984) Susceptibility of Culicidae to the beta-exotoxin of *Bacillus thuringiensis*. *Entomophaga*, **29**, 95–108.

Larrain, S.P. (1986) Effectiveness of insecticides and frequency of application based on critical population levels of *Scrobipalpula absoluta* (Meyrick), on tomatoes. *Agric. Tecnica*, **46**, 329–333.

LaSota, L.R. (1996) Evaluation by the United States Environmental Protection Agency of pesticidal substances produced in plants. *Field Crops Res.*, **45**, 181–186.

Latha, K., Balasubramanian, G., Sundarababu, P.C. & Gopalan, M. (1994) Field evaluation of insecticides alone and in combination against leaffolders and their effect on natural enemies in rice. *Pest Manag. Econ. Zool.*, **2**, 105–109.

Lattin, A., Grimes, J., Hoxter, K.A. & Smith, G.J. (1990a) Vectobac technical material (*Bacillus thuringiensis* variety *israelensis*): an avian oral toxicity and pathogenicity study in the mallard. Wildlife International Ltd. Easton, Maryland. Project No. 161-115. pp. 1–24 (unpublished).

Lattin, A., Hoxter, K.A. & Smith, G.J. (1990b) Vectobac technical material (*Bacillus thuringiensis* variety *israelensis*): an avian oral toxicity and pathogenicity study in the bobwhite. Wildlife International Ltd. Easton, Maryland. Project No. 161-114. pp. 1–27 (unpublished).

Laurent, P., Ripouteau, H., Dumanoir, V.C., Frachon, E. & Lecadet, M.M. (1996) A micromethod for serotyping *Bacillus thuringiensis*. *Lett. Appl. Microbiol.*, **22**, 259–261.

Lavrenyuk, N.M., Uzdenov, U.B. & Romasheva, L.F. (1977a) The effect of a heat stable exotoxin of the group *Bacillus thuringiensis*, isolated from strains obtained from blood-sucking arthropods, on the mite *Dermanyssus gallinae*. *Biologicheskie Metody Bor'by s Krovososushchimi Nasekomymi i Kleshchami* (recd. 1979), 32–34.

Lavrenyuk, N.M., Asylbaeva, N.S. & Romasheva, L.F. (1977b) Study of the effect of bacterial preparations from strains of the group *Bacillus thuringiensis*, isolated from blood-sucking mites, on the ectoparasites of birds on poultry farms. *Biological Methods of Control of Blood Sucking Insects and Acarina.: Biologicheskie Metody Bor'by s Krovososushchimi Nasekomymi i Kleshchami*, 35–37.

Law, S.E. & Mills, H.A. (1980) Electrostatic application of low-volume microbial insecticide spray on broccoli plants. *J. Amer. Soc. Hort. Sci*, **105**, 774–777.

Lawrence, D., Heitefuss, S. & Seifert, H.S.H. (1991) Differentiation of *Bacillus anthracis* from *Bacillus cereus* by gas chromatographic whole-cell fatty acid analysis. *J. Clinical Microbiol.*, **29**, 1508–1512.

Lebaron, P., Bauda, P., Lett, M.C., Duval Iflah, Y., Simonet, P., Jacq, E., Frank, N., Roux, B., Baleux, B., Faurle, G., Hubert, J.C., Normand, P., Prieur, D., Schmitt, S. & Block, J.C. (1997) Recombinant plasmid mobilization between *E. coli* strains in seven sterile microcosms. *Canad. J. Microbiol.*, **43**, 534–540.

Lebrun, P. & Vlayen, P. (1981) Comparative study of the bioactivity and secondary effects of *Bacillus thuringiensis* H 14. *Z. Angew. Entomol.*, **91**, 15–25.

Lecadet, M.M. & Martouret, D. (1987) Host specificity of the *Bacillus thuringiensis* delta-endotoxin toward lepidopteran species: *Spodoptera littoralis* Bdv. and *Pieris brassicae* L. *J. Invertebr. Pathol.*, **49**, 37–48.

Lecadet, M.M., Sanchis, V., Menou, G., Rabot, P., Lereclus, D., Chaufaux, J. & Martouret, D. (1988) Identification of a delta-endotoxin gene product specifically active against *Spodoptera littoralis* Bdv. among proteolysed fractions of the insecticidal crystals of *Bacillus thuringiensis* subsp. *aizawai* 7.29. *Appl. Environ. Microbiol.*, **54**, 2689–2698.

Lecadet, M.M., Frachon, E., Dumanoir, V.C., Ripouteau, H., Hamon, S., Laurent, P. & Thiery, I. (1999) Updating the H-antigen classification of *Bacillus thuringiensis*. *J. Appl. Microbiol.*, **86**, 660–672.

Lecheva, I. (1983) Geometrids in fruit plantations and their control. *Rast. Zashc.*, **31**, 42–43.

Lecheva, I. & Kuzmanova, I. (1980) The biological effect of bacterial-insecticidal combinations on larvae of the cherry spinner (*Biston hirtaria* Cl., Lepidoptera, Geometridae). *Rasteniev"dni Nauki*, **17**, 119–126.

Lee, B.M. & Scott, G.I. (1989) Acute toxicity of temephos, fenoxycarb, diflubenzuron, and methoprene and *Bacillus thuringiensis* var. *israelensis* to the mummichog (*Fundulus heteroclitus*). *Bull. Environ. Contam. Toxicol.*, **43**, 827–832.

Lee, K.L. & Kang, S.K. (1989) The microbial insecticide *Bacillus thuringiensis* and sericulture. 1. Comparative toxicities of *Bacillus thuringiensis* subspecies to silkworm larvae, *Bombyx mori* L., and persistence of insecticidal activity of parasporal crystal in mulberry fields. *Research Reports of the Rural Development Administration, Crop Prot.*, **31**, 24–28.

Lee, H.H., Jung, J.D., Yoon, M.S. & *et al.* (1995a) Distribution of *Bacillus thuringiensis* in Korea. *Bacillus thuringiensis Biotechnology and Environmental Benefits* (ed. T.Y. Feng), pp. 201–215. Hua Shiang Yuan Publishing Co., Taiwan.

Lee, H.H., Lee, J.A., Lee, K.Y., Chung, J.D., de Barjac, H., Charles, J.F., Cosmao Dumanoir, V. & Frachon, E. (1994) New serovars of *Bacillus thuringiensis*: *B. thuringiensis* ser. *coreanensis* (serotype H25), *B. thuringiensis* ser. *leesis* (serotype H33), and *B. thuringiensis* ser. *konkukian* (serotype H34). *J. Invertebr. Pathol.*, **63**, 217–219.

Lee, H.L. & Cheong, W.H. (1985) Laboratory evaluation of the potential efficacy of *Bacillus thuringiensis israelensis* for the control of mosquitoes in Malaysia. *Trop. Biomed.*, **2**, 133–137.

Lee, H.L. & Cheong, W.H. (1988) Isolation and evaluation of an entomopathogenic strain of *Bacillus thuringiensis* against mosquitoes of public health importance in Malaysia. *Trop. Biomed.*, **5**, 9–18.

Lee, H.L. & Seleena, P. (1990a) *Bacillus thuringiensis* ssp. *malaysianensis*: a new subspecies isolated from Malaysia. *Trop. Biomed*, **7**, 117–118.

Lee, H.L. & Seleena, P. (1990b) Preliminary field evaluation of a Malaysian isolate of *Bacillus thuringiensis* serotype H-14 against *Culex pseudovishnui*. *Southeast Asian J. Trop. Med. Public Health*, **21**, 143–144.

Lee, H.L., Seleena, P. & Lam, W.K. (1990) Preliminary field evaluation of indigenous (Malaysian) isolates and commercial preparations of *Bacillus thuringiensis* serotype H-14 and *Bacillus sphaericus* serotype H5a5b against *Anopheles karwari*. *Trop. Biomed.*, **7**, 49–57.

Lee, M.K., Curtiss, A., Alcantara, E. & Dean, D.H. (1996) Synergistic effect of the *Bacillus thuringiensis* toxins CryIAa and CryIAc on the gypsy moth, *Lymantria dispar*. *Appl. Environ. Microbiol.*, **62**, 583–586.

Lee, M.K., Rajamohan, F., Gould, F. & Dean, D.H. (1995b) Resistance to *Bacillus thuringiensis* CryIA delta-endotoxins in a laboratory-selected *Heliothis virescens* strain is related to receptor alteration. *Appl. Environ. Microbiol.*, **61**, 3836–3842.

Lentini, A. & Luciano, P. (1995) *Bacillus thuringiensis* in the management of gypsy moth (*Lymantria dispar* L.) in Sardinian cork-oak forests. *Bull. OILB-SROP*, **18**, 104–109.

Leong, K.L.H., Cano, R.J. & Kubinski, A.M. (1980) Factors affecting *Bacillus thuringiensis* total field persistence. *Environ. Entomol.*, **9**, 593–599.

Leong, K.L.H., Yoshimura, M.A. & Kaya, H.K. (1992) Low susceptibility of overwintering monarch butterflies to *Bacillus thuringiensis* Berliner. *Pan Pacific Entomol.*, **68**, 66–68.

Lereclus, D., Delecluse, A. & Lecadet, M.M. (1993) Diversity of *Bacillus thuringiensis* toxins and genes. *Bacillus thuringiensis, an Environmental Biopesticide: Theory and Practice* (eds. Entwistle, P.F., Cory, J.S., Bailey, M.J. & Higgs, S), John Wiley & Sons, Chichester UK, 37–70.

Lereclus, D., Vallade, M., Chaufaux, J., Arantes, O. & Rambaud, S. (1992) Expansion of insecticidal host range of *Bacillus thuringiensis* by *in vivo* genetic recombination. *BioTechnology*, **10**, 418–421.

Letendre, M. & McNeil, J.N. (1979) Field evaluation of insecticides for the control of the European skipper, *Thymelicus lineola* (Lepidoptera: Hesperiidae), in Quebec. *Canad. Entomol.*, **111**, 1313–1317.

Levinson, B.L., Kasyan, K.J., Chiu, S.S., Currier, T.C. & Gonzalez, J.M., Jr. (1990) Identification of beta-exotoxin production, plasmids encoding beta-exotoxin, and a new exotoxin in *Bacillus thuringiensis* by using high-performance liquid chromatography. *J. Bacteriol.*, **172**, 3172–3179.

Lewis, L.C. & Bing, L.A. (1991) *Bacillus thuringiensis* Berliner and *Beauveria bassiana* (Balsamo) Vuillimen for European corn borer control: program for immediate and season-long suppression. *Canad. Entomol.*, **123**, 387–393.

Li, H.K. (1987) Control of *Laelia coenosa candida* Leeh and *Chilo hyrax* Bls., two major pests of reed, with *Bacillus thuringiensis*. *Chin. J. Biol. Cont.*, **3**, 127–128.

Li, K.H., Xu, X., Li, Y.F., Meng, Q.Z. & Zhou, L.C. (1986) Determination of toxicity of 29 chemicals to *Trichogramma japonicum* at various developmental stages. *Natural Enemies of Insects*, **8**, 187–194, 199.

Li, L.Z., Zhou, X.S., Yan, J., Cui, Y.S., Yang, H.P., Zhao, M.G., Shi, J.L., Pan, H.Y. & Wang, Y.Y. (1992) Studies on the biological control technique for the white grub *Blitopertha pallidipennis* Reitter at the Bureau of Shangganling. *Collection of achievements on the technique cooperation project of P.R. China and F. R. Germany.*, 136–140.

Li, R.S. & Luo, C. (1989) Composition of proteins and polypeptides in crystals and spore coats of *Bacillus thuringiensis* and their toxicity to insects. *Acta Entomol. Sin.*, **32**, 149–157.

Li, R.S. & Sheng, Z.M. (1990) Host specificity of *Bacillus thuringiensis* delta-endotoxin proteolysed by proteinases of larval gut juice. *Proc. Abst. Vth Intern. Coll. Invertebr. Pathol. Microb. Con., Adelaide, Australia*, 296.

Li, R.S., Jarrett, P. & Burges, H.D. (1987) Importance of spores, crystals, and delta-endotoxins in the pathogenicity of different varieties of *Bacillus thuringiensis* in *Galleria mellonella* and *Pieris brassicae*. *J. Invertebr. Pathol.*, **50**, 277–284.

Li, S.Y. & Fitzpatrick, S.M. (1997) Responses of larval *Choristoneura rosaceana* (Harris) (Lepidoptera: Tortricidae) to a feeding stimulant. *Canad. Entomol.*, **129**, 363–369.

Li, S.Y., Fitzpatrick, S.M. & Isman, M.B. (1995) Effect of temperature on toxicity of *Bacillus thuringiensis* to the obliquebanded leafroller (Lepidoptera: Tortricidae). *Canad. Entomol.*, **127**, 271–273.

Li, Y.P., Wu, T.L., Pan, L.S. & Zhang, J.P. (1984) Control of second-generation masson's pine moth with *Bacillus thuringiensis* var. *dendrolimus*. *For. Sci.Tech. Linye Keji Tongxun*, **9**, 19–20.

Liang, Z.Q., Xiong, J.W. & Chen, H.Y. (1983) Effectiveness of a microsporidian parasite against *Prodenia litura* Fabricius. *Natural Enemies of Insects*, **5**, 79–81.

Liguori, M. (1988) Effect of different pest control treatments on the population of the phytoseiid predator *Typhlodromus exhilaratus* Ragusa and on phytophagous mites in a vineyard in Siena. *Redia*, **71**, 455–466.

Lim, G.S., Sivapragasam, A. & Ruwaida, M. (1986) Impact assessment of *Apanteles plutellae* on diamondback moth using an insecticide-check method. *Diamondback Moth Management*, Tainan, Taiwan, 11-15 March, 1985, 195–204.

Lima, M.M., dos Santos, L.M.M., da Silva, M.H.L. & Rabinovitch, L. (1994) Effects of the spore-endotoxin complex of a strain of *Bacillus thuringiensis* serovar *morrisoni* upon *Triatoma vitticeps* (Hemiptera: Reduviidae) under laboratory conditions. *Mem. Inst. Oswaldo Cruz*, **89**, 403–405.

Lin, K.C., Chen, Y.H., Peng, S.A., Zhu, H.M., Yu, H.P. & Yu, Z.N. (1990a) Applications of 'Shachongjin' in the control of pests of horticulture I. The effect of 'Shachongjin', a new *Bacillus thuringiensis* agent, against the larvae of *Apriona germari* (Coleoptera). *Proc. Abst. Vth Intern. Coll. Invertebr. Pathol. Microb. Con.*, Adelaide, Australia, 393.

Lin, K.C., Chen, Y.H., Peng, S.A., Zu, H.M. & Yu, Z.N. (1990b) Applications of 'Shachongjin' in the control of pests of horticulture II. The effect of 'Shachongjin' against larvae of *Contarinia citri* (Diptera). *Proc. Abst. Vth Intern. Coll. Invertebr. Pathol. Microb. Con.*, Adelaide, Australia, 394.

Lin, W., Guo, S., Song, H. (1998) Population fluctuation of *Apanteles glomeratus* and toxicity of insecticides. *Natural Enemies of Insects*, **20**, 150–155.

Lindquist, R.K. (1972) *Bacillus thuringiensis* formulations for cabbage looper control on greenhouse lettuce. *Res. Sum. Ohio Agr. Res. Develop. Cen.*, **No. 58**, 33–34.

Lindquist, R.K. (1975) Insecticides and insecticide combinations for control of tomato pinworm larvae on greenhouse tomatoes: A progress report. *Res. Sum. Ohio Agr. Res. Develop. Cent.*, **No. 82**, 37–39.

Lindquist, R.K. (1977) Cutworm control trials on greenhouse crops in 1976. *Ohio Florists Assoc. Bull.*, **568**, 8–9.

Liotta, G., Mineo, G. & Ragusa, S. (1976) On the current state of knowledge concerning certain arthropods injurious to citrus in Sicily. *Bol. Ist. Entomol. Agrar. Osse. Fitopatol. Palermo*, **10**, 29–67.

Lipa, J.J. (1976) Results of investigations on the use of microbial preparations containing *Bacillus thuringiensis* for the protection of cabbage during 1969-1974. *Pr. Nauk. Inst. Ochr. Rosl.*, **18**, 89–108.

Lipa, J.J. & Bakowski, G. (1979) Chemical and biological control of four orchard pests: the winter moth (*Cheimatobia brumata* L.), the gypsy moth (*Lymantria dispar* L.), the brown-tail (*Euproctis chrysorrhoea* L.) and the European tussock moth (*Orgyia antiqua* L.). *Rocz. Nauk Roln. E*, **9**, 159–168.

Lipa, J.J., Kowalska, T. & Szczepanska, K. (1969) The results of laboratory studies of the susceptibility of *Agrotis c-nigrum* L. (Lepidoptera, Noctuidae) to commercial biopreparations of *Bacillus thuringiensis* Berliner. *Biul. Inst. Ochr. Rosl.*, **45**, 115–130.

Lipa, J.J., Pruszynski, S. & Bartkowski, J. (1970) The use of biopreparations for the protection of brassica crops. *Biul. Inst. Ochr. Rosl.*, **47**, 347–354.

Lipa, J.J., Slizynski, K., Ziemnicka, J. & Bartkowski, J. (1975) Interaction of *Bacillus thuringiensis* and nuclear polyhedrosis virus in *Spodoptera exigua*. *Environ. Qual. Safety Suppl.*, **3**, 668–671.

Lipa, J.J., Bakowski, G. & Rychlewska, M. (1977) The effectiveness of the microbial insecticides Bactospeine, Dipel, Selectzin and Thuricide against the rusty tussock moth *Orgyia antiqua* L. (Lepidoptera, Lymantriidae) on apple trees. *Pr. Nauk. Inst. Ochr. Rosl.*, **19**, 183–190.

Litvinov, B.M. & Bondarenko, A.G. (1987) Bitoxibacillin against the raspberry-strawberry weevil. *Zaschc. Rast. Moskva,* **12**, 35–36.

Liu, C.H. & Tan, B. (1984) A strain of *Bacillus thuringiensis* highly toxic to *Polychrosis cunninghamiacola*. *For. Sci. Tech. Linye Keji Tongxun,* **5**, 27–28.

Liu, S., Yang, Z., Wu, T., Wang, Z. & Zhou, S. (1993) The production and application in large areas of GrNPV pesticide. *Acta Prataculturae Sin.*, **2**, 47–50.

Liu, Y. & Tabashnik, B.E. (1997a) Genetic basis of diamondback moth resistance to *Bacillus thuringiensis* toxin Cry1C. *Resist. Pest Manag.*, **9**, 21–22.

Liu, Y. & Tabashnik, B.E. (1997b) Inheritance of resistance to the *Bacillus thuringiensis* toxin Cry1C in the diamondback moth. *Appl. Environ. Microbiol.*, **63**, 2218–2223.

Liu, Y., Tabashnik, B.E. & Pusztai Carey, M. (1996) Field-evolved resistance to *Bacillus thuringiensis* toxin CryIC in diamondback moth (Lepidoptera: Plutellidae). *J. Econ. Entomol.*, **89**, 798–804.

Liu, Y.B. & Tabashnik, B.E. (1996) Diamondback moth resistance to *Bacillus thuringiensis* toxin Cry1C in the field. *Resist. Pest Manag.*, **8**, 44–46.

Liu, Y.T., Sui, M.J., Ji, D.D., Wu, I.H., Chou, C.C. & Chen, C.C. (1993) Protection from ultraviolet irradiation by melanin of mosquitocidal activity of *Bacillus thuringiensis* var. *israelensis*. *J. Invertebr. Pathol.*, **62**, 131–136.

Logan, N.A. & Berkeley, R.C.W. (1984) Identification of *Bacillus* strains using the API system. *J. Gen. Microbiol.*, **130**, 1871–1882.

Lonc, E. & Lachowicz, T.M. (1987) Insecticidal activity of *Bacillus thuringiensis* subspecies against *Menopon gallinae* (Mallophaga: Menoponidae). *Angew. Parasitol.*, **28**, 173–176.

Lonc, E., Mazurkiewicz, M. & Szewczuk, V. (1986) Susceptibility of poultry biting lice (Mallophaga) to Dipel and Bacilan (*Bacillus thuringiensis*). *Angew. Parasitol.*, **27**, 35–37.

Lonc, E., Lachowicz, T.M. & Mazurkiewicz, M. (1988) Efficacy of *Bacillus thuringiensis* against poultry ectoparasites. *Zesz. Nauk. Akad. Roln. Wroclaw. Weterynaria,* **45**, 63–66.

Lonc, E., Lachowicz, T.M. & Jedryka, U. (1991) Insecticidal activity of some *Bacillus* strains against larvae of house flies (*Musca domestica*). *Wiad. Parazyt.*, **37**, 357–365.

Lonc, E., Lecadet, M.M., Lachowicz, T.M. & Panek, E. (1997) Description of *Bacillus thuringiensis wratislaviensis* (H-47), a new serotype originating from Wroclaw (Poland), and other *Bt* soil isolates from the same area. *Lett. Appl. Microbiol.*, **24**, 467–473.

Longhini, L. & Busoli, A.C. (1993) Integrated control of *Brevicoryne brassicae* (L., 1758) (Homoptera, Aphididae) and *Ascia monuste orseis* (Latr., 1819) (Lepidoptera, Pieridae), in kale (*Brassica oleracea* var. *acephala*). *Cientifica Jaboticabal,* **21**, 231–237.

Lopatina, V.V. (1981) The apple leaf-curling moth. *Zaschc. Rast.*, **4**, 30.

Lopez Meza, J.E. & Ibarra, J.E. (1996) Characterization of a novel strain of *Bacillus thuringiensis*. *Appl. Environ. Microbiol.*, **62**, 1306–1310.

Lopez Meza, J., Federici, B.A., Poehner, W.J., Martinez Castillo, A. & Ibarra, J.E. (1995) Highly mosquitocidal isolates of *Bacillus thuringiensis* subspecies *kenyae* and *entomocidus* from Mexico. *Biochem. Syst. Ecol.*, **23**, 461–468.

Lopez, R. & Ferro, D.N. (1995) Larviposition response of *Myiopharus doryphorae* (Diptera: Tachinidae) to Colorado potato beetle (Coleoptera: Chrysomelidae) larvae treated with lethal and sublethal doses of *Bacillus thuringiensis* Berliner subsp. *tenebrionis*. *J. Econ. Entomol.*, **88**, 870–874.

Lord, J.C. & Fukuda, T. (1990) Relative potency of *Bacillus thuringiensis* var. *israelensis* and *Bacillus sphaericus* 2362 for *Mansonia titillans* and *Mansonia dyari*. *J. Amer. Mosq. Con. Assoc.*, **6**, 325–327.

Lord, J.C. & Undeen, A.H. (1990) Inhibition of the *Bacillus thuringiensis* var. *israelensis* toxin by dissolved tannins. *Environ. Entomol.*, **19**, 1547–1551.

Lorenzato, D. (1984) Laboratory study on the control of the apple moth *Phtheocroa cranaodes* Meyrick, 1937 with *Bacillus thuringiensis* Berliner and chemical insecticides. *Agron. Sulriogr.*, **20**, 157–163.

Lorenzato, D. & Corseuil, E. (1982) Effects of different control measures on the main pests of soyabean (*Glycine max* (L.) Merrill) and its predators. *Agron. Sulriogr.*, **18**, 61–84.

Lorenzato, D., Chouene, E.C., Medeiros, J., Rodrigues, A.E.C. & Pederzolli, R.C.D. (1997) Occurrence and control of the pineapple fruit borer *Thecla basalides* (Geyer, 1847). *Pesqui. Agropecu. Gaucha,* **3**, 15–19.

Losey, J.E., Fleischer, J., Calvin, D.D., Harkness, W.L. & Leahy, T. (1995) Evaluation of *Trichogramma nubilale* and *Bacillus thuringiensis* in management of *Ostrinia nubilalis* (Lepidoptera: Pyralidae) in sweet corn. *Environ. Entomol.*, **24**, 436–445.

Losey, J.E., Rayor, L.S. & Carter, M.E. (1999) Transgenic pollen harms monarch larvae. *Nature,* **399**, 214.

Lovgren, A., Carlson, C.R., Eskils, K. & Kolsto, A.B. (1998) Localization of putative virulence genes on a physical map of the *Bacillus thuringiensis* subsp. *gelechiae* chromosome. *Curr. Microbiol.*, **37**, 245–250.

Lu, Q.G., Zhang, N.X., Jiang, Y.Z. & Zhang, S.F. (1993) Laboratory and field tests on the efficacy of a Chinese produced *Bacillus thuringiensis* wettable powder against peach fruit borer, *Carposina niponensis* (Lep.: Carposinidae). *Chin. J. Biol. Cont.*, **9**, 156–159.

Lu, Z.Q., Lin, G.L., Chen, L.F., Zhu, Z.L. & Zhu, S.D. (1986) A study on the natural enemies of aphids on vegetable crops and their control. *Natural Enemies of Insects,* **8**, 63–71.

Lublinkhof, J. & Lewis, L.C. (1980) Virulence of *Nosema pyraustae* to the European corn borer when used in combination with insecticides. *Environ. Entomol.*, **9**, 67–71.

Lublinkhof, J., Lewis, L.C. & Berry, E.C. (1979) Effectiveness of integrating insecticides with *Nosema pyrausta* for suppressing populations of the European corn borer. *J. Econ. Entomol.*, **72**, 880–883.

Lumaban, M.D. & Raros, R.S. (1973) Yield responses of cabbage and mungo to injury by important insect pests in relation to insecticidal control efficiency. *Philipp. Entomol.*, **2**, 445–452.

Luo, S.B., Yan, J.P., Chai, C.J., Liang, S.P., Zhang, Y.M., Zhang, Y. & Le, G.K. (1986) Control of pink bollworm, *Pectinophora gossypiella* with *Bacillus thuringiensis* in cotton fields. *Chin. J. Biol. Cont.*, **2**, 167–169.

Luthy, P., Hofmann, C., Jaquet, F. & Hutter, R. (1985) Inactivation of delta-endotoxin of *Bacillus thuringiensis* by plant extracts. *Experientia,* **41**, 540.

Luthy, P., Raboud, G., Delucchi, V. & Kuenzi, M. (1980) Field efficacy of *Bacillus thuringiensis* var. *israelensis*. *Mitt. Schweiz. Entomol. Ges.*, **53**, 3–9.

Luttrell, R.G., Yearian, W.C. & Young, S.Y. (1981) Microbial and chemical insecticides against the cotton leafworm. *Arkansas Farm Res.*, **30**, 10.

Luttrell, R.G., Young, S.Y., Yearian, W.C. & Horton, D.L. (1982) Evaluation of *Bacillus thuringiensis*-spray adjuvant-viral insecticide combinations against *Heliothis* spp. (Lepidoptera: Noctuidae). *Environ. Entomol.*, **11**, 783–787.

Luttrell, R.G., Mascarenhas, V.J., Schneider, J.C., Parker, C.D. & Bullock, P.D. (1995) Effect of transgenic cotton expressing endotoxin protein on arthropod populations in Mississippi cotton. *Proc. Beltwide Cotton Conf.*, Texas, USA, January 4-7, 1995: Volume 2., 760–763.

Lynch, R.E., Lewis, L.C. & Brindley, T.A. (1976) Bacteria associated with eggs and first-instar larvae of the European corn borer: identification and frequency of occurrence. *J. Invertebr. Pathol.*, **27**, 229–237.

Lynch, R.E., Lewis, L.C., Berry, E.C. & Robinson, J.F. (1977) European corn borer: granular formulations of *Bacillus thuringiensis* for control. *J. Econ. Entomol.*, **70**, 389–391.

Lyness, E.W., Pinnock, D.E. & Cooper, D.J. (1994) Microbial ecology of sheep fleece. *Proceedings of a Workshop on Bacillus thuringiensis,* (ed. R. Milner), pp. 103–112, 24-26 September 1991, Canberra, A.C.T., Australia.

Lyutikova, L.I. & Yudina, T.G. (1995) The susceptibility of the termite *Anacanthotermes ahngerianus* Jacobs (Isoptera) to some subspecies of *Bacillus thuringiensis* Berliner. *Moscow Univ. Biol. Sci. Bull.*, **50**, 43–48.

Ma, C.Z., Gu, Z.R., Zhou, A.N., Qian, Z.G., Wang, D.S., Qian, G.C., Zhang, F.Q., Han, C.G. & Qi, D.G. (1994) Screening high-toxic strains of *Bacillus thuringiensis* against rice leaf roller (*Cnaphalo crocis midinalis* Guenee). *Acta Agric. Shanghai,* **10**, 52–56.

Macedo, N. (1976) Main points from a study on pests of *Eucalyptus* spp. *FAO Report, No. PNUD-FAO-IBDF-BRA-45, Comunicacao Tecnica* **4**, 11 p.

Maciejewska, J., Chamberlain, W.F. & Temeyer, K.B. (1988) Toxic and morphological effects of *Bacillus thuringiensis* preparations on larval stages of the oriental rat flea (Siphonaptera: Pulicidae). *J. Econ. Entomol.*, **81**, 1656–1661.

MacIntosh, S.C., Kishore, G.M., Perlak, F.J., Marrone, P.G., Stone, T.B., Sims, S.R. & Fuchs, R.L. (1990a) Potentiation of *Bacillus thuringiensis* insecticidal activity by serine protease inhibitors. *J. Agric. Food Chem.*, **38**, 1145–1152.

MacIntosh, S.C., McPherson, S.L., Perlak, F.J., Marrone, P.G. & Fuchs, R.L. (1990b) Purification and characterization of *Bacillus thuringiensis* var. *tenebrionis* insecticidal proteins produced in *E. coli*. *Biochem. Biophys. Res. Commun.*, **170**, 665–672.

MacIntosh, S.C., Stone, T.B., Sims, S.R., Hunst, P.L., Greenplate, J.T., Marrone, P.G., Perlak, F.J., Fischhoff, D.A. & Fuchs, R.L. (1990c) Specificity and efficacy of purified *Bacillus thuringiensis* proteins against agronomically important insects. *J. Invertebr. Pathol.*, **56**, 258–266.

MacQuillan, M.J. (1976) Insecticides for control of light brown apple moth and codling moth in Australia. *Aust. J. Exp. Agric. Animal Husb.*, **16**, 135–139.

Madatyan, A.V. (1985) Determining the times for control measures against the codling moth. *Zaschc. Rast.*, **11**, 39.

Madatyan, A.V. & Sharipov, D.D. (1984) Determining times for control of the grape leaf-roller. *Zaschc. Rast.*, **10**, 23.

Madsen, H.F. & Carty, B.E. (1979) Organic pest control: two years experience in a commercial apple orchard. *J. Entomol. Soc. British Columbia,* **76**, 3–5.

Madsen, H.F., Potter, S.A. & Peters, F.E. (1977) Pest management: control of *Archips argyrospilus* and *Archips rosanus* (Lepidoptera: Tortricidae) on apple. *Canad. Entomol.*, **109**, 171–174.

Magnoler, A. (1974) Ground application of a *Bacillus thuringiensis* preparation for gypsy moth control. *Z. Pflanzenkrankheiten Pflanzenschutz,* **81**, 575–583.

Magrini, E.A., Botelho, P.S.M. & Negrin, S.G. (1997) Control of *Anticarsia gemmatalis* Hubner on soybean, *Glycine max* (L.), with biological insecticides. *Rev. Agric. Piracicaba,* **72**, 319–329.

Mahmood, F. (1998) Laboratory bioassay to compare susceptibilities of *Aedes aegypti* and *Anopheles albimanus* to *Bacillus thuringiensis* var. *israelensis* as affected by their feeding rates. *J. Amer. Mosq. Con. Assoc.,* **14**, 69–71.

Mahmoud, E.A., Ali, A.S.A. & Abdulla, H.E. (1988) Influence of *Bacillus thuringiensis* Berliner on survival and development of great wax moth *Galleria mellonella*. *J. Biol. Sci. Res.,* **19**, 17–30.

Maillard, J.C. & Provost, A. (1975) Investigations on the pathogenicity of *Bacillus thuringiensis* for *Glossina* (Diptera - Muscidae). Study on *Glossina tachinoides* in Chad. *Rev. Elev. Med. Vet. Pays Trop.,* **28**, 61–65.

Maini, S. & Burgio, G. (1990) Biological control of *Ostrinia nubilalis* (Hb.) (Lepidoptera, Pyralidae) on protected pepper. *Boll. Ist. Entomol. 'Guido Grandi' Univ. Studi Bologna,* **44**, 23–36.

Maini, S. & Burgio, G. (1991) Biological control of the European corn borer in protected pepper by *Trichogramma maidis* Pint & Voeg. and *Bacillus thuringiensis* Berl. subsp. *kurstaki*. *Colloq. INRA,* **56**, 213–215.

Maini, S., Cappai, A. & Burgio, G. (1989) Laboratory experiments using different subspecies of *Bacillus thuringiensis* Berl. against *Ostrinia nubilalis* (Hb.). *Boll. Ist. Entomol. 'Guido Grandi' Univ. Studi Bologna,* **43**, 187–193.

Makarov, M. (1971) Noctuids - dangerous pests of plants. The lucerne Noctuid. *Rast. Zashc.,* **19**, 5–6.

Makhmudov, D., Azimov, A., Abdulagatov, A.Z. & Ataev, K.G. (1977) Control of the grape moth. *Zaschc. Rast.,* **7**, 24–25.

Maksymiuk, B. & Orchard, R.D. (1974) Techniques for evaluating *Bacillus thuringiensis* and spray equipment for aerial application against forest defoliating insects. *USDA For. Ser. Res. Paper, Pacific Northwest For. Range Exp. St.,* PNW-183, 13 p.

Maksymov, J.K. (1980) Biological control of the satin moth *Stilpnotia salicis* L. (Lep., Lymantriidae) with *Bacillus thuringiensis* Berliner. *Anz. Schadlingskd. Pflanzenschutz Umweltschutz,* **53**, 52–56.

Malokvasova, T.S. (1979) Bacterial preparations against the Siberian silkmoth. *Lesnoe Khoz.,* **4**, 59–61.

Malvar, T. & Baum, J.A. (1994) Tn5401 disruption of the spo0F gene. identified by direct chromosomal sequencing, results in CryIIIA overproduction in *Bacillus thuringiensis*. *J. Bacteriol.,* **176**, 4750–4753.

Malyi, L.P., Krushchev, L.T., Likhovidov, V.E., Kuksenkov, V.M. & Sinchuk, I.V. (1978) The use of bacterial preparations against leaf-eating pests of oak. *Lesnoe Khoz.,* **11**, 84–85.

Mamedova, S.R., Guseinov, D.G., Arakcheeva, L.I. & Kulieva, G.D. (1990) The economic threshold of injuriousness in control of *Chloridea obsoleta*. *Zaschc. Rast. Moskva,* **3**, 42–43.

Manasherob, R., Ben Dov, E., Zaritsky, A. & Barak, Z. (1998) Germination, growth, and sporulation of *Bacillus thuringiensis* subsp. *israelensis* in excreted food vacuoles of the protozoan *Tetrahymena pyriformis*. *Appl. Environ. Microbiol.,* **64**, 1750–1758.

Manchev, M. (1980) Induction of spontaneous nuclear polyhedrosis in the silkworm. *Rast. Zashc.,* **28**, 13–14.

Manjula, K. & Padmavathamma, K. (1996) Effect of microbial insecticides on the control of *Maruca testulalis* and on the predators of redgram pest complex. *Entomon,* **21**, 269–271.

Manonmani, A.M. & Hoti, S.L. (1995) Field efficacy of indigenous strains of *Bacillus thuringiensis* H-14 and *Bacillus sphaericus* H-5a5b against *Anopheles subpictus* larvae. *Trop. Biomed.,* **12**, 141–146.

Manonmani, A.M., Hoti, S.L. & Balaraman, K. (1987) Isolation of mosquito pathogenic *Bacillus thuringiensis* strains from mosquito breeding habitats in Tamil Nadu. *Indian J. Med. Res.,* **86**, 462–468.

Marchal Segault, D. (1974) Susceptibility of the hymenopterous braconids *Apanteles glomeratus* and *Phanerotoma flavitestacea* to the spore-crystal complex of *Bacillus thuringiensis* Berliner. *Ann. Zool. Ecol. Anim.,* **6**, 521–528.

Marchal Segault, D. (1975) Larval development of the parasitic Hymenoptera *Apanteles glomeratus* L. and *Phanerotoma flavitestacea* F. in caterpillars infected by *Bacillus thuringiensis* Berliner. *Ann. Parasitol. Hum. Comp.,* **50**, 223–232.

Marchenko Ya, I., Fokin, A.S., Kirienko, N.G. & Chupriyanov Yu, M. (1982) The effectiveness of bacterial preparations against *Lymantria monacha* and *Bupalus piniarius*. *Lesnoe Khoz.,* **11**, 48–51.

Marchenko Ya, I., Krushev, L.T. & Vitola, R.P. (1983) Experience in using bacterial preparations against *Bupalus piniarius* in Latvia. *Lesnoe Khoz.,* **10**, 72–75.

Marchenko, L.I. (1983) The quality of *Trichogramma* and its effectiveness. *Zaschc. Rast.,* **12**, 17.

Margalit, J. & Bobroglo, H. (1984) The effect of organic materials and solids in water on the persistence of *Bacillus thuringiensis* var. *israelensis* serotype H-14. *Z. Angew. Entomol.,* **97**, 516–520.

Mariau, D. (1982) Phyllophagous oil palm and coconut pests. Importance of entomopathogenic parasites for population regulation. *Oleagineux,* **37**, 3–7.

Marques, I.M.R. & Alves, S.B. (1995) Influence of *Bacillus thuringiensis* Berliner var. *kurstaki* on parasitism of *Scrobipalpuloides absoluta* Meyer (Lepidoptera: Gelechiidae) by *Trichogramma pretiosum* R. (Hymenoptera: Trichogrammatidae). *Arquivos de Biologia e Tecnologia,* **38**, 317–325.

Marques, I.M.R. & Alves, S.B. (1996) Effect of *Bacillus thuringiensis* Berl. var. *kurstaki* against *Scrobipalpuloides absoluta* Meyer. (Lepidoptera: Gelechiidae). *An. Soc. Entomol. Bras.,* **25**, 39–45.

Marston, N.L., Thomas, G.D., Ignoffo, C.M., Gebhardt, M.R., Hostetter, D.L. & Dickerson, W.A. (1979) Seasonal cycles of soybean arthropods in Missouri: effect of pesticidal and cultural practices. *Environ. Entomol.,* **8**, 165–173.

Mart, C. & Kilincer, N. (1994) Comparison of biopreparations with chemical and mechanical methods of controlling carob moth *Ectomyelois ceratoniae* Zell. (Lepidoptera: Pyralidae). *Turkiye III,* 25-28 Ocak 1994, 511–518.

Marten, G.G., Che, W., Bordes, E.S. & Che, W.Y. (1993) Compatibility of cyclopoid copepods with mosquito insecticides. *J. Amer. Mosq. Con. Assoc.*, **9**, 150–154.

Martin, N.A. & Workman, P. (1986) Greenlooper caterpillar control on greenhouse tomatoes with *Bacillus thuringiensis. Proc. 39th NZ Weed Pest Cont. Conf.* (ed. A.J. Popay), 130–132.

Martin, P.A.W. & Travers, R.S. (1989) Worldwide abundance and distribution of *Bacillus thuringiensis* isolates. *Appl. Environ. Microbiol.*, **55**, 2437–2442.

Martinez, A.J., Robacker, D.C. & Garcia, J.A. (1997) Toxicity of an isolate of *Bacillus thuringiensis* subspecies *darmstadiensis* to adults of the Mexican fruit fly (Diptera: Tephritidae) in the laboratory. *J. Econ. Entomol.*, **90**, 130–134.

Martouret, D. & Auer, C. (1977) Effects of *Bacillus thuringiensis* on a population of the larch tortrix, *Zeiraphera diniana* (Lep.: Tortricidae) during an outbreak period. *Entomophaga,* **22**, 37–44.

Martouret, D. & Euverte, G. (1964) The effect of *Bacillus thuringiensis* Berliner preparations on the honeybee under conditions of forced feeding. *J. Insect Pathol.*, **6**, 198–203.

Mascarenhas, R.N., Boethel, D.J., Leonard, B.R., Boyd, M.L. & Clemens, C.G. (1998) Resistance monitoring to *Bacillus thuringiensis* insecticides for soybean loopers (Lepidoptera: Noctuidae) collected from soybean and transgenic *Bt*-cotton. *J. Econ. Entomol.*, **91**, 1044–1050.

Mashanov, A.I., Gukasyan, A.B., Kobzar, V.F. & Olekh, S.A. (1976) Aerial bacteriological control of the Siberian silkworm in mountain forests in Tuva. *Izvestiya Sibirskogo Otdeleniya Akademii Nauk SSR, Biologicheskikh Nauk,* **10**, 90–95.

Mashanov, A.I., Baranovskii, V.I. & Pakhtuev, A.I. (1980) New bacterial preparations and their use in forest protection. *Izv. Sib. Otd. Akad. Nauk SSSR Biol,* **10**, 63–68.

Mashchenko, N.V. (1983) The rose budworm - a pest of soyabean. *Zaschc. Rast.*, **12**, 31.

Masson, L., Mazza, A., Gringorten, L., Baines, D., Aneliunas, V. & Brousseau, R. (1994) Specificity domain localization of *Bacillus thuringiensis* insecticidal toxins is highly dependent on the bioassay system. *Mol. Microbiol.*, **14**, 851–860.

Mata, V., Weiser, J., Olejnicek, J., Tonner, M. & Kluzak, Z. (1986) Possibilities of application of preparations based on *Bacillus thuringiensis* in the biological control of black fly larvae. *Dipterologica Bohemoslovaca IV. Sbornik referatu z VIII. celostatniho dipterologickeho seminare,* 135–138.

Matthew, J., Durr, H.J.R., Giliomee, J.H. & Neser, S. (1974) The orange tortrix, *Tortrix capensana* Wlk. (Lepidoptera: Tortricidae), as a pest of citrus, with special reference to its significance in orchards under integrated biological control. *Proc. 1st Congr. Entomol. Soc. Southern Africa,* 221–234.

Mathur, Y.K., Alam, M.A. & Kumar, J. (1994) Effectiveness of different formulations of *Bacillus thuringiensis* Berliner against *Pericallia ricini* Fabricius (Lepidoptera: Arctiidae). *J. Entomol. Res.*, **18**, 95–104.

Matsuyama, J., Yamamoto, K., Miwatani, T. & Honda, T. (1995) Monoclonal antibody developed against a hemolysin of *Bacillus thuringiensis. Microbiol. Immunol.*, **39**, 619–22.

Matter, M.M. & Keddis, M.E. (1988) Effect of *Bacillus thuringiensis* on field strains of the cotton leafworm, *Spodoptera littoralis* Boisd. *Agric. Res. Rev.*, **66**, 1–7.

Mayes, M.E., Held, G.A., Lau, C., Seely, J.C., Roe, R.M., Dauterman, W.C. & Kawanishi, C.Y. (1989) Characterization of the mammalian toxicity of the crystal polypeptides of *Bacillus thuringiensis* subsp. *israelensis. Fund. Appl. Toxicol.*, **13**, 310–322.

McClintock, J.T., Schaffer, C.R. & Sjoblad, R.D. (1995) A comparative review of the mammalian toxicity of *Bacillus thuringiensis*-based pesticides. *Pestic. Sci.*, **45**, 95–105.

McCracken, I.R. & Matthews, S.L. (1997) Effects of *Bacillus thuringiensis* subsp. *israelensis* (*B.t.i.*) applications on invertebrates from two streams on Prince Edward Island. *Bull. Environ. Contam. Toxicol.*, **58**, 291–298.

McDonald, R.C., Kok, L.T. & Yousten, A.A. (1990) Response of fourth instar *Pieris rapae* parasitized by the braconid *Cotesia rubecula* to *Bacillus thuringiensis* subsp. *kurstaki* delta-endotoxin. *J. Invertebr. Pathol.*, **56**, 422–423.

McGaughey, W.H. (1975) Compatability of *Bacillus thuringiensis* and granulosis virus treatments of stored grain with four fumigants. *J. Invertebr. Pathol.*, **26**, 247–250.

McGaughey, W.H. (1976) *Bacillus thuringiensis* for controlling three species of moths in stored grain. *Canad. Entomol.*, **108**, 105–112.

McGaughey, W.H. (1978) Response of *Plodia interpunctella* and *Ephestia cautella* larvae to spores and parasporal crystals of *Bacillus thuringiensis. J. Econ. Entomol.*, **71**, 687–688.

McGaughey, W.H. (1985) Evaluation of *Bacillus thuringiensis* for controlling Indianmeal moths (Lepidoptera: Pyralidae) in farm grain bins and elevator silos. *J. Econ. Entomol.*, **78**, 1089–1094.

McGaughey, W.H. (1990) Insect resistance to *Bacillus thuringiensis* delta-endotoxin. *New directions in biological control,* (eds. R.R. Baker & P.E. Dunn) *UCLA Symp. Mol. Cell. Biol.*, **112**, 583–598.

McGaughey, W.H. (1994a) Implications of cross-resistance among *Bacillus thuringiensis* toxins in resistance management. *Proc. OECD Worksh. Ecol. Implic. Transgenic Crops containing Bt Toxin Genes,* (eds. H.M.T. Hokkanen & J. Deacon), New Zealand on 10-14 January 1994, **4**, 427–435.

McGaughey, W.H. (1994b) Problems of insect resistance to *Bacillus thuringiensis*. *Agric. Ecosys. Environ.*, **49**, 95–102.

McGaughey, W.H. & Beeman, R.W. (1988) Resistance to *Bacillus thuringiensis* in colonies of Indianmeal moth and almond moth (Lepidoptera: Pyralidae). *J. Econ. Entomol.*, **81**, 28–33.

McGaughey, W.H. & Johnson, D.E. (1987) Toxicity of different serotypes and toxins of *Bacillus thuringiensis* to resistant and susceptible Indianmeal moths (Lepidoptera: Pyralidae). *J. Econ. Entomol.*, **80**, 1122–1126.

McGaughey, W.H. & Johnson, D.E. (1992) Indianmeal moth (Lepidoptera: Pyralidae) resistance to different strains and mixtures of *Bacillus thuringiensis*. *J. Econ. Entomol.*, **85**, 1594–1600.

McGaughey, W.H. & Johnson, D.E. (1994) Influence of crystal protein composition of *Bacillus thuringiensis* strains on cross-resistance in Indianmeal moths (Lepidoptera: Pyralidae). *J. Econ. Entomol.*, **87**, 535–540.

McGaughey, W.H. & Kinsinger, R.A. (1978) Susceptibility of *Angoumois* grain moths to *Bacillus thuringiensis*. *J. Econ. Entomol.*, **71**, 435–436.

McGaughey, W.H., Donahaye, E. & Navarro, S. (1987) *Bacillus thuringiensis*: a critical review. *Proc. 4th Intern. Work. Conf. Stored Product Prot.*, Israel, 21-26 September 1986, 14–23.

McGaughey, W.H., Kinsinger, R.A. & Dicke, E.B. (1975) Dispersal of *Bacillus thuringiensis* spores by nonsusceptible species of stored-grain beetles. *Environ. Entomol.*, **4**, 1007–1010.

McGaughey, W.H., Dicke, E.B., Finney, K.F., Bolte, L.C. & Shogren, M.D. (1980) Spores in dockage and mill fractions of wheat treated with *Bacillus thuringiensis*. *J. Econ. Entomol.*, **73**, 775–778.

McGuire, M.R. & Shasha, B.S. (1989) Sprayable self-encapsulating starch formulations for *Bacillus thuringiensis*. *J. Econ. Entomol.*, **83**, 1813–1817.

McGuire, M.R., Shasha, B.S., Lewis, L.C. & Nelsen, T.C. (1994) Residual activity of granular starch-encapsulated *Bacillus thuringiensis*. *J. Econ. Entomol.*, **87**, 631–637.

McGuire, M.R., Shasha, B.S., Eastman, C.E. & Oloumi Sadeghi, H. (1996) Starch- and flour-based sprayable formulations: effect on rainfastness and solar stability of *Bacillus thuringiensis*. *J. Econ. Entomol.*, **89**, 863–869.

McKeen, W.D., Mullins, B.A., Rodriguez, J.L. & Mandeville, J.D. (1988) *Bacillus thuringiensis* exotoxin for northern fowl mite control. *Proc. Western Poultry Dis. Conf.*, **37**, 140–141.

McKenna, C.E., Stevens, P.S. & Steven, D. (1995) A new *Bacillus thuringiensis* product for use on kiwifruit. *Proc. 48th NZ Plant Prot. Conf.* (ed. A.J. Popay), 135–138.

McKilling, S.C. & Brown, D.G. (1991) Evaluation of a formulation of *Bacillus thuringiensis* against waxmoths in stored honeycombs. *Aust. J. Exp. Agr.*, **31**, 709–711.

McLaughlin, R.E. & Vidrine, M.F. (1984) Distribution of a flowable concentrate formulation of *Bacillus thuringiensis* serotype H-14 during irrigation of rice fields as a function of the quantity of formulation. *Mosq. News*, **44**, 330–335.

McLaughlin, R.E., Fukuda, T., Willis, O.R. & Billodeaux, J. (1982) Effectiveness of *Bacillus thuringiensis* serotype H-14 against *Anopheles crucians*. *Mosq. News*, **42**, 370–374.

McLaughlin, R.E., Dulmage, H.T., Alls, R., Couch, T.L., Dame, D.A., Hall, I.M., Rose, R.I. & Versoi, P.L. (1984) U.S. Standard bioassay for the potency assessment of *Bacillus thuringiensis* serotype H-14 against mosquito larvae. *Bull. Entomol. Soc. Amer.*, **30**, 26–29.

McLeod, P.J., Yearian, W.C. & Young, S.Y. (1982) Effectiveness of *Bacillus thuringiensis* against the southern pine coneworm, *Dioryctria amatella* (Lepidoptera: Pyralidae). *Environ. Entomol.*, **11**, 1305–1306.

McLeod, P.J., Yearian, W.C. & Young, S.Y. (1983) Persistence of *Bacillus thuringiensis* on second-year loblolly pine cones. *Environ. Entomol.*, **12**, 1190–1192.

McLeod, P.J., Yearian, W.C. & Young, S.Y. (1984) Evaluation of *Bacillus thuringiensis* for coneworm, *Dioryctria* spp., control in southern pine seed orchards. *J. Georgia Entomol. Soc.*, **19**, 408–413.

McNeil, J.N., Smirnoff, W.A. & Letendre, M. (1977) *Bacillus thuringiensis* as a means of control for the European skipper, *Thymelicus lineola* (Lepidoptera: Hesperiidae). *Canad. Entomol.*, **109**, 37–38.

McVay, J.R., Gudauskas, R.T. & Harper, J.D. (1977) Effects of *Bacillus thuringiensis* nuclear-polyhedrosis virus mixtures on *Trichoplusia ni* larvae. *J. Invertebr. Pathol.*, **29**, 367–372.

McWhorter, G.M., Berry, E.C. & Lewis, L.C. (1972) Control of the European corn borer with two varieties of *Bacillus thuringiensis*. *J. Econ. Entomol.*, **65**, 1414–1417.

Meade, T. & Hare, J.D. (1994) Effects of genetic and environmental host plant variation on the susceptibility of two noctuids to *Bacillus thuringiensis*. *Entomol. Exp. Appl.*, **70**, 165–178.

Meadows, J.R., Gill, S.S. & Bone, L.W. (1989) Lethality of *Bacillus thuringiensis morrisoni* for eggs of *Trichostrongylus colubriformis* (Nematoda). *Intern. J. Invertebr. Reprod. Develop.*, **15**, 159–161.

Meadows, J., Gill, S.S. & Bone, L.W. (1990) *Bacillus thuringiensis* strains affect population growth of the free-living nematode *Turbatrix aceti*. *Invertebr. Reprod. Develop.*, **17**, 73–76.

Meadows, M.P. (1993) *Bacillus thuringiensis* in the environment: ecology and risk assessment. *Bacillus thuringiensis, an environmental biopesticide: Theory and Practice* (eds. P.F. Entwistle, J.S. Cory, M.J. Bailey & S. Higgs), pp. 193–220. John Wiley and Sons, Chichester.

Medvecky, B.A. & Zalom, F.G. (1992) Conventional and alternative insecticides, including a granular formulation of *Bacillus thuringiensis* var. *kurstaki*, for the control of *Busseola fusca* (Fuller) (Lepidoptera: Noctuidae) in Kenya. *Trop. Pest Manag.*, **38**, 186–189.

Meisner, J., Hadar, D., Wysoki, M. & Harpaz, I. (1990) Phagostimulants enhancing the efficacy of *Bacillus thuringiensis* formulations against the giant looper, *Boarmia* (*Ascotis*) *selenaria*, in avocado. *Phytoparasitica*, **18**, 107–115.

Men, U.B. & Thakare, H.S. (1998) Efficacy of biocides alone and in combination with insecticides against sunflower semilooper, *Thysanoplusia orichalcea* F. *PKV Res. J.*, **22**, 57–59.

Mena, J., Vazquez, R., Fernandez, M., Perez, L., Garcia, M., Pimentel, E., Lopez, A., Mencho, J.D., Zaldua, Z., Garcia, R., Somontes, D. & Moran, R. (1996) Use of *Bacillus thuringiensis* var. *kurstaki* to control *Meloidogyne incognita* and *Radopholus similis*. *Cent. Agric.*, **23**, 31–38.

Mena, J., Pimentel, E., Vazquez, R., Garcia, R., Fernandez, M., Moran, R., Perez, L., Garcia, M., Zaldua, Z., Somontes, D., Lopez, A., Gomez, M. & Mencho, J.D. (1997) Results of the use of *Bacillus thuringiensis* var. *kurstaki* in the control of *Radopholus similis* in banana and plantain plantations. *Cent. Agric.*, **24**, 41–49.

Mendel, Z. (1987) Major pests of man-made forests in Israel: origin, biology, damage and control. *Phytoparasitica*, **15**, 131–137.

Menendez, J.M., Echevarria, E., Garcia, A., Berrios, C., Fernandez, A., Valdes, H. & Ramos, R. (1986) Use of *Bacillus thuringiensis* in the control of *Rhyacionia frustrana*. *Rev. For. Baracoa*, **16**, 101–111.

Meneses Carbonell, R. (1983) Pathogens and nematodes for control of rice water weevil in Cuba. *Intern. Rice Res. Newsl.*, **8**, 16–17.

Meng, F., Jiang, W.C. & Bing, J.C. (1988) Preliminary observations on *Larerannis orthogrammaria* (Wehrli). *Insect Knowl.*, **25**, 32–34.

Menon, A.S. & De Mestral, J. (1985) Survival of *Bacillus thuringiensis* var. *kurstaki* in waters. *Water Air and Soil Pollut.*, **25**, 265–274.

Menon, K.K.R., Rao, A.S. & Amonkar, S.V. (1982) A new multicrystalliferous *Bacillus thuringiensis* isolate from diseased mosquito larvae. *Curr. Sci.*, **51**, 794–795.

Merca, F.E. & de los Reyes, A.M. (1997) Isolation and characterization of a lectin from *Bacillus thuringiensis* subsp. *morrisoni* (serotype H8a:8b) PG-14. *Philipp. J. Sci.*, **126**, 1–9.

Mercer, D.R., Nicolas, L. & Thiery, I. (1995) Evaluation of entomopathogenic bacteria against *Aedes polynesiensis*, the vector of lymphatic filariasis in French Polynesia. *J. Amer. Mosq. Con. Assoc.*, **11**, 485–488.

Merdan, A., Abdel Rahman, H. & Soliman, A. (1975) On the influence of host plants on insect resistance to bacterial diseases. *Z. Angew. Entomol.*, **78**, 280–285.

Merdan, A.I. (1977) Response of the common house-fly, *Musca domestica* (Linn.) to 3 commercial formulations of *Bacillus thuringiensis* Berliner. *Bull. High Inst. Public Health*, **7**, 27–33.

Merdan, A.I., Hilmy, N.M. & Ibrahim, A.A. (1986) Effectiveness of *Bacillus thuringiensis* serotype H-14 on certain Egyptian mosquito species in small ditches. *J. Egypt. Soc. Parasit.*, **16**, 171–83.

Meredith, W.R., Jr., Dugger, P. & Richter, D. (1998) The role of host plant resistance in *Lygus* management. *1998 Proc. Beltwide Cotton Conf.*, California, USA, 5-9 January 1998, 940–944.

Meretoja, T., Carlberg, G., Gripenberg, U., Linnainmaa, K. & Sorsa, M. (1977) Mutagenicity of *Bacillus thuringiensis* exotoxin. I. Mammalian tests. *Hereditas*, **85**, 105–112.

Merritt, R.W., Walker, E.D., Wilzbach, M.A., Cummins, K.W. & Morgan, W.T. (1989) A broad evaluation of *B.t.i.* for black fly (Diptera: Simuliidae) control in a Michigan river: efficacy, carry and nontarget effects on invertebrates and fish. *J. Amer. Mosq. Con. Assoc.*, **5**, 397–415.

Mertz, B.P., Fleischer, S.J., Calvin, D.D. & Ridgway, R.L. (1995) Field assessment of *Trichogramma brassicae* (Hymenoptera: Trichogrammatidae) and *Bacillus thuringiensis* for control of *Ostrinia nubilalis* (Lepidoptera: Pyralidae) in sweet corn. *J. Econ. Entomol.*, **88**, 1616–1625.

Meshram, P.B., Bisaria, A.K., Shamila, K. & Kalia, S. (1997) Efficacy of Bioasp and Biolep - a microbial insecticide against teak skeletonizer, *Eutectona machaeralis* Walk. *Indian For.*, **123**, 1202–1204.

Mielitz, L.R. & da Cruz, F.Z. (1980) Behaviour of three commercial products based on *Bacillus thuringiensis* Berliner against the soyabean larva (*Anticarsia gemmatalis* Hubner, 1818). *Agron. Sulriogr.*, **16**, 193–204.

Mihalache, G., Arsenescu, M. & Pirvescu, D. (1972) Effectiveness of the bacterial preparation Dipel for the control of some forest defoliators. *Rev. Padurilor*, **87**, 362–365.

Mihalache, G., Pirvescu, D. & Simionescu, A. (1974) Trials of the bacterial preparation Dipel for the control of some forest pests. *Rev. Padurilor Ind. Lemnului Silvic. Exploatarea Padurilor*, **89**, 134–138.

Mike, A., Ohba, M. & Aizawa, K. (1991) Spontaneous occurrence of heat-stable somatic antigen variants in *Bacillus thuringiensis*. *J. Appl. Bacteriol.*, **70**, 408–413.

Mikolajewicz, M., Studzinski, A., Wysakowska, I. & Wysocki, W. (1992) Attempts to use biochemicals for caterpillar control in medicinal plants. *Materialy Sesji Instytutu Ochrony Roslin*, **32**, 104–108.

Miller, J.C. (1990) Field assessment of the effects of a microbial pest control agent on nontarget Lepidoptera. *Bull. Entomol. Soc. Amer.*, **36**, 135–139.

Miller, J.C. (1992) Effects of a microbial insecticide, *Bacillus thuringiensis kurstaki*, on nontarget Lepidoptera in a spruce budworm-infested forest. *J. Res. Lepidoptera,* **29**, 267–276.

Miller, J.C. & West, K.J. (1987) Efficacy of *Bacillus thuringiensis* and diflubenzuron on Douglas-fir and oak for gypsy moth control in Oregon. *J. Arboric.,* **13**, 240–242.

Milner, R.J. & Akhurst, R.J. (1995) Effect of the biotic insecticide *Bacillus thuringiensis* on predators of *Chrysophtharta bimaculata*: a comment. *J. Aust. Entomol. Soc.,* **34**, 97–98.

Milstead, J.E., Odom, D. & Kirby, M. (1980) Control of early larval stages of the California oakworm by low concentrations of *Bacillus thuringiensis* applied to lower leaf surfaces. *J. Econ. Entomol.,* **73**, 344–345.

Mineo, G. (1970) Control tests against *Prays citri* Mill. by means of *Bacillus thuringiensis* (2nd note). *Bol. Ist. Entomol. Agrar. Osse. Fitopatol. Palermo,* **7**, 125–133.

Mink, J.S. & Boethel, D.J. (1993) Feeding activity and mortality of permethrin-resistant and susceptible soybean looper (Lepidoptera: Noctuidae) larvae on soybean foliage treated with insecticide. *J. Econ. Entomol.,* **86**, 265–274.

Miranpuri, G.S., Bidochka, M.J. & Khachatourians, G.G. (1993) Control of painted lady caterpillars, *Vanessa cardui* L. (Lep., Nymphalidae), on borage by *Bacillus thuringiensis* var. *kurstaki*. *J. Appl. Entomol.,* **116**, 156–162.

Misch, D.W., Anderson, L.M. & Boobar, L.R. (1987) The relative toxicity of a spore preparation of *Bacillus thuringiensis* var. *israelensis* against fourth instar larvae of *Aedes aegypti* and *Toxorhynchites amboinensis*: suspension feeding compared with enemas and forced feeding. *Entomol. Exp. Appl.,* **44**, 151–154.

Mistric, W.J. & Smith, F.D. (1973a) Tobacco budworm: control on flue-cured tobacco with certain microbial pesticides. *J. Econ. Entomol.,* **66**, 979–982.

Mistric, W.J., Jr. & Smith, F.D. (1973b) Tobacco hornworm: methomyl, monocrotophos, and other insecticides for control on flue-cured tobacco. *J. Econ. Entomol.,* **66**, 581–583.

Mitkov, A. & Raicheva, B. (1983) Achievements and problems in biological control and integrated plant protection. *Rast. Zashc.,* **31**, 12–15.

Mitrofanov, V.B. (1982) The role of microorganisms in the dynamics of abundance of the meadow moth and the use of microbiological preparations for its control. *Proceedings of the All Union Research Institute for Plant Protection,* (ed. K.V. Novozhilov), All Union Lenin Academy of Agricultural Sciences.; Leningrad; USSR, 94–99.

Mittal, P.K., Adak, T. & Sharma, V.P. (1994) Comparative toxicity of certain mosquitocidal compounds to larvivorous fish, *Poecilia reticulata*. *Indian J. Malariol.,* **31**, 43–47.

Miura, T., Takahashi, R.M. & Mulligan, F.S., III (1980) Effects of the bacterial mosquito larvicide, *Bacillus thuringiensis* serotype H–14 on selected aquatic organisms. *Mosq. News,* **40**, 619–622.

Miyamoto, K. & Aizawa, K. (1982) Effect of the existence of *Bacillus thuringiensis* AF101 spores (serotype 4a:4b) on toxicity toward the fall webworm, *Hyphantria cunea*. *Proc. Assoc. Plant Prot. Kyushu,* **28**, 207–209.

Miyasono, M., Inagaki, S., Yamamoto, M., Ohba, K., Ishiguro, T., Takeda, R. & Hayashi, Y. (1994) Enhancement of delta-endotoxin activity by toxin-free spore of *Bacillus thuringiensis* against the diamondback moth, *Plutella xylostella*. *J. Invertebr. Pathol.,* **63**, 111–112.

Mizell, R.F., III & Schiffhauer, D.E. (1987) Evaluation of insecticides for control of *Glyphidocera juniperella* (Lepidoptera: Blastobasidae) in container-grown juniper. *Florida Entomol.,* **70**, 316–320.

Moar, W.J. & Trumble, J.T. (1987a) Biologically derived insecticides for use against beet armyworm. *Calif. Agr.,* **41**, 13–15.

Moar, W.J. & Trumble, J.T. (1987b) Toxicity, joint action, and mean time of mortality of Dipel 2X, avermectin B1, neem, and thuringiensin against beet armyworms (Lepidoptera: Noctuidae). *J. Econ. Entomol.,* **80**, 588–592.

Moar, W.J. & Trumble, J.T. (1990) Comparative toxicity of five *Bacillus thuringiensis* strains and formulations against *Spodoptera exigua* (Lepidoptera: Noctuidae). *Florida Entomol.,* **73**, 195–197.

Moar, W.J., Osbrink, W.L.A. & Trumble, J.T. (1986) Potentiation of *Bacillus thuringiensis* var. *kurstaki* with thuringiensin on beet armyworm (Lepidoptera: Noctuidae). *J. Econ. Entomol.,* **79**, 1443–1446.

Moar, W.J., Trumble, J.T. & Federici, B.A. (1989) Comparative toxicity of spores and crystals from the NRD-12 and HD-1 strains of *Bacillus thuringiensis* subsp. *kurstaki* to neonate beet armyworm (Lepidoptera: Noctuidae). *J. Econ. Entomol.,* **82**, 1593–1603.

Moar, W.J., Pusztai Carey, M. & Mack, T.P. (1995a) Toxicity of purified proteins and the HD-1 strain from *Bacillus thuringiensis* against lesser cornstalk borer (Lepidoptera: Pyralidae). *J. Econ. Entomol.,* **88**, 606–609.

Moar, W.J., Pusztai Carey, M., van Faassen, H., Bosch, D., Frutos, R., Rang, C., Luo, K. & Adang, M.J. (1995b) Development of *Bacillus thuringiensis* CryIC resistance by *Spodoptera exigua* (Hubner) (Lepidoptera: Noctuidae). *Appl. Environ. Microbiol.,* **61**, 2086–2092.

Moawad, G.M., Shalaby, F.F., Metwally, A.G. and El Gemeily, M.M. (1982) Laboratory pathogenicity tests with two commercial preparations of *Bacillus thuringiensis* (Berliner) on the first instar larvae of the spiny bollworm. *Bull. Soc. Entomol. Egypt.,* **64**, 137–144.

Mohamed, A.I., Young, S.Y. & Yearian, W.C. (1983) Effects of microbial agent-chemical pesticide mixtures on *Heliothis virescens* (F.) (Lepidoptera: Noctuidae). *Environ. Entomol.,* **12**, 478–481.

Mohamed, S.H. (1993) Effect of some chemical and microbial insecticides on the lesser cotton leafworm, *Spodoptera exigua* (Hb.), and the associated predators. *Assiut J. Agr. Sci.*, **24**, 3–11.
Mohd Salleh, M.B. & Lewis, L.C. (1982a) Feeding deterrent response of corn insects to beta-exotoxin of *Bacillus thuringiensis*. *J. Invertebr. Pathol.*, **39**, 323–328.
Mohd Salleh, M.B. & Lewis, L.C. (1982b) Toxic effects of spore/crystal ratios of *Bacillus thuringiensis* on European corn borer larvae. *J. Invertebr. Pathol.*, **39**, 290–297.
Mohd Salleh, M.B. & Lewis, L.C. (1983) Comparative effects of spore-crystal complexes and thermostable exotoxins of six subspecies of *Bacillus thuringiensis* on *Ostrinia nubilalis* (Lepidoptera: Pyralidae). *J. Invertebr. Pathol.*, **41**, 336–340.
Mohd Salleh, M.B., Beegle, C.C. & Lewis, L.C. (1980) Fermentation media and production of exotoxin by three varieties of *Bacillus thuringiensis*. *J. Invertebr. Pathol.*, **35**, 75–83.
Mohrig, W., Schittek, D. & Hanschke, R. (1979) Investigations on cellular defense reactions with *Galleria mellonella* against *Bacillus thuringiensis*. *J. Invertebr. Pathol.*, **34**, 207–212.
Moiseenko, A.I. & Barybkina, L.V. (1977) Bitoxibacillin against the Colorado beetle. *Zaschc. Rast.*, **10**, 40.
Moiseeva, N.V., Kalashnikova, G.I. & Mashkina, L.G. (1979) The effectiveness of trichogrammatids against pests of cabbage at various rates of release. *Entomol. Issled. Kirgizii*, **13**, 79–81.
Moiseeva, N.V., Mashkina, L.G., Kalashnikova, G.I., Ratomskaya, A.A. & Protsenko, A.I. (1975) The effectiveness of the use of *Trichogramma* for control of certain cabbage pests in the Chu Valley, Kirgizia. *Entomol. Issled. Kirgizii*, **10**, 44–46.
Moldenke, A.F., Berry, R.E., Miller, J.C., Wernz, J.G. & Li, X.H. (1994) Toxicity of *Bacillus thuringiensis* subsp. *kurstaki* to gypsy moth, *Lymantria dispar*, fed with alder or Douglas-fir. *J. Invertebr. Pathol.*, **64**, 143–145.
Molloy, D. & Jamnback, H. (1981) Field evaluation of *Bacillus thuringiensis* var. *israelensis* as a black fly biocontrol agent and its effect on nontarget stream insects. *J. Econ. Entomol.*, **74**, 314–318.
Molloy, D., Gaugler, R. & Jamnback, H. (1981) Factors influencing efficacy of *Bacillus thuringiensis* var. *israelensis* as a biological control agent of black fly larvae. *J. Econ. Entomol.*, **74**, 61–64.
Molloy, D.P. (1992) Impact of the black fly (Diptera: Simuliidae) control agent *Bacillus thuringiensis* var. *israelensis* on chironomids (Diptera: Chironomidae) and other nontarget insects: results of ten field trials. *J. Amer. Mosq. Con. Assoc.*, **8**, 24–31.
Monnerat, R.G. & Bordat, D. (1998) Influence of HD1 (*Bacillus thuringiensis* ssp. *kurstaki*) on the developmental stages of *Diadegma* sp. (Hym., Ichneumonidae) parasitoid of *Plutella xylostella* (Lep., Yponomeutidae). *J. Appl. Entomol.*, **122**, 49–51.
Moore, G.E. (1972) Pathogenicity of ten strains of bacteria to larvae of the southern pine beetle. *J. Invertebr. Pathol.*, **20**, 41–45.
Moore, I. & Navon, A. (1973) Studies of the susceptibility of the cotton leafworm, *Spodoptera littoralis* (Boisduval), to various strains of *Bacillus thuringiensis*. *Phytoparasitica*, **1**, 23–32.
Moore, J.E. (1983) Control of tomato leafminer (*Scrobipalpula absoluta*) in Bolivia. *Trop. Pest Manag.*, **29**, 231–238.
Moraes, G.J., Macedo, N. & de Carvalho, J.C.B. (1975) The occurrence of *Dirphiopsis trisignata* Walker, 1855 and its control with *Bacillus thuringiensis* Berliner. *IPEF Instituto de Pesquisas e Estudos Florestais*, **11**, 43–47.
Morales, L., Moscardi, F., Kastelic, J.G., Sosa Gomez, D.R., Paro, F.E. & Soldorio, I.L. (1995) Susceptibility of *Anticarsia gemmatalis* Hubner and *Chrysodeixis includens* (Walker) (Lepidoptera: Noctuidae) to *Bacillus thuringiensis* (Berliner). *An. Soc. Entomol. Bras.*, **24**, 593–598.
Morallo Rejesus, B. (1988) Techniques for control of the Asian corn borer in the Philippines. *Proc. Asian Regional Maize Worksh.*, **3**, 105–116.
Morallo Rejesus, B. & Javier, P.A. (1985) Detasseling technique for the control of corn borer (*Ostrinia furnacalis* Guenee). *Philipp. Entomol.*, **6**, 287–306.
Moran Rodriguez, C. & Sandoval y Trujillo, H. (1991) Control of the greater wax moth *Galleria mellonella* by strains of *Bacillus thuringiensis* in the municipality of Tecoman, Colima, Mexico. *Rev. Latinoam. Microbiol.*, **33**, 203–207.
Morel, G. (1974) Study of the action of *Bacillus thuringiensis* in the scorpion *Buthus occitanus*. *Entomophaga*, **19**, 85–95.
Moreno Vazquez, R., Zorrilla Manas, J.A., Flores Cortinas, A., Canovas Martinez, I. & Rojo Sanchez, E. (1972) Tests against *Prays citri* on lemon, 'Berna' variety. *Bol. Inform. Plagas*, **100**, 9–11.
Moroz, A.A. (1979) The effect of bacterial preparations of the group *Bacillus thuringiensis* on larvae of cattle warble-flies. *Entomol. Issled. Kirgizii*, **13**, 105–106.
Morosini, S. (1979) Effect of different insecticides for the control of soyabean larvae. *Agron. Sulriogr.*, **15**, 77–82.
Morris, O.N. (1972a) Inhibitory effects of foliage extracts of some forest trees on commercial *Bacillus thuringiensis*. *Canad. Entomol.*, **104**, 1357–1361.
Morris, O.N. (1972b) Susceptibility of some forest insects to mixtures of commercial *Bacillus thuringiensis* and chemical insecticides, and sensitivity of the pathogen to the insecticides. *Canad. Entomol.*, **104**, 1419–1425.
Morris, O.N. (1973a) Dosage-mortality studies with commercial *Bacillus thuringiensis* sprayed in a modified Potter's tower against some forest insects. *J. Invertebr. Pathol.*, **22**, 108–114.

Morris, O.N. (1973b) A method of visualizing and assessing deposits of aerially sprayed insect microbes. *J. Invertebr. Pathol.*, **22**, 115–121.

Morris, O.N. (1975) Effect of some chemical insecticides on the germination and replication of commercial *Bacillus thuringiensis*. *J. Invertebr. Pathol.*, **26**, 199–204

Morris, O.N. (1977a) Long term study of the effectiveness of aerial application of *Bacillus thuringiensis* - acephate combinations against the spruce budworm, *Choristoneura fumiferana* (Lepidoptera: Tortricidae). *Canad. Entomol.*, **109**, 1239–1248.

Morris, O.N. (1977b) Compatibility of 27 chemical insecticides with *Bacillus thuringiensis* var. *kurstaki*. *Canad. Entomol.*, **109**, 855–864.

Morris, O.N. (1980) Report of the 1979 CANUSA cooperative *Bacillus thuringiensis* B.T. spray trials. *Rep. Canad. For. Ser.*, FPM-X-40.

Morris, O.N. (1986) Susceptibility of the bertha armyworm, *Mamestra configurata* (Lepidoptera: Noctuidae), to commercial formulations of *Bacillus thuringiensis* var. *kurstaki*. *Canad. Entomol.*, **118**, 473–478.

Morris, O.N. (1988) Comparative toxicity of delta-endotoxin and thuringiensin of *Bacillus thuringiensis* and mixtures of the two for the Bertha armyworm (Lepidoptera: Noctuidae). *J. Econ. Entomol.*, **81**, 135–141.

Morris, O.N. & Armstrong, J.A. (1975) Preliminary trials with *Bacillus thuringiensis* - chemical insecticide combinations in the integrated control of the spruce budworm, *Choristoneura fumiferana* (Lepidoptera: Tortricidae). *Canad. Entomol.*, **107**, 1281–1288.

Morris, O.N. & Moore, A. (1975) Studies on the protection of insect pathogens from sunlight inactivation. II. Preliminary field trials. *Chemical Research Institute, Ottawa, Ontario, Canada. Report CC-X-113*.

Morris, O.N. & Moore, A. (1983) Relative potencies of 50 isolates of *Bacillus thuringiensis* for larvae of the spruce budworm, *Choristoneura fumiferana* (Lepidoptera: Tortricidae). *Canad. Entomol.*, **115**, 815–822.

Morris, O.N., Hildebrand, M.J. & Moore, A. (1982) Response of the spruce budworm, *Choristoneura fumiferana* (Clem.), to graded dosage rates and single versus double applications of *Bacillus thuringiensis* var. *kurstaki*. *Rep. Canad. For. Ser.*, No. FPM-X-53.

Morris, O.N., Trottier, M., McLaughlin, N.B. & Converse, V. (1994) Interaction of caffeine and related compounds with *Bacillus thuringiensis* ssp. *kurstaki* in bertha armyworm (Lepidoptera: Noctuidae). *J. Econ. Entomol.*, **87**, 610–617.

Morris, O.N., Converse, V., Kanagaratnam, P. & Cote, J.C. (1998) Isolation, characterization, and culture of *Bacillus thuringiensis* from soil and dust from grain storage bins and their toxicity for *Mamestra configurata* (Lepidoptera: Noctuidae). *Canad. Entomol.*, **130**, 515–537.

Mosievskaya, L.M. & Makarov, E.M. (1974) The effect of bacterial preparations on the parasites of the codling moth. *Zaschc. Rast.*, **11**, 25.

Mottram, P., Madill, B., Ebsworth, P. (1989) Efficiency of *Bacillus thuringiensis* var. *israelensis* for control of saltmarsh mosquitoes. *Arbovirus Research in Australia* (eds. M.F. Uren, J. Blok & L.H Manderson), pp. 185–188. Aug. 28-Sept. 1 1989, CSIRO Division of Tropical Animal Production; Indooroopilly, Queensland; Australia.

Mowat, D.J. & Clawson, S. (1988) Insecticide treatments for the control of small ermine moth, *Yponomeuta padella* (L.) in hawthorn hedges. *Agric. Ecosys. Environ.*, **21**, 245–253.

Moyal, P. (1988) The borers of maize in the savannah area of Ivory Coast. Morphological, biological and ecological data. Control trials and plant-insect relations, 367 pp. Thesis. Office de la Recherche Scientifiques et Technique Outre-Mer, France.

Muck, O., Hassan, S., Huger, A.M. & Krieg, A. (1981) The effect of *Bacillus thuringiensis* Berliner on the parasitic hymenopterans *Apanteles glomeratus* L. (Braconidae) and *Pimpla turionellae* (L.) (Ichneumonidae). *Z. Angew. Entomol.*, **92**, 303–314.

Muckenfuss, A.E. & Shepard, B.M. (1994) Seasonal abundance and response of diamondback moth, *Plutella xylostella* (L.) (Lepidoptera: Plutellidae), and natural enemies to esfenvalerate and *Bacillus thuringiensis* subsp. *kurstaki* Berliner in coastal South Carolina. *J. Agric. Entomol.*, **11**, 361–373.

Mueller, M.D. & Harper, J.D. (1987) Interactions between a commercial preparation of *Bacillus thuringiensis* and beta-exotoxin in the fall armyworm, *Spodoptera frugiperda*. *J. Invertebr. Pathol.*, **50**, 201–206.

Mulla, M.S. (1990) Activity, field efficacy, and use of *Bacillus thuringiensis israelensis* against mosquitoes. *Bacterial control of mosquitoes and black flies: biochemistry, genetics and applications of Bacillus thuringiensis israelensis and Bacillus sphaericus* (eds. H. de Barjac & D.J. Sutherland), pp. 131–160. Unwin Hyman Ltd., London.

Mulla, M.S., Federici, B.A. & Darwazeh, H.A. (1980) Effectiveness of the bacterial pathogen *Bacillus thuringiensis* serotype H-14 against mosquito larvae. *Proc. Papers 48th Ann. Conf. Calif. Mosq. Vector Cont. Assoc.* (ed. C.D. Grant), Anaheim, California, 25–27.

Mulla, M.S., Federici, B.A. & Darwazeh, H.A. (1982) Larvicidal efficacy of *Bacillus thuringiensis* serotype H-14 against stagnant-water mosquitoes and its effects on nontarget organisms. *Environ. Entomol.*, **11**, 788–795.

Mulla, M.S., Darwazeh, H.A., Ede, L., Kennedy, B. & Dulmage, H.T. (1985) Efficacy and field evaluation of *Bacillus thuringiensis* (H-14) and *B. sphaericus* against floodwater mosquitoes in California. *J. Amer. Mosq. Con. Assoc.*, **1**, 310–315.

Mulla, M.S., Darwazeh, H.A. & Aly, C. (1986) Laboratory and field studies on new formulations of two microbial control agents against mosquitoes. *Bull. Soc. Vector Ecol.*, **11**, 255–263.

Mulla, M.S., Chaney, J.D. & Rodcharoen, J. (1990a) Control of nuisance aquatic midges (Diptera: Chironomidae) with the microbial larvicide *Bacillus thuringiensis* var. *israelensis* in a man-made lake in southern California. *Bull. Soc. Vector Ecol.*, **15**, 176–184.

Mulla, M.S., Darwazeh, H.A. & Zgomba, M. (1990b) Effect of some environmental factors on the efficacy of *Bacillus sphaericus* 2362 and *Bacillus thuringiensis* (H-14) against mosquitoes. *Bull. Soc. Vector Ecol.*, **15**, 166–175.

Mulla, M.S., Chaney, J.D. & Rodcharoen, J. (1993) Elevated dosages of *Bacillus thuringiensis* var. *israelensis* fail to extend control of *Culex* larvae. *Bull. Soc. Vector Ecol.*, **18**, 125–132.

Muller-Cohn, J., Marchal, M., Chaufaux, J., Gilois, N. & Lereclus, D. (1994) Segregational stability and conjugation of a *Bacillus thuringiensis* plasmid in artificial media, soil microcosms and insects. *Proc. 6th Intern. Coll. Invertebr. Pathol.*, Montpellier, France. pp. 36.

Muller Cohn, J., Chaufaux, J., Buisson, C., Gilois, N., Sanchis, V. & Lereclus, D. (1996) *Spodoptera littoralis* (Lepidoptera: Noctuidae) resistance to CryIC and cross-resistance to other *Bacillus thuringiensis* crystal toxins. *J. Econ. Entomol.*, **89**, 791–797.

Mulligan, F.S., III & Schaefer, C.H. (1981) Integration of selective mosquito control agent *Bacillus thuringiensis* serotype H-14, with natural predator populations in pesticide-sensitive habitats. *Proc. Papers 49th Ann. Conf. California Mosq. Vector Cont. Assoc.*, Redding, California. 1982, 19–22.

Muresan, F. & Mustea, D. (1995) Results obtained in European corn borer control *Ostrinia nubilalis* Hbn. at the agricultural research station Turda. *Probl. Prot. Plantelor,* **23**, 23–34.

Mushtaque, M., Irshad, M. & Mohyuddin, A.I. (1993) Studies on mortality factors in immature stages of *Pieris brassicae* (L.) (Pieridae; Lepidoptera) and effect of Bactospeine on its larvae and parasites. *Pakistan J. Zool.*, **25**, 165–167.

Muthukrishnan, P. & Rangarajan, M. (1974) Laboratory studies on the control of black headed caterpillar *Nephantis serinopa* Meyr, by microorganisms. *Labdev J. Sci. Tech. B,* **12**, 106–108.

Nagy, L.R. & Smith, K.G. (1997) Effects of insecticide-induction reduction in lepidopteran larvae on reproductive success of hooded warblers. *Auk,* **114**, 619–627.

Naidenov, G.P. (1977) Protection of seed lucerne. *Zaschc. Rast.*, **12**, 55.

Nair, K.S.S. (1998) KFRI's tryst with the teak defoliator. *Evergreen Trichur,* **40**, 1–7.

Nakamura, L.K. & Dulmage, H.T. (1988) *Bacillus thuringiensis* cultures available from the US Department of Agriculture. *Tech. Bull. USDA,* **1738**, 38pp.

Nakamura, Y., Kanayama, A., Sato, H., Fujita, K., Tabaru, Y. & Shimada, A. (1985) Field trials of *Bacillus thuringiensis* var. *israelensis* against blackfly larvae in Japan. *Jap. J. Sanitary Zool.*, **36**, 371–373.

Nakano, O., Sampaio, V.R. & Paro, L.A., Jr. (1973) The pecan borer, *Timocratica albella. Solo,* **65**, 41–47.

Narayanan, K. & Jayaraj, S. (1974a) Mode of action of *Bacillus thuringiensis* Berliner in citrus leaf caterpillar, *Papilio demoleus* L. (Papilionidae: Lepidoptera). *Indian J. Exp. Biol.*, **12**, 89–91.

Narayanan, K. & Jayaraj, S. (1974b) The effect of *Bacillus thuringiensis* endotoxin on hemolymph cation levels in the citrus leaf caterpillar, *Papilio demoleus. J. Invertebr. Pathol.*, **23**, 125–126.

Narayanan, K., Jayaraj, S. & Subramanian, T.R. (1970) Control of three species of lepidopteran insects with the pathogen, *Bacillus thuringiensis* Berliner. *Madras Agr. J.*, **57**, 665–673.

Narayanan, K., Jayaraj, S. & Govindarajan, R. (1976a) Further observations on the mode of action of *Bacillus thuringiensis* on *Papilio demoleus* and *Spodoptera litura*. *J. Invertebr. Pathol.*, **28,** 269–270.

Narayanan, K., Govindarajan, R., Jayaraj, S. & Muthu, M. (1976b) X-ray studies on the effect of *Bacillus thuringiensis* Berliner on the feeding activity in three species of Lepidoptera. *Curr. Sci.*, **45**, 772.

Narayanasamy, P. & Baskaran, P. (1979) Field efficiency of a fungus, a bacterium and organic insecticides against rice pests. *Intern. Rice Res. Newsl.*, **4**, 19.

Narayanasamy, P., Ragunathan, V. & Baskaran, P. (1979) Incidence of brinjal mosaic virus and brinjal little leaf diseases in Dipel-organic insecticide combination treatments. *Madras Agr. J.*, **66**, 170–173.

Nashnosh, I.M., Baraka, M.M., Ismail, W. & Maayuf, M.M. (1993) Laboratory evaluation of natural and commercial preparations of the entomopathogenic fungi and bacteria on leopard moth *Zeuzera pyrina* L. (Lepidoptera: Cossidae). *Arab J. Plant Prot.*, **11**, 73–76.

Nasrullaev, D.N. (1974) Not chemical protection alone. *Zaschc. Rast.*, **5**, 6–7.

Nataraju, B., Baig, M., Subbaiah, M.B., Reddy, S.V., Singh, B.D. & Noamani, M.K.R. (1993) Comparative toxicity and infectivity titer of *Bacillus thuringiensis* strains to silkworm, *Bombyx mori* L. *Indian J. Sericult.*, **32**, 103–105.

Naton, E. (1978) On testing the side-effects of pesticides on beneficial arthropods on living plants. Test with *Phygadeuon trichops* Thomson (Ichneumonidae). *Anz. Schaedlingskd. Pflanzenschutz Umweltschutz,* **51**, 136–139.

Navarro, M.A. (1988) Biological control of *Scrobipalpula absoluta* (Meyrick) by *Trichogramma* sp. in the tomato (*Lycopersicon esculentum* Mill.). *Colloq. INRA,* **43**, 453–458.

Navon, A. (1993) Control of lepidopteran pests with *Bacillus thuringiensis*. In: *Bacillus thuringiensis, an Environmental Biopesticide: Theory and Practice* (eds. P.F. Entwistle, J.S. Cory, M.J. Bailey & S. Higgs), pp. 125–146. John Wiley and Sons, Chichester.

Navon, A. & Melamed Madjar, V. (1986) Screening of *Bacillus thuringiensis* preparations for microbial control of *Ostrinia nubilalis* in sweet corn. *Phytoparasitica,* **14**, 111–117.

Nayak, P., Rao, P.S. & Padmanabhan, S.Y. (1978) Effect of Thuricide on rice stem borers. *Proc. Indian Acad. Sci. B,* **87**, 59–62.

NDoye, M. & Gahukar, R. (1986) Insect pests of pearl millet in West Africa and their control. *Proc. Intern. Pearl Millet Worksh.,* 183–205.

Nealis, V. & van Frankenhuyzen, K. (1990) Interactions between *Bacillus thuringiensis* Berliner and *Apanteles fumiferanae* Vier. (Hymenoptera: Braconidae), a parasitoid of the spruce budworm, *Choristoneura fumiferana* (Clem.) (Lepidoptera: Tortricidae). *Canad. Entomol.,* **122**, 585–594.

Nealis, V.G., van Frankenhuyzen, K. & Cadogan, B.L. (1992) Conservation of spruce budworm parasitoids following application of *Bacillus thuringiensis* var. *kurstaki* Berliner. *Canad. Entomol.,* **124**, 1085–1092.

Nef, L. (1972) Influence of chemical and microbial treatments on a population of *Stilpnotia* (=*Leucoma*) *salicis* L. and on its parasites. *Z. Angew. Entomol.,* **69**, 357–367.

Nef, L. (1975) Microorganismes pathogenes dans la lutte contre les Lymantriides defoliateurs. *Semaine d'etude. Agriculture et hygiene des plantes.* 8-12 Sept. 1975, 325–330.

Neisess, J. (1980) Effect of pH and chlorine concentration on activity of *Bacillus thuringiensis* tank mixes. *J. Econ. Entomol.,* **73**, 186–188.

Neisess, J.A., Hubbard, H.B., Ignoffo, C.M. & Falcon, L.A. (1978) Application of microbial insecticides on forests. *Misc. Publ. Entomol. Soc. Amer.,* **10**, 27–43.

Neumann, F.G. & Collett, N.G. (1997a) Insecticide trials for control of the steelblue sawfly (*Perga affinis affinis*), a primary defoliator in young commercial eucalypt plantations of south-eastern Australia. *Aust. For.,* **60**, 75–83.

Neumann, F.G. & Collett, N.G. (1997b) Insecticide trials for control of the autumn gum moth (*Mnesampela privata*), a primary defoliator in commercial eucalypt plantations prior to canopy closure. *Aust. For.,* **60**, 130–137.

Nguyen, V.D. (1971) Results of the use of bacterial preparations in the control of leaf-eating fruit-tree pests. *Beitr. Entomol.,* **21**, 397–402.

Niccoli, A. & Pelagatti, O. (1986) Persistence of products based on *Bacillus thuringiensis* Berl. used in the control of some species of Lepidoptera. *Redia,* **69**, 329–339.

Nicolescu, G. (1982) The action of two formulations containing *Bacillus thuringiensis* (serotype H-14) on some autochthonous mosquito species (Diptera: Culicidae). *Arch. Roum. Pathol. Exp. Microbiol.,* **41**, 67–72.

Nicoli, G., Cornale, R., Corazza, L. & Marzocchi, L. (1989) Activity of *Anthocoris nemoralis* (F.) (Rhyn. Anthocoridae) against *Psylla pyri* (L.) (Rhyn. Psyllidae) in pear orchards using various pest control strategies. *Boll. Ist. Entomol. 'Guido Grandi' Univ. Studi Bologna,* **43**, 171–186.

Nicoli, G., Corazza, L. & Cornale, R. (1990) Biological control of lepidopterous tortricid leafrollers on pear with *Bacillus thuringiensis* Berl. ssp. *kurstaki*. *Inf. Fitopatol.,* **40**, 55–62.

Niemczyk, E. (1971) The duration of the toxicity of the bacterial preparations Thuricide 90 TS and Entobakterin 3 for the control of three species of orchard pests. *Rocz. Nauk Roln. Seria E,* **1**, 7–16.

Niemczyk, E. (1997) The occurrence of different groups of phytophagous and predatory mites on apple plots sprayed according to different programs. *Zahradnictvi,* **24**, 45–52.

Niemczyk, E. (1980) Applying bacterial preparations against orchard pests. *Proceedings International symposium of IOBC/WPRS on integrated control in agriculture* (eds. K. Russ & H. Berger), pp. 416–419. International Organization for Biological Control of Noxious Animals and Plants, West Palearctic Regional Section.; Vienna; Austria.

Niemczyk, E. (1997) The occurrence of different groups of phytophagous and predatory mites on apple plots sprayed according to different programs. *Zahradnictvi,* **24**, 45–52.

Niemczyk, E. & Bakowski, G. (1970) The effectiveness of Thuricide 90TS and Entobakterin 3 for control of the rose Tortricid (*Cacoecia rosana* L.), the small ermine moth (*Hyponomeuta malinellus* Zell.) and the apple leaf skeletoniser (*Simaethis pariana* Clerk.). *Rocz. Nauk Roln. E,* **1**, 103–117.

Niemczyk, E. & Dronka, K. (1976) The effectiveness of the bacterial preparation Dipel for the control of leaf-rollers (Tortricidae) in apple orchards. *Rocz. Nauk Roln. E,* **6**, 143–149.

Niemczyk, E., Dadaj, J. & Dronka, K. (1973) The control of the winter moth (*Operophthera brumata* L.) with some bacterial preparations and various insecticidal products. *Rocz. Nauk Roln. E,* **3**, 7–19.

Niemczyk, E., Dronka, K., Dawydko, B. & Predki, S. (1975) Effectiveness of bacterial preparation Thuricide 90TS to control leaf rollers (Tortricidae) in apple-orchards. *Rocz. Nauk Roln. E,* **5**, 75–85.

Niemczyk, E., Olszak, R. & Miszczak, M. (1976) The effectiveness of the bacterial preparation Dipel in controlling the codling moth (*Laspeyresia pomonella* L.) and leaf rollers (Tortricidae) during the summer period. *Rocz. Nauk Roln. E,* **6**, 151–157.

Nishiitsutsuji Uwo, J., & Endo, Y. (1980) Mode of action of *Bacillus thuringiensis* delta-endotoxin: relative role of spores and crystals in toxicity to *Pieris, Lymantria* and *Ephestia* larvae. *Appl. Entomol. Zool.,* **15**, 416–424.

Nishiitsutsuji Uwo, J., & Endo, Y. (1981) Mode of action of *Bacillus thuringiensis* delta-entotoxin: changes in hemolymph pH and ions of *Pieris, Lymantria* and *Ephestia* larvae. *Appl. Entomol. Zool.,* **16**, 225–230.

Niwa, C.G., Stelzer, M.J. & Beckwith, R.C. (1987) Effects of *Bacillus thuringiensis* on parasites of western spruce budworm (Lepidoptera: Tortricidae). *J. Econ. Entomol.,* **80**, 750–753.

Nolting, S.P. & Poston, F.L. (1982) Application of *Bacillus thuringiensis* through center-pivot irrigation systems for control of the southwestern corn borer and European corn borer (Lepidoptera: Pyralidae). *J. Econ. Entomol.,* **75**, 1069–1073.

Nora, I., Reis Filho, W. & Stuker, H. (1989) Larval damage in fruits and leaves of apple trees: changes in the agroecosystem cause appearance of unwanted insects in the orchards. *Agropecu. Catarinense,* **2**, 54–55.

Norris, J.R. (1964) The classification of *Bacillus thuringiensis*. *J. Appl. Bacteriol.,* **27**, 439–447.

Norton, M. (1991) Timing *Bacillus thuringiensis* insecticides for omnivorous leafroller control in grapes. *Components,* **2**, 5–6.

Noteborn, H.P.J.M., Kuiper, H.A. & Jones, D.D. (1994) Safety assessment strategies for genetically modified plant products: a case study of *Bacillus thuringiensis*-toxin tomato. *The Biosafety Results of Field Tests of Genetically Modified Plants and Microorganisms,* Monterey, California, USA, 13-16 November, 1994, 199–207.

Noteborn, H.P.J.M., Bienenmann Ploum, M.E., Alink, G.M., Zolla, L., Reynaerts, A., Pensa, M. & Kuiper, H.A. (1996) Safety assessment of the *Bacillus thuringiensis* insecticidal crystal protein CryIA(b) expressed in transgenic tomatoes. *Agri food quality: An Interdisciplinary Approach* (eds. G.R. Fenwick, C. Hedley, R.L. Richards & S. Khokhar), pp. 23–26; Special Publication No. 179. Royal Society of Chemistry; Cambridge; UK.

Novinskii Yu, S. (1977) No relaxation of attention to the meadow moth. *Zaschc. Rast.,* **4**, 20–21.

Novitskaya, T.N. & Abzianidze, N.V. (1985) Studies of the effect of biopreparation Bitoxibacillin on citrus scale insects. *Subtrop. Kul't.,* **6**, 142–144.

Novitskaya, T.N. & Dzholiya, D.G. (1980) The use of bacterial preparations from cultures of *Bacillus thuringinesis* var. *caucasicus* in the control of orchard pests. *Biol. Z. Armenii,* **33**, 429–431.

Novotny, J. (1988a) Susceptibility of *Lymantria dispar* (Lepidoptera, Lymantriidae) larvae to *Bacillus thuringiensis* and 2 moulting inhibitors, alone and in combination. *Anz. Schaedlingskd. Pflanzenschutz Umweltschutz,* **61**, 11–14.

Novotny, J. (1988b) Effectiveness of tank-mix Nomolt and microbial preparation in control of the caterpillars of *Lymantria dispar* L. (Lepidoptera: Lymantriidae). *Lesnictvi,* **34**, 525–536.

Novotny, J. (1988c) Time distribution of the effectiveness of bacterial insecticides on caterpillars of *Lymantria dispar*. *Lesnicky Casopis,* **34**, 325–335.

Novotny, J. & Svestka, M. (1986) Synergism of biological and chemical insecticides against larvae of the gypsy moth *Lymantria dispar* (L.). *Lesnictvi,* **32**, 1115–1128.

Novozhilov, K.V., Gusev, G.V., Sazonova, I.N., Koval, A.G., Moralev, S.N. & Patrashku, F.I. (1991) Prospect for a complex system of egg plant protection against Colorado beetle in Moldova. *Ekologicheskie osnovy primeneniya insektoakaritsidov,* 78–89.

Nugud, A.D. & White, G.B. (1982) Evaluation of *Bacillus thuringiensis* serotype H-14 formulations as larvicides for *Anopheles arabiensis* (species B of the *An. gambiae* complex). *Mosq. News,* **42**, 36–40.

Nwanze, K.F., Partida, G.J. & McGaughey, W.H. (1975) Susceptibility of *Cadra cautella* and *Plodia interpunctella* to *Bacillus thuringiensis* on wheat. *J. Econ. Entomol.,* **68**, 751–752.

Nyirady, S.A. (1973) The germfree culture of three species of Triatominae: *Triatoma protracta* (Uhler), *Triatoma rubida* (Uhler) and *Rhodnius prolixus* Stal. *J. Med. Entomol.,* **10**, 417–448.

Nyouki, F.F.R. & Fuxa, J.R. (1994) Persistence of natural and genetically engineered insecticides based on *Bacillus thuringiensis*. *J. Entomol. Sci.,* **29**, 347–356.

Nyouki, F.F.R., Fuxa, J.R. & Richter, A.R. (1996) Spore-toxin interactions and sublethal effects of *Bacillus thuringiensis* in *Spodoptera frugiperda* and *Pseudoplusia includens* (Lepidoptera: Noctuidae). *J. Entomol. Sci.,* **31**, 52–62.

Oatman, E.R., Wyman, J.A., van Steenwyk, R.A., Johnson, M.W. (1983) Integrated control of the tomato fruitworm (Lepidoptera: Noctuidae) and other lepidopterous pests on fresh-market tomatoes in southern California. *J. Econ. Entomol.,* **76**, 1363–1369.

Obadofin, A.A. & Finlayson, D.G. (1977) Interactions of several insecticides and a carabid predator (*Bembidion lampros* (Hrbst.)) and their effects on *Hylemya brassicae* (Bouche). *Canad. J. Plant Sci.,* **57**, 1121–1126.

Obama, E., Soria, S. & Tome de la Vega, F., (1988) Serious attack of *Malacosoma neustria* (Linnaeus, 1758) (Lepidoptera: Lasiocampidae) and other lepidopterans on holly oak of Monte del Pardo (Madrid); laboratory trials for its control and evaluation of the chemical control campaign. *Boletin de Sanidad Vegetal Plagas,* **14**, 27–38.

Obra, J.B. & Morallo Rejesus, B. (1997) Sensitivity of *Cotesia plutellae* and *Diadegma semiclausum*, parasitoids of diamondback moth, *Plutella xylostella* (L.) to several insecticides. *Philipp. Entomol.*, **11**, 49–56.

Ochoa, R. & von Lindeman, G. (1988) Importance of Acari in the cultivation of tomato (*Lycopersicon esculentum*) and sweet pepper (*Capsicum annuum*) in Panama. *Manejo Integrado de Plagas,* **7**, 29–36.

Ogiwara, K., Indrasith, L.S., Asano, S. & Hori, H. (1992) Processing of delta-endotoxin from *Bacillus thuringiensis* subsp. *kurstaki* HD-1 and HD-73 by gut juices of various insect larvae. *J. Invertebr. Pathol.*, **60**, 121–126.

Ohba, M. (1996) *Bacillus thuringiensis* populations naturally occurring on mulberry leaves: a possible source of the populations associated with silkworm-rearing insectaries. *J. Appl. Bacteriol.*, **80**, 56–64.

Ohba, M. & Aizawa, K. (1979a) Properties of *Bacillus thuringiensis* subsp. *ostriniae* and subsp. *wuhanensis*. *Proc. Assoc. Plant Prot. Kyushu,* **25**, 125–128.

Ohba, M. & Aizawa, K. (1979b) A new subspecies of *Bacillus thuringiensis* possessing 11a:11c flagellar antigenic structure: *Bacillus thuringiensis* subsp. *kyushuensis*. *J. Invertebr. Pathol.*, **33**, 387–388.

Ohba, M. & Aizawa, K. (1986a) *Bacillus thuringiensis* subsp. *japonensis* (flagellar serotype 23): a new subspecies of *Bacillus thuringiensis* with a novel flagellar antigen. *J. Invertebr. Pathol.*, **48**, 129–130.

Ohba, M. & Aizawa, K. (1986b) Frequency of Cry- spore forming *B. cereus* possessing flagellar antigens of *Bt*. *J. Basic Microbiol.*, **26**, 185–188.

Ohba, M. & Aizawa, K. (1989a) Distribution of the four flagellar (H) antigenic subserotypes of *Bacillus thuringiensis* H serotype 3 in Japan. *J. Appl. Bacteriol.*, **67**, 505–509.

Ohba, M. & Aizawa, K. (1989b) New flagellar (H) antigenic subfactors in *Bacillus thuringiensis* H serotype 3 with description of two new subspecies, *Bacillus thuringiensis* subsp. *sumiyoshiensis* (H serotype 3a:3d) and *Bacillus thuringiensis* subsp. *fukuokaensis* (H serotype 3a:3d:3e). *J. Invertebr. Pathol.*, **54**, 208–210.

Ohba, M. & Aizawa, K. (1990) Occurrence of two pathotypes in *Bacillus thuringiensis* subsp. *fukuokaensis* (flagellar serotype 3a:3d:3e). *J. Invertebr. Pathol.*, **55**, 293–294.

Ohba, M., Aizawa, K. & Furusawa, T. (1979) Distribution of *Bacillus thuringiensis* serotypes in Ehime Prefecture, Japan. *Appl. Entomol. Zool.*, **14**, 340–345.

Ohba, M., Aizawa, K. & Shimizu, S. (1981a) A new subspecies of *Bacillus thuringiensis* isolated in Japan: *Bacillus thuringiensis* subsp. *tohokuensis* (serotype 17). *J. Invertebr. Pathol.*, **38**, 307–309.

Ohba, M., Ono, K., Aizawa, K. & Iwanami, S. (1981b) Two new subspecies of *Bacillus thuringiensis* isolated in Japan: *Bacillus thuringiensis* subsp. *kumamotoensis* (serotype 18) and *Bacillus thuringiensis* subsp. *tochigiensis* (serotype 19). *J. Invertebr. Pathol.*, **38**, 184–190.

Ohba, M., Tantichodok, A. & Aizawa, K. (1981c) Production of heat-stable exotoxin by *Bacillus thuringiensis* and related bacteria. *J. Invertebr. Pathol.*, **38**, 26–32.

Ohba, M., Aizawa, K. & Sudo, S.I. (1984) Distribution of *Bacillus thuringiensis* in sericultural farms of Fukuoka Prefecture, Japan. *Proc. Assoc. Plant Prot. Kyushu,* **30**, 152–155.

Ohba, M., Iwahana, H., Asano, S., Suzuki, N., Sato, R. & Hori, H. (1992) A unique isolate of *Bacillus thuringiensis* serovar *japonensis* with a high larvicidal activity specific for scarabaeid beetles. *Lett. Appl. Microbiol.*, **14**, 54–57.

Ohba, M., Saitoh, H., Miyamoto, K., Higuchi, K. & Mizuki, E. (1995) *Bacillus thuringiensis* serovar *higo* (flagellar serotype 44), a new serogroup with larvicidal activity preferential for the anopheline mosquito. *Lett. Appl. Microbiol.*, **21**, 316–318.

Okada, T., Imamura, K., Matsutani, S. & Sone, K. (1977) Effect of the particle size of mulberry leaf powder in an artificial diet on the potency of a formulation of *Bacillus thuringiensis*. *Bull. Agr. Chem. Inspect. St.*, **17**, 28–33.

Olejnicek, J. (1986) The use of *Bacillus thuringiensis* var. *israelensis* in the biological control of blackflies in Czechoslovakia. *Wiad. Parazyt.*, **32**, 539–542.

Olszak, R. (1982) Impact of different pesticides on ladybird beetles (Coccinellidae, Col.). *Rocz. Nauk Roln. E Ochr. Rosl.*, **12**, 141–149.

Ono, K., Ohba, M., Aizawa, K. & Iwanami, S. (1988) Flagellar subserotype in *Bacillus thuringiensis* serotype 6: description of *Bacillus thuringiensis* subsp. *oyamensis* (serotype 6a:6c). *J. Invertebr. Pathol.*, **51**, 296–297.

Ooi, P.A.C. (1980) The pathogenicity of *Bacillus thuringiensis* for *Crocidolomia binotalis*. *MAPPS Newsl.*, **3**, 4.

Orduz, S., Realpe, M., Arango, R., Murillo, L.A. & Delecluse, A. (1998) Sequence of the *cry*11Bb11 gene from *Bacillus thuringiensis* subsp. *medellin* and toxicity analysis of its encoded protein. *Biochim. Biophys. Acta,* **1388**, 267–272.

Orekhov, D.A. (1978) Pathogenesis of septicaemia in larvae of *Dendrolimus pini*. *Lesovedenie,* **4**, 84–89.

Orekhov, D.A., Grimal' skii, V.I. & Lozinskii, V.A. (1978) Ants - vectors of infection. *Zaschc. Rast.*, **9**, 25.

Orr, D.B. & Landis, D.A. (1997) Oviposition of European corn borer (Lepidoptera: Pyralidae) and impact of natural enemy populations in transgenic versus isogenic corn. *J. Econ. Entomol.*, **90**, 905–909.

Ortegon Martinez, J.J. & Quiroz Martinez, H. (1990) Effect of the GM-10 strain of *Bacillus thuringiensis* on the predatory capacity of *Buenoa* sp. (Hemiptera: Notonectidae) on larvae of *Culex pipiens quinquefasciatus*. *Folia Entomol. Mex.*, **79**, 197–206.

Osborne, L.S., Boucias, D.G. & Lindquist, R.K. (1985) Activity of *Bacillus thuringiensis* var. *israelensis* on *Bradysia coprophila* (Diptera: Sciaridae). *J. Econ. Entomol.*, **78**, 922–925.

Osintseva, L.A. (1996) Ecologically safe means of controlling flea beetles in fields of white cabbage. *Agrokhimiya*, **8–9**, 112–116.

Osman, G.Y. & Mohamed, A.M. (1991) Bio-efficacy of bacterial insecticide, *Bacillus thuringiensis* Berl. as biological control agent against snail vectors of schistosomiasis in Egypt. *Anz. Schaedlingskd. Pflanzenschutz Umweltschutz*, **64**, 136–139.

Osman, G.Y., Salem, F.M. & Ghattas, A. (1988) Bio-efficacy of two bacterial insecticide strains of *Bacillus thuringiensis* as a biological control agent in comparison with a nematicide, Nemacur, on certain parasitic Nematoda. *Anz. Schaedlingskd. Pflanzenschutz Umweltschutz*, **61**, 35–37.

Osman, G.Y., Mohamed, A.M. & Jamel Al Layl, K. (1992) Studies on molluscicidal activity of different preparations of *Bacillus thuringiensis* as biocidal agents on *Biomphalaria alexandrina* snails as vectors of schistosomiasis (bilharziasis) in Saudi Arabia. *Anz. Schaedlingskd. Pflanzenschutz Umweltschutz*, **65**, 67–70.

Otvos, I.S. & Raske, A.G. (1980) The effects of fenitrothion, Matacil, and *Bacillus thuringiensis* plus Orthene on larval parasites of the spruce budworm, *Choristoneura fumiferana* (Lepidoptera: Tortricidae). *Inf. Rep. Canad. For. Ser.*, N-X-184, 24 pp.

Ouzounis, K. & Samanidou Voyadjoglou, A. (1993) Preliminary research on *Bacillus thuringiensis* var. *israelensis* use for mosquito control in northeast Greece. *Bull. Soc. Vector Ecol.*, **18**, 147–151.

Ozawa, A., Saito, T. & Ikeda, F. (1998) Effects of pesticides on *Diglyphus isaea* (Walker) and *Dacnusa sibirica* Telenga, parasitoids of *Liriomyza trifolii* (Burgess). *Jap. J. Appl. Entomol. Zool.*, **42**, 149–161.

Ozino Marletto, O.I., Arzone, A.& Marletto, F.(1972) Tests of infection of *Apis mellifera ligustica* Spinola with increasing doses of *Bacillus thuringiensis dendrolimus* Talalaev. *Ann. Fac. Sci. Agrar. Univ. Studi Torino*, **8**, 157–172.

Padua, L.E., Gabriel, B.P., Aizawa, K. & Ohba, M. (1981) *Bacillus thuringiensis* isolated from the Philippines. *Philipp. Entomol.*, **5**, 199–208.

Padua, L.E., Ohba, M. & Aizawa, K. (1980) The isolates of *Bacillus thuringiensis* serotype 10 with a highly preferential toxicity to mosquito larvae. *J. Invertebr. Pathol.*, **36**, 180–186.

Padua, L.E., Ebora, R.V. & Moran, D.G. (1987) Screening of *Bacillus thuringiensis* against the Asian corn borer, *Ostrinia furnacalis* (Guenee) and the diamond back moth, *Plutella xylostella* (L.). *Ext.Bull., ASPAC Food Fert. Tech. Center Asian Pacific Region Taiwan*, **No. 257**, 28–29.

Painter, M.K., Tennessen, K.J. & Richardson, T.D. (1996) Effects of repeated applications of *Bacillus thuringiensis israelensis* on the mosquito predator *Erythemis simplicicollis* (Odonata: Libellulidae) from hatching to final instar. *Environ. Entomol.*, **25**, 184–191.

Pal, S.K. (1977) Relative effectiveness of some chemical insecticides and bacterial insecticides against castor semilooper (*Achoea janata* L.). *Indian J. Plant Prot.*, **5**, 195–198.

Palanichamy, M. & Arunachalam, A. (1995) Temperature dependent feeding rate and the larval mortality of *Spodoptera litura* (Fabricius) treated with *Bacillus thuringiensis kurstaki* (Berliner). *New Agricult.*, **6**, 55–58.

Palm, C.J., Schaller, D.L., Donegan, K.K. & Seidler, R.J. (1996) Persistence in soil of transgenic plant produced *Bacillus thuringiensis* var. *kurstaki* delta-endotoxin. *Canad. J. Microbiol.*, **42**, 1258–1262.

Palmer, R.W. (1993) Short-term impacts of formulations of *Bacillus thuringiensis* var. *israelensis* de Barjac and the organophosphate temephos used in blackfly (Diptera: Simuliidae) control on rheophilic benthic macroinvertebrates in the middle Orange River, South Africa. *Southern African J. Aquat. Sci.*, **19**, 14–33.

Palmer, R.W., Edwardes, M. & Nevill, E.M. (1996) Control of pest blackflies (Diptera: Simuliidae) along the Orange River, South Africa: 1990-1995. *Onderstepoort J. Vet. Res.*, **63**, 289–304.

Palmisano, C.T. (1987) Overview of the larvicide program of the St. Tammany parish mosquito abatement district No. 2. *Proc. Ann. Meet. Utah Mosq. Abatement Assoc.*, **40**, 9–10.

Pana Beratlief, Z. (1968) The behaviour of different strains of *Bacillus thuringiensis* in deep culture and their entomopathogenic effect. *Anal. Inst. Cerc. Prot. Plantelor*, **6**, 407–414.

Panbangred, W., Pantuwatana, S. & Bhumiratana, A. (1979) Toxicity of *Bacillus thuringiensis* toward *Aedes aegypti* larva. *J. Invertebr. Pathol.*, **33**, 340–347.

Pandey, B.N., Singh, B.P., Dayal, R. & Sanehi, R. (1977) Efficacy of quinalphos and chlorpyriphos emulsions in controlling black bug of sugarcane. *Indian Sugar Crops J.*, **4**, 89–91.

Panetta, J.D. (1989) M-one insecticide (active ingredient *Bacillus thuringiensis* subsp. *san diego*)- a case study in the successful registration of a new biological product. *AgBiotech '89.* Proceedings of a conference held in Arlington, Virginia, USA, 28-30 March 1989, 50–56.

Panis, A. (1979) Integrated control in olive cultivation. *Inf. Fitopatol.*, **29**, 27–28.

Panizzi, L. & Pinzauti, M. (1988) Proliferation of pathogenic bacteria in the nest of the *Apis mellifera* following attack by *Varroa jacobsoni* Oud. *Apiacta*, **23**, 74–78.

Pardede, D. (1986) Integrated control of cocoa tussock moth *Orgyia postica* Wlk in North Sumatera. *Bul. Perkebunan,* **17**, 131–138.

Pardede, D. (1992) Study of integrated pest control of *Darna trima* Moore (Lepidoptera: Limacodidae) in oil palm (*Elaeis guineensis* Jacq.). *Bul. Perkebunan,* **23**, 103–114.

Pari, P., Carli, G., Molinari, F. & Cravedi, P. (1993) Evaluations of the efficacy of *Bacillus thuringiensis* Berliner against *Cydia molesta* (Busck). *Bull. OILB-SROP,* **16**, 38–41.

Park, H.W., Roh, J.Y., Je, Y.H., Jin, B.R., Oh, H.W., Park, H.Y. & Kang, S.K. (1998) Isolation of a non-insecticidal *Bacillus thuringiensis* strain belonging to serotype H8a8b. *Lett. Appl. Microbiol.*, **27**, 62–66.

Parnata, Y., Muchtar, A. & Subardjo (1981) A Boarmia pest on cocoa and *Moghania macrophylla* in the Deli-Serdang district. *Bull. Balai Penelitian Perkebunan Medan,* **12**, 49–54.

Pascovici, V., Mihalache, G., Pirvescu, D. & Simionescu, D. (1978) Experiments on the control of pest *Lymantria dispar* L. and other defoliators by means of bacterial preparations. *Zast. Bilja,* **29**, 69–76.

Pasqualini, E., Antropoli, A., Pari, P. & Faccioli, G. (1992) Biological control in integrated pest management systems for apple and pear orchards. *Acta Phytopathol. Entomol. Hung.*, **27**, 507–512.

Pasqualini, E., Antropoli, A. & Civolani, S. (1996) Recent experiments in the control of *Zeuzera pyrina* with chemical and microbiological products. *Inform. Agar.*, **52**, 67–69.

Patti, J.H. & Carner, G.R. (1974) *Bacillus thuringiensis* investigations for control of *Heliothis* spp. on cotton. *J. Econ. Entomol.*, **67**, 415–418.

Paulov, S. (1985) Interactions of *Bacillus thuringiensis* var. *israelensis* with developmental stages of amphibians (*Rana temporaria* L.). *Biol. Czech. B Zool.*, **40**, 133–138.

Pavan, M. (1979) Utilisation of ants of the group *Formica rufa* for the protection of forests by means of biological control. *Bull. SROP,* **2,** 135–159.

Pawar, V.M. & Thombre, U.T. (1990) Biological activity of *Bacillus thuringiensis* (B.T.) formulation against some crop pests. *Proc. Abst. Vth Intern. Coll. Invertebr. Pathol. Microb. Con.*, Adelaide, Australia, 496.

Peacock, J.W., Schweitzer, D.F., Carter, J.L. & Dubois, N.R. (1998) Laboratory assessment of the effects of *Bacillus thuringiensis* on native Lepidoptera. *Environ. Entomol.*, **27**, 450–457.

Pedersen, A.F., & van Frankenhuyzen, K. (1996) Debilitating effects on spruce budworm, *Choristoneura fumiferana* (Clemens), caused by treatments with sublethal doses of *Bacillus thuringiensis* var. *kurstaki*. *Bull. OILB-SROP.* **19,** 69–74.

Pedersen, A., Dedes, J., Gauthier, D. & Van Frankenhuyzen, K. (1997) Sublethal effects of *Bacillus thuringiensis* on the spruce budworm, *Choristoneura fumiferana*. *Entomol. Exp. Appl.*, **83**, 253–262.

Pedersen, J.C., Damgaard, P.H., Eilenberg, J. & Hansen, B.M. (1995) Dispersal of *Bacillus thuringiensis* var. *kurstaki* in an experimental cabbage field. *Canad. J. Microbiol.*, **41**, 118–125.

Peferoen, M. (1997) Progress and prospects for field use of *Bt* genes in crops. *Trends Biotech.*, **15**, 173–177.

Perani, M., Bishop, A.H. & Vaid, A. (1998) Prevalence of beta-exotoxin, diarrhoeal toxin and specific delta-endotoxin in natural isolates of *Bacillus thuringiensis*. *FEMS Microbiol. Lett.*, **160**, 55–60.

Perez, C.J. & Shelton, A.M. (1997) Resistance of *Plutella xylostella* (Lepidoptera: Plutellidae) to *Bacillus thuringiensis* Berliner in central America. *J. Econ. Entomol.*, **90**, 87–93.

Perez, O.Y., Rodriguez, A. & Cotes, A.M. (1997) A rapid and reliable method for the evaluation of *Bacillus thuringiensis* native strains against *Tecia solanivora* (Povolny) (Lepidoptera: Gelechiidae). *Rev. Colombiana Entomol.*, **23**, 113–118.

Perez, T., Jimenez, J. & Pazos, R. (1991) Effectiveness of *Bacillus thuringiensis* obtained in static liquid medium against *Mocis* spp. in grassland. *Proteccion de Plantas,* **1**, 15–20.

Perez Ibanez, T., Alberti Maurici, J., Calderon Forns, E., Martinez Canales Murcia, G. & Lozano Conejero, M.(1973) Study of the effectiveness of various insecticidal products against the pine processionary. *Boletin Informativo de Plagas,* **112**, 87–93.

Petcu, I.P. & Nastase, I.G. (1974) Trials of the bacterial agent Dipel to control the defoliator *Leucoma salicis* (Lep.). *Rev. Padurilor Ind. Lemnului Silvic. Exploatarea Padurilor,* **89**, 246–248.

Peteanu, S. (1980) Contributions to the study of the biological and integrated control of the hemp moth (*Grapholitha delineana* Walker). *Cereale si Plante Tehnice, Productia Vegetala,* **32**, 39–43.

Petras, S.F. & Casida, L.E., Jr. (1985) Survival of *Bacillus thuringiensis* spores in soil. *Appl. Environ. Microbiol.*, **50**, 1496–1501.

Petrova, V.I. (1987) Study of the mechanism of action of Bitoxibacillin on spider mites. *Tr. Latv. Sel'skokhozy. Akad.*, **237**, 20–31.

Petrova, V.I. & Khrameeva, A.V. (1989) Regulatory effect of Bitoxibacillin in limiting the numbers of two-spotted spider mites and its impact on *Phytoseiulus persimilis*. *Aspekty biologicheskoi regulyatsii chislennosti vreditelei rastenii* (ed. M.T. Shternbergs), Zinatne; Riga; USSR , 5–38.

Petrushov, A.Z., Barishvili, M.K. & Kasradze, D.T. (1983) Experience in mass rearing of *Metaseiulus*. *Zaschc. Rast.*, **9**, 23.

Pfrimmer, T.R. (1979) *Heliothis* spp.: control on cotton with pyrethroids, carbamates, organophosphates, and biological insecticides. *J. Econ. Entomol.*, **72**, 593–598.

Phillips, J.D. & Griffins, M.W. (1986) Factors contributing to the seasonal variation of *Bacillus* spp. in pasteurized dairy products. *J. Appl. Bacteriol.*, **61**, 275–285.
Pietrantonio, P.V. & Gill, S.S. (1992) The parasporal inclusion of *Bacillus thuringiensis* subsp. *shandongiensis*: characterization and screening for insecticidal activity. *J. Invertebr. Pathol.*, **59**, 295–302.
Pilcher, C.D., Obrycki, J.J., Rice, M.E. & Lewis, L.C. (1997) Preimaginal development, survival, and field abundance of insect predators on transgenic *Bacillus thuringiensis* corn. *Environ. Entomol.*, **26**, 446–454.
Pinkham, J.D., Frye, R.D. & Carlson, R.B. (1984) Toxicities of *Bacillus thuringiensis* isolates against the forest tent caterpillar (Lepidoptera: Lasiocampidae). *J. Kansas Entomol. Soc.*, **57**, 672–674.
Pinnock, D.E. & Milstead, J.E. (1972) Evaluation of *Bacillus thuringiensis* for suppression of navel orangeworm infestation of almonds. *J. Econ. Entomol.*, **65**, 1747–1749.
Pinnock, D.E. & Milstead, J.E. (1978) Microbial control of the fruit tree leafroller, *Archips argyrospila* (Lep.: Tortricidae) in California. *Entomophaga*, **23**, 203–206.
Pinnock, D.E. (1994) The use of *Bacillus thuringiensis* for control of pests of livestock. *Agric. Ecosys. Environ.*, **49**, 59–63.
Pinnock, D.E., Milstead, J.E., Coe, N.F. & Stegmiller, F. (1973) Evaluation of *Bacillus thuringiensis* formulations for control of larvae of the western grapeleaf skeletonizer. *J. Econ. Entomol.*, **66**, 194–197.
Pinnock, D.E., Milstead, J.E., Coe, N.F. & Brand, R.J. (1974) The effectiveness of *Bacillus thuringiensis* formulations for the control of larvae of *Schizura concinna* on *Cercis occidentalis* trees in California. *Entomophaga*, **19**, 221–227.
Pinnock, D.E., Brand, R.J., Milstead, J.E. & Jackson, K.L. (1975) Effect of tree species on the coverage and field persistence of *Bacillus thuringiensis* spores. *J. Invertebr. Pathol.*, **25**, 209–214.
Pinnock, D.E., Hagen, K.S., Cassidy, D.V., Brand, R.J., Milstead, J.E. & Tasson, R.L. (1978a) Integrated pest management in highway landscapes. *Calif. Agr.*, **32**, 33–34.
Pinnock, D.E., Brand, R.J., Milstead, J.E., Kirby, M.E. & Coe, N.F. (1978b) Development of a model for prediction of target insect mortality following field application of a *Bacillus thuringiensis* formulation. *J. Invertebr. Pathol.*, **31**, 31–36.
Pirvescu, D. (1973) The control of *Drymonia ruficornis* by the use of mixtures of *Bacillus thuringiensis* preparations with insecticides. *Rev. Padurilor*, **88**, 381–384.
Pishokha, N.P. (1985) Persistence of several bacterial preparations in the forest. *Lesovods. Agrolesomelior.*, **70**, 51–54.
Pishokha, N.P. & Timchenko, G.A. (1981) Spore retention time on trees, in litter and in the soil following stand infection with *Bacillus dendrolimus*. *Lesovods. Agrolesomelior.*, **61**, 21–25.
Pistrang, L.A. & Burger, J.F. (1984) Effect of *Bacillus thuringiensis* var. *israelensis* on a genetically-defined population of black flies (Diptera: Simuliidae) and associated insects in a montane New Hampshire stream. *Canad. Entomol.*, **116**, 975–981.
Plaut, H.N. (1976) Development and behavior of young larvae of *Zeuzera pyrina* L. on apple and pear seedlings, and pesticide tests in 1975. *Special Publication, Agricultural Research Organization, Bet Dagan*, **No. 61**, 15pp.
Poe, S.L. & Everett, P.H. (1974) Comparison of single and combined insecticides for control of tomato pinworm in Florida. *J. Econ. Entomol.*, **67**, 671–674.
Poinar, G.O. Jr., Thomas, G.M. & Lighthart, B. (1990) Bioassay to determine the effect of commercial preparation of *Bacillus thuringiensis* on entomogenous rhabditoid nematodes. *Agric. Ecosys. Environ.*, **30**, 195–202.
Pojananuwong, S., Catling, D. & Chareondham, P. (1988) Aerial application of insecticides against yellow stem borer in deepwater rice. *Thai J. Agr. Sci.*, **21**, 23–27.
Polles, S.G. (1974) Evaluation of foliar sprays for control of two webworms and the walnut caterpillar on pecan. *J. Georgia Entomol. Soc.*, **9**, 182–186.
Polles, S.G. & Payne, J.A. (1975) Webworm, walnut caterpillar can be controlled by spraying. *Pecan Quarterly*, **9**, 6–7.
Pollini, A. (1979) Diprionidae injurious to pines. *Inf. Fitopatol.*, **29**, 19–21.
Poncet, S., Delecluse, A., Klier, A. & Rapoport, G. (1995) Evaluation of synergistic interactions among the CryIVA, CryIVB, and CryIVD toxic components of *B. thuringiensis* subsp. *israelensis* crystals. *J. Invertebr. Pathol.*, **66**, 131–135.
Poplavskii, V.V. (1984) The serious danger presented by the meadow moth. *Zaschc. Rast.*, **5**, 43.
Popova, V. (1971) The dynamics of the development of the cotton bollworm in 1969-70. *Rast. Zashc.*, **19**, 23–25.
Porto Santos, G., Vinha Zanuncio, T., Neto, H.F. & Cola Zanuncio, J. (1995) Susceptibility of *Eustema sericea* (Lepidoptera; Notodontidae) to *Bacillus thuringiensis* var. *kurstaki*. *Rev. Ceres*, **42**, 423–430.
Powell, K.A. & Jutsum, A.R. (1993) Technical and commercial aspects of biocontrol products. *Pestic. Sci.*, **37**, 315–321.
Powell, K.A., Faull, J.L. & Renwick, A. (1990) The commercial and regulatory challenge. *Biological Control of Soil Borne Plant Pathogens* (ed. D. Hornby), pp. 445–463. CABI, Wallingford, Oxon; UK.
Pozsgay, M., Fast, P., Kaplan, H. & Carey, P.R. (1987) The effect of sunlight on the protein crystals from *Bacillus thuringiensis* var. *kurstaki* HD1 and NRD12: a Raman spectroscopic study. *J. Invertebr. Pathol.*, **50**, 246–253.
Prada, R.M. & Gutierrez, P.J. (1974) Preliminary contribution to the microbiological control of *Scrobipalpula absoluta* (Meyrick), with *Neoaplectana carpocapsae* Weiser and *Bacillus thuringiensis* Berl. on tomato *Lycopersicum esculentum* Mill. *Acta Agron.*, **24**, 116–137.

Pradhan, S., Jotwani, M.G. & Young, W.R. (eds.)(1972) Other approaches for the control of shoot fly. *Proc. Intern. Symp. Control of sorghum shoot fly,* Rockefeller Foundation, Oxford and IBH Publishing Co., New Delhi, India. 324 p.
Prando, H.F. & Silva, A.A., Jr. (1990) Efficacy of six insecticides for the control of *Neoleucinodes elegantalis* (Guenee, 1854) (Lepidoptera: Pyralidae) on tomatoes. *An. Soc. Entomol. Bras.,* **19**, 59–65.
Prasad, S., Tilak, K. & Gollakota, K.G. (1972) Role of *Bacillus thuringiensis* var. *thuringiensis* on the larval survivability and egg hatching of *Meloidogyne* spp., the causative agent of root knot disease. *J. Invertebr. Pathol.,* **20**, 377–378.
Prasertphon, S. (1975) Development of production and application of *Bacillus thuringiensis* Berliner in Thailand. *Plant Prot. Ser. Tech. Bull.,* **34**, 26 p.
Prasertphon, S., Areekul, P. & Tanada, Y. (1973) Sporulation of *Bacillus thuringiensis* in host cadavers. *J. Invertebr. Pathol.,* **21**, 205–207.
Pree, D.J. & Daly, J.C. (1996) Toxicity of mixtures of *Bacillus thuringiensis* with endosulfan and other insecticides to the cotton boll worm *Helicoverpa armigera. Pestic. Sci.,* **48**, 199–204.
Prieditis, A. & Rituma, I. (1974) The possibility of using microbiological preparations in the integrated control of apple tree pests. *Agronomija Lauksaimnieciskajai Razosanai,* **79**, 68–75.
Priest, F.G., Kaji, D.A., Rosato, Y.B. & Canhos, V.P. (1994) Characterization of *Bacillus thuringiensis* and related bacteria by ribosomal RNA gene restriction fragment length polymorphisms. *Microbiol. Reading,* **140**, 1015–1022.
Prishchepa, L.I. & Mikul' skaya, N.I. (1998) Biopreparations on rape and clover. *Zashc. Karantin Rast.,* **6**, 30.
Prishchepa, L.I. & Vanyushina, N.V. (1997) Biological protection of rape. *Zashc. Karantin Rast.,* **5**, 11.
Pruett, C.J.H., Burges, H.D. & Wyborn, C.H. (1980) Effect of exposure to soil on potency and spore viability of *Bacillus thuringiensis. J. Invertebr. Pathol.,* **35**, 168–174.
Pukhaev, R.V. (1981) How best to prepare a working liquid of Bitoxibacillin. *Zaschc. Rast.,* **6**, 62.
Puntambekar, U.S., Mukherjee, S.N. & Ranjekar, P.K. (1997) Laboratory screening of different *Bacillus thuringiensis* strains against certain lepidopteran pests and subsequent field evaluation on the pod boring pest complex of pigeonpea (*Cajanus cajan*). *Antonie van Leeuwenhoek,* **71**, 319–323.
Purcell, B.H. (1981) Effects of *Bacillus thuringiensis* var. *israelensis* on *Aedes taeniorhynchus* and some non-target organisms in the salt marsh. *Mosq. News,* **41**, 476–484.
Purcell, M. & Granett, J. (1985) Toxicity of benzoylphenyl ureas and thuringiensin to *Trioxys pallidus* (Hymenoptera: Braconidae) and the walnut aphid (Homoptera: Aphididae). *J. Econ. Entomol.,* **78**, 1133–1137.
Pusztai, M., Fast, P., Gringorten, L., Kaplan, H., Lessard, T. & Carey, P.R. (1991) The mechanism of sunlight-mediated inactivation of *Bacillus thuringiensis* crystals. *Biochem. J.,* **273**, 43–47.
Pye, A.E. & Burman, M. (1977) Pathogenicity of the nematode *Neoaplectana carpocapsae* (Rhabditida, Steinernematidae) and certain microorganisms towards the large pine weevil, *Hylobius abietis* (Coleoptera, Curculionidae). *Ann. Entomol. Fennici,* **43**, 115–119.
Quiroz Martinez, H., Herrera Delgadillo, M.A. & Badii, M.H. (1996) Effect of *Bacillus thuringiensis* on the predation of *Buenoa antigone* on larvae of *Aedes aegypti. Southwestern Entomol.,* **21**, 483–484.
Rabinovitch, L., de Jesus, F.F., Cavados, C.F.G., Zahner, V., Momen, H., de Silva, M.H.L., Dumanoir, V.C., Frachon, E. & Lecadet, M.M. (1995) *Bacillus thuringiensis* subsp. *oswaldocruzi* and *Bacillus thuringiensis* subsp. *brasiliensis*, two novel Brazilian strains which determine new serotype H38 and H39, respectively. *Mem. Inst. Oswaldo Cruz,* **90**, 41–42.
Radhika, P. (1998) Effect of *Bacillus thuringiensis* on *Trichogramma* spp. *Insect Environment,* **4**, 22.
Radwan, H.S.A., Ammar, I.M.A., Eisa, A.A., Omar, H.I.H. & Moftah, E.A.M. (1986) Latent effects of certain *Bacillus* preparations on the biology of the cotton whitefly, *Bemisia tabaci. Minufiya J. Agr. Res.,* **8**, 417–429.
Rady, M.H., Merdan, A.I. & Salem, A. (1991) Water soluble toxins of *Bacillus thuringiensis* serotype H-14 demonstrating lethal action on *Culex pipiens* larvae. *Egyptian J. Micro.,* **26**, 147-156.
Ragaei, M. (1998) Laboratory evaluation of shellac as ultraviolet screen for the *Bacillus thuringiensis* var. *entomocidus* against *Spodoptera littoralis* larvae. *Anz. Schaedlingskd. Pflanzenschutz Umweltschutz,* **71**, 132–134.
Ragni, A., Thiery, I. & Delecluse, A. (1996) Characterization of six highly mosquitocidal *Bacillus thuringiensis* strains that do not belong to H-14 serotype. *Curr. Microbiol.,* **32**, 48–54.
Rahman, W.U. & Chaudhury, M.I. (1987) Efficacy of Alsystin, Dimilin and Bactospeine against babul defoliator, *Euproctis lunata* Walk. *Pakistan J. Zool.,* **19**, 307–311.
Rahman, A. & Faruki, S.I. (1993) Effect of *Bacillus thuringiensis* var. *kurstaki* on the tropical warehouse moth, *Cadra cautella* (Walker) (Lepidoptera: Phycitidae). *Pakistan J. Zool.,* **25**, 45–48.
Rahman, S.M.M., Gupta, C.P., Sidik, M., Rejesus, B.M., Garcia, R.P., Champ, B.R., Bengston, M., Dharmaputa, O.S. & Halid, H. (1997) Application of neem oil and *Bacillus thuringiensis* preparations to control insects in stored paddy. *Proc. Symp. Pest Manag. Stored Food Feed,* Bogor, Indonesia, 5-7 September 1995, 173–197.
Rafes, P.M., Gninenko Yu, I. & Sokolov, V.K. (1976) Population dynamics of competing leaf-feeding pests of birch. *Byull. Mosk. Obshch. Ispyt. Prir. Otd. Biol.,* **81**, 48–55.
Rai, G.P. & Rana, R.S. (1979) Effectiveness of beta-exotoxin of *Bacillus thuringiensis* var. *thuringiensis* on the ability of *Meloidogyne* sp. from brinjal (*Solanum melongenea* L.) to survive. *J. Biosciences,* **1**, 271–278.

Rajagopal, D. & Trivedi, T.P. (1989) Status, bioecology and management of *Epilachna* beetle, *Epilachna vigintioctopunctata* (Fab.) (Coleoptera: Coccinellidae) on potato in India: a review. *Trop. Pest Manag.*, **35**, 410–413.

Rajagopal, D., Mallikarjunappa, S. & Gowda, J. (1988) Occurrence of natural enemies of the groundnut leaf miner, *Aproaerema modicella* Deventer (Lepidoptera: Gelechiidae). *J. Biol. Control*, **2**, 129–130.

Rajamohan, F., Lee, M.K. & Dean, D.H. (1998) *Bacillus thuringiensis* insecticidal proteins: molecular mode of action. *Prog. Nucleic Acid Res. Mol. Biol.*, **60**, 1–27.

Rajamohan, N. & Jayaraj, S. (1978) Field efficacy of *Bacillus thuringiensis* and some other insecticides against pests of cabbage. *Indian J. Agr. Sci.*, **48**, 672–675.

Ram, S. & Gupta, M.P. (1990) Integrated pest management in lucerne (*Medicago sativa* L.) and its economics in India. *Trop. Pest Manag.*, **36**, 258–262.

Ramachandran, R., Raffa, K.F., Bradley, D., Miller, M., Ellis, D.D. & McCown, B.H. (1993a) Activity of an insecticidal protein from *Bacillus thuringiensis* subsp. *thuringiensis* HD-290-1 strain to coleopteran and lepidopteran defoliators of poplars. *Environ. Entomol.*, **22**, 190–196.

Ramachandran, R., Raffa, K.F., Miller, M.J., Ellis, D.D. & McCown, B.H. (1993b) Behavioral responses and sublethal effects of spruce budworm (Lepidoptera: Tortricidae) and fall webworm (Lepidoptera: Arctiidae) larvae to *Bacillus thuringiensis* Cry1A(a) toxin in diet. *Environ. Entomol.*, **22**, 197–211.

Ramakrishnan, N. & Pant, N.C. (1971) Effect of *Bacillus thuringiensis* Berliner on the free amino acids and oxygen consumption in *Earias fabia* (Stoll.). *Indian J. Entomol.*, **33**, 312–316.

Ramaprasad, G., Joshi, B.G. & Satyanarayana, S.V.V. (1982) Efficacy of some insecticides and *Bacillus thuringiensis* Berliner in the control of *Spodoptera litura* Fabricius. *Tob. Res.*, **8**, 13–18.

Ramos, L.H.M., McGuire, M.R. & Wong, L.J.G. (1998) Utilization of several biopolymers for granular formulations of *Bacillus thuringiensis*. *J. Econ. Entomol.*, **91**, 1109–1113.

Ramoska, W.A., Watts, S. & Rodriguez, R.E. (1982) Influence of suspended particulates on the activity of *Bacillus thuringiensis* serotype H-14 against mosquito larvae. *J. Econ. Entomol.*, **75**, 1–4.

Ramzan, M., Mann, G.S. & Darshan, S. (1979) Preliminary studies on the life history and control of *Metanastria* sp. *Indian J. Plant Prot.*, **7**, 223–225.

Randall, A.P., Moore, A. & Carter, N. (1979) Cessna Ag-truck calibration trials for the dispersal of aqueous formulations of *Bacillus thuringiensis kurstaki* in Newfoundland, 1979. *Rep. Canad. For. Ser.*, FPM-X-38, 33 p.

Rani, S.S. & Balaraman, K. (1996) Effect of insecticidal crystal proteins of *Bacillus thuringiensis* on human erythrocytes *in vitro*. *Indian J. Exp. Biol.*, **34**, 1241–1244.

Rao, A.S. & Amonkar, S.V. (1980) Nucleopolyhedrosis of *Amsacta moorei* Buttler (Lepidoptera: Arctiidae). *Curr. Sci.*, **49**, 709–710.

Rao, B.M. & Krishnayya, P.V. (1996) Effect of diflubenzuron and *Bacillus thuringiensis* var. *kurstaki* baits on the growth and development of *Spodoptera litura* (Fab.) larvae. *Pesticide Res. J.*, **8**, 80–83.

Rao, P.R.M. & Rao, R.S.P. (1979) Note on the use of *Bacillus thuringiensis* and extracts of *Eclipta alba* (Linn.) Hassak. and *Azadirachta indica* A. Juss. for the control of rice brown planthopper. *Indian J. Agr. Sci.*, **49**, 905–906.

Rao, P.R.M. & Rao, R.S.P.(1979) Effects of biocides on brown planthopper adults on rice. *Intern. Rice Res. Newsl.*, **4**, 20.

Rao, V.P., Datta, B. & Ramaseshiah, G. (1970) Natural enemy complex of flushworm and phytophagous mites on tea in India. *Sci. Publ. Ser. Tea Board India*, **5**, 53 p.

Rashid, H. & Khan, A.M. (1995) Susceptibility of *Sitophilus granarius* L. to various commercial formulations of bacterial insecticides. *Bioved*, **6**, 111–114.

Raske, A.G., Clarke, L.J. & Sutton, W.J. (1986) The status of the spruce budworm in Newfoundland and Labrador from 1983 to 1985. *Inf. Rep. Newfoundland For. Centre Canad. For. Ser.*, No. N-X-235, 51 p.

Raske, A.G., Retnakaran, A., West, R.J., Hudak, J. & Lim, K.P. (1986) The effectiveness of *Bacillus thuringiensis*, Dimilin, Sumithion and Matacil against the hemlock looper, *Lambdina fiscellaria fiscellaria*, in Newfoundland in 1985. *Inf. Rep. Newfoundland For. Centre Canad. For. Ser.*, No. N-X-238, 56 p.

Rasnitsyn, S.P., Voitsik, A.A. & Skidan, K.B. (1993) Possible use of bacterial mosquito control preparations in salt water. *Med. Parazitol. Parazit. Bolezni*, **3**, 33–34.

Rassmann, W. (1986) Studies on the effectiveness of a preparation of *Bacillus thuringiensis* against stored-products moths in grain stores. *Nachrichtenbl. Dtsch. Pflanzenschutzdienst*, **38**, 61–63.

Ravensberg, W.J. & Berger, H.K. (1988) Biological control of the European corn borer (*Ostrinia nubilalis* Hbn, Pyralidae) with *Trichogramma maidis* Pintureau and Voegele in Austria in 1980-1985. *Colloq. INRA*, **43**, 557–564.

Ray, D.E. (1991) Pesticides derived from plants and other organisms. *Handbook of Pesticide Toxicology* (eds. W.J. Hayes & E.R. Laws), pp. 585–636. Academic Press, San Diego.

Reanney, D.C., Roberts, W.P. & Kelly, W.J. (1982) Genetic interactions among microbial communities. *Microbial Interactions and Communities* (eds. A.T. Bull & J.H. Slater), pp. 287–322. Academic Press, New York.

Reardon, R., Metterhouse, W. & Balaam, R. (1979) Impact of aerially applied *Bacillus thuringiensis* and carbaryl on gypsy moth (Lep.: Lymantriidae) and adult parasites. *Entomophaga*, **24**, 305–310.

Reardon, R.C. & Haissig, K. (1984) Efficacy and field persistence of *Bacillus thuringiensis* after ground application to balsam fir and white spruce in Wisconsin. *Canad. Entomol.*, **116**, 153–158.

Reardon, R.C. & Wagner, D.L. (1995) Impact of *Bacillus thuringiensis* on nontarget lepidopteran species in broad-leaved forests. *Biorational Pest Control Agents: Formulation and Delivery. 207th National Meeting of the American Chemical Society, San Diego, California, March 13-17, 1994.* (eds. F.R. Hall & J.W. Barry), pp. 284–292. American Chemical Society, Washington; USA.

Reddy, S.T., Kumar, N.S. & Venkateswerlu, G. (1998) Comparative analysis of intracellular proteases in sporulated *Bacillus thuringiensis* strains. *Biotechn. Lett.*, **20**, 279–281.

Reeves, E.L. & Garcia, C., Jr. (1970) Pathogenicity of bicrystalliferous *Bacillus* isolate for *Aedes aegypti* and other Aedine mosquito larvae. *Proc. Intern. Coll. Insect Pathol., Soc. Invertebr. Pathol.*, College Park, Maryland USA, 219–228.

Reeves, E.L., Garcia, C. & Peck, T.D. (1971) Susceptibility of *Aedes* mosquito larvae to certain crystalliferous *Bacillus* pathogens. *Proc. Papers 39th Ann. Conf. Calif. Mosq. Cont. Assoc.*, Oakland, California, 18–120.

Regev, A., Keller, M., Strizhov, N., Sneh, B., Prudovsky, E., Chet, I., Ginzberg, I., Koncz Kalman, Z., Koncz, C., Schell, J. & Zilberstein, A. (1996) Synergistic activity of a *Bacillus thuringiensis* delta-endotoxin and a bacterial endochitinase against *Spodoptera littoralis* larvae. *Appl. Environ. Microbiol.*, **62**, 3581–3586.

Reigart, J.R. & Roberts, J.R. (eds.) (1999) Recognition and Management of Pesticide Poisonings, 5th Edition. *EPA document* 735-R-98-003.

Reinert, J.A. (1973) Sod webworm control in Florida turfgrass. *Florida Entomol.*, **56**, 333–337.

Reinert, J.A. (1974) Tropical sod webworm and southern chinch bud control in Florida. *Florida Entomol.*, **57**, 275–279.

Reinert, J.A. (1981) Control of the oleander caterpillar on oleander. *Proc. Florida State Hort. Soc.*, **93**, 168–169.

Reis, P.R. & Souza, J.C. (1996) Control of *Neoleucinodes elegantalis* (Guenee) (Lepidoptera: Pyralidae) with physiological insecticides in staked tomato plants. *An. Soc. Entomol. Bras.*, **25**, 65–69.

Reis, P.R. & de Souza, J.C. (1998) Chemical control of *Tuta absoluta* (Meyrick) in staked tomato plants. *Cienc. Agrotecnol.*, **22**, 13–21.

Reisner, W.M., Feir, D.J., Lavrik, P.B. & Ryerse, J.S. (1989) Effect of *Bacillus thuringiensis kurstaki* alpha-endotoxin on insect Malpighian tubule structure and function. *J. Invertebr. Pathol.*, **54**, 175–190.

Remund, U. & Siegfried, W. (1982) *Eupoecilia ambiguella—Botrytis* relations. *Schweiz. Z. Obst Weinb.*, **118**, 277–285.

Ren, G., Li, K., Yang, M. & Yi, X. (1975) The classification of the strains of *Bacillus thuringiensis* group. *Acta Microbiol. Sinica*, **15**, 292–301.

Retnakaran, A., Lauzon, H. & Fast, P. (1983) *Bacillus thuringiensis* induced anorexia in the spruce budworm, *Choristoneura fumiferana*. *Entomol. Exp. Appl.*, **34**, 233–239.

Rettich, F. (1983) Effects of *Bacillus thuringiensis* serotype H-14 on mosquito larvae in the Elbe lowland. *Acta Entomol. Bohemoslov.*, **80**, 21–28.

Reynolds, R.B., Reddy, A. & Thorne, C.B. (1988) Five unique temperate phages from a polylysogenic strain of *Bacillus thuringiensis* subsp. *aizawai*. *J. Gen. Microbiol.*, **134**, 1577–1585.

Rezk, G.N., Azmy, N.M. & Hamed, A.A. (1981) A laboratory study on the use of *Bacillus thuringiensis* against lepidopterous cotton pests. *Res. Bull. Fac. Agric. Ain Shams Univ.*, **1672**, 8 p.

Ribeiro, B.M. & Crook, N.E. (1998) Construction of occluded recombinant baculoviruses containing the full- length *cry*1Ab and *cry*1Ac genes from *Bacillus thuringiensis*. *Braz. J. Med. Biol. Res.*, **31**, 763–769.

Richardson, J.S. & Perrin, C.J. (1994) Effects of the bacterial insecticide *Bacillus thuringiensis* var. *kurstaki* (*Btk*) on a stream benthic community. *Canad. J. Fish. Aquatic Sci.*, **51**, 1037–1045.

Richter, A.R. & Fuxa, J.R. (1984) Pathogen-pathogen and pathogen-insecticide interactions in velvetbean caterpillar (Lepidoptera: Noctuidae). *J. Econ. Entomol.*, **77**, 1559–1564.

Riddick, E.W. & Mills, N.J. (1995) Seasonal activity of carabids (Coleoptera: Carabidae) affected by microbial and oil insecticides in an apple orchard in California. *Environ. Entomol.*, **24**, 361–366.

Riddick, E.W., Dively, G. & Barbosa, P. (1998) Effect of a seed-mix deployment of Cry3A-transgenic and nontransgenic potato on the abundance of *Lebia grandis* (Coleoptera: Carabidae) and *Coleomegilla maculata* (Coleoptera: Coccinellidae). *Ann. Entomol. Soc. Amer.*, **91**, 647–653.

Ridet, J.M. (1973) Study of the sensitivity of *Lymantria dispar* L. to the crystal-spore complex of *Bacillus thuringiensis* Berliner. *Zast. Bilja*, **24,** 205–218.

Riethmuller, U. & Langenbruch, G.A. (1989) Two bioassay methods to test the efficacy of *Bacillus thuringiensis* subsp. *tenebrionis* against larvae of the Colorado potato beetle (*Leptinotarsus decemlineata*). *Entomophaga*, **34**, 237–245.

Riffiod, G. (1976) Control of the maize pyralid. *Phytoma*, **28**, 3–7.

Riggin Bucci, T.M. & Gould, F. (1997) Impact of intraplot mixtures of toxic and nontoxic plants on population dynamics of diamondback moth (Lepidoptera: Plutellidae) and its natural enemies. *J. Econ. Entomol.*, **90**, 241–251.

Riley, C.M. & Fusco, R. (1990) Field efficacy of Vectobac-12AS and Vectobac-24AS against black fly larvae in New Brunswick streams (Diptera: Simuliidae). *J. Amer. Mosq. Con. Assoc.*, **6**, 43–46.

Rimmington, A. (1989) The production and use of microbial pesticides in the USSR. *Intern. Indust. Biotech.*, **9**, 10–14.

Ripa, S.R. (1981) Advances in the control of the tomato fruit moth *Scrobipalpula absoluta* (Meyr.) II. Tests on chemical control. *Agric. Tecnica,* **41**, 113–119.

Rituma, I. (1985) Possibilities of using microbiological preparations for the control of the winter geometrid. *Tr. Latv. Sel'skokhozy. Akad.*, **222**, 21–27.

Rituma, I. (1987) Effectiveness of using microbiological methods of plant protection against the cabbage white butterfly and the winter moth. *Tr. Latv. Sel'skokhozy. Akad.,* **236**, 42–47.

Rizvi, N.H. & Talhouk, A.S. (1984) Further studies on the chemical control of the alfalfa weevil *Phytonomus variablis* (Herbst) in Lebanon. *Proc. Entomol. Soc. Karachi,* **14–15**, 81–92.

Robert, P., Chaufaux, J. & Marchal, M. (1994) Sensitivity of larval *Oxythyrea funesta* (Coleoptera: Scarabaeidae, Cetoniinae) to three strains of *Bacillus thuringiensis* (subsp. *tenebrionis*). *J. Invertebr. Pathol.*, **63**, 99–100.

Robert, P., Chaufaux, J. & Marchal, M. (1996) Sensitivity of larval *Oxythyrea funesta* Poda (Coleoptera: Scarabaeidae, Cetoniinae) to different strains of *Bacillus thuringiensis*. *J. Invertebr. Pathol.*, **67**, 187–189.

Roberts, G.M. (1995) Salt-marsh crustaceans, *Gammarus duebeni* and *Palaemonetes varians* as predators of mosquito larvae and their reaction to *Bacillus thuringiensis* subsp. *israelensis*. *Biocontr. Sci. Technol.*, **5**, 379–385.

Robertson, J.L., Preisler, H.K., Ng, S.S., Hickle, L.A. & Gelernter, W.D. (1995) Natural variation: a complicating factor in bioassays with chemical and microbial pesticides. *J. Econ. Entomol.*, **88**, 1–10.

Rock, G.C. & Monroe, R.J. (1983) Interaction of larval age and dietary formaldehyde on the susceptibility of tufted apple budmoth (Lepidoptera: Tortricidae) to *Bacillus thuringiensis*. *J. Invertebr. Pathol.*, **42**, 71–76.

Rodcharoen, J., Mulla, M.S. & Chaney, J.D. (1991) Microbial larvicides for the control of nuisance aquatic midges (Diptera: Chironomidae) inhabiting mesocosms and man-made lakes in California. *J. Amer. Mosq. Con. Assoc.*, **7**, 56–62.

Rodenhouse, N.L. & Holmes, R.T. (1992) Results of experimental and natural food reductions for breeding black-throated blue warblers. *Ecology* **73**, 357–372.

Rodriguez Padilla, C., Galan Wong, L., de Barjac, H., Roman Calderon, E., Tamez Guerra, R. & Dulmage, H. (1990) *Bacillus thuringiensis* subspecies *neoleonensis* serotype H-24, a new subspecies which produces a triangular crystal. *J. Invertebr. Pathol.*, **56**, 280–282.

Rogers, C.E., Archer, T.L. & Bynum, E.D., Jr. (1984) *Bacillus thuringiensis* for controlling larvae of *Homoeosoma electellum* on sunflower. *J. Agric. Entomol.,* **1**, 323–329.

Rogoff, M.H. & Yousten, A.A. (1969) *Bacillus thuringiensis*: microbiological considerations. *Ann. Rev. Microbiol.*, **23**, 357–386.

Rohl, L.J. & Woods, W. (1994) Biological and insecticidal control of *Arotrophora arcuatalis* (Walker) (Lepidoptera: Tortricidae): an important pest of banksias in Western Australia. *Plant Prot. Quarterly,* **9**, 20–23.

Roltsch, W.J., Zalom, F.G., Barry, J.W., Kirfman, G.W. & Edstrom, J.P. (1995) Ultra-low volume aerial applications of *Bacillus thuringiensis* variety *kurstaki* for control of peach twig borer in almonds trees. *Appl. Engineer. Agr.*, **11**, 25–30.

Romasheva, L.F. & Uzdenov, U.B. (1979) Distribution of bacteria of the group *Bacillus thuringiensis* in arthropods and wild birds. *Entomol. Issled. Kirgizii,* **13**, 86–90.

Romasheva, L.F., Dubitskii, A.M., Shcherbak, V.P., Lavrenyuk, N.M. & Protsenko, A.I. (1977) The effect of Bitoxibacillin and thermostable exotoxin on blood-sucking arthropods. *Entomol. Issled. Kirgizii,* **10**, 120–122.

Romasheva, L.F., Uzdenov, U.V., Krylova, V.N. & Verchagina, V.V. (1978) The relationship between bacteria of the *Bacillus thuringiensis* group and their hosts. *First All-Union Conference of Parasitocoenologists (Poltava, September 1978)* (ed. A. P. Markevich). *Tezisy Dokladov*, 64–65.

Rombach, M.C., Aguda, R.M., Picard, L. & Roberts, D.W. (1989) Arrested feeding of the Asiatic rice borer (Lepidoptera: Pyralidae) by *Bacillus thuringiensis*. *J. Econ. Entomol.*, **82**, 416–419.

Romi, R., Ravoniharimelina, B., Ramiakajato, M. & Majori, G. (1993) Field trials of *Bacillus thuringiensis* H-14 and *Bacillus sphaericus* (strain 2362) formulations against *Anopheles arabiensis* in the central highlands of Madagascar. *J. Amer. Mosq. Con. Assoc.*, **9**, 325–329.

Ros, M.A., Casadevall, M., Ferrer, X. & Sorribas, R. (1987) Trial of pesticide products for the control of the carnation tortricids (*Epichoristodes acerbella* Wlk) (*Cacoecimorpha pronubana* Mbn). *Fulls d'Informacio Tecnica,* **131**, 4 p.

Roshchin, N.T., Shestopal, A.V. & Dzhafarov Sh, M. (1986) Lepidocide against tortricids. *Zaschc. Rast.*, **5**, 33–34.

Rosley, M. & Jusoh, M.M. (1982) The efficacy and economics of several insecticides against major insect pests of Virginia tobacco. *Malay. Agr. J.*, **53**, 248–252.

Rosset, P.M. (1988) Control of insect pests in tomato: some considerations on the Central American experience. *Manejo Integrado de Plagas,* **7**, 1–12.

Rossiter, M., Yendol, W.G. & Dubois, N.R. (1990) Resistance to *Bacillus thuringiensis* in gypsy moth (Lepidoptera: Lymantriidae): genetic and environmental causes. *J. Econ. Entomol.*, **83**, 2211–2218.

Rotundo, G. & Giacometti, R. (1986) Reality and prospects of control of chestnut tortricids. *Inform. Agr.*, **42**, 69–72.

Roush, R.T. (1997) *Bt*-transgenic crops: just another pretty insecticide or a chance for a new start in resistance management? *Pestic. Sci.*, **51**, 328–334.

Roversi, P.F., Boretti, R., Merendi, G.A., Bartolozzi, L. & Toccafondi, P. (1997) Oak processionary caterpillar: damage and control criteria. *Sherwood Foreste ed Alberi Oggi,* **3**, 13–16.

Ruelle, P., Nef, L. & Lebrun, P. (1977) Laboratory study of the long-term efficacity of *Bacillus thuringiensis* on *Euproctis chrysorrhoea*. *Parasitica,* **33**, 127–137.

Rupar, M.J., Donovan, W.P., Groat, R.G., Slaney, A.C., Mattison, J.W., Johnson, T.B., Charles, J.F., Dumanoir, V.C. & de Barjac, H. (1991) Two novel strains of *Bacillus thuringiensis* toxic to coleopterans. *Appl. Environ. Microbiol.,* **57**, 3337–3344.

Rusul, G. & Yaacob, N.H. (1995) Prevalence of *Bacillus thuringiensis* in selected foods and detection of enterotoxin in selected foods using TECRA-VIA and BCET-RPLA. *Intern. J. Food Microbiol.,* **25**, 131–139.

Ryan, M. & Nicholas, W.L. (1972) The reaction of the cockroach *Periplaneta americana* to the injection of foreign particulate material. *J. Invertebr. Pathol.,* **19**, 299–307.

Saad, A., El Bahrawi, A. & Dabbas, A. (1977) Pesticidal efficiency of different insecticidal groups against cotton boll worm *Earias insulana* infesting cotton in Iraq. *Meded. Fac. Landbouwwet. Rijksuniv. Gent,* **42,** 937–942.

Saade, F.E., Dunphy, G.B. & Bernier, R.L. (1996) Response of the carrot weevil, *Listronotus oregonensis* (Coleoptera: Curculionidae), to strains of *Bacillus thuringiensis*. *Biological Control,* **7**, 293–298.

Sachan, G.C. & Rathore, Y.S. (1983) Studies on the complete protection of sorghum crop against insect pests by chemicals and Thuricide at various stages of crop growth. *Pesticides,* **17**, 15–16.

Safronov, A.N. (1996) Protection of Scots pine stands against defoliators. *Lesnoe Khoz.,* **5**, 49–50.

Saitoh, H., Higuchi, K., Mizuki, E. & Ohba, M. (1996) Larvicidal activity of *Bacillus thuringiensis* natural isolates, indigenous to Japan, against two nematoceran insect pests occurring in urban sewage environments. *Microbiol. Res.,* **151**, 263–271.

Saitoh, H., Higuchi, K., Mizuki, E. & Ohba, M. (1998) Larvicidal toxicity of Japanese *Bacillus thuringiensis* against the mosquito *Anopheles stephensi*. *Med. Vet. Entomol.,* **12**, 98–102.

Salama, H.S. (1991) Potency of *Bacillus thuringiensis* Berliner strains against major pests of oil-seed crops. *J. Appl. Entomol.,* **111**, 418–424.

Salama, H.S. & Abdel Razek, A. (1992) Effect of different kinds of food on susceptibility of some stored products insects to *Bacillus thuringiensis*. *J. Appl. Entomol.,* **113**, 107–110.

Salama, H.S. & Foda, M.S. (1984) Studies on the susceptibility of some cotton pests to various strains of *Bacillus thuringiensis*. *Z. Pflanzenkrankheiten Pflanzenschutz,* **91**, 65–70.

Salama, H.S. & Sharaby, A. (1985) Histopathological changes in *Heliothis armigera* infected with *Bacillus thuringiensis* as detected by electron microscopy. *Insect Sci. Its Applic.,* **6**, 503–511.

Salama, H.S. & Sharaby, A.F. (1988) Effects of exposure to sublethal doses of *Bacillus thuringiensis* (Berl.) on the development of the greasy cutworm *Agrotis ypsilon* (Hufn.). *J. Appl. Entomol.,* **106**, 396–401.

Salama, H.S. & Zaki, F.N. (1983) Interaction between *Bacillus thuringiensis* Berliner and the parasites and predators of *Spodoptera littoralis* in Egypt. *Z. Angew. Entomol.,* **95**, 425–429.

Salama, H.S. & Zaki, F.N. (1984) Impact of *Bacillus thuringiensis* Berl. on the predator complex of *Spodoptera littoralis* (Boisd.) in cotton fields. *Z. Angew. Entomol.,* **97**, 485–490.

Salama, H.S. & Zaki, F.N. (1985) Biological effects of *Bacillus thuringiensis* on the egg parasitoid, *Trichogramma evanescens*. *Insect Sci. Its Applic.,* **6**, 145–148.

Salama, H.S., Foda, M.S., El Sharaby, A.M. & Sharaby, A.M.E. (1981a) Potency of spore- delta -endotoxin complexes of *Bacillus thuringiensis* against some cotton pests. *Z. Angew. Entomol.,* **91**, 388–398.

Salama, H.S., Foda, M.S., El Sharaby, A., Matter, M. & Khalafallah, M. (1981b) Development of some lepidopterous cotton pests as affected by exposure to sublethal levels of endotoxins of *Bacillus thuringiensis* for different periods. *J. Invertebr. Pathol.,* **38**, 220–229.

Salama, H.S., Zaki, F.N. & Sharaby, A.F. (1982) Effect of *Bacillus thuringiensis* Berl. on parasites and predators of the cotton leafworm *Spodoptera littoralis* (Boisd.). *Z. Angew. Entomol.,* **94**, 498–504.

Salama, H.S., Foda, M.S. & Sharaby, A. (1983a) Biological activity of mixtures of *Bacillus thuringiensis* varieties against some cotton pests. *Z. Angew. Entomol.,* **95**, 69–74.

Salama, H.S., Foda, M. & Sharaby, A. (1983b) *Bacillus thuringiensis* as a biological control agent vs. cotton pests in Egypt. *Proceedings of the 10th International Congress of Plant Protection Volume 2*. Brighton, England, 20-25 November, 1983, 781.

Salama, H.S., Foda, M.S., Zaki, F.N. & Khalafallah, A. (1983c) Persistence of *Bacillus thuringiensis* Berliner spores in cotton cultivations. *Z. Angew. Entomol.,* **95**, 321–326.

Salama, H.S., Foda, M.S., Zaki, F.N. & Moawad, S. (1984) Potency of combinations of *Bacillus thuringiensis* and chemical insecticides on *Spodoptera littoralis* (Lepidoptera: Noctuidae). *J. Econ. Entomol.,* **77**, 885–890.

Salama, H.S., Foda, S. & Sharaby, A. (1985) Role of feeding stimulants in increasing the efficiency of *Bacillus thuringiensis* versus *Spodoptera littoralis* (Lepidoptera: Noctuidae). *Entomol. Gen.,* **10**, 111–119.

Salama, H.S., Moawed, S.M. & Zaki, F.N. (1987) Effects of nuclear polyhedrosis-*Bacillus thuringiensis* combinations on *Spodoptera littoralis* (Boisd.). *J. Appl. Entomol.,* **104**, 23–27.

Salama, H.S., Moawed, S., Saleh, R. & Ragaei, M. (1990a) Field tests on the efficacy of baits based on *Bacillus thuringiensis* and chemical insecticides against the greasy cutworm *Agrotis ypsilon* Rdf. in Egypt. *Anz. Schaedlingskd. Pflanzenschutz Umweltschutz,* **63**, 33–36.

Salama, H.S., Salem, S., Zaki, F.N. & Matter, M. (1990b) Control of *Agrotis ypsilon* (Hufn.) (Lep., Noctuidae) on some vegetable crops in Eygpt using the microbial agent *Bacillus thuringiensis. Anz. Schaedlingskd. Pflanzenschutz Umweltschutz,* **63**, 147–151.

Salama, H.S., Aboul Ela, R., El Moursy, A. & Abdel Razek, A. (1991a) Biology and development of some stored grain pests as affected by delta-endotoxin and beta-exotoxin of *Bacillus thuringiensis. Biocontr. Sci. Technol.,* **1**, 281–287.

Salama, H.S., El Moursy, A., Aboul Ela, R. & Abdel Razek, A. (1991b) Potency of different varieties of *Bacillus thuringiensis* (Berl.) against some lepidopterous stored product pests. *J. Appl. Entomol.,* **112**, 19–26.

Salama, H.S., El Moursy, A., Zaki, F.N., Aboul Ela, R. & Abdel Razek, A. (1991c) Parasites and predators of the meal moth *Plodia interpunctella* Hbn. as affected by *Bacillus thuringiensis* Berl. *J. Appl. Entomol.,* **112**, 244–253.

Salama, H.S., Salem, S. & Matter, M. (1991d) Field evaluation of the potency of *Bacillus thuringiensis* on lepidopterous insects infesting some field crops in Egypt. *Anz. Schaedlingskd. Pflanzenschutz Umweltschutz,* **64**, 150–154.

Salama, H.S., Ragaei, M. & Sabbour, M. (1995a) Larvae of *Phthorimaea operculella* (Zell.) as affected by various strains of *Bacillus thuringiensis. J. Appl. Entomol.,* **119**, 241–243.

Salama, H.S., Zaki, F.N., Ragaei, M. & Sabbour, M. (1995b) Persistence and potency of *Bacillus thuringiensis* against *Phthorimaea operculella* (Zell.) (Lep., Gelechiidae) in potato stores. *J. Appl. Entomol.,* **119**, 493–494.

Salama, H.S., Aboul Ela, R. & Abdel Razek, A. (1996a) Persistence of *Bacillus thuringiensis* delta-endotoxin and beta-exotoxin on stored grains and their potency against *Plodia interpunctella* and *Sitotroga cerealella. J. Appl. Entomol.,* **120**, 249–254.

Salama, H.S., Zaki, F.N. & Sabbour, M.M. (1996b) Effect of *Bacillus thuringiensis* endotoxin on *Apanteles litae* Nixon and *Bracon instabilis* Marsh. (Hym., Braconidae), two parasitoids of the potato tuber moth *Phthorimia operculella* Zeller (Lep., Gelishiidae). *J. Appl. Entomol.,* **120**, 565–568.

Sale, P.R., Sawden, D. & Steven, D. (1985) Trials with *Bacillus thuringiensis* on kiwifruit 1982-1984. *Proc. 38th NZ Weed Pest Cont. Conf.* (ed. A.J. Popay), 162–164.

Saleem, M.A. & Shakoori, A.R. (1996) Synergistic effects of permethrin and cypermethrin on the toxicity of *Bacillus thuringiensis* in the adult beetles of *Tribolium castaneum. Pakistan J. Zool.,* **28**, 191–198.

Saleem, M.A., Tufail, N. & Shakoori, A.R. (1995) Synergistic effect of synthetic pyrethroids on the toxicity of *Bacillus thuringiensis* as shown by the biochemical changes in the sixth instar larvae of *Tribolium castaneum. Pakistan J. Zool.,* **27**, 317–323.

Saleh, S.M., Harris, R.F. & Allen, O.N. (1969) Method for determining *Bacillus thuringiensis* var. *thuringiensis* Berliner in soil. *Canad. J. Microbiol.,* **15**, 1101–1104.

Saleh, S.M., Kelada, N.L. & Shaker, N. (1991) Control of European house dust mite *Dermatophagoides pteronyssinus* (Trouessart) with *Bacillus* spp. *Acarologia,* **32**, 257–260.

Sample, B.E., Butler, L., Zivkovich, C., Whitmore, R.C. & Reardon, R. (1996) Effects of *Bacillus thuringiensis* Berliner var. *kurstaki* and defoliation by the gypsy moth (*Lymantria dispar* (L.) (Lepidoptera: Lymantriidae)) on native arthropods in West Virginia. *Canad. Entomol.,* **128**, 573–592.

Samples, J.R. & Buettner, H. (1983) Corneal ulcer caused by a biologic insecticide (*Bacillus thuringiensis*). *Amer. J. Ophthalmol.,* **95**, 258–260.

Sampson, M.N. & Gooday, G.W. (1998) Involvement of chitinases of *Bacillus thuringiensis* during pathogenesis in insects. *Microbiol. Reading,* **144**, 2189–2194.

Sandhu, G.S. & Varma, G.C. (1980) Control of an infestation by the German cockroach, *Blattella germanica* (L.), in a parasite rearing laboratory. *Intern. Pest Cont.,* **22**, 58, 60, 65.

Sandler, H.A. & Mason, J. (1997) Evaluation of three bioinsecticides for control of lepidopteran pests in cranberries. *Acta Horticulturae,* **446**, 447–455.

Sandner, H. & Szczypiorska, M. (1972) An integrated use of pathogens of insects and insecticides in plant protection. *Biul. Inst. Ochr. Rosl.,* **52**, 45–54.

Santharam, G., Victoria, D.R., Rabindra, R.J. & Jayaraj, S. (1994) Studies on biological control of pigeonpea pod borers in India. *Anz. Schaedlingskd. Pflanzenschutz Umweltschutz,* **67**, 103–106.

Sanzhimitupova, R.D. (1984) For the protection of buckthorn. *Zaschc. Rast.,* **12**, 32.

Sarag, D.G. & Satpute, U.S. (1988) Efficacy of some synthetic insecticides and *Bacillus thuringiensis* var. *kenyae* against bollworms of cotton. *PKV Res. J.,* **12**, 119–122.

Saraswathi, A. & Ranganathan, L.S. (1996) Larvicidal effect of *Bacillus thuringiensis* var. *israelensis* on *Tabanus triceps* (Thunberg) (Diptera: Tabanidae). *Indian J. Exp. Biol.,* **34**, 1155–1157.

Sareen, V., Rathore, Y.S. & Bhattacharya, A.K. (1983) Response of *Spodoptera litura* (Fab.) to various concentrations of *Bacillus thuringiensis* var. *thuringiensis. Sci Cult.,* **49**, 186–187.

Sarmiento, M.J. & Razuri, R.V. (1978) *Bacillus thuringiensis* for the control of *Spodoptera frugiperda* and of *Diatraea saccharalis* on maize. *Rev. Peruana Entomol.*, **21**, 121–124.

Sarnthoy, O., Li, T., Keinmeesuke, P., Sinchaisri, N., Miyata, T. & Saito, T. (1997) Cross-resistance of *Bacillus thuringiensis* resistant population of diamondback moth *Plutella xylostella* (Lepidoptera: Yponomeutidae). *Resist. Pest Manag.*, **9**, 11–13.

Sastrosiswojo, S. (1996) Biological control of the diamondback moth in IPM systems: case study from Asia (Indonesia). *Biological Control Introductions Opportunities for Improved Crop Production,* UK 18 November 1996, 13–32.

Satapathy, C.R. & Panda, S.K. (1997) Effect of commercial *Bt* formulations against fruit borers of okra. *Insect Environ.*, **3**, 54.

Satpute, U.S., Mishra, M.B., Bhalerao, P.D. & Sarag, D.G. (1988) Utility of modern insecticides in relation to fibre properties of cotton. *Cotton Develop.*, **17**, 53–54.

Satpute, U.S., Peshkar, L.N., Bhalerao, P.D. & Sarag, D.G. (1989) Effect of modern insecticides on sucking pests of cotton when used for control of bollworms. *Pestology,* **13**, 5–6.

Saubenova, O.G. (1973) Influence of biotic and abiotic factors on the toxicity of Entobakterin to larvae of blood-sucking mosquitos and midges. *Izv. Akad. Nauk Kazakhskoi SSR Biol.*, **5**, 26–30.

Saubenova, O.G., Sadovnikova, T.P., Dubitskii, A.M. & Sinitsina, L.P. (1973) Tests of the effect of microbial preparations on larvae of mosquitos of the genus *Culex* in south-eastern Kazakhstan. *Parazitologiya,* **7**, 227–230.

Sauma, S.Y. & Strand, M. (1990) Identification and characterization of glycosylphosphatidylinositol-linked *Schistosoma mansoni* adult worm immunogens. *Mol. Biochem. Parasitol.*, **38**, 199–209.

Savel' ev, V.N. & Kozlov, M.P. (1974) Comparative studies of the effect of certain bacterial insecticides on flea larvae. *Especially dangerous diseases in the Caucasus: Abstracts of papers of the Third Scientific Practical Conference of Anti plague Institutions of the Caucasus on the Natural Foci, Epidemiology and Prophylaxis of Especially Dangerous Diseases* (eds. I.F. Taran & A.I. Dyatlov) 14-16 May 1974, 179–180.

Savel' ev, V.N., Kozlov, M.P., Nadeina, V.P. (1974) A study of the susceptibility of fleas to the combined effect of Entobakterin and DDT. *Especially dangerous diseases in the Caucasus: abstracts of papers of the Third Scientific Practical Conference of Anti plague Institutions of the Caucasus on the Natural Foci, Epidemiology and Prophylaxis of Especially Dangerous Diseases* (eds. I.F. Taran & A.I. Dyatlov), 14-16 May 1974, 180–181.

Savel' ev, V.N., Kozlov, M.P., Kandybin, N.V. & Markevich, A.P. (1977) The effect of Bitoxibacillin on adult fleas in experiments. *Probl. Parasitol.*, 146–147.

Saxena, R.C. (1978) Efficiency of *Bacillus thuringiensis* Berliner for the control of *Athalia proxima* Klug. *Indian J. Plant Prot.*, **6**, 33–35.

Saxena, H. & Ahmad, R. (1998) Evaluation of *Bacillus thuringiensis* formulations against gram semilooper *Autographa nigrisigna* Walker (Lepidoptera: Noctuidae). *J. Biol. Con.*, **11**, 89–91.

Sayaboc, A.S., Raros, R.S. & Raros, L.C. (1973) The abundance of predatory and saprophagous acarines associated with decomposing rice stubble with a consideration of the effects of insecticide residues. *Philipp. Entomol.*, **2**, 375–383.

Sayed, A.A. & Ali, A.G. (1995) Timing of application of certain organophosphates versus a biocide to control the cosmopterigid, *Batrachedra amydraula* Meyr infesting date palm fruits in New Valley. *Assiut J. Agric. Sci.*, **26**, 253–259.

Sayed, A.A. & Temerak, S.A. (1995) Mechanical, chemical and biological control of *Cadra* spp. in date palm trees at Kharga Oasis, New Valley Governorate. *Assiut J. Agric. Sci.*, **26**, 51–58.

Scalco, A., Charmillot, P.J., Pasquier, D. & Antonin, P. (1997) Comparison of *Bacillus thuringiensis* (BT) based products to control grape and wine moth: from laboratory to vineyard. *Rev. Suisse Vitic. Arboric. Hortic.*, **29**, 345–350.

Scarpellini, J.R. & dos Santos, J.C.C. (1997) Chemical control of *Ecdytolopha aurantiana* Lima, 1927 (Lepidoptera - Olethreutidae) in a citrus orchard. *Ecossistema,* **22**, 27–28.

Schaefer, C.H. & Kirnowardoyo, S. (1984) An operational evaluation of *Bacillus thuringiensis* serotype H-14 against *Anopheles sundaicus* in west Java, Indonesia. *1984,* (WHO-vbc-84.896), unpublished.

Schesser, J.H. & Bulla, L.A., Jr. (1978) Toxicity of *Bacillus thu*ringiensis spores to the tobacco hornworm, *Manduca sexta*. *Appl. Environ. Microbiol.*, **35**, 121–123.

Schmid, A. (1975) Interaction between the specific granulosis virus and two bacterial preparations in larvae of *Zeiraphera diniana*. *Mitt. Schweiz. Entomol. Ges.*, **48**, 173–179.

Schmid, A. & Antonin, P. (1977) *Bacillus thuringiensis* in the control of the grape moths *Lobesia botrana* and *Clysia ambiguella* in French Switzerland. *Rev. Suisse Vitic. Arboric. Hortic.*, **9**, 119–126.

Schmidt, H.U. (1979) Investigations on the effectiveness of *Bacillus thuringiensis* Berliner on the dried-fruit moth (*Plodia interpunctella* Hbn.). *Anz. Schadlingskd. Pflanzenschutz Umweltschutz,* **52**, 36–39.

Schmidt, N.C.J. & Kirfman, G.W. (1992) NovoBtt - a novel *Bacillus thuringiensis* ssp. *tenebrionis* for superior control of Colorado potato beetle, and other leaf-feeding Chrysomelidae. *Proceedings, Pests and Diseases*, 1992 Brighton, November 23-26, 1992, 381–386.

Schmidt, R.F. (1988) Helicopter calibration of *Bti* granules. *Proc. Ann. Meet. New Jersey Mosq. Cont. Assoc.*, **75**, 22–25.

Schnepf, H.E., Wong, H.C. & Whiteley, H.R. (1985) The amino acid sequence of a crystal protein from *Bacillus thuringiensis* deduced from the DNA base sequence. *J. Biol. Chem.*, **260**, 6264–6272.

Schnetter, W., Engler, S., Morawcsik, J. & Becker, N. (1981) The effectiveness of *Bacillus thuringiensis* var. *israelensis* against mosquito larvae and non-target organisms. *Mitt. Dtsch. Ges. Allg. Angew. Entomol.*, **2**, 195–202.

Schnetter, W., Engler Fritz, S., Aly, C. & Becker, N. (1983) Application of preparations of *Bacillus thuringiensis* var. *israelensis* against mosquitoes in the Upper Rhine Valley. *Mitt. Dtsch. Ges. Allg. Angew. Entomol.*, **4**, 18–25.

Schonherr, J. & Ketterer, R. (1979) On combined applications of polyhedrosis virus and *Bacillus thuringiensis* for control of the nun moth, *Lymantria monacha* L. (Lepidoptera). *Z. Pflanzenkrankheiten Pflanzenschutz*, **86**, 483–488.

Schreiner, I.H. (1991) Damage threshold for *Diaphania indica* Saunders (Lepidoptera: Pyralidae) on cucumbers in Guam. *Trop. Pest Manag.*, **37**, 17–20.

Schreiner, I.H. & Nafus, D.M. (1987) Detasselling and insecticides for control of *Ostrinia furnacalis* (Lepidoptera: Pyralidae) on sweet corn. *J. Econ. Entomol.*, **80**, 263–267.

Schreiner, I.H. & Nafus, D.M. (1988) No-tillage and detasselling: effect on the Asian corn borer *Ostrinia furnacalis* and ants. *Philipp. Entomol.*, **7**, 435–442.

Schroder, R.F.W. & Athanas, M.M. (1989) Use of the egg parasite *Edovum puttleri* (Hym.: Eulophidae) in an IPM system developed for Colorado potato beetle (Col.: Chrysomelidae) control on potatoes, Beltsville, Maryland. *Entomophaga*, **34**, 193–199.

Schubert, G. & Stengel, M. (1974) Comparison of the insecticidal activity of four formulations based on *Bacillus thuringiensis* and of a phosphoric ester on the European corn borer (*Ostrinia nubilalis* Hubn.). *Rev. Zool. Agri. Pathol. Veg.*, **73**, 47–52.

Schuster, D.J. (1994) Life-stage specific toxicity of insecticides to parasitoids of *Liriomyza trifolii* (Burgess) (Diptera: Agromyzidae). *Intern. J. Pest Manag.*, **40**, 191–194.

Schuster, D.J. & Engelhard, A.W. (1979) Insecticide-fungicide combinations for control of arthropods and *Ascochyta* blight on chrysanthemum. *Phytoprotection*, **60**, 125–135.

Schuster, D.J. & Harbaugh, B.K. (1977) Effect of insecticides on poinsettia and *Erinnyis ello*. *Proc. Florida State Hort. Soc.*, **90**, 326–327.

Schwartz, A. (1976) A new moth problem makes its appearance. *Citrus Subtrop. Fruit J.*, **510**, 12–14.

Scopes, N.E.A. & Biggerstaff, S.M. (1974) Progress towards integrated pest control on year round chrysanthemums. *Proceedings of the Seventh British Insecticide and Fungicide Conference*, 19-22 Nov. 1973, Brighton, England, 227–234.

Scriber, J.M. (1998) Non-target Lepidoptera impacted by *Btk* pesticide prays, induced phytochemical defenses, and generalised parasitoids and predators of *Lymantria dispar*. *Proc. 6th Aust. Appl. Entomol. Res. Conf.*, Brisbane, Australia, pp. 236–245.

Sears, M.K., Jaques, R.P. & Laing, J.E. (1983) Utilization of action thresholds for microbial and chemical control of lepidopterous pests (Lepidoptera: Noctuidae, Pieridae) on cabbage. *J. Econ. Entomol.*, **76**, 368–374.

Sebesta, K., Farkas, J., Horska, K. & Vankova, J. (1981) Thuringiensin, the beta-exotoxin of *Bacillus thuringiensis*. *Microbial Control of Pests and Plant Diseases 1970-1980* (ed. H.D. Burges), pp. 249–281. Academic Press, London.

Sekar, P. & Baskaran, P. (1976) Investigation on Dipel-organic insecticide combinations against insect pests of brinjal. *Madras Agric. J.*, **63,** 542–544.

Seleena, P., Lee, H.L. & Lecadet, M.M. (1995) A new serovar of *Bacillus thuringiensis* possessing 28a28c flagellar antigenic structure: *Bacillus thuringiensis* serovar *jegathesan*, selectively toxic against mosquito larvae. *J. Amer. Mosq. Con. Assoc.*, **11**, 471–473.

Selinger, L.B., Khachatourians, G.G., Byers, J.R. & Hynes, M.F. (1998) Expression of a *Bacillus thuringiensis* delta-endotoxin gene by *Bacillus pumilus*. *Canad. J. Microbiol.*, **44**, 259–269.

Selvanarayanan, V. & Baskaran, P. (1996) Efficacy of non-conventional insecticides against the sesame shoot webber and capsule borer *Antigastra catalaunalis*. *Indian J. Entomol.*, **58**, 326–336.

Sem' yanov, V.P. & Tsybul' ko, V.I. (1983) Some characteristics of the control of cabbage pests in the Kharkov region. *Noveishie dostizheniya sel'skokhozyaistvennoi entomologii po materialam USh s"ezda VEO, Vil'nyus,* 9-13 Oktyabrya 1979, 192–194.

Sem' yanov, V.P., Miselyunene, I.S. & Valyukas Yu, B. (1983) The effect of microbial preparations on the most important pests of orchards. *Noveishie dostizheniya sel'skokhozyaistvennoi entomologii po materialam USh s"ezda VEO, Vil'nyus,* 9-13 Oktyabrya 1979, 117–119.

Sen, A.K., Bhattacharya, A. & Naqvi, A.H. (1993) Effect of *Bacillus thuringiensis* Berliner infection on the larval haemolymph of the lac predator, *Eublemma amabilis* Moore. *J. Insect Sci.*, **6**, 290–291.

Seskevicius, A. (1977) The effectiveness of *Trichogramma* against orchard pests. *Liet. Zemdir. Moks. Tyrimo Inst. Darb.*, **21**, 77–82.

Sevryukova, M.V. (1979) The vapourer moth - a pest of orchards. *Zaschc. Rast.*, **10**, 45–46.

Seyoum, A. & Abate, D. (1997) Larvicidal efficacy of *Bacillus thuringiensis* var. *israelensis* and *Bacillus sphaericus* on *Anopheles arabiensis* in Ethiopia. *World J. Microbiol. Biotech.*, **13**, 21–24.

Sha, C.Y. & Xie, Q.J. (1992) Changes of haemocyte number, protein and esterase in the haemolymph of *Mythimna separata* (Walker) infected with *Bacillus thuringiensis* Berliner. *Entomol. Knowl.*, **29**, 215–217.

Shah, B.H., Gul, H. & Chaudhry, M.I. (1979) Efficacy of microbial insecticides against poplar defoliator, *Ichthyura anastomosis* Steph (Notodontidae: Lepidoptera). *Pakistan J. For.*, **29**, 129–133.

Shah, P.A. & Goettell, M.S. (eds.) (1999) *Directory of Microbial Control Products and Services*. Microbial Control Division, Society for Invertebrate Pathology, 31 p.

Shakoori, A.R., Iftikhar, A. & Majeed, S. (1991a) Effect of heptachlor, MNNG and thioacetamide on the growth of *Bacillus thuringiensis* Berliner, strain HD-1 (*kurstaki*) and its toxicity to dipterous fly *Zaprionus indiana*. *Proc. 11th Pakistan Congr. Zool.*, **11**, 57–73.

Shakoori, A.R., Majeed, S. & Iftikhar, A. (1991b) Effect of synthetic pyrethroids on the growth of *Bacillus thuringiensis* subsp. *kurstaki* (strain HD1) and its toxicity to dipterous fly, *Zaprionis indiana*. *Pakistan J. Zool.*, **23**, 239–249.

Shalaby, F.F., Moawad, G.M., El Lakwah, F.A. & El Gemeiy, H.M. (1986) Laboratory pathogenecity tests with a commercial product of *Bacillus thuringiensis* (Ber.) against active and resting larvae of the pink bollworm. *Agric. Res. Rev.*, **61**, 23–43.

Shamiyeh, N.B., Mullins, C.A., Southards, C.J., Straw, R.A. & Roberts, C.H. (1993) Control of major insect pests of cole crops in Tennessee: 1988-1991. *Tenn. Farm Home Sci.*, **165**, 37–42.

Shamseldean, M.M.M. & Ismail, A.A. (1997) Effect of the nematode *Heterorhabditis bacteriophora* and the bacterium *Bacillus thuringiensis* as integrated biocontrol agents of the black cutworm. *Anz. Schaedlingskd. Pflanzenschutz Umweltschutz*, **70**, 77–79.

Shapiro, J.P., Schroeder, W.J. & Stansly, P.A. (1998) Bioassay and efficacy of *Bacillus thuringiensis* and an organosilicone surfactant against the citrus leafminer (Lepidoptera: Phyllocnistidae). *Florida Entomol.*, **81**, 201–210.

Sharafutdinov Sh, A. & Salikhov, R.R. (1975) Control of the cotton bollworm. *Zaschc. Rast.*, 20.

Sharma, C.B.S., Panigrahi, S. & Sahu, R.K. (1977) Mitodepressive activity of two bacterial toxins in the root meristems of *Allium cepa*. *Z. Pflanzenphysiol.*, **84**, 163–166.

Sharma, M.L. & Odak, S.C. (1996) Efficacy of *Bacillus thuringiensis* separately and in combination with a chemical insecticide against sorghum stem borer. *Indian J. Entomol.*, **58**, 354–358.

Sharma, M.L., Odak, S.C. & Pathak, S.C. (1998) Effect of weather parameters on the efficiency of biotic factors in regulating larval population of *Chilo partellus* (Swinhoe). *Shashpa*, **5**, 159–164.

Sharma, R.D. (1994) *Bacillus thuringiensis*: a biocontrol agent of *Meloidogyne incognita* on barley. *Nematol. Bras.*, **18**, 79–84.

Sharma, R.D. & Gomes, A.C. (1996) Effect of seed treatment with delta-endotoxins of *Bacillus* spp. on the multiplication of *Heterodera glycines* on soybean and corn. *Nematol. Bras.*, **20**, 21–29.

Sharpe, E.S. & Detroy, R.W. (1979) Susceptibility of Japanese beetle larvae to *Bacillus thuringiensis*: associated effects of diapause, midgut pH, and milk disease. *J. Invertebr. Pathol.*, **34**, 90–91.

Sheeran, W. & Fisher, S.W. (1992) The effects of agitation, sediment, and competition on the persistence and efficacy of *Bacillus thuringiensis* var. *israelensis* (*Bti*). *Ecotoxicol. Environ. Safety*, **24**, 338–346.

Shelton, A.M., Robertson, J.L., Tang, J.D., Perez, C., Eigenbrode, S.D., Priesler, H.K., Wilsey, W.T. & Cooley, R.J. (1993) Resistance of diamondback moth (Lepidoptera: Plutellidae) to *Bacillus thuringiensis* subspecies in the field. *J. Econ. Entomol.*, **86**, 697–705.

Shek, G.K. & Telepa, N.G. (1981) The meadow moth in Kazakhstan. *Zaschc. Rast.*, **6**, 23–24.

Shen, J.L., Zhou, W.J., Wu, Y.D., Lin, X.W. & Zhu, X.F. (1998) Early resistance of *Helicoverpa armigera* (Hubner) to *Bacillus thuringiensis* and its relation to the effect of transgenic cotton lines expressing BT toxin on the insect. *Acta Entomol. Sin.*, **41**, 8–14.

Shetata, W.A. & Nasr, F.N. (1998) Laboratory evaluation and field application of bacterial and fungal insecticides on the citrus flower moth, *Prays citri* Miller (Lep., Hyponomeutidae) in lime orchards in Egypt. *Anz. Schaedlingskd. Pflanzenschutz Umweltschutz*, **71**, 57–60.

Shevchenko, S.F., Ivanova, V.S. & Bul'ba, N.P. (1983) The effect of bacterial preparations of the *thuringiensis* group and their components on adult fleas. *Med. Parazitol. Parazit. Bolezni*, **52**, 53–57.

Shevtsov, V., Schyolokova, E., Krainova, O., Jigletsova, S. & Ichtchenko, V. (1996) Application horizons of crystalliferous bacilli for control of pest insects, nematodes and mosquitoes. *Bull. OILB-SROP*, **19**, 289–292.

Shi, Z.H. & Liu, S.S. (1998) Toxicity of insecticides commonly used in vegetable fields to the diamondback moth, *Plutella xylostella* and its parasite, *Cotesia plutellae*. *Chin. J. Biol. Cont.*, **14**, 53–57.

Shibano, Y., Yamagata, A., Nakamura, N., Iizuka, T., Sugisaki, H. & Takanami, M. (1985) Nucleotide sequence coding for the insecticidal fragment of the *Bacillus thuringiensis* crystal protein. *Gene*, **34**, 243–251.

Shimizu, M., Oshie, K., Nakamura, K., Takada, Y., Oeda, K. & Ohkawa, H. (1988) Cloning and expression in *Escherichia coli* of the 135-kDa insecticidal protein gene from *Bacillus thuringiensis* subsp. *aizawai* IPL7. *Agric. Biol. Chem.*, **52**, 1565–1573.

Shin, B., Park, S., Choi, S., Koo, B., Lee, S. & Kim, J. (1995) Distribution of *cry*V-type insecticidal protein genes in *Bacillus thuringiensis* and cloning of *cry*V-type genes from *Bacillus thuringiensis* subsp. *kurstaki* and *Bacillus thuringiensis* subsp. *entomocidus*. *Appl. Environ. Microbiol.*, **61**, 2402–2407.

Shternshis, M.V. & Zurabova, E.R. (1987) A new form of Lepidocide. *Zaschc. Rast. Moskva*, **9**, 39.

Shurovenkov Yu, B., Alekhin, V.T. & Lysenko, N.N. (1988) A system of measures for controlling the meadow moth. *Zaschc. Rast. Moskva,* **5**, 42–43.

Siegel, J.P. & Shadduck, J.A. (1990) Clearance of *Bacillus sphaericus* and *Bacillus thuringiensis* ssp. *israelensis* from mammals. *J. Econ. Entomol.*, **83**, 347–55.

Siegel, J.P., Shadduck, J.A. & Szabo, J. (1987) Safety of the entomopathogen *Bacillus thuringiensis* var. *israelensis* for mammals. *J. Econ. Entomol.*, **80**, 717–23.

Siegfried, B.D., Marcon, P., Witkowski, J.F., Wright, R.J. & Warren, G.W. (1995) Susceptibility of field populations of the European corn borer, *Ostrinia nubilalis* (Hubner) (Lepidoptera: Pyralidae), to *Bacillus thuringiensis* (Berliner). *J. Agric. Entomol.*, **12**, 267–273.

Sierpinska, A. (1996) The insecticidal activity of some *Bacillus thuringiensis* strains against forest Lepidoptera larvae at different temperatures. *Bull. OILB-SROP*, **19**, 75–78.

Sierpinska, A. (1997) The use of *Bacillus thuringiensis* in the control of forest defoliating insects - the present state and perspectives. *Sylwan,* **141**, 63–70.

Sierpinska, A., Sierpinski, A. & Pruszynski, S. (1995) The susceptibility of larvae of the pine moth *Dendrolimus pini* L. to autumnal control treatments with Foray 02.2 UL. *Proc. 35th Sci. Meet. Inst. Plant Prot.*, **35**, 153–156.

Sikoura, A.J. & Sikura, A.I. (1975) Utilisation of microorganisms against Noctuids. *Rev. Zool. Agri. Pathol. Veg.*, **74**, 54–60.

Sikura, A.I. (1976) A test of microbiological control of the American white butterfly. *Tr. Vses. Nauchno issled. Inst. Zashch. Rast.*, **47**, 45–51.

Sikura, A.I. & Simchuk, P.A. (1970) The microbiological method for the control of the leopard moth. *Zaschc. Rast.*, **15**, 16.

Sikura, A.I. & Simchuk, P.A. (1981) Microbiological protection from the American white butterfly. *Zashch. Rast.*, **9**, 39.

Sims, S.R. (1995) *Bacillus thuringiensis* var. *kurstaki* (CryIA (C)) protein expressed in transgenic cotton: effects on beneficial and other non-target insects. *Southwestern Entomol.*, **20**, 493–500.

Sims, S.R. (1997) Host activity spectrum of the CryIIA *Bacillus thuringiensis* subsp. *kurstaki* protein: effects on Lepidoptera, Diptera, and non-target arthropods. *Southwestern Entomol.*, **22**, 395–404.

Sims, S.R. & Martin, J.W. (1997) Effect of the *Bacillus thuringiensis* insecticidal proteins CryIA(b), CryIA(c), CryIIA, and CryIIIA on *Folsomia candida* and *Xenylla grisea* (Insecta: Collembola). *Pedobiologia,* **41**, 412–416.

Sims, S.R. & Ream, J.E. (1997) Soil inactivation of the *Bacillus thuringiensis* subsp. *kurstaki* cryIIA insecticidal protein within transgenic cotton tissue: laboratory microcosm and field studies. *J. Agric. Food Chem.*, **45**, 1502–1505.

Sims, S.R. & Stone, T.B. (1991) Genetic basis of tobacco budworm resistance to an engineered *Pseudomonas fluorescens* expressing the delta-endotoxin of *Bacillus thuringiensis kurstaki*. *J. Invertebr. Pathol.*, **57**, 206–210.

Sinegre, G., Gaven, B. & Jullien, J.L. (1980) Safety of application of *Bacillus thuringiensis* serotype H-14 for the non-target fauna of the mosquito breeding-sites on the Mediterranean coast of France. *Parassitologia,* **22**, 205–211.

Sinegre, G., Gaven, B. & Vigo, G. (1981a) Contribution to the standardisation of laboratory tests on experimental and commercial formulations of the serotype H-14 of *Bacillus thuringiensis* II - Influence of temperature, free chlorine, pH and water depth on the biological activity or a primary powder. *Cah. ORSTROM Entomol. Med. Parasitol.*, **19**, 149–155.

Sinegre, G., Gaven, B. & Jullien, J.L. (1981b) Contribution to the standardisation of laboratory tests on experimental and commercial formulations of the serotype H-14 of *Bacillus thuringiensis*. III — Separate or combined influence of larval density, of the volume or depth of water, and of the presence of earth on the effectiveness and residual larvicidal action of a primary powder. *Cah. ORSTOM Entomol. Med. Parasitol.*, **19**, 157–163.

Singer, S. (1974) Entomogenous bacilli against mosquito larvae. *Develop. Indust. Microbiol.*, **15**, 187–194.

Singer, S. & Smirnoff, W.A. (1974) *Bacillus thuringiensis* and its possible use in a programme of mosquito control. *Proceedings of the International Seminar on Mosquito Contro,* University of Quebec at Trois-Rivieres, Quebec, Canada, May 8-10, 1973, 137–142.

Singh, S., Kumar, R. & Chhabra, M.B. (1992) Efficacy of Dipel-8L and Sumicidin 20-EC against mammalian lice. *J. Vet. Parasit.*, **6**, 1–5.

Sinitsyna, L.P., Ostrogskaya, N.A., Biryukova, A.P. & Novikova, L.K. (1980) Method for the biological evaluation of the preparation BIP. *Biol. Z. Armenii,* **33**, 379–384.

Sivamani, E., Rajendran, N., Senrayan, R., Ananthakrishnan, T.N. & Jayaraman, K. (1992) Influence of some plant phenolics on the activity of delta-endotoxin of *Bacillus thuringiensis* var. *galleriae* on *Heliothis armigera*. *Entomol. Exp. Appl.*, **63**, 243–248.

Skatulla, U. (1973) Differential effects of *Bacillus thuringiensis* on *Orgyia antiqua* in relation to the food plant. *Anz. Schadlingskund. Planzen Umweltschutz,* **46**, 46–47.

Skatulla, U. (1974) On the outbreak of the vapourer moth, *Orgyia antique* (L.) in 1971/72 in Bavaria. *Anz. Schadlingskd. Pflanzen Umweltschutz,* **47**, 89–93.

Sklodowski, J.J.W. (1996) Communities of epigeic insects (Col. Carabidae) one year after spraying the nun moth with the preparations Trebon, Decis, Foray and Dimilin. *Sylwan,* **140**, 83–97.

Sklyarov, N.A. (1983) From continuous treatments to integrated protection. *Zaschc. Rast.*, **9**, 17–18.

Skovmand, O. & Eriksen, A.G. (1993) Field trials of a fizzy tablet with *Bacillus thuringiensis* subsp. *israelensis* in forest spring ponds in Denmark. *Bull. Soc. Vector Ecol.*, **18**, 160–163.

Skovmand, O., Hoegh, D., Pedersen, H.S. & Rasmussen, T. (1997) Parameters influencing potency of *Bacillus thuringiensis* var. *israelensis* products. *J. Econ. Entomol.*, **90**, 361–369.

Skvortsova, M.M., Burtseva, L.I., Shashkina, N.I., Robertus, L.A., Filipova, E.A. & Zhuravetskaya, N.I. (1976) Comparative study of two virulent phages of *Bacillus thuringiensis* var. *galleriae*. *Izv. Sib. Otd. Akad. Nauk SSSR Biologicheskikh Nauk*, **5**, 49–55.

Slabospitskii, A.G. & Yakulov, F.Y. (1979) Effects of BTB on pests of cabbage. *Zaschc. Rast.*, **12**, 28.

Slama, K. & Lysenko, O. (1981) Monitoring the course of bacterial infections by hemolymph pressure pulses in insects. *J. Invertebr. Pathol.*, **37**, 11–21.

Sleesman, J.P. (1971) Biological insecticides control cabbage looper. *Ohio Rep. Res. Develop. Agric. Home Econ. Nat. Res.*, **56**, 41–43.

Smirnoff, W.A. (1972) The susceptibility of *Archips cerasivoranus* (Lepidoptera-Tortricidae) to infection by *Bacillus thuringiensis*. *Canad. Entomol.*, **104**, 1153–1159.

Smirnoff, W.A. (1973) The possible use of *Bacillus thuringiensis* plus chitinase formulation for the control of spruce budworm outbreaks. *J. New York Entomol. Soc.*, **81**, 196–200.

Smirnoff, W.A. (1974) Susceptibility of *Lambdina fiscellaria fiscellaria* (Lepidoptera: Geometridae) to infection by *Bacillus thuringiensis* Berliner alone or in the presence of chitinase. *Canad. Entomol.*, **106**, 429–432.

Smirnoff, W.A. (1983) Microbial control of the European skipper, *Thymelicus lineola* Ochs. *Crop Prot.*, **2**, 353–360.

Smirnoff, W.A. & Fast, P.G. (1978) Laboratory bioassays of mixtures of *Bacillus thuringiensis* and chitinase. *Canad. Entomol.*, **110**, 201–203.

Smirnoff, W.A. & Heimpel, A.M. (1961) Notes on the pathogenicity of *Bacillus thuringiensis* Berliner for the earthworm *Lumbricus terrestris* Linnaeus. *J. Insect Pathol.*, **3**, 403–408.

Smirnoff, W.A. & Hutchison, P.M. (1965) Bacteriostatic and bacteriocidal effects of extracts of foliage from various plant species on *Bacillus thuringiensis* var. *thuringiensis* Berliner. *J. Invertebr. Pathol.*, **7**, 273–80.

Smirnoff, W.A. & MacLeod, C.F. (1961) Study of the survival of *Bacillus thuringiensis* var. *thuringiensis* Berliner in the digestive tracts and in feces of a small mammal and birds. *J. Insect Pathol.*, **3**, 266–270.

Smirnoff, W.A. & Valero, J. (1977) Determination of the chitinolytic activity of nine subspecies of *Bacillus thuringiensis*. *J. Invertebr. Pathol.*, **30**, 265–266.

Smith, D.B. & Hostetter, D.L. (1982) Laboratory and field evaluations of pathogen-adjuvant treatments. *J. Econ. Entomol.*, **75**, 472–476.

Smith, R.A. & Barry, J.W. (1998) Environmental persistence of *Bacillus thuringiensis* spores following aerial application. *J. Invertebr. Pathol.*, **71**, 263–267.

Smith, R.A. & Couche, G.A. (1991) The phylloplane as a source of *Bacillus thuringiensis* variants. *Appl. Environ. Microbiol.*, **57**, 311–315.

Smith, S.M., Wallace, D.R., Howse, G. & Meating, J. (1990) Suppression of spruce budworm populations by *Trichogramma minutum* Riley, 1982-1986. *Mem. Entomol. Soc. Canad.*, **153**, 56–81.

Smits, P.H. & Vlug, H.J. (1990) Control of tipulid larvae with *Bacillus thuringiensis* var. *israelensis*. *Proc. Abst. Vth Intern. Coll. Invertebr. Pathol. Microb. Con.*, Adelaide, Australia, 343.

Smits, P.H., Vlug, H.J. & Wiegers, G.L. (1993) Biological control of leatherjackets with *Bacillus thuringiensis*. *Proc. Section Exp. Appl. Entomol. Netherlands Entomol. Soc.*, 187–192.

Snarski, V.M. (1990) Interactions between *Bacillus thuringiensis* subsp. *israelensis* and fathead minnows, *Pimephales promelas rafinesque*, under laboratory conditions. *Appl. Environ. Micro.*, **56**, 2618–2622.

Sneath, P.H.A., Mair, N.S., Sharpe, M.E. & Holt, J.G. (eds.)(1986) *Bergey's Manual of Systematic Bacteriology*, Volume 2. Willams & Wilkins. 1599 pp.

Sneh, B. & Gross, S. (1983) Biological control of the Egyptian cotton leafworm *Spodoptera littoralis* (Boisd.) (Lep., Noctuidae) in cotton and alfalfa fields using a preparation of *Bacillus thuringiensis* ssp. *entomocidus*, supplemented with adjuvants. *Z. Angew. Entomol.*, **95**, 418–424.

Sneh, B., Gross, S. & Gasith, A. (1983) Biological control of *Spodoptera littoralis* (Boisd.) (Lep., Noctuidae) by *Bacillus thuringiensis* subsp. *entomocidus* and *Bracon hebetor* Say (Hym., Braconidae). *Z. Angew. Entomol.*, **96**, 408–412.

Sneh, B., Schuster, S. & Broza, M. (1981) Insecticidal activity of *Bacillus thuringiensis* strains against the Egyptian cotton leaf worm *Spodoptera littoralis* (Lep.: Noctuidae). *Entomophaga*, **26**, 179–190.

Snodgrass, G.L. & Stadelbacher, E.A. (1994) Population levels of tarnished plant bugs (Heteroptera: Miridae) and beneficial arthropods following early-season treatments of *Geranium dissectum* for control of bollworms and tobacco budworms (Lepidoptera: Noctuidae). *Environ. Entomol.*, **23**, 1091–1096.

Snow, A. (1997) Potential for gene flow between transgenic crops and wild relatives. *Transgenic plants: Bacillus thuringiensis in Mesoamerican Agriculture*, (eds. A.J. Hruska & M.L. Pavon) pp. 53–57. Escuela Agricola Panamericana Zamorano; Tegucigalpa, Central America.

Snow, K.R. (1984) Evaluation of *Bacillus thuringiensis*, serotype H-14 formulations as larvicides for *Aedes punctor* (Kirby). *Intern. Pest Cont.*, **26**, 12–14.
Sokolov, G.I. (1985) The harmfulness of birch sawflies. *Zaschc. Rast.*, **3**, 27–28.
Sokolova, E.I. & Ganushkina, L.A. (1982) The combined effect of bacterial preparations and insect development regulators on larvae of blood-sucking mosquitoes. *Med. Parazitol. Parazit. Bolezni*, **51**, 42–45.
Sokolova, E.I., Ganushkina, L.A., Kosovskikh, V.L. & Gavrilyuk, L.A. (1982) The results of laboratory tests of preparations of *Bacillus thuringiensis* Berliner on larvae of blood-sucking mosquitoes of the family Culicidae. *Med. Parazitol. Parazit. Bolezni*, **51**, 44–48.
Sokolova, E.I., Kulieva, N.M., Erlikh, V.D. & Shatalova, I.A. (1986) Results of tests of the preparation *Bac. thuringiensis* H-14 strain BTS-393. *Med. Parazitol. Parazit. Bolezni*, **56**, 25–28.
Sokolova, E.I., Makarova, G.Y., Kulieva, N.M., Pavlova Ivanova, L.K. & Ulyanova, E.A. (1985) Natural strains of *Bacillus thuringiensis* Berliner pathogenic for blood-sucking mosquitoes. *Med. Parazitol. Parazit. Bolezni*, **3**, 35–41.
Solanes, R., Mallea, A.R., Macola, G.S., Garcia, S.J., Bahamondes, L.A., Suarez, J.H. & Aranda, D. (1974) Biological insecticide against the common bagworm. *Rev. Fac. Cienc. Agrar.*, 3–10.
Soliman, A.A., Afify, A.M., Abdel Rahman, H.A. & Atwa, W.A. (1970) Effect of *Bacillus thuringiensis* on the biological potency of *Pieris rapae* (Lep., Pieridae). *Z. Angew. Entomol.*, **66**, 399–403.
Solomon, J.D. (1985) Comparative effectiveness of gallery-injected insecticides and fumigants to control carpenterworms (Lepidoptera: Cossidae) and oak clearwing borers (Lepidoptera: Sesiidae). *J. Econ. Entomol.*, **78**, 485–488.
Somermaa, K. (1970) Control experiments in the laboratory against the Delphacid *Javesella* (= *Calligypona*) *pellucida* (F.). *Vaxtskyddsnotiser*, **34**, 59–64.
Somerville, H.J., Tanada, Y. & Omi, E.M. (1970) Lethal effect of purified spore and crystalline endotoxin preparations of *Bacillus thuringiensis* on several Lepidopterous insects. *J. Invertebr. Pathol.*, **16**, 241–248.
Song, S.S. (1991) Resistance of diamondback moth (*Plutella xylostella* L.: Yponomeutidae: Lepidoptera) against *Bacillus thuringiensis* Berliner. *Korean J. Appl. Entomol.*, **30**, 291–293.
Sorensen, A.A. & Falcon, L.A. (1980) Microdroplet application of *Bacillus thuringiensis*: methods to increase coverage on field crops. *J. Econ. Entomol.*, **73**, 252–257.
Spackman, E.W. (1985) Palisades reservoir - problem - control. *Proc. Ann. Meet. Utah Mosq. Abatement Assoc.*, **39**, 5–6.
Spektor, M.R. & Vykhovets, A.N. (1975) A promising method of control of insect pests. *Lesnoe Khoz.*, **1**, 74–75.
Srivastava, K.L. & Ramakrishnan, N. (1980) Potency of Bactospeine and Dipel, two commercial formulations of *Bacillus thuringiensis* Berliner against castor semilooper, *Achaea janata* Linn. *Indian J. Entomol.*, **42**, 769–772.
Srivastava, R.P. & Nayak, P. (1978) Laboratory evaluation of four formulations of *Bacillus thuringiensis* against *Cnaphalocrocis medinalis* Gn. (Pyralidae: Lepidoptera), the rice leafroller. *Z. Pflanzenkrankheiten Pflanzenschutz*, **85**, 641–644.
Standaert, J.Y. (1981) Persistence and effectiveness of *Bacillus thuringiensis* H 14 against larvae of *Anopheles stephensi*. *Z. Angew. Entomol.*, **91**, 292–300.
Stansly, P.A., Schuster, D.J., McAuslane, H.J. & Bottcher, A.B. (1994) Biological control of silverleaf whitefly: an evolving sustainable technology. *Proc. 2nd Conf. Environ. Sound Agric.*, (eds. K.L. Campbell & W.D. Graham), 20-22 April 1994, Orlando, Florida, 484–491.
Stapel, J.O., Waters, D.J., Ruberson, J.R. & Lewis, W.J. (1998) Development and behavior of *Spodoptera exigua* (Lepidoptera: Noctuidae) larvae in choice tests with food substrates containing toxins of *Bacillus thuringiensis*. *Biological Control*, **11**, 29–37.
Stark, P.M. & Meisch, M.V. (1983) Efficacy of *Bacillus thuringiensis* serotype H-14 against *Psorophora columbiae* and *Anopheles quadrimaculatus* in Arkansas ricelands. *Mosq. News*, **43**, 59–62.
Starnes, R.L., Liu, C.L. & Marrone, P.G. (1993) History, use, and future of microbial insecticides. *Amer. Entomol.*, **39**, 83–91.
Stelzer, M.J. & Beckwith, R.C. (1988) Comparison of two isolates of *Bacillus thuringiensis* in a field test on western spruce budworm (Lepidoptera: Tortricidae). *J. Econ. Entomol.*, **81**, 880–886.
Stelzer, M.J., Neisess, J. & Thompson, C.G. (1975) Aerial applications of a nucleopolyhedrosis virus and *Bacillus thuringiensis* against the Douglas fir tussock moth. *J. Econ. Entomol.*, **68**, 269–272.
Stengel, M. (1972) The maize borer (*Ostrinia nubilalis* Hubn.) in Alsace. Biology, influence of attack, forecasting of damage and control methods. *Rev. Suisse Agric.*, **4**, 100–106.
Stengel, M. & Atak, E.D. (1971) Tests of treatments against the maize borer (*Ostrinia nubilalis* Hubn. Lep. Pyralidae) in maize crops of the Colmar region (France) in 1971. *Rev. Zool. Agri. Pathol. Veg.*, **70**, 67–74.
Stephenson, J.C., Williams, M.L. & Miller, G.L. (1987) A false spider mite on liriope. *Res. Rep. Ser. Alabama Agric. Exp. St. Auburn Univ.*, **5**, 23–24.
Stern, V.M., Flaherty, D.L. & Peacock, W.L. (1980) Control of the grapeleaf skeletonizer. *Calif. Agr.*, **34**, 17–19.
Stern, V.M., Flaherty, D.L. & Peacock, W.L. (1983) Control of the western grapeleaf skeletonizer (Lepidoptera: Zygaenidae), a new grape pest in the San Joaquin Valley, California. *J. Econ. Entomol.*, **76**, 192–195.

Sternlicht, M. (1979) Improving control of the citrus flower moth, *Prays citri*, by mass trapping of males. *Alon Hanotea,* **34**, 189–192.

Steven, D. & Sale, P.R. (1985) Insect control trials on persimmons. *Proc. 38th NZ Weed Pest Cont. Conf.* (ed. A.J. Popay), 203–206.

Steven, D., Barnett, S.W., Stevens, P.S. & McKenna, C.E. (1997) Changing pest control on New Zealand kiwifruit. *Proc. 3rd Intern. Symp. Kiwifruit,* (eds. E. Sfakiotakis & J. Porlingis) Thessaloniki, Greece, 19-22 September 1995, **444**, 765–771.

Stevens, P.J.G., Walker, J.T.S., Shaw, P.W. & Suckling, D.M. (1994) Organosilicone surfactants: tools for horticultural crop protection. *Proc. Brighton Crop Prot. Conference,* Brighton, UK, 21-24 Nov. 1994, Volume 2, 755–760.

Stewart, J.G., Lund, J.E. & Thompson, L.S. (1991) Factors affecting the efficacy of *Bacillus thuringiensis* var. *san diego* against larvae of the Colorado potato beetle. *Proc. Entomol. Soc. Ontario,* **122**, 21–25.

Stoakley, J.T. (1978) The pine beauty moth - its distribution, life cycle and importance as a pest in Scottish forests. *Control of Pine Beauty Moth by Fenitrothion in Scotland 1978,* (eds. A.V. Holden & D. Bevan), Forestry Commission, Edinburgh; UK, 7–12.

Stone, T.B., Sims, S.R. & Marrone, P.G. (1989) Selection of tobacco budworm for resistance to a genetically engineered *Pseudomonas fluorescens* containing the alpha-endotoxin of *Bacillus thuringiensis* subsp. *kurstaki*. *J. Invertebr. Pathol.,* **53**, 228–234.

Strapazzon, A., & Dalla Monta, L. (1986) *Bacillus thuringiensis* Berliner and traditional products in an apple orchard. Summer control of the fruit feeders and side-effects on mites. *Atti Giornate Fitopatol.,* **1**, 41–51.

Stroeva, I.A., Vainovska, E.Y., Sniedze, Z.Y., Yarkulov, F.M., Chilingaryan, V.A., Galetenko, S.M., Kholchenkov, V.A. & Pastukh, G.I. (1980) The effectiveness of the bacterial preparation BIP in the control of leaf-gnawing Lepidoptera. *Biol. Z. Armenii,* **33**, 416–419.

Stus, A.A. (1977) The Colorado beetle in the Crimea. *Zaschc. Rast.,* **No. 10**, 39.

Stus, A.A. (1979) Protection of older orchards. *Zaschc Rast.,* **No. 12**, 28.

Stus, A.A. (1980) The gold-tail. *Zaschc. Rast.,* **No. 7**, 64.

Stus, A.A. & Mikherskaya, N.I. (1986) The effectiveness of the application of Bactoculicide for the control of larvae of mosquitoes in the Crimea. *Med. Parazitol. Parazit. Bolezni,* **1**, 28–32.

Sudarsan, N., Suma, N.R., Vennison, S.J., Vaithilingam, S. & Sekar, V. (1994) Survival of a strain of *Bacillus megaterium* carrying a lepidopteran-specific gene of *Bacillus thuringiensis* in the phyllospheres of various economically important plants. *Plant and Soil,* **167**, 321–324.

Sukhoruchenko, G.I., Niyazov, O.D., Daricheva, M.A., Alekseev Yu, I., Kamalov, K., Zavodchikova, V.V. & Tashlieva, A.O. (1976) The effect of chemical treatments on the harmful and useful fauna of cotton. *Ecology and economic importance of the insects of Turkmenia (Ekologiya i khozyaistvennoe znachenie nasekomykh Turkmenii),* 46–61.

Sulaiman, S., Jeffery, J., Sohadi, A.R. & Abdul Rahim, S. (1991) Residual efficacy of Altosid and Bactimos briquets for control of dengue/dengue haemorrhagic fever vector *Aedes aegypti* (L.). *Mosq. Borne Dis. Bull.,* **8**, 123–126.

Sun, F., Wang, H., Ji, M., Sun, F.R., Wang, H.K. & Ji, M. (1993) Studies on the nuclear polyhedrosis virus disease of *Clanis bilineata tsingtauica* Mell. *Acta Agric. Boreali Sin.,* **8**, 59–63.

Sun, Q.N. (1985) Epizootiology of a baculovirus disease of the bagworm, *Cryptothelea variegata*, and its application for the control of its insect host. *Chin. J. Biol. Cont.,* **3**, 13–16.

Sundaram, A. & Sundaram, K.M.S. (1996) Effect of sunlight radiation, rainfall and droplet spectra of sprays on persistence of *Bacillus thuringiensis* deposits after application of Dipel R 76AF formulation onto conifers. *J. Environ. Sci. Health B,* **31**, 1119–1154.

Sundaram, A., Leung, J.W. & Devisetty, B.N. (1993) *Pesticide Formulations and Application Systems.* (eds. P.D. Berger, B.N. Devisetty & F.R. Hall), pp. 227–241. American Society for Testing and Materials, Philadelphia.

Sundaram, A., Sundaram, K.M.S. & Sloane, L. (1996) Spray deposition and persistence of a *Bacillus thuringiensis* formulation (Foray R76B) on spruce foliage, following aerial application over a northern Ontario forest. *J. Environ. Sci. Health B,* **31**, 763–813.

Sundaram, K.M.S. (1996) Sorptive interactions and binding of delta-endotoxin protein from *Bacillus thuringiensis* subsp. *kurstaki* in forest soils. *J. Environ. Sci. Health B,* **31**, 1321–1340.

Sundaram, K.M.S. & Sundaram, A. (1996) Influence of droplet sizes of deposits on persistence of *Bacillus thuringiensis* applied as an aqueous flowable, and azadirachtin applied as a nonaqueous solution, onto oak foliage. *Pesticide Formulations and Application systems* (eds. H.M. Collins, F.R. Hall & M. Hopkinson) Volume 15, ASTM; West Conshohocken, USA, 29–41.

Sundaram, K.M.S., Sundaram, A., Huddleston, E., Nott, R., Sloane, L., Ross, J. & Ledson, M. (1997) Deposition, distribution, persistence and fate of *Bacillus thuringiensis* variety *kurstaki* (*Btk*) in pecan orchards following aerial and ground applications to control pecan nut casebearer larvae. *J. Environ. Sci. Health B,* **32**, 741–788.

Supatashvili, A.S. (1990) Biological agents controlling *Diprion pini*. *Zaschc. Rast. Moskva,* **6**, 23–24.

Surany, P. (1970) Hyperactive toxigenesis by sporeless variants of *Bacillus thuringiensis*. *Intern. Pest Cont.,* **12**, 25–27.

Suzuki, N., Hori, H., Ogiwara, K., Asano, S., Sato, R., Ohba, M. & Iwahana, H. (1992) Insecticidal spectrum of a novel isolate of *Bacillus thuringiensis* serovar *japonensis*. *Biological Control*, **2**, 138–142.

Suzuki, N., Hori, H. & Asano, S. (1993) Sensitivity of cupreous chafer, *Anomala cuprea* (Coleoptera: Scarabaeidae), in different larval stages to *Bacillus thuringiensis* serovar *japonensis* strain Buibui. *Appl. Entomol. Zool.*, **28**, 403–405.

Suzuki, N., Hori, H., Tachibana, M., Indrasith, L.S. & Fujiie, A. (1994) Residual effects of the toxin from *Bacillus thuringiensis* serovar *japonensis* strain Buibui in the soil. *Appl. Entomol. Zool.*, **29**, 610–613.

Svestka, M. (1974) The use of *Bacillus thuringiensis* for the biological control of leaf-eating pests of flood-plain forests in S. Moravia. *Lesnictvi*, **20**, 439–464.

Svestka, M. (1977) The possibility of the use of bacteria *Bacillus thuringiensis* Berl. for the biological control of *Operophtera brumata* and *Tortrix viridana*. *Lesnictvi*, **23**, 875–892.

Svestka, M. & Vankova, J. (1976) Effects of artificial rain on the persistence of spores of *Bacillus thuringiensis* Berl. *Lesnictvi*, **22**, 829–838.

Svestka, M. & Vankova, J. (1978) On the effect of *Bacillus thuringiensis* combined with a synthetic pyrethroid on *Orgyia antiqua* L. *Anz. Schadlingskd. Pflanzenschutz Umweltschutz*, **51**, 5–9.

Svestka, M. & Vankova, J. (1980) On the effect of *Bacillus thuringiensis* in combination with the synthetic pyrethroid Ambush on *Operophthera brumata*, *Tortrix viridana* and the insect fauna of an oak stand. *Anz. Schadlingskd. Pflanzenschutz Umweltschutz*, **53**, 6–10.

Svestka, M. & Vankova, J. (1984) Action of preparations of *Bacillus thuringiensis* Berl. against the population of larch bud moth *Zeiraphera diniana* Gn. in spruce growths of the Krkonose (Giant Mountains) region. *Z. Angew. Entomol.*, **98**, 164–173.

Swadener, C. (1994) *Bacillus thuringiensis* (*B.t.*). *J. Pestic. Reform*, **14**, 13–20.

Swezey, S.L. & Salamanca, M.L. (1988) Trials of *Bacillus thuringiensis* var. *israelensis* for the control of larvae of *Anopheles albimanus* Wiedemann, in Leon department, Nicaragua, 1982. *Rev. Nicaraguense Entomol.*, **4**, 11–28.

Swiezynska, H. & Glowacka Pilot, B. (1973) Susceptibility of *Panolis flammea* larvae to the action of *Bacillus thuringiensis* and reduced rates of Nogos. *Sylwan*, **117**, 38–44.

Swirski, E., Wysoki, M., Yizhar, J., Gurevitz, E. & Greenberg, S. (1979) The honeydew moth (*Cryptoblabes gnidiella*) and the long-tailed mealybug (*Pseudococcus longispinus*) in avocado plantations. *Alon Hanotea*, **34**, 83–86.

Szalay Marzso, L., Halmagyi, L. & Fodor, S. (1981) Microbial control experiment against *Stilpnotia salicis* L., pest of poplar stands in northwest Hungary. *Acta Phytopathol. Acad. Sci. Hung.*, **16**, 189–197.

Tabashnik, B.E. (1992) Evaluation of synergism among *Bacillus thuringiensis* toxins. *Appl. Environ. Microbiol.*, **58**, 3343–3346.

Tabashnik, B.E. (1994) Evolution of resistance to *Bacillus thuringiensis*. *Ann. Rev. Entomol.*, **39**, 47–79.

Tabashnik, B.E., Finson, N. & Johnson, M.W. (1991) Managing resistance to *Bacillus thuringiensis*: lessons from the diamondback moth (Lepidoptera: Plutellidae). *J. Econ. Entomol.*, **84**, 49–55.

Tabashnik, B.E., Cushing, N.L., Finson, N. & Johnson, M.W. (1990) Field development of resistance to *Bacillus thuringiensis* in diamondback moth (Lepidoptera: Plutellidae). *J. Econ. Entomol.*, **83**, 1671–1676.

Tabashnik, B.E., Liu, Y.B., Finson, N., Masson, L. & Heckel, D.G. (1997) One gene in diamondback moth confers resistance to four *Bacillus thuringiensis* toxins. *Proc. Nat. Acad. Sci. USA*, **94**, 1640–1644.

Tabashnik, B.E., Finson, N., Groeters, F.R., Moar, W.J., Johnson, M.W., Luo, K. & Adang, M.J. (1994) Reversal of resistance to *Bacillus thuringiensis* in *Plutella xylostella*. *Proc. Nat. Acad. Sci. USA*, **91**, 4120–4124.

Tabashnik, B.E., Malvar, T., Liu, Y., Finson, N., Borthakur, D., Shin, B., Park, S., Masson, L., Maagd, R.A.d. & Bosch, D. (1996) Cross-resistance of the diamondback moth indicates altered interactions with domain II of *Bacillus thuringiensis* toxins. *Appl. Environ. Microbiol.*, **62**, 2839–2844.

Tabashnik, B.E., Liu, Y., Malvar, T., Heckel, D.G., Masson, L., Ballester, V., Granero, F., Mensua, J.L. & Ferre, J. (1997) Global variation in the genetic and biochemical basis of diamondback moth resistance to *Bacillus thuringiensis*. *Proc. Nat. Acad. Sci. USA*, **94**, 12780–12785.

Tailor, R., Tippett, J., Gibb, G., Pells, S., Pike, D., Jordan, L. & Ely, S. (1992) Identification and characterization of a novel *Bacillus thuringiensis* delta-endotoxin entomocidal to coleopteran and lepidopteran larvae. *Mol. Microbiol.*, **6**, 1211–1217.

Takagi, K. (1974) Monitoring of hymenopterous parasite in tea field. *Bull. Nat. Res. Inst. Tea*, **10**, 91–131.

Takahashi, Y., Hori, H., Furuno, H., Kawano, T., Takahashi, M. & Wada, Y. (1998) Enzyme-linked immunosorbent assays for rapid and quantitative detection of insecticidal crystal proteins of *Bt* pesticides. *J. Pestic. Sci.*, **23**, 386–391.

Takaki, S. (1975) *Bt* preparations in Japan. *Jap. Pestic. Inform.*, **25**, 23–26.

Takatsuka, J. & Kunimi, Y. (1998) Replication of *Bacillus thuringiensis* in larvae of the Mediterranean flour moth, *Ephestia kuehniella* (Lepidoptera: Pyralidae): growth, sporulation and insecticidal activity of parasporal crystals. *Appl. Entomol. Zool.*, **33**, 479–486.

Talalaev, E.V. (1976) Forest sanitation by bacteria. *Zaschc. Rast.*, **7**, 32–34.

Talekar, N.S. & Yang, J.C. (1991) Characteristic of parasitism of diamondback moth by two larval parasites. *Entomophaga*, **36**, 95–104.

Tamez Guerra, P., Castro Franco, R., Medranno Roldan, H., McGuire, M.R., Galan Wong, L.J.G. & Luna Olvera, H.A. (1998) Laboratory and field comparisons of strains of *Bacillus thuringiensis* for activity against noctuid larvae using granular formulations (Lepidoptera). *J. Econ. Entomol.*, **91**, 86–93.

Tan, A., Heaton, S., Farr, L. & Bates, J. (1997) The use of *Bacillus* diarrhoeal enterotoxin (BDE) detection using an ELISA technique in the confeirmation of the aetiology of *Bacillus*-mediated diarrhoea. *J. Appl. Microbiol.*, **82**, 677–682.

Tanada, Y. & Kaya, H.K. (1993) *Insect Pathology*, Academic Press, Inc., London.

Tanaka, H. & Kimura, Y. (1991) Resistance to BT formulation in diamondback moth, *Plutella xylostella* L., on watercress. *Jap. J. Appl. Entomol. Zool.*, **35**, 253–255.

Tanaka, K., Sadakane, H., Murakami, H., Hatano, S. & Watanabe, T. (1972) On the exotoxin of *Bacillus thuringiensis* var. *thuringiensis* Berliner. (1) Separation and the effect of toxin on silk worms. *Sci. Bull. Fac. Agric. Kyushu Univ.*, **26**, 537–544.

Tandan, J.S. & Nillama, N.C. (1987) Biological control of Asiatic corn borer (*Ostrinia furnacalis* Guenee) and corn earworm (*Helicoverpa armigera* Hubner). *CMU J. Agric. Food Nutrition*, **9**, 33–48.

Tang, J.D. & Shelton, A.M. (1995) Stable resistance to *Bacillus thuringiensis* in *Plutella xylostella*. *Resist. Pest Manag.*, **7**, 8–9.

Tang, J.D., Gilboa, S., Roush, R.T. & Shelton, A.M. (1997) Inheritance, stability, and lack-of-fitness costs of field-selected resistance to *Bacillus thuringiensis* in diamondback moth (Lepidoptera: Plutellidae) from Florida. *J. Econ. Entomol.*, **90**, 732–741.

Tanigoshi, L.K., Mayer, D.F., Babcock, J.M. & Lunden, J.D. (1990) Efficacy of the beta-endotoxin of *Bacillus thuringiensis* to *Lygus hesperus* (Heteroptera: Miridae): laboratory and field responses. *J. Econ. Entomol.*, **83**, 2200–2206.

Tapp, H. & Stotzky, G. (1995) Insecticidal activity of the toxins from *Bacillus thuringiensis* subspecies *kurstaki* and *tenebrionis* adsorbed and bound on pure and soil clays. *Appl. Environ. Microbiol.*, **61**, 1786–1790.

Tapp, H. & Stotzky, G. (1997) Monitoring the insecticidal toxins from *Bacillus thuringiensis* in soil with flow cytometry. *Canad. J. Microbiol.*, **43**, 1074–1078.

Tapp, H. & Stotzky, G. (1998) Persistence of the insecticidal toxin from *Bacillus thuringiensis* subsp. *kurstaki* in soil. *Soil Biol. Biochem.*, **30**, 471–476.

Tapp, H., Calamai, L. & Stotzky, G. (1994) Adsorption and binding of the insecticidal proteins from *Bacillus thuringiensis* subsp. *kurstaki* and subsp. *tenebrionis* on clay minerals. *Soil Biol. Biochem.*, **26**, 663–679.

Tavamaishvili, L.E. (1988) The nut weevil and its control. *Subtrop. Kul't.*, **5**, 128–131.

Tavamaishvili, L.E. (1990) The chief pests of hazelnuts in subtropical western Georgia and their control. *Subtrop. Kul't.*, **5**, 119–125.

Taylor, T.A. (1974) Evaluation of Dipel for control of lepidopterous pests of okra. *J. Econ. Entomol.*, **67**, 690–691.

Te Giffel, M.C., Beumer, R.R., Klijn, N., Wagendorp, A., & Rombouts, F.M. (1997) Discrimination between *Bacillus cereus* and *Bacillus thuringiensis* using specific DNA probes based on variable regions of 16S rRNA. *FEMS Microbiol. Lett.*, **146**, 47–51.

Teakle, R.E. (1994) Present use of, and problems with, *Bacillus thuringiensis* in Australia. *Agric. Ecosys. Environ.*, **49**, 39–44.

Teakle, R.E., Caon, G., Grimshaw, J.F. & Byrne, V.S. (1992) Larvicidal activity of strains of *Bacillus thuringiensis* for *Helicoverpa armigera* (Hubner) and *Helicoverpa punctigera* Wallengren (Lepidoptera: Noctuidae) estimates using a droplet-feeding method. *J. Aust. Entomol. Soc.*, **31**, 209–213.

Tedders, W.L. & Ellis, H.C. (1977) Aerial application of *Bacillus thuringiensis* var. *kurstaki* (HD-1) to shade and ornamental pecan trees against *Hyphantria cunea* and *Datana integerrima*. *J. Georgia Entomol. Soc.*, **12**, 248–250.

Tedders, W.L. & Gottwald, T.R. (1986) Evaluation of an insect collecting system and an ultra-low-volume spray system on a remotely piloted vehicle. *J. Econ. Entomol.*, **79**, 709–713.

Tedders, W.L., Gottwald, T. & Kaniuka, R. (1985) Biocontrol takes off in a pilotless miniplane. *Agric. Res. USA*, **33**, 8–9.

Temerak, S.A. (1980) Detrimental effects of rearing a braconid parasitoid on the pink borer larvae inoculated by different concentrations of the bacterium, *Bacillus thuringiensis* Berliner. *Z. Angew. Entomol.*, **89**, 315–319.

Temerak, S.A. (1982) Transmission of two bacterial pathogens by means of the ovipositor of *Bracon brevicornis* Wesm. (Hym., Braconidae) into the body of *Sesamia cretica* Led. (Lep., Tortricidae) *Anz. Schadlingskd. Pflanzenschutz Umweltschutz*, **55**, 89–92.

Temerak, S.A. (1984) On the ability of *Bracon brevicornis* Wesm. (Hym., Braconidae) to distinguish between uninfected and bacteria-infected larvae of *Sesamia cretica* (Lep., Noctuidae). *Anz. Schaedlingskd. Pflanzenschutz Umweltschutz*, **57**, 54–56.

Temeyer, K.B. (1984) Larvicidal activity of *Bacillus thuringiensis* subsp. *israelensis* in the dipteran *Haematobia irritans*. *Appl. Environ. Microbiol.*, **47**, 952–955.

Terytze, K. & Terytze, H. (1987) The use of bacterial preparations (*Bacillus thuringiensis* Berliner) for the control of harmful Lepidoptera species on cabbage (Eubacterales: Bacillaceae). *Arch. Phytopath. Pflschutz.*, **23**, 377–387.

Theoduloz, C., Roman, P., Bravo, J., Padilla, C., Vasquez, C., MezaZepeda, L. & MezaBasso, L. (1997) Relative toxicity of native Chilean *Bacillus thuringiensis* strains against *Scrobipalpuloides absoluta* (Lepidoptera: Gelechiidae). *J. Appl. Microbiol.*, **82**, 462–468.

Thiery, I., Delecluse, A., Tamayo, M.C. & Orduz, S. (1997) Identification of a gene for Cyt1A-like hemolysin from *Bacillus thuringiensis* subsp. *medellin* and expression in a crystal-negative *B. thuringiensis* strain. *Appl. Environ. Microbiol.*, **63**, 468–473.

Thomas, W.E. & Ellar, D.J. (1983) *Bacillus thuringiensis* var. *israelensis* crystal delta -endotoxin: effects on insect and mammalian cells *in vitro* and *in vivo*. *J. Cell Sci.*, **60**, 181–197.

Thompson, L.S. (1977) Field tests with chemical and biological insecticides for control of *Thymelicus lineola* on timothy. *J. Econ. Entomol.*, **70**, 324–326.

Thoms, E.M. & Watson, T.F. (1986) Effect of Dipel (*Bacillus thuringiensis*) on the survival of immature and adult *Hyposoter exiguae* (Hymenoptera: Ichneumonidae). *J. Invertebr. Pathol.*, **47**, 178–183.

Thomsen, L., Damgaard, P.H., Eilenberg, J. & Smits, P.H. (1998) Screening of selected *Bacillus thuringiensis* strains against *Agrotis segetum* larvae. *Bull. OILB-SROP*, **21**, 235–239.

Thomson, C., Tomkins, A.R. & Wilson, D.J. (1996) Effect of insecticides on immature and mature stages of *Encarsia citrina*, an armoured scale parasitoid. *Proc. 49th NZ Plant Prot. Conf* (ed. M. O'Callaghan), 1–5.

Ticehurst, M., Fusco, R.A. & Blumenthal, E.M. (1982) Effects of reduced rates of Dipel 4L, Dylox 1.5 Oil, and Dimilin W-25 on *Lymantria dispar* (L.) (Lepidoptera: Lymantriidae), parasitism, and defoliation. *Environ. Entomol.*, **11**, 1058–1062.

Tipping, P.W. & Burbutis, P.P. (1983) Some effects of pesticide residues on *Trichogramma nubilale* (Hymenoptera: Trichogrammatidae). *J. Econ. Entomol.*, **76**, 892–896.

Tkachev, V.M. (1972) The role of *Tachina larvarum* in the control of leaf-biting pests of apple trees. *Sadovod. Vinograd. Vinodel. Mold.*, **9**, 37–39.

Tkachev, V.M. (1974) Insect enemies of the codling moth. *Zaschc. Rast.*, **8**, 26.

Tollo, B. & Chougourou, D.C. (1997) Feeding rate and survival of *Plutella xylostella* (L.) larvae (Lepidoptera: Plutellidae) after intoxication by *Bacillus thuringiensis* Berliner var. *kurstaki* and var. *aizawai*. *Arch. Phytopath. Plant Prot.*, **31**, 201–206.

Tolstova, Y., S. & Ionova, Z.A. (1976) Toxicity of pesticides to *Trichogramma*. *Zaschc. Rast.*, **9**, 21.

Tomar, R.K.S. (1998a) Efficacy of some insecticides against okra petiole maggot *Melanagromyza hibisci*. *Indian J. Entomol.*, **60**, 22–24.

Tomar, R.K.S. (1998b) Efficacy and economics of biopesticide and insecticide combinations against okra shoot and fruit borer. *Indian J. Entomol.*, **60**, 25–28.

Tomkins, A.R. (1996) Pest control on kiwifruit with an insecticidal soap. *Proc. 49th NZ Plant Prot. Conf.* (ed. M. O'Callaghan), 6–11.

Tompkins, G.J., Linduska, J.J., Young, J.M. & Dougherty, E.M. (1986) Effectiveness of microbial and chemical insecticides for controlling cabbage looper (Lepidoptera: Noctuidae) and imported cabbageworm (Lepidoptera: Pieridae) on collards in Maryland. *J. Econ. Entomol.*, **79**, 497–501.

Tonkonozhenko, A.P. (1981) The new biological insecticide thuringin (from *Bacillus thuringiensis*) for controlling *Oestrus ovis* infestation. *Toksikologiya i zashchita sel'skokhozyaistvennykh zhivotnykh ot ektoparazitov* (ed. V.S. Yarnykh). pp. 11–17; Series: Trudy VNIIVS.

Tonkonozhenko, A.P., Alekseenok, A.Y., Simetskii, M.A. & Kudryavtsev, E.N. (1977) Biological preparation "miazol" (*Bacillus thuringiensis* exotoxin) for myiasis in animals. *Veterinariya, Moscow* **6**, 41–42.

Tonks, N.V., Everson, P.R. & Theaker, T.L. (1978) Efficacy of insecticides against geometrid larvae, *Operophtera* spp., on southern Vancouver Island, British Columbia. *J. Entomol. Soc. British Columbia*, **75**, 6–9.

Toumanoff, C. & Vago, C. (1951) L'agent pathogene de la flacherie du Ver endemique dans la region des Cevennes: *Bacillus cereus* var. *alesti*. *Compt. Rend. Acad. Sci.*, **233**, 1504–1506.

Tousignant, M.E., Boisvert, J.L. & Chalifour, A. (1992) Reduction of mortality rates of *Bacillus thuringiensis* var. *israelensis* aqueous suspensions due to freezing and thawing. *J. Amer. Mosq. Con. Assoc.*, **8**, 149–155.

Travers, R.S., Faust, R.M. & Reichelderfer, C.F. (1976) Effects of *Bacillus thuringiensis* var. *kurstaki* delta -endotoxin on isolated lepidopteran mitochondria. *J. Invertebr. Pathol.*, **28**, 351–356.

Tremblay, F.L.J., Huot, L. & Perron, J.M. (1972) Penetration of the thermostable exotoxin of *Bacillus thuringiensis* through the chorion of *Acheta domesticus*. *Entomol. Exp. Appl.*, **15**, 397–398.

Trenchev, G. & Pavlov, A. (1982) Morphological characteristics of *Orgyia antiqua* L. and *Orgyia gonostigma* F. (Lepidoptera, Lymantriidae) and methods of controlling the larvae. *Gradinar. Lozar. Nauka*, **19**, 53–58.

Treverrow, N. (1985) Susceptibility of *Chironomus tepperi* (Diptera: Chironomidae) to *Bacillus thuringiensis* serovar *israelensis*. *J. Aust. Entomol. Soc.*, **24**, 303–304.

Triggiani, O. (1980) Tests of the susceptibility of the larvae of *Lymantria dispar* L. (Lep. Lymantriidae) to various concentrations of *Bacillus thuringiensis* Berl. var. *kurstaki* and Baculovirus (subgroup A) combined. *Entomologica*, **16**, 5–12.

Triggiani, O. & Sidor, C. (1982) Microbiological control tests against the pine processionary (*Thaumetopoea pityocampa* Schiff., Lepid. Thaumetopoeidae) in pinewoods in Puglia. *Entomologica,* **17**, 91–102.

Tripp, H.A. (1972) Field trials to control spruce budworm, *Choristoneura fumiferana* (Clem.), through aerial application of *Bacillus thuringiensis*. *Proc. Entomol. Soc. Ontario,* **103**, 64–69.

Trisyono, A. & Whalon, M.E. (1997) Fitness costs of resistance to *Bacillus thuringiensis* in Colorado potato beetle (Coleoptera: Chrysomelidae). *J. Econ. Entomol.*, **90**, 267–271.

Trottier, M.R., Morris, O.N. & Dulmage, H.T. (1988) Susceptibility of the bertha armyworm, *Mamestra configurata* (Lepidoptera, Noctuidae), to sixty-one strains from ten varieties of *Bacillus thuringiensis*. *J. Invertebr. Pathol.*, **51**, 242–249.

Trudel, R., Bauce, E., Cabana, J. & Guertin, C. (1997) Vulnerability of the fir coneworm, *Dioryctria abietivorella* (Grote) (Lepidoptera: Pyralidae), in different larval stages to the HD-1 strain of *Bacillus thuringiensis*. *Canad. Entomol.*, **129**, 197–198.

Trumble, J.T. & Alvarado Rodriguez, B. (1993) Development and economic evaluation of an IPM program for fresh market tomato production in Mexico. *Agric. Ecosys. Environ.*, **43**, 267–284.

Tsai, S.F., Liao, J.W. & Wang, S.C. (1997) Clearance and effects of intratracheal instillation to spores of *Bacillus thuringiensis* or *Metarhizium anisopliae* in rats. *J. Chin. Soc. Vet. Sci.*, **23**, 515–522.

Tsankov, G. & Mirchev, P. (1983) The effect of some plant protection measures on the egg parasite complex of the pine processionary (*Thaumetopoea pityocampa*). *Gorskostop. Nauka,* **20**, 84–89.

Tsankov, G. & Ovcharov, D. (1986) The chestnut fruit worm (*Pammene fasciana*), a new pest of chestnut fruits in Bulgaria. *Gorskostop. Nauka,* **23**, 41–45.

Tseng, C.T. (1990) Use of *Trichogramma ostriniae* (Hym., Trichogrammatidae) for controlling the oriental corn borer, *Ostrinia furnacalis* (Lep., Pyralidae) in Taiwan, China. *FFTC-NARC International Seminar on 'The use of parasitoids and predators to control agricultural pests,* Ibaraki-ken, Japan, 2-7 October, 15 p.

Tseng, C.T. & Wu, Y.Z. (1990) The integrated control of the Asian corn borer, *Ostrinia furnacalis* Guen'ee, on sweet corn. *Plant Prot. Bull. Taipei,* **32**, 177–182.

Tsilosani, G.A., Shoniya, D.I. & Butliashvili, R.N. (1976) Effectiveness of using bacterial preparations. *Lesnoe Khoz.*, **11**, 30–32.

Tsinitis, R.Y. (1980) Results of the application of biopreparations of *Bacillus thuringiensis* var. *caucasicus* in Latvia. *Biol. Z. Armenii,* **33**, 436–438.

Tulisalo, U. & Rautapaa, J. (1983) The control of flies in cowsheds and piggeries with a preparation of *Bacillus thuringiensis*. *Vaxtskyddsnotiser,* **47**, 15–22.

Turnipseed, S.G., Habib, M.E.M. & Amaral, M.E.C. (1985) Aerial application of *Bacillus thuringiensis* against the velvetbean caterpillar, *Anticarsia gemmatalis* Huebner, in soybean fields. *Rev. Agr. Piracicaba Brazil,* **60**, 141–149.

Tutkun, E., Inci, A. & Yilmaz, B. (1987) Investigations on the effectiveness of preparations of *Bacillus thuringiensis* against larvae of the greater wax moth (*Galleria mellonella* L.) causing damage in the hives of the honeybee (*Apis mellifera* L.). *Turk. Entomol. Kongr. Bildirileri,* 585–594.

Twine, P.H. & Lloyd, R.J. (1982) Observations on the effect of regular releases of *Trichogramma* spp. in controlling *Heliothis* spp. and other insects in cotton. *Queensland J. Agric. Anim. Sci.*, **39**, 159–167.

Tyrell, D.J., Bulla, L.A., Jr. & Davidson, L.I. (1981a) Characterization of spore coat proteins of *Bacillus thuringiensis* and *Bacillus cereus*. *Comp. Biochem. Physiol. B,* **70**, 535–539.

Tyrell, D.J., Bulla, L.A., Jr., Andrews, R.E., Jr., Kramer, K.J., Davidson, L.I. & Nordin, P. (1981b) Comparative biochemistry of entomocidal parasporal crystals of selected *Bacillus thuringiensis* strains. *J. Bacteriol.*, **145**, 1052–1062.

Udayasuriyan, V., Nakamura, A., Mori, H., Masaki, H. & Uozumi, T. (1994) Cloning of a new *cry*IA(a) gene from *Bacillus thuringiensis* strain FU-2-7 and analysis of chimaeric CryIA(a) proteins for toxicity. *Biosci. Biotech. Biochem.*, **58**, 830–835.

Ulewicz, K. & Bakowski, S. (1975) Investigations on effective methods of controlling cockroaches in ships. *Anz. Schaedlingskd. Pflanzenschutz Umweltschutz,* **48**, 49–52.

Ulpah, S. & Kok, L.T. (1996) Interrelationship of *Bacillus thuringiensis* Berliner to the diamondback moth (Lepidoptera: Noctuidae) and its primary parasitoid, *Diadegma insulare* (Hymenoptera: Ichneumonidae). *J. Entomol. Sci.*, **31**, 371–377.

Umarov Sh, A., Nilova, G.N. & Davlyatov, I.D. (1975) The effect of Entobakterin and Dendrobacillin on beneficial arthropods. *Zaschc. Rast.*, **3**, 25–26.

Undeen, A.H., Takaoka, H. & Hansen, K. (1981) A test of *Bacillus thuringiensis* var. *israelensis* de Barjac as a larvicide for *Simulium ochraceum*, the Central American vector of onchocerciasis. *Mosq. News,* **41**, 37–40.

Unnamalai, N. & Vaithilingam, S. (1995) *Bacillus thuringiensis*, a biocontrol agent for major tea pests. *Curr. Sci.*, **69**, 939–940.

Urias Lopez, M.A., Bellotti, A.C., Bravo Mojica, H. & Carrillo Sanchez, J.L. (1987) Effect of insecticides on 3 parasitoids of the yucca pest *Erinnyis ello* (L.). *Agrociencia,* **67**, 137–146.

Utami, R., Whalon, M.E. & Rahardja, U. (1995) Inheritance of resistance to *Bacillus thuringiensis* subsp. *tenebrionis* CryIIIA delta-endotoxin in Colorado potato beetle, (Coleoptera: Chrysomelidae). *J. Econ. Entomol.*, **88**, 21–26.

Vago, C. & Burges, H.D. (1964) International symposium on the identification and assay of viruses and *Bacillus thuringiensis* Berliner used for insect control. *J. Insect Pathol.* **6**, 544–547.

Vail, P.V., Soo Hoo, C.F., Seay, R.S., Killinen, R.G. & Wolf, W.W. (1972) Microbial control of lepidopterous pests of fall lettuce in Arizona and effects of chemical and microbial pesticides on parasitoids. *Environ. Entomol.*, **1**, 780–785.

Vakenti, J.M., Campbell, C.J. & Madsen, H.F. (1984) A strain of fruittree leafroller, *Archips argyrospilus* (Lepidoptera: Tortricidae), tolerant to azinphos-methyl in an apple orchard region of the Okanagan Valley of British Columbia. *Canad. Entomol.*, **116**, 69–73.

Valli, G. (1977) Integrated control in vineyards. Studies and preliminary assessments of the vine moths. *Not. Mal. Piante*, **92–93**, 407–419.

van der Geest, L.P.S. (1971) Use of *Bacillus thuringiensis* for the control of orchard pests. *Z. Angew. Entomol.*, **69**, 263–266.

van der Geest, L.P.S. & de Barjac, H. (1982) Pathogenicity of *Bacillus thuringiensis* towards the tsetse fly *Glossina pallidipes*. *Z. Angew. Entomol.*, **93**, 504–507.

van Epenhuijsen, C.W. & Carpenter, A. (1990) Pear slug and apple leaf curling midge: observations on their biology and control in an organic orchard. *Proc. 43rd NZ Weed Pest Cont. Conf.* (ed. A.J. Popay), 119–122.

van Essen, F.W. & Hembree, S.C. (1982) Simulated field studies with four formulations of *Bacillus thuringiensis* var. *israelensis* against mosquitoes: residual activity and effect of soil constituents. *Mosq. News*, **42**, 66–72.

van Frankenhuyzen, K. (1987) Effect of wet foliage on efficacy of *Bacillus thuringiensis* spray against the spruce budworm, *Choristoneura fumiferana* Clem. (Lepidoptera: Tortricidae). *Canad. Entomol.*, **119**, 955–956.

van Frankenhuyzen, K. (1990a) Development and current status of *Bacillus thuringiensis* for control of defoliating forest insects. *For. Chron.*, **66**, 498–507.

van Frankenhuyzen, K. (1990b) Effect of temperature and exposure time on toxicity of *Bacillus thuringiensis* Berliner spray deposits to spruce budworm, *Choristoneura fumiferana* Clemens (Lepidoptera: Tortricidae). *Canad. Entomol.*, **122**, 69–75.

van Frankenhuyzen, K. (1993) The challenge of *Bacillus thuringiensis*. *Bacillus thuringiensis, an Environmental Biopesticide: Theory and Practice* (eds. P.F. Entwistle, J.S. Cory, M.J. Bailey & S. Higgs), pp. 311. John Wiley and Sons Ltd, Chichester, UK.

van Frankenhuyzen, K. (1994) Effect of temperature on the pathogenesis of *Bacillus thuringiensis* Berliner in larvae of the spruce budworm, *Choristoneura fumiferana* Clem. (Lepidoptera: Tortricidae). *Canad. Entomol.*, **126**, 1061–1065.

van Frankenhuyzen, K. & Fast, P.G. (1989) Susceptibility of three coniferophagous *Choristoneura* species (Lepidoptera: Tortricidae) to *Bacillus thuringiensis* var. *kurstaki*. *J. Econ. Entomol.*, **82**, 193–196.

van Frankenhuyzen, K. & Gringorten, J.L. (1991) Frass failure and pupation failure as quantal measurements of *Bacillus thuringiensis* toxicity to Lepidoptera. *J. Invertebr. Pathol.*, **58**, 465–467.

van Frankenhuyzen, K. & Nystrom, C. (1989) Residual toxicity of a high-potency formulation of *Bacillus thuringiensis* to spruce budworm (Lepidoptera: Tortricidae). *J. Econ. Entomol.*, **82**, 868–872.

van Frankenhuyzen, K. & Nystrom, C. (1998). The *Bacillus thuringiensis* toxin specificity database (1998). http://www.glfc.forestry.ca/english/res/ Bt_HomePage /netintro.htm (February 2000).

van Frankenhuyzen, K., Gringorten, J.L., Milne, R.E., Gauthier, D., Pusztai, M., Brousseau, R. & Masson, L. (1991) Specificity of activated CryIA proteins from *Bacillus thuringiensis* subsp. *kurstaki* HD-1 for defoliating forest Lepidoptera. *Appl. Environ. Microbiol.*, **57**, 1650–1655.

van Frankenhuyzen, K., Milne, R., Brousseau, R. & Masson, L. (1992) Comparative toxicity of the HD-1 and NRD-12 strains of *Bacillus thuringiensis* subsp. *kurstaki* to defoliating forest Lepidoptera. *J. Invertebr. Pathol.*, **59**, 149–154.

van Rie, J., Jansens, S., Hofte, H., Degheele, D. & Van Mellaert, H. (1990a) Receptors on the brushborder membranes of the insect midgut as determinants of the specificity of *Bacillus thuringiensis* delta-endotoxins. *Appl. Environ. Microbiol.*, **56**, 1378–1385.

van Rie, J., McGaughey, W.H., Johnson, D.E., Barnett, B.D. & van Mellaert, H. (1990b) Mechanism of insect resistance to the microbial insecticide *Bacillus thuringiensis*. *Science*, **247**, 72–74.

Vandenberg, J.D. (1990) Safety of four entomopathogens for caged adult honey bees (Hymenoptera: Apidae). *J. Econ. Entomol.*, **83**, 755–759.

Vandenberg, J.D. & Shimanuki, H. (1986) Two commercial preparations of the beta-exotoxin of *Bacillus thuringiensis* influence the mortality of caged adult honey bees, *Apis mellifera* (Hymenoptera: Apidae). *Environ. Entomol.*, **15**, 166–169.

Vankova, J. (1973) Comparison of the insecticidal effects of two *Bacillus thuringiensis* preparations, Bathurin and Dipel. *Acta Entomol. Bohemoslov.*, **70**, 328–333.

Vankova, J. (1978) The heat-stable exotoxin of *Bacillus thuringiensis*. *Folia Microbiol.*, **23**, 162–174.

Vankova, J. (1981) House-fly susceptibility to *Bacillus thuringiensis* var. *israelensis* and a comparison with the activity of other insecticidal bacterial preparations. *Acta Entomol. Bohemoslov.*, **78**, 358–362.

Vankova, J. & Purrini, K. (1979) Natural epizootics caused by bacilli of the species *Bacillus thuringiensis* and *Bacillus cereus*. *Z. Angew. Entomol.*, **88**, 216–221.

Vankova, J., Horska, K. & Sebesta, K. (1974) The fate of exotoxin of *Bacillus thuringiensis* in *Galleria mellonella* caterpillars. *J. Invertebr. Pathol.*, **23**, 209–212.

Vargas Osuna, E., Merino, R., Aldebis, H.K. & Santiago Alvarez, C. (1994) Field application of a baculovirus (ObGV) for the control of *Ocnogyna baetica* Rambur (Lepidoptera: Arctiidae) on beans. *Bol. Sanidad Vegetal Plagas*, **20**, 487–493.

Varlez, S., Jervis, M.A., Kidd, N.A.C., Campos, M. & McEwen, P.K. (1993) Effects of *Bacillus thuringiensis* on parasitoids of the olive moth, *Prays oleae* Bern. (Lep., Yponomeutidae). *J. Appl. Entomol.*, **116**, 267–272.

Vasquez, M., Parra, C., Hubert, E., Espinoza, P., Theoduloz, C. & Meza Basso, L. (1995) Specificity and insecticidal activity of Chilean strains of *Bacillus thuringiensis*. *J. Invertebr. Pathol.*, **66**, 143–148.

Vasudevan, G. & Baskaran, P. (1979) Efficacy of certain foliar spray of insecticides against important insect pests of tobacco. *Indian J. Plant Prot.*, **7**, 150–155.

Venezian, A. & Blumberg, D. (1982) Phenology, damage and control of the lesser date moth, *Batrachedra amydraula*, in date palms in Israel. *Alon Hanotea*, **36**, 785–788.

Verigin, V.F. (1975) Entobakterin in the fields. *Zaschc. Rast.*, **5**, 30.

Verma, S.K. (1995) Studies on the control of greater wax moth, *Galleria mellonella* L. in *Apis cerana* F. colonies with the biological insecticide, Dipel. *Indian Bee J.*, **57**, 121–123.

Vervelle, C. (1975) Contribution to the study of the effects of *Bacillus thuringiensis* Berliner on the reproduction of *Laspeyresia pomonella* L. *Rev. Zool. Agri. Pathol. Veg.*, **74**, 108–115.

Videnova, E., Tsankov, G. & Chernev, T. (1972) Study on the effect of *Bacillus thuringiensis* on larvae of *Thaumetopoea pityocampa*. *Gorskostop. Nauka*, **9**, 59–65.

Videnova, E., Matev, T., Ganchev, K. & Pulev, V. (1980) Dipel - a highly effective biopreparation. Results of experiments. *Zashc. Rast.*, **28**, 5–6.

Viggiani, G. (1981) Recenti acquisizioni sulla lotta integrata nell'oliveto (Recent findings on integrated control in olive groves). *Inf. Fitopatol.*, **31**, 37–43.

Viggiani, G. & Tranfaglia, A. (1975) Indicative control tests against the vine moth (*Lobesia botrana* - Schiff.) in Campania. *Boll. Lab. Entomol. Agrar. Portici*, **32**, 140–144.

Viggiani, G., Bernardo, U. & Giorgini, M. (1998) Contact effects of pesticides on some entomophagous insects. *Inf. Fitopatol.*, **48**, 76–78.

Viggiani, G., Castronuovo, N. & Borrelli, C. (1972) Secondary effects of 40 pesticides on *Leptomastidea abnormis* Grlt. (Hym. Encyrtidae) and *Scymnus includens* Kirsch (Col. Coccinellidae), important natural enemies of *Planococcus citri* (Risso). *Boll. Lab. Entomol. Agrar. Portici*.**30,** 88–103.

Vilas-Bôas, G.F.L.T, Vilas-Boas, L.A., Lereclus, D. & Arantes, O.M.N. (1998). *Bacillus thuringiensis* conjugation under environmental conditions. *FEMS Microbiol. Ecol.*, **25**, 369–374.

Villani, H.C., Campos, A.R. & Gravena, S. (1980) Effectiveness of *Bacillus thuringiensis* Berliner and fenitrothion + fenvalerate for the control of the passion-fruit larva *Dione juno juno* (Cramer, 1779) (Lepidoptera, Heliconidae). *An. Soc. Entomol. Bras.*, **9**, 255–260.

Villas-Boas, G.L. & Franca, F.H. (1996) Use of the parasitoid *Trichogramma pretiosum* for control of Brazilian tomato pinworm in tomato grown in the greenhouse. *Hort. Bras.*, **14**, 223–225.

Visser, B., Salm, T.v.d., Brink, W.v.d. & Folkers, G. (1988) Genes from *Bacillus thuringiensis entomocidus* 60.5 coding for insect-specific crystal proteins. *Mol. Gen. Genet.*, **212**, 219–224.

Visser, S., Addison, J.A. & Holmes, S.B. (1994) Effects of Dipel 176, a *Bacillus thuringiensis* subsp. *kurstaki* (*B.t.k.*) formulation, on the soil microflora and the fate of *B.t.k.* in an acid forest soil: a laboratory study. *Canad. J. For. Res.*, **24**, 462–471.

Vlayen, P., van Impe, G. & Semaille, R. (1978) Activity of a commercial preparation of *Bacillus thuringiensis* against the common spider mite *Tetranychus urticae* Koch (Acari: Tetranychidae). *Meded. Fac. Landbouwwet. Rijksuniv. Gent*, **43**, 471–479.

Vlug, H.J., Smits, P.H. & Grijpma, P. (1988) The occurrence, biology and control with *Bacillus thuringiensis* of the antler moth, *Cerapteryx graminis*. *Rapport Rijksinstituut voor Onderzoek in de Bos en Landschapsbouw 'De Dorschkamp', Wageningen*, **516**, 45–49.

Volzhinskii, D.V., Sokolova, E.I., Kosovskikh, V.L., Kulieva, N.M., Bikunova, A.N., Ganushkina, L.A. & Erlikh, V.D. (1984) Application of *Bacillus thuringiensis* Berl. of the 14th serotype against larvae of blood-sucking mosquitoes. *Med. Parazitol. Parazit. Bolezni*, **3**, 69–73.

von Arx, R, Goueder, J.; Cheikh, M & Temime, A.B. (1987) Integrated control of potato tubermoth *Phthorimaea operculella* (Zeller) in Tunisia. *Insect Sci. Its Applic.* **8,** 989–994.

Wabiko, H., Held, G.A. & Bulla, L.A., Jr. (1985) Only part of the protoxin gene of *Bacillus thuringiensis* subsp. *berliner* 1715 is necessary for insecticidal activity. *Appl. Environ. Microbiol.*, **49**, 706–708.

Wagner, D.L., Peacock, J.W., Carter, J.L. & Talley, S.E. (1996) Field assessment of *Bacillus thuringiensis* on nontarget Lepidoptera. *Environ. Entomol.*, **25**, 1444–1454.

Wahba, M.E., El Gemeiy, H.M., Naguib, S.M., Rofail, M.F. & El Gogary, O.A. (1996) Studies on the combined effect of sabadilla and *Bacillus thuringiensis* on the 1st instar larvae of *Pectinophora gossypiella* and *Earias insulana*. *Arab Univ. J. Agric. Sci.*, **4**, 155–161.

Waikwa, J.W. & Mathenge, W.M. (1977) Field studies on the effects of *Bacillus thuringiensis* (Berliner) on the larvae of the giant coffee looper, *Ascotis selenaria reciprocaria*, (Lepidoptera: Geometridae) and its side effects on the larval parasites of the leaf miner (*Leucoptera* spp.). *Kenya Coffee*, **42**, 95–101.

Waites, R.E., Gouger, R.J. & Habeck, D.H. (1978) Synthetic pyrethroids for control of caterpillars on cabbage and brussel sprouts and Colorado potato beetle on Irish potatoes. *J. Georgia Entomol. Soc.*, **13**, 247–250.

Walgenbach, J.F., Leidy, R.B. & Sheets, T.J. (1991) Persistence of insecticides on tomato foliage and implications for control of tomato fruitworm (Lepidoptera: Noctuidae). *J. Econ. Entomol.*, **84**, 978–986.

Walker, E.D. (1995) Effect of low temperature on feeding rate of *Aedes stimulans* larvae and efficacy of *Bacillus thuringiensis* var. *israelensis* (H-14). *J. Amer. Mosq. Con. Assoc.*, **11**, 107–110.

Wallner, W.E., Dubois, N.R. & Grinberg, P.S. (1983) Alteration of parasitism by *Rogas lymantriae* (Hymenoptera: Braconidae) in *Bacillus thuringiensis*-stressed gypsy moth (Lepidoptera: Lymantriidae) hosts. *J. Econ. Entomol.*, **76**, 275–277.

Walters, P.J. (1976) Susceptibility of three *Stethorus* spp. (Coleoptera: Coccinellidae) to selected chemicals used in N.S.W. apple orchards. *J. Aust. Entomol. Soc.*, **15**, 49–52.

Wang, A.F., Huang, J.N. & Han, P.E. (1989) A study on the control of insects on vegetables with preparations of *Bacillus thuringiensis*. *Zhejiang Agric. Sci.*, **2**, 79–81.

Wang, D.S., Wu, S.C., Yuan, Y.D. & Shen, B. (1997) Testing residual toxicity of insecticides on vegetables with susceptible house fly (*Musca domestica*). *Acta Agric. Shanghai*, **13**, 52–55.

Wang, J.B. & Cheung, W.W.K. (1994) Histopathological effects of *Bacillus thuringiensis* delta-endotoxin on the Malpighian tubules of *Pieris canidia* larva. *Zool. Stud.*, **33**, 192–199.

Wang, L.C., Chen, J., Wang, Y., Liu, X.F., Wang, Z.G., Fu, G.B., Huang, Z.F. & Jiang, R.H. (1988) A study on chemical control of *Quadraspidiotus gigas*. *J. Northeast For. Univer.*, **16**, 1–6.

Wang, X.G. & Liu, S.S. (1998) Bionomics of *Diadromus collaris* (Hymenoptera: Ichneumonidae), a major pupal parasitoid of *Plutella xylostella*. *Acta Entomol. Sin.*, **41**, 389–395.

Wang, Y., Wen, J. & Feng, X.C. (1986) A new serovar of *Bacillus thuringiensis*. *Acta Microbiol. Sinica*, **26**, 1–6.

Wang, Y.Z., Zhang, L.H. & Huang, C.Y. (1990) Pathological modifications of haemocytes of indian meal moth after infection by *Bacillus thuringiensis*. *Insect Knowl.*, **27**, 20–21.

Wang, Z., Jia, C., Sun, S., Huang, Y., Hong, J., Sun, B. & Zhang, Z. (1996) Effects of *Bacillus thuringiensis* on parasitic wasps in the egg period of *Dendrolimus superans*. *J. Northeast For. Univ.*, **24**, 51–55.

Wang, Z.X. & Chen, H.Z. (1984) A note on the research and application of cytoplasmic polyhedrosis virus of pine moth in Japan. *For. Sci.Tech. Linye Keji Tongxun*, **4**, 29–31.

Waquil, J.M., Viana, P.A., Lordello, A.I., Cruz, I. & de Oliveira, A.C. (1982) Control of the armyworm on maize with chemical and biological insecticides. *Pesqu. Agropecu. Bras.*, **17**, 163–166.

Ward, R.V. & Redovan, S.S. (1983) Operational applications of *Bacillus thuringiensis* var. *israelensis*. *J. Florida Anti Mosq. Assoc.*, **54**, 45–47.

Warelas, D. (1976) Contribution to the study of *Pandemis dumetana* (Treitschke) (Lep.: Tortricidae) a new pest of strawberry in Switzerland. Thesis, Ecole Polytechnique Federale, Zurich.

Warren, R.E., Rubenstein, D., Ellar, D.J., Kramer, J.M. & Gilbert, R.J. (1984) *Bacillus thuringiensis* var. *israelensis* protoxin activation and safety. *Lancet*, **8378**, 678–679.

Wasano, N. & Ohba, M. (1998) Assignment of delta-endotoxin genes of the four Lepidoptera-specific *Bacillus thuringiensis* strains that produce spherical parasporal inclusions. *Curr. Microbiol.*, **37**, 408–411.

Wasano, N., Kim, K.H. & Ohba, M. (1998) Delta-endotoxin proteins associated with spherical parasporal inclusions of the four Lepidoptera-specific *Bacillus thuringiensis* strains. *J. Appl. Microbiol.*, **84**, 501–508.

Watson, T.F. & Kelly Johnson, S. (1995) A bioassay to assess pink bollworm, *Pectinophora gossypiella* (Saunders), susceptibility to *B.t.* toxins. *Proc. Beltwide Cotton Conf.*, San Antonio, Texas, USA, January 4-7, 1995, **2**, 878–879.

Wawrzyniak, M. (1987) The susceptibility of *Apanteles glomeratus* L. to insecticides applied against its host. *Rocz. Nauk Roln. Ser. E Ochr. Rosl.*, **17**, 225–235.

Wearing, C.H. & Thomas, W.P. (1978) Integrated control of apple pests in New Zealand. 13. Selective insect control using diflubenzuron and *Bacillus thuringiensis*. *Proc. 31st NZ Weed Pest Cont. Conf.* (ed. M.J. Hartley), 221–228.

Webb, J.P. & Dhillon, M.S. (1984) The effect of *Bacillus thuringiensis israelensis* (serotype H-14) on *Aedes squamiger* at the Bolsa Chica Marsh, Orange County, California. *Mosq. News*, **44**, 412–414.

Webb, R.E., Shapiro, M., Podgwaite, J.D., Reardon, R.C., Tatman, K.M., Venables, L. & Kolodny Hirsch, D.M. (1989) Effect of aerial spraying with Dimilin, Dipel, or Gypchek on two natural enemies of the gypsy moth (Lepidoptera: Lymantriidae). *J. Econ. Entomol.*, **82**, 1695–1701.

Webb, S.E. (1994) Management of insect pests of squash. *Proc. Florida State Hort. Soc.*, **106**, 165–168.

Weigand, S., Lateef, S.S., El Din, N.E.S., Mahmoud, S.F., Ahmed, K., Ali, K., Muehlbauer, F.J. & Kaiser, W.J. (1994) Integrated control of insect pests of cool season food legumes. *Expanding the production and use of cool season food legumes, lentil, faba bean, chickpea, and grasspea,* Cairo, Egypt, 12-16 April 1992, 679–694.

Weiser, J. & Prasertphon, S. (1984) Entomopathogenic spore-formers from soil samples of mosquito habitats in Northern Nigeria. *Zentralbl. Microbiol.*, **139**, 49–55.

Weiser, J., Tonka, T., Weiser, J. Jr. & Horak, P. (1992) A water soluble molluscicidal metabolite of *Bacillus thuringiensis. Proc. 25th Ann. Meet. Soc. Invertebr. Pathol.*, Heidelberg, Germany, 204.

Welton, J.S. & Ladle, M. (1993) The experimental treatment of the blackfly, *Simulium posticatum* in the Dorset Stour using the biologically produced insecticide *Bacillus thuringiensis* var. *israelensis. J. Appl. Ecol.*, **30**, 772–782.

Wernicke, K. & Funke, W. (1995) Impact of Dipel (*Bacillus thuringiensis* var. *kurstaki*) and BIO1020 (*Metarhizium anisopliae*) on arthropods with soil-living developmental stages). *Mitt. Dtsch. Ges. Allg. Angew. Entomol.*, **10**, 207–210.

Weseloh, R.M. & Andreadis, T.G. (1982) Possible mechanism for synergism between *Bacillus thuringiensis* and the gypsy moth (Lepidoptera: Lymantriidae) parasitoid, *Apanteles melanoscelus* (Hymenoptera: Braconidae). *Ann. Entomol. Soc. Amer.*, **75**, 435–438.

West, A.W. (1984) Fate of the insecticidal, proteinaceous parasporal crystal of *Bacillus thuringiensis* in soil. *Soil Biol. Biochem.*, **16**, 357–360.

West, A.W. & Burges, H.D. (1985) Persistence of *Bacillus thuringiensis* and *Bacillus cereus* in soil supplemented with grass or manure. *Plant and Soil,* **83**, 389–398.

West, R.J. & Carter, J. (1992) Aerial applications of *Bacillus thuringiensis* formulations against eastern blackheaded budworm in Newfoundland in 1990. *Inf. Rep. Newfoundl. Labrador Region For. Canad.*, N-X-282, 10 pp.

West, A.W., Crook, N.E. & Burges, H.D. (1984) Detection of *Bacillus thuringiensis* in soil by immunoflurescence. *J. Invertebr. Pathol.*, **43**, 150–155.

West, A.W., Burges, H.D., Dixon, T.J. & Wyborn, C.H. (1985) Survival of *Bacillus thuringiensis* and *Bacillus cereus* spore inocula in soil: effects of pH, moisture, nutrient availability and indigenous microorganisms. *Soil Biol. Biochem.*, **17**, 657–665.

West, R.J., Raske, A.G., Retnakaran, A. & Lim, K.P. (1987) Efficacy of various *Bacillus thuringiensis* Berliner var. *kurstaki* formulations and dosages in the field against the hemlock looper, *Lambdina fiscellaria fiscellaria* (Guen.) (Lepidoptera: Geometridae), in Newfoundland. *Canad. Entomol.*, **119**, 449–458.

West, R.J., Raske, A.G. & Sundaram, A. (1989) Efficacy of oil-based formulations of *Bacillus thuringiensis* Berliner var. *kurstaki* against the hemlock looper, *Lambdina fiscellaria fiscellaria* (Guen.) (Lepidoptera: Geometridae). *Canad. Entomol.*, **121**, 55–63.

Whaley, W.H., Anhold, J. & Schaalje, G.B. (1998) Canyon drift and dispersion of *Bacillus thuringiensis* and its effects on select nontarget lepidopterans in Utah. *Environ. Entomol.*, **27**, 539–548.

Whalon, M.E., Miller, D.L., Hollingworth, R.M., Grafius, E.J. & Miller, J.R. (1993) Selection of a Colorado potato beetle (Coleoptera: Chrysomelidae) strain resistant to *Bacillus thuringiensis. J. Econ. Entomol.*, **86**, 226–233.

Wharton, D.A. & Bone, L.W. (1989) *Bacillus thuringiensis israelensis* toxin affects egg-shell ultrastructure of *Trichostrongylus colubriformis* (Nematoda). *Intern. J. Invertebr. Reprod. Develop.*, **15**, 155–158.

White, P.F., Butt, J., Pethybridge, N.K., Jarrett, P. & Elliott, T.J. (1995) The story of a strain: development of GC327, a dipteran-active strain of *Bacillus thuringiensis* effective against the mushroom sciarid, *Lycoriella auripila. Science and cultivation of edible fungi,* Oxford, UK, 17-22 September 1995, 499–506.

Widiastuti, H., Darmono, T.W. & Hadioetomo, R.S. (1996) Characteristics of selected *Bacillus thuringiensis* isolates from Indonesia and their toxicity to *Hyposidra talaca. Menara Perkebunan,* **64**, 65–78.

Wigley, P. & Chilcott, C. (1990) *Bacillus thuringiensis* isolates active against the New Zealand pasture pest, *Costelytra zealandica* (Coleoptera: Scarabaeidae). *Proc. Abst. Vth Intern. Coll. Invertebr. Pathol. Microb. Con., Adelaide, Australia,* 344.

Wilkinson, J.D., Biever, K.D. & Ignoffo, C.M. (1975) Contact toxicity of some chemical and biological pesticides to several insect parasitoids and predators. *Entomophaga,* **20**, 113–120.

Williams, M.L., Miller, G.L. & Hendricks, H.J. (1987) The holly looper, a new pest of holly in the southern landscape. *Res. Rep. Ser. Alabama Agric. Exp. St. Auburn University,* **5**, 30.

Willoughby, B., Prestidge, R.A., Seay, C. & Gray, J. (1994) Evaluation of two strains of *Bacillus thuringiensis* var. *kurstaki* against early instar grass grub (*Costelytra zealandica*) larvae. *Proc. 47th N.Z. Plant Prot. Conf.* (ed. A.J. Popay), 277–278.

Wilson, A.G.L. (1981) Field evaluation of formamidine insecticides and *Bacillus thuringiensis* for selective control of *Heliothis* spp. on cotton. *Gen. Appl. Entomol.*, **13**, 105–111.

Wilson, A.G.L., Desmarchelier, J.M. & Malafant, K. (1983) Persistence on cotton foliage of insecticide residues toxic to *Heliothis* larvae. *Pestic. Sci.*, **14**, 623–633.

Wilson, B.H. & Burns, E.C. (1968) Induction of resistance to *Bacillus thuringiensis* in a laboratory strain of house flies. *J. Econ. Entomol.*, **61**, 1747–1748.

Wilson, M.C., Chen, F.C. & Shaw, M.C. (1984) Susceptibility of the alfalfa weevil to a *Bacillus thuringiensis* exotoxin. *J. Georgia Entomol. Soc.*, **19**, 366–371.

Wipfli, M.S. & Merritt, R.W. (1994) Effects of *Bacillus thuringiensis* var. *israelensis* on nontarget benthic insects through direct and indirect exposure. *J. North Amer. Benthological Soc.*, **13**, 190–205.

Wipfli, M.S., Merritt, R.W. & Taylor, W.W. (1994) Low toxicity of the black fly larvicide *Bacillus thuringiensis* var. *israelensis* to early stages of brook trout (*Salvelinus fontinalis*), brown trout, (*Salmo trutta*), and steelhead trout (*Oncorhynchus mykiss*) following direct and indirect exposure. *Canad. J. Fish. Aquat. Sci.*, **51**, 1451–1458.

Wirth, M.C., Georghiou, G.P. & Federici, B.A. (1997) CytA enables CryIV endotoxins of *Bacillus thuringiensis* to overcome high levels of CryIV resistance in the mosquito, *Culex quinquefasciatus*. *Proc. Nat. Acad. Sci. USA*, **94**, 10536–10540.

Wirth, M.C., Delecluse, A., Federici, B.A. & Walton, W.E. (1998) Variable cross-resistance to Cry11B from *Bacillus thuringiensis* subsp. *jegathesan* in *Culex quinquefasciatus* (Diptera: Culicidae) resistant to single or multiple toxins of *Bacillus thuringiensis* subsp. *israelensis*. *Appl. Environ. Microbiol.*, **64**, 4174–4179.

Wiwat, C., Panbangred, W. & Bhumiratana, A. (1990) Transfer of plasmids and chromosomal genes amongst subspecies of *Bacillus thuringiensis*. *J. Indust. Microbiol.*, **6**, 19–27.

Wolfenbarger, D.O. & Poe, S.L. (1973) Tomato pinworm control. *Proc. Florida State Hort. Soc.*, **86**, 139–143.

Wolfenbarger, D.A., Guerra, A.A., Dulmage, H.T. & Garcia, R.D. (1972) Properties of the beta-exotoxin of *Bacillus thuringiensis* IMC 10,001 against the tobacco budworm. *J. Econ. Entomol.*, **65**, 1245–1248.

Wolfenbarger, D.A., Hamed, A.A. & Luttrell, R.G. (1997) Toxicity of *Bacillus thuringiensis* var. *tenebrionis* and CA-thuringiensin against the boll weevil *Anthonomus grandis* (Boh.) (Coleoptera: Curculionidae). *1997 Proc. Beltwide Cotton Conf.*, USA, 6-10 January, 1997, Volume 2, 1296–1300.

Woods, W. (1981) Controlling cotton pests with egg parasites. *J. Agric. Western Australia*, **22**, 63–64.

World Health Organisation (1992) Fourteenth report of the WHO Expert Committee on Vector Biology and Control, Safe Use of Pesticides. *Tech Rep. Ser. WHO Geneva*, 813 pp.

Wraight, S.P., Molloy, D.P. & Singer, S. (1987) Studies on the culicine mosquito host range of *Bacillus sphaericus* and *Bacillus thuringiensis* var. *israelensis* with notes on the effects of temperature and instar on bacterial efficacy. *J. Invertebr. Pathol.*, **49**, 291–302.

Wraight, S.P., Molloy, D., Jamnback, H. & McCoy, P. (1981) Effects of temperature and instar on the efficacy of *Bacillus thuringiensis* var. *israelensis* and *Bacillus sphaericus* strain 1593 against *Aedes stimulans* larvae. *J. Invertebr. Pathol.*, **38**, 78–87.

Wu, F.Y. & Tang, K.F. (1981) Isolation and identification of an insect-pathogenic strain, CW-I. *Weishengwuxue Tongbao*, **8**, 101–102.

Wu, H.F. & Yang, X.M. (1988) Control of *Brachmia macrosocopa* with B.t. (*Bacillus thuringiensis*) emulsion. *Chin. J. Biol. Cont.*, **4**, 144.

Wu, J.X., Chen, Z. & Zhang, H.Z. (1998) Studies on the effects of temperature and humidity on bacteriophage infection in fermentation of *Bacillus thuringiensis*. *Chin. J. Biol. Cont.*, **14**, 101–104.

Wu, Q.L. (1986) Investigation on the fluctuations of dominant natural enemy populations in different cotton habitats and integrated application with biological agents to control cotton pests. *Natural Enemies of Insects*, **8**, 29–34.

Wysoki, M. (1989) *Bacillus thuringiensis* preparations as a means for the control of lepidopterous avocado pests in Israel. *Israel J. Entomol.*, **23**, 119–129.

Wysoki, M. & Jarvinen, L. (1986) Evaluation of nine strains of *Bacillus thuringiensis* Berliner against the giant looper, *Boarmia* (*Ascotis*) *selenaria* Schiffermuller (Lepidoptera, Geometridae). *Anz. Schaedlingskd. Pflanzenschutz Umweltschutz*, **59**, 74–77.

Wysoki, M. & Scheepens, M.H.M. (1988) Effect of 20 strains of *Bacillus thuringiensis* Berliner on larvae of *Boarmia selenaria* Schiffermuller (Lepidoptera: Geometridae). *Insect Sci. Its Applic.*, **9**, 433–436.

Wysoki, M., Izhar, Y., Gurevitz, E., Swirski, E. & Greenberg, S. (1975) Control of the honeydew moth, *Cryptoblabes gnidiella* Mill. (Lepidoptera: Phycitidae), with *Bacillus thuringiensis* Berliner in avocado plantations. *Phytoparasitica*, **3**, 103–111.

Wysoki, M., Swirski, E. & Izhar, Y. (1981) Biological control of avocado pests in Israel. *Prot. Ecol.*, **3**, 25–28.

Wysoki, M., de Haan, P. & Izhar, Y. (1986) Influence of a spore killed *Bacillus thuringiensis* preparation (Toarow CT) on mortality of two avocado pests, the giant looper *Boarmia selenaria* and the honeydew moth *Cryptoblabes gnidiella*. *Alon Hanotea*, **40**, 1043–1049.

Xin, Y.F., Liu, Z.X., Ma, S.Y. & Xu, W.G. (1989) Experiment on controlling Asiatic corn borer (*Ostrinia furnacalis*) by aerospraying of *Bt* (*Bacillus thuringiensis*) and *Bt* mixture. *J. Shenyang Agric. Univ.*, **20**, 77–80.

Xu, J., Zhang, Q.W., Tian, H.Y. & Cheo, M.T. (1998) *In vitro* and planta bioassays of engineered endophytic bacterium in corn with corn borer. *Acta Entomol. Sin.*, **41**, 126–131.

Xu, R.M., Chang, J.S. & Lu, B.L. (1986) Susceptibility of eight Chinese mosquitoes to *Bacillus thuringiensis* H-14. *Chin. J. Biol. Cont.*, **2**, 20–22.

Xue, R., Xue, R.D., Zhang, W. & Xiao, A. (1990) Status of study on the biological control of filth flies in China. *Biological Control of Filth Flies*, **6**, 6–9.

Yabas, C. & Zeren, O. (1992) Studies on the chemical control of *Hellula undalis* F. in the eastern Mediterranean region. *Proc. 2nd Turk. Nat. Cong. Entomol.*, 335–341.

Yadava, C.P. (1978) Toxicity of *Bacillus thuringiensis* to the larvae of *Sesamia inferens* Walker (Noctuidae: Lepidoptera) the pink borer of rice. *Oryza*, **15**, 105.

Yakunin, B.M. (1977) The influence of thermostable exotoxin on the fertility of fleas of the daughter generation. *Med. Parazitol. Parazit. Bolezni*, **4**, 491–493.

Yakunin, B.M., Prokop' ev, V.N. & Dubitskii, A.M. (1974) The insecticidal action of the exotoxin of *Bacillus thuringiensis* on rodent flea larvae. *Probl. Osobo Opasnykh Infekts.*, **1**, 95–99.

Yamaguchi, H., Furuta, K. & Akita, Y. (1971) Preliminary studies on the control of forest defoliators with *Bacillus thuringiensis* in Hokkaido. *Ann. Rep. Hokkaido Branch Govern. For. Exp. St.*, 29–34.

Yamamoto, P.T., Benetoli, I., Fernandes, O.D. & Gravena, S. (1990) Effect of *Bacillus thuringiensis* and insecticides on the cotton leafworm *Alabama argillacea* (Hubner) (Lepidoptera: Noctuidae) and predatory arthropods. *Ecossistema*, **15**, 36–44.

Yamamoto, T. & McLaughlin, R.E. (1981) Isolation of a protein from the parasporal crystal of *Bacillus thuringiensis* var. *kurstaki* toxic to the mosquito larva, *Aedes taeniorhynchus*. *Biochem. Biophys. Res. Commun.*, **103**, 414–421.

Yameogo, L., Leveque, C., Traore, K. & Fairhurst, C.P. (1988) Ten years of monitoring the aquatic fauna of West African rivers treated against blackflies (Diptera: Simuliidae), vectors of onchocerciasis. *Nat. Canad.*, **115**, 287–298.

Yamvrias, C. (1972) Tests with bacterial preparations against larvae of the olive moth (*Prays oleae* (Bern.)). *Ann. Inst. Phytopathol. Benaki*, **10**, 256–266.

Yamvrias, C. & Young, E.C. (1977) Trials using *Bacillus thuringiensis* to control the olive moth, *Prays oleae* in Greece in 1976. *Z. Angew. Entomol.*, **84**, 436–440.

Yang, F.S., Chen, Y.C. & Chen, X. (1984) A report of study on control of pine moth by low volume spray with *Bacillus thuringiensis* var. *dendrolimus*. *For. Sci.Tech. Linye Keji Tongxun*, **11**, 27–29.

Yang, L.C., Yen, D.F. & Hsu, E.L. (1985) Secondary effect of *Bacillus thuringiensis* var. *kurstaki* to larvae of *Spodoptera litura*. *Chin. J. Entomol.*, **5**, 19–21.

Yang, M.H. (1978) Research and application of *Bacillus thuringiensis* in China. *Biological insect control in China and Sweden*, Royal Swedish Academy of Sciences, Stockholm, Sweden, 13.

Yao, B.A., Wang, Q.L., Zhao, J.L., Ma, L.H. & Yu, Z.I. (1995a) Ovicidal activity of the parasporal crystal toxin of *Bacillus thuringiensis* to *Haemonchus contortus* eggs. *Chin. J. Vet. Sci.*, **15**, 332–334.

Yao, B.A., Wang, Q.L., Zhao, J.L., Ma, L.H. & Yu, Z.N. (1995b) Lethal effects of the companion cell crystal of *Bacillus thuringiensis* on *Strongyloides papillosus* larvae. *Chin. J. Vet. Sci. Tech.*, **25**, 39–40.

Yao, J., Dong, Y.S., Lang, J.H., Wang, Y.L., Chen, C.H. & Xia, X.G. (1998) Distribution and toxicity of *Bacillus thuringiensis* in soils in the north of Jiangsu Province. *Jiangsu J. Agric. Sci.*, **14**, 96–102.

Yara, K., Kunimi, Y. & Iwahana, H. (1997) Comparative studies of growth characteristic and competitive ability in *Bacillus thuringiensis* and *Bacillus cereus* in soil. *Appl. Entomol. Zool.*, **32**, 625–634.

Yarnykh, V.S. & Kats, M.B. (1981) Comparative evaluation of some strains of *Bacillus thuringiensis* as producers of thermostable exotoxin. *Toksikologiya i zashchita sel'skokhozyaistvennykh zhivotnykh ot ektoparazitov*, 17–20.

Yarnykh, V. & Tonkonozhenko, A. (1975) Turingin, a new biological insecticide for use against harmful Diptera. *Proc. 20th World Vet. Cong., Thessaloniki*, **1**, 567–568.

Yarnykh, V.S., Tonkonozhenko, A.P. & Ermakova, G.I. (1981a) Study of the pathogenicity of *Bacillus thuringiensis* to larvae of mosquitoes and house-flies. *Toksikologiya i zashchita sel'skokhozyaistvennykh zhivotnykh ot ektoparazitov*, 38–42.

Yarnykh, V.S., Tonkonozhenko, A.P. & Ermakova, G.I. (1981b) The insecticidal activity of *Bacillus thuringiensis* against blood-sucking mosquitoes. *Toksikologiya i zashchita sel'skokhozyaistvennykh zhivotnykh ot ektoparazitov*, 33–38.

Yastrebov, I.O. (1978) The role of certain ecological factors in the control of cabbage pests. *Vestn. Zool.*, **1**, 84–87.

Yates, M.M. (1985) Further trials with ultralow-volume aerial applications of *Bti* using the Beecomist system. *Proc. 72nd Ann. Meet. New Jersey Mosq. Cont. Assoc.*, Atlantic City, New Jersey, 157–160.

Ye, Z.C., Cheng, D.F., Wu, Y.Z., Li, M.Y. & Bai, G.J. (1985) Application of *Bacillus thuringiensis* var. *galleriae* 81-6 for controlling larvae of *Clanis bilineata* (Walker). *Chin. J. Biol. Con.*, **3**, 44.

Yearian, W.C., Luttrell, R.G., Stacy, A.L. & Young, S.Y. (1980) Efficacy of *Bacillus thuringiensis* and Baculovirus *heliothis*-chlordimeform spray mixtures against *Heliothis* spp. on cotton. *J. Georgia Entomol. Soc.*, **15**, 260–271.

Yen, F.C. (1988) Studies on the integrated control of key insect pests of supersweet corn. *Res. Bull. Tainan District Agric. Improv. St.*, **22**, 25–37.

Yendol, W.G. & Hedlund, R.C. (1972) Microbial insecticide example of bacterium to control gypsy moth. Gypsy moth potent threat to Pennsylvania forests. *Sc. Agric.*, **19**, 2–3.

Yendol, W.G., Hamlen, R.A. & Rosario, S.B. (1975) Feeding behavior of gypsy moth larvae on *Bacillus thuringiensis*-treated foliage. *J. Econ. Entomol.*, **68**, 25–27.

Yin, R.G. (1993) Bionomics of *Leucinodes orbonalis* Guenee and its control. *Entomol. Know.*, **30**, 91–92.

Ying, S.L. & Klimetzek, D. (1986) A decade of successful control of pine caterpillar, *Dendrolimus punctatus* Walker (Lepidoptera: Lasiocampidae), by microbial agents. *Forest Entomology, Hamburg. Forest Ecology and Management*, 69–74.

Young, J.M., Chilcott, C.N., Broadwell, A., Wigley, P.J. & Lecadet, M.M. (1998) Identification of serovars of *Bacillus thuringiensis* Berliner 1915 in New Zealand. *NZ J. Crop Hort. Sci.*, **26**, 63–68.

Young, S.Y., McCaul, L.A. & Yearian, W.C. (1980) Effect of *Bacillus thuringiensis-Trichoplusia* nuclear polyhedrosis virus mixtures on the cabbage looper, *Trichoplusia ni. J. Georgia Entomol. Soc.*, **15**, 1–8.

Young, S.Y., Kring, T.J., Johnson, D.R. & Klein, C.D. (1997) *Bacillus thuringiensis* alone and in mixtures with chemical insecticides against heliothines and effects on predator densities in cotton. *J. Entomol. Sci.*, **32**, 183–191.

Yousten, A.A. (1973) Effect of the *Bacillus thuringiensis* delta -endotoxin on an insect predator which has consumed intoxicated cabbage looper larvae. *J. Invertebr. Pathol.*, **21**, 312–314.

Yousten, A.A., Jones, M.E. & Benoit, R.E. (1982) Development of selective/differential bacteriological media for the enumeration of *Bacillus thuringiensis* serovar. *israelensis* (H14) and *Bacillus sphaericus* (WHO-VBC-82.844), Geneva, Switzerland 7 pp. (unpublished).

Yu, C.G., Mullins, M.A., Warren, G.W., Koziel, M.G. & Estruch, J.J. (1997b) The *Bacillus thuringiensis* vegetative insecticidal protein Vip3A lyses midgut epithelium cells of susceptible insects. *Appl. Environ. Microbiol.*, **63**, 532–536.

Yu, H.S., Lee, D.K., Lee, W.J. & Shim, J.C. (1982) Mosquito control evaluation of *Bacillus thuringiensis* var. *israelensis* in the laboratory, simulated rice paddies, and confined field trials in marsh and sewage effluent in South Korea. *Korean J. Entomol.*, **12**, 69–82.

Yu, J.H., He, C.N. & Wang, S.N. (1990) Observations on bionomics of *Dioryctria schutazeella. For. Pest Dis.*, **2**, 10–11.

Yu, L., Berry, R.E. & Croft, B.A. (1997a) Effect of *Bacillus thuringiensis* toxins in transgenic cotton and potato on *Folsomia candida* (Collembola: Isotomidae) and *Oppia nitens* (Acari: Oribatidae). *J. Econ. Entomol.*, **90**, 113–118.

Yu, Z.N., Dai, J.Y., Zhou, H.B. & Dong, Z.R. (1984) A new serotype of *Bacillus thuringiensis. Acta Microbiol. Sinica,* **24**, 117–121.

Yu, Z.N. & Dai, D.S. (1987) Impact of input quantity of Nong-Ru 100 on the quality of *Bacillus thuringiensis* wettable powder. *Natural Enemies of Insects,* **9**, 6–9.

Yuan, J.R., Lin, Z.M. & Zhang, X.X. (1983) A study on the screening and utilization of the armyworm *Bacillus. Shanxi Agricultural Science Shanxi Nongye Kexue,* **12**, 13–17.

Yudina, T.G. & Burtseva, L.I. (1997) Activity of delta-endotoxins of four *Bacillus thuringiensis* subspecies against prokaryotes. *Microbiol. New York,* **66**, 17–22.

Yunusov, I. (1974) Biopreparations destroy the lucerne bug. *Zaschc. Rast.*, **8**, 32.

Yuval, B. & Warburg, A. (1989) Susceptibility of adult phlebotomine sandflies (Diptera: Psychodidae) to *Bacillus thuringiensis* var. *israeliensis. Ann. Trop. Med. Parasitol.*, **83**, 195–196.

Zahner, V., Momen, H., Salles, C.A. & Rabinovitch, L. (1989) A comparative study of enzyme variation in *Bacillus cereus* and *Bacillus thuringiensis. J. Appl. Bacteriol.*, **67**, 275–282.

Zaritsky, A., Zalkinder, V., Ben Dov, E., Barak, Z. & Dov, E.B. (1991) Bioencapsulation and delivery of mosquito larvae of *Bacillus thuringiensis* H14 toxicity by *Tetrahymena pyriformis. J. Invertebr. Pathol.*, **58**, 455–457.

Zayats Yu, V., Suradeeva, A.N. & Rzhavina, E.K. (1976) Biological control of the Colorado beetle on egg-plant. *Zaschc. Rast.,* **9**, 51.

Zaz, G.M. (1989) Effectiveness of *Bacillus thuringiensis* Berliner against different instars of *Spodoptera litura* (Fabricius). *Indian J. Plant Prot.*, **17**, 119–121.

Zaz, G.M. & Kushwaha, K.S. (1983) Quantitative incidence of tobacco caterpillar, *Spodoptera litura* (F.) and related natural enemies in cole crops. *Indian J. Entomol.*, **45**, 201–202.

Zethner, O., Khan, B.M., Chaudhry, M.I., Bolet, B., Khan, S., Khan, H., Gul, H., Ogaard, L., Zaman, M. & Nawaz, G. (1987) *Agrotis segetum* granulosis virus as a control agent against field populations of *Agrotis ipsilon* and *A. segetum* (Lep: Noctuidae) on tobacco, okra, potato and sugar beet in northern Pakistan. *Entomophaga,* **32**, 449–455.

Zelazny, B. & Welling, M. (1994) Isolation of *Bacillus thuringiensis* from tropical and subtropical soil samples. *Nachrichtenbl. Dtsch. Pflanzenschutzdienst,* **46**, 192–194.

Zelenev, N.N. (1982) Natural enemies of the juniper argent. *Lesnoe Khoz.*, **2**, 48–50.

Zeng, H.F., Jiao, L.J. & Sun, L. (1984) The life history, habits and control measures of *Sparganothis pilleriana. Sci. Silvae Sin.*, **20**, 100–103.

Zethner, O., Khan, B.M., Chaudhry, M.I., Bolet, B., Khan, S., Khan, H., Gul, H., Ogaard, L., Zaman, M. & Nawaz, G. (1987) *Agrotis segetum granulosis* virus as a control agent against field populations of *Agrotis ipsilon* and *A. segetum* (Lep: Noctuidae) on tobacco, okra, potato and sugar beet in northern Pakistan. *Entomophaga,* **32**, 449–455.

Zhang, B.Y., Lei, J.Y. & Jin, X.L. (1987) A new strain of *Bacillus thuringiensis* found in Gansu. *Chin. J. Biol. Cont.*, **3**, 30–32.

Zhang, F. (1998) Reaction of *Trichogramma chilonis* to some pesticides commonly used in cotton fields. *J. Jilin Agric. Univ.*, **20**, 43–45.

Zhang, M.L. (1997) Effects of 14 insecticides on adults, larvae, eggs, and pupae of *Trichogramma confusum*. *Natural Enemies of Insects*, **19**, 11–14.

Zhang, M.Y., Lovgren, A. & Landen, R. (1995) Adhesion and cytotoxicity of *Bacillus thuringiensis* to cultured *Spodoptera* and *Drosophila* cells. *J. Invertebr. Pathol.*, **66**, 46–51.

Zhang, X.H., Zhang, C.Z., Ma, Z.Q., Gao, F.K. & Meng, F.S. (1996) The application of Junduwei, a mixture of *Bt* and HaNPV, against the cotton bollworm. *Chin. J. Biol. Cont.*, **12**, 1–4.

Zhao, J.Z., Zhu, G.R., Ju, Z.L. & Wang, W.Z. (1993) Resistance of diamondback moth to *Bacillus thuringiensis* in China. *Resist. Pest Manag.*, **5**, 11–12.

Zhao, Q., Zhao, X., Liu, G., Wu, B., Zhao, Q.Y., Zhao, X.H., Liu, G.P. & Wu, B.G. (1998) Isolation and determination of a plant pathogen of *Dendrolimus superans*. *J. Northeast For. Univer.*, **26**, 70–71.

Zhou, A., Ma, C. & Ma, X. (1994) Co-toxicity of diazinon and *Bacillus thuringiensis* formulations on the diamondback moth (*Plutella xylostella* L.). *Acta Agric. Shanghai*, **10**, 75–78.

Zhou, X.S. (1992) Biological effect of M-one (containing 5% *Bacillus thuringiensis* var. *san diego*) on the lady beetle *Olla abdominalis* Casey. (eds. M.G. Zhao, J.L. Shi, H.Y. Pan & Y.Y. Wang), *Collection of Achievements on the Technique Cooperation Project of P.R. China and F.R. Germany*, 175–178, Ministry of Forestry; China.

Zhumanov, B.Z., Dzhumaev Sh, B. & Yakhishev, E.B. (1988) Preserving lacewings in the field. *Zaschc. Rast. Moskva*, **8**, 13.

Ziogas, A., Gedminas, A. & Bartninkaite, I. (1997) Foray-48B against insect pests in Lithuania. *Integrated Plant Protection: Achievements and Problems*, Dotnuva-Akademija, Lithuania, 7-9 September 1997, 172–174.

Zivanovic, V. & Stamenkovic, S. (1976) The insecticide Dipel for control of the larvae of some Lepidoptera. *Zast. Bilja*, **27**, 381–387.

Zlatanova, A.A. & Lukin, V.A. (1971) The conservation of parasites of the codling moth during integrated protection of orchards. *Zaschc. Rast.*, **16**, 17.

Znamenskii, V.S. & Kupriyanova, V.A. (1973) Practical use of 'biopreparations' in forest protection. *Lesnoe Khoz.*, **12**, 63–64.

Zoebelein, G. (1990) Twenty-three year surveillance of development of insecticide resistance in diamondback moth from Thailand (*Plutella xylostella* L., Lepidoptera, Plutellidae). *Med. Fac. Landbouww. Rijk. Gent*, **55**, 313–322.

Zohdy, N.Z.M. & Matter, M.M. (1982) Effect of *Bacillus thuringiensis* var. *israelensis* on some Egyptian mosquito larvae. *J. Egypt. Soc. Parasit.*, **12**, 349–357.

Zubkov, A.F., Titova, R.P., Nesterova, O.A., Zakladnaya, A.G., Volkava, L.D. & Novozhilov, K.V. (1986) Biocoenological after-effects of using pesticides in sowings of fodder pea cultivated under intensive conditions. *Ekologicheskie osnovy predotvrashcheniya poter' urozhaya ot vreditelei, boleznei i sornyakov*, 80–91.

Zuckerman, B.M., Dicklow, M.B. & Acosta, N. (1993) A strain of *Bacillus thuringiensis* for the control of plant-parasitic nematodes. *Biocontr. Sci. Technol.*, **3**, 41–46.

Zukauskiene, J. & Petrauskas, V. (1973) The destruction of leaf-eating pests of berry fruits by entomopathogenic microorganisms and entomophagous insects in natural conditions. *Acta Entomol. Lit.*, **2**, 153–160.

Zukowski, K. (1994) Testing the effectiveness of selected bioinsecticides in reduction of the population of cockroaches (*Blattella germanica* L.). *Rocz. Panstw. Zakl. Hig.*, **45**, 139–144.

Zukowski, K. (1995) Testing the effectiveness of the new bioinsecticides proposed as reductants of the population of cockroaches (*Blattella germanica* L.). *Rocz. Panstw. Zakl. Hig.*, **46**, 293–297.

Zuo, G.S., Guo, Y.J. & Wang, N.Y. (1994a) Effects of thuringiensin on cotton bollworm and its main natural enemies. *Plant Prot.*, **20**, 2–4.

Zuo, G.S., Guo, Y.J., Wang, N.Y. & Guo, Y.Y. (1994b) Impact of thuringiensin on the predation of *Orius sauteri* nymph on *Tetranychus urticae* eggs. *Chin. J. Biol. Cont.*, **10**, 126–130.

Appendices

APPENDIX 1: DELTA-ENDOTOXIN GENES

(Modified from the website, Crickmore *et al.* 2000 [February 2000] and Crickmore *et al.* 1998. For full list of references, see the website)

Name	Previous Name	Subspecies/variety in original description	Accession Number
cry1Aa1	CryIA(a)	*kurstaki*	M11250
cry1Aa2		*sotto*	M10917
cry1Aa3		*aizawai*	D00348
cry1Aa4		*entomocidus*	X13535
cry1Aa5		FU-2-7	D17518
cry1Aa6		*kurstaki*	U43605
cry1Aa7		unspecified	AF081790
cry1Aa8		unspecified	I26149
cry1Aa9		*dendrolimus*	AB026261
cry1Ab1	CryIA(b)	*berliner*	M13898
cry1Ab2		*kurstaki*	M12661
cry1Ab3		*kurstaki*	M15271
cry1Ab4		*kurstaki*	D00117
cry1Ab5		*berliner*	X04698
cry1Ab6		*kurstaki*	M37263
cry1Ab7		*aizawai*	X13233
cry1Ab8		*aizawai*	M16463
cry1Ab9		*aizawai*	X54939
cry1Ab10		*kurstaki*	A29125
cry1Ab11		*kurstaki*	I12419
cry1Ab12		?	AF057670
cry1Ac1	CryIA(c)	*kurstaki*	M11068
cry1Ac2		*kenyae*	M35524
cry1Ac3		unspecified	X54159
cry1Ac4		*kurstaki*	M73249
cry1Ac5		*kurstaki*	M73248
cry1Ac6		*kurstaki*	U43606
cry1Ac7		*kurstaki*	U87793
cry1Ac8		*kurstaki*	U87397
cry1Ac9		*kurstaki*	U89872
cry1Ac10		*kurstaki*	AJ002514
cry1Ac11		unspecified	AJ130970
cry1Ac12		unspecified	I12418
cry1Ad1	CryIA(d)	*aizawai*	M73250
cry1Ad2		unspecified	A27531

APPENDIX 1 (*Continued*)

Name	Previous Name	Subspecies/variety in original description	Accession Number
cry1Ae1	CryIA(e)	*alesti*	M65252
cry1Af1		unspecified	U82003
cry1Ag1		unspecified	AF081248
cry1Ba1	CryIB	*thuringiensis*	X06711
cry1Ba2		unspecified	X95704
cry1Bb1	ET5	EG 5847	L32020
cry1Bc1	PEG5	*morrisoni*	Z46442
cry1Bd1	CryE1	*wuhanensis*	U70726
cry1Be1			
cry1Ca1	CryIC	*entomocidus*	X07518
cry1Ca2		*aizawai*	X13620
cry1Ca3		*aizawai*	M73251
cry1Ca4		*entomocidus*	A27642
cry1Ca5		*aizawai*	X96682
cry1Cb1	CryIC(b)	*galleria*	M97880
cry1Da1	CryID	*aizawai*	X54160
cry1Da2		unspecified	I76415
cry1Db1	PrtB	unspecified	Z22511
cry1Ea1	CryIE	unspecified	X53985
cry1Ea2		*kenyae*	X56144
cry1Ea3		*kenyae*	M73252
cry1Ea4		*kenyae*	U94323
cry1Ea5		unspecified	A15535
cry1Eb1	CryIE(b)	*aizawai*	M73253
cry1Fa1	CryIF	*aizawa*	M63897
cry1Fa2		*aizawai*	M73254
cry1Fb1	PrtD	unspecified	Z22512
cry1Fb2		*morrisoni*	AB012288
cry1Fb3		*morrisoni*	AF062350
cry1Fb4		NA	I73895
cry1Ga1	PrtA	unspecified	Z22510
cry1Ga2		*wuhanensis*	Y09326
cry1Gb1	CryH2	*wuhanensis*	U70725
cry1Ha1	PrtC	unspecified	Z22513
cry1Hb1		*morrisoni*	U35780
cry1Ia1	CryV	*kurstaki*	X62821
cry1Ia2		*kurstaki*	M98544
cry1Ia3		*kurstaki*	L36338
cry1Ia4		unspecified	L49391
cry1Ia5		unspecified	Y08920
cry1Ia6		*kurstaki*	AF076953
cry1Ib1	CryV	*kurstaki* or *entomocidus*	U07642
cry1Ic1		unspecified	AF056933
cry1Id1	NRcryV	unspecified	AF047579
cry1Ja1	ET4	EG 5847	L32019
cry1Jb1	ET1	EG5092	U31527

APPENDIX 1 (*Continued*)

Name	Previous Name	Subspecies/variety in original description	Accession Number
cry1Jc1		NA	I90730
cry1Ka1		*morrisoni*	U28801
cry2Aa1	CryIIA	*kurstaki*	M31738
cry2Aa2		*kurstaki*	M23723
cry2Aa3		*sotto*	D86064
cry2Aa4		*kenyae*	AF047038
cry2Aa5		unspecified	AJ132464
cry2Aa6		unspecified	AJ132465
cry2Aa7		unspecified	AJ132463
cry2Ab1	CryIIB	*kurstaki*	M23724
cry2Ab2		*kurstaki*	X55416
cry2Ac1	CryIIC	non motile strain	X57252
cry3Aa1	CryIIIA	*tenebrionis (=san diego)*	M22472
cry3Aa2		*tenebrionis*	J02978
cry3Aa3		*tenebrionis*	Y00420
cry3Aa4		*tenebrionis*	M30503
cry3Aa5		*morrisoni*	M37207
cry3Aa6		*tenebrionis*	U10985
cry3Aa7		unspecified	AJ237900
cry3Ba1	CryIIIB	*tolworthi*	X17123
cry3Ba2		unspecified	A07234
cry3Bb1	CryIIIBb	EG4961	M89794
cry3Bb2		EG5144	U31633
cry3Bb3		unspecified	I15475
cry3Ca1	CryIIID	*kurstaki*	X59797
cry4Aa1	CryIVA	*israelensis*	Y00423
cry4Aa2		*israelensis*	D00248
cry4Ba1	CryIVB	*israelensis*	X07423
cry4Ba2		*israelensis*	X07082
cry4Ba3		*israelensis?*	M20242
cry4Ba4		*israelensis*	D00247
cry5Aa1	CryVA(a)	*darmstadiensis*	L07025
cry5Ab1	CryVA(b)	*darmstadiensis*	L07026
cry5Ac1		unspecified	I34543
cry5Ba1		unspecified.	U19725
cry6Aa1	CryVIA	unspecified.	L07022
cry6Ba1	CryVIB	unspecified.	L07024
cry7Aa1	CryIIIC	*galleriae*	M64478
cry7Ab1	CryIIICb	*dakota*	U04367
cry7Ab2		*kumamotoensis*	U04368
cry8Aa1	CryIIIE	*kumamotoensis*	U04364
cry8Ba1	CryIIIG	*kumamotoensis*	U04365
cry8Ca1	CryIIIF	*japonensis (buibui)*	U04366
cry9Ba1	CryIX	*galleriae*	X75019
cry9Aa1	CryIG	*galleriae*	X58120
cry9Aa2	CryIG	DSIR517	X58534

APPENDIX 1 (*Continued*)

Name	Previous Name	Subspecies/variety in original description	Accession Number
cry9Ca1	CryIH	unspecified	Z37527
cry9Da1		*japonensis*	D85560
cry9Da2		*japonensis*	AF042733
cry9Ea1		*aizawai*	AB011496
cry10Aa1	CryIVC	*israelensis*	M12662
cry10Aa2		*israelensis*	E00614
cry11Aa1	CryIVD	*israelensis*	M31737
cry11Aa2		*israelensis*	M22860
cry11Ba1	Jeg80	*jegathesan*	X86902
cry11Bb1		*medellin*	AF017416
cry12Aa1	CryVB	unspecified	L07027
cry13Aa1	CryVC	unspecified	L07023
cry14Aa1	CryVD	*sotto*	U13955
cry15Aa1	34kDa	*thompsoni*	M76442
cry16Aa1	cbm71	*Clostridium bifermentans* subsp. *malaysia*	X94146
cry17Aa1	cbm72	*Clostridium bifermentans* subsp. *malaysia*	X99478
cry18Aa1	CryBP1	*Bacillus popilliae melolontha*	X99049
cry18Ba1		NA	AF169250
cry18Ca1		NA	AF169251
cry19Aa1	Jeg65	*jegathesan*	Y07603
cry19Ba1		*higo*	D88381
cry20Aa1		*fukuokaensis*	U82518
cry21Aa1		(nematode active)	I32932
cry21Aa2		(nematode active)	I66477
cry22Aa1		NA	I34547
cry23Aa1		NA	AF03048
cry24Aa1	Jeg72	*jegathesan*	U88188
cry25Aa1	Jeg74	*jegathesan*	U88189
cry26Aa1		*finitimus*	AF122897
cry27Aa1			AB023293
cry28Aa1		*finitimus*	AF132928
cyt1Aa1	CytA	*israelensis*	X03182
cyt1Aa2		*israelensis*	X04338
cyt1Aa3		*morrisoni*	Y00135
cyt1Aa4		*morrisoni*	M35968
cyt1Ab1	CytM	*medellin*	X98793
cyt1Ba1		*neoleoensis*	U37196
cyt2Aa1	CytB	*kyushuensis*	Z14147
cyt2Ba1	"CytB"	*israelensis*	U52043
cyt2Ba2		*israelensis*	AF020789
cyt2Ba3		*fuokukaensis*	AF022884
cyt2Ba4		*israelensis*	AF022885
cyt2Ba5		*israelensis*	AF022886
cyt2Ba6		*tenebrionis*	AF034926
cyt2Bb1		*jegathesan*	U82519

NA: Genebank number not accessible.

APPENDIX 1 (*Continued*)

The following toxins or toxin-like proteins have not been assigned a name or entered into the nomenclature for the reasons given.

Name	Subspecies or variety of orignal description	Accession	Reference	Reason
Cry1I-like	NA	I90732	Payne *et al.* 1998	Insufficient sequence data
Cry1-like	NA	I90729	Payne *et al.* 1998	Insufficient sequence data
Cry9-like	*galleriae*	AF093107	Wasano & Ohba 1998	Insufficient sequence data
40kDa	*thompsonii*	M76442	Brown & Whiteley 1992	No reported toxicity
cryC35	*cameroun*	X92691	Juarez-Perez *et al.* 1995	No reported toxicity
cryTDK	*mexicanensis*	D86346	Hashimoto 1996	No reported toxicity
cryC53	*cameroun*	X98616	Juarez-Perez *et al.* 1996	No reported toxicity
p21med	*medellin*	X98794	Thiery *et al.* 1997	No reported toxicity
ET34	NA	AF038049	Donovan & Slaney 1998	No reported toxicity
vip3A(a)	unspecified	L48811	Estruch *et al.* 1996	Not a crystal protein
vip3A(b)	unspecified	L48812	Estruch *et al.* 1996	Not a crystal protein

NA: Genebank number not accessible.

APPENDIX 2: SUSCEPTIBILITY OF INVERTEBRATES TO *BT*

Tables include records from previously published lists of susceptible invertebrates compiled by Krieg and Langenbruch (1981); Keller and Langenbruch (1993) and Glare and O'Callaghan (1998).

APPENDIX 2.1 INVERTEBRATE SPECIES REPORTED AS SUSCEPTIBLE TO *BT*

Up to three publications listed for each entry. * - serovar unidentified; L - laboratory; F - field; GH - glasshouse; NS - site not stated;
† Report from the list of Krieg and Langenbruch (1981); full references in that publication.

Subspecies/ Serovar	Order: Family	Genus and Species	Lab/Field	References
aizawai	Diptera: Culicidae	*Aedes aegypti*	L	Hall *et al.* 1977†; Ohba *et al.* 1984
		Aedes triseriatus	L	Hall *et al.* 1977†
		Culex pipiens	L	Hall *et al.* 1977†
		Culex pipiens molestus	L	Saitoh *et al.* 1996
		Culex tarsalis	L	Hall *et al.* 1977†
	Hemiptera: Pentatomidae	*Mucanum* sp.	F	Intari 1996
	Lepidoptera: Arctiidae	*Hyphantria cunea*	L/F	Takaki 1975†; Akiba 1986b
	Lepidoptera: Bombycidae	*Bombyx mori*	L	Angus & Norris 1968†; Galowalia *et al.* 1973†
	Lepidoptera: Gelechiidae	*Pectinophora gossypiella*	L	Salama & Foda 1984; Wahba *et al.* 1996
		Phthorimaea operculella	L	Salama *et al.* 1995a
		Sitotroga cerealella	L	Salama *et al.* 1991b
	Lepidoptera: Geometridae	*Boarmia (Ascotis) selenaria*	L	Cohen *et al.* 1983; Wysoki & Jarvinen 1986
		Operophtera brumata	F	Sierpinska 1997
	Lepidoptera: Gracillariidae	*Caloptilia (Gracillaria) theivora*	L/F	Takaki 1975†; Unnamalai & Vaithilingam 1995
	Lepidoptera: Hesperiidae	*Parnara guttata*	F	Takaki 1975†
	Lepidoptera: Lasiocampidae	*Dendrolimus spectabilis*	L	Katagiri *et al.* 1978; Yao *et al.* 1998
		Malacosoma disstria	L	Pinkham *et al.* 1984
	Lepidoptera: Limacodidae	*Monema flavescens*	F	Takaki 1975†
	Lepidoptera: Lymantriidae	*Euproctis pseudoconspersa*	F	Takaki 1975†
		Lymantria dispar	L	Dubois & Squires 1971†; Dubois *et al.* 1989
		Orgyia antiqua	F	Lipa *et al.* 1977†
		Orgyia leucostigma	L	Rossmoore *et al.* 1970†
		Orgyia thyellina	F	Takaki 1975†
	Lepidoptera: Noctuidae	*Agrotis ipsilon (ypsilon)*	L	Salama 1991; Aboul Ela *et al.* 1993
		Agrotis segetum	L	Jarrett & Burges 1986
		Anticarsia gemmatalis	F	Magrini *et al.* 1997
		Earias biplaga	L	Frutos *et al.* 1987
		Earias insulana	L	Salama & Foda 1984; Wahba *et al.* 1996; Frutos *et al.* 1987
		Helicoverpa (Heliothis) armigera	L	Salama *et al.* 1981a, 1983a, b
		Helicoverpa (Heliothis) punctigera	L	Teakle *et al.* 1992
		Helicoverpa (Heliothis) zea	L	Ameen *et al.* 1998; Rogoff *et al.* 1969†

APPENDIX 2.1 (*continued*)

Subspecies/ Serovar	Order: Family	Genus and Species	Lab/Field	References
aizawai	Lepidoptera: Noctuidae	*Heliothis virescens*	L/F	Dulmage 1973†; Ameen *et al.* 1998; Tamez Guerra *et al.* 1998
		Hellula undalis	F	Anon. 1987b
		Lacanobia (Polia) oleracea	L	Jarrett & Burges 1986
		Mamestra (Barathra) brassicae	L/F	Takaki 1975†; Jarrett & Burges 1986
		Mamestra (Barathra) configurata	L	Trottier *et al.* 1988
		Pseudoplusia (Chrysodeixis) includens	L	Mascarenhas *et al.* 1998
		Spodoptera exempta	L/F	Broza *et al.* 1991b; Bai *et al.* 1992
		Spodoptera exigua	L/F	Salama *et al.* 1983a; Baki *et al.* 1988
		Spodoptera frugiperda	L	Hernandez 1988; Alfonso *et al.* 1994; Castro Franco *et al.* 1998
		Spodoptera littoralis	L	Salama *et al.* 1983a; Salama *et al.* 1984; Kalfon & de Barjac 1985
		Spodoptera litura	L/F	Aizawa 1975; Takaki 1975†; Li & Sheng 1990
		Trichoplusia ni	L/F	Rogoff *et al.* 1969†; Ramos *et al.* 1998; Tamez Guerra *et al.* 1998
	Lepidoptera: Phyllocnistidae	*Phyllocnistis citrella*	L	Shapiro *et al.* 1998
	Lepidoptera: Pieridae	*Pieris brassicae*	L	Galowalia *et al.* 1973†; Lecadet & Martouret 1987
		Pieris rapae	L/F	Takaki 1975†; Nishiitsutsuji Uwo & Endo 1980, 1981
	Lepidoptera: Psychidae	*Canephora asiatica*	F	Takaki 1975†
	Lepidoptera: Pyralidae	*Chilo partellus*	L	Brownbridge & Onyango 1992a, b
		Chilo suppressalis	F	Takaki 1975†
		Ephestia cautella	L	Kinsinger *et al.* 1980; Nishiitsutsuji Uwo & Endo 1980, 1981
		Ephestia kuehniella	L	Takatsuka & Kunimi 1998
		Galleria mellonella	L	Nishiitsutsuji Uwo & Endo 1981; Goodwin 1985
		Ostrinia furnacalis	L	Yao *et al.* 1998
		Ostrinia nubilalis	L/F	Navon & Melamed Madjar 1986; Curto 1996
		Plodia interpunctella	L	Kinsinger *et al.* 1980; Salama *et al.* 1991b
	Lepidoptera: Stathmopodidae	*Kakivoria flavofasciata*	F	Takaki 1975†
	Lepidoptera: Tortricidae	*Adoxophyes orana*	L	Ioriatti *et al.* 1996
		Autographa nigrisigna	F	Takaki 1975†
		Choristoneura fumiferana	L	Morris & Moore 1983
		Lobesia botrana	F	Fougeroux & Lacroze 1996
	Lepidoptera: Yponomeutidae	*Plutella xylostella (maculipennis)*	L/F	Takaki 1975†
		Yponomeuta malinella	F	Takaki 1975†
	Phthiraptera: Menonponidae	*Menopon gallinae*	L	Lonc & Lachowicz 1987
	Phthiraptera: Trichodectidae	*Damalinia (Bovicola) ovis*	L	Drummond *et al.* 1992
aizawai?	Lepidoptera: Carposinidae	*Carposina niponensis*	F	Takaki 1975†
aizawai?	Lepidoptera: Gracillariidae	*Phyllonorycter (Lithocolletis) ringoniella*	F	Takaki 1975†
aizawai?	Lepidoptera: Lymantriidae	*Lymantria dispar*	F	Takaki 1975†
aizawai?	Lepidoptera: Tortricidae	*Adoxophyes orana*	F	Takaki 1975†
alesti	Diptera: Chloropidae	*Hippelates collusor*	L	Hall *et al.* 1977†
	Diptera: Culicidae	*Culex tarsalis*	L	Hall *et al.* 1977†
	Diptera: Muscidae	*Musca domestica*	L	Rogoff *et al.* 1969†
	Hymenoptera: Apidae	*Apis mellifera*	L	Cantwell *et al.* 1966†; Haragsim & Vankova 1968†
	Hymenoptera: Diprionidae	*Neodiprion sertifer*	F	Donaubauer & Schmutzenhofer 1973†

Lepidoptera: Arctiidae	*Arctia caja*	L/F	Martouret 1959†; Burgerjon & Biache 1967b†
	Hyphantria cunea	L	Vankova 1962†; Akiba 1986b
	Spilosoma lubricipeda (as *menthastri*)	L	Burgerjon & Biache 1967b†
	Spilosoma sp.	L	Burgerjon & Grison 1959†
Lepidoptera: Bombycidae	*Bombyx mori*	L	Burgerjon & Biache 1967b†; Angus & Norris 1968†; Akiba 1986b
Lepidoptera: Gelechiidae	*Phthorimaea* (*Gnorimoschema*) *operculella*	L	Toumanoff & Grison 1954†
	Scrobipalpula absoluta	L	Attathom *et al.* 1995
Lepidoptera: Geometridae	*Alsophila aescularia*	L	Burgerjon & Grison 1959†
	Alsophila pometaria	F	Hildahl & Peterson 1974; Frye *et al.* 1976
	Apocheima pilosaria (*pedaria*)	L	Burgerjon & Grison 1959†
	Bupalus piniarius	F	Kochanov *et al.* 1976; Marchenko *et al.* 1983
	Colotois pennaria	L/F	Biliotti 1956†; Balinski *et al.* 1969†
	Erannis defoliaria	L	Burgerjon & Grison 1959†
	Operophtera brumata	L	Biliotti 1956†
Lepidoptera: Lasiocampidae	*Malacosoma disstria*	L	Pinkham *et al.* 1984
	Malacosoma neustria	L	Burgerjon & Biache 1967b†; Balinski *et al.* 1969†
Lepidoptera: Lymantriidae	*Euproctis chrysorrhoea*	L	Burgerjon & Biache 1967a†
	Leucoma wiltshirei	L	Alizadeh 1977†
	Leucoma (*Stilpnotia*) *salicis*	L/F	Burgerjon & Biache 1967b†; Donaubauer & Schmutzenhofer 1973
	Lymantria dispar	L/F	Burgerjon & Biache 1967b†; Dubois & Squires 1971†; Ridet 1973
	Lymantria monacha	L/F	Balinski *et al.* 1969†; Marchenko Ya *et al.* 1982
	Orgyia leucostigma	L	Rossmoore *et al.* 1970†
	Orgyia antiqua	L	Balinski *et al.* 1969†
Lepidoptera: Noctuidae	*Agrotis ipsilon* (*ypsilon*)	L	Burgerjon & Grisson 1959†; Salama 1991
	Brachionycha sphinx	L	Burgerion & Grison 1959†
	Chrysodeixis chalcites (*eriosoma*)	F	Daricheva *et al.* 1983
	Earias insulana	L	Burgerjon & Grison 1959†
	Helicoverpa (*Heliothis*) *armigera*	L	Salama *et al.* 1983a; Attathom *et al.* 1995
	Helicoverpa (*Heliothis*) *zea*	F	Cowan & Davis 1972
	Heliothis peltigera	F	Martouret 1959†
	Heliothis spp.	F	Patti & Carner 1974
	Heliothis virescens	F	Cowan & Davis 1972
	Loxagrotis albicosta	L	Helms & Wedberg 1976
	Mamestra (*Barathra*) *brassicae*	L	Burgerjon & Biache 1967b†
	Orthosia gothica	L	Burgerjon & Biache 1967b†
	Spodoptera exigua	L	Salama *et al.* 1983a; Attathom *et al.* 1995
	Spodoptera frugiperda	L/F	Creighton *et al.* 1972; Hernandez 1988
	Spodoptera littoralis	L	Salama *et al.* 1983a
	Trichoplusia ni	F	Creighton *et al.* 1972; Cowan & Davis 1972
	Xestia (*Amathes*) *c-nigrum*	L	Burgerjon & Biache 1967b†
Lepidoptera: Notodontidae	*Clostera* (*Pygaera*) *anastomosis*	L	Burgerjon & Biache 1967b†
	Drymonia sp.	L	Balinski *et al.* 1969†
	Heterocampa manteo	L	Ignoffo *et al.* 1973
	Ptilophora plumigera	F	Kulikovskii Yu 1984
Lepidoptera: Nymphalidae	*Vanessa io*	L	Burgerjon & Biache 1967b†
Lepidoptera: Phycitidae	*Homeosoma vagella*	F	Ironside & Giles 1981
Lepidoptera: Pieridae	*Pieris brassicae*	L/F	Lemoigne *et al.* 1956†; Burgerjon & Biache 1967b†

APPENDIX 2.1 *(continued)*

Subspecies/ Serovar	Order: Family	Genus and Species	Lab/Field	References
alesti	Lepidoptera: Pieridae	*Pieris rapae*	F	Creighton *et al.* 1972; Creighton & McFadden 1974
	Lepidoptera: Psychidae	*Oiketicus moyanoi*	L/F	Solanes *et al.* 1980
		Thyridopteryx ephemeraeformis	F	Kearby *et al.* 1972
	Lepidoptera: Pyralidae	*Chilo suppressalis*	L	Attathom *et al.* 1995
		Ephestia kuehniella	L	Burgerjon & Grison 1959†; Vankova 1962†
		Galleria mellonella	L	Vankova 1966†; Jarrett & Burges 1982b
		Loxostege stricticalis	F	Mitrofanov 1982
		Ostrinia nubilalis	F	Martouret 1959†
	Lepidoptera: Saturnidae	*Antheraea pernyi*	L	Burgerjon & Biache 1967b†
		Anisota senatoria	L/F	Frye *et al.* 1973; Ignoffo *et al.* 1974; Kaya 1974
		Hylesia nigricans	F	Anon. 1972
		Samia cynthia	L	Burgerjon & Biache 1967b†
		Saturnia pavonia	L	Burgerjon & Biache 1967b†
	Lepidoptera: Sphingidae	*Manduca sexta*	L	Tyrell *et al.* 1981b
	Lepidoptera: Thaumetopoeidae	*Thaumetopoea pityocampa*	L/F	Grison & Beguin 1954†
		Thaumetopoea processionea	L	Burgerjon & Biache 1967b†
	Lepidoptera: Tortricidae	*Agriopis (Erannis) aurantiaria*	L	Ceianu *et al.* 1970†
		Agriopis (Erannis) marginaria	L	Burgerjon & Grison 1959†
		Choristoneura fumiferana	L	Smirnoff 1965†; Yamvrias & Angus 1970†; Morris & Moore 1983
		Tortrix viridana	L	Burgerjon & Klinger 1959†; Burgerjon & Biache 1967b†
	Lepidoptera: Yponomeutidae	*Acrolepiopsis (Acrolepia) assectella*	L/F	Burgerjon & Grison 1959†; Goix 1959†
		Plutella xylostella (maculipennis)	L/F	Burgerjon & Biache 1967b†; Creighton *et al.* 1972
		Yponomeuta cognatella (evonymi)	L	Burgerjon & Biache 1967b†
		Yponomeuta mahalebella	L	Ceianu *et al.* 1970†
		Yponomeuta malinella	L/F	Kuchly 1959†; Burgerjon & Biache 1967b†
		Yponomeuta padella	L	Toumanoff 1955†
	Orthoptera: Phasmatidae	*Diapheromera femorata*	L	Ignoffo *et al.* 1973†
	Phthiraptera: Trichodectidae	*Damalinia (Bovicola) ovis*	L	Drummond *et al.* 1992
alesti?	Lepidoptera: Geometridae	*Paleacrita vernata*	F	Hildahl & Peterson 1974
alesti?	Lepidoptera: Noctuidae	*Spodoptera eridania*	F	Creighton *et al.* 1971†
amagiensis	Diptera: Culicidae	*Aedes aegypti*	L	Ishii & Ohba 1993a
		Culex pipiens molestus	L	Saitoh *et al.* 1996
		Culex pipiens pallens	L	Ishii & Ohba 1993a
	Isoptera: Hodotermitidae	*Anacanthotermes ahngerianus*	L	Lyutikova & Yudina 1995
	Lepidoptera: Yponomeutidae	*Plutella xylostella*	L	Iriarte *et al.* 1998
brasiliensis	Diptera: Culicidae	*Aedes aegypti*	L	Rabinovitch *et al.* 1995
		Anopheles stephensi	L	Rabinovitch *et al.* 1995
canadensis	Diptera: Culicidae	*Aedes aegypti*	L	Ragni *et al.* 1996; Ishii & Ohba 1997
		Anopheles stephensi	L	Ragni *et al.* 1996
		Culex pipiens	L	Ragni *et al.* 1996

		Helicoverpa (Heliothis) armigera	L	Teakle *et al.* 1992
		Spodoptera littoralis	L	Salama *et al.* 1991b
	Phthiraptera: Trichodectidae	*Damalinia (Bovicola) ovis*	L	Drummond *et al.* 1992
caucasicus	Lepidoptera: Arctiidae	*Hyphantria cunea*	F	Kondrya *et al.* 1980
	Lepidoptera: Geometridae	*Operophtera brumata*	F	Rituma 1985
	Lepidoptera: Lasiocampidae	*Dendrolimus sibiricus*	L	Gukasyan & Gukasyan 1980
	Lepidoptera: Lymantriidae	*Lymantria dispar*	L/F	Sinitsyna *et al.* 1980; Kondrya *et al.* 1980
	Lepidoptera: Noctuidae	*Mamestra (Barathra) brassicae*	F	Stroeva *et al.* 1980
	Lepidoptera: Pieridae	*Pieris brassicae*	L/F	Tsinitis 1980
		Pieris rapae	L/F	Stroeva *et al.* 1980; Tsinitis 1980
	Lepidoptera: Tortricidae	*Archips (Cacoecia) rosanus*	F	Stroeva *et al.* 1980
		Tortrix viridana	F	Gamayunova & Timchenko 1985
	Lepidoptera: Yponomeutidae	*Yponomeuta (padellus) malinellus*	L/F	Novitskaya & Dzholiya 1980; Tsinitis 1980; Kondrya *et al.* 1980
	Siphonaptera: Pulicidae	*Xenopsylla cheopis*	L	Savel' ev & Kozlov 1974
colmeri	Lepidoptera: Noctuidae	*Spodoptera frugiperda*	L	Hernandez 1988
coreanensis	Diptera: Culicidae	*Culex pipiens*	L	Lee *et al.* 1994
darmstadiensis	Coleoptera: Bostrichidae	*Rhyzopertha dominica*	L	Beegle 1996
	Coleoptera: Chrysomelidae	*Diabrotica undecimpunctata*	L	Abdel-Hameed & Landen 1994
	Diptera: Agromyzidae	*Liriomyza trifolii*	L	patent No. 05298245 & 05436002, J.M. Payne 1994, 1995
	Diptera: Chironomidae	*Paratanytarsus grimmii*	L	Kondo *et al.* 1995a
	Diptera: Culicidae	*Aedes aegypti*	L	Padua *et al.* 1980; Finney & Harding 1982; Lacey & Oldacre 1983
		Anopheles albimanus	L	Lacey & Oldacre 1983
		Anopheles quadrimaculatus	L	Lacey & Oldacre 1983
		Culex molestus	L	Padua et al. 1980
		Culex pipiens	L	Hall *et al.* 1977†
		Culex pipiens molestus	L	Saitoh *et al.* 1996
		Culex quinquefasciatus	L	Hall *et al.* 1977†; Lacey & Oldacre 1983
		Culex tritaeniorhynchus	L	Padua *et al.* 1980
	Diptera: Muscidae	*Musca domestica*	L	McGaughey & Johnson 1987; Lonc *et al.* 1991
	Diptera: Psychodidae	*Telmatoscopus albipunctatus* (*Clogmia albipunctata*)	L	Saitoh *et al.* 1996
	Diptera: Simuliidae	*Simulium verecundum*	L	Finney & Harding 1982
	Diptera: Tephritidae	*Anastrepha ludens*	L	Martinez *et al.* 1997
	Lepidoptera: Bombycidae	*Bombyx mori*	L	Padua *et al.* 1981
	Lepidoptera: Lymantriidae	*Lymantria dispar*	L	Dubois *et al.* 1989
	Lepidoptera: Noctuidae	*Heliothis virescens*	L	Dulmage 1973†
		Spodoptera frugiperda	L	Dulmage 1973†
	Lepidoptera: Pieridae	*Pieris rapae*	L	Yao *et al.* 1998
	Lepidoptera: Pyralidae	*Chilo suppressalis*	F	Goarant *et al.* 1995
		Galleria mellonella	L	Krieg unpubl. obs.†
		Ostrinia furnacalis	L	Yao *et al.* 1998
		Ostrinia nubilalis	L	Mohd Salleh & Lewis 1983
	Lepidoptera: Tortricidae	*Choristoneura fumiferana*	L	Morris & Moore 1983

APPENDIX 2.1 *(continued)*

Subspecies/ Serovar	Order: Family	Genus and Species	Lab/Field	References
darmstadiensis	Lepidoptera: Tortricidae	*Cydia (Laspeyresia) pomonella*	L	Andermatt *et al.* 1988
	Phthiraptera: Trichodectidae	*Damalinia (Bovicola) ovis*	L	Drummond *et al.* 1992
dendrolimus	Acari: Argasidae	*Argas persicus*	F	Frolov *et al.* 1979
	Acari: Dermanyssidae	*Dermanyssus gallinae*	F	Lavrenyuk *et al.* 1977; Frolov 1977
	Coleoptera: Chrysomelidae	*Diabrotica undecimpunctata*	L	Abdel-Hameed & Landen 1994
		Leptinotarsa decemlineata	L	Bartninkaite & Babonas 1985
		Phaedon cochleariae	L	Kamenek 1988
		Pyrrhalta (Xanthogaleruca) luteola	F	Kalyuzhnaya *et al.* 1995
	Diptera: Chironomidae	*Chironomus plumosus*	L	Lavrentyev *et al.* 1965†
	Diptera: Culicidae	*Aedes* sp.	L	Lavrentyev *et al.* 1965†
		Aedes caspius	L/F	Kadyrova *et al.* 1977
		Anopheles sp.	L	Lavrentyev *et al.* 1965†
		Culex sp.	L	Lavrentyev *et al.* 1965†
		Culex modestus	L/F	Kadyrova *et al.* 1977
		Culex pipiens	L/F	Kadyrova *et al.* 1977
	Diptera: Oestridae	*Hypoderma bovis*	L	Moroz 1979
	Diptera: Tabanidae	*Tabanus autumnalis brunnescens*	L/F	Kadyrova *et al.* 1977
	Hemiptera: Cimicidae	*Cimex lectularis*	F	Frolov *et al.* 1979
	Hemiptera: Miridae	*Adelphocoris lineolatus*	F	Yunusov 1974
	Hymenoptera: Apidae	*Apis mellifera*	L	Haragsim & Vankova 1968†
	Hymenoptera: Diprionidae	*Neodiprion sertifer*	F	Safronov 1996
	Hymenoptera: Formicidae	*Monomorium pharaonis*	L	Brikman *et al.* 1967†
	Hymenoptera: Tenthredinidae	*Macrophya punctumalbum*	F	Kulikovskii Yu 1984
		Nematus (Pteronidea) ribesii	F	Zukauskiene & Petrauskas 1973
	Hymenoptera: Trichogrammatidae	*Trichogramma* sp.	L	Tolstova Yu & Ionova 1976
	Lepidoptera: Arctiidae	*Arctia caja*	L	Burgerjon & Biache 1967b†
		Hyphantria cunea	L/F	Vankova 1962†; Sikura & Simchuk 1981
		Spilosoma lubricipeda (as *menthastri*)	L	Burgerjon & Biache 1967b†
	Lepidoptera: Bombycidae	*Bombyx mori*	L/F	Angus & Norris 1968†; Ozino Marletto *et al.* 1972
	Lepidoptera: Gelechiidae	*Gelechia hippophaella*	F	Sanzhimtupova 1984
		Pectinophora gossypiella	L	Salama & Foda 1984
		Phthorimaea operculella	L	Salama *et al.* 1995a
		Sitotroga cerealella	L	Salama *et al.* 1991b
	Lepidoptera: Geometridae	*Bupalus piniarius*	F	Fadeev 1974; Kochanov *et al.* 1976
		Calospilos (Abraxas) pantaria	F	Kosenko & Anferov 1996
		Chloroclystis rectangulata	F	Bolotnikova 1984
		Erannis defoliaria	F	Dariichuk 1981; Aukshtikal'nene & Imnadze 1981; Bolotnikova 1984
		Lycia (Biston) hirtaria	F	Kholchenkov & Galetenko 1979
		Operophtera brumata	F	Znamenskii & Kupriyanova 1973
	Lepidoptera: Lasiocampidae	*Dendrolimus pini*	F	Gorokhov & Kaplenko 1980; Safronov 1996
		Dendrolimus punctatus	F	Li *et al.* 1984; Yang *et al.* 1984
		Dendrolimus sibiricus	L/F	Talalaev 1957†, 1959†; Gukasyan 1962†

	Malacosoma neustria	L/F	Burgerjon & Biache 1967b†
Lepidoptera: Lymantriidae	*Euproctis chrysorrhoea*	L/F	Lappa 1964†; Znamenskii & Kupriyanova 1973
	Leucoma (Stilpnotia) salicis	L/F	Burgerjon & Biache 1967b†; Tsilosani *et al.* 1976
	Leucoma wiltshirei	L	Alizadeh 1977†
	Lymantria dispar	L	Burgerjon & Biache 1967b†; Dubois & Squires 1971†; Ridet 1973
	Lymantria monacha	L/F	Szmidt & Slizynski 1965†; Marchenko Ya *et al.* 1982
	Orgyia leucostigma	L	Rossmoore *et al.* 1970†
	Orgyia prisca	F	Akhmedov 1982
	Selenephera lunigera	L/F	Firstov 1965†
Lepidoptera: Noctuidae	*Agrotis ipsilon (ypsilon)*	L	Salama & Foda 1984; Aboul Ela *et al.* 1993
	Agrotis segetum	F	Sikoura & Tkatsch 1974†; Kashkarova 1975
	Apamea anceps	F	Dusenko 1986
	Helicoverpa (Heliothis) armigera	L/F	Nasrullaev 1974; Sharafutdinov & Salikhov 1975;
	Helicoverpa (Heliothis) zea	L	Rogoff *et al.* 1969†
	Heliothis virescens	F	Jimenez *et al.* 1983
	Heliothis (Chloridea) viriplaca	L	Charafoutdinof 1970†; Sikoura & Sikura 1975
	Mamestra (Barathra) brassicae	L/F	Burgerjon & Biache 1967b†; Yastrebov 1978; Korol 1986
	Mythimna (Pseudaletia) unipuncta	L	Burgerjon & Biache 1967b†
	Orthosia gothica	L	Burgerjon & Biache 1967b†
	Panolis flammea	F	Fedoryak 1985
	Spodoptera exigua	L/F	Nasrullaev 1974; Sikoura & Sikura 1975; Agzamova Kh *et al.* 1988
	Spodoptera littoralis	L	Merdan *et al.* 1975†; Salama 1991
	Syngrapha (Cornutiplusia) circumflexa	F	Kadamshoev 1996
	Xestia (Amathes) c-nigrum	L	Burgerjon & Biache 1967b†
Lepidoptera: Notodontidae	*Clostera (Pygaera) anastomosis*	L	Burgerjon & Biache 1967b†
	Ptilophora plumigera	F	Kulikovskii Yu 1984
Lepidoptera: Nymphalidae	*Vanessa io*	L	Burgerjon & Biache 1967b†
	Vanessa turkestanica	F	Guzeev 1986
Lepidoptera: Pieridae	*Aporia crataegi*	L/F	Korchagin 1983; Grigoryan *et al.* 1988
	Pieris brassicae	F	Burgerjon & Biache 1967b†; Yastrebov 1978; Rituma 1987
	Pieris rapae	L/F	Merdan *et al.* 1975†; Goral *et al.* 1984
	Pieris spp.	F	Dzhivaladze 1979; Goral *et al.* 1984
Lepidoptera: Pyralidae	*Ephestia kuehniella*	L	Vankova 1962†
	Galleria mellonella	L	Vankova 1966†; Jarrett & Burges 1982b
	Loxostege stricticalis	F	Shek & Telepa 1981; Kornilov *et al.* 1982
	Noorda blitealis	F	Kalia & Joshi 1997
	Plodia interpunctella	L	Salama *et al.* 1991b
	Zophodia convolutella	L	Sem'yanov *et al.* 1983
Lepidoptera: Pyralidae?	*Mesographe forficalis*	F	Issi 1978†
Lepidoptera: Saturnidae	*Antheraea pernyi*	L	Burgerjon & Biache 1967b†
	Neoris huttoni	F	Guzeev & Mansurov 1983
Lepidoptera: Thaumetopoeidae	*Thaumetopoea processionea*	L	Burgerjon & Biache 1967b†
Lepidoptera: Tortricidae	*Archips* spp.	F	Gamayunova 1987
	Archips crataegana	L/F	Kudler *et al.* 1965†
	Archips (Cacoecia) rosana	L	Zukauskiene 1973†; Zukauskiene & Petrauskas 1973
	Choristoneura fumiferana	L	Smirnoff 1965†; Yamvrias & Angus 1970†; Morris & Moore 1983
	Cydia (Laspeyresia) pomonella	F	Madatyan 1985; Korol 1986

APPENDIX 2.1 (*continued*)

Subspecies/ Serovar	Order: Family	Genus and Species	Lab/Field	References
dendrolimus	Lepidoptera: Tortricidae	*Lobesia botrana*	F	Dzhivaladze 1979; Madatyan & Sharipov 1984
		Rhyacionia frustrana	F	Menendez *et al.* 1986
		Tortrix viridana	L/F	Burgerjon & Biache 1967b†; Znamenskii & Kupriyanova 1973
		Zeiraphera diniana	L	Kudler 1984
	Lepidoptera: Yponomeutidae	*Atteva fabriciella*	L	Joshi *et al.* 1996
		Plutella xylostella (*maculipennis*)	F	Burgerjon & Biache 1967b†
		Yponomeuta cognatella (*evonymi*)	L	Burgerjon & Biache 1967b†
		Yponomeuta malinella	L/F	Rybina 1966†; Burgerjon & Biache 1967b†
	Phthiraptera:	Mallophaga spp.	F	Frolov 1974
	Phthiraptera: Trichodectidae	*Damalinia* (*Bovicola*) *ovis*	L	Drummond *et al.* 1992
	Siphonaptera: Pulicidae	*Xenopsylla cheopis*	L	Ershova *et al.* 1982
entomocidus	Coleoptera: Coccinellidae	*Coccinella undecimpunctata*	L	Salama *et al.* 1982
	Coleoptera: Scolytidae	*Scolytus multistriatus*	L	Jassim *et al.* 1990
		Scolytus scolytus	L	Jassim *et al.* 1990
	Diptera: Chloropidae	*Hippelates collusor*	L	Hall *et al.* 1977†
	Diptera: Culicidae	*Aedes aegypti*	L	Hall *et al.* 1977†; Lopez Meza *et al.* 1995
		Aedes triseriatus	L	Hall *et al.* 1977†
		Culex pipiens	L	Hall *et al.* 1977†
		Culex pipiens pipiens	L	Larget & de Barjac 1981a
		Culex quinquefasciatus	L	Lopez Meza *et al.* 1995
		Culex tarsalis	L	Hall *et al.* 1977†
	Diptera: Simuliidae	*Simulium vittatum*	L	Lacey & Mulla 1977†
	Hymenoptera: Apidae	*Apis mellifera*	L	Haragsim & Vankova 1968†
	Hymenoptera: Braconidae	*Microgaster* (*Microplitis*) *demolitor*	L	Salama *et al.* 1982
		Zele chlorophthalma	L	Salama & Zaki 1983
	Lepidoptera: Arctiidae	*Arctia caja*	L	Burgerjon & Biache 1967b†
		Datana ministra	F	Angus & Heimpel 1959†
		Hyphantria cunea	L	Angus & Heimpel 1959†; Vankova 1962†
		Spilosoma lubricipeda (*menthastri*)	L	Burgerjon & Biache 1967b†
	Lepidoptera: Bombycidae	*Bombyx mori*	L	Burgerjon & Biache 1967b†; Angus & Norris 1968†
	Lepidoptera: Cosmopterigidae	*Batrachedra amydraula*	F	Venezian & Blumberg 1982
	Lepidoptera: Gelechiidae	*Pectinophora gossypiella*	L	Salama & Foda 1984
		Phthorimaea operculella	L	Salama *et al.* 1995a
		Sitotroga cerealella	L	Salama *et al.* 1991b
	Lepidoptera: Geometridae	*Boarmia* (*Ascotis*) *selenaria*	L	Cohen *et al.* 1983; Wysoki & Jarvinen 1986
	Lepidoptera: Lasiocampidae	*Malacosoma neustria*	L	Burgerjon & Biache 1967b†
	Lepidoptera: Lymantriidae	*Euproctis chrysorrhoea*	L	Burgerjon & Biache 1967a†
		Leucoma (*Stilpnotia*) *salicis*	L	Burgerjon & Biache 1967b†
		Lymantria dispar	L	Burgerjon & Biache 1967b†; Dubois & Squires 1971†; Ridet 1973
		Orgyia leucostigma	L	Rossmoore *et al.* 1970†
	Lepidoptera: Noctuidae	*Agrotis ipsilon* (*ypsilon*)	L	Salama & Foda 1984; Salama *et al.* 1991b
		Agrotis (*Euxoa*) *segetum*	L	Jarrett & Burges 1986; Feng *et al.* 1995

		Helicoverpa (Heliothis) zea	L	Rogoff *et al.* 1969†
		Heliothis virescens	L	Visser *et al.* 1988
		Lacanobia (Polia) oleracea	L	Jarrett & Burges 1986
		Mamestra (Barathra) brassicae	L	Burgerjon & Biache 1967b†; Jarrett & Burges 1986
		Mamestra (Barathra) configurata	L	Trottier *et al.* 1988
		Mythimna (Pseudaletia) unipuncta	L	Burgerjon & Biache 1967b†
		Orthosia gothica	L	Burgerjon & Biache 1967b†
		Spodoptera exempta	F	Broza *et al.* 1991b
		Spodoptera exigua	L	Salama *et al.* 1981a, 1983a; Sneh *et al.* 1983
		Spodoptera littoralis	L/F	Sneh *et al.* 1981, 1983; Sneh & Gross 1983
		Spodoptera praefica	L	Steinhaus 1951a†
		Trichoplusia ni	L	Rogoff *et al.* 1969†
		Xestia (Amathes) c-nigrum	L	Burgerjon & Biache 1967b†
	Lepidoptera: Notodontidae	*Clostera (Pygaera) anastomosis*	L	Burgerjon & Biache 1967b†
	Lepidoptera: Nymphalidae	*Junonia coenia*	L	Steinhaus 1951a†
		Vanessa io	L	Burgerjon & Biache 1967b†
	Lepidoptera: Pieridae	*Pieris brassicae*	L	Burgerjon & Biache 1967b†; Lecadet & Martouret 1987
		Pieris rapae	F	Jaques & Fox 1960†
	Lepidoptera: Pyralidae	*Chilo partellus*	L	Brownbridge & Onyango 1992b
		Ephestia kuehniella	L	Vankova 1962†
		Galleria mellonella	L	Vankova 1966†; Burgerjon & Biache 1967b†
		Hypsipyla grandella	L	Hidalgo Salvatierra & Palm 1972
		Mythimna separata	L	Feng *et al.* 1995
		Ostrinia furnacalis	L	Feng *et al.* 1995
		Plodia interpunctella	L	Salama *et al.* 1991b; Johnson *et al.* 1998
	Lepidoptera: Saturnidae	*Anisota senatoria*	L	Angus 1956a†; Angus & Heimpel 1959†
		Antheraea pernyi	L	Burgerjon & Biache 1967b†
		Samia cynthia	L	Burgerjon & Biache 1967b†
		Saturnia pavonia	L	Burgerjon & Biache 1967b†
	Lepidoptera: Thaumetopoeidae	*Thaumetopoea processionea*	L	Burgerjon & Biache 1967b†
	Lepidoptera: Tortricidae	*Choristoneura fumiferana*	L	Smirnoff 1965†; Yamvrias & Angus 1970†; Morris & Moore 1983
		Cydia (Laspeyresia) pomonella	L	Andermatt *et al.* 1988
		Tortrix viridana	L	Burgerjon & Biache 1967b†
	Lepidoptera: Yponomeutidae	*Plutella xylostella (maculipennis)*	L	Burgerjon & Biache 1967b†; Kim *et al.* 1998b
		Yponomeuta cognatella (evonymi)	L	Burgerjon & Biache 1967b†
		Yponomeuta malinella	L	Burgerjon & Biache 1967b†
		Yponomeuta padella	L	Burgerjon & Biache 1967b†
	Neuroptera: Chrysopidae	*Chrysoperla (Chrysopa) carnea*	L	Salama *et al.* 1982
	Phthiraptera: Trichodectidae	*Damalinia (Bovicola) ovis*	L	Drummond *et al.* 1992
finitimus	Diptera: Chloropidae	*Hippelates collusor*	L	Hall *et al.* 1977†
	Lepidoptera: Arctiidae	*Hyphantria cunea*	L	Lee & Kang 1989
	Lepidoptera: Bombycidae	*Bombyx mori*	L/F	Lee & Kang 1989
	Lepidoptera: Gelechiidae	*Pectinophora gossypiella*	L/F	Abul Nasr *et al.* 1979, 1983; Salama & Foda 1984
		Phthorimaea operculella	L	Puntambekar *et al.* 1997
	Lepidoptera: Lymantriidae	*Lymantria dispar*	L	Dubois *et al.* 1989
	Lepidoptera: Noctuidae	*Agrotis ipsilon (ypsilon)*	L	Salama & Foda 1984

APPENDIX 2.1 *(continued)*

Subspecies/ Serovar	Order: Family	Genus and Species	Lab/Field	References
finitimus	Lepidoptera: Noctuidae	*Earias insulana*	F	Abul Nasr *et al.* 1983
		Spodoptera litura	L	Puntambekar *et al.* 1997
	Lepidoptera: Pyralidae	*Galleria mellonella*	L	Krieg 1982
	Phthiraptera: Menonponidae	*Menopon gallinae*	L	Lonc & Lachowicz 1987; Lonc *et al.* 1983
finitimus (?)	Lepidoptera: Geometridae	*Buzura suppressaria* (*benescripta*)	L	Wu & Tang 1981
fukuokaensis	Diptera: Culicidae	*Aedes aegypti*	L	Ohba & Aizawa 1990
		Aedes albopictus	L	Ohba & Aizawa 1990
		Anopheles stephensi	L	Saitoh *et al.* 1998
		Culex pipiens molestus	L	Saitoh *et al.* 1996
		Culex tritaeniorhynchus	L	Ohba & Aizawa 1990
	Diptera: Psychodidae	*Telmatoscopus albipunctatus*	L	Saitoh *et al.* 1996
	Diptera: Tephritidae	*Dacus* (*Bactrocera*) *oleae*	L	Dimitriadis & Domouhtsidou 1996
	Lepidoptera: Bombycidae	*Bombyx mori*	L	Ohba & Aizawa 1989b; Hastowo *et al.* 1992
	Lepidoptera: Notodontidae	*Clostera* (*Pygaera*) *anastomosis*	L	Ohba & Aizawa 1989b
galleriae	Acari: Argasidae	*Argas persicus*	F	Frolov *et al.* 1979; Romasheva *et al.* 1978
	Acari: Dermanyssidae	*Dermanyssus gallinae*	F	Lavrenyuk *et al.* 1977a; Frolov 1977
	Acari: Phytoseiidae	*Amblyseius* (*Neoseiulus*) *fallacis*	GH	Dong & Niu 1991
		Phytoseiulus persimilis	GH	Dong & Niu 1991
	Acari: Tenthredinidae	*Hoplocampa testudinea*	F	Prieditis & Rituma 1974
	Acari: Tetranychidae	*Panonychus ulmi*	F	G'onev 1975
	Coleoptera: Chrysomelidae	*Leptinotarsa decemlineata*	L/F	Kiselek & Zayats 1976; Bartninkaite & Babonas 1985
		Phyllotreta spp.	F	Sem'yanov & Tsybul'ko 1983
	Coleoptera: Curculionidae	*Tychius flavus*	F	Naidenov 1977
	Coleoptera: Scarabaeidae	*Popillia japonica*	L	Sharpe 1976†
	Diptera: Anthomyiidae	*Delia radicum* (*brassicae*)	F	Sem'yanov & Tsybul'ko 1983
	Diptera: Calliphoridae	*Lucilia cuprina*	F	Cooper *et al.* 1985
	Diptera: Chironomidae	*Chironomus plumosus*	L	Lavrentyev *et al.* 1965†
	Diptera: Culicidae	*Aedes aegypti*	L	Saubenova 1973; Hall *et al.* 1977†; Ignoffo *et al.* 1980a
		Aedes caspius	L/F	Saubenova 1973; Kadyrova *et al.* 1977
		Aedes flavescens	L	Saubenova 1973
		Aedes triseriatus	L	Hall *et al.* 1977†
		Aedes sp.	L	Lavrentyev *et al.* 1965†
		Anopheles sp.	L	Lavrentyev *et al.* 1965†
		Culex alaskaensis	L	Saubenova 1973
		Culex modestus (or *molestus*?)	L/F	Kadyrova *et al.* 1977
		Culex molestus	L	Saubenova *et al.* 1973; Saubenova 1973; Sokolova & Ganushkina 1982
		Culex pipiens	L	Saubenova *et al.* 1973; Saubenova 1973; Hall *et al.* 1977†
		Culex tarsalis	L	Hall *et al.* 1977†
		Culex sp.	L	Lavrentyev *et al.* 1965†
		Culicoides sp.	L	Saubenova 1973

Diptera: Oestridae	*Hypoderma bovis*	L	Moroz 1979
	Oestrus ovis	F	Sartbaev 1979
Diptera: Tephritidae	*Rhagoletis batava obscuriosa*	L	Andrashchuk 1981
Hemiptera: Aphididae	*Acyrthosiphon pisum*	F	Grechkanev & Maksimova 1980
	Brevicoryne brassicae	F	Sem'yanov & Tsybul'ko 1983
Hemiptera: Cimicidae	*Cimex lectularis*	F	Frolov *et al.* 1979
Hemiptera: Margarodidae	*Matsucoccus matsumurae*	F	Cheng *et al.* 1983
Hemiptera: Miridae	*Adelphocoris lineolatus*	F	Yunusov 1974
Hymenoptera: Diprionidae	*Neodiprion sertifer*	F	Kulikovskii Yu 1984
Hymenoptera: Formicidae	*Monomorium pharaonis*	L	Brikman *et al.* 1967†; Berndt *et al.* 1974
Hymenoptera: Tenthredinidae	*Athalia rosae* (*colibri*)	L	Charpentier 1971†; Johansson 1971
	Cimbex femoratus	F	Sokolov 1985
	Croesus septentrionalis	F	Sokolov 1985
	Nematus (*Pteronidea*) *ribesii*	L/F	Fedorinchik 1963†; Khrameeva 1972; Zukauskiene & Petrauskas 1973
Isoptera: Hodotermitidae	*Anacanthotermes ahngerianus*	L/F	Kakaliev & Saparliev 1975
Lepidoptera: ?	*Diaphora mendica*	L	Isakova 1958†
Lepidoptera: Arctiidae	*Arctia caja*	L	Burgerjon & Biache 1967b†
	Halisidota argentata	L	Morris 1969†
	Hyphantria cunea	L	Vankova 1962†; Videnova 1970†; Takaki 1975†
	Spilosoma lubricipeda (as *menthastri*)	L	Burgerjon & Biache 1967b†
Lepidoptera: Blastobasidae	*Holcocera pulverea*	L/F	Malhotra & Choudhary 1968†
Lepidoptera: Bombycidae	*Bombyx mori*	L	Burgerjon & Biache 1967b†; Angus & Norris 1968†; Okada *et al.* 1977
Lepidoptera: Choreutidae	*Choreutis* (*Simaethis*) *pariana*	F	Lopatina 1981
Lepidoptera: Cossidae	*Zeuzera pyrina*	F	Sikura & Simchuk 1970
Lepidoptera: Dioptidae	*Phryganidia californica*	F	Pinnock & Milstead 1972†
Lepidoptera: Gelechiidae	*Coleotechnites nanella*	F	Rafal'skii 1974†
	Phthorimaea operculella	L	Salama *et al.* 1995a
	Scrobipalpula absoluta	L	Attathom *et al.* 1995
	Sitotroga cerealella	L	Salama *et al.* 1991b
Lepidoptera: Geometridae	*Alsophila pometaria*	L	Larson & Ignoffo 1971†
	Apocheima (*Biston*) *hispidaria*	L	Fedorinchik 1969†
	Bupalus piniarius	L/F	Fankhanel 1962†; Skatulla 1971†; Fadeev 1974
	Chloroclystis rectangulata	F	Bolotnikova 1984
	Colotois pennaria	L	Balinski *et al.* 1969†
	Erannis defoliaria	L/F	Isakova 1958†; Tsilosani *et al.* 1976; Bolotnikova 1984
	Itame (*Thamnonoma*) *wauaria*	F	Isakova 1958†
	Lambdina fiscellaria lugubrosa	F	Harper 1974†
	Lycia (*Biston*) *hirtaria*	L/F	Dirimanov & Lecheva 1980; Kuzmanova & Lecheva 1981
	Melanolophia imitata	F	Morris 1969†
	Operophtera brumata	L/F	Isakova 1958†; Ceianu *et al.* 1970†; Niemczyk *et al.* 1973
Lepidoptera: Glyphipterigyidae	*Eutromula* (*Simaethis*) *pariana*	L/F	Niemczyk & Bakowski 1970; Niemczyk & Bakowski 1971†
Lepidoptera: Gracillariidae	*Phyllonorycter* (*Lithocolletis*) *blancardella*	F	Jaques 1965†
	Phyllonorycter (*Lithocolletis*) *pyrifoliella*	F	Boldyrev 1970
	Caloptilia (*Gracillaria*) *invariabilis*	L	Morris 1969†
	Caloptilia (*Gracillaria*) *syringella*	L	Fedorinchik 1969†

APPENDIX 2.1 *(continued)*

Subspecies/ Serovar	Order: Family	Genus and Species	Lab/Field	References
galleriae	Lepidoptera: Hepialidae	*Maculella* spp.	F	Carrillo & Mundaca 1975
	Lepidoptera: Hesperiidae	*Parnara guttata*	F	Takaki 1975†
		Thymelicus lineola	F	Arthur 1968†
	Lepidoptera: Lasiocampidae	*Dendrolimus pini*	L/F	Fedorinchik 1969†; Krushev *et al.* 1972
		Dendrolimus punctatus	F	Anon. 1985a; Anon. 1987a
		Dendrolimus sibiricus	L	Sem'yanov *et al.* 1983
		Eriogaster henkei	F	Alimdzhanov & Anufrieva 1966†
		Malacosoma fragilis (fragile)	F	Stelzer 1965†
		Malacosoma neustria	L/F	Shvecova 1959†; Burgerjon & Biache 1967b†; Nguyen 1971
		Malacosoma parallela	F	Aukshtikal'nene *et al.* 1984
		Malacosoma pluviale	L/F	Morris 1969†
	Lepidoptera: Limacodidae	*Monema flavescens*	F	Takaki 1975†
	Lepidoptera: Lymantriidae	*Euproctis chrysorrhoea*	L/F	Kondrja 1966b†; Burgerjon & Biache 1967a†; Erfurth & Motte 1971
		Euproctis similis	F	Korchagin 1975; Stus 1980
		Hypogymna morio	L	Dobrivojevic *et al.* 1969†
		Leucoma (Stilpnotia) salicis	L/F	Burgerjon & Biache 1967b†; Morris 1969†; Szalay Marzso *et al.* 1981
		Lymantria dispar	L/F	Doane & Hitchcock 1964†; Ceianu *et al.* 1970†; Dubois & Squires 1971†
		Lymantria monacha	L	Balinski *et al.* 1969†; Johansson 1971†; Krushev & Marchenko Ya 1981
		Orgyia antiqua	L/F	Balinski *et al.* 1969†; Niemczyk 1971†; Lipa & Bakowski 1979
		Orgyia gonostigma	F	Sevryukova 1979; Ivanova 1984
		Orgyia thyellina	F	Takaki 1975†
	Lepidoptera: Lyonetiidae	*Bucculatrix thurberiella*	F	Aragon 1964†
	Lepidoptera: Noctuidae	*Achaea janata*	L	Deshpande & Ramakrishnan 1982
		Achaea janata	L	Govindaraian *et al.* 1976†; Deshpande & Ramakrishnan 1982
		Acronicta leporina	F	Rafes *et al.* 1976
		Acronicta psi	F	Rafes *et al.* 1976
		Agrotis ipsilon (ypsilon)	L	Salama & Sharaby 1988; Aboul Ela *et al.* 1993
		Agrotis segetum	L/F	Onishchenko 1966†; Rassoulof *et al.* 1966†
		Apamea anceps	F	Dusenko 1986
		Bena (Hylophila) capucina (prasinana)	F	Rafes *et al.* 1976
		Cosmia trapezina	L	Baganich 1976
		Earias biplaga	L	Frutos *et al.* 1987
		Earias insulana	L	El Husseini & Afifi 1984a, 1984b; Salama & Foda 1984
		Eublemma amabilis	L/F	Malhotra & Choudhary 1968†
		Euxoa messoria	L/F	Cheng 1973†
		Helicoverpa (Heliothis) armigera	L/F	Isakova 1958†; Popova 1971; Sukhoruchenko *et al.* 1976
		Helicoverpa (Heliothis) punctigera	L	Teakle *et al.* 1992
		Helicoverpa (Heliothis) zea	L/F	Rogoff *et al.* 1969†; Creighton *et al.* 1971†
		Heliothis virescens	L/F	Mistric & Smith 1973†; Tamez Guerra *et al.* 1998
		Hydraecia micacea	F	Derecha *et al.* 1981

	Mamestra (Barathra) brassicae	L/F	Burgerjon & Biache 1967b†; Vergin 1975; Voskresenskaya [illegible]
	Mythimna (Pseudaletia) unipuncta	L	Burgerjon & Biache 1967b†
	Noctua pronuba	L	Burges & Jarrett pers. com.†
	Orthosia gothica	L	Burgerjon & Biache 1967b†
	Phytometra gamma	F	Isakova & Mogilevskaya 1975
	Pseudoplusia includens	L	Chalfant 1969†
	Ptilodon (Lophopteryx) capucina (camelina)	F	Rafes *et al.* 1976
	Spodoptera exigua	L/F	Creighton *et al.* 1970; Sukhoruchenko *et al.* 1976; Tamez Guerra *et al.* 1998
	Spodoptera littoralis	L	Altahtawy & Abaless 1973†; Salama *et al.* 1984, 1987
	Spodoptera liturna	F	Takaki 1975†
	Trichoplusia ni	L/F	Rogoff *et al.* 1969†; Creighton *et al.* 1971†; Jaques 1972†
	Triphaena (Noctua) pronuba	L	Burges unpubl. obs.†
	Xestia (Amathes) c-nigrum	L	Burgerjon & Biache 1967b†; Lipa *et al.* 1969
Lepidoptera: Notodontidae	*Clostera (Pygaera) anastomosis*	L	Burgerjon & Biache 1967b†
	Drymonia ruficornis (as *chaonia*)	F	Pirvescu 1973†
	Drymonia sp.	L	Balinski *et al.* 1969†
	Phalera bucephala	L	Isakova 1958†
	Schizura concinna	F	Pinnock *et al.* 1974†
Lepidoptera: Nymphalidae	*Vanessa cardui*	L	Morris 1969†
	Vanessa io	L	Burgerjon & Biache 1967b†
	Vanessa turkestanica	F	Guzeev 1986
Lepidoptera: Olethreutidae	*Epinotia tsugana*	L	Morris 1969†
Lepidoptera: Papilionidae	*Papilio demoleus*	L	Narayanan *et al.* 1976a†
Lepidoptera: Pieridae	*Aporia crataegi*	L/F	Isakova 1958†; Kondrja 1966a†; Korchagin 1983
	Pieris brassicae	L/F	Isakova 1958†; Burgerjon & Biache 1967b†; Lipa *et al.* 1970
	Pieris rapae	L/F	Isakova 1958†; Jaques 1972†; Takaki 1975†
	Pieris spp.	F	Dzhivaladze 1979
	Pontia (Synchloe) daplidice	F	Effremova 1976; Moiseeva *et al.* 1975
Lepidoptera: Psychidae	*Canephora asiatica*	F	Takaki 1975†
Lepidoptera: Pyralidae	*Acrobasis* sp.	L	Fedorinchik 1969†
	Chilo suppressalis	L	Attathom *et al.* 1995
	Cnaphalocrocis medinalis	F	Takaki 1975†
	Diaphania (Margaronia) nitidalis	F	Canerday 1967†
	Ephestia cautella	L	Kinsinger *et al.* 1980; Burges & Hurst 1977
	Ephestia elutella	L	Burges & Hurst 1977
	Ephestia kuehniella	L	Vankova 1962†; Burges & Hurst 1977
	Galleria mellonella	L/F	Vankova 1966†; Burgerjon & Biache 1967b†; Burges *et al.* 1976b
	Loxostege stricticalis	F	Dyadechko *et al.* 1976; Novinskii 1977; Endakov 1979
	Ostrinia nubilalis	L/F	Gerginov 1978; El Husseini 1980; Mohd Salleh & Lewis 1982b
	Plodia interpunctella	L	Burges & Hurst 1977; Kinsinger *et al.* 1980; Wang *et al.* 1990
	Schoenobius bipunctifer	F	Kwangsi Kweishien, Biol. Contr. Stat 1974†
	Zophodia convolutella	L	Sem'yanov *et al.* 1983
Lepidoptera: Pyralidae?	*Mesographe forficalis*	L	Isakova 1958†; Issi 1978†
Lepidoptera: Saturnidae	*Antheraea pernyi*	L	Burgerjon & Biache 1967b†
	Dictyoploca japonica	F	Kharchenko & Chelysheva 1975
	Samia cynthia	L	Burgerjon & Biache 1967b†
	Saturnia pavonia	L	Burgerjon & Biache 1967b†

APPENDIX 2.1 *(continued)*

Subspecies/ Serovar	Order: Family	Genus and Species	Lab/Field	References
galleriae	Lepidoptera: Sphingidae	*Clanis bilineata*	F	Ye *et al.* 1985
	Lepidoptera: Stathmopodidae	*Kakivoria flavofasciata*	F	Takaki 1975†
	Lepidoptera: Thaumetopoeidae	*Thaumetopoea pityocampa*	L/F	Videnova *et al.* 1972†
		Thaumetopoea processionea	L	Burgerjon & Biache 1967b†
	Lepidoptera: Tortricidae	*Achroia grisella*	L	Karabash 1974
		Acleris variana	L	Morris 1963†
		Adoxophyes orana	F	Takaki 1975†
		Agriopis (*Erannis*) *aurantiaria*	F	Ceianu *et al.* 1970†
		Agriopis (*Erannis*) *leucophaearia*	F	Ceianu *et al.* 1970†
		Archips podana	L	Fedorinchik 1969†
		Archips (*Cacoecia*) *rosana*	L	Zukauskiene 1973†
		Archips (*Cacoecia*) *rosanus*	L	Niemczyk & Bakowski 1970; Zukauskiene & Petrauskas 1973;
		Archips spp.	F	Gamayunova 1987
		Argyrotaenia pulchellana	F	Gaprindashvili 1975
		Argyrotaenia velutinana	F	Dolphin *et al.* 1967†
		Autographa gamma	L	Burges & Jarrett pers. com.†
		Cacoecimorpha pronubana	L	Burges unpubl. obs.†
		Choristoneura fumiferana	L/F	Yamvrias & Angus 1970†; Smirnoff 1973; Morris & Moore 1983
		Cydia funebrana	F	Ilieva 1973; G'onev 1975
		Cydia (*Grapholita, Laspeyresia*) *molesta*	F	Fedorinchik & Sogoyan 1975
		Cydia (*Laspeyresia*) *pomonella*	F	Dolphin *et al.* 1967†; Batalova 1970; Kleimenova 1970
		Desmia funeralis	F	Jensen 1969†
		Hedya nubiferana	L	Johansson 1971†
		Lobesia botrana	L/F	Roehrich 1970†; Makhmudov *et al.* 1977
		Phlogophora meticulosa	L	Burges & Jarrett unpubl. obs.†
		Rhyacionia buoliana	L	Morris 1969†
		Spilonota (*Tmetocera*) *ocellana*	F	Jaques 1965†; Prieditis & Rituma 1974
		Tortrix viridana	L/F	Fankhanel 1962†; Burgerjon & Biache 1967b†; Svestka 1974, 1977
	Lepidoptera: Yponomeutidae	*Argyresthia conjugella*	F	Gerasimovich 1971
		Plutella xylostella (*maculipennis*)	L/F	Leskova 1960†; Burgerjon & Biache 1967b†; Johansson 1971
		Prays oleae	F	Yamvrias 1972†
		Yponomeuta cognatella (*evonymi*)	L/F	Kondrja 1966a†; Burgerjon & Biache 1967b†
		Yponomeuta euonymellus	L	Khrameeva 1972
		Yponomeuta mahalebella	L	Ceianu *et al.* 1970†
		Yponomeuta malinella	L/F	Isakova 1958†; Burgerjon & Biache 1967b†
		Yponomeuta (*Hyponomeuta*) *malinellus*	F	Prieditis & Rituma 1974
		Yponomeuta (*padellus*) *malinellus*	L	Niemczyk & Bakowski 1970; Niemczyk 1971
		Yponomeuta padella	L/F	Kondrja 1966a†; Burgerjon & Biache 1967b†
		Yponomeuta rorella	L/F	Karasev 1968†
	Neuroptera: Chrysopidae	*Chrysopa* (*Chrysoperla*) *carnea*	L	Babrikova *et al.* 1982
		Chrysopa formosa	L	Babrikova & Kuzmanova 1984
		Chrysopa perla	L	Babrikova & Kuzmanova 1984
		Chrysopa septempunctata	L	Babrikova & Kuzmanova 1984
	Phthiraptera:	Mallophaga spp.	F	Frolov 1974

	Phthiraptera: Trichodectidae	*Damalinia (Bovicola) ovis*	L	Drummond *et al.* 1992
	Prostigmata: Tetranychidae	*Tetranychus urticae*	F	Sukhoruchenko *et al.* 1976
	Siphonaptera: Ceratophyllidae	*Citellophilus (Ceratophyllus) tesquorum*	L	Savel' ev *et al.* 1974
		Nosopsyllus (Ceratophyllus) consimilis	L	Savel' ev *et al.* 1974
		Nosopsyllus (Ceratophyllus) fasciatus	L	Shevchenko *et al.* 1983
		Nosopsyllus (Ceratophyllus) laeviceps	L	Savel' ev *et al.* 1974
	Siphonaptera: Pulicidae	*Xenopsylla cheopis*	L	Savel'ev & Kozlov 1974; Kozlov *et al.* 1975, 1977
galleriae?	Lepidoptera: Geometridae	*Erannis tiliaria*	F	Doane & Hitchcock 1964†
galleriae?	Lepidoptera: Lasiocampidae	*Malacosoma disstria*	L	Morris 1972a†
galleriae?	Lepidoptera: Lymantriidae	*Orgyia psuedotsugata*	L	Morris 1972a†
galleriae?	Lepidoptera: Pyralidae	*Maruca testulalis*	L	Taylor 1968†
galleriae?	Lepidoptera: Pyraustidae	*Dichomeris marginalla*	F	Nordin & Appleby 1969†
galleriae?	Lepidoptera: Thaumetopoeidae	*Thaumetopoea pityocampa*	F	Kailidis *et al.* 1971†
galleriae?	Lepidoptera: Tortricidae	*Archips (Cacoecia) rosana*	L/F	Niemczyk & Bakowski 1971†
galleriae?		*Zeiraphera diniana*	F	Benz 1975†
guiyaniensis	Lepidoptera: Pieridae	*Pieris rapae*	L	Yao *et al.* 1998
higo	Diptera: Culicidae	*Anopheles stephensi*	L	Ohba *et al.* 1995; Saitoh *et al.* 1998
		Culex pipiens	L	Ohba *et al.* 1995
		Culex tritaeniorhynchus	L	Jung *et al.* 1998
	Diptera: Psychodidae	*Telmatoscopus (Clogmia) albipunctatus*	L	Ohba *et al.* 1995
huazhongensis	Diptera: Culicidae	*Anopheles stephensi*	L	Dai *et al.* 1996
		Culex pipiens	L	Dai *et al.* 1996
indiana	Lepidoptera: Gelechiidae	*Phthorimaea operculella*	L	Salama *et al.* 1995a
		Sitotroga cerealella	L	Salama *et al.* 1991b
	Lepidoptera: Pyralidae	*Plodia interpunctella*	L	Salama *et al.* 1991b
	Phthiraptera: Trichodectidae	*Damalinia (Bovicola) ovis*	L	Drummond *et al.* 1992
israelensis	Acari: Argasidae	*Argas persicus*	L	Hassanain *et al.* 1997b
	Acari: Ixodidae	*Hyalomma dromedarii*	L	Hassanain *et al.* 1997b
	Acari: Pyroglyphidae	*Dermatophagoides pteronyssinus*	L	Saleh *et al.* 1991
	Crustacea: Anostraca	*Chirocephalus grubei*	L	Morawczik & Schnetter unpubl. obs.†
	Crustacea: Cyclopidae	*Cyclops* sp.	L	Weiser & Vankova 1978†
		Megacyclops sp.	L	Weiser & Vankova 1978†
	Diptera: ?	*Diamessa* sp.	L	Weiser & Vankova 1978†
	Diptera: Anisopodidae	*Sylvicola fenestralis*	L	Houston *et al.* 1989a; Coombs *et al.* 1991
	Diptera: Chironomidae	*Chironomus attenuatus*	L	Garcia *et al.* 1982
		Chironomus crassicaudatus	L	Ali *et al.* 1981
		Chironomus decorus	F	Mulla *et al.* 1990a; Rodcharoen *et al.* 1991
		Chironomus fulvipilus	F	Rodcharoen *et al.* 1991
		Chironomus kiiensis	L	Kondo *et al.* 1995b
		Chironomus riparius	F	Cilek & Knapp 1992
		Chironomus tentans	L	Benz & Joeressen 1994
		Chironomus tepperi	L	Treverrow 1985

APPENDIX 2.1 (*continued*)

Subspecies/ Serovar	Order: Family	Genus and Species	Lab/Field	References
israelensis	Diptera: Chironomidae	*Chironomus tummi*	L	Rieger & Schnetter unpubl. obs.†
		Chironomus yoshimatsui	L	Kondo *et al.* 1992; 1995b
		Chironomus sp.	L	Weiser & Vankova 1978†
		Dicrotendipes pelochloris	L	Kondo *et al.* 1995b
		Dicrotendipes sp.	F	Rodcharoen *et al.* 1991
		Glyptotendipes paripes	L	Ali *et al.* 1981
		Glyptotendipes tokunagai	L	Kondo *et al.* 1995b
		Hydrobaenus kondoi	L	Kondo *et al.* 1992
		Limnophyes minimus	L/F	Houston *et al.* 1989a, b
		Metriocnemus hygropetricus	L/F	Houston *et al.* 1989a, b
		Orthocladius fuscimanus	L	Houston *et al.* 1989b
		Paralauterborniella elachista	F	Rodcharoen *et al.* 1991
		Paratanytarsus grimmii	L	Kondo *et al.* 1995a
		Paratanytarsus sp.	L	Kondo *et al.* 1995b
		Pentapedilum tigrinum	L	Kondo *et al.* 1995b
		Rheotanytarsus fuscus	F	Palmer 1993
		Rheotanytarsus spp.	F	Molloy 1992; Merritt *et al.* 1989
		Stictochironomus akizukii	L	Kondo *et al.* 1995b
		Tanytarsus spp.	L	Ali *et al.* 1981
		Tokunagayusurika akamusi	L	Kondo *et al.* 1992
	Diptera: Culicidae	*Aedes aegypti*	L	Goldberg & Margalit 1977†; Lee & Cheong 1985; Larget & de Barjac 1981b
		Aedes albopictus	L	Lee & Cheong 1985; 1988; Huang *et al.* 1993
		Aedes atlanticus	F	Palmisano 1987
		Aedes atropalpus	L	Tousignant *et al.* 1992
		Aedes campestris	F	Spackman 1985
		Aedes canadensis	L	Knepper & Walker 1989
		Aedes cantans	L/F	Vankova & Weiser 1978†; Schnetter *et al.* 1983; Rettich 1983
		Aedes cantator	F	Christie 1990
		Aedes caspius	L/F	Sinegre *et al.* 1979b†; Zohdy & Matter 1982; Chabanenko *et al.* 1992
		Aedes cataphylla	F	Luthy *et al.* 1980
		Aedes cinereus	F	Rettich 1983
		Aedes communis	L	Rieger & Schnetter unpubl. obs.†; Luthy *et al.* 1980; Rettich 1983
		Aedes detritus	F	Sinegre *et al.* 1979b†, 1980; Merdan *et al.* 1986
		Aedes dorsalis	F	Garcia *et al.* 1980; Spackman 1985
		Aedes dupreei	NS	Beck 1982
		Aedes fitchii	L	Knepper & Walker 1989
		Aedes flavopictus	L	Xu *et al.* 1986
		Aedes hexodontus	F	Eldridge *et al.* 1985
		Aedes implicatus	F	Spackman 1985
		Aedes melanimon	L/F	Spackman 1985; Mulla *et al.* 1986
		Aedes mercurator	F	Spackman 1985
		Aedes nigromaculis	L/F	Mulla *et al.* 1980, 1982, 1986

Aedes pseudoscutellaris	L	Goettel *et al.* 1982
Aedes pullatus	F	Spackman 1985
Aedes punctor	L	Rieger & Schnetter unpubl. obs.†; Rettich 1983; Snow 1984
Aedes rusticus	F	Schnetter *et al.* 1983
Aedes sierrensis	NS	Beck 1982
Aedes sollicitan	F	Christie 1990; Ward & Redovan 1983; Lake *et al.* 1982
Aedes spencerii (& varieties)	F	Spackman 1985
Aedes squamiger	F	Webb & Dhillon 1984
Aedes sticticus	F	Schnetter *et al.* 1983; Rettich 1983
Aedes stimulans	L	Knepper & Walker 1989
Aedes taeniorhynchus	L/F	Purcell 1981; Mulla *et al.* 1980; Ward & Redovan 1983
Aedes togoi	L	Lee & Cheong 1985; Huang *et al.* 1993
Aedes tomentor	NS	Beck 1982
Aedes triseriatus	L/F	Lacey & Singer 1982; Gharib & Hilsenhoff 1988
Aedes ventrovittis	F	Eldridge *et al.* 1985
Aedes vexans	L/F	Luthy *et al.* 1980; Schnetter *et al.* 1981; Gharib & Hilsenhoff 1988
Aedes vigilax	L/F	Mottram *et al.* 1989; Goettel *et al.* 1982
Anopheles albimanus	L/F	Lacey & Singer 1982; Mulla *et al.* 1986; Swezey & Salamanca 1988
Anopheles annulipes	F	Davidson *et al.* 1981
Anopheles anthropophagus	L	Xu *et al.* 1986
Anopheles arabiensis	L/F	Romi *et al.* 1993; Seyoum & Abate 1997; Nugud & White 1982
Anopheles atroparvus	L	Nicolescu 1982; Sokolova *et al.* 1986
Anopheles balabacensis	L	Lee & Cheong 1985; Foo & Yap 1982
Anopheles crucians	F	McLaughlin *et al.* 1982
Anopheles culicifacies	L	Balaraman *et al.* 1981; Bhateshwar & Mandal 1991; Dua *et al.* 1993
Anopheles dthali	F	Ladoni *et al.* 1986
Anopheles franciscanus	L/F	Kramer 1989; Garcia *et al.* 1982
Anopheles freeborni	F	Kimball *et al.* 1986
Anopheles gambiae complex	F	Hougard *et al.* 1983
Anopheles hyrcanus	F	Dubitskii *et al.* 1981
Anopheles karwari	F	Lee *et al.* 1990
Anopheles maculatus	L/F	Lee & Cheong 1988; Lee *et al.* 1990
Anopheles maculipennis	F	Ouzounis & Samanidou Voyadjoglou 1993; Stus & Mikherskaya 1986
Anopheles multicolor	L	Zohdy & Matter 1982
Anopheles nigerrimus	F	Kramer 1984
Anopheles pharoensis	L	Bekheit 1984
Anopheles pulcherrimus	L/F	Chabanenko *et al.* 1992
Anopheles quadrimaculatus	L/F	Mulla *et al.* 1980; Lacey & Singer 1982; Stark & Meisch 1983
Anopheles sacharovi	L/F	Volzhinskii *et al.* 1984; Ceber 1992
Anopheles sergentii	L	Goldberg & Margalit 1977†
Anopheles sinensis	L/F	Xu *et al.* 1986; Yu *et al.* 1982
Anopheles stephensi	L	de Barjac 1978†; Balaraman *et al.* 1981, 1983a; Larget & de Barjac 1981b
Anopheles subpictus	L/F	Dua *et al.* 1993; Balaraman *et al.* 1983a; MAnonmani & Hoti 1995
Anopheles sundaicus	F	Schaefer & Kirnowardoyo 1984
Anopheles superpictus	F	Ladoni *et al.* 1986
Anopheles vagus	F	Kramer 1984

APPENDIX 2.1 (*continued*)

Subspecies/ Serovar	Order: Family	Genus and Species	Lab/Field	References
israelensis	Diptera: Culicidae	*Armigeres durhami*	L	Lee & Cheong 1985
		Armigeres kesseli	L	Lee & Cheong 1988
		Armigeres subalbatus	L	Huang *et al.* 1993; Kang & Chen 1986
		Culex annulirostris	F	Davidson *et al.* 1981
		Culex antennatus	F	Merdan *et al.* 1986; Bekheit 1984
		Culex declarator	L	Habib 1983
		Culex erraticus	NS	Beck 1982
		Culex fuscanus	F	Kramer 1984
		Culex laticinctus	F	Alten & Bosgelmez 1990
		Culex nigripalpu	NS	Beck 1982
		Culex orientalis	L/F	Yu *et al.* 1982
		Culex peus	L/F	Mulla *et al.* 1980; Eldridge & Callicrate 1982
		Culex pipiens	L/F	Goldberg & Margalit 1977†; Sinegre *et al.* 1979b†
		Culex pipiens complex	L	Farghal 1982a; Zohdy & Matter 1982; Schnetter *et al.* 1983
		Culex pseudovishnui	L	Lee & Seleena 1990b
		Culex quinquefasciatus	L/F	Mulla *et al.* 1980; Balaraman *et al.* 1981; Davidson *et al.* 1981
		Culex restuans	L	Wraight *et al.* 1987
		Culex salinarius	L	Holck & Meek 1987
		Culex sitiens	L	Goettel *et al.* 1982; Balaraman *et al.* 1983a; Mottram *et al.* 1989
		Culex tarsalis	L/F	Mulla *et al.* 1980; Garcia *et al.* 1980, 1982
		Culex theileri	L/F	Chabanenko *et al.* 1992
		Culex tritaeniorhynchus	L/F	Balaraman *et al.* 1981, 1983a
		Culex univittatus	L	Goldberg & Margalit 1977†
		Culex vishnui	F	Kramer 1984
		Culiseta alaskaensis	F	Dubitskii *et al.* 1981
		Culiseta incidens	NS	Beck 1982
		Culiseta inornata	L/F	Mulla *et al.* 1980; Garcia *et al.* 1982; Spackman 1985
		Culiseta longiareolata	L	Zohdy & Matter 1982; Merdan *et al.* 1986; Farghal 1982b
		Culiseta melanura	L	Wraight *et al.* 1987
		Culiseta nigripalpus	L	Ali *et al.* 1989
		Limatus durhamii	L	Lacey & Lacey 1981
		Limatus flavisetosus	L	Lacey & Lacey 1981
		Mansonia bonneae	F	Chang *et al.* 1990
		Mansonia dyari	L	Lord & Fukuda 1990
		Mansonia indiana	L	Foo & Yap 1982
		Mansonia richardii	L	Sinegre *et al.* 1979b†
		Mansonia titillans	L	Lord & Fukuda 1990
		Mansonia spp.	F	Chang *et al.* 1990
		Psorophora columbiae	L/F	Mulla *et al.* 1980; Holck & Meek 1987; McLaughlin & Vidrine 1984
		Psorophora ferox	F	Palmisano 1987
		Psorophora varipes	F	Palmisano 1987
		Toxorhynchites amboinensis	L	Larget & Charles 1982; Misch *et al.* 1987
		Toxorhynchites splendens	L	Lee & Cheong 1988
		Trichoprosopon digitatum	L	Lacey & Lacey 1981

	Uranotaenia inguiculata	L	Goldberg & Margalit 1977†
	Uranotaenia lowii	NS	Beck 1982
Diptera: Glossinidae	*Glossina pallidipes*	L	van der Geest & de Barjac 1982
Diptera: Muscidae	*Haematobia irritans*	L	Temeyer 1984
Diptera: Phlebotominae	*Phlebotomus argentipes*	L	Yuval & Warburg 1989
	Phlebotomus papatasi	L	de Barjac *et al.* 1981; Yuval & Warburg 1989
	Phlebotomus perniciosus	L	Yuval & Warburg 1989
Diptera: Phoridae	*Megaselia halterata*	L/F	Kiel 1991
Diptera: Psychodidae	*Lutzomyia longipalpis*	L	de Barjac *et al.* 1981; Yuval & Warburg 1989
	Psychoda alternata	L	Houston *et al.* 1989b
	Psychoda severini	L/F	Houston *et al.* 1989a, b
Diptera: Sciaridae	*Bradysia coprophila*	L	Osborne *et al.* 1985
	Bradysia spp.	F	Harris 1993
	Lycoriella mali	L/F	Cantwell & Cantelo 1984; Kiel 1991
Diptera: Simuliidae	*Austrosimulium multicorne*	F	Chilcott *et al.* 1983
	Cleitosimulium argenteostriatum	F	Car & Kutzer 1988
	Cnephia mutata	L	Goldberg *et al.* 1978
	Cnephia ornithophilia	L	Goldberg *et al.* 1978
	Cnephia pecuarum	L	Atwood *et al.* 1992
	Cnetha verna (*Simulium vernum*)	F	Olejnicek 1986
	Eusimulium vernum (*Simulium vernum*)	F	Colbo & O'Brien 1984
	Eusimulium latipes	L	Weiser & Vankova 1978†
	Odagmia ornata (*Simulium ornatum*)	L	Weiser & Vankova 1978†; Olejnicek 1986; Deschle *et al.* 1988
	Prosimulium mixtum	L	Goldberg *et al.* 1978; Colbo & O'Brien 1984
	Prosimulium tomosvaryi	F	Olejnicek 1986
	Simulium alcocki	F	Guillet & Escaffre 1979b†
	Simulium aokii	F	Nakamura *et al.* 1985
	Simulium aureum	L	Barton *et al.* 1991
	Simulium cervicornutum	F	Guillet & Escaffre 1979b†
	Simulium chutteri	F	Palmer 1993; Palmer *et al.* 1996
	Simulium damnosum	L/F	Guillet & Escaffre 1979b†
	Simulium damnosum	F	Guillet *et al.* 1982; Palmer 1993
	Simulium gariepense	F	Palmer 1993
	Simulium goeldii	L	Habib 1983
	Simulium japonicum	F	Nakamura *et al.* 1985
	Simulium jenningsi	F	Merritt *et al.* 1989
	Simulium mcmahoni	F	Palmer 1993
	Simulium monticola	F	Car & Kutzer 1988
	Simulium noelleri	F	Car & Kutzer 1988
	Simulium notiale	L	Barton *et al.* 1991
	Simulium ochraceum	F	Undeen *et al.* 1981
	Simulium pertinax	F	Araujo Coutinho & Lacey 1990; Castello-Branco & Andrade 1992
	Simulium posticatum	F	Welton & Ladle 1993
	Simulium pugetense	F	Molloy & Jamnback 1981
	Simulium reptans	L/F	Car & Kutzer 1988; Coupland 1993
	Simulium rorotaense	L	Habib 1983
	Simulium tuberosum complex	F	Molloy & Jamnback 1981; Car & Kutzer 1988; Merritt *et al.* 1989

APPENDIX 2.1 (*continued*)

Subspecies/ Serovar	Order: Family	Genus and Species	Lab/Field	References
israelensis	Diptera: Simuliidae	*Simulium uchidai*	F	Nakamura *et al.* 1985
		Simulium unicornutum	F	Guillet & Escaffre 1979b†
		Simulium variegatum group (*variegatum/argyreatum*)	L/F	Car & Kutzer 1988; Coupland 1993
		Simulium venustum/verecundum complex	L	Undeen unpubl. obs.†; Molloy & Jamnback 1981; Barton *et al.* 1991
		Simulium verecundum	L	Goldberg *et al.* 1978
		Simulium vernum	F	Riley & Fusco 1990
		Simulium vittatum	L	Goldberg *et al.* 1978; Lacey *et al.* 1978; Lacey & Federici 1979
		Stegopterna mutata	F	Colbo & O'Brien 1984
	Diptera: Tabanidae	*Tabanus triceps*	L	Saraswathi & Ranganathan 1996
	Diptera: Tephritidae	*Ceratitis capitata*	L	El-Sebae & Komeil 1990
	Diptera: Tipulidae	*Tipula oleracea*	L/F	Smits & Vlug 1990; Smits *et al.* 1993
		Tipula paludosa	L/F	Smits & Vlug 1990
	Hymenoptera: Diprionidae	*Gilpinia hercyniae*	L	Benz & Joeressen 1994
		Neodiprion sertifer	L	Benz & Joeressen 1994
	Hymenoptera: Pamphiliidae	*Acantholyda erythrocephala*	L	Benz & Joeressen 1994
		Cephalcia albietis	L	Benz & Joeressen 1994
		Cephalcia falleni	L	Benz & Joeressen 1994
	Hymenoptera: Tenthredinidae	*Athalia rosae*	L	Benz & Joeressen 1994
		Nematus melanaspis	L	Benz & Joeressen 1994
		Pristiphora abietina	L	Benz & Joeressen 1994
	Hymenoptera: Vespidae	*Vespula* sp.	L	Gambino 1988
	Lepidoptera: Geometridae	*Ennomos subsignarius*	F	Dunbar *et al.* 1973†
	Lepidoptera: Lasiocampidae	*Malacosoma neustria*	L	Vankova & Weiser 1978†
	Lepidoptera: Lymantriidae	*Euproctis chrysorrhoea*	L	Vankova & Weiser 1978†
	Lepidoptera: Noctuidae	*Spodoptera litura*	L	Puntambekar *et al.* 1997
	Lepidoptera: Pieridae	*Aporia crataegi*	L	Vankova & Weiser 1978†
	Nematoda: Meloidogynidae	*Meloidogyne incognita*	GH	Sharma 1994
	Nematoda: Trichostrongylidae	*Trichostrongylus colubriformis*	L	Bottjer *et al.* 1985 (in patent No. 05281530, A.J. Sick)
	Phthiraptera: Menoponidae	*Menopon gallinae*	L	Lonc & Lachowicz 1987
	Phthiraptera: Trichodectidae	*Damalinia (Bovicola) ovis*	L	Drummond *et al.* 1992
	Trichoptera: Hydropsychidae	*Hydropsyche pellucida*	L	Weiser & Vankova 1978†
		Potamophylax rotundipennis	L	Weiser & Vankova 1978†
japonensis	Coleoptera: Chrysomelidae	*Pyrrhalta tibialis*	L	Ohba & Aizawa 1986a
	Coleoptera: Scarabaeidae	*Anomala albopilosa*	L	Suzuki *et al.* 1992
		Anomala cuprea	L	Ohba *et al.* 1992; Suzuki *et al.* 1992, 1993
		Anomala daimiana	L	Suzuki *et al.* 1992
		Anomala orientalis	F	Koppenhofer *et al.* 1999
		Anomala rufocuprea	L	Ohba *et al.* 1992; Suzuki *et al.* 1992; Hori *et al.* 1994
		Anomala schonfeldti	L	Suzuki *et al.* 1992
		Cyclocephala hirta	L/F	Koppenhofer & Kaya 1997; Koppenhofer *et al.* 1999
		Cyclocephala pasadenae	L/F	Koppenhofer & Kaya 1997; Koppenhofer *et al.* 1999

		Popillia japonica	L/F	Suzuki *et al.* 1992; Ohba *et al.* 1992; Alm *et al.* 1997
	Lepidoptera: Bombycidae	*Bombyx mori*	L	Ohba & Aizawa 1986a
	Lepidoptera: Geometridae	*Larerannis orthogrammaria*	F	Meng *et al.* 1988
	Lepidoptera: Limacodidae	*Latoia (Parasa) lepida*	L	Ohba & Aizawa 1986a
	Lepidoptera: Noctuidae	*Spodoptera litura*	L	Asano & Hori 1995
	Lepidoptera: Notodontidae	*Clostera (Pygaera) anastomosis*	L	Ohba & Aizawa 1986a
	Lepidoptera: Pieridae	*Pieris rapae crucivora*	L	Ohba & Aizawa 1986a
jegathesan	Diptera: Culicidae	*Aedes aegypti*	L	Kawalek *et al.* 1995; Seleena *et al.* 1995; Ragni *et al.* 1996
		Aedes albopictus	L	Kawalek *et al.* 1995
		Aedes togoi	L	Kawalek *et al.* 1995
		Anopheles maculatus	L	Kawalek *et al.* 1995; Seleena *et al.* 1995
		Anopheles stephensi	L	Ragni *et al.* 1996
		Culex pipiens	L	Ragni *et al.* 1996
		Culex quinquefasciatus	L	Kawalek *et al.* 1995; Seleena *et al.* 1995; Wirth *et al.* 1998
		Mansonia uniformis	L	Kawalek *et al.* 1995
kenyae	Coleoptera: Scolytidae	*Dendroctonus frontalis*	L	Moore 1972
	Diptera: Culicidae	*Aedes aegypti*	L	Hall *et al.* 1977†; Lopez Meza *et al.* 1995
		Aedes triseriatus	L	Hall *et al.* 1977†
		Culex pipiens molestus	L	Saitoh *et al.* 1996
		Culex quinquefasciatus	L	Lopez Meza *et al.* 1995
		Culex tarsalis	L	Hall *et al.* 1977†
	Diptera: Simuliidae	*Simulium vittatum*	L	Lacey & Mulla 1977†
	Lepidoptera: Arctiidae	*Amsacta moorei*	L	Rao & Amonkar 1980
	Lepidoptera: Bombycidae	*Bombyx mori*	L	Angus & Norris 1968†; Ohba *et al.* 1984; Hastowo *et al.* 1992
	Lepidoptera: Gelechiidae	*Phthorimaea operculella*	L	Amonkar *et al.* 1979; Salama *et al.* 1995a
		Sitotroga cerealella	L	Salama *et al.* 1991b
	Lepidoptera: Geometridae	*Boarmia (Ascotis) selenaria*	L	Wysoki & Jarvinen 1986; Wysoki & Scheepens 1988
		Lycia (Biston) hirtaria	L	Kuzmanova & Lecheva 1981
	Lepidoptera: Lymantriidae	*Lymantria dispar*	F	Cabral 1973; Dubois *et al.* 1989
	Lepidoptera: Noctuidae	*Agrotis ipsilon (ypsilon)*	L	Salama *et al.* 1991b
		Earias vittella	F	Sarag & Satpute 1988
		Helicoverpa (Heliothis) armigera	L	Kulkarni & Amonkar 1988a, b; Teakle *et al.* 1992
		Helicoverpa (Heliothis) punctigera	L	Teakle *et al.* 1992
		Mamestra (Barathra) brassicae	L	Trottier *et al.* 1988
		Mamestra (Barathra) configurata	L	Trottier *et al.* 1988
		Spodoptera exigua	L	patent No. 05336492, J.M. Payne 1994
		Spodoptera frugiperda	L	Hernandez 1988
		Spodoptera littoralis	L	Kalfon & de Barjac 1985
		Spodoptera litura	L	Amonkar *et al.* 1985
	Lepidoptera: Pieridae	*Pieris rapae*	L	Yuan *et al.* 1983
	Lepidoptera: Pyralidae	*Chilo partellus*	L	Brownbridge & Onyango 1992b
		Galleria mellonella	L	Jarrett & Burges 1982b
		Loxostege sticticalis	L	Yuan *et al.* 1983
		Ostrinia nubilalis	L	Mohd Salleh & Lewis 1982b, 1983; Maini *et al.* 1989
		Plodia interpunctella	L	Salama *et al.* 1991b

APPENDIX 2.1 (*continued*)

Subspecies/ Serovar	Order: Family	Genus and Species	Lab/Field	References
kenyae	Lepidoptera: Tortricidae	*Choristoneura fumiferana*	L	Yamvrias & Angus 1970†; Morris & Moore 1983
		Choristoneura occidentalis	L	Patent No. 05336492, J.M. Payne 1994
		Cydia (Laspeyresia) pomonella	L	Andermatt *et al.* 1988
	Lepidoptera: Yponomeutidae	*Plutella xylostella (maculipennis)*	L	Kim *et al.* 1998b
	Phthiraptera: Menonponidae	*Menopon gallinae*	L	Lonc & Lachowicz 1987; Lonc *et al.* 1988
	Phthiraptera: Trichodectidae	*Damalinia (Bovicola) ovis*	L	Drummond *et al.* 1992
konkukian	Lepidoptera: Bombycidae	*Bombyx mori*	L	Lee *et al.* 1994
kumamotoensis	Coleoptera: Chrysomelidae	*Diabrotica undecimpunctata howardi*	L	Johnson *et al.* 1993
	Diptera: Psychodidae	*Telmatoscopus albipunctatus*	L	Saitoh *et al.* 1996
	Lepidoptera: Bombycidae	*Bombyx mori*	L	Ohba *et al.* 1981b
	Lepidoptera: Yponomeutidae	*Plutella xylostella (maculipennis)*	L	patent No. 05286485, K.A. Uyeda 1994
kurstaki	Acari: Argasidae	*Argas persicus*	L/F	Abdel Megeed *et al.* 1997; Hassanain *et al.* 1997
	Acari: Ixodidae	*Hyalomma dromedarii*	L	Hassanain *et al.* 1997b
	Acari: Mesostigmata	*Boophilus annulatus (calcaratus)*	F	Abdel Megeed *et al.* 1997
	Coleoptera: Anobiidae	*Lasioderma serricorne*	F	Eberhardt 1997
	Coleoptera: Chrysomelidae	*Diabrotica unidecimpunctata howardi*	L	Sims 1997
		Epitrix hirtipennis	F	Deseo *et al.* 1993
		Leptinotarsa decemlineata	L/F	Kiselek & Zayats 1976; Ignoffo *et al.* 1982a; Sims 1997
		Phaedon cochleariae	L	Kamenek 1988
		Phyllotreta striolata	F	Shamiyeh *et al.* 1993
	Coleoptera: Coccinellidae	*Henosepilachna (Epilachna) vigintioctopunctata*	F	Sekar & Baskaran 1976; Baskaran & Kumar 1980
		Hippodamia convergens	L	Haverty 1982
	Coleoptera: Curculionidae	*Anthonomus grandis*	L	MacIntosh *et al.* 1990c; Sims 1997
		Sitophilus granarius	F	Sandner & Szczpiorska 1972 ; McGaughey *et al.* 1975; Rashid & Khan 1995
	Coleoptera: Scarabaeidae	*Costelytra zealandica*	F	Willoughby *et al.* 1994
		Oryctes rhinoceros	L	Babu *et al.* 1971
	Coleoptera: Scolytidae	*Scolytus multistriatus*	L	Jassim *et al.* 1990
		Scolytus scolytus	L	Jassim *et al.* 1990
	Coleoptera: Tenebrionidae	*Tribolium castaneum*	L/F	McGaughey *et al.* 1975; Al Hafidh 1985; Arthur & Brown 1994
	Dictyoptera: Blattellidae	*Blattella germanica*	L	Sandhu & Varma 1980; Zukowski 1994, 1995
	Diptera: Agromyzidae	*Liriomyza sativae*	F	Schuster & Engelhard 1979; Johnson *et al.* 1980, 1984
		Liriomyza trifolii	F	Jones *et al.* 1986
		Melanagromyza hibisci	F	Tomar 1998a, 1998b
		Melanagromyza obtusa	F	Puntambekar *et al.* 1997
	Diptera: Anthomyiidae	*Delia (Hylemya) brassicae*	F	Obadofin & Finlayson 1977
	Diptera: Calliphoridae	*Lucilia cuprina*	L	Lyness *et al.* 1994
	Diptera: Cecidomyiidae	*Dasineura mali*	F	van Epenhuijsen & Carpenter 1990
	Diptera: Culicidae	*Aedes aegypti*	L	Hall *et al.* 1977†; Panbangred *et al.* 1979; Ignoffo *et al.* 1980a,1982b

	Aedes triseriatus	L	Hall *et al.* 1977†; Sims 1997
	Anopheles quadrimaculatus	L	Sims 1997
	Culex molestus	L	Sokolova & Ganushkina 1982
	Culex pipiens	L	Hall *et al.* 1977†
	Culex pipiens molestus	L	Saitoh *et al.* 1996
	Culex quinquefasciatus	L	Ignoffo *et al.* 1981; Menon *et al.* 1982
	Culex tarsalis	L	Hall *et al.* 1977†; Goldberg *et al.* 1974
Diptera: Drosophilidae	*Drosophila melanogaster*	L	Sims 1997
Diptera: Ephydridae	*Hydrellia* sp.	F	Narayanasamy & Baskaran 1979
Diptera: Muscidae	*Musca domestica*	L	Indrasith *et al.* 1992
	Haematobia irritans	L	Gingrich unpubl. data†
Diptera: Simuliidae	*Prosimulium fuscum* (*mixtum*)	L	Eidt 1984
	Simulium argus	L	Lacey *et al.* 1978†
	Simulium piperi	L	Lacey *et al.* 1978†
	Simulium tescorum	L	Lacey *et al.* 1978†
	Simulium vittatum	L	Lacey & Mulla 1977†, 1978; Lacey *et al.* 1978†
	Simulium (*Eusimulium*) *aureum*	L	Lacey *et al.* 1978†
	Simulium (*Hemicnetha*) *virgatum*	L	Lacey *et al.* 1978†
Diptera: Syrphidae	*Syrphus vitripennis*	L	Hassan *et al.* 1983
Diptera: Tachinidae	*Cermoasia auricaudata*	F	Hamel 1977†
	Madremyia saundersii	F	Hamel 1977†
	Pales pavida	L	Franz *et al.* 1980
Diptera: Tephritidae	*Rhagoletis cerasi*	F	Cardei & Rominger 1997
Hemiptera: Aleyrodidae	*Bemisia tabaci*	L	Radwan *et al.* 1986
Hemiptera: Aphididae	*Aphis citricola* (*spiraecola*)	F	Schuster & Engelhard 1979
	Aphis gossypii	F	Schuster & Engelhard 1979; Salama & Zaki 1984; Shevtsov *et al.* 1996
	Brevicoryne brassicae	F	Kennedy & Oatman 1976; Chalfant 1978
	Dysaphis plantaginea	F	Pasqualini *et al.* 1992
	Macrosiphoniella sanborni	F	Schuster & Engelhard 1979
	Myzus persicae	F	Kennedy & Oatman 1976; Chalfant 1978; Vasudevan & Baskaran 1979
Hemiptera: Cicadellidae	*Hishimonus phycitis*	F	Naryanaswamy *et al.* 1979
	Amrasca biguttula	F	Baskaran & Kumar 1980
	Amrasca devastans	F	Naryanaswamy *et al.* 1979
Hemiptera: Coccidae	*Sphaerolecanium prunastri*	F	Chepurnaya 1994
Hemiptera: Delphacidae	*Nilaparvata lugens*	L/F	Rao & Rao 1979; Rao & Prakasa Rao 1979
Hemiptera: Diaspididae	*Quadraspidiotus perniciosus*	F	Pasqualini *et al.* 1992
Hemiptera: Lygaeidae	*Dimorphopterus gibbus*	F	Pandey *et al.* 1977
Hemiptera: Miridae	*Adelphocoris lineolatus*	F	Yunusov 1974
Hemiptera: Pentatomidae	*Nezara viridula*	F	Turnipseed *et al.* 1985
Hemiptera: Pseudococcidae	*Pseudococcus longispinus*	L	Wysoki *et al.* 1975
Hemiptera: Psyllidae	*Psylla* (*Cacopsylla*) *pyri*	F	Nicoli *et al.* 1989; Pasqualini *et al.* 1992
Hymenoptera: Aphelinidae	*Aphytis melinus*	L	Haverty 1982
Hymenoptera: Braconidae	*Apanteles* (*Glyptapanteles*) *africanus*	F	Chari *et al.* 1996
	Apanieles fumiferanae	F	Cadogan *et al.* 1995
	Cotesia marginiventris	L	Atwood *et al.* 1997a

APPENDIX 2.1 *(continued)*

Subspecies/ Serovar	Order: Family	Genus and Species	Lab/Field	References
kurstaki	Hymenoptera: Braconidae	*Cotesia melanoscela*	L	Chenot & Raffa 1998
	Hymenoptera: Diprionidae	*Diprion pini*	F	Ziogas *et al.* 1997
		Neodiprion sertifer	F	Safronov 1996
	Hymenoptera: Encrytidae	*Ageniaspis fuscicollis*	L	Hamed 1979
		Leptomastix dactylopii	L	Franz *et al.* 1980
	Hymenoptera: Eulophidae	*Tetrastichus evynomellae*	L	Hamed 1978/1979†
	Hymenoptera: Formicidae	*Pheidole megacephala*	L	Castineiras & Calderon 1985
	Hymenoptera: Ichneumonidae	*Coccygomimus turionellae*	L	Franz *et al.* 1980
		Diadegma armillata	L	Hamed 1978/1979†
		Hyposoter exiguae	L	Thoms & Watson 1986
		Phaeogenes hariolus	F	Hamel 1977†
		Phygadeuon trichops	L	Franz *et al.* 1980
		Pimpla turionellae	L	Hamed 1978/1979†
		Trichionotus sp.	L	Hamed 1978/1979†
	Hymenoptera: Pamphilidae	*Acantholyda posticalis posticalis*	F	Chung & Shin 1986
	Hymenoptera: Pergidae	*Perga affinis affinis*	F	Neumann & Collett 1997a
	Hymenoptera: Tenthredinidae	*Athalia lugens proxima*	L	Saxena 1978
		Pristiphora abietina	F	Konig 1975
	Hymenoptera: Trichogrammatidae	*Trichogramma cacoeciae*	L	Franz *et al.* 1980
		Trichogramma pretiosum	L	Marques & Alves 1995
	Hymenoptera: Vespidae	*Vespula* sp.	L	Gambino 1988
	Isoptera: Rhinotermitidae	*Psammotermes hybostoma*	L	Farghal *et al.* 1987
	Isoptera: Termitidae	*Amitermes desertorum*	L	Farghal *et al.* 1987
	Lepidoptera: ?	*Epiglaea apiata*	F	Sandler & Mason 1997
	Lepidoptera: Arctiidae	*Datana integerrima*	F	Polles 1974†; Tedders & Ellis 1977; Farris & Appleby 1980
		Ergolis (Ariadne) merione	L	Pawar & Thombre 1990
		Halisidota argentata	F	Morris 1972b
		Hyphantria cunea	L/F	Ko & Lee 1972; Morris 1972b†; Takaki 1975†
		Ocnogyna baetica	L	Barreiro & Santiago Alvarez 1985
		Pericallia ricini	L	Kumar & Jayaraj 1978†; Mathur *et al.* 1994
		Spilosoma (Spilarctia) obliqua	L	Dabi *et al.* 1980b; Pawar & Thombre 1990; Biswas *et al.* 1996a, b, c
		Tyria jacobaeae	L/F	James *et al.* 1993b
		Utetheisa pulchella	L	Gangwar *et al.* 1980
	Lepidoptera: Bombycidae	*Bombyx mori*	L	Travers *et al.* 1976; Okada *et al.* 1977; Morris *et al.* 1982
	Lepidoptera: Brassolidae	*Brassolis astyra astyra*	L	Berti Filho & Gallo 1977
	Lepidoptera: Carposinidae	*Carposina niponensis*	F	Takaki 1975†
	Lepidoptera: Cosmopterigidae	*Batrachedra amydraula*	F	Venezian & Blumberg 1982; Sayed & Ali 1995
	Lepidoptera: Cossidae	*Prionoxystus robinae*	F	Solomon 1985
		Zeuzera pyrina	F	Plaut 1976; Pasqualini *et al.* 1996
	Lepidoptera: Crambidae	*Uresiphita reversalis*	F	Crosswhite 1985
	Lepidoptera: Dioptidae	*Phryganidia californica*	L/F	Milstead *et al.* 1980
	Lepidoptera: Gelechiidae	*Anarsia lineatella*	F	Chepurnaya 1994; Roltsch *et al.* 1995
		Aproaerema modicella	F	Jayanthi & Padmavathamma 1996a

	Keiferia (Gnorimoschema) lycopersicella	F	Poe & Everett 1974
	Pectinophora gossypiella	L/F	Dulmage 1970a†; Luo *et al.* 1986; Wu 1986
	Phthorimaea operculella	L/F	Ali 1974/1976†, El Sayed *et al.* 1979; Gravena *et al.* 1980
	Scrobipalpuloides (Tuta) absoluta	F	Navarro 1988; Marques & Alves 1996; Reis & de Souza 1998
	Sitotroga cerealella	L/F	McGaughey 1976†; Salama *et al.* 1991b; Salama & Abdel Razek 1992
Lepidoptera: Geometridae	*Alsophila pometaria*	F	Harper 1974†; Appleby *et al.* 1975; Frye *et al.* 1977, 1983
	Apocheima pilosaria	F	Lecheva 1983
	Boarmia (Ascotis) selenaria	L/F	Takaki 1975†; Izhar *et al.* 1979; Cohen *et al.* 1983
	Calospilos (Abraxas) pantaria	F	Kosenko & Anferov 1996
	Colotois (Himera) pennaria	F	Donaubauer & Schmutzenhofer 1973†; Donaubauer 1976
	Ematurga amitaria	F	Sandler & Mason 1997
	Ennomos magnaria	L	Peacock *et al.* 1998
	Ennomos subsignarius	F	Dunbar & Kaya 1972; Dunbar *et al.* 1973
	Erannis (Hibernia) defoliaria	F	Glowacka Pilot 1974; Glowacka Pilot *et al.* 1974
	Eutrapela clemataria	L	Peacock *et al.* 1998
	Itame sulphurea	F	Sandler & Mason 1997
	Lambdina fevidaria	L	Peacock *et al.* 1998
	Lambdina fiscellaria fiscellaria	F	Raske *et al.* 1986; West *et al.* 1987, 1989
	Lycia (Biston) hirtaria	L/F	Lecheva & Kuzmanova 1980; Kuzmanova & Lecheva 1981
	Operophtera bruceata	F	Tonks *et al.* 1978
	Operophtera brumata	F	van der Geest 1971†; Svestka 1974, 1977; Glowacka Pilot *et al.* 1974
	Operophtera fagata	L	Vankova 1973†
	Operophtera sp.	F	Glowacka Pilot 1974
	Paleacrita vernata	F	Harper 1974†; Frye *et al.* 1977, 1983
	Peribatodes rhomboidaria	F	Filip & Blaise 1998
	Prochoerodes transversata	L	Peacock *et al.* 1998
	Mnesampela privata	L/F	Harcourt *et al.* 1996; Neumann & Collett 1997b
	Sabulodes aegrotata	F	Bailey & Olsen 1990
Lepidoptera: Glyphipterigyidae	*Eutromula (Simaethis) pariana*	F	Dronka *et al.* 1976†
Lepidoptera: Gracillariidae	*Caloptilia (Gracillaria) theivora*	F	Takaki 1975†; Kariya 1977; Unnamalai & Vaithilingam 1995
	Phyllonorycter (Lithocolletis) ringoniella	F	Takaki 1975†
Lepidoptera: Heliconidae	*Dione juno juno*	L/F	Villani *et al.* 1980
Lepidoptera: Hesperiidae	*Calpodes ethlius*	L	Reisner *et al.* 1989
	Parnara guttata	F	Takaki 1975†
	Thymelicus lineola	L/F	Thompson 1977; McNeil *et al.* 1977†; Letendre & McNeil 1979
Lepidoptera: Lasiocampidae	*Dendrolimus pini*	L/F	Glowacka Pilot 1974; Kuteev *et al.* 1983; Ziogas *et al.* 1997
	Dendrolimus punctatus	F	Baranovskii *et al.* 1986
	Dendrolimus sibiricus	F	Baranovskii *et al.* 1988
	Dendrolimus spectabilis	F	Ko & Lee 1972; Asano *et al.* 1976
	Eriogaster lanestris	L	Vankova 1973†
	Malacosoma disstria	F	Abrahamson & Harper 1973; Harper 1974†; Johnson & Morris 1981
	Malacosoma neustria	L/F	Mihalache *et al.* 1972, 1974; Vankova 1973†
	Malacosoma neustria var. *testacea*	F	Takaki 1975†
Lepidoptera: Limacodidae	*Darna trima*	F	Pardede 1992
	Monema flavescens	F	Takaki 1975†
	Sibine apicalis	L	Jaramillo Celis *et al.* 1974†

APPENDIX 2.1 (*continued*)

Subspecies/ Serovar	Order: Family	Genus and Species	Lab/Field	References
kurstaki	Lepidoptera: Limacodidae	*Thosea (Setothosea) asigna*	F	Pushparaja *et al.* 1981
	Lepidoptera: Lycaenidae	*Incisalia fotis*	F	Whaley *et al.* 1998
		Lampides boeticus	F	Takaki 1975†
		Lycaeides melissa samuelis	L/F	Herms *et al.* 1997
		Satyrium calanus	F	Wagner *et al.* 1996
	Lepidoptera: Lymantriidae	*Euproctis chrysorrhoea*	L/F	Vankova 1973†; Gurses & Doganay 1976†; Ruelle *et al.* 1977
		Euproctis fraterna	L	Kumar & Jayaraj 1978†; Gangwar *et al.* 1980
		Euproctis lunata	F	Dabi *et al.* 1980a
		Euproctis melania	L/F	El Bahrawi *et al.* 1979
		Euproctis phaeorrhoea	F	Kneifl 1977
		Euproctis pseudoconspersa	F	Takaki 1975†
		Hemerocampa (Orgyia) pseudotsugata	L/F	Morris 1973a; Stelzer *et al.* 1975
		Hypogymna morio	L	Dobrivojevic & Injac 1975†
		Lambdina athasaria pellucidaria	F	Sorensen & Barbosa 1975†
		Leucoma wiltshirei	F	Abai 1981
		Leucoma (Stilpnotia) salicis	F	Donaubauer & Schmutzenhofer 1973†; Szalay Marzso *et al.* 1981
		Lymantria dispar	L/F	Yendol & Hedlund 1972; Dunbar *et al.* 1973; Yendol *et al.* 1973†
		Lymantria monacha	L/F	Bejer Petersen 1974†; Glowacka Pilot 1974; Schonherr & Ketterer 1979
		Orgyia antiqua	F	Lipa *et al.* 1977†; Svestka & Vankova 1978; Niemczyk 1980
		Orgyia gonostigma	F	Trenchev & Pavlov 1982; Ivanova 1984
		Orgyia leucostigma	L	Morris 1973a†; van Frankenhuyzen *et al.* 1992; Bernier *et al.* 1990
		Orgyia pseudotsugata	L/F	Morris 1973a†; Stelzer *et al.* 1975†
		Orgyia thyellina	F	Takaki 1975†
	Lepidoptera: Lyonetiidae	*Bucclatrix thurberiella*	F	Bell & Romine 1982
		Leucoptera malifoliella	F	Pasqualini *et al.* 1992
	Lepidoptera: Noctuidae	*Achaea janata*	L/F	Pal 1977; Deshpande & Ramakrishnan 1982
		Agrochola lychnidis	L	Benuzzi & Antoniacci 1995
		Agrotis ipsilon (ypsilon)	L/F	Salama *et al.* 1990a, b, 1991d
		Agrotis segetum	L/F	Langenbruch 1977†; Agzamova Kh *et al.* 1988
		Alabama argillacea	L/F	Luttrell *et al.* 1981; Gravena *et al.* 1983; Habib & de Andrade 1984
		Amphipyra pyramidoides	L	Peacock *et al.* 1998
		Anomis flava	F	Chen *et al.* 1991b
		Anomis leona	F	Taylor 1974
		Anticarsia gemmatalis	L/F	Yearian *et al.* 1973†; da Silva & Heinrichs 1975; Ignoffo *et al.* 1977†
		Apamea anceps	F	Shternshis & Zurabova 1987
		Autoba olivacea	F	Baskaran & Kumar 1980
		Busseola fusca	F	Medvecky & Zalom 1992
		Catocala coccinata	L	Peacock *et al.* 1998
		Catocala ilia	L	Peacock *et al.* 1998
		Catocala lineella	L	Peacock *et al.* 1998
		Catocala nymphaea	L	Obama *et al.* 1988
		Catocala nymphogoga	L	Obama *et al.* 1988

Catocala vidua	L	Peacock *et al.* 1998
Cerapteryx graminis	L	Vlug *et al.* 1988
Chrysodeixis chalcites (*eriosoma*)	F	Martin & Workman 1986; Wysoki 1989; Broza & Sneh 1994
Earias biplaga	L/F	Taylor 1974; Frutos *et al.* 1987
Earias huegeli	F	Wilson 1981
Earias insulana	L/F	Taylor 1974; Saad *et al.* 1977; Moawad *et al.* 1982
Earias vittella	F	Krishnaiah *et al.* 1981; Satapathy & Panda 1997
Egira alterans	L	Peacock *et al.* 1998
Eublemma amabilis	L	Sen *et al.* 1993
Euxoa auxiliaris	F	Bauernfeind & Wilde 1993
Euxoa messoria	L	Cheng 1973†
Evergestis forficalis	F	Langenbruch 1984b
Helicoverpa (*Heliothis*) *armigera*	F	Taylor 1974; Roome 1975†
Helicoverpa (*Heliothis*) *punctigera*	L/F	Cooper 1984; Teakle *et al.* 1992
Helicoverpa (*Heliothis*) *zea*	L/F	Rogoff *et al.* 1969†; Creighton *et al.* 1971†; Janes 1973
Heliothis virescens	L/F	Mistric & Smith 1973a; Dulmage & Martinez 1973 ; Johnson 1974
Heliothis (*Chloridea*) *assulta*	L/F	Takaki 1975†; Chung & Hyun 1980
Heliothis (*Protoschinia*) *scutosa* (*nervosa*)	F	Mikolajewicz *et al.* 1992
Hellula rogatalis	F	Waites *et al.* 1978
Hellula undalis	F	Anon. 1987b
Lacanobia oleracea	L/F	Chan Tkho 1973†; Hussey *et al.* 1976; Burges & Jarrett 1978†
Lithophane grotei	L	Peacock *et al.* 1998
Mamestra (*Barathra*) *brassicae*	L/F	Asano *et al.* 1973a; Vankova 1973†; Takaki 1975†
Mamestra (*Barathra*) *configurata*	L/F	Morris 1986; Trottier *et al.* 1988; Morris *et al.* 1998
Mamestra (*Barathra*) *suasa*	F	Langenbruch 1984b
Noctua pronuba	L/F	Burges & Jarrett pers. com.†
Orthosia rubescens	F	Wagner *et al.* 1996
Panolis flammea	L/F	Swiezynska & Glowacka Pilot 1973; Adomas 1994
Peridroma saucia	F	Lindquist 1977
Phoberia atomaris	F	Wagner *et al.* 1996
Plathypena scabra	L/F	Yearian *et al.* 1973†; Ignoffo *et al.* 1977†
Psaphida resumens	L	Peacock *et al.* 1998
Psaphida rolandi	L	Peacock *et al.* 1998
Pseudoplusia (*Chrysodeixis*) *includens*	L/F	Yearian *et al.* 1973†; Ignoffo *et al.* 1977†; Morales *et al.* 1995
Sesamia cretica	F	Abul Nasr *et al.* 1968
Sesamia inferens	L	Nayak *et al.* 1978†; Yadava 1978
Spodoptera exempta	L/F	Bai & Degheele 1993
Spodoptera exigua	L/F	Vail *et al.* 1972†; Ignoffo *et al.* 1977†; Moar & Trumble 1987a
Spodoptera frugiperda	F	Creighton *et al.* 1972†; Janes 1973; Law & Mills 1980
Spodoptera latifascia	L	Habib 1986
Spodoptera littoralis	L/F	Abul Nasr & Abdallah 1970; Narayanan *et al.* 1970; Asano *et al.* 1973a
Spodoptera litura	L/F	Babu & Subramaniam 1973; Takaki 1975†; Yang *et al.* 1985
Trichoplusia ni	L/F	Rogoff *et al.* 1969†; Narayanan *et al.* 1970; Jaques 1972†
Xylena nupera	F	Sandler & Mason 1997
Xystopeplus rugago	L	Peacock *et al.* 1998
Zale aeruginosa	L	Peacock *et al.* 1998

APPENDIX 2.1 *(continued)*

Subspecies/ Serovar	Order: Family	Genus and Species	Lab/Field	References
kurstaki	Lepidoptera: Notodontidae	*Botyoides asialis*	L	Kalia & Joshi 1996
		Clostera cupreata	L	Kalia & Joshi 1996
		Drymonia chaonia (*ruficornis*)	F	Pirvescu 1973; Mihalache *et al.* 1974
		Drymonia ruficornis (as *chaonia*)	F	Pirvescu 1973†
		Eustema sericea	L	Porto Santos *et al.* 1995
		Heterocampa gutticitta	F	Wallner 1971†
		Heterocampa manteo	L	Ignoffo *et al.* 1973†
		Ichthyura anastomosis	L/F	Chaudhry & Shah 1977; Shah *et al.* 1979
		Phalantha phalantha	L	Kalia & Joshi 1996
		Schizura concinna	F	Pinnock *et al.* 1974†; Band *et al.* 1976
	Lepidoptera: Nymphalidae	*Astercampa clyton*	L	Peacock *et al.* 1998
		Brassolis sophorae	L	Habib *et al.* 1986
		Limenitis arthemis astyanax	L	Peacock *et al.* 1998
		Speyeria diana	L	Peacock *et al.* 1998
		Vanessa (*Cynthia*) *cardui*	F	Miranpuri *et al.* 1993
		Vanessa turkestanica	F	Guzeev 1986
	Lepidoptera: Oecophoridae	*Depressaria nervosa* (*daucella*)	F	Mikolajewicz *et al.* 1992; Cate 1997
	Lepidoptera: Olethreutidae	*Ecdytolopha aurantiana*	F	Scarpellini & dos Santos 1997
		Epinotia tsugana	L	Morris 1969†
	Lepidoptera: Papilionidae	*Papilio canadensis*	F	Johnson *et al.* 1995; Scriber 1998
		Papilio glaucus	L/F	Johnson *et al.* 1995; Haas & Scriber 1998; Scriber 1998
		Papilio palamedes	L	Scriber 1998
		Papilio troilus	L	Scriber 1998
		Papilio xuthus	L	Gao *et al.* 1985
	Lepidoptera: Phycitidae	*Elasmopalpus lignosellus*	L	Gilreath & Funderburk 1987
		Homeosoma electellum	L/F	Carlson 1975; Rogers *et al.* 1984; Brewer 1991
	Lepidoptera: Phyllocnistidae	*Phyllocnistis citrella*	L	Shapiro *et al.* 1998
	Lepidoptera: Pieridae	*Aporia crataegi*	F	Langenbruch unpubl. obs.†; Niccoli & Pelagatti 1986
		Pieris brassicae	F	Galowalia *et al.* 1973†; Langenbruch 1984b
		Pieris canidia	L	Cheung & Lam 1993; Wang & Cheung 1994
		Pieris rapae	F	Jaques 1972†; Kouskolekas & Harper 1973; Takaki 1975†
		Pieris rapae crucivora	F	Asano *et al.* 1973a
	Lepidoptera: Plutellidae	*Leptophobia aripa*	F	Garcia 1991
	Lepidoptera: Psychidae	*Canephora asiatica*	F	Takaki 1975†
		Oiketicus kirbyi	L	Almela Pons *et al.* 1972; Escobar *et al.* 1979
		Oiketicus moyanoi	L	Almela Pons *et al.* 1972†
		Thyridopteryx ephemeraeformis	L/F	Bishop *et al.* 1973; Kearby *et al.* 1975; Gill & Raupp 1994
	Lepidoptera: Pterophoridae	*Exelastia atomosa*	F	Puntambekar *et al.* 1997
		Platyptilia carduidactyla	F	Bari & Kaya 1984
	Lepidoptera: Pyralidae	*Acrobasis nuxvorella*	F	Sundaram *et al.* 1997
		Aganais (*Hypsa*) *ficus*	L	Gangwar *et al.* 1980
		Antigastra catalaunalis	F	Selvanarayanan & Baskaran 1996
		Cadra spp.	F	Mohamed 1993; Sayed & Temerak 1995

	Chilo infuscatellus	F	Kennedy 1993
	Chilo partellus	F	Khan & Khan 1968; Brownbridge 1990
	Chilo suppressalis	L	Rombach *et al.* 1989
	Cnaphalocrocis medinalis	L/F	Srivastava & Nayak 1978†; Aguda *et al.* 1988; Latha *et al.* 1994
	Corcyra cephalonica	L	El Moursy *et al.* 1996; do Amaral Filho & de Carvalho 1996
	Crocidolomia binotalis	F	Krishnaiah *et al.* 1981; Babu & Krishnayya 1998
	Cryptoblabes gnidiella	F	Wysoki *et al.* 1975†; Sternlicht 1979; Wysoki *et al.* 1986
	Diaphania (Palpita) hyalinata	L/F	Delplanque & Gruner 1975; Delphanque *et al.* 1974
	Diaphania nitidalis	F	Dougherty & Schuster 1985
	Diatraea grandiosella	F	Nolting & Poston 1982
	Diatraea saccharalis	F	Al Badry *et al.* 1972; Charpentier *et al.* 1973†; Sarmiento & Razuri 1978
	Dioryctria amatella	L/F	McLeod *et al.* 1982, 1984
	Ectomyelois (Apomyelois) ceratoniae	L/F	Harpaz & Wysoki 1984; Dhouibi 1992; Mart & Kilincer 1992
	Eldana saccharina	L/F	Moyal 1988; Jacobs 1989
	Ephestia cautella	L/F	Dunbar *et al.* 1973†; McGaughey 1979†; Faruki & Khan 1996
	Ephestia elutella	F	Rassmann 1986; Eberhardt 1997
	Ephestia kuehniella	L	del Bene & Melis Porcinai 1980; Ignatowicz 1997
	Etiella behrii	L	Cooper 1983
	Eutectona machaeralis	F	Meshram *et al.* 1997
	Evergestis rimosalis	F	Waites *et al.* 1978
	Galleria mellonella	L/F	Ali *et al.* 1973†
	Herpetogramma phaeopteralis	F	Reinert 1974†
	Herpetogramma (Psara) bipunctalis	F	Delplanque & Gruner 1975
	Hymenia (Kinckenia) recurvalis (fascialis)	L/F	Delplanque & Gruner 1975; Gangwar *et al.* 1980
	Leucinodes orbonalis	F	Sekar & Baskaran 1976; Baskaran & Kumar 1980
	Loxostege sticticalis	L	Iacob & Iacob 1977†; Kornilov *et al.* 1982
	Macalla thyrsisalis	L/F	Howard 1990
	Marasmia patnalis	F	Latha *et al.* 1994
	Mussidia nigrivenella	F	Moyal 1988
	Mythimna separata	L	Sha & Xie 1992; Feng *et al.* 1995
	Noorda blitealis	F	Kalia & Joshi 1997
	Ostrinia furnacalis	F	Morallo Rejesus & Javier 1985; Morallo Rejesus 1988; Yen 1988
	Ostrinia nubilalis	L/F	McWhorter *et al.* 1972; Schubert & Stengel 1974†; Lynch *et al.* 1977
	Paramyelois (Amyelois) transitella	F	Pinnock & Milstead 1972†; Kellen *et al.* 1977
	Plodia interpunctella	L/F	Nwanze *et al.* 1975; Kinsinger & McGaughey 1976; McGaughey 1978†
	Polytela gloriosae	L	Gangwar *et al.* 1980
	Schoenobius bipunctifer	L	Nayak *et al.* 1978†
	Scirpophaga (Tryporyza) incertulas	L/F	Nayak *et al.* 1978; Israel & Padmanabhan 1978
	Sylepta (Syllepte) balteata	F	Kalia *et al.* 1997
	Sylepta (Syllepte) derogata	L/F	Taylor 1974†
	Symmerista canicosta	L	Millars *et al.* 1974†
	Tetralopha subcanalis	F	Polles & Payne 1975
Lepidoptera: Pyralidae?	*Mesographe forficalis*	L	Langenbruch unpubl. obs.†
Lepidoptera: Saturnidae	*Antheraea polphemus*	L	Peacock *et al.* 1998

APPENDIX 2.1 *(continued)*

Subspecies/ Serovar	Order: Family	Genus and Species	Lab/Field	References
kurstaki	Lepidoptera: Saturnidae	*Actias luna*	L	Peacock *et al.* 1998
		Anisota senatoria	L/F	Kaya 1974†
		Hemileuca maia	F	Wagner *et al.* 1996
		Callophrys sheridanii	F	Whaley *et al.* 1998
		Callosamia primethea	F	Johnson *et al.* 1995
	Lepidoptera: Satyridae	*Neominois ridingsii*	F	Whaley *et al.* 1998
	Lepidoptera: Sesiidae	*Synanthedon myopaeformis*	F	Balazs *et al.* 1996
	Lepidoptera: Sphingidae	*Erinnyis ello*	L/F	Abreu 1974†, 1984; Cruz 1977†
		Manduca quinquemaculata	L/F	Cheng & Hanlon 1990
		Manduca sexta	L/F	Mistric & Smith 1973b; Tyrell *et al.* 1981a, b
	Lepidoptera: Stathmopodidae	*Kakivoria flavofasciata*	F	Takaki 1975†
		Stathmopoda spp.	F	McKenna *et al.* 1995
	Lepidoptera: Thaumetopoeidae	*Thaumetopoea pityocampa*	L/F	Videnova *et al.* 1972†; Perez Ibanez *et al.* 1973
		Thaumetopoea processionea	F	Niccoli & Pelagatti 1986; Roversi *et al.* 1997
	Lepidoptera: Tineidae	*Opogona sacchari*	F	Heppner *et al.* 1987
	Lepidoptera: Tortricidae	*Acleris variana*	F	West & Carter 1992
		Adoxophyes orana	L/F	van der Geest & Velterop 1971†; Kreidl 1974; Takaki 1975
		Adoxophyes sp.	L/F	Kariya 1977; Asano *et al.* 1994
		Agriopis (Erannis) marginaria	L	Kuzmanova & Lecheva 1984
		Archips argyrospilus	F	Madsen *et al.* 1977; Pinnock & Milstead 1978†; Roshchin *et al.* 1986
		Archips podanus (podana)	F	Pasqualini *et al.* 1992
		Archips xylosteana	F	Milalache *et al.* 1972†; Niemczyk & Dronka 1976; Niemczyk 1980
		Archips spp.	F	Gamayunova 1987
		Archips (Cacoecia) rosana	L/F	Vankova 1973†; Ali Niazee 1974†
		Archips (Cacoecia) rosanus	F	AliNiazee 1974; Vankova 1973; Niemczyk & Dronka 1976
		Argyrotaenia pulchellana	F	Nicoli *et al.* 1990
		Arotrophora arcuatalis	F	Rohl & Woods 1994
		Autographa confusa	F	Mikolajewicz *et al.* 1992
		Autographa gamma	L/F	Burges & Jarrett pers. com.†; Scopes & Biggerstaff 1974
		Autographa nigrisigna	L/F	Takaki 1975†
		Cacoecimorpha pronubana	L/F	Scopes & Biggerstaff 1974; Burges & Jarrett 1978; Wysoki 1989
		Choristoneura conflictana	F	Holsten & Hard 1985
		Choristoneura fumiferana	L/F	Smirnoff 1965†; Morris 1973a
		Choristoneura occidentalis	F	Maksymiuk & Orchard 1974; Kaya & Reardon 1982
		Choristoneura pinus	L	van Frankenhuyzen & Fast 1989
		Choristoneura pinus pinus	F	Cadogan *et al.* 1986; Cadogan 1993
		Choristoneura rosaceana	L	Li *et al.* 1995; Li & Fitzpatrick 1997
		Clepsis spectrana	L/F	Burges & Jarrett unpubl. obs.†
		Cnephasia longana	F	Norton 1991
		Cochylis (Hysterosia) hospes	L	Barker 1998
		Cryptophlebia leucotreta	F	Moyal 1988
		Ctenopseustis obliquana	F	Wearing & Thomas 1978
		Cydia (Grapholita) delineana	F	Peteanu 1980

			Dirimanov *et al.* 1980
	Desmia funeralis	F	AliNiazee & Jensen 1973†; Biever & Hostetter 1975
	Dioryctria abietivorella	L	Trudel *et al.* 1997
	Epichoristodes acerbella	F	del Bene 1981
	Epiphyas postvittana	L/F	Buchanan 1977†; Wearing & Thomas 1978; Bailey *et al.* 1996
	Eupoecilia ambiguella	F	Schmid & Antonin 1977†
	Hedya (Argyroploce) nubiferana (variegana)	F	Niemczyk *et al.* 1976; Niemczyk & Dronka 1976
	Homona magnanima	L/F	Takaki 1975†; Kariya 1977
	Lobesia botrana	F	Buess & Bassand 1976; Schmid & Antonin 1977†; El Hakim *et al.* 1988
	Pammene fasciana	F	Tsankov & Ovcharov 1986
	Pandemis cerasana	F	Nicoli *et al.* 1990; Pasqualini *et al.* 1992
	Pandemis heparana	F	Dirimanov *et al.* 1980; Injac & Dulic 1982
	Paralobesia viteana	F	Biever & Hostetter 1975†
	Phlogophora meticulosa	L/F	Burges & Jarrett pers. com.†; Scopes & Biggerstaff 1974
	Planotortrix excessana	F	Wearing & Thomas 1978
	Platynota idaeusalis	L	Rock & Monroe 1983
	Rhyacionia frustrana	F	Menendez *et al.* 1986
	Spilonota ocellana	F	Niemczyk & Dronka 1976; Niemczyk *et al.* 1976; Niemczyk 1980
	Tortrix viridana	F	Svestka 1974†, 1977; Kuteev *et al.* 1983
	Zeiraphera canadensis	F	Eidt & Dunphy 1991
	Zeiraphera diniana	L/F	Schmid 1975; Svestka & Vankova 1984; Kudler 1984
Lepidoptera: Yponomeutidae	*Acrolepia alliella*	F	Takaki 1975†
	Argyresthia thuiella	F	Langenbruch 1984a
	Atteva fabriciella	L	Joshi *et al.* 1996
	Plutella xylostella (maculipennis)	L/F	Narayanan *et al.* 1970; Asano *et al.* 1973a; Takaki 1975†
	Prays citri	F	Moreno Vazquez *et al.* 1972
	Prays oleae	F	Yamvrias 1972†; Yamvrias & Young 1977
	Swammerdamia pyrella	F	Niemczyk 1980
	Yponomeuta evonymella	L/F	Hamed 1978†
	Yponomeuta malinella	L	Vankova 1973†
	Yponomeuta (padellus) malinellus	F	Vankova 1973; Niemczyk 1980
	Yponomeuta padella	L	Hamed 1978†
Lepidoptera: Zygaenidae	*Harrisina brillians*	L/F	Pinnock *et al.* 1973†; Stern *et al.* 1980; 1983
	Theresimima (Ino) ampelophaga	F	Kharizanov & Kharizanov 1983
Mestostigmata: Tetranychidae	*Panonychus ulmi*	F	Dirimanov *et al.* 1980
Mollusca: Planorbidae	*Biomphalaria alexandrina*	L	Osman & Mohamed 1991; Osman *et al.* 1992
Nematoda: Aphelenchidae	*Aphelenchus avenae*	L	Shevtsov *et al.* 1996
Nematoda: Heteroderidae	*Heterodera glycines*	GH/F	Sharma & Gomes 1996
Nematoda: Meloidgynidae	*Meloidogyne hapla*	L	Shevtsov *et al.* 1996
	Meloidogyne incognita	L/F	Chahal & Chahal 1993; Mena *et al.* 1996
	Meloidogyne javanica	F	Osman *et al.* 1988
	Tylenchulus semipenetrans	F	Osman *et al.* 1988
Nematoda: Pratylenchidae	*Radopholus similis*	F	Mena *et al.* 1996; 1997
Nematoda: Steinernematidae	*Steinernema (Neoaplectana) feltiae (carpocapsae)*	L	Kaya & Burlando 1989
Nematoda: Trichostrongylidae	*Trichostrongylus colubriformis*	L	Bottjer *et al.* 1985 (in patent No. 05281530, A.J. Sick)

APPENDIX 2.1 (*continued*)

Subspecies/ Serovar	Order: Family	Genus and Species	Lab/Field	References
kurstaki	Nematoda: Trichostrongylidae	?		Meadows *et al.* 1989
	Neuroptera: Chrysopidae	*Chrysopa formosa*	L	Babrikova & Kuzmanova 1984
		Chrysopa perla	L	Babrikova & Kuzmanova 1984
		Chrysopa septempunctata	L	Babrikova & Kuzmanova 1984
		Chrysoperla (*Chrysopa*) *carnea*	L/F	Franz *et al.* 1980; Haverty 1982; Jayanthi & Padmavathamma 1996
	Phthiraptera : Linognathidae	*Linognathus stenopsis*	F	Singh *et al.* 1992
	Phthiraptera : Menonponidae	*Menopon gallinae*	L	Gingrich *et al.* 1974†; Lonc *et al.* 1986; Lonc & Lachowicz 1987
		Eomenacanthus (*Menacanthus*) *stramineus*	L	Lonc *et al.* 1986
	Phthiraptera : Trichodectidae	*Damalinia* (*Bovicola*) *caprae*	F	Singh *et al.* 1992
		Damalinia (*Bovicola*) *ovis*	L	Drummond *et al.* 1992; 1995
		Heterodoxus spiniger	F	Singh *et al.* 1992
	Plecoptera: ?	*Taeniopteryx nivalis*	L	Kreutzweiser *et al.* 1992
	Plecoptera: Leuctridae	*Leuctra tenuis*	L/F	Kreutzweiser *et al.* 1994
	Prostigmata: Tetranychidae	*Tetranychus urticae turkistani*	F	Saad *et al.* 1977
		Tetranychus viennensis	F	Chepurnaya 1994
	Siphonaptera: Pulicidae	*Xenopsylla cheopis*	L	Maciejewska *et al.* 1988
	Thysanoptera: Thripidae	*Frankliniella* spp.	F	Schuster & Engelhard 1979
kurstaki?	Lepidoptera: Lasiocampidae	*Dendrolimus spectabilis*	L	Yao *et al.* 1998
kurstaki? (ser 3)		*Dendrolimus superans*	L/F	Zhao *et al.* 1998
kurstaki? (ser 3)	Lepidoptera: Lymantriidae	*Lymantria dispar*	L	Dubois & Squires 1971†
kurstaki? (ser 3)	Lepidoptera: Noctuidae	*Hellula undalis*	F	Anon. 1987b
kurstaki?	Lepidoptera: Pieridae	*Pieris rapae*	L	Yao *et al.* 1998
kurstaki?	Lepidoptera: Pyralidae	*Ostrinia furnacalis*	L	Yao *et al.* 1998
kurstaki? (ser 3)	Lepidoptera: Tortricidae	*Archips argyrospilus*	F	Pinnock & Milstead 1978
kyushuensis	Diptera: Chironomidae	*Paratanytarsus grimmii*	L	Kondo *et al.* 1995a
	Diptera: Culicidae	*Aedes aegypti*	L	Ishii & Ohba 1993a, 1994
		Culex pipiens molestus	L	Saitoh *et al.* 1996
		Culex pipiens pallens	L	Ishii & Ohba 1993a
		Culex tritaeniorhynchus	L	Ohba & Aizawa 1979†
	Diptera: Psychodidae	*Telmatoscopus albipunctatus*	L	Saitoh *et al.* 1996
	Lepidoptera: Gelechiidae	*Phthorimaea operculella*	L	Salama *et al.* 1995a
		Sitotroga cerealella	L	Salama *et al.* 1991b
	Lepidoptera: Pyralidae	*Plodia interpunctella*	L	Salama *et al.* 1991b
	Phthiraptera: Trichodectidae	*Damalinia* (*Bovicola*) *ovis*	L	Drummond *et al.* 1992
leesis	Diptera: Psychodidae	*Telmatoscopus* (*Clogmia*) *albipunctatus*	L	Higuchi *et al.* 1998b
	Lepidoptera: Bombycidae	*Bombyx mori*	L	Lee *et al.* 1994
	Lepidoptera: Yponomeutidae	*Plutella xylostella*	L	Higuchi *et al.* 1998b
malaysiensis	Diptera: Culicidae	*Aedes aegypti*	L	Lee & Seleena 1990a; de Barjac *et al.* 1990; Ragni *et al.* 1996
		Anopheles maculatus	L	de Barjac *et al.* 1990
		Anopheles stephensi	L	Ragni *et al.* 1996

medellin	Diptera: Culicidae	*Aedes aegypti*	L	Orduz *et al.* 1992; Diaz *et al.* 1993; Mercer *et al.*1995
		Aedes polynesiensis	L	Mercer *et al.* 1995
		Anopheles stephensi	L	Orduz *et al.* 1992; Diaz *et al.* 1993
		Culex pipiens	L	Orduz *et al.* 1992; Diaz *et al.* 1993
		Culex quinquefasciatus	L	Orduz *et al.* 1992; Mercer *et al.* 1995
mexicanensis	Diptera: Culicidae	*Culex pipiens molestus*	L	Saitoh *et al.* 1996
	Diptera: Psychodidae	*Telmatoscopus albipunctatus*	L	Saitoh *et al.* 1996
morrisoni	Coleoptera: Chrysomelidae	*Colaspidema atrum*	L	Iriarte *et al.* 1998
	Diptera: Calliphoridae	*Lucilia cuprina*	L	Lyness *et al.* 1994
	Diptera: Culicidae	*Aedes aegypti*	L	Padua *et al.* 1981; patent No. 05298245, J.M. Payne 1994
		Aedes albopictus	L	Merca *et al.* 1997
		Aedes taeniorhynchus	L	Lacey *et al.* 1988
		Anopheles aibimanus	L	Lacey *et al.* 1988
		Anopheles gambiae	L	Granum *et al.* 1988
		Anopheles quadrimaculatus	L	Lacey *et al.* 1988
		Culex pipiens pallens	L	Park *et al.* 1998
		Culex quinquefasciatus	L	Hall *et al.* 1977†; Granum *et al.* 1988
		Culex salinarius	L	Lacey *et al.* 1988
		Culex tarsalis	L	Hall *et al.* 1977†
		Psorophora columbiae	L	Lacey *et al.* 1988
	Diptera: Muscidae	*Haematobia irritans*	L	Ames *et al.* 1982; Haufler & Kunz 1985
		Musca domestica	L	Lonc *et al.* 1991
	Hemiptera: Reduviidae	*Triatoma vitticeps*	L	Lima *et al.* 1994
	Lepidoptera: Arctiidae	*Hyphantria cunea*	L	Akiba 1986b
	Lepidoptera: Bombycidae	*Bombyx mori*	L	Angus & Norris 1968†; Akiba 1986b; Hastowo *et al.* 1992
	Lepidoptera: Gelechiidae	*Pectinophora gossypiella*	F	Luo *et al.* 1986
	Lepidoptera: Lymantriidae	*Euproctis chrysorrhoea*	F	Cabral 1978
		Lymantria dispar	L/F	Dubois & Squires 1971†; Cabral 1973, 1978, 1980
	Lepidoptera: Noctuidae	*Helicoverpa* (*Heliothis*) *armigera*	L	Iriarte *et al.* 1998
		Spodoptera littoralis	L	Kalfon & de Barjac 1985
		Spodoptera litura	L	Asano & Hori 1995
	Lepidoptera: Pieridae	*Pieris brassicae*	L	Galowalia *et al.* 1973†; Granum *et al.* 1988
	Lepidoptera: Pyralidae	*Ostrinia furnacalis*	L	Padua *et al.* 1987; Yao *et al.* 1998; Cheng *et al.* 1998
	Lepidoptera: Tortricidae	*Choristoneura fumiferana*	L	Granum *et al.* 1988
		Cydia (*Laspeyresia*) *pomonella*	L	Andermatt *et al.* 1988
	Nematoda: Cephalobidae	*Turbatrix aceti*	F	Meadows *et al.* 1990
	Nematoda: Trichostrongylidae	*Trichostrongylus colubriformis*	L	Meadows *et al.* 1989
	Phthiraptera: Menonponidae	*Menopon gallinae*	L	Lonc *et al.* 1988
	Phthiraptera: Trichodectidae	*Damalinia* (*Bovicola*) *ovis*	L	Drummond *et al.* 1992
neoleonensis	Diptera: Agromyzidae	*Liriomyza trifolii*	L	Patent No. 05298245 & 05436002, J.M. Payne 1994, 1995
	Diptera: Culicidae	*Aedes aegypti*	L	Patent No. 05298245 & 05436002, J.M. Payne 1994, 1995
	Diptera: Muscidae	*Musca domestica*	L	Patent No. 05298245 & 05436002, J.M. Payne 1994, 1995
	Lepidoptera: Noctuidae	*Heliothis virescens*	L	Rodriguez Padilla *et al.* 1990

APPENDIX 2.1 *(continued)*

Subspecies/ Serovar	Order: Family	Genus and Species	Lab/Field	References
neoleonensis	Lepidoptera: Pyralidae	*Ephestia kuehniella*	L	Rodriguez Padilla *et al.* 1990
nigeriensis	Lepidoptera: Noctuidae	*Helicoverpa (Heliothis) armigera*	L	Iriarte *et al.* 1998
novosibirsk	Lepidoptera: Noctuidae	*Helicoverpa (Heliothis) armigera*	L	Iriarte *et al.* 1998
ostriniae	Lepidoptera: Geometridae	*Boarmia (Ascotis) selenaria*	L	Wysoki & Scheepens 1988
	Phthiraptera: Trichodectidae	*Damalinia (Bovicola) ovis*	L	Drummond *et al.* 1992
oswaldocruzi	Lepidoptera: Noctuidae	*Spodoptera littoralis*	L	Rabinovitch *et al.* 1995
oyamensis	Coleoptera: Bruchidae	*Callosobruchus maculatus*	L	Lopez Meza & Ibarra 1996
	Diptera: Culicidae	*Anopheles pseudopunctipennis*	L	Lopez Meza & Ibarra 1996
		Culex pipiens pallens	L	Ono *et al.* 1988
	Lepidoptera: Noctuidae	*Spodoptera frugiperda*	L	Lopez Meza & Ibarra 1996
	Orthoptera: Gryllidae	*Acheta* sp.	L	Lopez Meza & Ibarra 1996
pakistani	Diptera: Culicidae	*Aedes aegypti*	L	Yarnykh *et al.* 1981b
	Lepidoptera: Gelechiidae	*Phthorimaea operculella*	L	Salama *et al.* 1995a
		Sitotroga cerealella	L	Salama *et al.* 1991b
	Lepidoptera: Pieridae	*Pieris brassicae*	L	de Barjac *et al.* 1977†
	Lepidoptera: Pyralidae	*Ephestia kuehniella*	L	de Barjac *et al.* 1977†
		Plodia interpunctella	L	Salama *et al.* 1991b
	Phthiraptera: Menonponidae	*Menopon gallinae*	L	Lonc & Lachowicz 1987
	Phthiraptera: Trichodectidae	*Damalinia (Bovicola) ovis*	L	Drummond *et al.* 1992
pondicheriensis	Diptera: Culicidae	*Culex quinquefasciatus*	L	Manonmani *et al.* 1987
	Lepidoptera: Yponomeutidae	*Plutella xylostella*	L	Kim *et al.* 1998b
shandongiensis	Diptera: Culicidae	*Culex pipiens molestus*	L	Saitoh *et al.* 1996
sotto	Diptera: Fanniidae	*Fannia canicularis*	L	Hall *et al.* 1972
		Fannia femoralis	L	Hall *et al.* 1972
	Hymenoptera: Apidae	*Apis mellifera*	L	Cantwell *et al.* 1966†
	Hymenoptera: Diprionidae	*Neodiprion lecontei*	L	Heimpel 1961†
	Hymenoptera: Tenthredinidae	*Pristiphora erichsonii*	L	Heimpel 1961†
	Lepidoptera: Arctiidae	*Arctia caja*	L	Burgerjon & Biache 1967b†
		Datana integerrima	L	Angus 1956b†
		Datana ministra	L/F	Angus 1956b†; Angus & Heimpel 1959†
		Halisidota caryae	L	Angus & Heimpel 1959†
		Hyphantria cunea	L	Angus & Heimpel 1959†; Miyamoto & Aizawa 1982
		Hyphantria sp.	L	Aizawa 1975
		Spilosoma lubricipeda (as *menthastri*)	L	Burgerjon & Biache 1967b†

	Sitotroga cerealella	L	Salama *et al.* 1991b
Lepidoptera: Geometridae	*Erannis tiliaria*	L	Angus 1956a†
Lepidoptera: Lasiocampidae	*Malacosoma americana*	F	Angus & Heimpel 1959†
	Malacosoma disstria	L/F	Angus & Heimpel 1959†
	Malacosoma neustria	L	Burgerjon & Biache 1967b†
Lepidoptera: Lymantriidae	*Euproctis chrysorrhoea*	L	Burgerjon & Biache 1967a†
	Leucoma (Stilpnotia) salicis	L	Burgerjon & Biache 1967b†
	Lymantria dispar	L	Burgerjon & Biache 1967b†; Dubois & Squires 1971†; Ridet 1973
	Orgyia leucostigma	L	Rossmoore *et al.* 1970†
Lepidoptera: Noctuidae	*Achaea janata*	L	Deshpande & Ramakrishnan 1982
	Agrotis ipsilon (ypsilon)	L	Salama & Foda 1984
	Earias insulana	L	Salama & Foda 1984
	Mamestra (Barathra) brassicae	L/F	Burgerjon & Biache 1967b†; Takaki 1975†
	Mythimna (Pseudaletia) unipuncta	L	Burgerjon & Biache 1967b†
	Spodoptera exigua	L	Iriarte *et al.* 1998
	Spodoptera frugiperda	L	Hernandez 1988
	Spodoptera litura	L	Mike *et al.* 1991
	Xestia (Amathes) c-nigrum	L	Burgerjon & Biache 1967b†
Lepidoptera: Notodontidae	*Clostera (Pygaera) anastomosis*	L	Burgerjon & Biache 1967b†
Lepidoptera: Nymphalidae	*Nymphalis antiopa*	L	Angus 1956b†
	Vanessa cardui	L	Angus & Heimpel 1959†
	Vanessa io	L	Burgerjon & Biache 1967b†
Lepidoptera: Pieridae	*Pieris brassicae*	L	Burgerjon & Biache 1967b†; Galowalia *et al.* 1973†
	Pieris rapae	F	Takaki 1975†; Aizawa 1975
Lepidoptera: Pyralidae	*Galleria mellonella*	L	Vankova 1966†
	Hypsipyla grandella	L	Hidalgo Salvatierra & Grijpma 1973
	Ostrinia nubilalis	L	Maini *et al.* 1989
	Plodia interpunctella	L	Salama *et al.* 1991b
Lepidoptera: Saturnidae	*Antheraea pernyi*	L	Burgerjon & Biache 1967b†
	Anisota senatoria	L/F	Angus & Heimpel 1959†
	Dryocampa rubicunda	L/F	Angus 1956a†; Angus & Heimpel 1959†
	Samia cynthia	L	Burgerjon & Biache 1967b†
	Saturnia pavonia	L	Burgerjon & Biache 1967b†
Lepidoptera: Sphingidae	*Manduca quinquemaculata*	L	Angus 1956a†
	Manduca sexta	L/F	Guthrie *et al.* 1959†
Lepidoptera: Thaumetopoeidae	*Thaumetopoea processionea*	L	Burgerjon & Biache 1967b†
Lepidoptera: Tineidae	*Tineola bisselliella*	L	Yamvrias & Angus 1969†
Lepidoptera: Tortricidae	*Autographa nigrisigna*	F	Takaki 1975†
	Choristoneura fumiferana	L	Smirnoff 1965†; Yamvrias & Angus 1970†; Morris & Moore 1983
	Cydia (Laspeyresia) pomonella	L	Andermatt *et al.* 1988
	Tortrix viridana	L	Burgerjon & Biache 1967b†
Lepidoptera: Yponomeutidae	*Plutella xylostella (maculipennis)*	L	Burgerjon & Biache 1967b†; Aizawa 1975; Takaki 1975†
	Yponomeuta cognatella (evonymi)	L	Burgerjon & Biache 1967b†
	Yponomeuta malinella	L	Burgerjon & Biache 1967b†
	Yponomeuta padella	L	Burgerjon & Biache 1967b†

APPENDIX 2.1 (*continued*)

Subspecies/ Serovar	Order: Family	Genus and Species	Lab/Field	References
subtoxicus	Lepidoptera: Gelechiidae	*Pectinophora gossypiella*	L	Salama & Foda 1984
		Phthorimaea operculella	L	Salama *et al.* 1995a
		Sitotroga cerealella	L	Salama *et al.* 1991b
	Lepidoptera: Lymantriidae	*Euproctis chrysorrhoea*	L	Cabral 1980
		Lymantria dispar	F	Cabral 1973; Ridet 1973
	Lepidoptera: Noctuidae	*Earias insulana*	L	Salama & Foda 1984
		Helicoverpa (*Heliothis*) *armigera*	L	Salama *et al.* 1983a
		Spodoptera exigua	L	Salama *et al.* 1983a
		Spodoptera littoralis	L	Salama *et al.* 1983a
		Spodoptera litura	L	Puntambekar *et al.* 1997
	Lepidoptera: Psychidae	*Oiketicus moyanoi*	L/F	Solanes *et al.* 1974
	Lepidoptera: Pyralidae	*Galleria mellonella*	L	Mohrig *et al.* 1979
		Plodia interpunctella	L	Salama *et al.* 1991b
	Lepidoptera: Tortricidae	*Cydia* (*Laspeyresia*) *pomonella*	L	Andermatt *et al.* 1988
	Phthiraptera: Trichodectidae	*Damalinia* (*Bovicola*) *ovis*	L	Drummond *et al.* 1992
sumiyoshiensis	Diptera: Culicidae	*Culex tritaeniorhynchus*	L	Ohba & Aizawa 1989b
	Lepidoptera: Bombycidae	*Bombyx mori*	L	Ohba & Aizawa 1989b
	Lepidoptera: Noctuidae	*Spodoptera exigua*	L	Iriarte *et al.* 1998
	Lepidoptera: Notodontidae	*Clostera* (*Pygaera*) *anastomosis*	L	Ohba & Aizawa 1989b
	Lepidoptera: Yponomeutidae	*Plutella xylostella*	L	Iriarte *et al.* 1998
tenebrionis	Acari: Phytoseiidae	*Metaseiulus occidentalis*	L/F	Chapman & Hoy 1991
	Coleoptera: Cantharidae	*Chauliognathus lugubris*	L	Greener & Candy 1994
		Chrysophtharta bimaculata	L/F	Schmidt & Kirfman 1992; Elek 1997; Beveridge & Elek 1999
	Coleoptera: Carabidae	*Amara* sp.	L	Ferrari & Maini 1992
	Coleoptera: Chrysomelidae	*Agelastica alni*	L	Krieg *et al.* 1983; Ferrari & Maini 1992
		Calligrapha scalaris	L	W. Gelernter in Keller & Langenbruch 1993
		Chrysolima (*Chrysomela*) *fastuosa*	L	Meyer 1989 in Keller & Langenbruch 1993
		Chrysomela herbacea	L	Riethmuller 1990 in Keller & Langenbruch 1993; Ferrari & Maini 1992
		Chrysomela populi	L	Ferrari & Maini 1992
		Chrysomela scripta	L/F	Bauer 1989; Ferrari & Maini 1992
		Crioceris asparagi	L/F	Riethmuller 1990 in Keller & Langenbruch 1993; Ferrari & Maini 1992
		Crioceris duodecimpunctata	L	Riethmuller 1990 in Keller & Langenbruch 1993
		Diabrotica longicornis	L	Gelernter 1990
		Diabrotica undecimpunctata	L	Herrnstadt *et al.* 1986
		Epitrix hirtipennis	F	Deseo *et al.* 1993
		Galeruca tanaceti	L	Riethmuller 1990 in Keller & Langenbruch 1993
		Galerucella (*Pyrrhalta*) *viburni*	L/F	Riethmuller 1990 in Keller & Langenbruch 1993
		Gastroidea viridula	L	Ferrari & Maini 1992; C. Balser in Keller & Langenbruch 1993
		Gonioctena fornicata	L	Burgio *et al.* 1992

	Leptinotarsa jucta	L	Ferrari & Maini 1992
	Lilioceris lilii	L	Riethmuller 1990 in Keller & Langenbruch 1993; Ferrari & Maini 1992
	Melasoma vigintipuncta	L	Riethmuller 1990 in Keller & Langenbruch 1993
	Phaedon cochleariae	L	Carroll *et al.* 1989; Ferrari & Maini 1992
	Phyllodecta vulgatissima	L	Riethmuller 1990 in Keller & Langenbruch 1993
	Phyllotreta armoraciae	L	MacIntosh *et al.* 1990c
	Plagiodera versicolora	L	Meyer 1989 in Keller & Langenbruch 1993; Ferrari & Maini 1992; Bauer 1992
	Podagrica fuscicornis	L	Riethmuller 1990 in Keller & Langenbruch 1993
	Pyrrhalta (Xanthogaleruca) exclamalionis	L	W. Gelernter in Keller & Langenbruch 1993
	Pyrrhalta (Xanthogaleruca) luteola	L/F	Herrnstadt *et al.* 1986; Cranshaw *et al.* 1989; Francardi 1990
	Pyrrhalta (Xanthogaleruca) viburni	L	Ferrari & Maini 1992
	Pyrrhalta (Xanthogaleruca) zygogramma	L	W. Gelernter in Keller & Langenbruch 1993
	Trirhabda nitidicollis	F	Eckberg & Cranshaw 1994
	Chrysophtharta agricola	L	Harcourt *et al.* 1996
	Oulema melanopus	L	Meyer 1989 in Keller & Langenbruch 1993; Ferrari & Maini 1992
	Oulema spp.	L	G. A. Langenbruch in Keller & Langenbruch 1993
	Paropsis charybdis	L	Jackson & Poinar 1989
	Phyllotreta undulata	L	Krieg *et al.* 1983; Ferrari & Maini 1992
Coleoptera: Coccinellidae	*Epilachna varivestis*	L	C. Balser in Keller & Langenbruch 1993; Ferrari & Maini 1992
Coleoptera: Curculionidae	*Anthonomus grandis*	L	Herrnstadt *et al.* 1986; MacIntosh *et al.* 1990b; Ferrari & Maini 1992
	Diabrotica virgifera	L	Herrnstadt & Soares US patent 4797276, 1989; Ferrari & Maini 1992
	Hypera brunneipennis (postica)	L	Herrnstadt *et al.* 1987; Herrnstadt & Soares US patent 4797276, 1989
	Listronotus oregonensis	L	Saade *et al.* 1996
	Othiorrhynchus salicicola	L	Ferrari & Maini 1992
	Othiorrhynchus sulcatus	L/F	Herrnstadt *et al.* 1986; Landi 1990
Coleoptera: Dermestidae	*Attagenus unicolor*	L	Herrnstadt *et al.* 1986
Coleoptera: Elateridae	*Agriotes lineatus*	L	Riethmuller 1990 in Keller & Langenbruch 1993
Coleoptera: Nitidulidae	*Meligethes aeneus*	L	Ferrari & Maini 1992; Prishchepa & Mikul'skaya 1998
Coleoptera: Scarabaeidae	*Blitopertha pallidipennis*	F	Li *et al.* 1992
	Melolontha hippocastani	L	Meyer 1989 in Keller & Langenbruch 1993
	Melolontha melolontha	NS	Riethmuller 1990 in Keller & Langenbruch 1993
	Oxythyrea funesta	L	Robert *et al.* 1994
	Popillia japonica	L	patent No. 05378625, W.P. Donovan 1995
Dictyoptera: Blattellidae	*Blattella germanica*	L	MacIntosh *et al.* 1990c; Ferrari & Maini 1992; Zukowski 1995
	Periplaneta americana	L	Ferrari & Maini 1992
Diptera: Muscidae	*Musca domestica*	L	Ferrari & Maini 1992; Indrasith *et al.* 1992
Hemiptera: Aphididae	*Apion apricans*	F	Prishchepa & Mikul'skaya 1998
	Apion flavipes (fulvipes)	F	Prishchepa & Mikul'skaya 1998
	Myzus persicae	L	MacIntosh *et al.* 1990c
Hemiptera: Miridae	*Lygus pratensis*	F	Prishchepa & Mikul'skaya 1998
Isoptera: Rhinotermitidae	*Coptotermes formosanus*	L	Grace & Ewart 1996
	Reticulitermes flavipes	L	MacIntosh *et al.* 1990c
Lepidoptera: Lymantridae	*Euproctis chrysorrhoea*	NS	Meyer 1989 in Keller & Langenbruch 1993
Lepidoptera: Noctuidae	*Helicoverpa (Heliothis) zea*	L	MacIntosh *et al.* 1990c; Ferrari & Maini 1992
	Heliothis virescens	L	MacIntosh *et al.* 1990c; Ferrari & Maini 1992

APPENDIX 2.1 *(continued)*

Subspecies/ Serovar	Order: Family	Genus and Species	Lab/Field	References
tenebrionis	Lepidoptera: Tenebrionidae	*Tenebrio molitor*	L	Krieg *et al.* 1983; Carroll *et al.* 1989; Ferrari & Maini 1992
	Lepidoptera: Yponomeutidae	*Plutella xylostella*	L	Ferrari & Maini 1992
	Orthoptera : Acrididae	*Melanoplus sanguinipes*	L	Ferrari & Maini 1992
		Locusta migratoria	L	Ferrari & Maini 1992
	Prostigmata: Tetranychidae	*Tetranychus urticae*	L/F	MacIntosh *et al.* 1990c; Chapman & Hoy 1991
thompsoni	Diptera: Culicidae	*Aedes aegypti*	L	Ragni *et al.* 1996
		Anopheles stephensi	L	Ragni *et al.* 1996
		Culex pipiens	L	Ragni *et al.* 1996
	Lepidoptera: Gelechiidae	*Phthorimaea operculella*	L	Salama *et al.* 1995a
		Sitotroga cerealella	L	Salama *et al.* 1991b
	Lepidoptera: Lymantriidae	*Lymantria dispar*	L	Dubois *et al.* 1989
	Lepidoptera: Noctuidae	*Earias insulana*	L	Salama & Foda 1984
		Trichoplusia ni	L	de Barjac & Thompson 1970†
	Lepidoptera: Pieridae	*Pieris brassicae*	L	de Barjac & Thompson 1970†
	Lepidoptera: Pyralidae	*Galleria mellonella*	L	de Barjac & Thompson 1970†
		Plodia interpunctella	L	Salama *et al.* 1991b
	Phthiraptera: Trichodectidae	*Damalinia* (*Bovicola*) *ovis*	L	Drummond *et al.* 1992
thuringiensis	Acari: Acaridae	*Proctolaelaps scolyti*	L	Bolanos *et al.* 1991
		Tyrophagus putrescentiae	L	Bolanos *et al.* 1991
	Acari: Argasidae	*Argas persicus*	L	Hassanain *et al.* 1997b
	Acari: Ixodidae	*Hyalomma dromedarii*	L	Hassanain *et al.* 1997b
	Acari: Phytoseiidae	*Phytoseiulus persimilis*	L	Petrova & Khrameeva 1989; Guo *et al.* 1993
	Acari: Prostigmata	*Phyllocoptruta oleivora*	F	Novitskaya & Abzianidze 1985
	Acari: Tenuipalpidae	*Brevipalpus* sp.	F	Stephenson *et al.* 1987
	Coleoptera: Bostrichidae	*Rhyzopertha dominica*	L	Keever 1994
	Coleoptera: Bruchidae	*Acanthoscelides obtectus*	L	Sander & Cichy 1967†
	Coleoptera: Chrysomelidae	*Altica ambiens*	L	Hall & Dunn 1958†
		Chrysomela scripta	L	Ramachandran *et al.* 1993a
		Diabrotica undecimpunctata	L	Abdel-Hameed & Landen 1994
		Leptinotarsa decemlineata	L/F	Lipa 1976; Stus 1977; Moiseenko & Barybkina 1977
		Phyllotreta striolata	F	Chang & Pegn 1971
		Oulema oryzae	F	Buryi *et al.* 1991
	Coleoptera: Coccinellidae	*Coccinella undecimpunctata*	F	Kazakova *et al.* 1977
		Epilachna varivestis	F	Cantwell *et al.* 1985
	Coleoptera: Curculionidae	*Anthonomus grandis*	L	Wolfenbarger *et al.* 1997
		Anthonomus rubi	L/F	Litvinov & Bondarenko 1987
		Hypera brunneipennis	L	Hall 1957†
		Hypera (*Phytonomus*) spp.	F	Kovtun 1984
		Sitona lineatus	F	Zubkov *et al.* 1986
		Sitophilus granarius	L/F	Steinhaus & Bell 1953†; Sander & Cichy 1967†
		Sitophilus oryzae	L	Steinhaus & Bell 1953†

Coleoptera: Nitidulidae	*Meligethes aeneus*	F	Prishchepa & Vanyushina 1997; Prishchepa & Mikul'skaya 1998
Coleoptera: Scarabaeidae	*Oryctes rhinoceros*	L	Steinhaus 1951b†; Babu *et al.* 1971
Coleoptera: Scolytidae	*Dendroctonus frontalis*	L	Moore 1972
	Scolytus multistriatus	L	Jassim *et al.* 1990
	Scolytus scolytus	L	Jassim *et al.* 1990
	Ips subelongatus	F	Gusteleva 1980, 1982
Dictyoptera: Blattellidae	*Blattella germanica*	L	Zukowski 1994, 1995
	Schelfordella tartarica	L	Brikman *et al.* 1966†
Diptera: Anthomyiidae	*Hylemyia* (*Strobilomyia*) *laricicola*	F	Banaszak & Szmidt 1987
Diptera: Calliphoridae	*Lucilia cuprina*	L	Lyness *et al.* 1994
Diptera: Chironomidae	*Chironomus plumosus*	L	Lavrentyev *et al.* 1965†
Diptera: Chloropidae	*Hippelates collusor*	L	Hall *et al.* 1972; Hall *et al.* 1977†
Diptera: Culicidae	*Aedes aegypti*	L	Reeves & Garcia 1970; Hall *et al.* 1977†; Panbangred *et al.* 1979
	Aedes caspius	L/F	Kadyrova *et al.* 1977
	Aedes nigromaculis	L	Reeves & Garcia 1970; Reeves *et al.* 1971
	Aedes triseriatus	L	Reeves & Garcia 1970; Reeves *et al.* 1971
	Aedes sp.	L	Lavrentyev *et al.* 1965†
	Anopheles atroparvus	L/F	Sokolova *et al.* 1982
	Anopheles stephensi	L	Larget Thiery *et al.* 1984
	Anopheles sp.	L	Lavrentyev *et al.* 1965†
	Culex modestus	L/F	Kadyrova *et al.* 1977
	Culex molestus	L	Sokolova & Ganushkina 1982; Sokolova *et al.* 1982
	Culex pipiens	L	Larget Thiery *et al.* 1984; Farghal *et al.* 1987
	Culex tarsalis	L	Hall *et al.* 1977†
	Culex sp.	L	Lavrentyev *et al.* 1965†
	Culiseta (*Theobaldia*) *longiareolata*	L	Farghal *et al.* 1987
Diptera: Drosophilidae	*Drosophila melanogaster*	L	Carlberg & Lindstrom 1987; Zhang *et al.* 1995
Diptera: Fanniidae	*Fannia canicularis*	L	Hall *et al.* 1971
	Fannia femoralis	L	Hall *et al.* 1972
Diptera: Glossinidae	*Glossinia pallidipes*	L	v. d. Geest & de Barjac 1982
Diptera: Muscidae	*Atherigona* (*varia*) *soccata*	F	Sachan & Rathore 1983
	Haematobia irritans	L	Gingrich & Eschle 1971
	Musca autumnalis	L	Korzh *et al.* 1975
	Musca domestica	L	Sherman *et al.* 1962†; Yarnykh & Kats 1981; Mohd Salleh & Lewis 1982a
	Stomoxys calcitrans	L/F	Korzh *et al.* 1975; Tulisalo & Rautapaa 1983
Diptera: Oestridae	*Oestrus ovis*	F	Tonkonozhenko 1981
Diptera: Simuliidae	*Simulium vittatum*	L	Lacey & Mulla 1977†
Diptera: Syrphidae	*Sphaerophoria scripta*	F	Kazakova *et al.* 1977
Diptera: Tabanidae	*Tabanus autumnalis brunnescens*	L/F	Kadyrova *et al.* 1977
Diptera: Tipulidae	*Tipula paludosa*	F	Lam & Webster 1972
Hemiptera: Aleyrodidae	*Bemisia tabaci*	L	Radwan *et al.* 1986
Hemiptera: Aphididae	*Aphis gossypii*	F	Begunov & Storozhkov Yu 1986
	Brevicoryne brassicae	F	Crooks 1975
	Myzus persicae	L/F/GH	Crooks 1975; Khrameeva *et al.* 1988
Hemiptera: Apionidae	*Apion apricans*	F	Prishchepa & Mikul'skaya 1998
	Apion flavipes (*fulvipes*)	F	Prishchepa & Mikul'skaya 1998

APPENDIX 2.1 (*continued*)

Subspecies/ Serovar	Order: Family	Genus and Species	Lab/Field	References
thuringiensis	Hemiptera: Apionidae	*Apion seniculus*	F	Prishchepa & Mikul'skaya 1998
	Hemiptera: Cimicidae	*Cimex lectularius*	F	Romasheva & Uzdenov 1979
	Hemiptera: Lygaeidae	*Geocoris punctipes*	L	Herbert & Harper 1986
	Hemiptera: Miridae	*Lygus hesperus*	F	Grau 1987; Tanigoshi *et al.* 1990
		Lygus pratensis	F	Prishchepa & Mikul'skaya 1998
	Hymenoptera: Apidae	*Apis mellifera*	L	Cantwell *et al.* 1966†; Haragsim & Vankova 1968†; Kazakova *et al.* 1977
	Hymenoptera: Braconidae	*Bracon hebetor*	L	Tamashiro 1968†
		Phanerotoma flavitestacea	L	Marchal Segault 1975†
	Hymenoptera: Diprionidae	*Diprion pini*	F	Supatashvili 1990
		Neodiprion banksianae	L	Angus 1956†
	Hymenoptera: Formicidae	*Monomorium pharaonis*	L/F	Brikman *et al.* 1967†; Vankova *et al.* 1975†
		Pheidole megacephala	L	Castineiras & Calderon 1985
	Hymenoptera: Ichneumonidae	*Liotryphon punctulatus*	F	Kazakova *et al.* 1977
	Hymenoptera: Tenthredinidae	*Athalia rosae* (*colibri*)	L/F	Rautapaa 1967†; Prishchepa & Vanyushina 1997
		Macrophya punctumalbum	F	Kulikovskii Yu 1984
		Nematus (*Pteronidea*) *ribesii*	L	Rautapaa 1967†
		Priophorus tristis	L	Rautapaa 1967†
		Pristiphora erichsonii	L	Heimpel 1961†
	Hymenoptera: Trichogrammatidae	*Trichogramma* sp.	F	Kazakova *et al.* 1977
	Isoptera: Rhinotermitidae	*Psammotermes hybostoma*	L	Farghal *et al.* 1987
	Isoptera: Termitidae	*Amitermes desertorum*	L	Farghal *et al.* 1987
	Lepidoptera: ?	*Hypeuryntis coricopa*	F	Helson 1965†
	Lepidoptera: ?	*Molippa* (*Orthetrum?*) *sabina*	L/F	Fugueiredo *et al.* 1960†
	Lepidoptera: ?	*Ommatopteryx texana*	L	White & Briggs 1964†
	Lepidoptera: Arctiidae	*Arctia caja*	L	Burgerjon & Biache 1967b†
		Datana integerrima	L/F	Angus 1956a†; Harris & Cutler 1975
		Datana ministra	L/F	Angus 1955†; Angus & Heimpel 1959†
		Estigmene acrea	L	Hall & Dunn 1958†
		Hyphantria cunea	L/F	Vankova 1962†; Krieg & Schmidt 1962†; Pana Beratlief 1968
		Ocnogyna baetica	L	Barreiro & Santiago Alvarez 1985
		Pericallia ricini	L	Kumar & Jayaraj 1978; Mathur *et al.* 1994
		Spilosoma lubricipeda (as *menthastri*)	L	Burgerjon & Biache 1967b†
		Spilosoma (*Diacrisia*) *obliqua*	L	Ahmed *et al.* 1988
		Spilosoma (*Spilarctia*) *obliqua*	L	Gupta & Rana 1991; Biswas *et al.* 1994
		Spilosoma virginica	L	Steinhaus 1957†
	Lepidoptera: Bombycidae	*Bombyx mori*	L	Burgerjon & Biache 1967b†; Angus & Norris 1968†; Tanaka *et al.* 1972
	Lepidoptera: Carposinidae	*Carposina niponensis*	F	Takaki 1975†
	Lepidoptera: Cossidae	*Zeuzera pyrina*	F	Deseo & Docci 1985
	Lepidoptera: Cryptophasidae	*Nephantis serinopa*	L/F	Ayyar 1961†
	Lepidoptera: Culicidae	*Aegeria* (*Sanninoidea*) *exitiosa*	F	Snapp 1962†
	Lepidoptera: Dioptidae	*Phryganidia californica*	L/F	Steinhaus 1951a†; Pinnock & Milstead 1972†

Lepidoptera: Gelechiidae	*Coleotechnites* (*Recurvaria*) *milleri*	L	Struble 1965†
	Gelechia hippophaella	F	Sanzhimtupova 1984
	Keiferia (*Gnorimoschema*) *lycopersicella*	F	Middlekauff *et al.* 1963†
	Pectinophora gossypiella	L/F	Graves & Watson 1970†; Abul Nasr *et al.* 1979; Rezk *et al.* 1981
	Phthorimaea operculella	L/F	El Sayed *et al.* 1979; Baklanova *et al.* 1990; Das *et al.* 1992
	Scrobipalpula absoluta	F	Prada & Gutierrez 1974; Ripa 1981
	Sitotroga cerealella	L	Steinhaus & Bell 1953†; Salama *et al.* 1991b
Lepidoptera: Geometridae	*Alsophila pometaria*	L/F	Larson & Ignoffo 1971†; Harper 1974†
	Boarmia (*Ascotis*) *selenaria*	L	Cohen *et al.* 1983
	Bupalus piniarius	L/F	Skatulla 1971†; Mashanov *et al.* 1980
	Buzura suppressaria (form *benescripta*)	F	Borthakur & Raghunathan 1987
	Colotois pennaria	L	Balinski *et al.* 1969†
	Ectropis (*Boarmia*) *crepuscularia*	L	Morris 1962†
	Ennomos subsignarius	F	Dunbar & Kaya 1972
	Erannis defoliaria	F	Aukshtikal'nene & Imnadze 1981
	Lambdina fiscellaria fiscellaria	L/F	Heimpel & Angus 1959†
	Lambdina fiscellaria lugubrosa	L/F	Heimpel & Angus 1959†; Carolin & Thompson 1967
	Lambdina fiscellaria somniaria	L/F	Heimpel & Angus 1959†; Morris 1962†
	Lycia (*Biston*) *hirtaria*	L	Dirimanov & Lecheva 1980; Kuzmanova & Lecheva 1981
	Melanolophia imitata	F	Morris 1962†, 1969†
	Operophtera brumata	L/F	Jaques 1961†; Ceianu *et al.* 1970†; Niemczyk *et al.* 1973
	Operophtera fagata	L	Vankova 1973†
	Paleacrita vernata	F	Harper 1974†
	Sabulodes caberata	L	Steinhaus 1957†
Lepidoptera: Glyphipterigyidae	*Eutromula* (*Simaethis*) *pariana*	L	Niemczyk & Bakowski 1970
Lepidoptera: Gracillariidae	*Caloptilia theivora*	L/F	Kariya 1977; Unnamalai & Vaithilingam 1995
	Caloptilia (*Gracillaria*) *invariabilis*	L	Morris 1969†
	Phyllonorycter (*Lithocolletis*) *blancardella*	F	Jaques 1965†
Lepidoptera: Hesperiidae	*Parnara guttata*	F	Takaki 1975†
	Thymelicus lineola	F	Arthur 1968†
Lepidoptera: Lasciocampidae	*Dendrolimus pini*	L	Sierpinska *et al.* 1995
	Dendrolimus sibiricus	F	Baranovskii *et al.* 1988
	Dendrolimus superans	F	Mashanov *et al.* 1976; Malokvasova 1979
	Eriogaster lanestris	L	Vankova 1973†
	Gastropacha quercifolia	F	Panait & Ciortan 1974†
	Malacosoma disstria	L/F	Angus & Heimpel 1959†; Pinkham *et al.* 1984
	Malacosoma neustria	L	Burgerjon & Biache 1967b†; v. d. Laan & Wassink 1962†;
	Malacosoma parallela	F	Aukshtikal'nene *et al.* 1984
	Malacosoma pluviale	L	Morris 1969†
Lepidoptera: Lymantriidae	*Dasychira pudibunda*	L	Krieg 1957b†
	Euproctis chrysorrhoea	L/F	Kondrja 1966b†; Burgerjon & Biache 1967a†; Danguy 1971
	Euproctis fraterna	L	Kumar & Jayaraj 1978
	Euproctis lunata	F	Rahman & Chaudhury 1987
	Euproctis similis	F	Stus 1979
	Hypogymna morio	L	Dobrivojevic *et al.* 1969†; Dobrivojevic & Injac 1975
	Leucoma wiltshirei	L/F	Alizadeh 1977†; Adeli 1980; Abai 1981

APPENDIX 2.1 (*continued*)

Subspecies/ Serovar	Order: Family	Genus and Species	Lab/Field	References
thuringiensis	Lepidoptera: Lymantriidae	*Leucoma* (*Stilpnotia*) *salicis*	L/F	Kudler & Lysenko 1963†; Burgerjon & Biache 1967b†; Tsilosani *et al.* 1976
		Lymantria dispar	L/F	Burgerjon & Biache 1967b†; Vankova 1973†; Magnoler 1974
		Lymantria monacha	L/F	Szmidt & Slizynski 1965†; Glowacka-Pilot 1968†; Balinski *et al.* 1969†
		Ocnerogyia amanda	F	Abai & Faseli 1986
		Orgyia antiqua	L/F	Balinski *et al.* 1969†; Niemczyk 1971†; Lipa *et al.* 1977
		Orgyia leucostigma	L	Rossmoore *et al.* 1970†
		Orgyia psuedotsugata	L	Morris 1963†
	Lepidoptera: Lyonetiidae	*Bucculatrix thurberiella*	L	Hall & Dunn 1958†
	Lepidoptera: Noctuidae	*Achaea janata*	L	Govindarajan *et al.* 1976†
		Agrotis ipsilon (*ypsilon*)	L/F	Chang 1972; Revelo 1973†
		Agrotis segetum	L/F	Agzamova Kh *et al.* 1988
		Alabama argillacea	L/F	Ignoffo *et al.* 1964†
		Anadevidia peponis	L/F	Ayyar 1961†
		Anomis (*Cosmophila*) *sabulifera*	F	Chatterjee 1965†
		Anticarsia gemmatalis	L/F	Mielitz & da Cruz. 1980; Bertoldo & Morosini 1982
		Autoplusia egena	F	Genung 1960†
		Cerapteryx graminis	F	Weiser 1960†; Vlug *et al.* 1988
		Chrysodeixis (*plusis*) *chalcites*	F	Helson 1965†
		Diachrysia orichalcea	F	Basu & Chatterjee 1969†
		Earias biplaga	L	Frutos *et al.* 1987
		Earias insulana	F	Al-Azawi 1964†; Abul Nasr *et al.* 1983; Moawad *et al.* 1982
		Euxoa lutescens	F	Ripa 1981
		Euxoa messoria	L/F	Cheng 1973†
		Helicoverpa (*Heliothis*) *armigera*	L/F	Daoust & Roome 1974†; El Sayed *et al.* 1979; Pukhaev 1981
		Helicoverpa (*Heliothis*) *zea*	L/F	Hall & Dunn 1958†; Shorey & Hall 1963†; Herbert & Harper 1986
		Heliothis assulta	F	Takaki 1975†
		Heliothis virescens	L/F	Wolfenbarger *et al.* 1972; Mistric & Smith 1973a; Patti & Carner 1974†
		Heliothis (*Chloridea*) *viriplaca*	F	Kovtun 1984
		Heliothis (*Protoschinia*) *scutosa* (*nervosa*)	F	Mikolajewicz *et al.* 1992
		Hellula undalis	L/F	Tanada 1956†; Chang & Pegn 1971
		Lacanobia (*Polia*) *oleracea*	L	Burgerjon & Barjac 1960†
		Mamestra (*Barathra*) *brassicae*	L/F	Burgerjon & Biache 1967b†; Duvlea *et al.* 1969; Burges & Jarrett 1976
		Mamestra (*Barathra*) *configurata*	L	Morris 1988
		Mocis latipes	L	Perez & Gonzales 1984
		Mocis repanda	L/F	Fugueiredo *et al.* 1960†
		Mythimna (*Pseudaletia*) *separata*	L	Sha & Xie 1992
		Mythimna (*Pseudaletia*) *unipuncta*	L	Burgerjon & Biache 1967b†
		Noctua pronuba	L	Burges & Jarrett pers. com.†
		Orthosia gothica	L	Burgerjon & Biache 1967b†

	Pseudoleucania (*Euxoa*) *bilitura*	F	Ripa 1981
	Pyrrhia umbra	F	Mashchenko 1983
	Spodoptera exigua	L/F	Hall & Dunn 1958†; Shorey & Hall 1963†; Merdan *et al.* 1975†
	Spodoptera frugiperda	L/F	Fugueiredo *et al.* 1960†; Dulmage 1973†; Zhang *et al.* 1995
	Spodoptera littoralis	L	Abul Nasr & Abdallah 1970; Lecadet & Martouret 1987
	Spodoptera litura	L/F	Chang 1972; Babu & Subramaniam 1973
	Spodoptera mauritia	L/F	Ayyar 1961†
	Spodoptera praefica	L/F	Steinhaus 1951a†; Middlekauff *et al.* 1963†
	Trichoplusia ni	L/F	Tanada 1956†; Rogoff *et al.* 1969†; Lindquist 1972
	Triphaena (*Noctua*) *pronuba*	L	Burges unpubl. obs.†
	Xanthopastis timais	L/F	Fugueiredo *et al.* 1960†
	Xestia (*Amathes*) *c-nigrum*	L	Burgerjon & Biache 1967b†; Lipa *et al.* 1969
Lepidoptera: Notodontidae	*Botyoides asialis*	L	Kalia & Joshi 1996
	Clostera cupreata	L	Kalia & Joshi 1996
	Clostera (*Pygaera*) *anastomosis*	L	Burgerjon & Biache 1967b†
	Drymonia ruficornis (as *chaonia*)	L/F	Ceianu *et al.* 1970†; Pirvescu 1973†
	Drymonia sp.	L	Balinski *et al.* 1969†
	Heterocampa gutticitta	F	Wallner 1971†
	Ichthyura anastomosis	L/F	Shah *et al.* 1979
	Phalantha phalantha	L	Kalia & Joshi 1996
	Phalera bucephala	L	Ceianu *et al.* 1970†
	Ptilophora plumigera	F	Kulikovskii Yu 1984
	Schizura concinna	L/F	White & Briggs 1964†; Pinnock *et al.* 1974†
Lepidoptera: Nymphalidae	*Aglais urticae*	L	Metalnikov & Chorine 1929†
	Euphydryas chalcedona	L	Steinhaus 1957†
	Junonia coenia	L	Steinhaus 1951a†
	Nymphalis antiopa	L	Angus 1956a†
	Vanessa cardui	L	Morris 1969†
	Vanessa io	L	Burgerjon & Biache 1967b†
	Vanessa turkestanica	F	Guzeev 1986
Lepidoptera: Oecophoridae	*Depressaria marcella*	L/F	Celli 1970†
	Depressaria nervosa (*daucella*)	F	Mikolajewicz *et al.* 1992
Lepidoptera: Papilionidae	*Papilio demoleus*	L	Narayanan & Jayaraj 1974b
	Papilio philenor	L	Steinhaus 1957†
Lepidoptera: Phycitidae	*Ectomyelois* (*Apomyelois*) *ceratoniae*	F	Dhouibi & Jemmazi 1996
Lepidoptera: Pieridae	*Aporia crataegi*	L/F	Krieg unpubl. obs.†; Kondrja 1966a†
	Ascia (*Pieris*) *monuste orseis*	L/F	Fugueitedo *et al.* 1960†
	Colias eurytheme	L/F	Steinhaus & Jerrel 1954†
	Colias lesbia	L/F	Faldini & Pastrana 1952†
	Pieris brassicae	L/F	Krieg 1957a†; Burgerjon & Biache 1967b†; Lipa *et al.* 1970
	Pieris canidia sordida	L	Yu-Chen Lee 1966†
	Pieris rapae	L/F	Tanada 1953†; Jaques & Fox 1960†; Takaki 1975†
	Pieris (*Artogeia*) *rapae crucivora*	F	Chang 1972
Lepidoptera: Psychidae	*Oiketicus moyanoi*	L/F	Solanes *et al.* 1974
Lepidoptera: Pterophoridae	*Platyptilia carduidactyla*	L/F	Tanada & Reiner 1960†
Lepidoptera: Pyralidae	*Acigona steniellus*	F	Irshad *et al.* 1982b
	Bissetia steniella	L/F	Atwal & Paul 1964†

APPENDIX 2.1 *(continued)*

Subspecies/ Serovar	Order: Family	Genus and Species	Lab/Field	References
thuringiensis	Lepidoptera: Pyralidae	*Cactoblastis cactorum*	L	Huang & Tamashiro 1966†
		Chilo auricilius	F	Kalra & Kumar 1963†
		Chilo infuscatellus	F	Irshad *et al.* 1982a
		Chilo partellus	F	Sachan & Rathore 1983
		Chilo sacchariphagus indicus	F	Kalra & Kumar 1963†
		Chilo suppressalis	L/F	Bounias & Guennelon 1974†; Takaki 1975†
		Cnaphalocrocis medinalis	L/F	Takaki 1975†; Srivastava & Nayak 1978†
		Corcyra cephalonica	L	El Husseini *et al.* 1987
		Crambus bonifatellus	L	Hall 1954†
		Crambus sperryellus	L/F	Hall 1957†
		Crambus spp.	F	Jefferson *et al.* 1964†
		Crocidolomia binotalis	L/F	Chang 1972; Ooi 1980
		Cryptoblabes sp.	F	Katanyukul & Bhudhasamai 1983
		Diaphania (*Margaronia*) *indica*	L/F	Ayyar 1961†
		Diaphania (*Palpita*) *hyalinata*	L/F	Delplanque & Gruner 1975; Delphanque *et al.* 1974
		Diatraea grandiosella	L	Sikorowski *et al.* 1970†
		Diatraea saccharalis	L/F	Revelo 1973†
		Ephestia cautella	L	Kinsinger *et al.* 1980
		Ephestia elutella	L	v.d. Laan & Wassink 1962†; Keever 1994
		Ephestia kuehniella	L/F	Berliner 1911†; Jacobs 1950†; Vankova 1962†
		Galleria mellonella	L	Burgerjon & Biache 1967b†; Karabash 1974; Burges *et al.* 1976b
		Herpetogramma (*Psara*) *bipunctalis*	F	Delplanque & Gruner 1975
		Herpetogramma phaeopteralis	F	Reinert 1974
		Homoeosoma electellum	F	Carlson 1967†
		Hymenia (*Kinckenia*) *recurvalis* (*fascialis*)	F	Delplanque & Gruner 1975
		Hypsipyla grandella	L	Hidalgo Salvatierra & Palm 1972
		Loxostege stricticalis	F	Mitrofanov 1982; Poplavskii 1984
		Nomophila noctuella	L	White & Briggs 1964†
		Noorda blitealis	F	Kalia & Joshi 1997
		Nymphula enixalis	F	Katanyukul & Bhudhasamai 1983
		Ostrinia nubilalis	L/F	McConnell & Cutkomp 1954†; McWhorter *et al.* 1972
		Paramyelois (*Amyelois*) *transitella*	F	Kellen *et al.* 1977
		Piesmopoda obliquifasciella	L	Khawaja *et al.* 1983
		Plodia interpunctella	L	Kinsinger *et al.* unpubl. obs.†
		Scirpophaga nivella	F	Irshad *et al.* 1982a
		Sylepta (*Syllepte*) *balteata*	F	Kalia *et al.* 1997
		Syllepte silicalis	L/F	Fugueiredo *et al.* 1960†
		Udea profundalis	L	White & Briggs 1964†
		Udea rubigalis	L	Hall & Dunn 1958†
	Lepidoptera: Pyraustidae	*Azochis gripusalis*	L/F	Fugueitedo *et al.* 1960†
	Lepidoptera: Saturnidae	*Anisota senatoria*	L/F	Angus 1956a†; Angus & Heimpel 1959†; Kaya 1974
		Antheraea pernyi	L	Burgerjon & Biache 1967b†
		Dryocampa rubicunda	L/F	Angus & Heimpel 1959†

Lepidoptera: Sphingidae	*Erinnyis ello*	L	Pigatti *et al.* 1960†
	Hyloicus pinastri	L	Glowacka-Pilot 1968†
	Manduca quinquemaculata	L/F	Angus 1955†; Rabb *et al.* 1957†; Cheng & Hanlon 1990
	Manduca sexta	L/F	Guthrie *et al.* 1959†; Tyrell *et al.* 1981b; Wabiko *et al.* 1985
Lepidoptera: Thaumetopoeidae	*Thaumetopoea pityocampa*	F	de Bellis & Cavalcaselle 1970; Kailides *et al.* 1971; Touzeau 1971†
	Thaumetopoea processionea	L	Burgerjon & Biache 1967b†
	Thaumetopoea wilkinsoni	L/F	Moore *et al.* 1962†
Lepidoptera: Thomisidae	*Dichomeris (Nothris) marginella*	F	Zelenev 1982
Lepidoptera: Tineidae	*Tineola bisselliella*	L	Yamvrias & Angus 1969†
Lepidoptera: Tortricidae	*Achroia grisella*	L	Burges & Bailey 1968†; Karabash 1974
	Acleris variana	F	Morris 1963†
	Adoxophyes orana	L/F	v.d. Geest & Velterop 1971†; Takaki 1975†
	Adoxophyes sp.	L/F	Kariya 1977
	Agriopis (Erannis) aurantiaria	F	Ceianu *et al.* 1970†
	Agriopis (Erannis) leucophaearia	F	Ceianu *et al.* 1970†
	Amorbia essigana	L	Hall & Dunn 1958†
	Archips argyropilus	F	Pinnock & Milstead 1978†
	Archips cerasivoranus	L	Smirnoff 1972†
	Archips crataegana	L/F	Kudler *et al.* 1965†
	Archips (Cacoecia) rosana	L	Vankova 1973†
	Archips rosanus	L/F	Niemczyk & Bakowski 1970; Niemczyk 1971; Vankova 1973
	Archips xylosteana	F	Niemczyk *et al.* 1975
	Argyrotaenia mariana	L/F	Jaques 1961†
	Argyrotaenia pulchellana	F	Strapazzon & Dalla Monta 1986
	Argyrotaenia velutinana	F	McEwen *et al.* 1960†
	Autographa californica	L	White & Briggs 1964†
	Autographa confusa	F	Mikolajewicz *et al.* 1992
	Autographa gamma	L/F	Burges & Jarrett pers. com.†; Slabospitskii & Yakulov 1979
	Cacoecimorpha pronubana	L	Burges unpubl. obs.†
	Choristoneura conflictana	F	Holsten & Hard 1985
	Choristoneura fumiferana	L	Smirnoff 1965†; Yamvrias & Angus 1970†
	Choristoneura murinana	L	Krieg 1956†; Weiser 1962†
	Choristoneura occidentalis	F	Stelzer & Beckwith 1988
	Cydia (Grapholita) delineana	F	Peteanu 1980
	Cydia funebrana	F	Wiackowski & Wiackowska 1966†
	Cydia (Grapholitha) molesta	L	Roehrich 1964†
	Cydia (Laspeyresia) pomonella	L/F	Jaques 1961†; Roehrich 1964†; Vervelle 1975
	Desmia funeralis	F	Jensen 1969†; AliNiazee & Jensen 1973
	Eupoecilia ambiguella	L/F	Metalnikov 1937†; Schmid & Antonin 1977†; Scalco *et al.* 1997
	Hedya (Argyroploce) nubiferana (variegana)	L	Johansson 1971†
	Homona magnanima	L/F	Kariya 1977
	Lobesia botrana	L/F	Roehrich 1970†; Roehrich 1970; Schmid & Antonin 1977†
	Pammene juliana	L	Muller 1957†
	Pandemis dumetana	F	Balazs Klara 1966†
	Phlogophora meticulosa	L	Burges & Jarrett pers. com.†
	Platynota stultana	L	Hall & Dunn 1958†
	Rhyacionia buoliana	L	Ponting 1962†

APPENDIX 2.1 (*continued*)

Subspecies/ Serovar	Order: Family	Genus and Species	Lab/Field	References
thuringiensis	Lepidoptera: Tortricidae	*Sparganothis pilleriana*	F	Harranger 1961†; Herfs 1964†
		Spilonota (Tmetocera) ocellana	L/F	Jaques 1961†
		Tortrix viridana	L/F	Franz *et al.* 1967†; Burgerjon & Biache 1967b†; Svestka 1977
		Zeiraphera diniana	L/F	Grison *et al.* 1971; Benz 1975†; Martouret & Auer 1977†
	Lepidoptera: Yponomeutidae	*Acrolepiopsis (Acrolepia) assectella*	F	Goix 1959†
		Atteva aurea	L	Hull & Onuoha 1962†
		Plutella xylostella (maculipennis)	L/F	Tanada 1956†; Takaki 1975†; Johansson 1971
		Plutella spp.	F	Crooks 1975
		Prays citri	F	Giammanco *et al.* 1966†; Shetata & Nasr 1998
		Prays oleae	F	Yamvrias 1972†; Jardak & Ksantini 1986
		Yponomeuta cognatella (evonymi)	L	Kondrja 1966a†; Burgerjon & Biache 1967b†
		Yponomeuta evonymella	L	Hamed 1978†
		Yponomeuta malinella	L/F	Wildbolz & Staub 1962†; Burgerjon & Biache 1967b†
		Yponomeuta padella	L/F	Kondrja 1966a†; Hamed 1978†
		Yponomeuta (padellus) malinellus	L	Niemczyk & Bakowski 1970; Niemczyk 1971; Sem'yanov *et al.* 1983
	Lepidoptera: Zygaenidae	*Harrisina brillians*	L/F	Hall 1955†; Pinnock *et al.* 1973†
	Mesostigmata: Dermanyssidae	*Dermanyssus gallinae*	F	Romasheva *et al.* 1977
	Mesostigmata: Tetranychidae	*Panonychus citri*	F	Novitskaya & Abzianidze 1985
	Nematoda: Heligmonellidae	*Nippostrongylus brasiliensis*	L	Bone 1989
	Nematoda: Meloidgynidae	*Meloidogyne* sp.	L	Prasad *et al.* 1972; Rai & Rana 1979
		Meloidogyne incognita	GH	Sharma 1994
	Nematoda: Trichostrongylidae	*Trichostrongylus colubriformis*	L	Bone 1989
	Neuroptera: Chrysopidae	*Chrysopa formosa*	L	Babrikova & Kuzmanova 1984
		Chrysopa perla	L	Babrikova & Kuzmanova 1984
		Chrysopa septempunctata	L	Babrikova & Kuzmanova 1984
		Chrysoperla (Chrysopa) carnea	L	Babrikova *et al.* 1982
	Orthoptera: Phasmatidae	*Diapheromera femorata*	L	Ignoffo *et al.* 1973†
	Phthiraptera	Mallophaga spp.	F	Frolov *et al.* 1984
	Phthiraptera: Menonponidae	*Lipeurus caponis*	F	Hoffman & Gingrich 1968†
		Menopon gallinae	L/F	Hoffman & Gingrich 1968†; Gingrich *et al.* 1974; Lonc & Lachowicz 1987
		Menacanthus stramineus	F	Hoffman & Gingrich 1968†
		Menacanthus sp.	F	Romasheva *et al.* 1977
	Phthiraptera: Trichodectidae	*Damalinia (Bovicola) bovis*	L	Gingrich *et al.* 1974†
		Damalinia (Bovicola) crassipes	L	Gingrich *et al.* 1974†
		Damalinia (Bovicola) limbata	L	Gingrich *et al.* 1974†
		Damalinia (Bovicola) ovis	L	Gingrich *et al.* 1974†; Drummond *et al.* 1992
	Prostigmata: Tetranychidae	*Tetranychus pacificus*	F	Grau 1987
		Tetranychus urticae	L/F	Krieg 1972; Chilingaryan *et al.* 1977; Petrushov *et al.* 1983
	Siphonaptera: Ceratophyllidae	*Nosopsyllus (Ceratophyllus) fasciatus*	L	Shevchenko *et al.* 1983
	Siphonaptera: Pulicidae	*Xenopsylla cheopis*	L	Ershova *et al.* 1976, 1980; Savel' ev *et al.* 1977
	Thysanoptera: Thripidae	*Thrips tabaci*	F	Begunov & Storozhkov Yu 1986
thuringiensis?	Lepidoptera: Geometridae	*Calospilos (Abraxas) pantaria*	F	Kosenko & Anferov 1996

tianmensis	Lepidoptera: Gelechiidae	*Pectinophora gossypiella*	L	Anon. 1980
	Lepidoptera: Pieridae	*Pieris rapae*	F	Yu & Dai 1987
	Lepidoptera: Pyralidae	*Cnaphalocrocis medinalis*	L	Anon. 1981
tochigiensis	Diptera: Culicidae	*Aedes aegypti*	L	Patent No. 05298245, J.M. Payne 1994
		Culex pipiens molestus	L	Saitoh *et al.* 1996
toguchini	Lepidoptera: Yponomeutidae	*Plutella xylostella*	L	Iriarte *et al.* 1998
tohokuensis	Diptera: Culicidae	*Culex pipiens molestus*	L	Saitoh *et al.* 1996
	Diptera: Psychodidae	*Telmatoscopus albipunctatus* (*Clogmia albipunctata*)	L	Saitoh *et al.* 1996
tolworthi	Coleoptera: Chrysomelidae	*Diabrotica undecimpunctata howardi*	L	Rupar *et al.* 1991
		Leptinotarsa decemlineata	L	Rupar *et al.* 1991
	Diptera: Culicidae	*Aedes aegypti*	L	Hall *et al.* 1977†; Aboul Ela *et al.* 1993
		Culex pipiens	L	Hall *et al.* 1977†
		Culex quinquefasciatus	L	Hall *et al.* 1977†
		Culex tarsalis	L	Hall *et al.* 1977†
	Diptera: Muscidae	*Musca domestica*	L/F	McGaughey & Johnson 1987; Xue *et al.* 1990
	Diptera: Psychodidae	*Telmatoscopus albipunctatus* (*Clogmia albipunctata*)	L	Saitoh *et al.* 1996
	Diptera: Simuliidae	*Simulium vittatum*	L	Lacey & Mulla 1977†
	Lepidoptera: Bombycidae	*Bombyx mori*	L	Angus & Norris 1968†
	Lepidoptera: Gelechiidae	*Phthorimaea operculella*	L	Salama *et al.* 1995a
		Sitotroga cerealella	L	Salama *et al.* 1991b
	Lepidoptera: Lasiocampidae	*Malacosoma disstria*	L	Pinkham *et al.* 1984
	Lepidoptera: Lymantriidae	*Lymantria dispar*	L	Dubois *et al.* 1989
	Lepidoptera: Noctuidae	*Agrotis ipsilon* (*ypsilon*)	L	Salama & Foda 1984
		Earias insulana	L	Salama & Foda 1984
		Spodoptera frugiperda	L	Hernandez 1988
		Spodoptera littoralis	L	Kalfon & De Barjac 1985; Salama 1991
		Spodoptera litura	L	Amonkar *et al.* 1985
	Lepidoptera: Pieridae	*Pieris brassicae*	L	Galowalia *et al.* 1973†
	Lepidoptera: Pyralidae	*Ephestia cautella*	L	Kinsinger *et al.* 1980
		Galleria mellonella	L	Jarrett & Burges 1982b
		Ostrinia nubilalis	L	Mohd Salleh & Lewis 1982b, 1983
		Plodia interpunctella	L	Kinsinger *et al.* 1980; Salama *et al.* 1991b
	Lepidoptera: Sphingidae	*Manduca sexta*	L	Tyrell *et al.* 1981b
	Lepidoptera: Tortricidae	*Choristoneura fumiferana*	L	Morris & Moore 1983
		Cydia (*Laspeyresia*) *pomonella*	L	Andermatt *et al.* 1988
	Phthiraptera: Menonponidae	*Menopon gallinae*	L	Lonc & Lachowicz 1987
	Phthiraptera: Trichodectidae	*Damalinia* (*Bovicola*) *ovis*	L	Drummond *et al.* 1992
toumanoffi	Diptera: Calliphoridae	*Lucilia cuprina*	L	Lyness *et al.* 1994
	Diptera: Culicidae	*Aedes triseriatus*	L	Hall *et al.* 1977†

APPENDIX 2.1 (*continued*)

Subspecies/ Serovar	Order: Family	Genus and Species	Lab/Field	References
toumanoffi	Diptera: Culicidae	*Culex quinquefasciatus*	L	Hall *et al.* 1977†
	Lepidoptera: Lasiocampidae	*Dendrolimus sibiricus*	F	Gukasyan & Rybakova 1964†
		Malacosoma disstria	L	Pinkham *et al.* 1984
	Lepidoptera: Lymantriidae	*Lymantria dispar*	L	Dubois *et al.* 1989
	Lepidoptera: Pyralidae	*Galleria mellonella*	L	Krieg unpubl. obs.†
	Phthiraptera: Trichodectidae	*Damalinia* (*Bovicola*) *ovis*	L	Drummond *et al.* 1992
wratislaviensis	Diptera: Drosophilidae	*Drosophila melanogaster*	L	Lonc *et al.* 1997
	Diptera: Muscidae	*Musca domestica*	L	Lonc *et al.* 1997
wuhanensis	Diptera: Calliphoridae	*Lucilia cuprina*	F	Cooper *et al.* 1985
	Lepidoptera: Gelechiidae	*Pectinophora gossypiella*	F	Luo *et al.* 1986
	Lepidoptera: Geometridae	*Apocheima cinerarium*	L	Zhang *et al.* 1987
	Lepidoptera: Hesperiidae	*Parnara guttatus*	L	Zhang *et al.* 1987
	Lepidoptera: Lasiocampidae	*Dendrolimus sibiricus*	L/F	Rishbeth 1978†
	Lepidoptera: Lymantriidae	*Euproctis pseudoconspersa*	L/F	Rishbeth 1978†
		Lymantria dispar	L	Dubois *et al.* 1989
	Lepidoptera: Noctuidae	*Anomis flava*	L/F	Rishbeth 1978†
		Helicoverpa (*Heliothis*) *armigera*	L/F	Rishbeth 1978†; Teakle *et al.* 1992
		Helicoverpa (*Heliothis*) *punctigera*	L	Teakle *et al.* 1992
	Lepidoptera: Pieridae	*Aporia crataegi*	L	Zhang *et al.* 1987
		Pieris brassicae	L	Li *et al.* 1987
		Pieris rapae	L/F	Hubei Inst. Res. Group 1976†; Zhang *et al.* 1987
	Lepidoptera: Pyralidae	*Cnaphalocrocis medinalis*	L/F	Rishbeth 1978†
		Galleria mellonella	L	Li *et al.* 1987
	Lepidoptera: Yponomeutidae	*Plutella xylostella* (*maculipennis*)	L/F	Rishbeth 1978†
	Phthiraptera: Trichodectidae	*Damalinia* (*Bovicola*) *ovis*	L	Drummond *et al.* 1992
Unknown	Acari: Goniodidae	*Goniocotes gallinae*	F	Lavrenyuk *et al.* 1977b
*		*Goniocotes gallinae hologaster*	F?	Asylbaeva & Fedorova 1977
*	Acari: Pyroglyphidae	*Dermatophagiodes pteronysinus*	L	Patent No. 05350576, J.M. Payne 1994
*	Acari: Tarsonemidae	*Polyphagotarsonemus latus*	F	Ochoa & von Lindeman 1988
*	Coleoptera: Bostrichidae	*Rhyzopertha dominica*	F	Rahman *et al.* 1997
*	Coleoptera: Carabidae	*Pterostichus coracinus*	F	Cameron & Reeves 1990
*		*Pterostichus mutus*	F	Cameron & Reeves 1990
*	Coleoptera: Cerambycidae	*Oberia linearis*	F	Tavamaishvili 1990
*		*Apriona germari*	L	Lin *et al.* 1990a
*	Coleoptera: Chrysomelidae	*Chrysomela populi*	F	Orekhov *et al.* 1978
*		*Chrysophtharta bimaculata*	F	Clarke *et al.* 1997
*		*Costalimaita ferruginea vulgata*	F	Macedo 1976
*		*Haltica* (*Altica*) *amphelophaga*	F	Tavamaishvili 1990
*		*Leptinotarsa decemlineata*	F	Zayats Yu *et al.* 1976
*		*Pyrrhalta* (*Xanthogaleruca*) *luteola*	F	Dahlsten *et al.* 1994

*		*Phyllotreta vittula*	F	Osintseva 1996
*	Coleoptera: Coccinellidae	*Epilachna varivestis*	L/F	Cantwell & Cantelo 1982
*		*Henosepilachna (Epilachna) vigintioctopunctata*	F	Rajagopal & Trivedi 1989
*	Coleoptera: Curculionidae	*Anthonomus pomorum*	F	Balazs *et al.* 1997
*		*Hylobius abietis*	L	Pye & Burman 1977
*		*Lissorhoptrus brevirostris*	F	Meneses Carbonell 1983
*		*Otiorhynchus sulcatus*	F	Blackshaw 1984a, 1984b
*		*Sitonia crinitus (macularius)*	F	Weigand *et al.* 1994
*		*Curculio nucum*	F	Tavamaishvili 1990
*	Coleoptera: Scarabaeidae	*Antitrogus consanguineus*	L	Allsopp *et al.* 1996
*		*Costelytra zealandica*	L	Chilcott & Wigley 1990; Wigley & Chilcott 1990
*		*Oryctes rhinoceros*	L	Hurpin 1974
*		*Oxythyrea funesta*	L	Robert *et al.* 1996
*		*Popillia japonica*	L	Sharpe & Detroy 1979
*	Coleoptera: Scolytidae	*Dendroctonus frontalis*	L	Cane *et al.* 1995
*	Coleoptera: Tenebrionidae	*Tenebrio molitor*	L	Benz & Lebrun 1976; Slama & Lysenko 1981; Chilcott & Wigley 1993
*	Dictyoptera: Blattellidae	*Blattella germanica*	L	Ulewicz & Bakowski 1975
*		*Periplaneta americana*	L	Ryan & Nicholas 1972
*	Diptera: Agromyzidae	*Liriomyza bryoniae*	F	Chang & Chen 1993
*		*Liriomyza sativae*	F	Trumble & Alvarado Rodriguez 1993
*		*Liriomyza trifolii*	F	Espino *et al.* 1988
*	Diptera: Anthomyiidae	*Delia floralis*	F	Havukkala 1982, 1988
*		*Delia radicum*	F	Havukkala 1982, 1988
*	Diptera: Calliphoridae	*Lucilia cuprina*	L	Arellano *et al.* 1990
*	Diptera: Cecidomyiidae	*Pachydiplosis oryzae*	F	Kulshreshtha *et al.* 1971
*	Diptera: Culicidae	*Aedes cantans*	F	Khazipov & Alekseev 1982
*		*Aedes dorsalis*	F	Khazipov & Alekseev 1982
*		*Anopheles hyrcanus*	F	Cristescu *et al.* 1975
*		*Anopheles hyrcanus*	F	Cristescu *et al.* 1975
*		*Culex pervigilana (pervigilans)*	L	Chilcott & Wigley 1993
*		*Culex pipiens*	L	Singer 1974; Theoduloz *et al.* 1997
*		*Culex quinquefasciatus*	L	Manonmani *et al.* 1987
*		*Culex tarsalis*	L	Singer & Smirnoff 1974
*	Diptera: Fanniidae	*Fannia canicularis*	L	Hall *et al.* 1972
*		*Fannia femoralis*	L	Hall *et al.* 1972
*	Diptera: Muscidae	*Atherigona soccata*	F	Pradhan *et al.* 1972
*		*Haematobia irritans*	L	Gingrich & Haufler 1978
*		*Musca domestica*	L	Burgerjon *et al.* 1974; Kudler & Lysenko 1976
*	Tephritidae	*Dacus (Bactrocera) oleae*	L	Karamanlidou *et al.* 1991
*	Gastropoda: Arionidae	*Arion distinctus*	L	Kienlen *et al.* 1996
*		*Limax valentianus*	L	Kienlen *et al.* 1996
*	Hemiptera: Aleyrodidae	*Bemisia tabaci*	F	Stansly *et al.* 1994
*		*Trialeurodes vaporariorum*	F	Kornilov & Ivanova 1987; Cardona 1995
*	Hemiptera: Anthocoridae	*Xylocoris flavipes*	L	Salama *et al.* 1991c
*	Hemiptera: Aphididae	*Aphis gossypii*	F	Satpute *et al.* 1989
*		*Aphis pomi*	F	Balazs *et al.* 1997

APPENDIX 2.1 (*continued*)

Subspecies/ Serovar	Order: Family	Genus and Species	Lab/Field	References
Unknown	Hemiptera: Aphididae	*Brevicoryne brassicae*	F	Longhini & Busoli 1993
*		*Dysaphis plantaginea*	F	Balazs *et al.* 1997
*		*Myzocallis coryli*	F	Tavamaishvili 1990
*		*Myzus persicae*	F	Imenes *et al.* 1990
*	Hemiptera: Cicadellidae	*Amrasca biguttula biguttula* (*devastans*)	F	Satpute *et al.* 1989
*		*Edwardsiana flaresens* (*flavescens*)	F	Chang & Chen 1993
*		*Empoasca* spp.	F	Ram & Gupta 1990
*	Hemiptera: Coccidae	*Saissetia oleae*	F	Panis 1979
*	Hemiptera: Delphacidae	*Javesella* (*Calligypona*) *pellucida*	F	Somermaa 1970
*	Hemiptera: Diaspididae	*Aspidiotus nerii*	F	Tomkins 1996
*		*Hemiberlesia rapax*	F	Sale *et al.* 1985; Tomkins 1996
*	Hemiptera: Miridae	*Lygus lineolaris*	F	Snodgrass & Stadelbacher 1994
*		*Lygus* spp.	F	Meredith *et al.* 1998
*	Hemiptera: Pseudococcidae	*Planococcus citri*	F	Liotta *et al.* 1976; Dolidze 1983
*		*Planococcus vitis* (*ficus*)	F	Baum 1986
*	Hemiptera: Reduviidae	*Rhodnius prolixus*	L	Nyirady 1973
*		*Triatoma protracta protracta*	L	Nyirady 1973
*	Hymenoptera: Aphelinidae	*Encarsia formosa*	L	Castaner & Garrido 1995
*	Hymenoptera: Apidae	*Apis mellifera*	L	Vandenberg & Shimanuki 1986
*	Hymenoptera: Braconidae	*Cotesia plutellae*	F	Chilcutt & Tabashnik 1997
*		*Lysiphlebus testaceipes*	L	Castaner & Garrido 1995
*		*Cardiochiles nigriceps*	L	Dunbar & Johnson 1975
*		*Microgaster* (*Microplitis*) *rufiventris*	L	El Maghraby *et al.* 1988
*		*Meteorus laeviventris*	L	Hafez *et al.* 1997b
*		*Chelonus blackburni*	L	Jadhav & Khaire 1987
*		*Bracon instabilis*	L	Salama *et al.* 1996b
*		*Apanteles* (*Dolichogenidea*) *litae*	L	Salama *et al.* 1996b
*		*Rogas lymantriae*	L	Wallner *et al.* 1983
*	Hymenoptera: Diprionidae	*Diprion pini*	F	Pollini 1979
*		*Neodiprion sertifer*	F	Pollini 1979
*	Hymenoptera: Tenthredinidae	*Pteronidea* (*Nematus*) sp.	F	Bulukhto & Korotkova 1988
*	Isoptera: Kalotermitidae	*Bifiditermes beesoni*	L	Khan *et al.* 1985
*	Isoptera: Termitidae	*Microcerotermes championi*	L	Khan *et al.* 1985
*	Lepidoptera: ?	*Furcivena rhodenuralis*	F	Mariau 1982
*	Lepidoptera: Arctiidae	*Datana integerrima*	F	Polles & Payne 1975; Tedders *et al.* 1985; Tedders & Gottwald 1986
*		*Eupseudosoma involuta*	F	Macedo 1976
*		*Hyphantria cunea*	L/F	Surany 1970; D'Aguilar *et al.* 1978
*		*Ocnogyna baetica*	F	Vargas Osuna *et al.* 1994
*		*Syntomeida epilais jucundissima*	F	Reinert 1981
*		*Halisidota* (*Lophocampa*) *argentata*	F	Antonelli & Collman 1993
*	Lepidoptera: Attacidae	*Dirphiopsis trisignata*	L/F	Moraes *et al.* 1975
*	Lepidoptera: Blastobasidae	*Glyphidocera juniperella*	F	Mizell & Schiffhauer 1987
*	Lepidoptera: Bombycidae	*Andraca bipunctata*	F	Hou 1987

	Lepidoptera: Brassolidae	*Opsiphanes cassina*	[illegible]	[illegible]
*		*Opsiphanes tamarindi*	F	Briceno 1997
*	Lepidoptera: Carposinidae	*Carposina niponensis (sasakii)*	L	Lu *et al.* 1993
*	Lepidoptera: Cossidae	*Zeuzera pyrina*	F	Deseo *et al.* 1984; Nashnosh *et al.* 1993; Delrio 1995
*	Lepidoptera: Dalceridae	*Dalcera* sp.	F	Alves *et al.* 1988
*	Lepidoptera: Dioptidae	*Phryganidia californica*	F	Pinnock *et al.* 1978a
*	Lepidoptera: Gelechiidae	*Anarsia lineatella*	F	Kheiri *et al.* 1974; Hendricks & Barbera 1994
*		*Brachmia macrosocopa*	L	Wu & Yang 1988
*		*Gnorimoschema (Keiferia) lycopersicella*	F	Rosset 1988; Trumble & Alvarado Rodriguez 1993
*		*Pectinophora gossypiella*	L	Watson & Kelly Johnson 1995
*		*Phthorimaea operculella*	F	von Arx *et al.* 1987; Bekheit *et al.* 1997
*		*Scrobipalpuloides absoluta*	L/F	Larrain 1986; Vilas Boas & Franca 1996
*		*Sitotroga cerealella*	L/F	McGaughey & Kinsinger 1978; McGaughey *et al.* 1987
*	Lepidoptera: Geometridae	*Anacamptodes pergracilus*	F	Dixon 1982
*		*Apocheima* spp.	F	Yang 1978
*		*Boarmia* sp.	F	Parnata *et al.* 1981
*		*Boarmia (Ascotis) selenaria*	L/F	Wysoki *et al.* 1981; Wysoki & Izhar 1986; Meisner *et al.* 1990
*		*Boarmia (Ascotis) selenaria reciprocaria*	F	Waikwa & Mathenge 1977
*		*Bupalus piniarius*	F	Fadeev 1974; Kudler & Lysenko 1977
*		*Fulgurodes* sp.	F	Macedo 1976
*		*Lambdina fiscellaria fiscellaria*	L/F	Smirnoff 1974
*		*Operophtera brumata*	L/F	Kudler & Lysenko 1975; Aukshtikal'nene & Imnadze 1981
*		*Thyrinteina arnobia*	F	Macedo 1976
*	Lepidoptera: Gracillariidae	*Phyllonorycter* sp.	F	Cortes *et al.* 1990
*		*Phyllonorycter corylifoliella*	F	Balazs *et al.* 1997
*		*Caloptilia theivora*	L/F	Kariya 1977
*	Lepidoptera: Hesperiidae	*Parnara guttata*	F	Yang 1978
*		*Thymelicus lineola*	F	Smirnoff 1983
*	Lepidoptera: Hyblaeidae	*Hyblaea puera*	F	Nair 1998
*	Lepidoptera: Lasiocampidae	*Dendrolimus pini*	L	Orekhov 1978
*		*Dendrolimus punctatus*	L/F	Ying & Klimetzek 1986
*		*Dendrolimus spectabilis*	L	Katagiri & Shimazu 1974; Wang & Chen 1984; Yao *et al.* 1998
*		*Dendrolimus* spp.	F	Yang 1978
*		*Malacosoma californicum*	F	Pinnock *et al.* 1978a; Milstead *et al.* 1980
*		*Malacosoma constrictum*	F	Pinnock *et al.* 1978a; Milstead *et al.* 1980
*		*Metanastria* sp.	L	Ramzan *et al.* 1979
*	Lepidoptera: Limacodidae	*Euprosterna eleasa*	F	Mariau 1982
*		*Latoia pallida*	F	Mariau 1982
*		*Latoia (Parasa) lepida*	F	de Chenon 1982
*	Lepidoptera: Lycaenidae	*Cacyreus marshalli*	F	Karnowski & Labanowski 1998
*		*Eumaeus atala florida*	F	Culbert 1995
*		*Thecla basalides*	F	Lorenzato *et al.* 1997
*	Lepidoptera: Lymantriidae	*Euproctis melania*	F	Abai 1976
*		*Gynaephora ruoergensis*	F	Liu *et al.* 1993
*		*Laelia coenosa candida*	L/F	Li 1987
*		*Leucoma (Stilpnotia) salicis*	F	Nef 1972, 1975
*		*Lymantria dispar*	F	Chernev 1976
*		*Lymantria monacha*	L/F	Anishchenko *et al.* 1982

APPENDIX 2.1 *(continued)*

Subspecies/ Serovar	Order: Family	Genus and Species	Lab/Field	References
*	Lepidoptera: Lymantriidae	*Orgyia antiqua*	L/F	Galani 1973; Skatulla 1974
*		*Orgyia postica*	F	Pardede 1986
*		*Orgyia pseudotsugata*	F	Neisess *et al.* 1978
*		*Sarsina violascens*	F	Macedo 1976
*	Lepidoptera: Lyonetiidae	*Leucoptera malifoliella*	F	Balazs *et al.* 1997
*		*Perileucooptera (Leucoptera) coffeella*	F	Benavides Gomez & Cardenas Murillo 1975; Gravena 1984
*	Lepidoptera: Noctuidae	*Achaea janata*	L	Narayanan *et al.* 1976b; Deshmukh & Deshpande 1989
*		*Agrotis ipsilon*	L/F	Mohd Salleh &Lewis 1982a; Zethner *et al.* 1987
*		*Agrotis segetum*	F	Zethner *et al.* 1987
*		*Alabama argillacea*	F	Bleicher *et al.* 1990; Yamamoto *et al.* 1990
*		*Anomis flava*	F	Twine & Lloyd 1982
*		*Anomis sabulifera*	F	Das & Singh 1998
*		*Anomis (Cosmophila) flava*	F	Angelini & Couilloud 1972; Delattre 1974
*		*Anticarsia gemmatalis*	L/F	Lorenzato & Corseuil 1982; Richter & Fuza 1984; da Silva 1995
*		*Athetis (Proxenus) ignava*	F	Carnegie & Dick 1972
*		*Athetis (Proxenus) xantholopha*	F	Carnegie & Dick 1972
*		*Busseola fusca*	F	Brownbridge 1990
*		*Chrysodeixis chalcites*	F	Hachler *et al.* 1998
*		*Diparopsis watersi*	F	Jacquemard 1982
*		*Earias insulana*	F	Hussain & Askari 1976; Ghobrial & Ali 1978; Ghobrial & Dittrich 1980
*		*Earias* spp.	F	Jacquemard 1982
*		*Earias vittella (fabia)*	L	Ramakrishnan & Pant 1971
*		*Elydna* sp.	F	Carnegie & Dick 1972
*		*Helicoverpa (Heliothis) armigera*	F	Woods 1981; Jacquemard 1982
*		*Helicoverpa (Heliothis) zea*	L/F	Sharma *et al.* 1977; Fuxa 1979; Smith & Hostetter 1982
*		*Heliothis* spp.	F	Broadley *et al.* 1979; Johnson 1982
*		*Heliothis (Chloridea)* sp.	F	Denver 1974
*		*Heliothis (Chloridea) maritima bulgarica*	F	Makarov 1971
*		*Hellula (Oebia) undalis*	F	Hou 1987; Yabas & Zeren 1992
*		*Lacanobia oleracea*	F	Burges & Jarrett 1976; Foster & Crook 1983
*		*Mamestra (Barathra) brassicae*	F	Groner 1977; Terytze & Terytze 1987
*		*Mocis* spp.	F	Perez *et al.* 1991
*		*Mythimna phaea*	F	Carnegie & Dick 1972
*		*Mythimna polyrabda*	F	Carnegie & Dick 1972
*		*Mythimna (Borolia) pinna (longirostris)*	F	Carnegie & Dick 1972
*		*Mythimna (Pseudaletia) unipuncta*	L	Somerville *et al.* 1970
*		*Nystalea nyseus*	F	Laranjeiro & Evans 1994
*		*Plathypena scabra*	F	Beegle *et al.* 1973; Marston *et al.* 1979
*		*Plusia* spp.	F	Lorenzato & Corseuil 1982
*		*Rachiplusia nu*	L/F	Ibarra *et al.* 1992
*		*Rachiplusia nubilalis*	F	da Silva 1987
*		*Raghuva (Heliocheilus) albipunctella*	F	NDoyle & Gahukar 1986
*		*Simplicia extinctalis*	F	Carnegie & Dick 1972
*		[illegible]	F	[illegible]

*		*Spodoptera exigua*	L/F	Lipa *et al.* 1975; Kheyri 1977; Heinz *et al.* 1988
*		*Spodoptera frugiperda*	L/F	Gardner & Fuxa 1980; Mohd Salleh & Lewis 1982b; Gardener 1988
*		*Spodoptera latifascia*	F	Nora *et al.* 1989
*		*Spodoptera litura*	L/F	Govindarajan *et al.* 1976†; Zaz & Kushwaha 1983; Liang *et al.* 1983
*		*Spodoptera praefica*	F	Anon. 1976
*		*Spodoptera (Prodenia) sunia*	F	Gloria 1975
*		*Thysanoplusia orichalcea*	F	Men & Thakare 1998
*		*Trichoplusia ni*	L/F	Somerville *et al.* 1970 ; Sleesman 1971
*		*Xestia (Agrotis) c-nigrum*	L	Lipa *et al.* 1969
*	Lepidoptera: Notodontidae	*Schizura concinna*	L/F	Pinnock *et al.* 1978a, 1978b
*		*Thaumetopoea pityocampa*	L/F	Chernev 1976; Gadais *et al.* 1978
*		*Thaumetopoea processionea*	F	Danguy 1971; Currado & Brussino 1985
*		*Thaumetopoea wilkinsoni*	F	Mendel 1987
*	Lepidoptera: Nymphalidae	*Danaus plexippus*	L	Leong *et al.* 1992
*	Lepidoptera: Olethreutidae	*Epinotia aporema*	L/F	Ibarra *et al.* 1992
*	Lepidoptera: Papilionidae	*Papilio demoleus*	L/F	Prasertphon 1975
*		*Papilio* spp.	F	Butani 1979
*	Lepidoptera: Pieridae	*Ascia monuste orseis*	F	Longhini & Busoli 1993
*		*Colias eurytheme*	L	Somerville *et al.* 1970
*		*Pieris brassicae*	L/F	Ebersold *et al.* 1977; Terytze & Terytze 1987
*		*Pieris canidia*	L	Cheung *et al.* 1990
*		*Pieris canidia sordida*	F	Hou 1987
*		*Pieris rapae*	F	Eckenrode *et al.* 1981; Creighton *et al.* 1981; Sears *et al.* 1983
*		*Pieris rapae crucivora*	F	Hou 1987
*	Lepidoptera: Psychidae	*Cryptothelea (Eumeta) variegata*	F	Sun 1985
*	Lepidoptera: Pyralidae	*Amyelois transitella*	F	Connell *et al.* 1998
*		*Chilo hyrax (niponella)*	L/F	Li 1987
*		*Chilo partellus*	F	Brownbridge 1990; Sharma & Odak 1996; Sharma *et al.* 1998
*		*Cnaphalocrocis medinalis*	L/F	Yang 1978; Srivastava & Nayak 1978;Joshi *et al.* 1987
*		*Corcyra cephalonica*	L	Chiang *et al.* 1986; Hasan *et al.* 1994
*		*Crambus* spp.	F	Reinert 1973
*		*Crocidolomia binotalis*	F	Krishnaiah *et al.* 1981
*		*Cryptoblabes gnidiella*	F	Swirski *et al.* 1979; Baum 1986
*		*Diaphania (Palpita) hyalinata*	L/F	Delplanque & Gruner 1975; Webb 1994
*		*Diaphania (Palpita) nitidalis*	F	Botelho *et al.* 1975; Webb 1994
*		*Diatraea grandiosella*	L	Bohorova *et al.* 1996
*		*Diatraea saccharalis*	L	Bohorova *et al.* 1996
*		*Dioryctria amatella*	L	McLeod *et al.* 1983
*		*Dioryctria rubella*	F	Halos *et al.* 1985
*		*Dioryctria schutazeella*	F	Yu *et al.* 1990
*		*Dioryctria* spp.	F	Cameron 1989; Fatzinger *et al.* 1992
*		*Ectomyelois ceratoniae*	L/F	Alrubeai 1988
*		*Ephestia cautella*	L/F	Hagstrum & Sharp 1975; McGaughey 1978; McGaughey *et al.* 1987
*		*Ephestia kuehniella*	L	Burgerjon *et al.* 1974
*		*Galleria mellonella*	L/F	Krieg 1971; Krieg 1973a; Jafri & Sabiha 1974
*		*Herpetogramma* spp.	F	Reinert 1973
*		*Homoeosoma electellum*	L	Brewer & Anderson 1990

APPENDIX 2.1 (*continued*)

Subspecies/ Serovar	Order: Family	Genus and Species	Lab/Field	References
*	Lepidoptera: Pyralidae	*Hulstia undulatella*	F	Corliaa 1994
*		*Hymenia recurvalis*	F	Johnson 1968
*		*Hypsipyla grandella*	L	Hidalgo Salvatierra & Grijpma 1973
*		*Leucinodes orbonalis*	F	Krishnaiah *et al.* 1981; Yin 1993
*		*Loxostege sticticalis*	F	Shurovenkov Yu *et al.* 1988
*		*Mussidia nigrivenella*	F	Xin *et al.* 1989
*		*Neoleucinodes elegantalis*	F	Prando & Silva 1990; Reis & Souza 1996
*		*Ostrinia nubilalis*	F	Stengel & Atak 1971; Cangardel 1971; Stengel 1972
*		*Ostrinia* spp.	F	Yang 1978
*		*Palpita (Margaronia) quadristigmalis*	F	Aguilera *et al.* 1992
*		*Palpita (Margaronia) unionalis*	L/F	El Hakim & Hanna 1982; Fodale & Mule 1990; Fodale *et al.* 1990
*		*Plodia interpunctella*	L/F	McGaughey 1978; 1985; McGaughey *et al.* 1987
*		*Pyrausta (Eutectona) machaeralis*	F	Chadhar 1996
*	Lepidoptera: Pyraustidae	*Dichocrocis diminutiva*	F	Fu *et al.* 1983
*	Lepidoptera: Riodinidae	*Euselasia eucerus*	F	Macedo 1976
*	Lepidoptera: Saturniidae	*Hyalophora cecropia*	L	Boman *et al.* 1978
*		*Samia cynthia*	L	Boman *et al.* 1978
*		*Callosamia primethea*	L	Boman *et al.* 1978
*	Lepidoptera: Sphingidae	*Erinnyis ello*	L/F	Bellotti & Arias 1978
*		*Herse (Agrius) convolvuli*	F	Feng *et al.* 1997
*		*Manduca sexta*	L	Griego *et al.* 1979
*		*Clanis bilineata tsingtauica*	NS	Sun *et al.* 1993
*	Lepidoptera: Tortricidae	*Achroia grisella*	F	Krieg 1974; Burges 1978
*		*Adoxophyes orana*	F	van der Geest 1971; Takagi 1974; Balazs *et al.* 1997
*		*Adoxophyes* sp.	L/F	Kariya 1977; Kosugi 1998
*		*Archips argyrospilus*	F	Pinnock *et al.* 1978a; Madsen & Carty 1979
*		*Archips rosanus*	F	Madsen & Carty 1979; Benfatto 1981
*		*Archips* spp.	F	de Reede 1985
*		*Archips (Cacoecia) occidentalis*	F	Schwartz 1976
*		*Autographa gamma*	F	Burges & Jarrett 1976, 1978
*		*Cacoecimorpha (Epichoristodes) pronubana (acerbella)*	F	Burges & Jarrett 1976, 1978; Ros *et al.* 1987
*		*Choristoneura conflictana*	F	Averill & Vandre 1982
*		*Choristoneura diversana*	L/F	Yamaguchi *et al.* 1971
*		*Choristoneura occidentalis*	F	Hodgkinson *et al.* 1979; Beckwith & Stelzer 1987
*		*Choristoneura pinus pinus*	F	Cadogan 1993
*		*Cnephasia jactatana*	F	Sale *et al.* 1985
*		*Cryptophlebia leucotreta*	F	Xin *et al.* 1989
*		*Ctenopseustis obliquana*	F	Sale *et al.* 1985; Steven & Sale 1985
*		*Cydia (Laspeyresia) pomonella*	L	Dobzhenok 1976; MacQuillan 1976
*		*Cydia (Laspeyresia) splendana*	F	Rotundo & Giacometti 1986
*		*Epichoristodes acerbella*	F	Ros *et al.* 1987
*		*Epiphyas postvittana*	F	MacQuillan 1976; Buchanan 1977; Sale *et al.* 1985

*		*Lobesia botrana*	F	Viggiani & Tranfaglia 1975; Baggiolini *et al.* 1976
*		*Pammene fasciana*	F	Rotundo & Giacometti 1986
		Pammene rhediella	F	Forti & Ioriatti 1990; 1992
*		*Pandemis dumetana*	F	Warelas 1976
*		*Pandemis heparana*	F	Balazs *et al.* 1997
*		*Petrova cristata*	F	Halos *et al.* 1985
*		*Phlogophora meticulosa*	F	Burges & Jarrett 1976, 1978
*		*Phtheocroa cranaodes*	L	Lorenzato 1984
*		*Planotortrix excessana*	F	Sale *et al.* 1985; Steven & Sale 1985
*		*Planotortrix octa (octo)*	L	Chilcott & Wigley 1993
*		*Platynota idaeusalis*	F	Hull *et al.* 1997
*		*Platynota stultana*	F	Helgesen & Zenner Polania 1974
*		*Polychrosis cunninghamiacola*	L	Liu & Tan 1984; McGaughey *et al.* 1987
*		*Sparganothis pilleriana*	F	Zeng *et al.* 1984
*		*Tortrix capensana*	F	Matthew *et al.* 1974
*		*Tortrix viridana*	F	Konig 1970
*		*Zeiraphera diniana*	F	Martouret & Auer 1977; Aeschlimann 1978
*	Lepidoptera: Xylorictidae	*Timocratica albella*	F	Nakano *et al.* 1973
*	Lepidoptera: Yponomeutidae	*Acrolepiopis (Acrolepia) assectella*	F	Gerst *et al.* 1977
*		*Plutella xylostella (maculipennis)*	F	Lumaban & Raros 1973
*		*Prays oleae*	F	Viggiani 1981
*		*Yponomeuta (padellus) malinellus*	F	Mowat & Clawson 1988; Antonelli *et al.* 1989
*	Mestostigmata: Tetranychidae	*Tetranychus urticae*	F	Englert & Kettner 1983
*	Nematoda: Meloidogynidae	*Meloidogyne* spp.	F	Ivanova *et al.* 1996
*	Nematoda: Strongyloididae	*Strongyloides papillosus*	L	Yao *et al.* 1995b
*	Nematoda: Trichostrongylidae	*Haemonchus contortus*	L	Yao *et al.* 1995a
*	Orthoptera: Acrididae	*Hieroglyphus nigrorepletus*	L	Khan *et al.* 1994
*		*Locusta migratoria*	L	Hoffman *et al.* 1974; Hoffmann 1980
*	Phthiraptera : Trichodectidae	*Damalinia (Bovicola) ovis*	L	Hill & Pinnock 1998
*	Prostigmata: Nalepellidae	*Phytoptus avellanae*	F	Tavamaishvili 1990
*	Prostigmata: Tetranychidae	*Panonychus citri*	L/GH	Hall *et al.* 1971
*		*Panonychus ulmi*	F	Corino *et al.* 1983
*		*Tetranychus urticae*	L/F	Berendt *et al.* 1973; Charles *et al.* 1985; Chang & Chen 1993
*	Pulmonata: Limacidae	*Deroceras reticulatum*	L	Ester & Nijenstein 1995
*	Thysanoptera: Thripidae	*Frankliniella intonsa*	F	Chang & Chen 1993
*		*Thrips tabaci*	F	Satpute *et al.* 1989; Borisevich 1998
*	Trematoda: Schistosomatidae	*Schistosoma mansoni*	L	Sauma & Strand 1990
not ser. 1-15	Diptera: Culicidae	*Aedes aegypti*	L	Sokolova *et al.* 1985
not ser. 1-15		*Anopheles atroparvus*	L	Sokolova *et al.* 1985
not ser. 1-15		*Anopheles messeae*	L	Sokolova *et al.* 1985
not ser. 1-15		*Anopheles sacharovi*	L	Sokolova *et al.* 1985
not ser. 1-15)		*Anopheles stephensi*	L	Sokolova *et al.* 1985
not ser. 1-15)		*Culex pipiens*	L	Sokolova *et al.* 1985
AC19	Lepidoptera: Bombycidae	*Bombyx (Theophila) mandarina*	L	Feng & Xing 1982
	Lepidoptera: Noctuidae	*Mythimna (Leucania) separata*	L	Feng & Xing 1982
auto-agglutinate	Lepidoptera: Pieridae	*Pieris rapae*	L	Yao *et al.* 1998
	Lepidoptera: Yponomeutidae	*Plutella xylostella*	L	Yao *et al.* 1998

APPENDIX 2.1 (*continued*)

Subspecies/ Serovar	Order: Family	Genus and Species	Lab/Field	References
BT1	Lepidoptera: Pyralidae	*Diaphania* (*Palpita*) *indica*	L/F	Wang *et al.* 1989; Schreiner 1991
BT1		*Maruca testulalis* (*vitrata*)	F	Karel & Schoonhoven 1986; Wang *et al.* 1989
BT2	Hemiptera: Diaspididae	*Quadraspidiotus* (*Diaspidiotus*) *gigas* (*gigax*)	F	Wang *et al.* 1988
Bt-40, Bt-119, Bt-285, Bt-536 & Bt-538	Lepidoptera: Gelechiidae	*Tecia solanivora*	L	Perez *et al.* 1997
BTSO2584B	Coleoptera: Chrysomelidae	*Leptinotarsa decemlineata*	L	patent No. 05369027, B.J. Lambert 1994
BTSO2584B	Coleoptera: Curculionidae	*Diabrotica virgifera*	L	patent No. 05369027, B.J. Lambert 1994
BTSO2584C	Coleoptera: Chrysomelidae	*Leptinotarsa decemlineata*	L	patent No. 05369027, B.J. Lambert 1994
BTSO2584C	Coleoptera: Curculionidae	*Diabrotica virgifera*	L	patent No. 05369027, B.J. Lambert 1994
CR-371	Nematoda: Hoplolaimidae	*Rotylenchulus reniformis*	F	patent No. 05378460, B.M. Zuckerman 1995
CR-371		*Rotylenchulus reniformis*	F	Zuckerman *et al.* 1993
CR-371	Nematoda: Meloidgynidae	*Meloidogyne incognita*	GH/F	Zuckerman *et al.* 1993; patent No. 05378460, B.M. Zuckerman 1995
CR-371	Nematoda: Pratylenchidae	*Pratylenchus penetrans*	F	patent No. 05378460, B.M. Zuckerman 1995
CR-371	Nematoda: Rhabditidae	*Caenorhabditis elegans*	L	Zuckerman *et al.* 1993; Borgonie *et al.* 1996a, b
CR-450	Nematoda: Meloidgynidae	*Meloidogyne incognita*	GH	patent No. 05378460, B.M. Zuckerman 1995
EG2838	Coleoptera: Scarabaeidae	*Popillia japonica*	L	patent No. 05378625, W.P. Donovan 1995
EG4961	Coleoptera: Chrysomelidae	*Diabrotica undecimpunctata howardi*	L	patent No. 05378625, W.P. Donovan 1995
EG5092	Lepidoptera: Yponomeutidae	*Plutella xylostella* (*maculipennis*)	L	patent No. 05356623, M.A. von Tersch 1994
EG5144	Coleoptera: Chrysomelidae	*Diabrotica undecimpunctata howardi*	L	patent No. 05378625, W.P. Donovan 1995
EG5144	Coleoptera: Scarabaeidae	*Popillia japonica*	L	patent No. 05378625, W.P. Donovan 1995
EG5847	Lepidoptera: Lymantriidae	*Lymantria dispar*	L	patent No. 05322687, W.P. Donovan 1994
EG5847	Lepidoptera: Noctuidae	*Helicoverpa* (*Heliothis*) *zea*	L	patent No. 05322687, W.P. Donovan 1994
EG5847		*Heliothis virescens*	L	patent No. 05322687, W.P. Donovan 1994
EG5847		*Pseudoplusia* (*Chrysodeixis*) *includens*	L	patent No. 05322687, W.P. Donovan 1994
EG5847		*Spodoptera exigua*	L	patent No. 05322687, W.P. Donovan 1994
EG5847		*Spodoptera frugiperda*	L	patent No. 05322687, W.P. Donovan 1994
EG5847		*Trichoplusia ni*	L	patent No. 05322687, W.P. Donovan 1994
EG5847	Lepidoptera: Pyralidae	*Ostrinia nubilalis*	L	patent No. 05322687, W.P. Donovan 1994
EG5847	Lepidoptera: Yponomeutidae	*Plutella xylostella* (*maculipennis*)	L	patent No. 05322687, W.P. Donovan 1994
EG7237	Coleoptera: Chrysomelidae	*Diabrotica undecimpunctata howardi*	L	patent No. 05378625, W.P. Donovan 1995
EG7237		*Leptinotarsa decemlineata*	L	patent No. 05378625, W.P. Donovan 1995
EG7237	Coleoptera: Coccinellidae	*Epilachna varivestis*	L	patent No. 05378625, W.P. Donovan 1995
exotoxin	Hemiptera: Anthocoridae	*Orius albidepennis*	L	Hafez *et al.* 1997a
exotoxin	Coleoptera: Curculionidae	*Hypera variablilis* (*postica*)	L/F	Wilson *et al.* 1984
exotoxin	Orthoptera: Gryllidae	*Acheta domesticus*	L	Tremblay *et al.* 1972
exotoxin	Lepidoptera: Noctuidae	*Spodoptera littoralis*	L/F	Moore & Navon 1973
exotoxin	Siphonaptera: Pulicidae	*Xenopsylla cheopis*	L	Yakunin 1977
exotoxin		*Xenopsylla skrjabini*	L	Yakunin 1977
exotoxin	Hymenoptera: Vespoidea	*Vespula vulgaris*	L	Haragsim & Vankova 1973
exotoxin	Hymenoptera: Cecidomyiidae	*Contarinia citri*	F	Lin *et al.* 1990b
(Shachongjin)	Hymenoptera: Culicidae	*Culex pipiens*	L	Wang *et al.* 1986
	Lepidoptera: Noctuidae	*Helicoverpa* (*Heliothis*) *armigera*	L	Wang *et al.* 1986

(strain 202)				
GC327	Diptera: Sciaridae	*Lycoriella auripila*	L	White *et al.* 1995
IMC	Lepidoptera: Lymantriidae	*Lymantria dispar*	L/F	Dunbar *et al.* 1973
IMC 90012	Lepidoptera: Geometridae	*Ennomos subsignarius*	F	Dunbar *et al.* 1973; Anderson & Kaya 1975
LBT 5 & 6	Hymenoptera: Formicidae	*Wasmannia (Ochetomyrmex) auropunctatus*	L	Castineiras *et al.* 1991
many	Diptera: Muscidae	*Haematobia irritans*	L	Bucher 1981†
many	Lepidoptera: Bombycidae	*Bombyx mori*	L	Bucher 1981†
many	Lepidoptera: Lymantriidae	*Hypogymna morio*	L	Dulmage *et al.* 1981†
many		*Lymantria dispar*	L	Dulmage *et al.* 1981†
many		*Orgyia psuedotsugata*	L	Dulmage *et al.* 1981†
many	Lepidoptera: Noctuidae	*Agrotis ipsilon*	L	Dulmage *et al.* 1981† (see App. 2.3)
many		*Heliothis virescens*	L	Dulmage *et al.* 1981† (see App. 2.3)
many		*Spodoptera exigua*	L	Dulmage *et al.* 1981† (see App. 2.3)
many		*Spodoptera liturna*	L	Dulmage *et al.* 1981† (see App. 2.3)
many		*Trichoplusia ni*	L	Dulmage *et al.* 1981† (see App. 2.3)
many	Lepidoptera: Pieridae	*Pieris brassicae*	L	Dulmage *et al.* 1981† (see App. 2.3)
many	Lepidoptera: Pyralidae	*Ephestia cautella*	L	Dulmage *et al.* 1981† (see App. 2.3)
many		*Ephestia elutella*	L	Dulmage *et al.* 1981†
many		*Galleria mellonella*	L	Dulmage *et al.* 1981† (see App. 2.3)
many		*Ostrinia nubilalis*	L	Dulmage *et al.* 1981† (see App. 2.3)
many		*Plodia interpunctella*	L	Dulmage *et al.* 1981† (see App. 2.3)
many	Phthiraptera : Trichodectidae	*Damalinia (Bovicola) crassipes*	L	Dulmage *et al.* 1981† (see App. 2.3)
"Plantibac"	Lepidoptera: Yponomeutidae	*Prays citri*	F	Mineo 1970; Carles 1984
"Subdu"	Diptera: Muscidae	*Stomoxys calcitrans*	F	Campbell & Wright 1976
PS17	Nematoda: Pratylenchidae	*Pratylenchus* spp.	L	A.J. Sick patent No. 05281530
PS17	Nematoda: Rhabditidae	*Caenorhabditis elegans*	L	A.J. Sick patent No. 05281530
PS201L1	Diptera: Agromyzidae	*Liriomyza trifolii*	L	patent No. 05436002, J.M. Payne 1995
PS201L1	Diptera: Culicidae	*Aedes aegypti*	L	patent No. 05436002, J.M. Payne 1995
PS92J	Diptera: Agromyzidae	*Liriomyza trifolii*	L	patent No. 05298245 & 05436002, J.M. Payne 1994, 1995
PS92J	Diptera: Culicidae	*Aedes aegypti*	L	patent No. 05436002, J.M. Payne 1995
SD 5	Lepidoptera: Noctuidae	*Plusia (Chrysodeixis) agnata*	F	Dai *et al.* 1996
ser. 20	Diptera: Culicidae	*Anopheles stephensi*	L	Saitoh *et al.* 1998
several subsp.	Diptera: Tephritidae	*Ceratitis capitata*	L	Gingrich 1987
several subsp.	Lepidoptera: Pyralidae	*Cnaphalocrocis medinalis*	L	Ma *et al.* 1994
T83016	Lepidoptera: Lasiocampidae	*Dendrolimus tabulaeformis*	L	Dai *et al.* 1989
T83017		*Malacosoma neustria*	L	Dai *et al.* 1989
T83018	Lepidoptera: Lymantriidae	*Leucoma (Stilpnotia) salicis*	L/F	Dai *et al.* 1989
T83019	Lepidoptera: Pyralidae	*Pyralis farinalis*	L/F	Dai *et al.* 1989
var. *anduze*	Lepidoptera: Arctiidae	*Hyphantria cunea*	L	Pana Beratlief 1968

APPENDIX 2.2 INVERTEBRATE SPECIES REPORTED AS NON-SUSCEPTIBLE TO *BT*

Species were recorded as non-susceptible where there was no reported difference between toxicity in the treatments and the control in laboratory assays or no effect against field populations reported. Note that some species have been recorded as both susceptible (Appendix 2.1) and non-susceptible. See Chapter 5 for discussion.

Up to three publications listed for each entry. * – serovar unidentified; L – laboratory; F – field; NS – site not stated;
† – Report from the list of Krieg and Langenbruch 1981; full references in that publication.

Subspecies/ Serovar	Order: Family	Genus and Species	Lab/Field	References
aizawai	Diptera: Culicidae	*Aedes aegypti*	L	Hall *et al.* 1977†; Padua *et al.* 1981
		Aedes triseriatus	L	Hall *et al.* 1977†
		Anopheles stephensi	L	Larget & de Barjac 1981a
		Culex pipiens molestus	L	Saitoh *et al.* 1996
		Culex quinquefasciatus	L	Hall *et al.* 1977†
		Culex tarsalis	L	Hall *et al.* 1977†
	Diptera: Muscidae	*Musca domestica*	L	Rogoff *et al.* 1969†; McGaughey & Johnson 1987
	Diptera: Psychodidae	*Telmatoscopus albipunctatus*	L	Saitoh *et al.* 1996
	Ephemeroptera: Ephemerallidae	*Ephemera dancia*	L	Weiser & Vankova 1978†
	Hymenoptera: Aphelinidae	*Encarsia formosa*	L	Hayashi 1996
	Lepidoptera: Bombycidae	*Bombyx mori*	L	Padua *et al.* 1981
	Lepidoptera: Geometridae	*Hyposidra talaca*	L	Widiastuti *et al.* 1996
	Lepidoptera: Lymantriidae	*Lymantria dispar*	L	Dubois *et al.* 1989
	Lepidoptera: Noctuidae	*Achaea janata*	L	Deshpande & Ramakrishnan 1982
aizawai?	Lepidoptera: Phalaenidae	*Hadena (Mamestra) illoba*	F	Takaki 1975†
aizawai?	Lepidoptera: Tortricidae	*Cydia (Grapholitha) molesta*	F	Takaki 1975†
alesti	Diptera: Culicidae	*Aedes aegypti*	L	Hall *et al.* 1977†; Panbangred *et al.* 1979
		Aedes triseriatus	L	Hall *et al.* 1977†
		Culex pipiens molestus	L	Saitoh *et al.* 1996
		Culex quinquefasciatus	L	Hall *et al.* 1977†
		Culex tarsalis	L	Hall *et al.* 1977†
	Diptera: Fanniidae	*Fannia canicularis*	L	Hall *et al.* 1972
		Fannia femoralis	L	Hall *et al.* 1972
	Diptera: Muscidae	*Musca domestica*	L	Rogoff *et al.* 1969†
	Diptera: Psychodidae	*Telmatoscopus albipunctatus*	L	Saitoh *et al.* 1996
	Hymenoptera: Apidae	*Apis mellifera*	L	Haragsim & Vankova 1968†; Celli 1974
	Hymenoptera: Braconidae	*Apanteles glomeratus*	L	Biliotti 1956†
		Ascogaster quadridentata	L	Mosievskaya & Makarov 1974
		Microdus (Agathis) rufipes	L	Mosievskaya & Makarov 1974
	Hymenoptera: Ichneumonidae	*Hyposoter ebeninus*	L	Biliotti 1965†
	Hymenoptera: Tenthredinidae	*Pristiphora abietina*	F	Donaubauer & Schmutzenhofer 1973†

APPENDIX 2.2 (*continued*)

Subspecies/ Serovar	Order: Family	Genus and Species	Lab/Field	References
alesti	Lepidoptera: Cochylidae	*Eupoecilia ambiguella*	F	Chaboussou 1959†
	Lepidoptera: Lymantriidae	*Lymantria dispar*	L	Dubois *et al.* 1989
	Lepidoptera: Noctuidae	*Agrotis ipsilon*	L	Burgerjon & Grison 1959†
		Heliothis virescens	L	Dulmage 1973†
		Heliothis zea	L	Rogoff *et al.* 1969†
		Mythimna (Pseudaletia) unipuncta	L	Burgerjon & Biache 1967b†
		Oria musculosa	L	Burgerjon & Grison 1959†
		Spodoptera littoralis	L	Moore & Navon 1973
		Trichoplusia ni	L	Rogoff *et al.* 1969†
		Sparganothis pilleriana	F	Chaboussou 1959†
amagiensis	Diptera: Culicidae	*Culex pipiens molestus*	L	Saitoh *et al.* 1996
	Diptera: Psychodidae	*Telmatoscopus albipunctatus*	L	Saitoh *et al.* 1996
	Lepidoptera: Arctiidae	*Hyphantria cunea*	L	Ishii & Ohba 1993a
	Lepidoptera: Bombycidae	*Bombyx mori*	L	Ishii & Ohba 1993a
andaluciensis	Coleoptera: Chrysomelidae	*Leptinotarsa decemlineata*	L	Iriarte *et al.* 1998
	Diptera: Tipulidae	*Tipula oleracea*	L	Iriarte *et al.* 1998
	Lepidoptera: Noctuidae	*Heliothis armigera*	L	Iriarte *et al.* 1998
		Spodoptera exigua	L	Iriarte *et al.* 1998
	Lepidoptera: Yponomeutidae	*Plutella xylostella*	L	Iriarte *et al.* 1998
brasiliensis	Lepidoptera: Noctuidae	*Spodoptera littoralis*	L	Rabinovitch *et al.* 1995
canadensis	Diptera: Culicidae	*Aedes aegypti*	L	Hall *et al.* 1977†; Hastowo *et al.* 1992
		Aedes triseriatus	L	Hall *et al.* 1977†
		Culex pipiens molestus	L	Saitoh *et al.* 1996
		Culex quinquefasciatus	L	Hall *et al.* 1977†; Hastowo *et al.* 1992
		Culex tarsalis	L	Hall *et al.* 1977†
	Diptera: Psychodidae	*Telmatoscopus albipunctatus*	L	Saitoh *et al.* 1996
	Lepidoptera: Bombycidae	*Bombyx mori*	L	Hastowo *et al.* 1992
	Lepidoptera: Lymantriidae	*Lymantria dispar*	L	Dubois *et al.* 1989
	Lepidoptera: Noctuidae	*Heliothis punctigera*	L	Teakle *et al.* 1992
colmeri	Diptera: Culicidae	*Culex pipiens molestus*	L	Saitoh *et al.* 1996
	Diptera: Psychodidae	*Telmatoscopus albpunctatus*	L	Saitoh *et al.* 1996
	Lepidoptera: Lymantriidae	*Lymantria dispar*	L	Dubois *et al.* 1989
coreanensis	Diptera: Culicidae	*Culex pipiens molestus*	L	Saitoh *et al.* 1996
	Diptera: Psychodidae	*Telmatoscopus albipunctatus*	L	Saitoh *et al.* 1996
	Lepidoptera: Bombycidae	*Bombyx mori*	L	Lee *et al.* 1994
[illegible]	Diptera: Culicidae	*Culex pipiens molestus*	L	Saitoh *et al.* 1996

	Lepidoptera: Noctuidae	*Trichoplusia ni*	L	de Lucca†
darmstadiensis	Coleoptera: Chrysomelidae	*Leptinotarsa decemlineata*	L	Iriarte *et al.* 1998
		Diabrotica undecimpunctata howardii	L	Abdel-Hameed & Landen 1994
	Diptera: Culicidae	*Aedes aegypti*	L	Hall *et al.* 1977†
		Aedes triseriatus	L	Hall *et al.* 1977†
		Culex pipiens molestus	L	Saitoh *et al.* 1996
		Culex quinquefasciatus	L	Hall *et al.* 1977†
		Culex tarsalis	L	Hall *et al.* 1977†
	Diptera: Muscidae	*Musca domestica*	L	Abdel-Hameed & Landen 1994
	Diptera: Psychodidae	*Telmatoscopus albipunctatus*	L	Saitoh *et al.* 1996
	Diptera: Tipulidae	*Tipula oleracea*	L	Iriarte *et al.* 1998
	Lepidoptera: Bombycidae	*Bombyx mori*	L	Nataraju *et al.* 1993
	Lepidoptera: Lymantriidae	*Lymantria dispar*	L	Dubois *et al.* 1989
	Lepidoptera: Noctuidae	*Heliothis armigera*	L	Iriarte *et al.* 1998
		Spodoptera exigua	L	Iriarte *et al.* 1998
		Heliothis virescens	L	Abdel-Hameed & Landen 1994
		Trichoplusia ni	L	Abdel-Hameed & Landen 1994
	Lepidoptera: Yponomeutidae	*Plutella xylostella*	L	Iriarte *et al.* 1998
dendrolimus	Coleoptera: Bruchidae	*Acanthoscelides obtectus*	L	Sander & Cichy 1967†
	Coleoptera: Chrysomelidae	*Diabrotica undecimpunctata howardii*	L	Abdel-Hameed & Landen 1994
	Coleoptera: Coccinellidae	*Adonia variegata*	F	Umarov Sh *et al.* 1975
		Coccinella septempunctata	F	Umarov Sh *et al.* 1975
		Coccinella undecimpunctata	F	Umarov Sh *et al.* 1975
		Scolothrips acariphagus	F	Umarov Sh *et al.* 1975
	Coleoptera: Curculionidae	*Sitophilus granarius*	L	Sander & Cichy 1967†; McGaughey *et al.* 1975†
	Diptera: Culicidae	*Culex molestus*	L	Saubenova *et al.* 1973
		Aedes aegypti	L	Hall *et al.* 1977†
		Aedes triseriatus	L	Hall *et al.* 1977†
		Culex pipiens	L	Rogoff *et al.* 1969†
		Culex quinquefasciatus	L	Hall *et al.* 1977†
		Culex tarsalis	L	Hall *et al.* 1977†
	Diptera: Muscidae	*Musca domestica*	L	Rogoff *et al.* 1969†
	Hemiptera: Anthocoridae	*Orius niger*	F	Umarov Sh *et al.* 1975
	Hemiptera: Miridae	*Campylomma diversicornis*	F	Umarov Sh *et al.* 1975
		Campylomma variegata	F	Umarov Sh *et al.* 1975
	Hymenoptera: Apidae	*Apis mellifera (ligustica)*	L	Haragsim & Vankova 1968†
	Hymenoptera: Braconidae	*Apanteles (Cotesia) glomeratus*	F	Yastrebov 1978
		Apanteles (Cotesia) tibialis	F	Kashkarova 1975
		Bracon hebetor	F	Kovalenkov 1983
		Eurithia (Ernestria) consobrina	F	Yastrebov 1978
	Hymenoptera: Trichogrammatidae	*Trichogramma euproctidus*	F	Kovalenko 1983; Adashkevich & Rashidov 1986
		Trichogramma evanescens	F	Marchenko 1983
		Trichogramma sp.	F	Sharafutdinov Sh & Salikhov 1975
	Lepidoptera: Bombycidae	*Bombyx mori*	L	Burgerjon & Biache 1967b†

APPENDIX 2.2 *(continued)*

Subspecies/ Serovar	Order: Family	Genus and Species	Lab/Field	References
dendrolimus	Lepidoptera: Lasiocampidae	*Dendrolimus superans (sibiricus)*	F	Mashanov *et al.* 1976
	Lepidoptera: Lymantriidae	*Lymantria dispar*	L	Dubois *et al.* 1989
	Lepidoptera: Noctuidae	*Heliothis virescens*	L	Abdel-Hameed & Landen 1994
		Trichoplusia ni	L	Rogoff *et al.* 1969†
	Lepidoptera: Pyralidae	*Galleria mellonella*	L	Burgerjon & Biache 1967b†
	Lepidoptera: Saturniidae	*Samia cynthia*	L	Burgerjon & Biache 1967b†
		Saturnia pavonia	L	Burgerjon & Biache 1967b†
	Lepidoptera: Tortricidae	*Cydia (Grapholitha) funebrana*	F	Fedorinchik & Sogoyan 1975
	Neuroptera: Chrysopidae	*Chrysoperla (Chrysopa) carnea*	F	Umarov Sh *et al.* 1975; Adylov *et al.* 1990
	Thysanoptera: Aeolothripidae	*Aeolothrips intermedius*	F	Umarov Sh *et al.* 1975
entomocidus	Coleoptera: Staphylinidae	*Paederus alfierii*	L	Salama & Zaki 1983
	Diptera: Culicidae	*Aedes aegypti*	L	Hall *et al.* 1977†; Padua *et al.* 1981
		Aedes triseriatus	L	Hall *et al.* 1977†
		Anopheles stephensi	L	Larget & de Barjac 1981a
		Culex pipiens	L	Hall *et al.* 1977†; Kim *et al.* 1998b
		Culex quinquefasciatus	L	Hall *et al.* 1977†; Hastowo *et al.* 1992
		Culex tarsalis	L	Hall *et al.* 1977†
	Diptera: Muscidae	*Musca domestica*	L	Rogoff *et al.* 1969†; McGaughey & Johnson 1987
	Hymenoptera: Apidae	*Apis mellifera*	L	Haragsim & Vankova 1968†
	Hymenoptera: Braconidae	*Bracon hebetor*	L	Sneh *et al.* 1983
	Lepidoptera: Bombycidae	*Bombyx mori*	L	Padua *et al.* 1981; Hastowo *et al.* 1992
	Lepidoptera: Noctuidae	*Achaea janata*	L	Deshpande & Ramakrishnan 1982
		Heliothis zea	L	Rogoff *et al.* 1969†
finitimus	Diptera: Culicidae	*Aedes aegypti*	L	Hall *et al.* 1977†
		Aedes triseriatus	L	Hall *et al.* 1977†
		Culex pipiens	L	Hall *et al.* 1977†
		Culex quinquefasciatus	L	Hall *et al.* 1977†
		Culex tarsalis	L	Hall *et al.* 1977†
	Lepidoptera: Lymantriidae	*Lymantria dispar*	L	Dubois *et al.* 1989
		Orgyia leucostigma	L	Rossmoore *et al.* 1970†
	Lepidoptera: Noctuidae	*Heliothis zea*	L	Rogoff *et al.* 1969†
		Spodoptera littoralis	L	Moore & Navon 1973
		Trichoplusia ni	L	Rogoff *et al.* 1969†
	Lepidoptera: Tortricidae	*Choristoneura fumiferana*	L	Yamvrias & Angus 1970†
fukuokaensis	Coleoptera: Chrysomelidae	*Leptinotarsa decemlineata*	L	Iriarte *et al.* 1998
	Diptera: Culicidae	*Aedes aegypti*	L	Hastowo *et al.* 1992
		Culex pipiens molestus	L	Saitoh *et al.* 1996
		Culex quinquefasciatus	L	Hastowo *et al.* 1992
		Culex tritaeniorhynchus	L	Ohba & Aizawa 1989
	Diptera: Muscidae	*Musca domestica*	L	Ohba & Aizawa 1989

	Lepidoptera: Bombycidae	*Bombyx mori*	L	Hastowo *et al.* 1992
	Lepidoptera: Noctuidae	*Heliothis armigera*	L	Iriarte *et al.* 1998
		Spodoptera exigua	L	Iriarte *et al.* 1998
	Lepidoptera: Yponomeutidae	*Plutella xylostella*	L	Iriarte *et al.* 1998
galleriae	Acari: Anthocoridae	*Anthocoris nemoralis*	F	Karadzhov 1973a,b
		Anthocoris nemorum	F	Karadzhov 1973a,b
	Acari: Phytoseiidae	*Amblyseius (Typhlodromus) aberrans*	F	Karadzhov 1973a,b
		Amblyseius (Typhlodromus) finlandicus	F	Karadzhov 1973
	Coleoptera: Chrysomelidae	*Diabrotica longicornis*	L	Sutter 1969†
		Diabrotica undecimpunctata	L	Sutter 1969†; Abdel-Hameed & Landen 1994
		Diabrotica virgifera	L	Sutter 1969†
		Leptinotarsa decemlineata	L	Iriarte *et al.* 1998
	Coleoptera: Coccinellidae	*Adonia variegata*	F	Umarov Sh *et al.* 1975
		Coccinella septempunctata	L/F	Kiselek 1975; Umarov Sh *et al.* 1975
		Coccinella undecimpunctata	F	Umarov Sh *et al.* 1975
		Cryptolaemus montrouzieri	L	Kiselek 1975
		Scolothrips acariphagus	F	Umarov Sh *et al.* 1975
	Coleoptera: Curculionidae	*Sitophilus granarius*	L	Sander & Cichy 1967†
	Diptera: Cecidomyiidae	*Oligotrophus* sp.	F	Cheng *et al.* 1983
	Diptera: Culicidae	*Aedes aegypti*	L	Hall *et al.* 1977†; Larget & de Barjac 1981a
		Aedes triseriatus	L	Hall *et al.* 1977†
		Anopheles hyrcanus	L	Saubenova 1973
		Culex pipiens	L/F	Hall *et al.* 1977†; Kim *et al.* 1998b
		Culex pipiens molestus	L	Saitoh *et al.* 1996
		Culex pipiens pipiens	L	Larget & de Barjac 1981a
		Culex quinquefasciatus	L	Hall *et al.* 1977†
		Culex tarsalis	L	Hall *et al.* 1977†
	Diptera: Muscidae	*Musca domestica*	L	Rogoff *et al.* 1969†
	Diptera: Psychodidae	*Telmatoscopus albipunctatus*	L	Saitoh *et al.* 1996
	Diptera: Tabanidae	*Tabanus autumnalis brunnescens*	L/F	Kadyrova *et al.* 1977
	Diptera: Tachinidae	*Tachina larvarum*	F	Tkachev 1972
	Diptera: Tipulidae	*Tipula oleracea*	L	Iriarte *et al.* 1998
	Hemiptera: Anthocoridae	*Orius niger*	F	Umarov Sh *et al.* 1975
	Hemiptera: Kerriidae	*Kerria lacca*	F	Malhotra & Choudhary 1968†
	Hemiptera: Miridae	*Campylomma diversicornis*	F	Umarov Sh *et al.* 1975
		Campylomma variegata	F	Umarov Sh *et al.* 1975
	Hemiptera: Scutelleridae	*Eurygaster integriceps*	L	Isakova 1958†
	Hymenoptera: Aphelinidae	*Elatophilus nipponensis*	F	Cheng *et al.* 1983
		Encarsia formosa	L	Beglyarov & Maslienko 1978
	Hymenoptera: Apidae	*Apis mellifera*	L	Haragsim & Vankova 1968†
	Hymenoptera: Braconidae	*Apanteles albipennis*	L	Adashkevich 1966†
		Apanteles circumscriptus	F	Boldyrev 1970
		Apanteles glomeratus	L	Isakova 1965†
		Apanteles rufricus	L	Adashkevich 1966†
		Apanteles sicarius	L	Adashkevich 1966†
		Apanteles (Cotesia) tibialis	F	Kashkarova 1975

APPENDIX 2.2 *(continued)*

Subspecies/ Serovar	Order: Family	Genus and Species	Lab/Field	References
galleriae	Hymenoptera: Braconidae	*Ascogaster quadridentata*	L/F	Tkachev 1974; Mosievskaya & Makarov 1974
		Meteorus versicolor	L	Isakova 1965†
		Microdus (Agathis) rufipes	L/F	Tkachev 1974; Mosievskaya & Makarov 1974
		Microplitis tuberculifera	L	Isakova 1965†
		Trichomma enecator	F	Tkachev 1974
	Hymenoptera: Encyrtidae	*Ageniaspis fuscicollis*	L	Isakova 1965†
	Hymenoptera: Ichneumonidae	*Diadegma cerophaga*	L	Adashkevich 1966†
		Diadegma fenestralis	L	Adashkevich 1966†
		Diadromus subtilicornis	L	Adashkevich 1966†
		Hyposoter ebeninus	L	Isakova 1965†
		Pimpla (Coccygomimus) turionellae	F	Isakova 1965†; Tkachev 1974
		Pristomerus vulnerator	F	Tkachev 1974
		Theronia atalantae	L	Isakova 1965†
	Hymenoptera: Pteromalidae	*Dibrachys cavus*	F	Tkachev 1974
		Pteromalus puparum	L	Isakova 1965†
	Hymenoptera: Trichogrammatidae	*Trichogramma cacoecia pallida*	L/F	Seskevicius 1977; Kapustina 1975
		Trichogramma embryophagum	F	Kolmakova 1971; Karadzhov 1975
		Trichogramma euproctidus	F	Moiseeva *et al.* 1979
		Trichogramma evanescens	L/F	Karadzhov 1975; Kim Chi 1978
		Trichogramma pallidum (cacoeciae pallida)	F	Kolmakova 1971; Karadzhov 1975
		Trichogramma sp.	F	Zlatanova & Lukin 1971; Moiseeva *et al.* 1975
		Trichogramma evanescens	F	Shchepetilnikova *et al.* 1968a, b†
	Isoptera: Hodotermitidae	*Anacanthotermes* sp.	F	Kakaliev & Saparliev 1975†
	Lepidoptera: Carposinidae	*Carposina niponensis*	F	Takaki 1975†
	Lepidoptera: Gelechiidae	*Gelechia hippophaella*	F	Sanzhimtupova 1984
	Lepidoptera: Lithocolletidae	*Phyllonorycter (Lithocolletis) ringoniella*	F	Takaki 1975†
	Lepidoptera: Lymantriidae	*Lymantria dispar*	L	Johansson 1971†; Dubois *et al.* 1989
	Lepidoptera: Noctuidae	*Heliothis virescens*	L	Abdel-Hameed & Landen 1994
		Heliothis armigera	L	Iriarte *et al.* 1998
		Spodoptera exigua	L	Iriarte *et al.* 1998
		Spodoptera litura	L	Narayanan *et al.* 1976†
		Trichoplusia ni	L	Abdel-Hameed & Landen 1994
	Lepidoptera: Tortricidae	*Cydia molesta*	F	Aliev & Yakubov 1971
		Cydia (Grapholitha) funebrana	F	Fedorinchik & Sogoyan 1975
		Tortrix viridana	L	Kal'vish & Krivtsova 1978
	Lepidoptera: Yponomeutidae	*Plutella xylostella*	L	Iriarte *et al.* 1998
	Neuroptera: Chrysopidae	*Chrysoperla (Chrysopa) carnea*	L/F	Kiselek 1975; Umarov Sh *et al.* 1975
	Orthoptera: Catantopidae	*Calliptamus italicus*	L	Isakova 1958†
	Prostigmata: Anystidae	*Anystis* sp.	F	Cheng *et al.* 1983
	Thysanoptera: Aeolothripidae	*Aeolothrips intermedius*	F	Umarov Sh *et al.* 1975
guiyangiensis	Coleoptera: Chrysomelidae	*Leptinotarsa decemlineata*	L	Iriarte *et al.* 1998

	Lepidoptera: Yponomeutidae	*Plutella xylostella*	L	Iriarte *et al.* 1998
higo	Coleoptera: Chrysomelidae	*Leptinotarsa decemlineata*	L	Jung *et al.* 1998
	Lepidoptera: Arctiidae	*Hyphantria cunea*	L	Jung *et al.* 1998
higo	Lepidoptera: Bombycidae	*Bombyx mori*	L	Jung *et al.* 1998
huazhongensis	Diptera: Culicidae	*Aedes aegypti*	L	Dai *et al.* 1996
		Culex quinquefasciatus (fatigans)	L	Dai *et al.* 1996
	Lepidoptera: Bombycidae	*Bombyx mori*	L	Dai *et al.* 1996
	Lepidoptera: Noctuidae	*Spodoptera exigua*	L	Dai *et al.* 1996
		Spodoptera littoralis	L	Dai *et al.* 1996
	Lepidoptera: Yponomeutidae	*Plutella xylostella*	L	Dai *et al.* 1996
indiana	Coleoptera: Chrysomelidae	*Diabrotica undecimpunctata howardii*	L	Abdel-Hameed & Landen 1994
	Diptera: Culicidae	*Aedes aegypti*	L	Abdel-Hameed & Landen 1994
	Lepidoptera: Lymantriidae	*Lymantria dispar*	L	Dubois *et al.* 1989
	Lepidoptera: Noctuidae	*Trichoplusia ni*	L	de Lucca†; Abdel-Hameed & Landen 1994
		Heliothis virescens	L	Abdel-Hameed & Landen 1994
israelensis	Acarina: Acari	*Hydrachna* sp.	L	Weiser & Vankova 1978†; Beck 1982
		Hydrachnella sp.	?	Becker & Margalit 1993
	Acarina: Hydracarina	?	NS	Beck 1982
	Annelida: Hirudinea	*Helobdella* sp.	L	Weiser & Vankova 1978†
		Helobdella stagnalis	?	Becker & Margalit 1993
	Annelida: Oligochaeta	*Tubifex* sp.	L	Morawczik & Schnetter†
	Arachnoidea: Acari	*Hydacarina* sp.	L	Morawczik & Schnetter†
	Cladocera: ?	*Simocephalus vetulus*	NS	Abbott Laboratories; Garcia *et al.* 1980
	Cnidaria	*Hydra* sp.	?	Becker & Margalit 1993
	Coleoptera: ?	*Anacaena globulus*	?	Becker & Margalit 1993
	Coleoptera: ?	*Baeosus* sp.	NS	Beck 1982
	Coleoptera: ?	*Coelambus impressopunctatus*	?	Becker & Margalit 1993
	Coleoptera: ?	*Coelambus* sp.	L	Beck 1982; Morawczik & Schnetter†
	Coleoptera: ?	*Guignotus pusillus*	?	Becker & Margalit 1993
	Coleoptera: ?	*Peltodytes edentulus*	L	Gharib & Hilsenhoff 1988
	Coleoptera: Chrysomelidae	*Diabrotica undecimpunctata howardii*	L	Abdel-Hameed & Landen 1994
		Gastrophysa virula	L	Benz & Joeresson 1994
		Melasoma aenea	L	Benz & Joeresson 1994
	Coleoptera: Curculionidae	*Sitophilus granarius*	L	Benz & Joeresson 1994
	Coleoptera: Dytiscidae	*Hydroporus palustris*	?	Becker & Margalit 1993
		Hydroporus undulatus	L	Gharib & Hilsenhoff 1988
		Hygrotus inaequalis	?	Becker & Margalit 1993
		Hyphydrus ovatus	?	Becker & Margalit 1993
		Ilybius fuliginosus	?	Becker & Margalit 1993
		Laccophilus maculosus	L	Gharib & Hilsenhoff 1988
		Laccophilus sp.	L	Morawczik & Schnetter†; Beck 1982
		Rhantus consputus	?	Becker & Margalit 1993

APPENDIX 2.2 (*continued*)

Subspecies/ Serovar	Order: Family	Genus and Species	Lab/Field	References
israelensis	Coleoptera: Dytiscidae	*Rhantus pulverosus*	?	Becker & Margalit 1993
	Coleoptera: Haliplidae	*Haliplus immaculicollis*	L	Gharib & Hilsenhoff 1988
	Coleoptera: Hydrophilidae	*Berosus signaticollis*	?	Becker & Margalit 1993
		Berosus sp.	L	Morawczik & Schnetter†
		Hydrobius fuscipes	?	Becker & Margalit 1993
		Hydrophilus caraboides	?	Becker & Margalit 1993
		Tropisternus sp.	NS	Abbott Laboratories; Garcia *et al.* 1980
		Tropisternus salsamentus	NS	Abbott Laboratories; Garcia *et al.* 1980
	Coleoptera: Tenebrionidae	*Tribolium confusum*	L	Benz & Joeresson 1994
	Coleoptera: Dytiscidae	?	NS	Abbott Laboratories; Garcia *et al.* 1980
	Coleoptera: Gyrinidae	?	L	Weiser & Vankova 1978†; Beck 1982
	Copepoda: Cyclopidae	*Acanthocyclops vernalis*	L/F	Marten *et al.* 1993
	Crustacea: ?	*Bathyomphalus contortus*	?	Becker & Margalit 1993
	Crustacea: Amphipoda	*Gammaridae* sp.	NS	Abbott Laboratories; Garcia *et al.* 1980
		Gammarus duebeni	L	Roberts 1995
		Gammarus pulex	?	Becker & Margalit 1993
		Hyallela azteca	NS	Abbott Laboratories; Garcia *et al.* 1980; Gharib & Hilsenhoff 1988
	Crustacea: Artemiidae	*Artemia salina*	NS	Abbott Laboratories; Garcia *et al.* 1980
	Crustacea: Cambaridae	*Orconectes limosus*	?	Becker & Margalit 1993
	Crustacea: Cladocera	*Chirocephalus grubei*	?	Becker & Margalit 1993
		Daphnia sp.	F	Ali 1981
		Daphnia magna	L	Sinegre *et al.* 1979a†; Lebrun & Vlayen 1981
		Daphnia pulex	L	Morawczik & Schnetter†;
			?	Becker & Margalit 1993
		Daphnia sp.	L	Krieg & Langenbruch 1981
		Moina macrocopa	NS	Beck 1982
		Moina rectirostris	L	Mulla 1990
	Crustacea: Conchostracans	*Eulimnadia texana*	L	Mulla 1990
	Crustacea: Copepoda	*Cyclops fuscus*	L	Sinegre *et al.* 1979a†
		Cyclops spp.	L	Ali 1981; Beck 1982; Krieg & Langenbruch 1981
		Cyclops strenuus	?	Becker & Margalit 1993
		Macrocyclops sp.	NS	Abbott Laboratories; Garcia *et al.* 1980
		Macrocyclops albidus	F	Marten *et al.* 1993
		Mesocyclops longisetus	F	Marten *et al.* 1993
		Mesocyclops ruttneri	F	Marten *et al.* 1993
	Crustacea: Decapoda	*Hemigrapsus* sp.	NS	Abbott Laboratories; Garcia *et al.* 1980
	Crustacea: Isopoda	*Asellus aquaticus*	?	Becker & Margalit 1993
	Crustacea: Palaemonidae	*Palaemonetes varians*	L	Roberts 1995
	Decapoda: Palaemonidae	*Leander tenuicornis*	L	Brown *et al.* 1996
	Diptera: ?	*Chelifera* sp.	NS	Abbott Laboratories; Garcia *et al.* 1980
	Diptera: Anthomyiidae	*Erioischia (Phorbia) brassicae*	L	Hassan & Krieg†

	Chironomus plumosus	NS	Beck 1982
	Procladius freemani	L	Mulla *et al.* 1990
	Procladius sublettei	L	Mulla *et al.* 1990
	Tanypus sp.	L	Mulla *et al.* 1990
Diptera: Culicidae	*Aedes aegypti*	L	Abdel-Hameed & Landen 1994
	Culex pipiens molestus	L	Saitoh *et al.* 1996
	Toxorhynchites sp.	?	Krieg & Miltenburger 1984
	Toxorhynchites splendens	NS	Beck 1982
Diptera: Culicoidea	*Mochlonyx* sp.	L	Krieg & Miltenburger 1984
Diptera: Drosophilidae	*Drosophila melanogaster*	L	de Barjac & Larget†; Benz & Joeresson 1994
Diptera: Ephydridae	*Ephydra riparia* complex	NS	Abbott Laboratories; Garcia *et al.* 1980
Diptera: Muscidae	*Musca domestica*	L	Vankova 1981; Benz & Joeresson 1994
Diptera: Psychodidae	*Telmatoscopus albipunctatus*	L	Saitoh *et al.* 1996
Diptera: Tipulidae	*Dicranota* sp.	L	Abbott Laboratories; Garcia *et al.* 1980
	Tilapia nilotica	NS	Lebrun & Vlayen 1981
Ephemeroptera: Baetidae	*Baetis* sp.	L	Ali 1981
	Callibaetis sp.	NS	Abbott Laboratories; Garcia *et al.* 1980
	Callibaetis pacificus	L	Mulla 1990
	Cloeon dipterum	?	Becker & Margalit 1993
	Cloeon sp.	L	Krieg & Langenbruch 1981
Ephemeroptera: Caenidae	*Caenis lactea*	L	Weiser & Vankova 1978†
Ephemeroptera: Ephemerellidae	*Ephemera danica*	NS	Beck 1982
Ephemeroptera: Heptageniidae	*Stenonema* sp.	F	Merrit *et al.* 1989
Ephemeroptera: Leptophlebiidae	*Leptophlebia* sp.	L	Weiser & Vankova 1978†; Beck 1982
Hemiptera: Corixidae	*Hesperocorixa leavigata*	NS	Abbott Laboratories; Garcia *et al.* 1980
	Micronecta meridionalis	L	Becker & Margalit 1993
	Sigara sp.	L	Krieg in Krieg & Langenbruch 1981
	Sigara lateralis	NS	Beck 1982; Becker & Margalit 1993
	Sigara striata	NS	Beck 1982; Becker & Margalit 1993
	Trichocorixa reticulata	NS	Abbott Laboratories; Garcia *et al.* 1980
Hemiptera: Hydrocorisae	*Notonecta glauca*	L	Krieg†; Beck 1982
	Notonecta kirby	NS	Abbott Laboratories; Garcia *et al.* 1980
	Notonecta undulata	L	Aly & Mulla 1987
	Ranatra sp.	NS	Beck 1982
Hemiptera: Naucoridae	*Ilyocoris cimicoides*	L	Becker & Margalit 1993
Hemiptera: Notonectidae	*Anisops varia*	L	Becker & Margalit 1993
	Buenoa antigone	L	Quiroz-Martinez *et al.* 1996
	Bueona scimitra	NS	Abbott Laboratories; Garcia *et al.* 1980
Hemiptera: Pleidae	*Plea leachi*	NS	Abbott Laboratories; Garcia *et al.* 1980; Becker & Margalit 1993
Hymenoptera: Apidae	*Apis mellifera*	L	Pinsdorf & Krieg†; Abbott Laboratories
Hymenoptera: Trichogrammatidae	*Trichogramma evanescens*	L	Hassan & Krieg†
Lepidoptera: Bombycidae	*Bombyx mori*	L	Nataraju *et al.* 1993
Lepidoptera: Lymantriidae	*Lymantria dispar*	L	Dubois *et al.* 1989
Lepidoptera: Noctuidae	*Heliothis virescens*	L	Abdel-Hameed & Landen 1994
	Spodoptera litura	L	de Barjac 1978†
	Trichoplusia ni	L	Abdel-Hameed & Landen 1994

APPENDIX 2.2 *(continued)*

Subspecies/ Serovar	Order: Family	Genus and Species	Lab/Field	References
israelensis	Lepidoptera: Noctuidae	*Scotia segetum*	L	Benz & Joeresson 1994
	Lepidoptera: Pieridae	*Pieris brassicae*	L	Benz & Joeresson 1994
	Lepidoptera: Pyralidae	*Chilo partellus*	L	Khanna *et al.* 1995
		Ephestia kuehniella	L	Benz & Joeresson 1994
	Lepidoptera: Tortricidae	*Adoxophyes orana*	L	Benz & Joeresson 1994
		Cydia pomonella	L	Benz & Joeresson 1994
	Lepidoptera: Yponomeutidae	*Plutella xylostella* as *maculipennis*	L	de Barjac 1978†
	Lumbricidae	*Tubifex* sp.	L	Becker & Margalit 1993
	Mollusca: ?	*Hippeutis complanatus*	L	Becker & Margalit 1993
	Mollusca: Bivalvia	*Ostrea edulis*	L	Sinegre *et al.* 1979b†; Beck 1982
	Mollusca: Bivalvia	*Pisidium* sp.	L	Becker & Margalit 1993
	Mollusca: Gastropoda	*Aplexa hypnorum*	L	Morawczik & Schnetter†
			NS	Beck 1982; Becker & Margalit 1993
		Bithynia tentaculata	L	Morawczik & Schnetter†; Beck 1982
		Galba palustris	L	Beck 1982; Becker & Margalit 1993
			L	Morawczik & Schnetter†
		Limnea stagnalis	L	Morawczik & Schnetter†
		Physa sp.	L	Weiser & Vankova 1978†; Abbott Laboratories; Garcia *et al.* 1980
		Physa acuta	L	Becker & Margalit 1993
		Planorbis planorbis	NS	Beck 1982; Morawczik & Schnetter†
		Radix sp.	L	Morawczik & Schnetter†; Beck 1982
		Viviparus contectus	L	Beck 1982; Morawczik & Schnetter†
	Mollusca: Mytilidae	*Mytilus edulis*	NS	Beck 1982
	Mollusca: Planorbidae	*Anisus leucostomus*	L	Becker & Margalit 1993
	Mollusca	*Taphius (Biomphalaria) glabratus*	L	Larget & de Barjac 1981b
	Mollusca: Gastropoda	?	NS	Abbott Laboratories; Garcia *et al.* 1980
	Nematoda: Steinernematidae	*Neoaplectana carpocapsae*	L	Poinar *et al.* 1990
	Nemotoda: Heterorhabditidae	*Heterorhabditis heliothidis*	L	Poinar *et al.* 1990
	Notostraca: Triopsidae	*Triops longicaudatus*	L	Fry-O'Brien & Mulla 1996
	Odonata: ?	*Cordulia* sp.	NS	Beck 1982
	Odonata: ?	*Symetrium striolatum*	L	Becker & Margalit 1993
	Odonata: Aeshnidae	*Anax* sp.	NS	Abbott Laboratories; Garcia *et al.* 1980
	Odonata: Coenagrionidae	*Enallagma civile*	L	Aly & Mulla 1987
		Ischnura sp.	NS	Abbott Laboratories; Garcia *et al.* 1980
		Ischnura elegans	L	Becker & Margalit 1993
	Odonata: Libellulidae	*Erythemis simplicicollis*	L	Painter *et al.* 1996
		Orthetrum brunneum	L	Becker & Margalit 1993
		Tarnetrum corruptum	L	Aly & Mulla 1987
	Oligochaetes: Turbellaria	*Bothromesostoma personatum*	L	Becker & Margalit 1993
	Ostracoda: Cypridae	?	NS	Abbott Laboratories; Garcia *et al.* 1980
	Plecoptera	Several species	L	Garcia *et al.* 1980; Kreig *et al.* 1980
	Rotifera	*Brachionus calyciflorus*	L	Becker & Margalit 1993

	Trichoptera: ?	*Mystacides alafunbriata*	NS	Abbott Laboratories; Garcia *et al.* 1980
	Trichoptera: Limnephilidae	*Limnephilus flavicornis*	?	Lebrun & Vlayen 1981
		Limnophilus sp.	L	Becker & Margalit 1993
	Trichoptera	*Several species*	NS	Abbott Laboratories; Garcia *et al.* 1980
	Tricladida: Turbellaria	*Dugesia tigrina*	L	Morawczik & Schnetter†; Beck 1982
			L	Becker & Margalit 1993
	Tubellaria	*Dugesia dorotocephala*	NS	Abbott Laboratories; Garcia *et al.* 1980
	Isopoda: Copepoda		NS	Abbott Laboratories; Garcia *et al.* 1980; Knepper & Walker 1989
	Ostracoda: Ostracoda		L	Becker & Margalit 1993; Ali 1981
	Oligochaeta		NS	Beck 1982
japonensis	Diptera: Culicidae	*Culex pipiens molestus*	L	Saitoh *et al.* 1996
		Aedes aegypti	L	Ohba & Aizawa 1986a
		Aedes albopictus	L	Ohba & Aizawa 1986a
	Diptera: Psychodidae	*Telmatoscopus albipunctatus*	L	Saitoh *et al.* 1995
kenyae	Diptera: Culicidae	*Aedes aegypti*	L	Hall *et al.* 1977†; Hastowo *et al.* 1992
		Aedes triseriatus	L	Hall *et al.* 1977†
		Culex pipiens	L	Hall *et al.* 1977†; Kim *et al.* 1998b
		Culex pipiens molestus	L	Saitoh *et al.* 1996
		Culex quinquefasciatus	L	Hall *et al.* 1977†; Hastowo *et al.* 1992
		Culex tarsalis	L	Hall *et al.* 1977†
	Diptera: Psychodidae	*Telmatoscopus albipunctatus*	L	Saitoh *et al.* 1996
	Lepidoptera: Bombycidae	*Bombyx mori*	L	Hastowo *et al.* 1992
	Lepidoptera: Gelechiidae	*Pectinophora gossypiella*	F	Sarag & Satpute 1988
	Lepidoptera: Lymantriidae	*Lymantria dispar*	L	Dubois *et al.* 1989
	Lepidoptera: Noctuidae	*Heliothis armigera*	F	Sarag & Satpute 1988
		Spodoptera littoralis	L	Moore & Navon 1973; Salama *et al.* 1991b
konkukian	Coleoptera: Chrysomelidae	*Leptinotarsa decemlineata*	L	Iriarte *et al.* 1998
	Diptera: Culicidae	*Culex pipiens*	L	Lee *et al.* 1994
	Diptera: Tipulidae	*Tipula oleracea*	L	Iriarte *et al.* 1998
	Lepidoptera: Noctuidae	*Heliothis armigera*	L	Iriarte *et al.* 1998
		Spodoptera exigua	L	Iriarte *et al.* 1998
	Lepidoptera: Yponomeutidae	*Plutella xylostella*	L	Iriarte *et al.* 1998
kumamotoensis	Coleoptera: Chrysomelidae	*Leptinotarsa decemlineata*	L	Iriarte *et al.* 1998
	Diptera: Culicidae	*Aedes aegypti*	L	Ohba *et al.* 1981b
	Diptera: Tipulidae	*Tipula oleracea*	L	Iriarte *et al.* 1998
	Lepidoptera: Noctuidae	*Heliothis armigera*	L	Iriarte *et al.* 1998
		Spodoptera exigua	L	Iriarte *et al.* 1998
	Lepidoptera: Yponomeutidae	*Plutella xylostella*	L	Iriarte *et al.* 1998
kurstaki	Acari: Anthocoridae	*Anthocoris nemoralis*	F	Nicoli *et al.* 1989; 1990
	Acari: Phytoseiidae	*Amblyseius potentillae*	L	Hassan *et al.* 1983
		Typhlodromus pyri	F	Hardman *et al.* 1995
	Annelida: Oligochaeta	*Lumbricus terrestris*	F	White 1960†; Benz & Altwegg 1975

APPENDIX 2.2 (*continued*)

Subspecies/ Serovar	Order: Family	Genus and Species	Lab/Field	References
kurstaki	Arachnida: Lycosidae	*Pardosa milvina*	F	Muckenfuss & Shepard 1994
	Arachnida: Araneae	NS	F	Hilburn & Jennings 1988
	Arachnoidea: Acari	*Blattisoccus tarsalis*	F	McGaughey 1979†
	Coleoptera: Bostridae	*Rhyzopertha dominica*	F	McGaughey *et al.* 1975†
	Coleoptera: Carabidae	*Agonum punctiforme*	F	Riddick & Mills 1995
		Anisodactylus californicus	F	Riddick & Mills 1995
		Bembidion lampros	F	Obadofin & Finlayson 1977; Finlayson 1979
		Calathus ruficolis	F	Riddick & Mills 1995
		Calosoma granulatum (alternans)	F	Turnipseed *et al.* 1985
		Chlaenius sp.	F	Riddick & Mills 1995
		Harpalus pennsylvanicus	F	Riddick & Mills 1995
		Pterostichus spp.	F	Riddick & Mills 1995
	Coleoptera: Chrysomelidae	*Diabrotica undecimpunctata howardii*	L	MacIntosh *et al.* 1990c; Abdel-Hameed & Landen 1994
		Diabrotica virgifera	L	Sims 1997
	Coleoptera: Coccinellidae	*Adalia bipunctata*	L	Olszak 1982
		Ceratomegilla (Coleomegilla)maculata	F	Johnson 1974
		Coccinella sp.	F	Mohamed 1993
		Coccinella undecimpunctata	F	Salama & Zaki 1984
		Coccinella septempunctata	L/F	Kiselek 1975; Baicu & Hussein 1984
		Hippodamia convergens	L/F	Johnson 1974; Sims 1997
		Menochilus (Cheilomenes) sexmaculatus	F	Jayanthi & Padmavathamma 1996a
		Rodolia cardinalis	L	Viggiani *et al.* 1998
		Scymnus interruptus	F	Salama & Zaki 1984
		Scymnus sp.	F	Mohamed 1993
		Scymnus syriacus	F	Salama & Zaki 1984
	Coleoptera: Cucujidae	*Cryptolestes pusillus*	F	McGaughey *et al.* 1975†
	Coleoptera: Curculionidae	*Hypera postica*	L	MacIntosh *et al.* 1990c
		Phytonomus variablis	F	Rizvi & Talhouk 1984
		Sitophilus oryzae	L/F	McGaughey *et al.* 1975†; Kramer *et al.* 1985
		Sitophilus granarius	L	McGaughey *et al.* 1975†
	Coleoptera: Elmidae	*Zaitzevia* sp.	F	Richardson & Perrin 1994
	Coleoptera: Halticinae	*Phyllotreta armoraciae*	L	MacIntosh *et al.* 1990c
	Coleoptera: Scarabaeidae	*Popillia japonica*	L	MacIntosh *et al.* 1990c
	Coleoptera: Silvanidae	*Oryzaephilus surinamensis*	F	McGaughey *et al.* 1975†
	Coleoptera: Staphylinidae	*Paederus alfierii*	F	Salama & Zaki 1984
	Coleoptera: Tenebrionidae	*Tribolium castaneum*	L	McGaughey *et al.* 1975†
		Tribolium confusum	F	McGaughey *et al.* 1975†; Kramer *et al.* 1985
	Collembola: Isotomidae	*Folsomia candida*	L/F	Addison & Holmes 1995; Sims 1997
	Collembola: ?	*Xenylla grisea*	L	Sims 1997
	Crustacea: Cladocera	*Daphnia* sp.	L	Krieg & Langenbruch 1981
	Crustacea: Copepoda	*Cyclops* sp.	L	Krieg & Langenbruch 1981
	Crustacea: Isopoda	*Porcellio scaber*	L	Sims 1997

Diptera: Chironomidae	*Chironomini* sp.	F	Richardson & Perrin 1994
	Corynoneura sp.	F	Richardson & Perrin 1994
	Diamesinae sp.	F	Richardson & Perrin 1994
	Natarsia sp.	F	Kreutzweiser *et al.* 1994
	Orthocladiinae	F	Richardson & Perrin 1994
	Paratanytarsus grimmii	L	Kondo *et al.* 1995
	Tanypodinae	F	Richardson & Perrin 1994
	Tanytarsini sp.	F	Richardson & Perrin 1994; Kreutzweiser *et al.* 1994
Diptera: Culicidae	*Aedes aegypti*	L	Hall *et al.* 1977†; Hastowo *et al.* 1992
	Aedes triseriatus	L	Hall *et al.* 1977†
	Anopheles gambiae	L	Johnson & Davidson 1984
	Culex pipiens	L	Hall *et al.* 1977†; Farghal *et al.* 1987
	Culex pipiens molestus	L	Saitoh *et al.* 1996
	Culex quinquefasciatus	L	Hall *et al.* 1977†; Hastowo *et al.* 1992
	Culex tarsalis	L	Hall *et al.* 1977†
	Theobaldia (Culiseta) longiareolata	L	Farghal *et al.* 1987
Diptera: Dixidae	NS	F	Richardson & Perrin 1994
Diptera: Drosophilidae	*Zaprionus indiana (indianus)*	L	Shakoori *et al.* 1991b
Diptera: Muscidae	*Musca domestica*	L	Ogiwara *et al.* 1992; Abdel-Hameed & Landen 1994; Sims 1997
Diptera: Psychodidae	*Telmatoscopus albipunctatus*	L	Saitoh *et al.* 1996
Diptera: Simuliidae	NS	F	Richardson & Perrin 1994
	Simulium sp.	F	Kreutzweiser *et al.* 1994
Diptera: Tachinidae	*Bessa fugax*	L	Hamed 1979†
	Blepharipa scutellata	F	Dunbar *et al.* 1973†
	Pales pavida	L	Huang 1979†; Hassan *et al.* 1983
	Parasetigena agilis	F	Dunbar *et al.* 1973†
	Zinellia dolosa	L	Hamed 1978 & 1979†
Diptera: Tipulidae	*Dicranota* sp.	F	Richardson & Perrin 1994
Ephemeroptera: ?	*Eurylophella* spp.	F	Kreutzweiser *et al.* 1994
Ephemeroptera: Baetidae	*Baetis* sp.	F	Richardson & Perrin 1994; Kreutzweiser *et al.* 1994
	Centroptilum sp.	F	Richardson & Perrin 1994
Ephemeroptera: Ephemerellidae	*Ephemerella sp.*	L/F	Kreutzweiser *et al.* 1992; Richardson & Perrin 1994
Ephemeroptera: Heptageniidae	*Epeorus vitrea*	L	Kreutzweiser *et al.* 1992, 1994
	Heptagenia sp.	F	Kreutzweiser *et al.* 1992, 1994; Richardson & Perrin 1994
	Rhithrogena sp.	F	Kreutzweiser *et al.* 1992; Richardson & Perrin 1994
	Stenonema vicarium	F	Kreutzweiser *et al.* 1992
Ephemeroptera: Isonychiidae	*Isonychia* sp.	F	Kreutzweiser *et al.* 1992
Ephemeroptera: Leptophlebiidae	*Paraleptophlebia ontario*	F	Kreutzweiser *et al.* 1992
	Paraleptophlebia sp.	F	Richardson & Perrin 1994
Ephemeroptera: Pentatomidae	*Picromerus bidens*	L	Hamed 1978 & 1979†
Ephemeroptera: Siphonuridae	*Ameletus* sp.	F	Richardson & Perrin 1994
Hemiptera: Aleyrodidae	*Bemisia tabaci*	L	Sims 1997
Hemiptera: Aphididae	*Aphis craccivora*	F	Jayanthi & Padmavathamma 1996a
	Rhopalosiphum padi	L	Sims 1997
Hemiptera: Anthocoridae	*Orius albidipennis*	F	Salama & Zaki 1984
	Orius insidiosus	F	Young *et al.* 1997
	Orius laevigatus	F	Salama & Zaki 1984

APPENDIX 2.2 *(continued)*

Subspecies/ Serovar	Order: Family	Genus and Species	Lab/Field	References
kurstaki	Hemiptera: Anthocoridae	*Orius* spp.	F	Mohamed 1993
	Hemiptera: Berytidae	*Jalysus spinosus*	F	Elsey 1973; Dunbar & Johnson 1975†
	Hemiptera: Cicadellidae	*Empoasca kerri*	F	Jayanthi & Padmavathamma 1996a
	Hemiptera: Lygaeidae	*Geocoris punctipes*	F	Boyd & Boethel 1998
		Geocoris sp.	F	Young *et al.* 1997
		Oncopeltus fasciatus	L	Sims 1997
	Hemiptera: Nabidae	*Nabis roseipennis*	F	Boyd & Boethel 1998
		Nabis capsiformis	F	Boyd & Boethel 1998
	Hemiptera: Pentatomidae	*Murgantia histrionica*	F	Chalfant 1978
		Podisus maculiventris	F	Boyd & Boethel 1998
	Hemiptera: Pseudococcidae	*Pseudococcus longispinus*	L	Wysoki *et al.* 1975†
	Hymenoptera: Agromyzidae	*Neochrysocharis punctiventris*	L	Schuster 1994
	Hymenoptera: Aphelinidae	*Encarsia* sp.	L	Hassan *et al.* 1983; Kadyrov *et al.* 1994
	Hymenoptera: Apidae	*Apis mellifera*	L	Ali *et al.* 1973†; Pinsdorf & Krieg†
	Hymenoptera: Braconidae	*Apanteles congregatus*	L	Urias Lopez *et al.* 1987
		Apanteles fumiferanae	F	Hamel 1977
		Apanteles (Cotesia) glomeratus	L	Muck *et al.* 1981
		Apanteles (Cotesia) melanoscelus	F	Dunbar *et al.* 1973†; Wollam & Yendol 1976†; Weseloh & Andreadis 1982
		Apanteles (Cotesia) plutellae	F	Krishnaiah *et al.* 1981; Lim *et al.* 1986
		Apanteles (Cotesia) sp.	F	Fernandez & Clavijo 1984
		Bracon gelechiae	F	Jayanthi & Padmavathamma 1996a
		Bracon hebetor	F	McGaughey 1979†
		Cardiochiles nigriceps	L	Dunbar & Johnson 1975†
		Chelonus sp.	F	Fernandez & Clavijo 1984
		Cotesia marginiventris	L/F	Atwood *et al.* 1996; 1998
		Cotesia melanoscelus	F	Webb *et al.* 1989
		Cotesia plutellae	L	Asokan & Mohan 1996; Obra & Morallo Rejesus 1997
		Cotesia rubecula	L	McDonald *et al.* 1990
		Macrocentrus ancylivorus	L	Sims 1997
		Meteorus sp.	F	Fernandez & Clavijo 1984
		Meteorus pulchricornis	L	Sims 1997
		Microplitis (Glabromicroplitis) croceipes	L	Katayama *et al.* 1987; Blumberg *et al.* 1997b
		Opius concolor	L	Jacas *et al.* 1992a,b
		Phanerotoma flavitestacea (ocularis)	F	Dhouibi 1992
		Telenomus remus	F	Chari *et al.* 1996
	Hymenoptera: Chalcidoidea	*Nasonia vitripennis*	L	Sims 1997
	Hymenoptera: Encyrtidae	*Anagyrus fusciventris*	L	Wysoki *et al.* 1975†
		Archopoides (Hungariella) peregrina	L	Wysoki *et al.* 1975†
		Hungariella peregrina	L	Wysoki *et al.* 1975†
		Leptomastix dactylopii	L	Viggiani & Tranfaglia 1979†; Hassan *et al.* 1983; Viggiani *et al.* 1998
	Hymenoptera: Eulophidae	*Chrysonotomyia punctiventris*	F	Johnson *et al.* 1980; 1984

Hymenoptera: Eupelmidae	*Anastatus bifasciatus*	F	Tsankov & Mirchev 1983
Hymenoptera: Formicidae	*Monomorium pharaonis*	L	Berndt *et al.* 1974
Hymenoptera: Ichneumonidae	*Agrypon (Trichionotus)* sp.	L	Hamed 1979
	Coccygomimus turionellae	L	Hassan *et al.* 1983
	Diadegma insulare	F	Carballo *et al.* 1989; Cordero & Cave 1990; Garcia 1991
	Diadegma semiclausum	L	Obra & Morallo Rejesus 1997
	Diadegma sp.	L/F	Jimenez & Fernandez 1985; Monnerat & Bordat 1998
	Glypta fumiferanae	F	Hamel 1977†; Niwa *et al.* 1987
	Macrocharops bimaculata	F	Turnipseed *et al.* 1985
	Phygadeuon trichops	L	Plattner 1979†
	Phygadeuon trichops	L	Naton 1978; Hassan *et al.* 1983
	Phytodietus fumiferanae	F	Niwa *et al.* 1987
	Pimpla turionellae	L	Boogenschutz 1979†
	Pimpla (Coccygomimus) turionelles	L	Muck *et al.* 1981
	Thyraeella collaris	L	Hamilton & Attia 1977†
Hymenoptera: Scelionidae	*Telenomus alsophilae*	L	Kaya & Dunbar 1972†
	Telenomus sphingis	L	Urias Lopez *et al.* 1987
Hymenoptera: Tenthredinidae	*Pristiphora abietina*	L/F	Donaubauer & Schmutzenhofer 1973†
Hymenoptera: Trichogrammatidae	*Trichogramma cacoeciae*	L	Hassan & Krieg 1975†; Hassan 1983
	Trichogramma carverae	F	Rohl & Woods 1994
	Trichogramma embryophagum	F	Tsankov & Mirchev 1983
	Trichogramma evanescens	F	Peteanu 1980; Tandan & Nillama 1987
	Trichogramma exiguum	L/F	Urias Lopez *et al.* 1987; Navarro 1988; Campbell *et al.* 1991
	Trichogramma japonicum	L	Borah & Basit 1996
	Trichogramma maidis	F	Maini & Burgio 1990; 1991
	Trichogramma nubillale	L/F	Tipping & Burbutis 1983; Losey *et al.* 1995
	Trichogramma pallidum (cacoeciae pallida)	L	Kiselek 1975
	Trichogramma platneri	L	Wellinga in Wysoki 1989
	Trichogramma pretiosum	F	Bull *et al.* 1979; Oatman *et al.* 1983; Campbell *et al.* 1991
	Trichogramma sp.	F	Grinberg Sh *et al.* 1992
	Trichogrammatoidea bactrae	F	Rohl & Woods 1994
Hymenoptera: Trichogrammatidae ?	*Ooencyrtus pityocampae*	F	Tsankov & Mirchev 1983
Hymenoptera: Trichogrammatidae ?	*Tetrastichus (Baryscapus) servadeii*	F	Tsankov & Mirchev 1983
Isopoda: Porcellionidae	*Porcellio scaber*	L	Sims 1997
Isoptera: Rhinotermitidae	*Coptotermes formosanus*	L	Grace & Ewart 1996
	Reticulitermes flavipes	L	MacIntosh *et al.* 1990a,b; Sims 1997
Lepidoptera: Cosmopterigidae	*Batrachedra amydraula*	F	Blumberg *et al.* 1977a
Lepidoptera: Danaidae	*Danaus plexippus*	L/F	Leong *et al.* 1992
Lepidoptera: Gelechiidae	*Chionodes* sp.	F	Miller 1992
Lepidoptera: Geometridae	*Drepanulatrix* sp.	F	Miller 1992
	Euchlaena obtusaria	L	Peacock *et al.* 1998
	Eupithecia sp.	F	Miller 1992
	Hyposidra talaca	L	Widiastuti *et al.* 1996
	Operophtera brumata	F	AliNiazee 1986
	Phiglia titea	L	Peacock *et al.* 1998

APPENDIX 2.2 (*continued*)

Subspecies/ Serovar	Order: Family	Genus and Species	Lab/Field	References
kurstaki	Lepidoptera: Lymantriidae	*Dasychira obliquata*	F	Wagner *et al.* 1996; Peacock *et al.* 1998
		Lymantria dispar	L	Dubois *et al.* 1989
	Lepidoptera: Noctuidae	*Abagrotis alternata*	L	Peacock *et al.* 1998
		Adisura atkinsoni	F	Krishnaiah *et al.* 1978
		Busseola fusca	F	Moyal 1988
		Catocalo similis	L	Peacock *et al.* 1998
		Catocalo sordida	L	Peacock *et al.* 1998
		Chaetaglaea sericea	L	Peacock *et al.* 1998
		Chilo auricilius	L	Nayak *et al.* 1978†
		Cnaphalocrocis medinalis	F	Narayanasamy & Baskaran 1979
		Earias insulana	F	Hussain & Askari 1976†
		Eupsilia vinulenta	L	Peacock *et al.* 1998
		Heliothis armigera	F	Krishnaiah *et al.* 1978; Collingwood & Bourdouxhe 1980
		Heliothis virescens	L	Abdel-Hameed & Landen 1994
		Leucinodes orbonalis	F	Krishnaiah *et al.* 1981
		Lithophane petulca	L	Peacock *et al.* 1998
		Lithophane unimoda	L	Peacock *et al.* 1998
		Metaxaglaea semitaria	L	Peacock *et al.* 1998
		Mythimna (Leucania) loreyimima	F	Hitchock 1974†
		Mythimna (Pseudaletia) convecta	F	Hitchock 1974†
		Mythimna (Pseudaletia) separata	F	Hitchock 1974†
		Orthosia alurina	L	Peacock *et al.* 1998
		Orthosia hibisci	L	Peacock *et al.* 1998
		Ostrinia furnacalis	F	Schreiner & Nafus 1987
		Sericaglaea signata	L	Peacock *et al.* 1998
		Sesamia calamistis	F	Moyal 1988
		Sesamia inferens	L	Nayak *et al.* 1978†
		Spodoptera frugiperda	L/F	Waquil *et al.* 1982
		Spodoptera litura	L	Pawar & Thombre 1990
		Sunira bicolorago	L	Peacock *et al.* 1998
		Trichoplusia ni	L	Abdel-Hameed & Landen 1994
		Xylotype capax	L	Peacock *et al.* 1998
	Lepidoptera: Sphingidae	*Cotesia congregata*	L/F	Cheng & Hanlon 1990
	Lepidoptera: Pyralidae	*Ephestia cautella*	F	McGaughey *et al.* 1980
	Lepidoptera: Tortricidae	*Archips argyropilus*	F	Madsen & Potter 1977†
		Archips rosanus	F	Madsen *et al.* 1977
		Archips (Cacoecia) rosana	F	Madsen & Potter 1977†
		Cacoecimorpha pronubana	F	Inserra *et al.* 1987
		Choristoneura fumiferana	F	Morris *et al.* 1977b†
		Cydia funebrana	F	Niemczyk 1975†
		Cydia (Laspeyresia) molesta	F	Zivanovic & Stamenkovic 1976†
		Cydia (Laspeyresia) pomonella	F	Zivanovic & Stamenkovic 1976†; Andermatt *et al.* 1988

		Rhyacionia frustrana	F	Dupree & Davis 1975†
	Nematoda: Panagrolaimidae	*Panagrellus* sp.	L	Shevtsov *et al.* 1996
	Nemotoda: Heterorhabditidae	*Heterorhabditis bacteriophora*	L	Shamseldean & Ismail 1997
		Heterorhabditis heliothidis	L	Poinar *et al.* 1990
	Nemotoda: Steinernematidae	*Neoaplectana carpocapsae*	L	Poinar *et al.* 1990
	Neuroptera: Chrysopidae	*Chrysoperla (Chrysopa) carnea*	L/F	Kiselek 1975; Adylov *et al.* 1990; Santharam *et al.* 1994
	Odonata: Aeshnidae	*Boyeria grafiana*	F	Kreutzweiser *et al.* 1992
		Ophiogomphus carolus	F	Kreutzweiser *et al.* 1992
	Oligochaeta: Lumbricidae	*Dendrobaena octaedra*	L	Addison & Holmes 1996
	Oligochaeta: Naididae	NS	F	Richardson & Perrin 1994
	Orthoptera: Gryllidae	*Acheta domesticus*	L	Sims 1997
	Phthiraptera: Boopiidae	*Hyalomma dromedarii*	F	Singh *et al.* 1992
	Plecoptera: ?	*Taeniopteryx nivalis*	F	Kreutzweiser *et al.* 1992; 1994
	Plecoptera: Chloroperlidae	*Suwallia* sp.	F	Richardson & Perrin 1994
	Plecoptera: Nemouridae	*Zapada* sp.	F	Richardson & Perrin 1994
	Plecoptera: Perlidae	*Acroneuria abnormis*	F	Kreutzweiser *et al.* 1992
	Plecoptera: Perlodidae	*Isogenoides* sp.	F	Kreutzweiser *et al.* 1992
		Isoperla sp.	F	Richardson & Perrin 1994
	Plecoptera: Pteronarcyidae	*Pteronarcys* sp.	F	Kreutzweiser *et al.* 1992
	Prostigmata: Tetranychidae	*Tetranychus urticae*	L	MacIntosh *et al.* 1990c
	Trichoptera: ?	*Hesperophylax designatus*	F	Kreutzweiser *et al.* 1992
	Trichoptera: Hydropsychidae	*Hydropsyche* sp.	F	Kreutzweiser *et al.* 1992
	Trichoptera: Hydroptilidae	*Agraylea* sp.	F	Richardson & Perrin 1994
		Oxyethira sp.	F	Richardson & Perrin 1994
	Trichoptera: Lepidostomatidae	*Lepidostoma* spp.	F	Kreutzweiser *et al.* 1992; Richardson & Perrin 1994
	Trichoptera: Limnephilidae	*Pycnopsyche guttifer*	L/F	Kreutzweiser *et al.* 1992; 1994
	Trichoptera: Polycentropodidae	*Polycentropus* sp.	F	Richardson & Perrin 1994
	Trichoptera	*Dolophilodes distinctus*	F	Kreutzweiser *et al.* 1994
	?	*Hydatophylax argus*	L	Kreutzweiser & Capell 1996
kyushuensis	Diptera: Culicidae	*Aedes aegypti*	L	Larget & de Barjac 1981a
		Anopheles stephensi	L	Larget & de Barjac 1981a
		Culex pipiens molestus	L	Saitoh *et al.* 1996
		Culex pipiens pipiens	L	Larget & de Barjac 1981a
	Diptera: Psychodidae	*Telmatoscopus albipunctatus*	L	Saitoh *et al.* 1996
	Lepidoptera: Arctiidae	*Hyphantria cunea*	L	Ishii & Ohba 1993a
	Lepidoptera: Bombycidae	*Bombyx mori*	L	Ohba & Aizawa 1979b†; Ishii & Ohba 1993a
	Lepidoptera: Lymantriidae	*Lymantria dispar*	L	Dubois *et al.* 1989
leesis	Diptera: Chironomidae	*Tokunagayusurika akamusi*	L	Higuchi *et al.* 1998
	Diptera: Culicidae	*Aedes aegypti*	L	Higuchi *et al.* 1998
		Culex pipiens molestus	L	Saitoh *et al.* 1996; Higuchi *et al.* 1998
		Culex pipiens	L	Lee *et al.* 1994
londrina	Coleoptera: Chrysomelidae	*Leptinotarsa decemlineata*	L	Iriarte *et al.* 1998
	Diptera: Psychodidae	*Telmatoscopus albipunctatus*	L	Saitoh *et al.* 1996
	Diptera: Tipulidae	*Tipula oleracea*	L	Iriarte *et al.* 1998

APPENDIX 2.2 *(continued)*

Subspecies/ Serovar	Order: Family	Genus and Species	Lab/Field	References
londrina	Lepidoptera: Bombycidae	*Bombyx mori*	L	Higuchi *et al.* 1998
	Lepidoptera: Noctuidae	*Heliothis armigera*	L	Iriarte *et al.* 1998
		Spodoptera exigua	L	Iriarte *et al.* 1998
mexicanensis	Coleoptera: Chrysomelidae	*Leptinotarsa decemlineata*	L	Iriarte *et al.* 1998
	Diptera: Culicidae	*Culex pipiens molestus*	L	Saitoh *et al.* 1996
	Diptera: Psychodidae	*Telmatoscopus albipunctatus*	L	Saitoh *et al.* 1996
	Diptera: Tipulidae	*Tipula oleracea*	L	Iriarte *et al.* 1998
	Lepidoptera: Noctuidae	*Heliothis armigera*	L	Iriarte *et al.* 1998
	Lepidoptera: Yponomeutidae	*Plutella xylostella*	L	Iriarte *et al.* 1998
		Spodoptera exigua	L	Iriarte *et al.* 1998
morrisoni	Coleoptera: Cerambycidae	*Apriona germari*	L	Park *et al.* 1998
	Coleoptera: Scarabaeidae	*Anomala rufocuprea*	L	Park *et al.* 1998
		Popillia mutans	L	Park *et al.* 1998
	Diptera: Culicidae	*Aedes aegypti*	L	Hall *et al.* 1977†; Padua *et al.* 1981; Granum *et al.* 1988
		Culex pipiens	L	Hall *et al.* 1977†
		Culex pipiens molestus	L	Saitoh *et al.* 1996
		Culex pipiens pallens	L	Park *et al.* 1998
		Culex quinquefasciatus	L	Hall *et al.* 1977†; Hastowo *et al.* 1992
		Culex tarsalis	L	Hall *et al.* 1977†
		Culex tritaeniorhynchus	L	Park *et al.* 1998
		Aedes triseriatus	L	Hall *et al.* 1977†
	Diptera: Psychodidae	*Telmatoscopus albipunctatus*	L	Saitoh *et al.* 1996
	Lepidoptera: ?	*Phrixolepia sericea*	L	Park *et al.* 1998
	Lepidoptera: ? (cochlid)	*Austrapoda dentatus*	L	Park *et al.* 1998
	Lepidoptera: Arctiidae	*Hyphantria cunea*	L	Park *et al.* 1998
		Spilosoma subcarneum	L	Park *et al.* 1998
	Lepidoptera: Bombycidae	*Bombyx mandarina*	L	Park *et al.* 1998
		Bombyx mori	L	Padua *et al.* 1981; Park *et al.* 1998
	Lepidoptera: Heterogeneidae	*Monema flavescens*	L	Park *et al.* 1998
	Lepidoptera: Lymantriidae	*Euproctis similis*	L	Park *et al.* 1998
		Lymantria dispar	L	Dubois *et al.* 1989
	Lepidoptera: Notodontidae	*Fentonia ocypete*	L	Park *et al.* 1998
	Lepidoptera: Noctuidae	*Helicoverpa assulta*	L	Park *et al.* 1998
		Spodoptera exigua	L	Park *et al.* 1998
		Spodoptera litura	L	Park *et al.* 1998
	Lepidoptera: Pyralidae	*Glyphodes pyloalis*	L	Park *et al.* 1998
neoleonensis	Diptera: Culicidae	*Aedes aegypti*	L	Rodriguez Padilla *et al.* 1990; Hastowo *et al.* 1992
		Culex pipiens	L	Rodriguez Padilla *et al.* 1990
		Culex pipiens molestus	L	Saitoh *et al.* 1996
		Culex quinquefasciatus	L	Hastowo *et al.* 1992
	Diptera: Psychodidae	*Telmatoscopus albipunctatus*	L	Saitoh *et al.* 1996

		Trichoplusia ni	L	Rodriguez Padilla *et al.* 1990
ostriniae	Diptera: Culicidae	*Culex pipiens molestus*	L	Saitoh *et al.* 1996
	Diptera: Psychodidae	*Telmatoscopus albpunctatus*	L	Saitoh *et al.* 1996
	Lepidoptera: Bombycidae	*Bombyx mori*	L	Ohba & Aizawa 1979a†
	Lepidoptera: Lymantriidae	*Lymantria dispar*	L	Dubois *et al.* 1989
oswaldocruzi	Diptera: Culicidae	*Aedes aegypti*	L	Rabinovitch *et al.* 1995
		Anopheles stephensi	L	Rabinovitch *et al.* 1995
		Culex pipiens	L	Rabinovitch *et al.* 1995
oyamensis	Coleoptera: Chrysomelidae	*Leptinotarsa texana*	L	Lopez Meza & Ibarra 1996
	Diptera: Culicidae	*Aedes aegypti*	L	Lopez Meza & Ibarra 1996
		Culex pipiens molestus	L	Saitoh *et al.* 1996
oyamensis		*Culex quinquefasciatus (fatigans)*	L	Lopez Meza & Ibarra 1996
	Diptera: Muscidae	*Musca domestica*	L	Ono *et al.* 1988
	Diptera: Psychodidae	*Telmatoscopus albipunctatus*	L	Saitoh *et al.* 1996
	Hymenoptera: Formicidae	*Atta* sp.	L	Lopez Meza & Ibarra 1996
	Lepidoptera: Bombycidae	*Bombyx mori*	L	Ono *et al.* 1988
	Lepidoptera: Noctuidae	*Heliothis (Heliocoverpa) zea*	L	Lopez Meza & Ibarra 1996
	Lepidoptera: Pyralidae	*Diatraea saccharalis*	L	Lopez Meza & Ibarra 1996
	Lepidoptera: Sphingidae	*Manduca sexta*	L	Lopez Meza & Ibarra 1996
	Nematoda: Steinernematidae	*Steinernema (Neoaplectana) feltiae (carpocapsae)*	L	Lopez Meza & Ibarra 1996
pakistani	Coleoptera: Chrysomelidae	*Leptinotarsa decemlineata*	L	Iriarte *et al.* 1998
	Diptera: Culicidae	*Culex pipiens molestus*	L	Saitoh *et al.* 1996
	Diptera: Muscidae	*Musca domestica*	L	Yarnykh *et al.* 1981a
	Diptera: Psychodidae	*Telmatoscopus albipunctatus*	L	Saitoh *et al.* 1996
	Diptera: Tipulidae	*Tipula oleracea*	L	Iriarte *et al.* 1998
	Lepidoptera: Noctuidae	*Heliothis armigera*	L	Iriarte *et al.* 1998
		Spodoptera exigua	L	Iriarte *et al.* 1998
	Lepidoptera: Lymantriidae	*Lymantria dispar*	L	Dubois *et al.* 1989
	Lepidoptera: Yponomeutidae	*Plutella xylostella*	L	Iriarte *et al.* 1998
pondicheriensis	Diptera: Culicidae	*Aedes aegypti*	L	Hastowo *et al.* 1992
		Culex pipiens	L	Kim *et al.* 1998b
		Culex quinquefasciatus	L	Hastowo *et al.* 1992
	Lepidoptera: Bombycidae	*Bombyx mori*	L	Hastowo *et al.* 1992
shandongiensis	Coleoptera: Chrysomelidae	*Leptinotarsa decemlineata*	L	Iriarte *et al.* 1998
	Diptera: Culicidae	*Aedes aegypti*	L	Pietrantonio & Gill 1992
		Culex pipiens molestus	L	Saitoh *et al.* 1996
		Culex quinquefasciatus	L	Pietrantonio & Gill 1992
	Diptera: Psychodidae	*Telmatoscopus albipunctatus*	L	Saitoh *et al.* 1996
	Diptera: Tipulidae	*Tipula oleracea*	L	Iriarte *et al.* 1998
	Lepidoptera: Noctuidae	*Heliothis armigera*	L	Iriarte *et al.* 1998
		Heliothis virescens	L	Pietrantonio & Gill 1992

APPENDIX 2.2 *(continued)*

Subspecies/ Serovar	Order: Family	Genus and Species	Lab/Field	References
shandongiensis	Lepidoptera: Noctuidae	*Spodoptera exigua*	L	Pietrantonio & Gill 1992; Iriarte *et al.* 1998
		Trichoplusia ni	L	Pietrantonio & Gill 1992
	Lepidoptera: Yponomeutidae	*Plutella xylostella*	L	Iriarte *et al.* 1998
silo	Coleoptera: Chrysomelidae	*Leptinotarsa decemlineata*	L	Iriarte *et al.* 1998
	Diptera: Tipulidae	*Tipula oleracea*	L	Iriarte *et al.* 1998
	Lepidoptera: Noctuidae	*Heliothis armigera*	L	Iriarte *et al.* 1998
		Spodoptera exigua	L	Iriarte *et al.* 1998
	Lepidoptera: Yponomeutidae	*Plutella xylostella*	L	Iriarte *et al.* 1998
sooncheon	Coleoptera: Chrysomelidae	*Leptinotarsa decemlineata*	L	Iriarte *et al.* 1998
sooncheon	Diptera: Tipulidae	*Tipula oleracea*	L	Iriarte *et al.* 1998
	Lepidoptera: Noctuidae	*Heliothis armigera*	L	Iriarte *et al.* 1998
	Lepidoptera: Noctuidae	*Spodoptera exigua*	L	Iriarte *et al.* 1998
	Lepidoptera: Yponomeutidae	*Plutella xylostella*	L	Iriarte *et al.* 1998
sotto	Diptera: Chloropidae	*Hippelates collusor*	L	Hall *et al.* 1977†
	Diptera: Culicidae	*Aedes aegypti*	L	Hall *et al.* 1977†; Mike *et al.* 1991
		Aedes triseriatus	L	Hall *et al.* 1977†
		Culex pipiens	L	Hall *et al.* 1977†
		Culex pipiens molestus	L	Saitoh *et al.* 1996
		Culex quinquefasciatus	L	Hall *et al.* 1977†
		Culex tarsalis	L	Hall *et al.* 1977†
	Diptera: Muscidae	*Musca domestica*	L	Rogoff *et al.* 1969†
	Diptera: Psychodidae	*Telmatoscopus albipunctatus*	L	Saitoh *et al.* 1996
	Hymenoptera: Apidae	*Apis mellifera*	L	Haragsim & Vankova 1968†
	Hymenoptera: Diprionidae	*Gilpinia hercyniae*	L	Angus 1956b†
		Neodiprion abietis	L	Angus 1956b†
		Neodiprion banksianae	L	Angus 1956b†
		Neodiprion sertifer	L	Angus 1956†
		Neodiprion swainei	L	Heimpel 1961†
	Lepidoptera: Geometridae	*Lambdina fiscellaria fiscellaria*	L	Angus 1956a†
	Lepidoptera: Lymantriidae	*Lymantria dispar*	L	Dubois *et al.* 1989
	Lepidoptera: Noctuidae	*Heliothis zea*	L	Rogoff *et al.* 1969†
		Orthosia gothica	L	Burgerjon & Biache 1967b†
		Spodoptera litura	F	Takaki 1975†
		Trichoplusia ni	L	Rogoff *et al.* 1969†
subtoxicus	Diptera: Culicidae	*Aedes aegypti*	L	Padua *et al.* 1981
	Lepidoptera: Bombycidae	*Bombyx mori*	L	Padua *et al.* 1981
sumiyoshiensis	Diptera: Culicidae	*Aedes aegypti*	L	Ohba & Aizawa 1989b
		Culex pipiens molestus	L	Saitoh *et al.* 1996

tenebrionis (*san diego*)	Coleoptera: Coccinellidae	*Coleomegilla maculata lengi*	L/F	Giroux *et al.* 1994b
		Olla abdominalis (v nigrum)	L	Zhou 1992
	Nematoda: Heterorhabditidae	*Heterorhabditis heliothidis*	L	Poinar *et al.* 1990
	Nematoda: Steinernematidae	*Neoaplectana carpocapsae*	L	Poinar *et al.* 1990
	Atheriniformes: Poeciliidae	*Lebistes reticulata*	?	Meyer 1989 in Keller & Langenbruch 1993
	Coleoptera: Cantharidae	*Chauliognathus lugubris*	L	Beveridge & Elek 1999
	Coleoptera: Carabidae	*Zabrus tenebrioides*	L	Meyer 1989 in Keller & Langenbruch 1993
	Coleoptera: Cerambycidae	*Strangalia maculata*	L	Meyer 1989 in Keller & Langenbruch 1993; Ferrari & Maini 1992
	Coleoptera: Chrysomelidae	*Diabrotica balteata*	L	Ferrari & Maini 1992
		Diabrotica undecimpunctata howardii	L	MacIntosh *et al.* 1990c
		Haltica oleracea	L	Schnetter in Keller & Langenbruch 1993
		Phyllotreta atra	L	Krieg *et al.* 1983; Ferrari & Maini 1992
tenebrionis	Coleoptera: Coccinellidae	*Cleobora mellyi*	L	Greener & Candy 1994
		Coccinella septempunctata	L	Ferrari & Maini 1992
		Cylas punticollis	L	Riethmuller 1990 in Keller & Langenbruch 1993
		Harmonia conformis	L	Greener & Candy 1994
		Otiorhynchus salicicola	F	Landi 1990
		Zacladus affinis	L	Meyer 1989 in Keller & Langenbruch 1993
		Hypera brunneipennis (postica)	L	MacIntosh *et al.* 1990; Ferrari & Maini 1992
	Coleoptera: Dermestidae	*Attagenus (megatoma, piceus)*	L	Herrnstadt & Soaers US patent 4797276, 1989
		Dermestes maculatus	L	Ferrari & Maini 1992
		Trogoderma granarium	L	Skovmand in Keller & Langenbruch 1993
	Coleoptera: Scarabaeidae	*Popillia japonica*	L	MacIntosh *et al.* 1990c
	Coleoptera: Tenebrionidae	*Tribolium castaneum*	L/F	Herrnstadt *et al.* 1986; Arthur & Brown 1994
	Dermaptera: Forficulidae	*Forficula auricularia*	?	Meyer 1989 in Keller & Langenbruch 1993; Ferrari & Maini 1992
	Dictyoptera: Blattidae	*Periplaneta americana*	L	Krieg *et al.* 1983
	Diptera: Culicidae	*Aedes aegypti*	L	Krieg *et al.* 1983; MacIntosh *et al.* 1990c; Ferrari & Maini 1992
		Culex pipiens	L	Krieg *et al.* 1983
	Diptera: Tachinidae	*Myiopharus doryphorae*	L	Lopez & Ferro 1995
	Diptera: Tipulidae	*Tipula* sp.	?	Meyer 1989 in Keller & Langenbruch 1993
	Hemiptera: Pentatomidae	*Perillus bioculatus*	F	Hough Goldstein & Whalen 1993; Hough Goldstein *et al.* 1996 Cloutier & Jean 1998
	Hymenoptera: Apidae	*Apis mellifera*	L	Ferrari & Maini 1992
	Hymenoptera: Tenthredinidae	*Athalia rosae*	L	Krieg *et al.* 1983; Ferrari & Maini 1992
	Isopoda: Porcellionidae	*Porcellio scaber*	?	Meyer 1989 in Keller & Langenbruch 1993
	Isoptera: Hodotermitidae	*Anacanthotermes ahngerianus*	L	Lyutikova & Yudina 1995
	Lepidoptera: Hyponomeutidae	*Hyponomeuta malinellus*	?	Meyer 1989 in Keller & Langenbruch 1993
	Lepidoptera: Noctuidae	*Agrotis ipsilon (ypsilon)*	L	MacIntosh *et al.* 1990c
		Spodoptera exigua	L	Hermastadt *et al.* 1986; Ferrari & Maini 1992
		Spodoptera frugiperda	L	Ferrari & Maini 1992
		Spodoptera littoralis	L	Ferrari & Maini 1992
		Trichoplusia ni	L	Herrnstadt *et al.* 1986; MacIntosh *et al.* 1990; Ferrari & Maini 1992
	Lepidoptera: Nymphalidae	*Inachis io*	?	Meyer 1989 in Keller & Langenbruch 1993

APPENDIX 2.2 (*continued*)

Subspecies/ Serovar	Order: Family	Genus and Species	Lab/Field	References
tenebrionis	Lepidoptera: Pyralidae	*Ephestia (Anagasta) kuehniella*	L	Krieg *et al.* 1983; Ferrari & Maini 1992
		Ostrinia nubilalis	L	MacIntosh *et al.* 1990c
	Lepidoptera: Sphingidae	*Manduca sexta*	L	MacIntosh *et al.* 1990c; Ferrari & Maini 1992
	Mestostigmata: Tetranychidae	*Tetranychus urticae*	L	Chapman & Hoy 1991
	Neuroptera: Chrysopidae	*Chrysoperla (Chrysopa) carnea*	?	Langenbruch in Keller & Langenbruch 1993; Ferrari & Maini 1992
	Orthoptera: Acrididae	*Locusta migratoria*	L	Meyer 1989 in Keller & Langenbruch 1993
	Orthoptera: Gryllidae	*Acheta domestica*	L	Meyer 1989 in Keller & Langenbruch 1993; Ferrari & Maini 1992
	Orthoptera: Phaneropteridae	*Leptophyes punctatissima*	L	Meyer 1989 in Keller & Langenbruch 1993; Ferrari & Maini 1992
thompsoni	Lepidoptera: Lymantriidae	*Lymantria dispar*	L	Dubois *et al.* 1989
	Lepidoptera: Noctuidae	*Spodoptera litura*	L	Puntambekar *et al.* 1997
	Lepidoptera: Pyralidae	*Plodia interpunctella*	L	Salama *et al.* 1991b
thuringiensis	Acari: Phytoseiidae	*Amblyseius (Typhlodromus) andersoni*	F	Strapazzon *et al.* 1986
		Metaseiulus (Typhlodromus) occidentalis	L	Petrushov *et al.* 1983
	Annelida: Oligochaeta	*Eisenia foetida*	L	Krieg & Langenbruch 1981
		Lumbricus terrestris	L/F	White 1960†; Benz & Altwegg 1975
	Arachnida: Scorpiones	*Buthus occitanus*	L	Morel 1974†
	Arachnoidea: Acari	*Anystis agilis*	F	Jaques 1965†
		Bryobia arborea	F	Jaques 1965†
		Panonychus ulmi	F	Oatman 1965†
		Pilophorous perplexus	F	Jaques 1965†
		Versates schlechtendali	F	Jaques 1965†
	Arachnoidea: Scorpioidae	*Buthus occitanus*	L	Morel 1974†
	Coleoptera: Anobiidae	*Lasioderma serricorne*	L	Thompson & Fletcher 1972†
	Coleoptera: Bostrichidae	*Rhyzopertha dominica*	L	Steinhaus & Bell 1953†
	Coleoptera: Carabidae	*Carabus hampei*	L/F	Koval 1985
	Coleoptera: Chrysomelidae	*Diabrotica undecimpunctata howardii*	L	Abdel-Hameed & Landen 1994
		Galerucella xanthomelaena	L	Hall & Dunn 1958†
		Leptinotarsa decemlineata	L	Krieg 1957b†; Drozdowicz 1964†
	Coleoptera: Coccinellidae	*Ceratomegilla (Coleomegilla) maculata*	F	Johnson 1974
		Coccinella septempunctata	L	Kiselek 1975; Baicu & Hussein 1984
		Coccinella sp.	L	Steiner 1960†
		Cryptolaemus montrouzieri	L	Kiselek 1975
		Epilachna varivestis	L	Shaikh & Morrison 1966a†
		Hippodamia convergens	F	Stern 1961†; Johnson 1974
		Scymnus sp.	L	Ayyar 1961†
	Coleoptera: Curculionidae	*Diaprepes abbreviatus*	F	Wong *et al.* 1975†
		Hypera brunneipennis	F	Stern 1961†
		Sitophilus granarius	L	Berliner 1915†; Shaikh & Morrison 1966a†

Coleoptera: Dermestidae	[illegible]	[illegible]	[illegible]
	Trogoderma granarium	L	De & Konar 1955†
	Dermestes lardarius	L	Berliner 1915†
Coleoptera: Scarabaeidae	*Melolontha* sp.	L	Krieg 1957b†
Coleoptera: Staphylinidae	*Oligota fageli*	L	Botha *et al.* 1994
Coleoptera: Tenebrionidae	*Gnatocerus cornutus*	L	Berliner 1915†
	Tenebrio molitor	L	Berliner 1915†
	Tribolium confusum	L	Steinhaus & Bell 1953†
Coleoptera: Trogositidae	*Tenebroides mauritanicus*	L	Shaikh & Morrison 1966a†
Dictyoptera: Blattidae	*Periplaneta americana*	L	Shaikh & Morrison 1966b†
Diptera: Agromyzidae	*Liriomyza* spp.	F	Shorey & Hall 1963†
Diptera: Calliphoridae	*Chrysomya megacephala*	L	Sherman *et al.* 1962†
Diptera: Culicidae	*Aedes aegypti*	L	Shaikh & Morrison 1966a†; Hall *et al.* 1977†
	Aedes nigromaculis	L	Kellen & Lewallen 1960†
	Aedes stimulans	L	Shaikh & Morrison 1966a†
	Aedes triseriatus	L	Hall *et al.* 1977†
	Culex pipiens	L/F	Kadyrova *et al.* 1977; Kim *et al.* 1998b
	Culex pipiens molestus	L	Saitoh *et al.* 1996
	Culex quinquefasciatus	L	Hall *et al.* 1977†
	Culex tarsalis	L	Hall *et al.* 1977†
Diptera: Fanniidae	*Fannia pusio*	L	Sherman *et al.* 1962†
Diptera: Muscidae	*Musca domestica*	L	Abdel-Hameed & Landen 1994
Diptera: Psychodidae	*Telmatoscopus albipunctatus*	L	Saitoh *et al.* 1996
Diptera: Sarcophagidae	*Parasarcophaga argyrostoma*	L	Sherman *et al.* 1962†
Diptera: Tachinidae	*Compsilura concinnata*	F	Adeli 1980; Mushtaque *et al.* 1993
Diptera: Tipulidae	*Tipula paludosa*	L	Krieg 1957†
Hemiptera: Aleyrodidae	*Trialeurodes vaporariorum*	F	Heungens & Pelerents 1977
Hemiptera: Aphididae	*Aphis pomi*	L	Oatman 1965†
	Chromaphis juglandicola	L	Purcell & Granett 1985
	Macrosiphum euphorbiae	F	Shorey & Hall 1963†
Hemiptera: Anthocoridae	*Orius sauteri*	L	Zuo *et al.* 1994b
	Orius tristicolor	F	Stern 1961†
	Xylocoris flavipes	L	El Husseini *et al.* 1987
Hemiptera: Miridae	*Atractotomus mali*	F	Jaques 1965†
	Pilophorus perplexus	F	Jaques 1965†
Hymenoptera: Apidae	*Apis mellifera*	L	Wilson 1962†; Haragsim & Vankova 1968†
Hymenoptera: Braconidae	*Apanteles glomeratus*	L	Marchal-Segault 1975†
	Apanteles (Cotesia) glomeratus (glomerata)	F	Mushtaque *et al.* 1993
	Apanteles melanoscelus	L	Wolliam & Yendol 1976†
	Trioxys utilis	F	Stern 1961†
	Bracon sp.	F	Mamedova *et al.* 1990
	Microdus (Agathis) rufipes	F	Kazakova *et al.* 1977
	Trioxys pallidus	L	Purcell & Granett 1985
Hymenoptera: Diprionidae	*Diprion pini*	L	Glowacka-Pilot 1968†
	Gilpinia hercyniae	L	Krieg 1957b†
	Neodiprion sertifer	L/F	Krieg 1957b†; Donaubauer & Schmutzenhofer 1973†

APPENDIX 2.2 (*continued*)

Subspecies/ Serovar	Order: Family	Genus and Species	Lab/Field	References
thuringiensis	Hymenoptera: Eulophidae	*Edovum puttleri*	F	Schroder & Athanas 1989
	Hymenoptera: Formicidae	*Formica polyctena*	L/F	Kneitz 1966†
		Formica pratensis/nigricans	L/F	Kneitz 1966†
		Formica rufa	L/F	Kneitz 1966†
		Monomorium pharaonis	L	Berndt *et al.* 1974
	Hymenoptera: Ichneumonidae	*Pimpla instigator*	L	Biache 1975†
		Venturia (=Nemeritis) canescens	L	Kurstak 1964†
		Diadegma pierisae	F	Mushtaque *et al.* 1993
		Diadegma sp.	F	Jimenez & Fernandez 1985
	Hymenoptera: Pamphiliidae	*Acantholyda nemoralis*	L	Glowacka-Pilot 1968†
	Hymenoptera: Pteromalidae	*Pteromalus puparum*	F	Mushtaque *et al.* 1993
	Hymenoptera: Tenthredinidae	*Croesus septentrionalis*	L	Krieg 1957b†
		Pristiphora abietina	L	Krieg and Langenbruch 1981
		Athalia rosae (colibri)	L	Johansson 1971
	Hymenoptera: Trichogrammatidae	*Trichogramma cacoeciae*	L	Hassan & Krieg 1975†; Krieg *et al.* 1980
		Trichogramma evanescens	F	Peteanu 1980
		Trichogramma pallidum (Cacoeciae pallida)	L	Kiselek 1975
		Trichogramma sp.	F	Kazakova *et al.* 1977; Mamedova *et al.* 1990
	Hymenoptera: Vespidae	*Polistes exclamans*	L	Guthrie *et al.* 1959†
	Isoptera: Rhinotermitidae	*Reticulitermes flavipes*	L	Smythe & Coppel 1965†
		Reticulitermes hesperus	L	Smythe & Coppel 1965†
		Reticulitermes virginicus	L	Smythe & Coppel 1965†
	Isoptera: Termopsidae	*Zootermopsis angusticollis*	L	Smythe & Coppel 1965†
	Lepidoptera: Arctiidae	*Hyphantria cunea*	L	Ramachandran *et al.* 1993b
	Lepidoptera: Bombycidae	*Bombyx mori*	L	Galowalia *et al.* 1973†
	Lepidoptera: Gelechiidae	*Pectinophora gossypiella*	L	Shalaby *et al.* 1986
	Lepidoptera: Geometridae	*Hyposidra talaca*	L	Widiastuti *et al.* 1996
	Lepidoptera: Lasiocampidae	*Dendrolimus superans (sibiricus)*	F	Mashanov *et al.* 1976
	Lepidoptera: Lymantriidae	*Lymantria dispar*	L	Dubois *et al.* 1989
		Lymantria monacha	L	Johansson 1971†
	Lepidoptera: Noctuidae	*Euxoa messoria*	L/F	Cheng 1973†
		Feltia subterranea	L	White & Briggs 1964†
		Heliothis (Chloridea) assulta	F	Rosley & Jusoh 1982
		Heliothis virescens	L	Abdel-Hameed & Landen 1994
		Heliothis zea	L	Jaques & Fox 1960†
		Mamestra brassicae	L	Vankova 1973†
		Trichoplusia ni	L	Abdel-Hameed & Landen 1994
	Lepidoptera: Pyralidae	*Diatraea saccharalis*	F	Hensley *et al.* 1961†
		Amyelois (Paramyelois) transitella	F	Pinnock & Milstead 1972†
		Scirpophaga incertulas	F	Pojananuwong *et al.* 1988
	Lepidoptera: Sphingidae	*Cotesia congregata*	L/F	Cheng & Hanlon 1990
		Erinnyis ello	L	Pigatti *et al.* 1960†
	Lepidoptera: Tortricidae	[illegible]	F	[illegible]

		Rhyacionia frustrana	F	Dupree & Davis 1975†
	Myriopoda: Diplopoda	*Poratophilus pretorianus*	L	Fiedler 1965†
		Poratophilus robustus	L	Fiedler 1965†
		Spinotarsus fiedleri	L	Fiedler 1965†
	Neuroptera: Chrysopidae	*Chrysopa carnea*	L	Hassan & Krieg†
		Chrysopa sp.	F	Stern 1961†
		Chrysoperla (Chrysopa) carnea	L/F	Kiselek 1975; Zhumanov *et al.* 1988; Adylov *et al.* 1990
	Orthoptera: Acrididae	*Chorthippus dorsatus*	L	Metalnikov & Chorine 1929†
		Chorthippus pulvinataus	L	Metalnikov & Chorine 1929†
		Stauroderus biguttulus	L	Metalnikov & Chorine 1929†
	Orthoptera: Gryllidae	*Gryllodes sigillatus*	L	Shaikh & Morrison 1966b†
		Teleogryllus commodus	L	Shaikh & Morrison 1966b†
	Orthoptera: Hydrophilidae	*Blatta orientalis*	L	Brikman *et al.* 1966†
	Orthoptera: Mantidae	*Tenodera aridifolia*	L	Yousten 1973†
	Orthoptera: Phasmatidae	*Carausius morosus*	L	Pendleton 1970†
	Thysanoptera: Thripidae	*Frankliniella occidentalis*	F	Shorey & Hall 1963†
tochigiensis	Diptera: Culicidae	*Aedes aegypti*	L	Ohba *et al.* 1981b
	Lepidoptera: Bombycidae	*Bombyx mori*	L	Ohba *et al.* 1981b
tohokuensis	Diptera: Culicidae	*Aedes aegypti*	L	Hastowo *et al.* 1992
		Culex pipiens molestus	L	Saitoh *et al.* 1996
	Diptera: Psychodidae	*Telmatoscopus albipunctatus*	L	Saitoh *et al.* 1996
	Lepidoptera: Bombycidae	*Bombyx mori*	L	Hastowo *et al.* 1992
tolworthi	Coleoptera: Chrysomelidae	*Leptinotarsa decemlineata*	L	Iriarte *et al.* 1998
	Diptera: Culicidae	*Aedes aegypti*	L	Hall *et al.* 1977; Panbangred *et al.* 1979; Padua *et al.* 1981
		Culex pipiens	L	Hall *et al.* 1977
		Culex pipiens molestus	L	Saitoh *et al.* 1996
		Culex quinquefasciatus	L	Hall *et al.* 1977†; Hastowo *et al.* 1992
		Culex tarsalis	L	Hall *et al.* 1977†
	Diptera: Psychodidae	*Telmatoscopus albipunctatus*	L	Saitoh *et al.* 1996
	Diptera: Tipulidae	*Tipula oleracea*	L	Iriarte *et al.* 1998
	Lepidoptera: Bombycidae	*Bombyx mori*	L	Galowalia *et al.* 1973†; Padua *et al.* 1981; Hastowo *et al.* 1992
	Lepidoptera: Noctuidae	*Heliothis armigera*	L	Iriarte *et al.* 1998
		Spodoptera exigua	L	Iriarte *et al.* 1998
	Lepidoptera: Yponomeutidae	*Plutella xylostella*	L	Iriarte *et al.* 1998
toumanoffi	Diptera: Culicidae	*Aedes aegypti*	L	Hall *et al.* 1977†
		Aedes triseriatus	L	Hall *et al.* 1977†
		Culex pipiens	L	Hall *et al.* 1977†
		Culex quinquefasciatus	L	Hall *et al.* 1977†
		Culex tarsalis	L	Hall *et al.* 1977†
wenguanensis	Coleoptera: Cerambycidae	*Anoplophora glabripennis*	L	Dai & Wang 1988
	Lepidoptera: Lasiocampidae	*Dendrolimus tabulaeformis*	L	Dai & Wang 1988

APPENDIX 2.2 (*continued*)

Subspecies/ Serovar	Order: Family	Genus and Species	Lab/Field	References
wenguanensis	Lepidoptera: Lasiocampidae	*Malacosoma neustria testacea*	L	Dai & Wang 1988
wratislaviensis	Dictyoptera: Blattidae	*Blatta orientalis*	L	Lonc *et al.* 1997
		Periplaneta americana	L	Lonc *et al.* 1997
wuhanensis	Isoptera: Hodotermitidae	*Anacanthotermes ahngerianus*	L	Lyutikova & Yudina 1995
	Lepidoptera: Lymantriidae	*Lymantria dispar*	L	Dubois *et al.* 1989
yunnanensis	Diptera: Culicidae	*Culex pipiens molestus*	L	Saitoh *et al.* 1996
		Culex pipiens quinquefasciatus	L	Yu *et al.* 1984
	Diptera: Psychodidae	*Telmatoscopus albipunctatus*	L	Saitoh *et al.* 1996
	Lepidoptera: Bombycidae	*Bombyx mori*	L	Yu *et al.* 1984
	Isoptera: Hodotermitidae	*Anacanthotermes ahngerianus*	L	Lyutikova & Yudina 1995
Unknown	Acarina: Phytoseiidae	*Euseius stipulatus*	F	Bellows *et al.* 1992
*		*Typhlodromus exhilaratus*	F	Liguori 1988
*		*Typhlodromus pyri*	F	Niemczyk 1997
*	Coleoptera: Carabidae	*Calosoma frigidum*	F	Cameron & Reeves 1990
*	Coleoptera: Cerambycidae	*Anoplophora glabripennis*	L	Dai *et al.* 1989
*	Coleoptera: Coccinellidae	*Coccinella septempunctata*	F	Manjula & Padmavathamma 1996
*		*Cryptolaemus montrouzieri*	L	Castaner & Garrido 1995
*		*Hippodamia convergens*	L	Wilkinson *et al.* 1975
*		*Menochilus (Cheilomenes) sexmaculatus*	F	Manjula & Padmavathamma 1996
*		*Phytoseiulus persimilis*	F	Charles *et al.* 1985
*		*Rhyzobius lophanthae*	L	Bellows & Morse 1993
*		*Scymnus includens*	L	Viggiani *et al.* 1972
*		*Stethorus loxtoni*	L	Walters 1976
*		*Stethorus nigripes*	L	Edwards & Hodgson 1973; Walters 1976
*		*Stethorus vagans*	L	Walters 1976
*	Coleoptera: Curculionidae	*Neochetina eichhorniae*	L	Haag & Boucias 1991
*		*Sitophilus oryzae*	F	Rahman *et al.* 1997
*	Coleoptera: Scarabaeidae	*Oryctes elegans*	L	Hurpin 1974
*		*Oryctes monoceros*	L	Hurpin 1974
*		*Oryctes nasicornis*	L	Hurpin 1974
*	Coleoptera: Scolytidae	*Ips calligraphus*	L	Cane *et al.* 1995
*	Coleoptera: Staphylinidae	*Oligota kashimirica benefica*	L	Gyoutoku & Kastio 1990
*		*Oligota yasumatsui*	L	Gyoutoku & Kastio 1990
*	Coleoptera: Tenebrionidae	*Tribolium castaneum*	L/F	Saleem & Shakoori 1996; Rahman *et al.* 1997
*	Dermaptera: Forficulidae	*Doru lineare*	F	Campos & Gravena 1984
*	Dictyoptera: Mantidae	*Tenodera sinensis*	L	Yousten 1973†
*	Diptera: Cecidomyiidae	*Aphidoletes aphidimyza*	L	Helyer 1991
*	Diptera: Culicidae	*Aedes aegypti*	L	Feng & Xing 1982

*		*Lydella thompsoni*	F	Anglade *et al.* 1980
*		*Parasetigena agilis*	F	Dunbar *et al.* 1973†
*		*Voria ruralis*	L	Wilkinson *et al.* 1975
*	Gastropoda: Limacidae	*Deroceras reticulatum*	L	Kienlen *et al.* 1996
*	Hemiptera: Aleyrodidae	*Bemisia argentifolii*	L	Davidson *et al.* 1996
*	Hemiptera: Anthocoridae	*Orius majusculus*	F	Angeli *et al.* 1998
*	Hemiptera: Lygaeidae	*Geocoris punctipes*	F	Greene *et al.* 1974
*	Hemiptera: Nabidae	*Nabis capsiformis*	F	Greene *et al.* 1974
*	Hemiptera: Notonectidae	*Buenoa* sp.	L	Ortegon Martinez & Quiroz Martinez 1990
*	Hemiptera: Pentatomidae	*Perillus bioculatus*	F	Hough Goldstein *et al.* 1996
*	Hemiptera: Reduviidae	*Rhinocoris (Rhynocoris) fuscipes*	F	Manjula & Padmavathamma 1996
*	Hymenoptera: Aphelinidae	*Aphytis melinus*	L/F	Davies & McLaren 1977; Bellows & Morse 1993; Bellows *et al.* 1993
*				
*		*Encarsia citrina*	L	Thomson *et al.* 1996
*		*Encarsia* sp.	F	Kornilov & Ivanova 1987
*	Hymenoptera: Apidae	*Apis mellifera*	L	Cantwell *et al.* 1972; Krieg 1973b
*	Hymenoptera: Braconidae	*Apanteles americanus*	F	Bellotti & Arias 1978
*		*Apanteles congregatus*	F	Bellotti & Arias 1978
*		*Apanteles fumiferanae*	F	Otvos & Raske 1980; Nealis & van Frankenhuyzen 1990
*		*Apanteles (Cotesia) glomeratus*	L	Wawrzyniak 1987; Lin *et al.* 1998
*		*Apanteles (Cotesia) melanoscelus*	F	Dunbar *et al.* 1973†; Ahmad *et al.* 1978
*		*Apanteles (Cotesia) plutellae*	F	Talekar & Yang 1991
*		*Bracon brevicornis*	L	Temerak 1984
*		*Chelonus blacksburni*	L	Wilkinson *et al.* 1975
*		*Cotesia marginiventris*	L	Atwood *et al.* 1995
*		*Cotesia plutellae*	L	Shi & Liu 1998; Chilcutt & Tabashnik 1997
*		*Dacnusa sibirica*	L	Ozawa *et al.* 1998
*		*Meteorus leviventris*	L	Wilkinson *et al.* 1975
*		*Microlpitis demolitor*	L	Hassan & Graham Smith 1995
*		*Opius concolor*	L	Jacas *et al.* 1992a,b
*	Hymenoptera: Chalcididae	*Brachymeria intermedia*	L	Wilkinson *et al.* 1975
*	Hymenoptera: Encyrtidae	*Leptomastidea abnormis*	L	Viggiani *et al.* 1972
*	Hymenoptera: Eulophidae	*Diglyphus isaea*	L/F	Beitia *et al.* 1991; Ozawa *et al.* 1998
*		*Elasmus albizziae*	F	Bastian & Hart 1989
*		*Oomyzus sokolowskii*	L	Guo *et al.* 1998
*	Hymenoptera: Eupelmidae	*Anastatus bifasciatus*	F	Jamaa *et al.* 1996
*	Hymenoptera: Formicidae	*Formica lugubris*	F	Finnegan 1979
*		*Formica polyctena*	F	Orekhov *et al.* 1978
*		*Formica rufa*	F	Pavan 1979
*		*Solenopsis invicta*	F	Jouvenaz *et al.* 1996
*		*Solenopsis* sp.	F	Gravena 1984
*	Hymenoptera: Ichneumonidae	*Campoletis sonorensis*	L	Wilkinson *et al.* 1975
*		*Diadegma eucerophaga*	F	Anon. 1987b; Talekar & Yang 1991
*		*Diadegma semiclausum*	F	Sastrosiswojo 1996; Amend & Basedow 1997
*		*Diadromus collaris*	L	Wang & Liu 1998
*		*Glypta fumiferanae*	F	Otvos & Raske 1980
*	Hymenoptera: Pteromalidae	*Pachyneuron solitatomus (solitarium)*	L	Wang *et al.* 1996

APPENDIX 2.2 *(continued)*

Subspecies/ Serovar	Order: Family	Genus and Species	Lab/Field	References
Unknown	Hymenoptera: Scelionidae	*Telenomus alsophilae*	F	Kaya & Dunbar 1972†
*		*Telenomus dilophonotae*	F	Bellotti & Arias 1978
*		*Telenomus tetratomus*	L	Wang *et al.* 1996
*	Hymenoptera: Trichogrammatidae	*Trichogramma brassicae (buesi)*	F	Mertz *et al.* 1995
*		*Trichogramma confusum (chilonis)*	L	Zhang 1997; Zhang 1998
*		*Trichogramma dendrolimi*	L/F	Feng *et al.* 1977; Wang *et al.* 1996
*		*Trichogramma embryophagum*	F	Jamaa *et al.* 1996
*		*Trichogramma japonicum*	L	Li *et al.* 1986
*		*Trichogramma maidis*	F	Ravensberg & Berger 1988; Muresan & Mustea 1995
*		*Trichogramma minutum*	F	Bellotti & Arias 1978; Smith *et al.* 1990
*		*Trichogramma ostriniae*	F	Feng *et al.* 1977; Tseng 1990; Tseng & Wu 1990
*		*Trichogramma pretiosum*	L/F	Trumble & Alvarado Rodriguez 1993; Villas Boas & Franca 1996
*		*Trichogramma* sp.	L/F	Wu 1986; Radhika 1998
*		*Trichogrammatoidea bactrae*	L	Hassan & Graham Smith 1995
*	Hymenoptera: Trichogrammatidae ?	*Ooencyrtus pityocampae*	F	Jamaa *et al.* 1996
*	Hymenoptera: Trichogrammatidae ?	*Ooencyrtus* sp.	L	Wang *et al.* 1996
*	Hymenoptera: Trichogrammatidae ?	*Tetrastichus (Baryscapus) servadeii*	F	Jamaa *et al.* 1996
*	Hymenoptera: Vespidae	*Polistes canadensis*	F	Bellotti & Arias 1978
*		*Polistes erythrocephalus*	F	Bellotti & Arias 1978
*	Lepidoptera: Elachistidae	*Stenoma cecropia*	L	Genty 1978
*	Lepidoptera: Gelechiidae	*Scrobipalpula absoluta*	F	Moore 1983
*		*Sitotroga cerealella*	F	Rahman *et al.* 1997
*	Lepidoptera: Geometridae	*Hyposidra talaca*	L	Widiastuti *et al.* 1996
*	Lepidoptera: Lymantriidae	*Lymantria dispar*	L	Dubois *et al.* 1989
*	Lepidoptera: Noctuidae	*Anticarsia gemmatalis*	F	Morosini 1979; Degaspari & Gomez 1982
*		*Heliothis virescens*	F	Pfrimmer 1979
*		*Panolis flammea*	F	Stoakley *et al.* 1978; Holden & Bevan 1979
*		*Sesamia nonagrioides*	F	Felip *et al.* 1987
*		*Spodoptera litura*	L	Narayanan *et al.* 1976†
*	Lepidoptera: Tortricidae	*Cydia leucostoma*	F	Rao *et al.* 1970
*		*Zeiraphera indiana*	F	Aeschlimann 1978
*	Lepidoptera: Xyloryctidae ?	*Nephantis serinopa*	F	Muthukrishnan & Rangarajan 1976
*	Lepidoptera: Yponomeutidae	*Plutella xylostella*	F	Krishnaiah *et al.* 1981
*	Nematoda: Steinernematidae	*Steinernema (Neoaplectana) feltiae (carpocapsae)*	L	Abe 1987
*	Orthoptera: Acrididae	*Locusta migratoria migratorioides*	L	Hoffmann & Brehelin 1976
*		*Melanoplus differentialis*	L	Bradley *et al.* 1989
*		*Melanoplus sanguinipes*	L	Bradley *et al.* 1989
*	Coleoptera: Coccinellidae	*Coccinella septempunctata*	L	Kiselek 1975
*		*Cryptolaemus montrouzieri*	L	Kiselek 1975
*	Diptera: Glossinidae	*Glossinia tachinoides*	L	Maillard & Provost 1975
*	Hymenoptera: Apidae	*Apis mellifera*	F	Buckner *et al.* 1976

many	Diptera: Muscidae	*Haematobia irritans*	L	Dulmage *et al.* 1981† (see App. 2.3)
many	Lepidoptera: Bombycidae	*Bombyx mori*	L	Dulmage *et al.* 1981† (see App. 2.3)
many	Lepidoptera: Lymantriidae	*Hypogymna morio*	L	Dulmage *et al.* 1981†
many		*Lymantria dispar*	L	Dulmage *et al.* 1981† (see App. 2.3)
many		*Orgyia pseudotsugata*	L	Dulmage *et al.* 1981†
many	Lepidoptera: Noctuidae	*Agrotis ipsilon*	L	Dulmage *et al.* 1981† (see App. 2.3)
many		*Heliothis virescens*	L	Dulmage *et al.* 1981† (see App. 2.3)
many		*Spodoptera exigua*	L	Dulmage *et al.* 1981† (see App. 2.3)
many		*Spodoptera litura*	L	Dulmage *et al.* 1981† (see App. 2.3)
many		*Trichoplusia ni*	L	Dulmage *et al.* 1981† (see App. 2.3)
many	Lepidoptera: Pieridae	*Pieris brassicae*	L	Dulmage *et al.* 1981† (see App. 2.3)
many	Lepidoptera: Pyralidae	*Ephestia cautella*	L	Dulmage *et al.* 1981† (see App. 2.3)
many		*Ephestia elutella*	L	Dulmage *et al.* 1981† (see App. 2.3)
many		*Galleria mellonella*	L	Dulmage *et al.* 1981† (see App. 2.3)
many		*Ostrinia nubilalis*	L	Dulmage *et al.* 1981† (see App. 2.3)
many	Phthiraptera: Trichodectidae	*Bovicola crassipes*	L	Dulmage *et al.* 1981† (see App. 2.3)
AF 101	Lepidoptera: Bombycidae	*Bombyx mori*	L	Aizawa *et al.* 1975
11 varieties	Hymenoptera: Apidae	*Apis mellifera*	L	Haragsim & Vankova 1973†

APPENDIX 2.3 SUSCEPTIBILITY OF INSECTS TO ISOLATES OF *BT* FROM THE USDA CULTURE COLLECTION.

The summarised results are previously unpublished but are the outcome of a study described in Dulmage *et al.* (1981).

The cooperators who produced the data below were: H.T. Dulmage, H.D. Burges, L.C. Lewis, K. Aizawa, N. Allen, K.Y. Arakawa, H. de Barjac, C.C. Beegle, E.B. Dick, N.R. Dubois, N. Fugiyoshi. L.P.S. van der Geest, R.E. Gingrich, I.M. Hall, M. Haufler, P. Jarrett, J. Krywienczyk, W.H. McGaughey, D.S. Needleman, M. Ohba, H. del var Petersen, C.G. Thompson and H.J.M. Wassink.

Note on the data from Dr H.D. Burges: In addition to the obvious usefulness of the potency values, these data help in the selection of strains for specific purposes. For example, a range of strains covering the whole collection can be assembled by taking some of the strains from each variety and from each crystal type within a variety. Workers wishing to search for strains for unique new endotoxins could look for strains within each group that appear different from the majority in some particular feature. H.T. Dulmage has further studied potency ratios from the data for many different insects and split the major groups shown in this appendix into a valuable hierarchy of subgroups, some of which contain only one or two strains, which thus appear unique. Copies of the details of these subgroups, including the statistical basis for their erection, can be obtained on request from Dr Leslie Lewis, Iowa State University, Genetics Laboratory c/o Insectary USDA-ARS, Ames, IA 50011-0001, USA; or Dr H. Denis Burges, 21 Withdean Ave., Goring-by-Sea, Worthing, West Sussex, UK.

SUSCEPTIBILITY OF INSECTS TO ISOLATES OF BT FROM THE USDA CULTURE COLLECTION

(IU= international units/mg *Bt* powder; LC_{50} values µg/ml agar based insect diet, or µg/g solid insect food depending on lepidoptera species and *Haemotobia irritans*, or % powder in solid food for *Galleria mellonella*, or µg/ml water for mosquitoes; the percentages are mortalities at the 500 µg/g or 500 µg/ml level except for *O. nubilalis* which are at the 25 µg/g level, mosquitoes 100 µg/ml and *B. crassipes* 100 mg/ml).

Bt subspecies/ Isolate code (HD-)	*Trichoplusia ni*	*Heliothis virescens*	*Hyphantria cunea*	*Bombyx mori*	*Ephestia cautella*	*Plodia interpunctella*	*Haematobia irrtans*	*Ostrinia nubilalis*	*Galleria mellonella*	*Aedes aegypti*	*Aedes triseriatus*	*Culx pipiens*	*Culex quinquefasiatus*	*Pieris brassicae*	*Spodoptera litura*	*Spodoptera exigua*	*Agrotis ipsilon*	*Bovicola crassipes*	XLT type**
	IU/mg	IU/mg	LC_{50}	LC_{50}	LC_{50}	LC_{50}	LC_{50}	LC_{50}	LC_{50}	LC_{50}	LC_{50}	LC_{50}	LC_{50}	LC_{50}	LC_{50}	LC_{50}	LC_{50}	LC_{50}	
Bt thuringiensis																			
54-R-586A	2,110	658	29	269	18.4	20.7	2%†	24%	0.069	3%	0%	31%	28%	2.25	20%	26%	4%	34.4†	Thu
57-R-657B	2,290	418	21.3	198	49.7	90%	29%†	8%		1%	9%	11%	8%		15%	24%	8%	10.3†	Thu
93-R-643C	2,890	781	29.5	148	21.8	90.4	11%†	32%	0.112	14%	2%	6%	5%	4.77		33%	2%	2.1.3†	Thu
26-R-619B	5,330	1,430	34.9	250	13.5	24.5		44%	0.205	4%	6%	64%	17%	35.9	15%	169	90%		Thu
60-R-630C	5,090	1,080	32.4	168	10.9	32.8	139	48%	0.074	7%	12%	12%	10%	2.89	30%	32%	1%	100%, 5.3†	Thu
92-R-643B	5,570	1,400	30.7	162	16.5	41		64%	0.082	3%	2%	4%	9%	1.8	40%	42%	4%	12.7†	Thu
138-R-581A	6,870	1,140	36.9	162	13.6	30.9	480	28%	0.263	10%	7%	3%	24%		0%	158	78%	4.9†	Thu
271-R-707A	5,990	1,230	8.79	73.8	16.1	25	96	30%	0.16	25%	2%	30%	34%		20%	157	32%	100%, 1.2†	Thu
25-R-619A	2,950	595	50.6	312	51.7	104	198	16%		2%	6%	4%	26%	2.54	15%	74%	18%	100,% 4.4†	Thu
55-R-629A	3,960	626	47.5	203	61.4	103	162	0%	1					2.84	0%	311	15%	100%, 2.1†	Thu
91-R-643A	3,870	966	39	195	38.7	70.5	445	12%	0.181	47%	16%	7%	9%		5%	273	0%	100%, 3.2†	Thu
26-R-672A	7,670	1,620	19.1	23.5	21.2	29.9	36.1	56%	0.046	47%	15%	36%	69%	15.6	20%	170	12%	100%, 2..3†	Thu
59-R-645A	7,960	1,440	46.9	250	19.2	23.4	52.1	44%	0.16	10	16%	38%	24%	2.62	35%	81.6	797	100%, 0.6†	Thu
288-R-679B	12,000	3,480	20.7	76.1	10.5	16.5	51.2	64%	0.14	9%	33%	26%	62%		30%	118	0%	100%, 3.0†	Thu
2-R-606A	8,980	1,410	38	69.6	13.9	23.3	43.1	64%	0.31	34%	12%	30%	79%	0.84	20%	127	28%	100%, 1.5†	Thu
264-R-709A	9,300	1,390	16.8	57.6	16.2	27.4	41.6	46%	0.16	13%	6%	15%	11%		15%	108	28%	100%, 1.2†	Thu
98-R-651C	4,890	1,040	26.6	67.2	36.1	70.2	136	28%	0.22	62%	16%	3%	5%	2.31	10%	159	8%	100%, 2.5†	Thu
300-R-715A	1,910	394	32.8	63.1	43.5	91.3	23%†	16%	0.44	14%	20%	20%	12%		0%	26%	2%	14.6†	Thu
90-R-642C	1,850	763	22.1	129	23.3	105		24%	0.31	2%	4%	5%	6%	2.7	10%	22%	0%	17.8†	Thu
96-R-644C	4,850	1,430	34.9	193	14.1	42.1	294	44%	0.23	8%	25%	30%	8%		15%	353	4%	100%, 8.0†	Thu

281-R-711B	5,290	2,580	125	9.32	11.3	144.6	216	56%	1.4	0.9	0.2	14%	9%		5%	163	4%	100%, 5.5†	Thu
20-R-617B	3,570	987	28.4	56.1				44%	0.068	4%	3%	7%	9%		10%	14%	2%		Thu
20-R-725A	5,000	1,370			8.7	65.5	238	44%	0.145	17%	1%	12%	80%	5.09		407	2%	100%, 4.6†	Thu
303-R-716C	1,480	377	38.1	45			21%†	24%	0.3	8%	21%	14%	6%		0%	22%	0%	61.0†	Thu
53-R-632C	2,140	316	0%	100	24.9	130.4	89.2	8%	1.3	51%	1%	15%	9%	1.51	5%			1.4†	Thu
307-R-719A	2,490	451	369	191	68%	6%	260†	12%	1.32	0%	11%	7%	5%		0%	46%	2%	100%, 7.3†	Thu
309-R-719C	6,700	1,190	35%	37.8	444	91.1	78.6	4%	2.6*	2%	19%	17%	11%		0%	48%	4%	100%, 2.3†	Thu
27-R-725B	1,890	218	0%	5%	12.7	203.1	130	12%	1.02	11%	2%	18%	84%	9.93	20%	52%		100%, 3.2†	Thu
94-R-644A	2,610	349	7%	15%	11.8	125	38.4	20%	0.69	68%	9%	15%	13%	6.69	0%	272	8%	100%, 1.6†	Thu
141-R-528A	1,810	255	0%	15%	40.5	92.9	164	0%	1.7	8%	3%	6%	48%	1.92	5%	235	33%	100%, 1.1†	Thu
308-R-719B	5,450	778	0%	30%	197	160	315	0%	0.62	5%	16%	49%	20%		0%	44%	12%	100%, 1.4†	Thu
28-R-622C	3,460	64%	0%	10%	14	91.5	261	4%	1.51	11%	23%	45%	27%		10%	373	10%	100%	Thu
46-R-604B	2,070	26%	223	204			305	4%	0.78	3%	8%	21%	10%		0%	315	8%	56%	Thu
139-R-581B	1,850	40%	0%	20%	23.8	208	341	20%	0.9	5%	5%	8%	9%	0.69	5%	64%	18%	100%, 2.7†	Thu
24-R-656B	55%	60%	70.9	82.3	54.5	233	5%†	13%	1.1	5%	5%	14%	7%	7.95	5%	10%	12%		Thu
58-R-630A	47%	56%	90.5	339	~5%	~5%	39%†	12%	1.35	3%	1%	2%	17%	3.02	0%	10%	11%	16.1†	Thu
120-R-660A	13,900	5,910	9.94	3.13	2	26.5	237	68%	0.049	1	0.1	25%	29%		26%	124	50%	100%, 1.9†	K-1
290-R-713B	8,980	5,680	11	4.11	14	18.3	36.9	67%	0.23	79%	0.45	9%	18%		20%	85.2	22%	100%, 1.0†	K-1
14-R-612C	4,440	1,520	27.9	5.16				64%	>15	0.7	0.1	53%	25%	92%	0%	22%	8%		K-1
15-R-622B	4,420	1,640	67.3	9.81	15.4	~5%	21%†	28%	4.0*	0.8	0.3	30%	20%	3.78	5%	23%	4%		K-1
95-R-644B	5,600	2,150	34.5	9.95	28.1	~5%	5%†	32%	0.75	0.75	0.07	16%	18%	3.3	10%	30%	0%		K-1
17-R-615B	1,430	652	40%	2.9	21%	~5%	0%†	30%	>15	51%	0.5	9%	16%	7.16	5%				K-1
22-R-655C	1,800	918	164	12.9	~5%	~5%	30%†	8%	>15	1	0.3	21%	24%	2.59	0%	38%	4%		K-1
103-R-647B	3,930	2,510	150	13.2	26.8	~5%	32%†	44%	4.0*	1	0.07	14%	30%	2.45	0%	10%	12%		K-1
117-R-653A	2,730	1,920	376	40.9	39.1	5%	36%†	20%	15*	1	0.06	14%	3%	1.2	0%	16%	8%		K-1
39-R-601C	6,350	1,220	173	16.6	17.8	~5%	2%†	32%	7.9*	0.8	0.3	52%	6%	2.48	0%	20%	4%		K-1
280-R-722B	2,280	1,820			11.7	256	44%†, 39%	36%	0.478	10	0.8	12%	6%		48%	48%	4%		Mxd or ?
316-R-720B	5,760	2,430	83.9	5.86	18.7	132	80.6	48%	>0.2	3	0.3	28%	36%		48%	129	4%	100%, 2.0†	Mxd or ?
317-R-720C	11,800	4,210	12	6.62	16.8	20.3	66.7	60%	0.094	2.3	0.1	48%	56%		326	57.4	8%	100%, 1.1†	Mxd or ?
318-R-723C	1,620	877			12.5	102	24%†	56%	>0.2	42%	0.7	60%	9%		40%	40%	4%	27.5†	Mxd or ?
319-R-721A	4,780	2,460	6.31	4.35	9	30.5	10%†	68%	>0.2	4.3	0.3	23%	8%		74%	74%	6%	9.9†	Mxd or ?
319-R-722C	4,230	3,120			2.7	50.5	35%†	68%		1.8	1	12%	5%					38.9†	Mxd or ?
65-R-635A	992	1,150	27%	23.2			23%†	28%	>10	76%	0.3	5%	2%	2.5	0%	2%	8%		Mxd or ?
225-R-692C	4,370	2,000	7.03	3.69	24.6	~5%	13%†	28%	1.95	2	0.05	5%	15%	1.6	0%	26%	6%		Mxd or ?

APPENDIX 2.3 (*continued*)

Bt subspecies/ Isolate code (HD-)	*Trichoplusia ni*	*Heliothis virescens*	*Hyphantria cunea*	*Bombyx mori*	*Ephestia cautella*	*Plodia interpunctella*	*Haematobia irrtans*	*Ostrinia nubilalis*	*Galleria mellonella*	*Aedes aegypti*	*Aedes triseriatus*	*Culx pipiens*	*Culex quinquefasiatus*	*Pieris brassicae*	*Spodoptera litura*	*Spodoptera exigua*	*Agrotis ipsilon*	*Bovicola crassipes*	XLT type**
18-R-615C	6%	8%	0%	5%	5%	5%	22%†	0%	>15	1%	28%	6%	12%	74%	10%	2%	4%		Mxd or ?
23-R-656A	2%	8%	7%	5%	~5%	~5%	29%†	0%	>15	4%	4%	23%	12%	25%	0%	20%	12%		Mxd or ?
32-R-623C	4%	10%	0%	10%	~5%	~5%	11%†	0%	<15	2%	1%	7%	7%	0%	0%	14%	8%		Mxd or ?
41-R-603B	82%	62%	0%	0%	~5%	~5%		0%	0.53*	18%	25%	14%	73%	69%	15%	121	8%		Mxd or ?
42-R-626C	2%	8%	0%	5%	~5%	~5%	30%†	4%	>15	1%	1%	5%	6%	4%	15%	2%	8%		Mxd or ?
56-R-629B	2%	4%	7%	0%	~5%	~5%	39%†	4%	>15	5%	9%	2%	10%	99.6	5%	4%	2%		Mxd or ?
140-R-581C	12%	8%	0%	0%	~5%	~5%	13%†	0%	>15	8%	3%	6%	48%	48%	10%	12%	6%	72.2†	Mxd or ?
311-R-696A	4%	7%	27%	0%	~5%	~5%	26%†	0%	1.52*	1%	2%	11%	33%	88%	0%	12%	4%	8.2†	Mxd or ?
Bt alesti																			
70-R-633C	18%	30%	11.7	7.17	51%	26%	0%†	52%	0.098	10%	2%	3%	1%	2.88	5%	4%	2%		AL or ?
72-R-636B	2%	16%	6.55	6.66	29%	32%	10%†	38%	0.072	2%	15%	7%	3%	1.41	5%	10%	0%		AL or ?
81-R-638C	22%	20%	13.1	8.94	24%	34%	0%†	40%	0.13	3%	4%	6%	6%	1.65	0%	14%	0%		AL or ?
86-R-651A	2%	18%	22.4	9.41	19%	0%	26%†	32%	0.21	13%	22%	1%	8%	7030	5%	14%	2%		AL or ?
104-R-647C	20%	44%	12	5.5	42%	50%	0%†	42%	0.046	20%	30%	11%	16%	2%	0%	12%	0%		AL or ?
69-R-633B	4%	12%	21.8	47	14%	10%	0%†	8%	0.43	13%	2%	4%	3%	7.83	5%	8%	6%		AL or ?
79-R-638A	4%	8%	34.6	19.2	5%	11%	39%†	40%	0.61	8%	7%	3%	8%	1.57	0%	2%	0%		AL or ?
80-R-638B	6%	2%	29.9	19.6	15%	4%	31%†	20%	0.82	5%	6%	6%	7%	2.2	0%	4%	4%		AL or ?
83-R-658A	14%	30%	5.8	8.46	60%	32%	35%†	24%	0.067	0%	1%	44%	23%	0.89	0%	32%	0%		AL or ?
4-R-607A	14%	14%	18.9	35.3	19%	4%	7%†	28%	0.6	4%	8%	5%	4%	5.61	5%	10%	6%		AL or ?
40-R-603A	8%	14%	19.7	40.7	12%	2%	38%†	48%	0.71	7%	9%	28%	4%	2.43	0%	6%	0%		AL or ?
45-R-604A	4%	10%	81.3	186				8%	1.8	11%	5%	7%	13%		10%	4%	0%		AL or ?
16-R-615A	0%	12%	39.5	54			24%†	32%	1.72	2%	27%	3%	9%	23.3	5%	4%	6%		AL or ?
45-R-632A	4%	10%	44.7	100	16%	0%	29%†	4%	1.8	5%	3%	1%	6%	340	0%	14%	4%		AL or ?
71-R-636A	2%	8%	26.2	37.7	7%	23%	7%†	28%	2.47	3%	2%	15%	2%	10	5%	16%	12%		AL or ?
76-R-637C	4%	16%	62.2	145	1%	3%	36%†	17%	2.2	11%	8%	5%	30%	14.2	0%	8%	10%		AL or ?
75-R-637B	1,330	1,070	40.2	19.1	22%	29%	9%†	29%	0.28	8%	13%	6%	13%	10.6	0%	16%	4%		AL or ?
84-R-666A	20,300	24,500	60	15%	27.9	72.8	13%†	80%	0.35	5%	1%	13%	6%	0.78	0%	10%	4%		K-73
88-R-642A	19,900	3,930	117	10.4	7	440.6	5%†	52.2	1.1	72%	0.1	15%	6%	0.97	5%	32%	4%		K-73

89-R-642B	32,800	11,600	42.7	5.94	2.9	108	9%†	8.7	0.48	0.4	0.03	19%	10%	0.46	0%	22%	10%		K-1
245-R-702C	13,700	5,210	37.6	6.93	16.1	74.5	10%†	44%	0.024	4.5	0.6	22%	10%		15%	199	5%	1.9†	K-1
250-R-600C	11,600	4,280	62.5	21.1					4.0*	59%	0.3	32%	4%	4.69	25%	10%	10%		K-1
259-R-714A	15,700	5,650	125	7.24	7.8	51%	28%†	64%	0.91	41%	0.47	6%	3%		0%	25%	6%		K-1
169-R-678B	12,800	9,460	96.4	22	6.1	43%	7%†	68%	0.47	67%	0.6	11%	18%	2.86	5%	4%	10%	76.9†	K-1
172-R-680B	25,500	21,400	45.9	11.8	5.3	48.4	14%†	5.99	0.4	8		12%	1%		5%	30%	2%		K-1
231-R-639B	52,400	28,600	51.4	8.67	1.4	26.8	17%†	6.72	0.23	0.4	0.45	22%	10%		5%	22%	2%		K-1
243-R-702C	43,300	34,200	10.4	4.35	2.9	31.1	14%†	9.75	0.45	1	0.3	20%	29%		5%	38%	7%		K-1
299-R-701B	18,700	7,070	53.9	8.22	3.7	294	22%†	10.5	2.1	1	0.4	9%	21%	5.47	0%	26%	2%		K-1
302-R-715C	15,600	8,690	81.9	8.33	6.1	47%	10%†	52%	0.41	55%	0.35	9%	0%		0%	22%	0%	43.1†	K-1
1-R-641A	39,800	15,400	15.8	5.68	2.6	16.4	0%†	5.94	0.056	0.5	0.06	15%	6%	0.78	224	22%	0%	62.6†	K-1
252-R-601B	14,100	4,660	34.1	14.9			12%†	11.2	2.3	2	0.5	49%	3%		0%	24%	2%		K-1
310-R-720A	20,500	9,960	16.1	4.83	2.5	63.6	5%†	10.1	0.42	78%	0.5	22%	5%		40%	24%	2%	32.5†	K-1
164-R-676C	42,000	35,000	14.7	13.3	4.5	24.2	10%†	4.27	0.23	53%	0.2	36%	46%	98%	0%	2%	1%	54.6†	K-1
265-R-705A	11,400	10,100	10.7	8.66	12.5	35.5	24%†	72%	0.24*	17	0.8	9%	13%		0%	12%	4%	38.7†	K-1
270-R-706C	35,200	20,400	2.99	4.84	8.3	16.7	1%†	5.2	>0.14	1	0.5	8%	22%		15%	10%	30%		K-1
203-R-589C	18,900	6,760	4.66	5.8	6.4	13.9		4.34	0.023	93%	0.25	9	3%		258	34%	4%		K-1
262-R-718A	9,820	3,850	6.47	5.84	10.4	55.4	12%†	60%	0.15	1	0.35	10%	1%		20%	18%	6%		K-1
267-R-705C	14,100	8,690	3.95	3.97	13.8	23.8		76%	>0.2	2	0.6	4%	16%		0%	30%	6%		K-1
247-R-704A	16,600	7,190	28.4	31.7	27%	14%	12%	64%	1.01	38%	0.5	1%	13%		0%	22%	15%		K-1
286-R-664A	4,280	2,400	69.7	52.3	11.6	93.9	25%†	50%	0.41	2%	0.9	9%	4%	11.8	0%	14%	4%		K-1
242-R-663B	25,400	33,600	100	34.2	4	59	18%†	68%	0.28	35%	0.3	40%	7%		5%	38%	5%		K-1
246-R-663C	24,900	31,700	15.5	4.36	7.2	41.9	20%†	3.31	0.24	76%	0.46	77%	8%		25%	34%	21%	28.4†	K-1
263-R-704B	39,600	54,600	13.5	7.12	2.1	12.1	28%†	9.76	0.19	1	0.25	7%	15%		0%	40%	19%	62.0†	K-1
73-R-636C	23,100	34,500	28.2	45%	9	38	28%†	7.34	0.41	0%	8%	7%	0%	1.43	15%	6%	2%		K-73
73-R-694A	29,300	34,500	18.3	20%	14.2	33		60%	0.28	1%	2%	18%	41%		5%	6%	8%		K-73
74-R-637A	11,100	16,400	91.4	20%	84%	88%	15%†	40%	1.9	3%	3%	11%	4%	1.38	0%	10%	4%		K-73
74-R-701A	20,700	33,900	22.7	40%	21.5	36.1	4%†	44%	0.33	12%	15%	14%	8%	1.93	5%	12%	5%		K-73
78-R-639A	9,460	22,500	60.4	20%	87%	91%	24%†	49.9	1.8	1%	6%	11%	4%	7.34	5%	16%	3%		K-73
162-R-676A	15,800	35,900	88.8	30%	18.2	72.8	15%†	56%	0.37	12%	8%	4%	5%		0%	6%	15%		K-73
171-R-680A	12,100	18,600	25.4	107	19.8	66.4	20%†	52%	0.74	13%	6%	6%	1%		5%	0%	4%		K-73
177-R-687B	15,700	30,900	36.3	15%	25	93.8	0%†	72%	0.46	2%	4%	31%	42%		5%	12%	17%		K-73
178-R-687C	12,500	27,700	48.2	35%	20.4	63.9	38%†	79%	0.28	2%	2%	27%	4%	6.66	0%	10%	8%		K-73
180-R-588A	13,800	21,200	113	20%	20.3	85.5	36%†	48%	0.65	4%	2%	4%	11%		0%	14%	2%		K-73

APPENDIX 2.3 (*continued*)

Bt subspecies/ Isolate code (HD-)	*Trichoplusia ni*	*Heliothis virescens*	*Hyphantria cunea*	*Bombyx mori*	*Ephestia cautella*	*Plodia interpunctella*	*Haematobia irrtans*	*Ostrinia nubilalis*	*Galleria mellonella*	*Aedes aegypti*	*Aedes triseriatus*	*Culx pipiens*	*Culex quinquefasiatus*	*Pieris brassicae*	*Spodoptera litura*	*Spodoptera exigua*	*Agrotis ipsilon*	*Bovicola crassipes*	XLT type**
181-R-588B	11,100	23,800	105	30%	15.2	83.8	13%†	7.29	0.99	8%	2%	10%	1%		10%	4%	9%		K-73
182-R-588C	25,100	46,800	14.1	30%	8.7	50.5	33%†	5.05	0.22	4%	7%	7%	16%		5%	14%	4%		K-73
183-R-685C	18,000	23,300	65.2	35%	90%	93%	21%†	56%	0.37	4%	6%	13%	9%		30%	2%	8%		K-73
187-R-587A	21,000	36,900	71.5	15%	12.4	53.7	38%†	8.48	0.29	2%	0%	16%	2%		5%	14%	2%		K-73
191-R-665A	26,100	61,000	58.7	30%	18.5	36.8	23%†	72%	0.18	3%	2%	39%	5%		0%	6%	6%		K-73
244-R-702B	42,000	70,600	10.5	111	6.1	26.9	20%†	9.18	0.2	8%	30%	21%	9%	2.23	5%	26%	8%		K-73
279-R-710C	21,500	24,200	7.51	8.69	10.6	31.6	12%†	60%	0.11	65%	0.5	11%	6%		0%	38%	2%	37.5†	K-73
313-R-696C	1,070	2,470	47.4	91.2	39.8	198	0%†	64%	2.4	3%	6%	16%	40%		20%	14%	18%		K-73
254-R-722A	5,740	3,160			4.2	32.1	15%†	52%	0.16	8%	4%	10%	8%			12%	4%		K-73
87-R-651B	18,700	12,500	9.08	9.7	4.3	10.1	281	88%	0.075	1.7	0.03	9%	18%	2.89	35%	242	30%	100	K-1 and K-73
255-R-699A	23,900	10,600	82.7	4.05	5	114	6%†	100%	0.75	0.5	0.05	16%	43%	1.04	0%	32%	4%	3.1†	K-1 and K-73
85-R-666B	20%	262	10.8	11.1	~5%	~5%	34%†	36%	0.17	2%	5%	7%	7%	26%	0%	30%	4%	77.8†	K-73
21-R-655B	4%	4%	0%	0%	~5%	~5%	0%†	~5%	>15	35%	6%	24%	15%	1%	0%	12%	8%		K-73
31-R-623B	0%	6%	0%	5%	~5%	~5%	0%†	~5%	>15	1%	1%	6%	9%	0%	5%	22%	4%		
102-R-647A	12%	4%	0%	0%	~5%	~5%		~5%	>15	5%	6%	16%	21%	1%	5%	14%	6%		
315-R-701C	0%	3%	0%	0%	~5%	~5%	32%†	~5%	0.9*	7%	11%	11%	5%		5%	20%	14%		
284-R-718B	7,880	5,710	17.7	6.45	11.7	64.2	20%†	40%		6	0.5	16%	3%		20%	20%	14%	31.7†	Aiz or ?
304-R-717A	9,510	4,150	12.5	38.2	12	22.4	7%†	43.8	0.15	3%	10%	18%	1%		30%	14%	0%	13.7†	Aiz or ?
306-R-717C	23,300	6,280	11.9	5.1	4.9		18%†	5.23	0.18	2	0.2	19%	15%		5%	40%	8%	18.6†	Aiz or ?
254-R-722A	5,740	3,160			4.2	32.1	15%†	52%	0.16	8%	4%	10%	8%			12%			?
258-R-733B	13,700	4,690					15%†	48%	>15							35%			?
266-R-705B	5,710	1,260	16.4	287	32.9	41.6	51.8	4%	0.37	4%	1%	7%	20%		0%	111	54%	100%, 1.7†	?
268-R-706A	4%	319	251	5%	~5%	~5%	24%†	0%	3.2*	1%	4%	7%	1%		0%	10%	6%		?
313-R-696C	1,070	2,470	47.4	91.2	39.8	198	0%†	64%	2.4*	3%	6%	16%	40%		20%	14%	18%		?
62-R-634A	30%	751	10%	0%	~5%	~5%	0%†	16%	>15	50%	7%	3%	1%	19.6	0%	8%	4%		Ken
296-R-681A	16%	1,200	0%	262	33%	~5%	20%†	20%	>15	35%	1	31%	8%	30%	0%	30%	2%		Ken.

Bt kenyae																			
278-R-711A	2,250	3,310	13%	108	29%	13%	4%†	7%	0.51	14%	1	7%	6%		0%	16%	12%		K-1
261-R-716B	3,090	1,870	148	5.5	12.1	0%	13%†	60%	0.85	1	0.5	26%	2%		0%	30%	4%	23.4†	Thu and K-1
5-R-607B	4,380	2,450	198	224	24%	~5%	10%†	64%	2.4*	57%	0.7	3%	10%	11	15%	6%	10%		Ken
136-R-580B	9,350	5,650	144	195	13.7	~5%	14%†	48%	0.78	9%	0.55	4%	11%	62%	23.5	20%	0%		Ken.
277-R-709B	2,660	2,440	198	167	23.9	~5%	9%†	4%	>15	14%	0.85	5%	15%	28.5	15%	42%	12%		Ken.
293-R-679C	8,560	2,440	41.5	30.2	26.8	172	5%†	14.6	2.1*	2.8	0.62	10%	4%		45%	48%	0%	19.4†	Ken.
123-R-667B	6,520	2,720	8.79	162	32.2	0.761	141†	4.7	0.34	0.4	30%	25%	1.8	25%		24%	100%	2.2†	Ken.
294-R-684C	2,840	1,050	33.2	20%	28%	0%	215	25.7	4.0*	2	0.45	7%	19%	44%	5%	2.9	10%	100%, 3.7†	Ken.
63-R-634B	4%	285	0%	25%	~5%	~5%	9%†	0%	>15	58%	2%	1%	0%	43.4	0%	4%	2%	50.6†	Atyp
63-R-654B	6,280	12,000																	Atyp
64-R-634C	1,730	1,520	0%	10%	~5%	~5%	9%†	13%	>15	77%	9%	0%	0%	21.4	0%	18%	4%		Atyp
291-R-713C	4,540	1,240	10.3	37.6	5.3	25.8	19%†	32%	0.098	7%	2%	23%	3%		0%	30%	2%		Atyp
292-R-679B	6%	4%	0%	5%	~5%	~5%	24%†	0%	2.9*	22%	24%	22%	50%		0%	6%	32%		Atyp
295-R-682A	1,190	352	243	15%	~5%	~5%	8%†	40%	>15	64%	0.85	27%	42%	30.2	45%	18%	6%		Atyp
Bt galleriae																			
8-R-662A	6,690	679	53.2	26.7	71%‡	67%‡	41%†, 5%	26.6	0.0012	32%	39%	35%	29%	3.93	0%	14%	8%		Gal
29-R-659A	6,760	341	22.8	22.1	85%‡	79%‡	63%†, 4%	52%	0.0009	12%	13%	33%	28%	1.75	0%	16%	12%		Gal
190-R-576A	3,800	364	91.1	47.1	2%	3%	66†, 39%	60%	0.0016	40%	27%	14%	50%	9.7	15%	6%	0%	33%	Gal
196-R-688B	8,490	674	13	9.41	70.6	16%	18%†	68%	0.0009	44%	5%	20%	7%		10%	26%	0%		Gal
220-R-684B	3,940	253	52.1	26.6	10%	0%	122†, 0%	64%	0.0022	37%	9%	43%	18%	80%	0%	26%	0%	7%	Gal
207-R-590C	5,870	24%	82.4	35.8	47% ‡	55%‡		32%	0.0057	56%	17%	40%	12%	10.4	285	10%	0%		Gal
208-R-591A	3,830	20%	50.6	23.2	38% ‡	51%‡	348†, 39%	28%	0.0098	60%	8%	79%	19%	4.76	10%	14%	0%	100%, 13†	Gal
212-R-592C	5,620	345	47.7	40.9	53% ‡	37%‡	68†, 13%	40%	0.0013	63%	12%	47%	30%	1.87	5%	16%	5%	7%	Gal
216-R-593A	3,630	30%	51.4	38.3	54% ‡	46%‡	14%†	28%	0.0017	73%	27%	4%	15%	71%	10%	28%	0%		Gal
216-R-493C	6,410	775	50.6	23.2	69% ‡	72%‡		44%	0.0011	45%	7%	7%	46%		5%	24%	0%		Gal
233-R-586C	4,060	24%	67	37	45%‡	54%‡	55	44%	0.0024	30%	24%	14%	21%	80%	5%	10%	0%	10%, 6.2†	Gal
149-R-583A	5,390	596	25	42.3	30%	1%	14%†	32%	0.0012	47%	7%	30%	54%	54%	0%	6%	10%		Gal
150-R-583B	5,080	541	16.8	19.8	25%	18%	88†, 30%	15.9	0.0014	53%	16%	37%	56%	78%	40%	14%	0%	27%	Gal
151-R-583C	4,910	571	19	38.5	16%	1%	156†, 33%	32%	0.002	45%	19%	15%	54%	82%	10%	12%	13%	27%	Gal
153-R-673C	4,100	789	18.8	25.8	62.2	65	127†, 13%	11.6	0.0013	44%	39%	62%	27%	84%	0%	18%		27%	Gal
155-R-671A	4,470	579	14.2	18	25%	6%	46†, 13%	50%	0.0014	13%	13%	56%	25%	80%	0%	14%	2%		Gal

APPENDIX 2.3 *(continued)*

Bt subspecies/ Isolate code (HD-)	*Trichoplusia ni*	*Heliothis virescens*	*Hyphantria cunea*	*Bombyx mori*	*Ephestia cautella*	*Plodia interpunctella*	*Haematobia irrtans*	*Ostrinia nubilalis*	*Galleria mellonella*	*Aedes aegypti*	*Aedes triseriatus*	*Culx pipiens*	*Culex quinquefasiatus*	*Pieris brassicae*	*Spodoptera litura*	*Spodoptera exigua*	*Agrotis ipsilon*	*Bovicola crassipes*	XLT type**
155-R-674B	5,780	657	39.6	36.6	28%	23%	21†, 23%	30.6	0.0011	43%	7%	63%	52%	68%	0%	12%	22%	13%	Gal
156-R-671B	4,120	479	18.9	19.2	25%	5%	126†, 4%	40%	0.0015	4%	16%	55%	21%	48%	0%	18%	15%	0%	Gal
157-R-671C	3,450	403	8.62	28.2	15%	16%	27%†	32%	0.0014	37%	2%	33%	28%	42%	5%	14%	4%		Gal
189-R-578A	3,080	203	23.5	51.5	27%	17%	15%†	32%	0.0015	57%	56%	10%	51%		15%	12%	6%		Gal
305-R-717B	6,060	943	7.45	9.06	45.5	4%	55†, 22%	60%	0.0037	79%	20%	22%	9%		0%	24%	0%	100%	Gal
160-R-675B	5,920	894	42.7	34.2	65.8	61.8	26†, 32%	76%	0.001	61%	18%	60%	14%		0%	12%	26%	40%	Gal
185-R-589B	1,610	22%	56.9	72.4	18%	15%	129†, 39%	12%	0.0017	3%	5%	35%	14%	24%	5%	20%	40%	27%	Gal
221-R-691B	2,400	42%	55	65.3	16%	1%	67.2	36%	0.0045	8%	3%	26%	10%	86%	0%	10%	6%	13%	Gal
222-R-691C	1,890	30%	27.8	45.7	11%	0%	211†, 36%	12%	0.0024	10%	5%	30%	8%	86%	0%	24%	15%	13%	Gal
129-R-672B	4,460	582	43.1	97.3	28%	14%	65†, 13%	40%	0.0013	39%	10%	63%	41%		5%	18%		0%, 14.6†	Gal
154-R-674A	2,350	394	15.5	35.5	20%	7%	389†, 0%	23.4	0.002	31%	22%	52%	13%	64%	0%	6%	4%	33%	Gal
179-R-685B	5,320	660	26.4	52.4	28%	9%	273†, 0%	64%	0.0016	43%	22%	54%	55%		5%	18%	17%	27%	Gal
186-R-686A	5,510	550	14.5	24.9	18%	12%	30.9	5.47	0.0014	46%	11%	49%	43%	76%	0%	30%	0%	20%	Gal
211-R-690C	2,200	42%	37.5	35.3	11%	0%	27%†	12%	0.0048	8%	2%	40%	6%		15%	10%	7%		Gal
215-R-593B	7,300	40%	37.5	57.2	~5%	~5%	56†, 13%	60%	0.0011	68%	20%	75%	28%		0%	28%	0%	0%	Gal
219-R-691A	2,140	36%	27.4	60.5	8%	0%	9%†	4%	0.0028	8%	3%	23%	9%	75%	0%	16%	0%		Gal
152-R-586B	1,880	24%	91.2	68	27%	1%	1.4†, 21%	36%	0.0021	43%	3%	6.8	10%	76%	5%			13%	Gal
170-R-678C	22,300	13,700	6.54	17.2	16.6	41.6	25%†	3	0.0014	65%	59%	57%	25%		5%	18%	25%		Gal
173-R-680C	12,100	7,490	16	12.5	21.4	100.5	10%†	6.73	0.0016	79%	49%	32%	40%		15%	22%	14%		Gal
176-R-683C	15,100	9,650	12.5	21	21	85.1	80†, 0%	3.99	0.002	56%	9%	39%	34%		10%	34%	15%	100%, 59†	Gal
236-R-703C	13,500	7,530	8.08	14.8	2%	21%	48%†, 24%	11.4	0.0101	25%	8%	30%	43%	92%	30%	24%	0%		Gal
238-R-661B	11,400	7,350	35.4	37	6%	21%	118†, 44%	76%	0.002	5%	4%	53%	43%	72%	0%	6%	4%	20%	Gal
165-R-677A	13,100	14,100	22	25.8	27.6	79.6	40†, 13%	8.22	0.0022	41%	41%	8%	7%	2.97	0%	18%	23%	7%	Gal
166-R-677B	18,000	13,400	24.1	33.4	38%	40%	39%†	9.42	0.0017	51%	39%	32%	15%	4.38	0%	22%	14%		Gal
174-R-687A	20,800	16,800	14.1	26.5	24.6	83.3	15%†	76%	0.0021	34%	23%	22%	19%		0%	22%	2%		Gal
775-R-683B	20,800	16,900	4.42	32.9	25.6	114	44†, 55%	44%	0.002	41%	5%	47%	35%		0%	47%	47%	40%	Gal

							49%												
193-R-665C	21,100	31,000	14.1	54.4	12.5	50.5	204†, 13%	7.05	0.0015	4%	16%	56%	20%	6.32	0%	14%	20%	40%	Gal
239-R-662A	25,800	15,300	10.5	25.8	19.8	86.4	74.9	67%	0.0019	26%	16%	10	10%	7852	5%	18%	11%	33%	Gal
240-R-662B	25,700	24,600	9.62	22.7	16.3	72.8	29%†	6.75	0.0015	18%	23%	10	17%		10%	22%	25%		Gal
184-R-589A	21,900	15,600	10.9	18.3	14.1	36.7	19†, 23%	4.49	0.001	26%	40%	17%	76%		35%	20%	8%	13%	Gal
235-R-703B	10,200	4,490	5.04	10.4	2%	18%	23%†	76%	0.027	45%	8%	20%	24%	88%	5%	28%	4%		Gal
167-R-677C	21,200	17,100	10.4	25.8	26.4	51.5	15†, 23%	10.1	0.0026	58%	42%	48%	26%	2.5	0%	8%	10%	33%	Gal
168-R-678B	16,300	10,100	10.4	24.8	58	66.4	0%†	5.26	0.0021	49%	31%	26%	20%	10.1	0%	14%	15%		Gal
161-R-675C	25,600	22,400	13.2	15.6	19.9	43.4	105†, 17%	4.18	0.0014	54%	46%	42%	11%		0%	26%	27%	33%	Gal
158-R-674C	6,070	5,280	48.6	40.9	46.2	86.9	460†, 77%	12.6	0.0013	50%	26%	64%	59%	88%	0%	16%	0%	87%, 35.8†	Gal
163-R-676B	20,500	15,400	20	36.6	21	65.9	26†, 9%	5.28	0.001	47%	42%	67%	13%	4.8	0%	16%	0%	27%	Gal
237-R-661A	3,020	1,740	60.4	67.8	~5%	7%	271†, 17%	32%	0.015	2%	2%	48%	39%	28%	0%	20%	0%	13%	Gal
107-R-648B	6%	44%	26.6	23.2	~5%	~5%	17.3	44%	0.0054	26%	33%	29%	11%	7.68	0%	4%	12%	67%	Gal
214-R-693B	5%	0%	60.4	36.2	~5%	~5%	13%†	8%	0.02	2%	1%	19%	13%	39%	10%	10%	6%		Gal
223-R-692A	12%	10%	76	36.2	~5%	~5%	0%†	4%	0.014	2%	0%	23%	16%	52%	0%	8%	54%		Gal
287-R-664B	18,700	11,500	14.1	4.22	2.6	14.9	16%†	84%	0.077	72%	0.2	10%	3%	3.36	30%	20%	10%	61.9†	K-1
188-R-686B	9,010	12,000	34.1	195	35%	21%	16%†	13	0.0035	12%	12%	51%	24%		0%	16%	8%		Aiz
206-R-590B	5,220	5,880	102	0%	22%	14%	5%†	56%	1.2	3%	4%	8%	2%		5%	12%	4%		Aiz
234-R-703A	15,600	15,800	25	12.5	27%	0%	18%†	11.5	0.002	9	40%	59%	52%	86%	0%	26%	4%	4.9†	Aiz
210-R-592B	7,410	3,080	289	31.7	5%	5%	35%†	68%	0.0016	70%	4%	20%	21%	1.92	5%	18%	0%	7.7†	Aiz
232-R-639C	52,200	29,400	15.4	8.24	3.1	10.7	5%†	3.84	0.17	1	0.1	12%	5%		205	26%	10%		Mxd K-1/K-73/Aiz
108-R-648C	3,200	40%	26.6	35.9	~5%	~5%	3.9	48%	0.0024	17%	26%	26%	6%	2.65	5%	22%	14%	13%	?
205-R-590A	1,490	14%	54	105	4%	0%	112†, 8%	20%	0.0023	12%	2%	68%	11%	52%	5%	16%	14%	0%	?
209-R-592A	1,370	6%	50.6	51.6	~5%	~5%	9%†	20%	0.0084	9%	2%	25%	27%	4.79	0%	2%	0%		?
217-R-594A	4%	2%	334	316	~5%	~5%		0%	0.009	1%	3%	2%	27%	45.8	0%	22%	0%		?
273-R-707C	36%	9%	29.2	12.1	0%	4%	11%†	16%	0.054	56%	2%	11%	31%		0%	10%	4%		?
66-R-635B	0%	6%	7%	25%	~5%	~5%	26%†	4%	0.22	0%	11%	8%	2%		0%	2%	0%		?
Bt canadensis																			
30-R-623A	1,140	22%	102	100	~5%	~5%	8%†	12%	0.0035	34%	3%	19%	8%		0%	10%	0%		K-1/Gal Mxd
224-R-692B	4%	2%	120	20%	~5%	~5%	7%†	~5%	>4.0	1%	3%	12%	3%	~5%	10%	12%	24%		K-1/Gal Mxd

APPENDIX 2.3 *(continued)*

Bt subspecies/ Isolate code (HD-)	*Trichoplusia ni*	*Heliothis virescens*	*Hyphantria cunea*	*Bombyx mori*	*Ephestia cautella*	*Plodia interpunctella*	*Haematobia irrtans*	*Ostrinia nubilalis*	*Galleria mellonella*	*Aedes aegypti*	*Aedes triseriatus*	*Culx pipiens*	*Culex quinquefasiatus*	*Pieris brassicae*	*Spodoptera litura*	*Spodoptera exigua*	*Agrotis ipsilon*	*Bovicola crassipes*	XLT type**
Bt subtoxicus																			
10-R-609B	40%	40%	102	319	19%, 90%*‡	~5%	26%†	28%	0.59	2%	6%	12%	5%	2.59	0%	6%	4%		Thu
109-R-649B	8%	8%	0%	0%	~ 5%, 87%‡	~5%	2%†, 59% ‡	4%	2.9	9%	9%	11%	16%	1.43	0%	16%	6%		?n
Bt entomocidus																			
198-R-689A	9,990	2,200	10	14.7	19.8	31.8	22%†	54%	0.0041	1	0.3	65%	77%	3.72	30%	405	17%	5.1†	Ent
110-R-694C	2,170	24%	33.4	5.67	~5%	30%	19%†	38%	0.02	5.5	77%	10	82%		10%	26%	0%		Ent
9-R-609A	28%	8%	0%	145	~5%	~5%	21%†	4%	0.105	27%	46%	7%	37%	31.1	0%	36%		39.2†	?
Bt aizawai																			
52-R-584B	9,810	890	9.94	10.6	18.4	20.7	34%†	68%	0.0011	7	40%	5	10	1.72	107	57%	10%	1.0†	Aiz
112-R-650B	8,040	817	12.1	8.62	32.3	19	5%†	72%	0.002	1.2	55%	1.1	6.5	1.32	10%	151	3%		Aiz
128-R-668C	9,740	723	7.14	7.27	67.3	20.8	38%†	40%	0.0026	8	75%	87%	6.8	8.55	5%	174	2%	1.9†	Aiz
132-R-577B	4,660	552	37.4	25	12%	50.9	15%†	44%	0.012	10	32%	71%	81%		292	34%	4%	2.3†	Aiz
133-R-579A	21,600	1,420	15.3	7.08	15.3	12.6	0%†	8.62	0.0008	2.6	3	8	10		182	178	9%	1.3†	Aiz
134-R-579B	13,200	1,660	10.2	10.1	16.3	21.8	0%†	52%	0.0017	5	53%	78%	10	0.81	178	223	12%	1.9†	Aiz
135-R-580A	18,000	1,460	7.53	6.95	4.2	29.9	22%†	10.8	0.0022	6	67%	9	10	99.6	235	189	0%	1.1†	Aiz
137-R-580C	13,000	1,600	8.01	5.68	15.7	19.5	19%†	52%	0.0012	6.5	66%	9	10		85.8	216	14%	1.2†	Aiz
228-R-697A	7,940	658	8.25	6.64	48.7	45.1	2%†	39.7	0.0032	3	9%	76%	70%	3.73	30%	155	0%	20.4†	Aiz
248-R-600A	12,700	850	14.3	18.7	27.3	23.1	0%†	15.8	0.0024	1	74%	8	72%		166	166	2%	2.5†	Aiz
113-R-650C	5,590	994	19.9	19.9	94.3	3.55	18%†	44%	0.004	10	24%	44%	55%		0%	28%	4%	5.6†	Aiz
127-R-668B	4,750	950	20	23.8	16%	7%	32%†	72%	0.0055	78%	43%	10	66%	14.9	0%	347	10%	2.5†	Aiz
131-R-577A	5,660	1,020	33.7	25	52.1	49.7	24%†	36%	0.0074	58%	9%	44%	69%		170	167	5%	2.7†	Aiz
229-R-695C	18,300	1,790	6.97	4.87	26.5	76.6	13%†	60%	0.002	10	20%	65%	10		45%	143	21%	1.9†	Aiz
282-R-723B	20,700	2,840			7.7	18.8	0%†	64%	0.002	10	69%	3.8	5			197	0%	1.4†	Aiz
115-R-652B	5,310	1,110	13.5	8.68	61.3	37	10%†	44%	0.0015	53%	1	26%	28%		5%	347	10%	16.9†	Aiz
130-R-669B	2,780	4%	49.2	13.6	10%	4%	39%†	4%	0.0067	5	61%	85%	5	44%	0%	132	14%		Aiz
283-R-711C	13,100	2,550	9.23	2.07	30.9	11.6	0%†	58%	0.0042	10	0.9	61%	75%		205	168	18%	0.6†	Aiz
111-R-650A	7,290	44%	18.4	11.1	~5%	~5%	5%†	0%	0.0017	1.6	57%	12%	10	13.92	5%	185	2%	1.9†	Aiz

275-R-708B	8,110	969	12.5	1.89	32.5	38.5	2%†	38%	0.0036	3	40%	72%	69%		40%	236	12%	1.9†	Aiz
276-R-708C	2,230	287	103	16.1	~5%	~5%	2%†	8%	0.017	36%	19%	20%	18%		15%	54%	4%	8.4†	Aiz
249-R-600B	13,100	1,420	12	14.5	15.2	29.4	0%†	76%	0.0018	5.5	40%	67%	58%		50%	38%	4%	2.7†	Aiz
122-R-723A	20,700	15,200			8.2	67	18%†	68%	0.0027	25%	25%	35%	39%			39%	4%	9.8†	Aiz
68-R-633A	9,660	12%	28	9.73	~5%	~5%	4%†	8%	0.0053	54%	5	74%	10	22.6	238	42%	0%	2.3†	Aiz
114-R-652A	34%	32%	47.4	14.4	~5%	~5%	19%†	28%	>15	5%	28%	11%	0%	27.8	5%	22%	10%		Aiz
143-R-670A	5,750	34%	28.8	14.1	116	95.3	27%†	16%	0.0024	9	60%	78%	79%	61.4	5%	211	18%	1.0†	Aiz
272-R-707B	4,120	4%	91.3	4.04	~5%	30%	33%†	5%	0.025*	10	53%	44%	69%		5%	191	12%	1.4†	Aiz or ?
11-R-609C	53%	42%	83.1	4.49	~5%	~5%	10%†	52%	>15	1%	8%	6%	3%	1.52	5%	8%	6%		Aiz or ?
314-R-697B	52%	15%	99.4	4.67	~5%	~5%	36%†	84%	0.016	74%	10%	39%	57%	25.6	0%	64%	0%		Aiz or ?
227-R-695A	2%	7%	57.8	25.1	~5%	~5%	33%†	12%	0.85*	2%	1%	16%	5%		0%	16%	26%		Aiz or ?
312-R-696B	38%	3%	91.4	29.7	~5%	~5%	19%†	20%	0.095*	38%	2%	43%	34%	4.18	0%	20%	6%		Aiz or ?
67-R-635C	8%	8%	0%	40%	~5%	~5%	0%†	5%	0.14*	0%	2%	4%	3%	67%	0%	8%	8%		Aiz or ?
126-R-668A	10%	10%	7%	224	~5%	~5%	27%†	~5%	>15	40%	9%	1%	3%		0%	4%	12%		Aiz or ?
142-R-669C	2%	6%	0%	0%	~5%	~5%	2%†	~5%	>15	2%	2%	5%	2%	50%	0%	8%	16%		Aiz or ?
Bt morrisoni																			
116-R-652C	2,320	36%	24	216	30%	0%	31.5	36%	1.1	14%	42%	20%	6%	2.38	5%	79%	34%	100%, 5.4†	?
12-R-612A	2,500	48%	40.1	27.2	~5%	~5%	52.7	40%	0.061	12%	65%	73%	7	13.78	0%	416	27%	100%, 1.6†	?
Bt tolworthi																			
121-R-666C	3,740	3,230	365	171	4.2	44.2	202	60%	1.3	1.6	0.3	26%	2%	42%	50	162	21%	100%, 4.3†	Tol
125-R-610C	11,700	5,030	201	108				72%	0.88	0.8	0.08	7	8	0.5	157	62.1	190		Tol
125-R-727C	7,160	6,600			2.4	30.6	108	48%	11.3*	0.8	0.6	13%	23%			199		100%, 9.0†	Tol
285-R-712B	21,500	10,000	10.1	4.54	8.2	19.5	28%†	8.02	0.095	78%	1	6%	10%		20%	32%	0%	95.2†	Tol
301-R-715B	18,800	11,100	6	6.49	7.9	10.6	79.1	68%	0.084	2.8	0.4	47%	17		2.66	77.5	35%	100%, 0.8†	Tol
13-R-612B	1,350	758	50.6	209				32%	2.5*	2.3	0.6	31%	22%		10%	60%	12%		Tol.
124-R-667C	4,170	6,970	75.4	85.8	3.6	18.1	29%†		0.36	1.4	0.08	42%	6%	1.2	35%	14%	10%		Tol.
Bt darmstadiensis																			
146-R-672C	9,390	752	0%	268	104.1	36.1	33.2	0%	0.46	62%	60%	10	5	98%	50%	81.8	67%	100%, 0.7†	?
147-R-673A	4,250	601	0%	193	123	41.6	64.9	4%	0.41	40%	47%	62%	83%	104	316	80.2	13%	100%, 1.0†	?
200-R-689C	2,750	525	0%	382			58.5	16%	0.11	2%	14%	39%	44%	47	15%	15%	13%	100%, 1.1†	?
199-R-693A	13,400	1,920	65.2	85	81	28.8	47.5	0%	0.062	10%	64%	85%	89%	95.7	244	51.1	578	100%, 0.3†	?
Bt toumanoffi																			
201-R-690A	30%	22%	20.4	19.4	4%	0%	24%†	40%	0.13	3%	2%	52%	5%	26.7	0%	18%	6%		Gal and Den

APPENDIX 2.3 *(continued)*

Bt *subspecies/* Isolate code (HD-)	*Trichoplusia ni*	*Heliothis virescens*	*Hyphantria cunea*	*Bombyx mori*	*Ephestia cautella*	*Plodia interpunctella*	*Haematobia irrtans*	*Ostrinia nubilalis*	*Galleria mellonella*	*Aedes aegypti*	*Aedes triseriatus*	*Culx pipiens*	*Culex quinquefasiatus*	*Pieris brassicae*	*Spodoptera litura*	*Spodoptera exigua*	*Agrotis ipsilon*	*Bovicola crassipes*	XLT type**
Bt dendrolimus																			
7-R-610B	3,620	36%	9.23	169	~5%	~5%	45%†, 5%	40%	>15	18%	18%	9%	7%	9.48	0%	12%	2%		Den
34-R-625A	56%	28%	24.3	183	~5%	~5%	14%†	14.3	>15	2%	2%	5%	6%	6.08	0%	8%	2%		Den
35-R-625B	48%	14%	29.9	275	~5%	~5%	18%†	19	>15	5%	2%	6%	10%	3.94	0%	8%	0%		Den
35-R-726A	1,600	6%			37.9	~5%	127†, 77%	44%	>15	8%	6%	33%	14%	7.58		28%		0%	Den
106-R-648A	1,930	486	20	175	~5%	~5%	63.8	50%	1.2	6%	12%	3%	4%	0.92	5%	38%	6%	73%	Den
226-R-693C	2,810	*395	4.24	76.9	~5%	~5%	0%†	52%	>15	26%	18%	12%	6%	3.28	5%	12%	4%		Den
37-R-631B	940	38%	108	4.64	5%	0%	7%†	24%	>15	1%	2%	18%	9%	25.3	5%	12%	0%		Den
48-R-604C	26%	42%	160	15				8%	1.6	2%	4%	18%	16%		5%	8%	4%		Den

† Preparation that was not autoclaved, assayed with *Haematobia* or *Bovicola*; unmarked results with this species were with autoclaved preparation.
‡ At the 500 mg/ level
* Estimated
** XLT = crystal type (see Dulmage *et al.* 1981). Endotoxin serotype obtained by using the endotoxins as antigens (abbreviations mxd = mixed; ? = unknown; atyp = atypical).

APPENDIX 3. A CASE STUDY: TOXICITY OF DIPEL (*BT KURSTAKI* PRODUCED BY ABBOTT LABORATORIES)

APPENDIX 3.1 INVERTEBRATES REPORTED AS SUSCEPTIBLE TO DIPEL

Species were recorded as susceptible where there was a reported difference between toxicity in the treatments and the control in laboratory assays or an effect against field populations was reported. Note that some species have been recorded as both susceptible and non-susceptible (Appendix 3.2). See Section 5.5.1 for discussion.

Order/family	Genus and species	Field (F) or Laboratory (L)	Sample references
Acari			
Argasidae	*Argas persicus*	L/F	Abdel Megeed *et al.* 1997; Hassanain *et al.* 1997b
Ixodidae	*Hyalomma dromedarii*	L	Hassanain *et al.* 1997b
Mesostigmata	*Boophilus annulatus (calcaratus)*	F	Abdel Megeed *et al.* 1997
	Panonychus ulmi	F	Dirimanov *et al.* 1980
Coleoptera			
Chrysomelidae	*Leptinotarsa decemlineata*	L/F	Kiselek & Zayats 1976
Coccinellidae	*Henosepilachna (Epilachna) vigintioctopunctata*	F	Sekar & Baskaran 1976; Baskaran & Kumar 1980
Curculionidae	*Sitophilus granarius*	F	Rashid & Khan 1995
Tenebrionidae	*Tribolium castaneum*	L/F	Al Hafidh 1985; Arthur & Brown 1994
Dictyoptera			
Blattellidae	*Blattella germanica*	L	Sandhu & Varma 1980
Diptera			
Agromyzidae	*Liriomyza sativae*	F	Johnson *et al.* 1980, 1984
	Liriomyza trifolii	F	Jones *et al.* 1986
	Melanagromyza hibisci	F	Tomar 1998a & b
Anthomyiidae	*Delia (Hylemya) brassicae*	F	Obadofin & Finlayson 1977
Culicidae	*Culex molestus*	L	Sokolova & Ganushkina 1982
	Culex quinquefasciatus	L	Ignoffo *et al.* 1981
Hemiptera			
Aleyrodidae	*Bemisia tabaci*	L	Radwan *et al.* 1984; Al Shayji *et al.* 1998
Aphididae	*Aphis gossypii*	F	Salama & Zaki 1984
	Brevicoryne brassicae	F	Kennedy & Oatman 1976; Chalfant 1978
	Myzus persicae	F	Kennedy & Oatman 1976; Chalfant 1978; Vasudevan & Baskaran 1979
Cicadellidae	*Amrasca biguttula*	F	Baskaran & Kumar 1980
Lygaeidae	*Dimorphopterus gibbus*	F	Pandey *et al.* 1977
Miridae	*Adelphocoris lineolatus*	F	Yunusov 1974
Pseudococcidae	*Pseudococcus longispinus*	L	Wysoki *et al.* 1975
Hymenoptera			
Formicidae	*Pheidole megacephala*	L	Castineiras & Calderon 1985
Pergidae	*Perga affinis affinis*	F	Neumann & Collett 1997a,b
Tenthredinidae	*Athalia lugens proxima*	L	Saxena 1978
	Pristiphora abietina	F	Konig 1975
Vespidae	*Vespula* sp.	L	Gambino 1988

APPENDIX 3.1 (*continued*)

Order/family	Genus and species	Field (F) or Laboratory (L)	Sample references
Isoptera			
Rhinotermitidae	*Psammotermes hybostoma*	L	Farghal *et al.* 1987
Termitidae	*Amitermes desertorum*	L	Farghal *et al.* 1987
Lepidoptera			
Arctiidae	*Datana integerrima*	F	Polles 1974; Tedders & Ellis 1977; Farris & Appleby 1980
	Hyphantria cunea	F	Polles 1974
	Pericallia ricini	L	Mathur *et al.* 1994
	Utetheisa pulchella	L	Gangwar *et al.* 1980
Brassolidae	*Brassolis astyra astyra*	L	Berti Filho & Gallo 1977
Cosmopterigidae	*Batrachedra amydraula*	F	Venezian & Blumberg 1982; Blumberg *et al.* 1977
Cossidae	*Zeuzera pyrina*	F	Plaut 1976
Dioptida	*Phryganidia californica*	L/F	Milstead *et al.* 1980
Gelechiidae	*Gnorimoschema (Keiferia) lycopersicella*	F	Wolfenbarger & Poe 1973; Poe & Everett 1974; Lindquist 1975
	Pectinophora gossypiella	L/F	Butter *et al.* 1995
	Phthorimaea operculella	L/F	El Sayed *et al.* 1979; Gravena *et al.* 1980
	Sitotroga cerealella	L/F	McGaughey 1976; Salama & Abdel 1992
Geometridae	*Alsophila pometaria*	F	Appleby *et al.* 1975; Frye *et al.* 1977; Frye *et al.* 1983
	Apocheima pilosaria	F	Lecheva 1983
	Boarmia (Ascotis) selenaria	L/F	Izhar *et al.* 1979; Cohen *et al.* 1983; Wysoki *et al.* 1986
	Ennomos subsignarius	F	Dunbar & Kaya 1972
	Erannis (Hibernia) defoliaria	F	Glowacka Pilot 1974; Glowacka Pilot *et al.* 1974
	Himera pennaria	F	Donaubauer 1976
	Itame sulphurea	F	Sandler & Mason 1997
	Lambdina fiscellaria fiscellaria	F	West *et al.* 1989
	Lycia (Biston) hirtaria	L/F	Lecheva & Kuzmanova 1980; Dirimanov & Lecheva 1980; Lecheva 1983
	Operophthera sp.	F	Glowacka Pilot 1974
	Operophthera bruceata	F	Svestka 1974; Glowacka Pilot *et al.* 1974; Tonks *et al.* 1978
	Operophthera fagata	F	Vankova 1973
	Mnesampela privata	L/F	Neumann & Collett 1997b
	Paleacrita vernata	F	Frye *et al.* 1977; Frye *et al.* 1983
	Peribatodes rhomboidaria	F	Filip & Blaise 1998
Glyphipterigyidae	*Eutromula (Simaethis) pariana*	F	Dronka *et al.* 1976
Heliconidae	*Dione juno juno*	L/F	Villani *et al.* 1980
Hesperiidae	*Thymelicus lineola*	F	Thompson 1977; McNeil *et al.* 1977; Letendre & McNeil 1979
Lasiocampidae	*Dendrolimus pini*	L/F	Glowacka Pilot 1974
	Eriogaster lanestris	F	Vankova 1973
	Malacosoma disstria	F	Abrahamson & Harper 1973; Harper & Abrahamson 1979; Johnson & Morris 1981
	Malacosoma neustria	F	Mihalache *et al.* 1972; Vankova 1973; Mihalache *et al.* 1974
Lymantriidae	*Euproctis chrysorrhoea*	L/F	Vankova 1973; Ruelle *et al.* 1977; Canivet *et al.* 1978
	Euproctis fraterna	L	Gangwar *et al.* 1980
	Euproctis phaeorrhoea	F	Kniefl 1977
	Hemerocampa leucostigma	L	Morris 1973a
	Hypogymna morio	L	Dobrivojevic & Injac 1975
	Leucoma (Stilpnotia) salicis	F	Petcu & Nastase 1974; Donaubauer 1976
	Leucoma wiltshirei	F	Abai 1981

APPENDIX 3.1 (*continued*)

Order/family	Genus and species	Field (F) or Laboratory (L)	Sample references
	Lymantria dispar	F	Mihalache *et al.* 1974
	Lymantria monacha	L/F	Bejer Petersen 1974; Glowacka Pilot 1974
	Orgyia antiqua	F	Lipa *et al.* 1977; Şvestka & Vankova 1978; Niemczyk 1980
	Orgyia gonostigma	F	Trenchev & Pavlov 1982; Ivanova 1984
	Orgyia pseudosugata	L/F	Morris 1973a; Maksymiuk & Orchard 1974; Stelzer *et al.* 1975
Noctuidae	*Achaea janata*	L/F	Pal 1977; Srivastava & Ramakrishnan 1980
	Agrotis ipsilon (ypsilon)	L/F	Salama *et al.* 1990a, b
	Agrotis segetum	L/F	Langenbruch 1977
	Alabama argillacea	L/F	Gravena *et al.* 1983; Habib & De Andrade 1984
	Anomis leona	F	Taylor 1974
	Anticarsia gemmatalis	L/F	Da Silva & Heinrichs 1975; Bertoldo & Morosini 1982
	Autoba olivacea	F	Baskaran & Kumar 1980
	Busseola fusca	F	Medvecky & Zalom 1992
	Cerapteryx graminis	L	Vlug *et al.* 1988
	Chrysodeixis chalcites (eriosoma)	F	Martin & Workman 1986; Wysoki 1989; Broza & Sneh 1994
	Earias biplaga	L/F	Taylor 1974
	Earias huegeli	F	Wilson 1981
	Earias insulana	L/F	Taylor 1974; Moawad *et al.* 1982
	Earias vittella	F	Krishnaiah *et al.* 1981
	Euxoa messoria	L	Cheng 1973
	Heliothis (Helicoverpa) armigera	F	Kabour & Sane 1972; Taylor 1974; El Sayed *et al.* 1979
	Heliothis punctigera	F	Cooper 1984; Teakle *et al.* 1992
	Heliothis (Protoschinia) scutosa (nervosa)	F	Mikolajewicz *et al.* 1992
	Heliothis (Helicoverpa) virescens	F	Mistric & Smith 1973a; Johnson 1974
	Heliothis (Helicoverpa) zea	L/F	Janes 1973; Creighton *et al.* 1973
	Hellula rogatalis	F	Waites *et al.* 1978
	Lacanobia oleracea	L/F	Hussey *et al.* 1976; Burges & Jarrett 1980
	Mamestra (Barathra) brassicae	F	Vankova 1973
	Mamestra (Barathra) configurata	L/F	Morris 1986
	Mamestra (Lacanobia) oleracea	F	Chan 1973; Jarrett & Burges 1982a; Langenbruch 1984b
	Peridroma saucia	F	Lindquist 1977
	Spodoptera exigua	L	Moar & Trumble 1987b
	Spodoptera frugiperda	F	Janes 1973; Law & Mills 1980; Sarmiento & Razuri 1978
	Spodoptera latifascia	L	Habib 1986
	Spodoptera litura	L/F	Ramaprasad *et al.* 1982; Yang *et al.* 1985
	Trichoplusia ni	L/F	Jaques 1973
	Xylena nupera	F	Sandler & Mason 1997
Notodontidae	*Botyoides asialis*	L	Kalia & Joshi 1996
	Clostera cupreata	L	Kalia & Joshi 1996
	Drymonia chaonia (ruficornis)	F	Pirvescu 1973; Mihalache *et al.* 1974
	Eustema sericea	L	Porto Santos *et al.* 1995
	Ichthyura anastomosis	L/F	Chaudhry & Shah 1977; Shah *et al.* 1979
	Phalantha phalantha	L	Kalia & Joshi 1996
	Schizura concinna	F	Pinnock *et al.* 1974; Band *et al.* 1976
Nymphalidae	*Vanessa (Cynthia) cardui*	F	Miranpuri *et al.* 1993

APPENDIX 3.1 (*continued*)

Order/family	Genus and species	Field (F) or Laboratory (L)	Sample references
Oecophoridae	*Depressaria nervosa (daucella)*	F	Mikolajewicz *et al.* 1992; Cate 1997
Olethreutidae	*Ecdytolopha aurantiana*	F	Scarpellini & Santos dos 1997
Phycitidae	*Homeosoma electellum*	L/F	Rogers *et al.* 1984
Pieridae	*Aporia crataegi*	F	Niccoli & Pelagatti 1986
	Artogeia (Pieris) rapae	F	Kirby 1978; Kouskolekas & Harper 1973; Waites *et al.* 1978
	Pieris brassicae	F	Langenbruch 1984b
Psychidae	*Oiketicus kirbyi*	L	Almela Pons *et al.* 1972
	Thyridopteryx ephemeraeformis	F	Bishop *et al.* 1973
Pyralidae	*Aganais (Hypsa) ficus*	L	Gangwar *et al.* 1980
	Antigastra catalaunalis	L	Selvanarayanan & Baskaran 1996
	Cadra cautella	L	Rahman & Faruki 1993; Faruki & Khan 1996
	Chilo agamemnon	L	Hafez *et al.* 1998a
	Chilo suppressalis	L	Rombach *et al.* 1989
	Cnaphalocrocis medinalis	L/F	Srivastava & Nayak 1978; Aguda *et al.* 1988
	Corcyra cephalonica	L	El Moursy *et al.* 1996
	Crocidolomia binotalis	F	Krishnaiah *et al.* 1981
	Cryptoblabes gnidiella	F	Wysoki *et al.* 1975; Sternlicht 1979; Wysoki *et al.* 1986
	Diaphania (Palpita) hyalinata	L/F	Delplanque & Gruner 1975
	Diaphania nitidalis	F	Dougherty & Schuster 1985
	Diatraea grandiosella	F	Nolting & Poston 1982
	Diatraea saccharalis	F	Al Badry *et al.* 1972; Sarmiento & Razuri 1978
	Dioryctria amatella	L/F	McLeod *et al.* 1982; McLeod *et al.* 1984
	Ephestia (Cadra) cautella	L/F	Nwanze *et al.* 1975
	Ephestia (Anagasta) kuehniella	L	Samples & Buettner 1983; Ignatowicz 1997
	Evergestis rimosalis	F	Waites *et al.* 1978
	Herpetogramma (Psara) bipunctalis	F	Delplanque & Gruner 1975
	Herpetogramma phaeopteralis	F	Reinert 1974
	Hymenia recurvalis	L/F	Delplanque & Gruner 1975; Gangwar *et al.* 1980
	Leucinodes orbonalis	F	Sekar & Baskaran 1976; Baskaran & Kumar 1980
	Loxostege stricticalis	F	Kornilov *et al.* 1982
	Macalla thyrsisalis	L/F	Howard 1990
	Ostrinia nubilalis	L/F	Lynch *et al.* 1977; Gerginov 1978
	Paramyelois (Amyelois) transitella	F	Pinnock & Milstead 1972; Kellen *et al.* 1977
	Plodia interpunctella	L/F	McGaughey *et al.* 1980; Nwanze *et al.* 1975; Kinsinger & McGaughey 1976
	Polytela gloriosae	L	Gangwar *et al.* 1980
	Sylepta (Syllepte) balteata	F	Kalia *et al.* 1997
	Sylepta (Syllepte) derogata	F	Taylor 1974
Sesiidae	*Synanthedon myopaeformis*	F	Balazs *et al.* 1996
Sphingidae	*Erinnyis ello*	F	Cruz 1977; de Abreu 1982
Stathmopodidae	*Stathmopoda* spp.	F	McKenna *et al.* 1995
Thaumetopoeidae	*Thaumetopoea pityocampa*	F	Videnova *et al.* 1972
	Thaumetopoea processionea	F	Niccoli & Pelagatti 1986
Tineidae	*Opogona sacchari*	F	Heppner *et al.* 1987
Tortricidae	*Acleris variana*	F	West & Carter 1992
	Adoxophyes orana	F	Kreidl 1974; Zivanovic & Stamenkovic 1976

APPENDIX 3.1 (*continued*)

Order/family	Genus and species	Field (F) or Laboratory (L)	Sample references
	Agriopis (Erannis) marginaria	L	Kuzmanova & Lecheva 1984
	Archips argyrospilus	F	Vakenti *et al.* 1984
	Archips (Cacoecia) rosanus	F	Vankova 1973; Niemczyk & Dronka 1976
	Archips xylosteana	F	Mihalache *et al.* 1972; Niemczyk & Dronka 1976; Niemczyk 1980
	Arotrophora arcuatalis	F	Rohl & Woods 1994
	Autographa confusa	F	Mikolajewicz *et al.* 1992
	Autographa nigrisigna	L	Saxena & Ahmad 1998
	Cacoecimorpha pronubana	F	Scopes & Biggerstaff 1974; Burges & Jarrett 1978; Burges 2000
	Choristoneura conflictana	F	Holsten & Hard 1985
	Choristoneura fumiferana	L/F	Tripp 1972; Morris 1973a; Morris 1977a
	Choristoneura occidentalis	F	Maksymiuk & Orchard 1974; Barry & Ekblad 1978; Kaya & Reardon 1982
	Choristoneura rosaceana	L	Li & Fitzpatrick 1997
	Ctenopseustis obliquana	F	Wearing & Thomas 1978
	Cydia (Grapholita) delineana	F	Peteanu 1980
	Cydia (Grapholita, Laspeyresia) molesta	F	Iacob *et al.* 1981
	Cydia (Laspeyresia) pomonella	F	Niemczyk *et al.* 1976; Balevski & Ivanov 1979; Dirimanov *et al.* 1980
	Desmia funeralis	F	AliNiazee & Jensen 1973
	Epiphyas postvittana	L/F	Wearing & Thomas 1978; Harris *et al.* 1997
	Eupoecilia (Clysia) ambiguella	L/F	Schmid and Antonin 1977
	Hedya (Argyroploce) nubiferana (variegana)	F	Niemczyk *et al.* 1976; Niemczyk & Dronka 1976
	Lobesia botrana	F	Buess & Bassand 1976; Mitkov & Raicheva 1983;
	Pammene fasciana	F	Tsankov & Ovcharov 1986
	Pandemis heparana	F	Dirimanov *et al.* 1980
	Planotortrix excessana	F	Wearing & Thomas 1978
	Rhyacionia frustrana	F	Menendez *et al.* 1986
	Spilonota ocellana	F	Niemczyk & Dronka 1976; Niemczyk *et al.* 1976; Niemczyk 1980
	Tortrix viridana	F	Svestka 1974; Svestka 1977
	Zeiraphera diniana	L/F	Schmid 1975; Svestka & Vankova 1984; Kudler 1984
Yponomeutidae	*Argyresthia thuiella*	F	Langenbruch 1984a
	Plutella xylostella (maculipennis)	L/F	Cadapan & Gabriel 1972
	Prays oleae	F	Yamvrias 1972; Yamvrias & Young 1977
	Swammerdamia pyrella	F	Niemczyk 1980
	Yponomeuta(padellus) malinellus	F	Vankova 1973; Niemczyk 1980
	Yponomeuta euonymellus	L/F	Hamed 1978
Zygaenidae	*Harrisina brillians*	F	Pinnock *et al.* 1973; Stern *et al.* 1980; Stern *et al.* 1983
	Theresimima (Ino) ampelophaga	F	Kharizanov & Kharizanov 1983
Nematoda			
Meloidogynidae	*Meloidogyne javanica*	GH[1]	Osman *et al.* 1988
	Tylenchulus semipenetrans	GH	Osman *et al.* 1988

APPENDIX 3.1 (*continued*)

Order/family	Genus and species	Field (F) or Laboratory (L)	Sample references
Phthiraptera			
Menoponidae	*Eomenacanthus (Menacanthus) stramineus*	L	Lonc *et al.* 1986
	Menopon gallinae	L	Gingrich *et al.* 1974; Lonc & Lachowicz 1987
Ixodidae	*Bovicola (Damalinia) caprae*	F	Singh *et al.* 1992
Linognathidae	*Linognathus stenopsis*	F	Singh *et al.* 1992
Trichodectidae	*Heterodoxus spiniger*	F	Singh *et al.* 1992

[1] Greenhouse trials

APPENDIX 3.2 INVERTEBRATES REPORTED AS NON-SUSCEPTIBLE TO DIPEL

Species were recorded as non-susceptible where there was no reported difference between toxicity in the treatments and the control in laboratory assays or no effect against field populations reported. Note that some species have been recorded as both susceptible (Appendix 3.1) and non-susceptible. See Section 5.5.1 for discussion.

Order/family	Genus and species	Field (F) or Laboratory (L)	Sample references
Acari			
Phytoseiidae	*Amblyseius potentillae*	L	Hassan *et al.* 1983
Annelida			
Lumbricidae	*Dendrobaena octaedra*	L	Addison & Holmes 1996
	Lumbricus terrestris	F	Benz & Altwegg 1975
Coleoptera			
Carabidae	*Bembidion lampros*	F	Finlayson 1979 (slight mortality in the lab- Obadofin & Finlayson 1977)
Coccinellidae	*Adalia bipunctata*	L	Olszak 1982
	Ceratomegilla (Coleomegilla) maculata	F	Johnson 1974
	Coccinella septempunctata	L/F	Kiselek 1975; Baicu & Hussein 1984; Babrikova & Lecheva 1986
	Coccinella undecimpunctata	F	Salama & Zaki 1984
	Coccinella sp.	F	Mohamed 1993
	Hippodamia convergens	L/F	Johnson 1974; Sims 1997
	Scymnus interruptus	F	Salama & Zaki 1984
	Scymnus syriacus	F	Salama & Zaki 1984
	Scymnus sp.	F	Mohamed 1993
Curculionidae	*Sitophilus oryzae*	F	Kramer *et al.* 1985
Staphylinidae	*Paederus alfierii*	F	Salama & Zaki 1984
Tenebrionidae	*Tribolium confusum*	F	Kramer *et al.* 1985
Collembola	*Folsomia candida*	L/F	Addison & Holmes 1995; Sims 1997
Diptera			
Chironomidae	*Natarsia* sp.	F	Kreutzweiser *et al.* 1994
	Tanytarsini sp.	F	Kreutzweiser *et al.* 1994
Culicidae	*Culex pipiens*	L	Farghal *et al.* 1987; Sims 1997; Kim *et al.* 1998a
	Theobaldia (Culiseta) longiareolata	L	Farghal *et al.* 1987
Muscidae	*Musca domestica*	L	Vankova 1981; Ogiwara *et al.* 1992; Sims 1997
Simuliidae	*Simulium* sp.	F	Kreutzweiser *et al.* 1994
Tachinidae	*Pales pavida*	L	Hassan *et al.* 1983
Ephemeroptera			
Baetidae	*Baetis* sp.	F	Kreutzweiser *et al.* 1994
Ephemerellidae	*Ephemerella* sp.	F	Kreutzweiser *et al.* 1992
Heptageniidae	*Epeorus vitrea*	L	Kreutzweiser *et al.* 1992; 1994
	Heptagenia sp.	F	Kreutzweiser *et al.* 1992; 1994
	Rhithrogena sp.	F	Kreutzweiser *et al.* 1992
	Stenonema vicarium	F	Kreutzweiser *et al.* 1992
Isonychiidae	*Isonychia* sp.	F	Kreutzweiser *et al.* 1992
Leptophlebiidae	*Eurylophella* spp.	F	Kreutzweiser *et al.* 1994
	Paraleptophlebia ontario	F	Kreutzweiser *et al.* 1992

APPENDIX 3.2 (*continued*)

Order/family	Genus and species	Field (F) or Laboratory (L)	Sample references
Hemiptera			
Anthocoridae	*Orius albidipennis*	F	Salama & Zaki 1984
	Orius laevigatus	F	Salama & Zaki 1984
	Orius spp.	F	Mohamed 1993
Berytidae	*Jalysus spinosus*	F	Elsey 1973
Pentatomidae	*Murgantia histrionica*	F	Chalfant 1978
Hymenoptera			
Aphelinidae	*Encarsia* sp.	L	Hassan *et al.* 1983; Kadyrov *et al.* 1994
Braconidae	*Apanteles congregatus*	L	Urias Lopez *et al.* 1987
	Apanteles glomeratus	L	Muck *et al.* 1981
	Apanteles (Cotesia) melanoscelus	L/F	Weseloh & Andreadis 1982; Webb *et al.* 1989
	Apanteles plutellae	F	Krishnaiah *et al.* 1981; Lim *et al.* 1986
	Cardiochiles nigriceps	L	Dunbar & Johnson 1975
	Microplitis (Glabromicroplitis) croceipes	L	Blumberg *et al.* 1997b
	Telenomus remus	F	Chari *et al.* 1996
Encyrtidae	*Anagyrus fusciventris*	L	Wysoki *et al.* 1975
	Archopoides (Hungariella) peregrina	L	Wysoki *et al.* 1975
Eulophidae	*Chrysonotomyia punctiventris*	F	Johnson *et al.* 1980; 1984
	Diglyphus begini	F	Johnson *et al.* 1980; 1984
Eupelmidae	*Anastatus bifasciatus*	F	Tsankov & Mirchev 1983
Ichneumonidae	*Coccygomimus turionellae*	L	Hassan *et al.* 1983; Muck *et al.* 1981
	Diadegma insulare	F	Cordero & Cave 1990
	Diadegma sp.	L/F	Jimenez & Fernandez 1985
	Phygadeuon trichops	L	Naton 1978; Hassan *et al.* 1983
	Thyraeella collaris	L	Hamilton & Attia 1977
Scelionidae	*Telenomus sphingis*	L	Urias Lopez *et al.* 1987
Trichogrammatidae	*Ooencyrtus pityocampae*	F	Tsankov & Mirchev 1983
	Tetrastichus (Baryscapus) servadeii	F	Tsankov & Mirchev 1983
	Trichogramma cacoeciae	L	Hassan & Krieg 1975; Hassan 1983
	Trichogramma carverae	F	Rohl & Woods 1994
	Trichogramma embryophagum	F	Tsankov & Mirchev 1983
	Trichogramma pallidum (Cacoeciae pallida)	L	Kiselek 1975
	Trichogramma platneri	L	S. Wellinga in Wysoki 1989
	Trichogramma pretiosum	F	Bull *et al.* 1979; Oatman *et al.* 1983; Campbell *et al.* 1991
	Trichogrammatoidea bactrae	F	Rohl & Woods 1994
Lepidoptera			
Geometridae	*Hyposidra talaca*	L	Widiastuti *et al.* 1996
Noctuidae	*Adisura atkinsoni*	F	Krishnaiah *et al.* 1978
	Heliothis armigera	F	Krishnaiah *et al.* 1978
	Spodoptera frugiperda	L/F	Waquil *et al.* 1982
Pyralidae	*Leucinodes orbonalis*	F	Krishnaiah *et al.* 1981
	Ostrinia furnacalis	F	Schreiner & Nafus 1988
Tortricidae	*Archips argyrospilus*	F	Madsen *et al.* 1977
	Archips rosanus	F	Madsen *et al.* 1977
	Cydia (Laspeyresia) molesta	F	Zivanovic & Stamenkovic 1976
	Cydia pomonella	F	Zivanovic & Stamenkovic 1976

APPENDIX 3.2 (*continued*)

Order/family	Genus and species	Field (F) or Laboratory (L)	Sample references
Odonata			
Aeshnidae	*Boyeria grafiana*	F	Kreutzweiser *et al.* 1992
Gomphidae	*Ophiogomphus carolus*	F	Kreutzweiser *et al.* 1992
Phthiraptera			
Boopiidae	*Hyalomma dromedarii*	F	Singh *et al.* 1992
Plecoptera			
Perlidae	*Acroneuria abnormis*	F	Kreutzweiser *et al.* 1992
Perlodidae	*Isogenoides* sp.	F	Kreutzweiser *et al.* 1992
Plecoptera	*Taeniopteryx nivalis*	F	Kreutzweiser *et al.* 1992; 1994
Pteronarcidae	*Pteronarcys* sp.	F	Kreutzweiser *et al.* 1992
Trichoptera			
Hydropsychidae	*Hydropsyche* sp.	F	Kreutzweiser *et al.* 1992
Lepidostomatidae	*Lepidostoma* spp.	F	Kreutzweiser *et al.* 1992
Limnephilidae	*Hesperophylax designatus*	F	Kreutzweiser *et al.* 1992
	Pycnopsyche guttifer	L/F	Kreutzweiser *et al.* 1992; 1994
Philopotamidae	*Dolophilodes distinctus*	F	Kreutzweiser *et al.* 1994
? ("aquatic insect")	*Hydatophylax argus*	L	Kreutzweiser & Capell 1996

APPENDIX 4. SUMMARY OF THE EFFECT OF *BT* ON PREDATORS OF INSECTS AND PARASITOIDS

APPENDIX 4.1 SUMMARY OF REPORTS ON THE EFFECT OF *BT* ON PREDATORS OF INSECTS

Order/Family	Predator	Target pest	*Bt* subspecies or product	Effect on predator	Reference
Acarina					
Anystidae	*Anystis* sp.	Pine blast scale	*galleriae*	*Bt* transmitted by predators to prey	Cheng *et al.* 1983
Phytoseiidae	Predatory mites	Tortricids (mainly *Lobesia botrana*)	*aizawai* and *kurstaki* (Biocillis)	Harmless	Fougeroux & Lacroze 1996
	Metaseiulus occidentalis	*Tetranychus urticae*	*tenebrionis*	More toxic to predator females than host	Chapman & Hoy 1991
	Typhlodromus exhilaratus		Unspecified	Population increased	Liguori 1988
	Phytoseiulus persimilis	*Tetranychus urticae*	*thuringiensis* (Bitoxibacillin)	Highly deleterious, reduced fertility and (temporary) sterility	Petrova & Khrameeva 1989
	Typhlodromus pyri	*Eupoecilia ambiguella*	*kurstaki* (Thuricide)	Least damaging to mite of several insecticides	Haas 1987
	Typhlodromus pyri	Unspecified	Unspecified	Harmless	Englert & Kettner 1983
	Typhlodromus pyri	Spider mite (tetranychid)	Unspecified	No effect	Niemczyk 1997
Coleoptera					
	Predacious beetle larvae	Mosquitoes	*israelensis*	No effect	Mulligan & Schaefer 1981
Cantharidae					
	Chauliognathus lugubris	*Chrysophtharta bimaculata*	*tenebrionis* (Novodor)	No toxicity	Beveridge & Elek 1999
	Chauliognathus lugubris	*Chrysophtharta bimaculata*	*tenebrionis*	Affected survival	Greener & Candy 1994
Carabidae	*Agonum punctiforme*	*Cydia pomonella*	*kurstaki*	No obvious effect	Riddick & Mills 1995
	Anisodactylus californicus	*Cydia pomonella*	*kurstaki*)	No obvious effect	Riddick & Mills 1995
	Bembidion lampros	*Delia brassicae* (*Hylemya brassicae*)	*kurstaki* (Dipel)	Slight mortality	Obadofin & Finlayson 1977
	Calathus ruficolis	*Delia brassicae*	*kurstaki* (Dipel)	No obvious effect	Riddick & Mills 1995
	Carabus hampei	*Leptinotarsa decemlineata*	*thuringiensis* (Bitoxibacillin)	No mortality	Koval 1985
	Chlaenius sp.	*Leptinotarsa decemlineata*	*thuringiensis* (Bitoxibacillin)	No obvious effect	Riddick & Mills 1995
	Harpalus pennsylvanicus	*Cydia pomonella*	*kurstaki*	No obvious effect	Riddick & Mills 1995
	Pterostichus spp.	*Cydia pomonella*	*kurstaki*	No obvious effect	Riddick & Mills 1995
Coccinellidae	*Adalia bipunctata*	*Cydia pomonella*	*kurstaki* (Dipel)	Low toxicity to eggs and neonates	Olszak 1982
	Adonia variegata	*Helicoverpa armigera*	*dendrolimus* (Dendrobacillin) and *galleriae* (Entobakterin)	No effect	Umarov Sh *et al.* 1975

APPENDIX 4.1 *(continued)*

Order/Family	Predator	Target pest	*Bt* subspecies or product	Effect on predator	Reference
Coccinellidae	*Ceratomegilla (Coleomegilla) maculata*	*Heliothis virescens*	Dipel, Biotrol XK and Thuricide HPC	Not harmful	Johnson 1974
	Coccinellidae Coccinellidae	*Helicoverpa zea* Lepidoptera	*kurstaki* *kurstaki*	No population effect No effect	Young *et al.* 1997 Jayanthi & Padmavathamma 1996a
	Coccinellidae	Lepidoptera	*thuringiensis* (Bitoxibacillin)	Little effect (<3% mortality)	Slabospitskii & Yakulov 1979
	Coccinellidae	*Helicoverpa armigera*	Thuringiensin (β-exotoxin)	Decreased slightly in 3-7 days following treatments but no significant variation	Zuo *et al.* 1994a
	two coccinellids	*Chrysophtharta bimaculata*	*tenebrionis*	No effect	Greener & Candy 1994
	Coccinella sp.	*Spodoptera exigua*	*kurstaki* (Dipel, SAN 415 I and Thuricide HP)	Least harmful of a number of insecticides	Krieg *et al.* 1984
	Coccinella septempunctata	*Helicoverpa armigera*	*dendrolimus* (Dendrobacillin) and *galleriae* (Entobakterin)	No effect	Umarov Sh *et al.* 1975
	Coccinella septempunctata	Aphids (Aphididae)	*tenebrionis*	Unaffected	Langenbruch 1992
	Coccinella septempunctata	*Helicoverpa armigera*	not specified	No effect	Manjula & Padmavathamma 1996
	Coccinella undecimpunctata	*Spodoptera littoralis*	*kurstaki* (Dipel)	Slightly affected population levels, due to host reduction?	Salama & Zaki 1984
	Coccinella undecimpunctata	*Helicoverpa armigera*	*dendrolimus* (Dendrobacillin) and *galleriae* (Entobakterin)	No effect	Umarov Sh *et al.* 1975
	Coccinella undecimpunctata	*Spodoptera littoralis*	*thuringinesis* (Bactospeine)	Harmless	El Husseini 1984
	Coleomegilla maculata lengi	*Leptinotarsa decemlineata*	*tenebrionis* (*san diego*) (M-One)	Level of predation on eggs reduced. Decreased food intake. Consumption of contaminated pollen had no lethal effects.	Giroux *et al.* 1994a
	Coleomegilla maculata lengi	*Leptinotarsa decemlineata*	*san diego* (M-One)	No mortality, egg attack lowered at 10x dosage	Giroux *et al.* 1994b
	Menochilus sexmaculatus	*Helicoverpa armigera*	Not specified	No effect	Manjula & Padmavathamma 1996
	Hippodamia convergens	*Heliothis virescens*	Dipel, Biotrol XK and Thuricide HPC	Not harmful	Johnson 1974
	Olla abdominalis (*O. a-nigrum*)	*Phthorimaea operculella*	Up to 10% of M-one (containing 5% *Bt san diego*)	No effect on longevity and oviposition rate. Detrimental effects on growth and oviposition of the adults	Zhou 1992
	Rodolia cardinalis	Unspecified	*kurstaki* (MVPR)	Appears to be relatively safe	Viggiani *et al.* 1998
	Scymnus sp.	*Spodoptera exigua*	*kurstaki* (Dipel, SAN 415 I and Thuricide HP)	Least harmful of a number of insecticides	

	Scymnus syriacus	*Spodoptera littoralis*	*kurstaki* (Dipel)	Slightly affected population levels, due to host reduction?	Salama & Zaki 1984
	Stethorus loxtoni	(Apple orchards)	Unspecified	Non-toxic	Walters 1976
	Stethorus nigripes	(Apple orchards)	Unspecified	Non-toxic	Walters 1976
	Stethorus vagans	(Apple orchards)	Unspecified	Non-toxic	Walters 1976
Staphylinidae	*Oligota fageli*	Unspecified	Thuringiensin (β-exotoxin)	15% mortality using residues of field dosages	Botha *et al.* 1994
	Oligota kashimirica benefica	*Panonychus ulmi*	Unspecified	No toxicity	Gyoutoku & Kastio 1990
	Oligota yasumatsui	*Panonychus ulmi*	Unspecified	No toxicity	Gyoutoku & Kastio 1990
	Paederus alfierii	*Spodoptera littoralis*	*kurstaki* (Dipel)	Slightly affected population levels, due to host reduction?	Salama & Zaki 1984
Diptera					
Cecidomyiidae	*Aphidoletes aphidimyza*	Aphids (Aphididae)	Unspecified	Safe	Helyer 1991
	Cecidomyiids	Lepidoptera	*thuringiensis* (Bitoxibacillin)	Little effect (<3% mortality)	Slabospitskii & Yakulov 1979
	Oligotrophus sp. and others	Pine blast scale	*galleriae*	*Bt* transmitted by predators	Cheng *et al.* 1983
Syrphidae	Syrphid larvae	Lepidoptera	*thuringiensis* (Bitoxibacillin)	Little effect (<3% mortality)	Slabospitskii & Yakulov 1979
Hemiptera					
Anthocoridae	Anthocorids	*Helicoverpa armigera*	thuringiensin (β-exotoxin)	Decreased slightly in 3-7 days following treatments but no significant variation	Zuo *et al.* 1994a
	Elatophilus nipponensis	Pine blast scale	*galleriae*	*Bt* transmitted by predators	Cheng *et al.* 1983
	Orius albidipennis	*Agrotis ypsilon*	Unspecified Thuringiensin (β-exotoxin)	Affect on nymphal duration, rate of food consumption and egg production Some effect	Hafez *et al.* 1997a
	Orius albidipennis	*Spodoptera littoralis*	*kurstaki* (Dipel)	Slightly affected population levels, due to host reduction?	Salama and Zaki 1984
	Orius insidiosus	*Helicoverpa zea*	*kurstaki*	No population effect	Young *et al.* 1997
	Orius laevigatus	*Spodoptera littoralis*	*kurstaki* (Dipel)	Slightly affected population levels, due to host reduction?	Salama and Zaki 1984
	Orius majusculus (5th instar)	Unspecified	Unspecified	No appreciable effect	Angeli *et al.* 1998
	Orius niger	*Helicoverpa armigera*	*dendrolimus* (Dendrobacillin) and *galleriae* (Entobakterin)	No effect	Umarov Sh *et al.* 1975
	Orius spp.	*Spodoptera exigua*	*kurstaki* (Dipel, SAN 415 I and Thuricide HP)	Least harmful of a number of insecticides	Mohamed 1993
	Xylocoris flavipes	*Corcyra cephalonica*	*kurstaki* (Bactospeine)	Feeding on infected prey by nymphs, little effect, by adults. Significantly longer incubation period and reduced hatch rate of their eggs, followed by unusually high mortality of the ensuing nymphs.	El-Husseini *et al.* 1987

APPENDIX 4.1 (*continued*)

Order/Family	Predator	Target pest	*Bt* subspecies or product	Effect on predator	Reference
Lygaeidae	*Geocoris punctipes*	*Pseudoplusia includens*	*kurstaki*	Lowest contact toxicity of several insecticides	Boyd and Boethel 1998
	Geocoris spp.	*Helicoverpa zea*	*kurstaki*	No population effect	Young *et al.* 1997
Miridae	*Campylomma diversicornis*	*Helicoverpa armigera*	*dendrolimus* (Dendrobacillin) and *galleriae* (Entobakterin)	No effect	Umarov Sh *et al.* 1975
	Campylomma verbasci	*Helicoverpa armigera*	*dendrolimus* (Dendrobacillin) and *galleriae* (Entobakterin)	No effect	Umarov Sh *et al.* 1975
Nabidae	*Nabis capsiformis*	*Pseudoplusia includens*	*kurstaki*	Lowest contact toxicity of several insecticides	Boyd and Boethel 1998
	Nabis roseipennis	*Pseudoplusia includens*	*kurstaki*	Lowest contact toxicity of several insecticides	Boyd and Boethel 1998
Notonectidae	*Buenoa* sp.	Mosquitoes	*aizawai* (GM-10)	No effect	Ortegon-Martinez & Quiroz Martinez 1990
Pentatomidae	*Perillus bioculatus*	*Leptinotarsa decemlineata*	*tenebrionis* (*san diego*)	Caused very little mortality at any life stage tested	Hough-Goldstein and Keil 1991
Pentatomidae	*Picromerus bidens*	*Yponomeuta evonymellus*	*kurstaki*	No effect	Hamed 1979
	Podisus maculiventris	*Pseudoplusia includens*	*kurstaki*	Lowest contact toxicity of several insecticides	Boyd and Boethel 1998
	Podisus maculiventris	*Leptinotarsa decemlineata*	*thuringiensis* (Bitoxibacillin)	Harmless	Gusev *et al.* 1983
Reduviidae	*Rhinocoris fuscipes*	*Helicoverpa armigera*	Unspecified	No effect	Manjula and Padmavathamma 1996
Hymenoptera					
Chalcididea	Chalcids	*Acyrthosiphon pisum*	*galleriae* (Entobakterin)	No effect	Grechkanev and Maksimova 1980
Encyrtidae	*Leptomastix dactylopii*	Unspecified	*kurstaki* (MVPR)	Appears to be relatively safe	Viggiani *et al.* 1998
Formicidae	*Formica* spp.	Lepidoptera	Dipel and Turingin	No adverse effects on activity	Pascovici *et al.* 1978
	Solenopsis sp	*Perileucoptera coffeella*	Unspecified	Less effect than chemicals, reduced the ant populations by 12% after 1 day and 0% after 7 days	Gravena 1984
Mantodea					
Mantidae	*Tenodera sinensis*	*Trichoplusia ni*	*kurstaki*	No effect from feeding on dead larvae	Yousten 1973
Neuroptera					
Chrysopidae	*Chrysopa carnea* (*Chrysoperla carnea*)	*Helicoverpa armigera*	*dendrolimus* (Dendrobacillin) and *galleriae* (Entobakterin)	No effect	Umarov Sh *et al.* 1975
	Chrysopa carnea	Aphids (Aphididae)	*tenebrionis*	Unaffected	Langenbruch 1992
	Chrysopa carnea	*Spodoptera littoralis*	*kurstaki* (Dipel)	Slightly affected population levels, due to host reduction?	Salama and Zaki 1984

	Chrysopa carnea	(On cotton)	*kurstaki* (Lepidocide) and *dendrolimus* (Dendrobacillin)	Had no adverse effect on the numbers of adults and immature stages	Adylov *et al.* 1990
	Chrysopa carnea	(On cotton)	*thuringiensis* (Bitoxibacillin)	Reduced populations within 15 days but not subsequently	Adylov *et al.* 1990
	Chrysopa carnea	Unspecified	*galleriae* (Entobacterin), *thuringiensis* (Bactospeine) and *kurstaki* (Dipel)	Practically non-toxic to larvae (up to 30%), highly toxic to adults, up to 75% mortality.	Babrikova *et al.* 1982
	Chrysopa septempunctata, C. formosa, C. perla	Unspecified	*galleriae* (Entobacterin), *thuringiensis* (Bactospeine) and *kurstaki* (Dipel)	20% to 40% mortality of larvae, 60-90% mortality of adults.	Babrikova and Kuzmanova 1984
	Chrysopids	*Helicoverpa armigera*	Thuringiensin (β-exotoxin)	Decreased slightly in 3-7 days following treatments but no significant variation	Zuo *et al.* 1994a
	Chrysopids	Lepidoptera	*thuringiensis* (Bitoxibacillin)	Little effect (<3% mortality)	Slabospitskii and Yakulov 1979
Scorpiones					
Buthidae	*Buthus occitanus occitanus*	*Acheta (Gryllus) domesticus*	*thuringiensis*	No effect of feeding, toxic at high dose when injected into haemolymph	Morel 1974
Thysanoptera					
Aeolothripidae	*Aeolothrips intermedius*	*Helicoverpa armigera*	*dendrolimus* (Dendrobacillin) and *galleriae* (Entobakterin)	No effect	
Thripidae	*Scolothrips acariphagus*	*Helicoverpa armigera*	*dendrolimus* (Dendrobacillin) and *galleriae* (Entobakterin)	No effect	Umarov Sh *et al.* 1975

APPENDIX 4.2 SUMMARY OF REPORTS ON THE EFFECTS OF *BT* ON INSECT PARASITOIDS

Parasite/parasitoid	Host and/or target pest	*Bt* subspecies/products	Effect of *Bt* on parasitism	Reference
Diptera: Tachinidae				
Tachinids	*Acyrthosiphon pisum*	*galleriae* (Entobakterin)	No effect	Grechkanev and Maksimova 1980
Bessa fugax	*Yponomeuta evonymellus*	*kurstaki*	Not susceptible	
Blepharipa scutellata	*Lymantria dispar*	Thuricide HPC and IMC 90012	No effect	Dunbar *et al.* 1973
Ceromasia auricaudata	*Choristoneura occidentalis*	*kurstaki* (Dipel)	Parasitism decreased	Hamel 1977
Compsilura concinnata	*Lymantria dispar*	*kurstaki* (Dipel)	Reduced populations of parasites in treated blocks	Reardon *et al.* 1979
Compsilura concinnata	*Pieris brassicae*	*thuringiensis* (Bactospeine)	No effect on adults	Mushtaque *et al.* 1993
Lydella thompsoni	*Ostrinia nubilalis*	Unspecified	Little effect	Anglade *et al.* 1980
Lydella thompsoni	*Ostrinia nubilalis*	Unspecified	Not injurious	Riffiod 1976
Madremyia saundersii	*Choristoneura occidentalis*	*kurstaki* (Dipel)	Parasitism decreased	
Myiopharus doryphorae	*Leptinotarsa decemlineata*	*tenebrionis*	Lower total number of parasitoid larvae in treated hosts	Lopez and Ferro 1995
Parasetigena agilis	*Lymnatria dispar*	Thuricide HPC and IMC 90012	No effect	Dunbar *et al.* 1973
Parasetigena silvestris	*Lymantria dispar*	*kurstaki* (Dipel)	Reduced populations of parasites in treated blocks	Reardon *et al.* 1979
Voria ruralis	(direct effect tested)	Thuricide HPC	Less than 4% mortality	Wilkinson *et al.* 1975
Zenillia dolosa	*Yponomeuta evonymellus*	*kurstaki*	Not susceptible	Hamed 1979
Hymenoptera: Aphelinidae				
Encarsia citrina	Scale insects (Diaspididae)	*kurstaki* (Delfin)	No effect	Thomson *et al.* 1996
Encarsia formosa	*Trialeurodes vaporariorum*	*aizawai*	Non-toxic to mummies and adults	Hayashi 1996
Braconidae				
Braconids	*Acyrthosiphon pisum*	*galleriae* (Entobakterin)	No effect	Grechkanev and Maksimova 1980
Agathis rufipes (Microdus rufipes)	*Cydia pomonella*	*galleriae* (Entobakterin) and *alesti*? (BIP)	No direct effect	Mosievskaya and Makarov 1974
Apanteles africanus (Glyptapanteles africanus)	*Spodoptera litura*	*kurstaki* (Dipel)	Slight reduction in parasitism of *A. africanus*, no effect after 7 days	Chari *et al.* 1996
Apanteles fumiferanae	*Choristoneura fumiferana*	*kurstaki*	Reduced parasitoid populations by 50-60% by killing their hosts before parasitoid emergence, offset by lack of direct effect	Nealis and van Frankenhuyzen 1990
Apanteles fumiferanae	*Choristoneura fumiferana*	*kurstaki*	Little effect in the field	Cadogan *et al.* 1995
Apanteles fumiferanae	*Choristoneura occidentalis*	*kurstaki* (Dipel)	Parasitism increased	Hamel 1977
Apanteles fumiferanae	*Choristoneura occidentalis*	*kurstaki* (Thuricide 32LV and SAN 415 I)	Little effect in field	Niwa *et al.* 1987
Apanteles fumiferanae	*Choristoneura fumiferana*	Unspecified	No significant effect on the rate of parasitism	Otvos and Raske 1980

APPENDIX 4.2 (*continued*)

Parasite/parasitoid	Host and/or target pest	*Bt* subspecies/products	Effect of *Bt* on parasitism	Reference
Apanteles fumiferanae	*Choristoneura fumiferana*	*kurstaki* (Foray and Dipel)	Parasitism increased	Nealis *et al.* 1992
Apanteles glomeratus (Cotesia glomerata)	*Pieris brassicae*	*thuringiensis* (Bactospeine)	No effect on adults	Mushtaque *et al.* 1993
Apanteles glomeratus	*Pieris brassicae*	*alesti* (*kurstaki?*)	Transmitted by parasitoid	Toumanoff 1959 in Forsberg *et al.* 1976
Apanteles glomeratus	*Pieris brassicae*	Unspecified	Least injurious of 7 pesticides	Wawrzyniak 1987
Apanteles glomeratus	*Pieris brassicae*	*kurstaki* (Dipel)	Some mortality at higher doses	Muck *et al.* 1981
Apanteles glomeratus		(Direct exposure)	Reduced adult life	Marchal-Segault 1974
Apanteles glomeratus	*Ephestia kuehniella and Pieris brassicae*	*kurstaki* (Bactospeine)	Reduced host food intake leading to reduced parasite survival	Marchal-Segault 1975
Apanteles litae (Dolichogenidea litae)	*Phthorimaea operculella*	*galleriae*	Some development lag, some adult life-span reduction	Salama *et al.* 1996b
Apanteles melanoscelus	*Lymantria dispar*	*kurstaki*	No direct effect, development lag	Weseloh and Andreadis 1982
Apanteles melanoscelus	*Lymantria dispar*	*kurstaki* (Dipel 4L)	Enhancement of parasitism in the field	Ticehurst *et al.* 1982
Apanteles melanoscelus	*Lymantria dispar*	Thuricide HPC and IMC 90012	No effect	Dunbar *et al.* 1973
Apanteles melanoscelus	*Lymantria dispar*	*kurstaki* (Dipel)	Reduced populations of parasites in treated blocks	Reardon *et al.* 1979
Apanteles plutellae	*Plutella xylostella*	*kurstaki* (Dipel)	Not toxic	Lim *et al.* 1986
Apanteles plutellae	*Plutella xylostella*	*kurstaki* (Dipel)	Not toxic	Talekar and Yang 1991
Apanteles tibialis (congestus)	*Agrotis segetum*	*galleriae* (Entobakterin) *dendrolimus* (Dendrobacillin)	Parasite larvae developed normally	Kashkarova 1975
Ascogaster quadridentata	*Cydia pomonella*	*galleriae* (Entobakterin) and *alesti*? (BIP)	No direct effect	Mosievskaya and Makarov 1974
Bracon brevicornis	*Plodia interpunctella*	Unspecified	Significant reduction in egg production, reduced cocoon formation and longevity of adults	Salama *et al.* 1991c
Bracon brevicornis	*Sesamia cretica*	Unspecified	Detrimental effect, reducing egg numbers, cocoons and emerged adults, adult life-span	Temerak 1980
Bracon brevicornis	*Sesamia cretica*	Unspecified	Transmitted *Bt* to healthy *Sesamia* larvae, some adverse effects on parasitods	Temerak 1982
Bracon brevicornis	*Sesamia cretica*	Unspecified	Detrimental effect, reducing egg numbers, cocoons and emerged adults, adult life-span	Temerak 1980
Bracon hebetor	*Spodoptera littoralis*	*entomocidus*	Sublethal *Bt* host inoculation increased mortality of host	Sneh *et al.* 1983
Bracon instabilis	*Phthorimaea operculella*	*galleriae*	Some development lag, some adult life-span reduction	Salama *et al.* 1996b
Cardiochiles nigriceps	*Heliothis virescens*	*kurstaki* (Dipel)	Significant decreases in life-span when fed *Bt*, no effect of *Bt*-treated hosts	Dunbar and Johnson 1975

Apanteles xanthostigmus				
Cotesia congregata	*Manduca quinquemaculata*	*kurstaki* (Bactospein-A and Thuricide-HPC)	No effect	Cheng and Hanlon 1990
Cotesia marginiventris	*Heliothis virescens*	*kurstaki*	Survival inversely related to concentration of *kurstaki*, related to the timing of exposure	Atwood *et al.* 1995
Cotesia marginiventris	*Heliothis virescens*	*kurstaki*	No detrimental effect	Atwood *et al.* 1998
Cotesia marginiventris	*Heliothis virescens*	*kurstaki*	Emergence inversely related to *Bt* concentration and directly related to timing	Atwood *et al.* 1997a
Cotesia melanoscelus	*Lymantria dispar*	*kurstaki* (Dipel)	Significant increase in field numbers	Webb *et al.* 1989
Cotesia plutellae	*Plutella xylostella*	Unspecified	In susceptible larvae, parasitoid did not affect performance of *Bt*, but *Bt* had a significant negative effect on the parasitoid	Chilcutt and Tabashnik 1997
Cotesia plutellae	Unspecified	*kurstaki*	Least toxic of 8 insecticides	Obra and Morallo-Rejesus 1997
Cotesia rubecula	*Pieris rapae*	Unspecified	High dosage stopped emergence	McDonald *et al.* 1990
Meteorus leviventris	*Agrotis ypsilon*	*kurstaki* (Dipel)	Significant reduction in reproductive potential and longevity, retarded developmen	Hafez *et al.* 1997b
Meteorus leviventris	*Agrotis ypsilon*	Thuringiensin (β-exotoxin)	Retarded development and emergence	Hafez *et al.* 1997b
Meteorus leviventris	(direct effect tested)	Thuricide HPC	Less than 4% mortality	Wilkinson *et al.* 1975
Microgaster demolitor (Microplitis demolitor)	*Spodoptera littoralis*	*entomocidus*	% emergence and reproductive potential reduced	Salama *et al.* 1982
Microplitis croceipes	*Helicoverpa armigera*	*kurstaki* (Dipel)	Some effect through premature host death, no direct effect on adults	Blumberg *et al.* 1997
Microplitis croceipes	*Heliothis virescens*	*kurstaki*	Emergence inversely related to *Bt*, but field compatible	Atwood *et al.* 1997b
Microplitis demolitor	Unspecified	Unspecified	No effect	Hassan and Graham Smith 1995
Microplitis rufiventris	*Spodoptera littoralis*	*kurstaki* (Dipel)	Delayed egg-larval stage, 10% mortality after emergence	El-Maghraby *et al.* 1988
Opius concolor	olive grove pests	*kurstaki* (Bactospeine)	Considered safe	Jacas *et al.* 1992a
Opius concolor	*Bactrocera oleae*	*kurstaki* (Bactospeine)	Harmless	Jacas *et al.* 1992b
Phanerotoma flavitestacea (P. ocularis)	*Ectomyelois ceratoniae (Apomyelois ceratoniae)*	*kurstaki* (Bactospeine)	No effect observed	Dhouibi 1992
Phanerotoma flavitestacea	(Direct exposure)	Unspecified	Reduced adult life	Marchal-Segault 1974
Phanerotoma flavitestacea	*Ephestia kuehniella and Pieris brassicae*	*kurstaki* (Bactospeine)	Reduced host food intake leading to reduced parasite survival	Marchal-Segault 1975
Rogas lymantriae	*Lymantria dispar*	*kurstaki*	Less females, more unfertilised eggs	Wallner *et al.* 1983
Zele chlorophthalma	*Spodoptera littoralis*	*entomocidus*	Reduction in parasitism, emergence and reproductive potential.	Salama and Zaki 1983
Chalcididae				
Brachymeria intermedia	*Lymantria dispar*	*kurstaki* (Dipel)	Reduced populations of parasites in treated blocks	Reardon *et al.* 1979
Brachymeria intermedia	Unspecified	Thuricide HPC	Less than 4% mortality	Wilkinson *et al.* 1975

APPENDIX 4.2 *(continued)*

Parasite/parasitoid	Host and/or target pest	*Bt* subspecies/products	Effect of *Bt* on parasitism	Reference
Pteromalus puparum	*Pieris brassicae*	*thuringiensis* (Bactospeine)	No effect on adults	Mushtaque *et al.* 1993
Encyrtidae				
Ageniaspis fuscicollis	*Yponomeuta evonymellus*	*kurstaki*	Sensitive if spores ingested	Hamed 1979
Anagyrus fusciventris	*Pseudococcus longispinus*	*kurstaki* (Dipel)	No adverse effect on parasitoids or primary host	Wysoki *et al.* 1975
Hungariella peregrina	*Pseudococcus longispinus*	*kurstaki* (Dipel)	No adverse effect on parasitoids or primary host	Wysoki *et al.* 1975
Ooencyrtus pityocampae	*Thaumetopoea pityocampa*	Unspecified	No direct effect on parasitoid eggs	Jamaa *et al.* 1996
Ooencyrtus sp.	*Dendrolimus superans*	? (treatment of eggs with *Bt*)	No effect	Wang *et al.* 1996
Eulophidae				
Elasmus albizziae	*Homadaula anisocentra*	Unspecified	Use to " minimize damage"	Bastian and Hart 1989
Chrysonotomyia punctiventris	*Liriomyza sativae*	*kurstaki* (Dipel)	Parasitism higher than with methomyl	Johnson *et al.* 1984
Chrysonotomyia punctiventris	*Liriomyza sativae*	*kurstaki*	No effect on host or parasites	Johnson *et al.* 1980
Diglyphus begini	*Liriomyza sativae*	*kurstaki* (Dipel)	Parasitism higher than with methomyl	Johnson *et al.* 1984
Diglyphus begini	*Liriomyza sativae*	*kurstaki*	No effect on host or parasites	Johnson *et al.* 1980
Oomyzus sokolowskii	Unspecified	Unspecified	Harmless	Guo *et al.* 1998
Tetrastichus evonymellae	*Yponomeuta evonymellus*	*kurstaki*	Sensitive if spores ingested	Hamed 1979
Tetrastichus servadeii (Baryscapus servadeii)	*Thaumetopoea pityocampa*	Unspecified	No direct effect on parasitoid eggs	Jamaa *et al.* 1996
Eupelmidae				
Anastatus bifasciatus	*Thaumetopoea pityocampa*	Unspecified	No direct effect on parasitoid eggs	Jamaa *et al.* 1996
Ichneumonidae				
Ichneumonids	*Acyrthosiphon pisum*	*galleriae* (Entobakterin)	No effect	Grechkanev and Maksimova 1980
Ichneumonid	*Yponomeuta evonymellus*	*kurstaki*	Not susceptible	Hamed 1979
Agrypon sp. (*Trichionotus* sp.)	*Yponomeuta evonymellus*	*kurstaki*	Not susceptible	Hamed 1979
Campoletis sonorensis	(direct effect tested)	Thuricide HPC	Less than 4% mortality	Wilkinson *et al.* 1975
Diadegma armillata	*Yponomeuta evonymellus*	*kurstaki*	Sensitive if spores ingested	Hamed 1979
Diadegma eurcerophaga	*Plutella xylostella*	*kurstaki* (Dipel)	Not toxic	Talekar and Yang 1991
Diadegma insulare	*Plutella xylostella*	*kurstaki*	No direct effect	Idris and Grafius 1993
Diadegma insulare	*Plutella xylostella*	*kurstaki*	Little effect on parasitism	Ulpah and Kok 1996
Diadegma insulare	*Plutella xylostella* and *Leptophobia aripa*	*kurstaki*	No adverse effect	Garcia 1991
Diadegma insulare	*Plutella xylostella*	*kurstaki*	No significant effect in the field	Carballo *et al.* 1989
Diadegma insulare	*Plutella xylostella*	*kurstaki* (Dipel)	Parasitism similar to untreated control	Cordero and Cave 1990
Diadegma pierisae	*Pieris brassicae*	*thuringiensis* (Bactospeine)	No effect on adults	Mushtaque *et al.* 1993
Diadegma semiclausum	*Plutella xylostella*	*kurstaki*	Least toxic of 8 insecticides	Obra and Morallo-Rejesus 1997
Diadegma sp.	*Heliothis virescens*	*dendrolimus* (Bitoxibacillin) and *kurstaki* (Dipel)	No effect	Jimenez and Fernandez 1985

(Horogenes punctoria)				
Glypta fumiferanae	*Choristoneura occidentalis*	*kurstaki* (Dipel)	Parasitism increased	Hamel 1977
Glypta fumiferanae	*Choristoneura occidentalis*	*kurstaki* (Thuricide and SAN 415I)	Little effect in field	Niwa *et al.* 1987
Glypta fumiferanae	*Choristoneura fumiferana*	Unspecified	No significant effect on the rate of parasitism	Otvos and Raske 1980
Macrocharops bimaculata	*Anticarsia gemmatalis*	*kurstaki*	No reduction in population	Turnipseed *et al.* 1985
Nemeritis canescens	*Ephestria kuehniella*	*thuringiensis*	Transmitted Bt, no direct mortality	Kustak 1964 in Forsberg *et al.* 1976
Phaeogenes hariolus	*Choristoneura occidentalis*	*kurstaki* (Dipel)	Parasitism decreased	Hamel 1977
Phygadeuon trichops	*Delia* spp.	*kurstaki* (Dipel)	Virtually harmless to the parasite	Naton 1978
Phytodietus fumiferanae	*Choristoneura occidentalis*	*kurstaki* (Thuricide and SAN 415)	Little effect in field	Niwa *et al.* 1987
Pimpla turionellae	*Yponomeuta evonymellus*	*kurstaki*	Sensitive if spores ingested	Hamed 1979
Pimpla turionellae	*Galleria mellonella*	*kurstaki* (Dipel)	Some mortality at higher doses	Muck *et al.* 1981
Pteromalidae				
Pachyneuron solitatomus (P. solitarium)	*Dendrolimus superans*	? (treatment of eggs with *Bt*)	No effect	Wang *et al.* 1996
Scelionidae				
Telenomus alsophilae	*Ennomos subsignarius*	Unspecified	No effect	Kaya and Dunbar 1972
Telenomus remus	*Spodoptera litura*	*kurstaki* (Dipel)	No effect after 7 days	Chari *et al.* 1996
Telenomus sphingis	*Erinnyis ello*	*kurstaki* (Dipel)	No effect on adults or emergence	Urias-Lopez *et al.* 1987
Telenomus tetratomus	*Dendrolimus superans*	? (treatment of eggs with *Bt*)	No effect	Wang *et al.* 1996
Trichogrammatidae				
Trichogrammatoidea bactrae	Unspecified	Unspecified	No effect on either parasitoid	Hassan and Graham Smith 1995
Trichogramma cacoeciae	*Sitotroga cerealella*	*kurstaki* (Thuricide-HP, Dipel and Bactospeine)	Did not significantly affect their ability to parasitise the eggs, repellent at high conc.	Hassan and Krieg 1975
Trichogramma cacoeciae	*Sitotroga cerealella*	*kurstaki* (Thuricide-HP, Dipel, and Bactospeine)	Tolerant, high concentration repellent	Hassan and Krieg 1975
Trichogramma cacoeciae	*Sitotroga cerealella*	*thuringiensis, kurstaki* and *israelensis*	No reduction in parasitism of eggs	Krieg *et al.* 1980
Trichogramma confusum (Trichogramma chilonis)	Unspecified	Unspecified	Safe to all stages	Zhang 1997
Trichogramma dendrolimi	*Dendrolimus superans*	? (treatment of eggs with *Bt*)	No effect	Wang *et al.* 1996
Trichogramma embryophagum	*Thaumetopoea pityocampa*	Unspecified	No direct effect on parasitoid eggs	Jamaa *et al.* 1996
Trichogramma embryophagum	*Cydia pomonella*	*galleriae* (Entobakterin)	Used together	Kolmakova 1971
Trichogramma embryophagum	*Cydia pomonella*	*galleriae* (Entobakterin)	Caused no mortality in 24 h	Kostadinov 1979
Trichogramma evanescens	*Ostrinia furnacalis* and *Helicoverpa armigera*	*kurstaki* (Thuricide or Dipel)	No effect on parastoid abundance	Tandan and Nillama 1987

APPENDIX 4.2 (*continued*)

Parasite/parasitoid	Host and/or target pest	*Bt* subspecies/products	Effect of *Bt* on parasitism	Reference
Trichogramma evanescens	*Spodoptera littoralis* or *Ephestia kuehniella*	*galleriae*	No effect on adults, decrease in parasitism of treated host eggs.	Salama and Zaki 1985
Trichogramma evanescens	Unspecified	*thuringiensis* (thermostable exotoxin)	36-62% mortality when intestinally applied, no effect when topically applied	Korostel' and Kapustina 1975
Trichogramma evanescens	Unspecified	*galleriae* (Entobakterin)	Very little effect on any of the stages	Kim Chi 1978
Trichogramma exiguum	*Erinnyis ello*	*kurstaki* (Dipel)	No effect on adults or emergence	Urias-Lopez *et al.* 1987
Trichogramma exiguum	*Helicoverpa zea, Manduca* spp.and *Trichoplusia ni*	*kurstaki*	Parasitism did not differ significantly from that in the untreated control	Campbell *et al.* 1991
Trichogramma japonicum	Unspecified	*Bt* (7216)	Safest of 29 insecticides	Li *et al.* 1986
Trichogramma japonicum	*Corcyra cephalonica*	*kurstaki*	Least effect of 10 insecticides on emergence of parasitoids	Borah and Basit 1996
Trichogramma maidis	*Ostrinia nubilalis*	*kurstaki* (Javelin)	Compatible	Maini and Burgio 1991
Trichogramma nubilale	*Ostrinia nubilalis*	*kurstaki*	Used together	Losey *et al.* 1995
Trichogramma nubilale	*Ostrinia nubilalis*	*kurstaki* (Thuricide)	Did not reduce emergence from or parasitism of eggs	Tipping and Burbutis 1983
Trichogramma ostriniae	*Ostrinia furnacalis*	*kurtsaki*	Used together	Tseng 1990
Trichogramma pallidum	*Cydia pomonella*	*galleriae* (Entobakterin)	Used together	Kolmakova 1971
Trichogramma pretiosum	*Heliothis virescens* or *Helicoverpa zea*	*kurstaki*	Some effect of combined insecticides	Kring and Smith 1995
Trichogramma pretiosum			Not highly toxic	Castelo-Branco and Franca 1995
Trichogramma pretiosum	*Helicoverpa zea, Manduca* spp., and *Trichoplusia ni*	*kurstaki*	Parasitism did not differ significantly from that in the untreated control	Campbell *et al.* 1991
Trichogramma pretiosum	*Scrobipalpuloides absoluta* (*Tuta absoluta*)	*kurstaki*	Not affect the ability to parasitize, but it did adversely affect emergence	Marques and Alves 1995
Trichogramma pretiosum	*Helicoverpa zea*	*kurstaki* (Dipel)	Did not adversely affect egg parasitism	Oatman *et al.* 1983
Trichogramma sp.	*Heliothis* spp.	Unspecified	No effect on emergence	Twine and Lloyd 1982
Trichogramma spp.	*Ostrinia nubilalis*	Unspecified	Used together	Muresan and Mustea 1995
Trichogramma spp.	Unspecified	*dendrolimus* (Dendrobacillin)	100% mortality on the first day	Tolstova and Ionova 1976
Trichogramma spp.	*Corcyra cephalonica*	Unspecified	Negligible effect	Radhika 1998
Unspecified				
not specified	*Trichoplusia ni* and *Autographa californica*	*kurstaki*	Reduced parasitism	Vail *et al.* 1972
aphid parasitoids	*Myzus persicae*	Unspecified	Did not affect the parasitoid population	Imenes *et al.* 1990

Index